Lecture Notes in Mathematics 2084

For further volumes:
http://www.springer.com/series/304

Dachun Yang • Dongyong Yang • Guoen Hu

The Hardy Space H^1 with Non-doubling Measures and Their Applications

Dachun Yang
School of Mathematical Sciences
Beijing Normal University
Laboratory of Mathematics
and Complex Systems
Ministry of Education
Beijing, People's Republic of China

Dongyong Yang
School of Mathematical Sciences
Xiamen University
Xiamen, People's Republic of China

Guoen Hu
Department of Applied Mathematics
Zhengzhou Information Science
and Technology Institute
Zhengzhou, People's Republic of China

ISBN 978-3-319-00824-0 ISBN 978-3-319-00825-7 (eBook)
DOI 10.1007/978-3-319-00825-7
Springer Cham Heidelberg New York Dordrecht London

Lecture Notes in Mathematics ISSN print edition: 0075-8434
ISSN electronic edition: 1617-9692

Library of Congress Control Number: 2013945073

Mathematics Subject Classification (2010): 42B35, 46E35, 42B25, 42B20, 42B30, 30L99

Printed on acid-free paper

Springer is part of Springer Science+Business Media (www.springer.com)

Preface

In many problems from analysis, the Hardy space, $H^1(\mathbb{R}^D)$, always appears as a suitable substitution for $L^1(\mathbb{R}^D)$. Thanks to the seminal papers of Charles Fefferman and Elias M. Stein, Ronald R. Coifman and Guido Weiss, Robert H. Latter and other mathematicians, the properties of the Hardy spaces $H^p(\mathbb{R}^D)$ with $p \in (0, 1]$, such as the endpoint spaces as the interpolation spaces, the characterizations in terms of various maximal functions, the atomic and the molecular decompositions, were established in the period 1970s to 1980s. Nowadays, the analysis relating to the Hardy spaces plays an important role in many fields of analysis, such as complex analysis, partial differential equations, functional analysis and geometrical analysis.

On the other hand, one of the most crucial assumptions in the classical harmonic analysis relating to the Hardy space is the doubling condition of the underlying measures. This is because the Vitali covering lemma and the Calderón–Zygmund decomposition lemma—two cornerstones of the classical harmonic analysis—essentially depend on the doubling condition of the underlying measures. For a long time, mathematicians tried to seek a theory about function spaces and the boundedness of operators which does not require the doubling condition on the underlying measures. The motivations for this come from partial differential equations, complex analysis and harmonic analysis itself. One typical example is the singular integral operators considered in an open domain $\Omega \subset \mathbb{R}^D$ with the usual D-dimensional Lebesgue measure, or on a surface with the usual surface area measure instead of the whole space. If the boundary of Ω is a Lipschitz surface, then the problem can be reduced to the related problem in spaces of homogeneous type in the sense of Ronald R. Coifman and Guido Weiss and can be solved by the standard argument. For the domain with extremely singular boundary (or called "wild" boundary), the results for singular integral operators with doubling measures are not suitable anymore. Another famous examples are the so-called Painlevé problem and Vitushkin's conjecture, in which the non-homogeneous Tb theorem plays a key role. To solve the Painlevé problem, in the 1990s, mathematicians made a great effort to establish the L^2 boundedness for the Cauchy integrals with the one-dimensional Hausdorff measure satisfying some linear growth condition on $\mathbb{R}$.

Due to the celebrated works concerning the boundedness of the Cauchy integrals with continuous measures, which were given by Guy David, Mark S. Melnikov and Joan Verdera, Xavier Tolsa, Fëdor Nazarov, Sergei Treil and Alexander Volberg, at the end of the last century, mathematicians realized that the doubling condition is superfluous for the boundedness of the Cauchy integral. Since then, the considerable attention in harmonic analysis has been paid to the study of various function spaces and the boundedness of operators on these function spaces over non-homogeneous spaces.

To the best of our knowledge, Alexander Volberg and Xavier Tolsa have already provided two interesting monographs containing the self-contained and unified full proofs of Vitushkin's conjecture and of the semiadditivity of analytic and Lipschitz harmonic capacities (Tolsa's solution of the Painlevé problem), two-weight estimates for the Hilbert transform, as well as some elements of the Calderón–Zygmund theory associated with non-negative Radon measures satisfying some polynomial growth conditions on $\mathbb{R}^D$, which are also called non-doubling measures on $\mathbb{R}^D$.

The purpose of this book is to give a detailed survey of the recent progress about the analysis relating to the Hardy space associated with non-doubling measures on $\mathbb{R}^D$ and with non-homogeneous metric measure spaces in the sense of Tuomas Hytönen. The content of the whole book is divided into two parts.

Part I of this book is concerned with the Hardy space $H^1(\mu)$ and its applications on $\mathbb{R}^D$ with non-doubling measures μ, which consists of six chapters. We begin Part I with briefly presenting the history of the development of the real-variable theory of the Hardy space on $\mathbb{R}^D$ and an overview on the main contents of Part I. Then, in Chap. 1, the necessary preliminaries, such as covering lemmas and the Calderón–Zygmund decomposition, are given. In Chap. 2, the approximation of the identity, which is used in the study of operators on functions spaces, is introduced. Chapter 3 is devoted to the Hardy space $H^1(\mu)$. The space $\mathrm{RBMO}(\mu)$, the dual space of $H^1(\mu)$, is also considered in Chap. 3. While in Chap. 4, we study $h^1(\mu)$ and $\mathrm{rbmo}(\mu)$—the local versions of $H^1(\mu)$ and $\mathrm{RBMO}(\mu)$, respectively. Chapters 5 and 6 are focused on the boundedness on function spaces for the Calderón–Zygmund operators and some classical operators—the Littlewood–Paley maximal operators.

As is well known, the metric space is a natural extension of the Euclidean space, and the analysis relating to metric spaces has its own interest. It is Tuomas Hytönen who overcame some essential difficulties and established a framework for the analysis on non-homogeneous spaces. This new framework turns out to be a geometrically doubling metric space $\mathcal{X}$ with a Borel measure ν satisfying the upper doubling condition. The second part, Part II, of this book is concerned with the analysis relating to the Hardy space $H^1(\mathcal{X}, \nu)$ and the boundedness of Calderón–Zygmund operators over $(\mathcal{X}, \nu)$, which consists of two chapters: Chaps. 7 and 8. Similar to Part I, we begin Part II with briefly presenting the history of the development of the theory of the Hardy space on spaces of homogeneous type and an overview on the main content of Part II. Then, in Chap. 7, we investigate basic properties of this framework, as well as the Hardy space $H^1(\mathcal{X}, \nu)$ and its dual space. While in Chap. 8, the boundedness of Calderón–Zygmund operators, commutators and some maximal operators in this setting are given. In some sense, the content of Part II is an extension of the results in Part I.

Besides the detailed and self-contained arguments for the main results, after introducing each important notion, we give at least one typical and easily explicable example, which further clarifies the relations between the known and the present notions. At the end of each chapter of this book, there exists a section, called Notes, in which we give the detailed references of the content of this chapter. Also, in Notes, we present more known related results and some unsolved interesting problems, which might be interesting to the reader.

Comparing with the monographs of Alexander Volberg and Xavier Tolsa, only Chaps. 1–3 of this book may partially have some overlaps with the monograph of Xavier Tolsa. The other parts of the present book mainly focus on the results obtained by the authors of this book and their collaborators throughout recent years.

We would like to thank Professor Shanzhen Lu for his consistent encouragement. Dachun Yang thanks Professor Hans Triebel for his consistent encouragement and Professor Yongsheng Han for his fruitful collaborations throughout many years. We also express our appreciations to Professor Xavier Tolsa for his helpful discussions on the subject of Part I. Dachun Yang and Dongyong Yang thank Professor Tuomas Hytönen for fruitful collaborations on the subject of Part II. Last but not least, we wish to thank all our colleagues and collaborators, in particular, we mention several of them: Wengu Chen, Yan Meng, Haibo Lin, Liguang Liu, Wen Yuan, Suile Liu and Xing Fu for their fruitful collaborations throughout these years. Without them, this book would not be presented by this final version.

Dachun Yang is supported by National Natural Science Foundation of China (Grant No. 11171027) and the Specialized Research Fund for the Doctoral Program of Higher Education of China (Grant No. 20120003110003). Dongyong Yang is supported by National Natural Science Foundation of China (Grant No. 11101339), Natural Science Foundation of Fujian Province of China (Grant No. 2013J01020) and Fundamental Research Funds for Central Universities of China (Grant No. 2013121004). Guoen Hu is supported by National Natural Science Foundation of China (Grant No. 10971228).

Beijing, People's Republic of China — Dachun Yang
Fujian, People's Republic of China — Dongyong Yang
Henan, People's Republic of China — Guoen Hu
August 2012

Abstract

The theory of the Hardy space is a fundamental tool of Fourier analysis, with applications and connections to complex analysis, partial differential equations, functional analysis and geometrical analysis. It extends to settings where the doubling condition of the underlying measures may fail. Beginning in the 1990s, it was discovered that the doubling condition is superfluous for most results of function spaces and the boundedness of operators. The present book tries to give a gentle introduction to those discoveries, the methods behind them, their consequences and some of their applications. The features of this book are that it provides the detailed and self-contained arguments, many typical and easily explicable examples and interesting unsolved problems.

Contents

Part I
$\mathbb{R}^D$ with Non-doubling Measures μ

The development of the theory of Hardy spaces in the D-dimensional Euclidean space $\mathbb{R}^D$ began with the remarkable paper of Stein and Weiss [122], which was connected closely to the theory of harmonic functions. Later, real variable methods were introduced into this subject by Fefferman and Stein [26], making possible a variety of important D-dimensional results and extensions; see [41, 121, 124]. Moreover, the advent of the methods of the atomic decomposition and molecular decomposition enabled the extension of the theory of Hardy spaces on $\mathbb{R}^D$ to various far more general settings; see [18]. In the development of the theory of Hardy spaces and Calderón–Zygmund operators during the period 1970s–1990s, the only thing that has remained unchanged was the doubling property of the underlying measure, which is a key assumption in the classical theory of harmonic analysis.

Let μ be a nonnegative Radon measure on $\mathbb{R}^D$ which only satisfies the following *polynomial growth condition*, namely, there exist positive constants C_0 and $n \in (0, D]$ such that, for all $x \in \mathbb{R}^D$ and $r \in (0, \infty)$,

$$\mu(B(x,r)) \le C_0 r^n, \tag{0.0.1}$$

where above and in what follows,

$$B(x,r) := \{y \in \mathbb{R}^D : |x-y| < r\}.$$

Such a measure is not necessary to be doubling. In the recent decades, it was shown that many results on the theory of Hardy spaces and Calderón–Zygmund operators remain valid for non-doubling measures; see [20, 21, 32, 33, 94, 102–106, 131, 132, 134, 141, 151] and references therein. We remark that the analysis in this context, especially, the $T(b)$ theorem and the boundedness of the Cauchy integral on $L^2(\mu)$, plays an essential role in solving the long open Vitushkin's conjecture and Painlevé's problem; see [136, 137, 139] or survey papers [138, 140, 142, 143, 145, 146] for more details. The purpose of this part is to introduce the theory of the Hardy space $H^1(\mu)$ and pay attention to its applications to the study of Calderón–Zygmund operators on $\mathbb{R}^D$ with the measure μ satisfying (0.0.1).

This part consists of six chapters. In Chap. 1, we introduce some basic covering lemmas on $\mathbb{R}^D$ and notions of doubling cubes, and we further establish the Lebesgue differentiation theorem and the Calderón–Zygmund decomposition. In Chap. 2, we introduce a notion of the coefficient for cubes in $\mathbb{R}^D$, which well describes the geometric properties of cubes and is a useful tool in the whole part. Using this notion, we also construct the approximations of the identity on $\mathbb{R}^D$ with μ satisfying (0.0.1). Chapter 3 is devoted to the Hardy space $H^1(\mu)$. We introduce the BMO-type space RBMO (μ), establish the John–Nirenberg inequality for functions in RBMO (μ). We then introduce the atomic Hardy space $H^1(\mu)$, obtain its basic properties, and prove that the dual space of $H^1(\mu)$ is RBMO (μ). A maximal function characterization of $H^1(\mu)$ is also presented. The contents in Chap. 4 involve the study of $h^1(\mu)$ and rbmo (μ), the local versions of $H^1(\mu)$ and RBMO (μ). After presenting some basic properties, corresponding to those of $H^1(\mu)$ and RBMO (μ), of these spaces, we also establish the relations between $H^1(\mu)$ and $h^1(\mu)$ and between RBMO (μ) and rbmo (μ). In addition, we also discuss a BLO-type space RBLO (μ) and its local version rblo (μ) in the present setting. Chapters 5 and 6 are devoted to the study of boundedness of operators. In Chap. 5, we first establish some weighted estimates for the local sharp maximal operators as well as several interpolation results which are useful. Then we investigate the boundedness of the singular integral operators on $L^p(\mu)$ and $H^1(\mu)$, and the boundedness of the maximal singular integral operators and commutators on $L^p(\mu)$ as well as their endpoint estimates. Weighted estimates for (maximal) singular integral operators are also presented. In Chap. 6, we discuss the boundedness of operators associated with approximations of the identity in Chap. 2, on Hardy-type spaces, BMO-type spaces and Morrey-type space, where these operators include Littlewood–Paley operators and maximal operators.

We now make some necessary conventions. Throughout the whole book, C, $\tilde{C}$, c and $\tilde{c}$ stand for *positive constants* which are independent of the main parameters, but they may vary from line to line. *Constants with subscripts*, such as C_0 and A_0, do not change in different occurrences throughout this part. Furthermore, we use $C_{(\rho,\gamma,\ldots)}$ to denote a positive constant depending on the parameter ρ, γ, ... Throughout this book, the *symbol* $Y \lesssim Z$ means that there exists a positive constant C such that $Y \leq CZ$, and $Y \sim Z$ means that $Y \lesssim Z \lesssim Y$.

In this part, our consideration always takes place in the D-dimensional Euclidean space $\mathbb{R}^D$ (*throughout this book, we use D to denote the dimension of the Euclidean space instead of d because d is used to denote the metric on the non-homogeneous space $\mathcal{X}$ in Part II*). By a *cube* $Q \subset \mathbb{R}^D$, we mean a closed cube whose sides are parallel to the axes and centered at some point of supp μ, and we denote its *side length* by $\ell(Q)$ and its *center* by z_Q. Given $\lambda \in (0,\infty)$ and any cube Q, λQ stands for the *cube concentric with Q and having side length $\lambda\ell(Q)$*. For any subset $E \subset \mathbb{R}^D$, we denote by χ_E the *characteristic function of E*.

Let μ be a nonnegative Radon measure on $\mathbb{R}^D$, we define $\|\mu\| := \mu(\mathbb{R}^D)$. For any $f \in L^1_{\mathrm{loc}}(\mu)$ and cube Q, $m_Q(f)$ denotes the *mean* of f over cube Q, that is,

$$m_Q(f) := \frac{1}{\mu(Q)} \int_Q f(x)\,d\mu(x).$$

For any $p \in [1, \infty]$, in this part, $L^p_c(\mu)$ stands for the *space of functions in $L^p(\mu)$ with compact support* and $L^p_{c,0}(\mu)$ the *space of functions in $L^p_c(\mu)$ having integral 0.*

Chapter 1
Preliminaries

In this chapter, we first recall some basic covering lemmas and notions of doubling cubes, using these we further establish the Lebesgue differentiation theorem and the Calderón–Zygmund decomposition.

1.1 Covering Lemmas

This section is devoted to some basic covering lemmas. We first recall the following Besicovitch covering theorem which is very important and useful in our context.

Theorem 1.1.1. *Let E be a bounded set in $\mathbb{R}^D$. If, for every $x \in E$, there exists a closed cube $Q(x)$ centered at x, then it is possible to choose, from among the given cubes $\{Q(x)\}_{x\in E}$, a subsequence $\{Q_k\}_k$ (possibly finite) such that*

(i) $E \subset \bigcup_k Q_k$;
(ii) *no point of $\mathbb{R}^D$ is in more than N_D (a number that only depends on D) cubes of the sequences $\{Q_k\}$, namely, for every $z \in \mathbb{R}^D$,*

$$\sum_k \chi_{Q_k}(z) \le N_D;$$

(iii) *the sequence $\{Q_k\}_k$ can be distributed in B_D (a natural number that only depends on D) families of disjoint cubes.*

Proof. For any set $\Omega \subset \mathbb{R}^D$, denote by d_Ω the *diameter* of Ω. Now let

$$a_0 := \sup\{d_{Q(x)} : x \in E\}.$$

If $a_0 = \infty$, then we can take a single cube $Q(x)$ to cover E and the conclusions of Theorem 1.1.1 hold true. Assume that $a_0 < \infty$. We choose $Q_1 \in \{Q(x)\}_{x\in E}$ with center $x_1 \in E$ such that $d_{Q_1} > a_0/2$. Let

$$a_1 := \sup\{d_{Q(x)} : x \in (E \setminus Q_1)\}.$$

D. Yang et al., *The Hardy Space H^1 with Non-doubling Measures and Their Applications*, Lecture Notes in Mathematics 2084, DOI 10.1007/978-3-319-00825-7_1,

We now choose Q_2 with center $x_2 \in (E \setminus Q_1)$ such that $d_{Q_2} > a_1/2$. Going on in this way, if there exists some $m \in \mathbb{N} := \{1, 2, \dots\}$ such that

$$E \setminus \left(\bigcup_{k=1}^{m} Q_k \right) = \emptyset, \tag{1.1.1}$$

then the selection process is finished. Otherwise, we go on our selection and obtain a sequence of points, $\{x_k\}_k$, and cubes, $\{Q_k\}_k$, such that, for all i, j with $i \neq j$,

$$\frac{1}{3} Q_i \bigcap \frac{1}{3} Q_j = \emptyset. \tag{1.1.2}$$

To see this, we first observe that, for all $k \in \mathbb{N}$, it holds true that $a_k \leq a_{k-1} < \infty$ and

$$a_{k-1}/2 < d_{Q_k} \leq a_{k-1}.$$

From this observation, we further deduce that, for all $0 \leq j < i$, $d_{Q_i}/2 < d_{Q_j}$, which is equivalent to the fact that $\ell(Q_i)/2 < \ell(Q_j)$. Combining this with the fact that $x_i \notin Q_j$, we obtain (1.1.2).

From (1.1.2) and the fact that E is a bounded set, it follows that the sequence $\{\ell(Q_k)\}_k$ is either finite or $\ell(Q_k) \to 0$ as $k \to \infty$ (For otherwise, (1.1.1) does not hold true for all $m \in \mathbb{N}$ and there exists $\epsilon \in (0, \infty)$ such that, for any $N \in \mathbb{N}$, there exists $k \in \mathbb{N}$ satisfying that $k > N$ and $\ell(Q_k) \geq \epsilon$. We then choose a subsequence $\{Q_{k_N}\}_{N=1}^{\infty}$ of $\{Q_k\}_{k=1}^{\infty}$ such that, for any k_N, $\ell(Q_{k_N}) \geq \epsilon$. This, together with (1.1.2) and the fact that E is bounded, further implies (1.1.1) for some $m \in \mathbb{N}$, which is impossible). If the selection process stops, the conclusion (i) is trivial. If the sequence $\{\ell(Q_k)\}_k$ is infinite and $\ell(Q_k) \to 0$, then $d_{Q_k} \to 0$ and hence $a_k \to 0$. Thus, there exists $x \in E \setminus (\cup_{k=1}^{\infty} Q_k)$ and hence there exists k_0 such that $a_{k_0} < d_{Q(x)}$, which is contradictory to our selection. Thus, $E \subset \cup_{k=1}^{\infty} Q_k$ and (i) holds true in this case.

To see (ii), fix $z \in \mathbb{R}^D$ and draw D hyperplanes through z and consider the 2^D closed "hyperquadrants" through z determined by them. Fix k with Q_k including z. Let

$$\mathcal{J} := \{j \in \mathbb{N} : \ Q_j \ni z \text{ and } x_j \text{ lies in the same "hyperquadrants" as } x_k\}.$$

By the fact that $x_i \notin Q_j$ and $\ell(Q_i)/2 < \ell(Q_j)$ for all $i, j \in \mathbb{N}$ with $i > j$, we see that

$$\ell(Q_k) < \ell(Q_j) < 2\ell(Q_k)$$

when $j \in \mathcal{J}$ and $j > k$, and

$$\ell(Q_j) < \ell(Q_k) < 2\ell(Q_j)$$

when $j \in \mathcal{J}$ and $j < k$. This further implies that $\frac{1}{3}Q_j \subset \frac{8}{3}Q_k$ for all $j \in \mathcal{J}$, which, together with (1.1.2), implies that there exists a positive constant N depending on D such that the cardinality of $\mathcal{J}$ is at most $N + 1$. Thus, the cardinality of cubes containing z is at most $N_D := 2^D(N + 1)$, which completes the proof of (ii).

In order to prove (iii), we rearrange the sequence $\{Q_k\}_k$ such that the side length of the new sequence, which is still denoted by $\{Q_k\}_k$, is decreasing in k. We fix a cube Q_j of the sequence $\{Q_k\}_k$. By (ii), at most N_D members of the sequence contain a fixed vertex of Q_j. Observe that every cube Q_k with $k < j$ is of a size not smaller than that of Q_j. Thus, if $Q_k \cap Q_j \neq \emptyset$ and $k < j$, then Q_k contains at least one of the 2^D vertices of Q_j. This implies that there exist at most $2^D N_D$ sets of the collection $\{Q_1, \ldots, Q_{j-1}\}$ with non empty intersection with Q_j. Consequently, we distribute the sequence $\{Q_k\}_k$ in $2^D N_D + 1$ disjoint sequences in the following way: we let $Q_i \in \mathcal{Q}_i$ for $i \in \{1, \ldots, 2^D N_D + 1\}$. Since $Q_{2^D N_D+2}$ is disjoint with Q_{k_0} for some $k_0 \le 2^D N_D + 1$, we let $Q_{2^D N_D+2} \in \mathcal{Q}_{k_0}$. In the same way, $Q_{2^D N_D+3}$ is disjoint with all sets in some $\mathcal{Q}_{\tilde{k}}$, and we let $Q_{2^D N_D+3} \in \mathcal{Q}_{\tilde{k}}$, and so on. This finishes the proof of (iii), and hence Theorem 1.1.1. □

Remark 1.1.2. (i) Theorem 1.1.1 is not valid anymore, if x can be in the boundary of $Q(x)$ or arbitrarily close to it. However, if the point x is "far" from the boundary of $Q(x)$ (for example, $x \in \rho^{-1}Q(x)$ for a fixed $\rho \in (1, \infty)$ and any point x and $Q(x)$), then Theorem 1.1.1 also holds true.[1]

(ii) We remark that, if E in Theorem 1.1.1 is not bounded, but

$$\sup_{x \in E}\{\ell(Q(x))\} =: M < \infty,$$

then Theorem 1.1.1 still holds true with N_D and B_D replaced by some positive constants $\tilde{N}_D$ and $\tilde{B}_D$. Indeed, it suffices to partition $\mathbb{R}^D$ in cubes of side length M and then apply Theorem 1.1.1 to the intersection of E with each one of these cubes. We omit the details.

Let $\rho \in (1, \infty)$. For any $f \in L^1_{\mathrm{loc}}(\mu)$ and $x \in \mathbb{R}^D$, let

$$\mathcal{M}^{(\rho)} f(x) := \sup_{\rho^{-1}Q \ni x} \frac{1}{\mu(Q)} \int_Q |f(y)|\, d\mu(y),$$

where the supremum is taken over all cubes Q satisfying that $\rho^{-1}Q \ni x$. As an application of Theorem 1.1.1, we obtain the boundedness of $\mathcal{M}^{(\rho)}$ from $L^1(\mu)$ to $L^{1,\infty}(\mu)$ and on $L^p(\mu)$ for $p \in (1, \infty]$ as follows.

Corollary 1.1.3. *Let $\rho \in (1, \infty)$ and $p \in (1, \infty]$. Then $\mathcal{M}^{(\rho)}$ is bounded from $L^1(\mu)$ to $L^{1,\infty}(\mu)$ and on $L^p(\mu)$.*

---

[1] See [23, p. 7].

Proof. Assume that $f \in L^1(\mu)$. For each $t \in (0, \infty)$, let

$$E_t := \{x \in \mathbb{R}^D : \mathcal{M}^{(\rho)} f(x) > t\}.$$

By applying Theorem 1.1.1 to E_t, it is not difficult to see that $\mathcal{M}^{(\rho)}$ is of weak type (1, 1). Observe that $\mathcal{M}^{(\rho)}$ is bounded on $L^\infty(\mu)$. These two facts, together with the Marcinkiewicz interpolation theorem, imply that $\mathcal{M}^{(\rho)}$ is also bounded on $L^p(\mu)$ for any $p \in (1, \infty)$, which completes the proof of Corollary 1.1.3. □

Also, we need the following Whitney decomposition.[2]

Proposition 1.1.4. *Let $\Omega \subset \mathbb{R}^D$ be open and $\Omega \neq \mathbb{R}^D$. Then Ω can be decomposed as*

$$\Omega = \bigcup_{i \in I} Q_i,$$

where $\{Q_i\}_{i\in I}$ are cubes with disjoint interiors, $20Q_i \subset \Omega$ for all $i \in I$, and there exist some constants $\beta \in (20, \infty)$ and $N_W \in \mathbb{N}$ such that, for all $k \in I$, $\beta Q_k \setminus \Omega \neq \emptyset$ and there are at most N_W cubes Q_i with $10Q_k \cap 10Q_i \neq \emptyset$ (in particular, the family of cubes $\{10Q_i\}_{i\in I}$ has finite overlapping).

1.2 Doubling Cubes

In this section, we aim to introduce the notion of doubling cubes. A *non-doubling measure* μ on $\mathbb{R}^D$ means that μ is a nonnegative Radon measure which only satisfies the *polynomial growth condition* (0.0.1). Also, let $Q(x, r)$ be the *cube* centered at x with side length r. Moreover, we always assume that the constant C_0 in (0.0.1) has been chosen big enough such that, for all cubes $Q \subset \mathbb{R}^D$,

$$\mu(Q) \le C_0[\ell(Q)]^n,$$

where $n \in (0, D]$. Observe that, if (0.0.1) holds true for any ball $B(x, r)$, then, for any cube $Q(x, r)$,

$$\mu(Q(x, r)) \le \mu\left(B\left(x, \frac{\sqrt{D}}{2} r\right)\right) \le C_0 \left(\frac{\sqrt{D}}{2}\right)^n r^n.$$

Conversely, if we have $\mu(Q(x, r)) \le C_0 r^n$ for any $x \in \mathbb{R}^D$ and $r \in (0, \infty)$, then, for any ball $B(x, r)$,

$$\mu(B(x, r)) \le \mu(Q(x, 2r)) \le C_0 2^n r^n.$$

[2]See [121, p. 15].

The measure in (0.0.1) is not necessary to satisfy the following *doubling condition* that there exists a positive constant C such that, for all balls B,

$$\mu(2B) \le C\mu(B), \tag{1.2.1}$$

where above and in what follows, for all balls $B := B(x,r)$ and positive constant λ, $\lambda B := B(x, \lambda r)$. Though (1.2.1) is not assumed uniformly for all balls, it turns out there exist some cubes satisfying such an inequality.

Definition 1.2.1. Let $\alpha \in (1,\infty)$ and $\beta \in (\alpha^n, \infty)$. A cube Q is called an *(α, β)-doubling cube* if $\mu(\alpha Q) \le \beta\mu(Q)$.

Proposition 1.2.2. *Let $\alpha \in (1,\infty)$ and $\beta \in (\alpha^n, \infty)$. Then the following two statements hold true:*

(i) *For any $x \in \operatorname{supp}\mu$ and $R \in (0,\infty)$, there exists some (α, β)-doubling cube Q centered at x with $\ell(Q) \ge R$;*
(ii) *If $\beta > \alpha^D$, then, for μ-almost every $x \in \mathbb{R}^D$, there exists a sequence of (α, β)-doubling cubes, $\{Q_k\}_{k\in\mathbb{N}}$, centered at x with $\ell(Q_k) \to 0$ as $k \to \infty$.*

Proof. We first prove (i). To this end, assume that (i) does not hold true. Then there exist some positive constant C and $x_0 \in \operatorname{supp}\mu$ such that, for any cube Q centered at x_0 with $\ell(Q) \ge C$, we have $\mu(\alpha Q) > \beta\mu(Q)$. Now we take Q_0 be such a cube with $\mu(Q_0) > 0$. Then, by our assumption and the growth condition, we see that, for any $k \in \mathbb{N}$,

$$\beta^k \mu(Q_0) < \mu(\alpha^k Q_0) \le C_0 \alpha^{kn} [\ell(Q_0)]^n,$$

which in turn implies that

$$\mu(Q_0) < C_0 \left(\frac{\alpha^n}{\beta}\right)^k [\ell(Q_0)]^n.$$

Letting $k \to \infty$, we have $\mu(Q_0) = 0$, which contracts to $\mu(Q_0) > 0$. This implies that there exists some (α, β)-doubling cube Q centered at x_0 with $\ell(Q) \ge C_0$. Thus, (i) holds true.

To prove (ii), for any fixed $\alpha \in (1,\infty)$ and $\beta \in (\alpha^D, \infty)$, let

$$\begin{aligned}\Omega := \{x \in \mathbb{R}^D : &\text{ there does not exist any sequence of } (\alpha, \beta) - \text{doubling}\\ &\text{cubes centered at } x \text{ whose side lengths tend to zero}\}.\end{aligned}$$

We show that $\mu(\Omega) = 0$. For any $m \in \mathbb{N}$, let

$$\begin{aligned}\Omega_m := \{x \in \mathbb{R}^D : &\text{ all cubes centered at } x \text{ with side lengths}\\ &\text{less than } 1/m \text{ are not } (\alpha, \beta) - \text{doubling cubes}\}.\end{aligned}$$

Observe that $\Omega = \cup_{m=1}^{\infty}\Omega_m$. It suffices to prove that $\mu(\Omega_m) = 0$ for any $m \in \mathbb{N}$. To this end, we fix a cube Q with $\ell(Q) \leq \frac{1}{2m}$ and denote by Q_x^N the cube centered at x whose side length is $\alpha^{-N}\ell(Q)$ for any $x \in \Omega_m \cap Q$ and $N \in \mathbb{N}$. By Theorem 1.1.1, there exists a sequence of cubes, $\{Q_k^N\}_{k\in I_N}$, such that

$$\Omega_m \bigcap Q \subset \bigcup_{k\in I_N} Q_k^N \quad \text{and} \quad \sum_{k\in I_N} \chi_{Q_k^N} \lesssim 1.$$

Since the center of Q_k^N is in Ω_m and $\ell(Q_k^N) \leq \frac{1}{2m}$, Q_k^N is not a (α, β)-doubling cube for each k. Therefore, from this and the fact that $\alpha^N Q_k^N \subset 3Q$, it follows that

$$\mu(Q_k^N) < \beta^{-1}\mu(\alpha Q_k^N) < \cdots < \beta^{-N}\mu(\alpha^N Q_k^N) \leq \beta^{-N}\mu(3Q). \tag{1.2.2}$$

On the other hand, by the facts $\sum_{k\in I_N} \chi_{Q_k^N} \lesssim 1$ and $Q_k^N \subset 3Q$, we conclude that

$$\sum_{k\in I_N} |Q_k^N| \lesssim |3Q|, \tag{1.2.3}$$

where $|\cdot|$ denotes the *D-dimensional Lebesgue measure*. The inequality (1.2.3) is equivalent to that

$$\#(I_N)\alpha^{-ND}[\ell(Q)]^D \lesssim 3^D[\ell(Q)]^D,$$

where above and in what follows, for any set E, $\#(E)$ denotes its *cardinality*. Then we have $\#(I_N) \lesssim \alpha^{ND}$, which, together with (1.2.2), in turn implies that

$$\mu\left(\Omega_m \bigcap Q\right) \leq \sum_{k\in I_N} \mu(Q_k^N) \lesssim \alpha^{ND}\beta^{-N}\mu(3Q).$$

Letting $N \to \infty$, we see that $\mu(\Omega_m \cap Q) = 0$.

Notice that, for each $m \in \mathbb{N}$, $\mathbb{R}^D = \cup_i Q_{m,i}$, where $\{Q_{m,i}\}_i$ are cubes with

$$\ell(Q_{m,i}) = \frac{1}{2m}$$

for all i. We then find that

$$\mu(\Omega_m) \leq \sum_i \mu\left(\Omega_m \bigcap Q_{m,i}\right) = 0.$$

This further implies that $\mu(\Omega) = 0$ and finishes the proof of (ii) and hence Proposition 1.2.2. $\square$

Let $\rho \in (1, \infty)$. In the following, we always take $\beta_\rho := \rho^{D+1}$. For any cube Q, let $\tilde{Q}^\rho$ be the *smallest* (ρ, β_ρ)*-doubling cube* which has the form $\rho^k Q$ with $k \in \mathbb{N} \cup \{0\} =: \mathbb{Z}_+$. If $\rho = 2$, we *denote the cube* $\tilde{Q}^\rho$ *simply by* $\tilde{Q}$. Moreover, by a doubling cube Q, we *always mean a* $(2, 2^{D+1})$*-doubling cube.*

Example 1.2.3. Let

$$\mu := \chi_Q \, dx \, dy + \chi_I \, dx,$$

where $Q := [-1, 1] \times [-1, 1]$ and $I := Q \cap \mathbb{R} = \{(x, 0) : -1 \le x \le 1\}$. If B is the disc centered at $(x, y) \in Q$, $y \in (0, \infty)$, of radius y, then $\mu(B) \sim y^2$ while $\mu(2B) \sim y$ with the implicit equivalent positive constants independent of x and y, and hence μ is a non-doubling measure.

Example 1.2.4. Let $E \subset \mathbb{C}$ be compact. Define the *capacity*

$$\alpha_+(E) := \sup\{\mu(E) : \mu \text{ is a positive Radon measure supported on } E \text{ such that } \mathcal{C}\mu \text{ is a continuous function on } \mathbb{C} \text{ and } \|\mathcal{C}\mu\|_{L^\infty(\mathbb{C})} \le 1\},$$

where $\mathcal{C}\mu$ is the *Cauchy transform* defined by setting, for all $x \notin \operatorname{supp} \mu$,

$$\mathcal{C}\mu(x) := \int_{\mathbb{C}} \frac{1}{z - x} \, d\mu(z).$$

Now let μ_0 be a Radon measure supported on E such that $\mathcal{C}\mu_0$ is a continuous function on $\mathbb{C}$, $\|\mathcal{C}\mu_0\|_{L^\infty(\mathbb{C})} \le 1$ and $\mu_0(E) \ge \alpha_+(E)/2$. Then we conclude that, for all $x \in \mathbb{C}$ and $r \in (0, \infty)$, $\mu_0(B(x, r)) \le r$.[3]

1.3 The Lebesgue Differentiation Theorem

In this section, we establish the Lebesgue differentiation theorem. To begin with, we recall the fact that continuous functions are dense in $L^p(\mu)$ for any $p \in [1, \infty)$.[4]

Lemma 1.3.1. *Let* $p \in [1, \infty)$ *and* $f \in L^p(\mu)$. *Then, for any* $\epsilon \in (0, \infty)$, *there exists a continuous function* g *with compact support on* $\mathbb{R}^D$ *such that* $\|f - g\|_{L^p(\mu)} < \epsilon$.

The main result of this section is as follows.

---

[3]See [137, p. 530] and [37, p. 40].

[4]See [111, p. 69].

Theorem 1.3.2. *Let $f \in L^1_{\mathrm{loc}}(\mu)$. Then, for μ-almost every $x \in \operatorname{supp}\mu$ and any sequence of cubes, $\{Q_k(x)\}_k$, centered at x with $\ell(Q_k(x)) \to 0$, $k \to \infty$,*

$$\lim_{k\to\infty} \frac{1}{\mu(Q_k(x))} \int_{Q_k(x)} f(y)\,d\mu(y) = f(x). \tag{1.3.1}$$

Proof. By a standard localization, it suffices to consider the case when $f \in L^1(\mu)$. We claim that (1.3.1) holds true for any continuous function g. To this end, for any $x \in \mathbb{R}^D$ and each k, let

$$\mathrm{I}_k(x) := \left| \frac{1}{\mu(Q_k(x))} \int_{Q_k(x)} g(y)\,d\mu(y) - g(x) \right|.$$

Since g is continuous, for any $\epsilon \in (0,\infty)$, there exists $K \in \mathbb{N}$, depending on x and ϵ, such that, for any $k > K$ and $y \in Q_k(x)$, $|g(y) - g(x)| < \epsilon$. From this fact, it follows that

$$\mathrm{I}_k(x) \le \frac{1}{\mu(Q_k(x))} \int_{Q_k(x)} |g(y) - g(x)|\,d\mu(y) \le \epsilon.$$

Since ϵ is arbitrary, we further conclude that $\mathrm{I}_k(x) \to 0$, $k \to \infty$. Thus, the claim holds true.

We now show that, for any $f \in L^1(\mu)$ and μ-almost every x,

$$\limsup_{k\to\infty} |m_{Q_k(x)}(f) - f(x)| = 0.$$

By Lemma 1.3.1, there exists a sequence of continuous functions, $\{f_n\}_n$, on $\mathbb{R}^D$ such that $\|f - f_n\|_{L^1(\mu)} \to 0$, $n \to \infty$. It then follows from the claim that, for each $n \in \mathbb{N}$,

$$\begin{aligned}
&\limsup_{k\to\infty} \left|m_{Q_k(x)}(f) - f(x)\right| \\
&\quad \le \limsup_{k\to\infty} \left[\left|m_{Q_k(x)}(f) - m_{Q_k(x)}(f_n)\right| + \left|m_{Q_k(x)}(f_n) - f_n(x)\right|\right] \\
&\qquad + |f_n(x) - f(x)| \\
&\quad \le \mathcal{M}^{(2)}(f - f_n)(x) + |f_n(x) - f(x)|.
\end{aligned}$$

For any $\epsilon \in (0,\infty)$, let

$$\mathrm{E}_\epsilon := \left\{ x \in \mathbb{R}^D : \limsup_{k\to\infty} \left|m_{Q_k(x)}(f) - f(x)\right| > \epsilon \right\}.$$

Then, by Corollary 1.1.3 and Lemma 1.3.1, we see that

$$\mu(\mathrm{E}_\epsilon) \leq \mu\left(\left\{x \in \mathbb{R}^D : \mathcal{M}^{(2)}(f - f_n)(x) > \frac{\epsilon}{2}\right\}\right)$$
$$+\mu\left(\left\{x \in \mathbb{R}^D : |f_n(x) - f(x)| > \frac{\epsilon}{2}\right\}\right)$$
$$\lesssim \frac{1}{\epsilon}\|f_n - f\|_{L^1(\mu)},$$

which tends to 0, as $n \to \infty$. Therefore, we obtain $\mu(\mathrm{E}_\epsilon) = 0$. This finishes the proof of Theorem 1.3.2. □

As a consequence of Theorem 1.3.2, we further obtain the following conclusion.

Corollary 1.3.3. *Let $p \in [1, \infty)$ and $f \in L^p_{\mathrm{loc}}(\mu)$. Then, for μ-almost every $x \in \operatorname{supp}\mu$ and $Q_k(x)$ as in Theorem 1.3.2,*

$$\lim_{k\to\infty} \frac{1}{\mu(Q_k(x))} \int_{Q_k(x)} |f(y) - f(x)|^p \, d\mu(y) = 0.$$

Proof. Let $\mathbb{Q} := \{r_i\}_{i\in\mathbb{N}}$ be the *set of all rational numbers* and, for each i,

$$Z_i := \left\{x \in \operatorname{supp}\mu : \limsup_{k\to\infty} \frac{1}{\mu(Q_k(x))} \int_{Q_k(x)} |f(y) - r_i|^p \, d\mu(y) \neq |f(x) - r_i|^p\right\}.$$

Since $|f(y) - r_i|^p \in L^1_{\mathrm{loc}}(\mu)$, it follows, from Theorem 1.3.2, that $\mu(Z_i) = 0$ for any $i \in \mathbb{N}$. Define

$$Z_0 := \{x \in \operatorname{supp}\mu : |f(x)| = \infty\}.$$

Then $\mu(\cup_{i=0}^\infty Z_i) = 0$ and, to show Corollary 1.3.3, it suffices to prove that, whenever $x \notin \cup_{i=0}^\infty Z_i$,

$$\limsup_{k\to\infty} \frac{1}{\mu(Q_k(x))} \int_{Q_k(x)} |f(y) - f(x)|^p \, d\mu(y) = 0. \tag{1.3.2}$$

Now, for any $\epsilon \in (0, \infty)$ and each x, we choose $r_i \in \mathbb{Q}$ such that $|f(x) - r_i|^p < \epsilon$. By the fact that $x \notin \cup_{i=0}^\infty Z_i$, we see that

$$\limsup_{k\to\infty} \frac{1}{\mu(Q_k(x))} \int_{Q_k(x)} |f(y) - f(x)|^p \, d\mu(y) = |f(x) - r_i|^p \leq \epsilon.$$

Since ϵ is arbitrary, it follows that (1.3.2) holds true, which completes the proof of Corollary 1.3.3. □

1.4 The Calderón–Zygmund Decomposition

This section is devoted to the Calderón–Zygmund decomposition.

Theorem 1.4.1. *Let $p \in [1, \infty)$. Then, for any $f \in L^p(\mu)$ and any $\lambda \in (0, \infty)$ (with $\lambda \in (2^{D+1}\|f\|_{L^p(\mu)}/\|\mu\|, \infty)$ if $\|\mu\| < \infty$),*

(a) *there exists a family $\{Q_i\}_i$ of almost disjoint cubes, that is, $\sum_i \chi_{Q_i} \le C$, such that*

$$\frac{1}{\mu(2Q_i)} \int_{Q_i} |f(x)|^p \, d\mu(x) > \frac{\lambda^p}{2^{D+1}}, \tag{1.4.1}$$

$$\frac{1}{\mu(2\eta Q_i)} \int_{\eta Q_i} |f(x)|^p \, d\mu(x) \le \frac{\lambda^p}{2^{D+1}} \text{ for all } \eta \in (2, \infty) \tag{1.4.2}$$

and

$$|f(x)| \le \lambda \text{ for } \mu\text{-almost every } x \in \mathbb{R}^D \setminus \left(\bigcup_i Q_i\right); \tag{1.4.3}$$

(b) *for each i, let R_i be a $(6, 6^{D+1})$-doubling cube concentric with Q_i with*

$$\ell(R_i) > 4\ell(Q_i) \quad \text{and} \quad \omega_i := \chi_{Q_i} / \left(\sum_k \chi_{Q_k}\right).$$

Then there exists a family $\{\varphi_i\}_i$ of functions such that, for each i and μ-almost every $x \in \mathbb{R}^D$, $\varphi_i(x) = 0$ if $x \notin R_i$, and φ_i has a constant sign on R_i,

$$\int_{\mathbb{R}^D} \varphi_i(x) \, d\mu(x) = \int_{Q_i} f(x)\omega_i(x) \, d\mu(x) \tag{1.4.4}$$

and

$$\sum_i |\varphi_i(x)| \le B\lambda \text{ for } \mu\text{-almost every } x \in \mathbb{R}^D, \tag{1.4.5}$$

where B is some positive constant and, when $p = 1$, it holds true that

$$\|\varphi_i\|_{L^\infty(\mu)}\mu(R_i) \le C \int_{Q_i} |f(x)| \, d\mu(x) \tag{1.4.6}$$

or, when $p \in (1, \infty)$, it holds true that

$$\left[\int_{R_i} |\varphi_i(x)|^p \, d\mu(x)\right]^{1/p} [\mu(R_i)]^{1/p'} \le \frac{C}{\lambda^{p-1}} \int_{Q_i} |f(x)|^p \, d\mu(x), \quad (1.4.7)$$

here above and in what follows, for $p \in [1, \infty]$, p' stands for the conjugate index of p, namely, $\frac{1}{p} + \frac{1}{p'} = 1$.

Proof. Since the proof in the case that $\|\mu\| < \infty$ is similar, we only consider the case that $\|\mu\| = \infty$. Taking into account Proposition 1.2.2 and Theorem 1.3.2, for μ-almost every $x \in \mathbb{R}^D$ such that $|f(x)|^p > \lambda^p$, there exists some cube Q_x satisfying that

$$\frac{1}{\mu(2Q_x)} \int_{Q_x} |f(x)|^p \, d\mu(x) > \frac{\lambda^p}{2^{D+1}} \quad (1.4.8)$$

and such that, if $\hat{Q}_x$ is centered at x with $\ell(\hat{Q}_x) > 2\ell(Q_x)$, then

$$\frac{1}{\mu(2\hat{Q}_x)} \int_{\hat{Q}_x} |f(x)|^p \, d\mu(x) \le \frac{\lambda^p}{2^{D+1}}.$$

Now we apply Theorem 1.1.1 to obtain an almost disjoint subfamily $\{Q_i\}_i$ of cubes satisfying (1.4.1), (1.4.2) and (1.4.3). Indeed, if

$$\Omega := \{x \in \mathbb{R}^D : |f(x)|^p > \lambda^p\}$$

is bounded, then the existence of $\{Q_i\}_i$ comes from Theorem 1.1.1 directly. Otherwise, we choose a cube Q_0 centered at the origin big enough such that

$$2^{D+1} \|f\|^p_{L^p(\mu)} / \mu(Q_0) < \lambda.$$

Then, for any cube Q containing Q_0, we have

$$2^{D+1} \|f\|^p_{L^p(\mu)} / \mu(Q) < \lambda. \quad (1.4.9)$$

For any $m \in \mathbb{Z}_+$, let $Q_m := (5/4)^m Q_0$. Now we apply Theorem 1.1.1 to

$$(Q_m \setminus Q_{m-1}) \bigcap \Omega$$

(if $m = 0$ then we apply Theorem 1.1.1 to $Q_0 \cap \Omega$) and Q_x centered at

$$x \in \operatorname{supp} \mu \bigcap (Q_m \setminus Q_{m-1}) \bigcap \Omega$$

to obtain a sequence $\{Q_{m_i}\}_{i \in \Lambda_m}$.

Now (a) is reduced to showing that the sequence $\{Q_{m_i}\}_{i\in\Lambda_m, m\in\mathbb{Z}_+}$ also has the finite overlapping property. To this end, we first claim that there exists some constant N_0 such that $Q_x \subset Q_{m+N_0}$ for all $m \in \mathbb{Z}_+$ and $x \in Q_m \setminus Q_{m-1}$. Indeed, for any $m \in \mathbb{Z}_+$ and $x \in Q_m$, we see that $Q_0 \subset Q(x, 2\ell(Q_m))$. Then, if $\ell(Q_x) > \ell(Q_m)$, we would have $Q_0 \subset 2Q_x$, which implies that $2Q_x$ satisfies (1.4.9). This contradicts (1.4.8). Thus, we conclude that $\ell(Q_x) \le \ell(Q_m)$, from which the claim follows. Furthermore, it is not difficult to see that there exist $\tilde{N}_0$ and M which is big enough and depends on $\tilde{N}_0$ such that, for all $m \ge M$ and $x \in Q_m \setminus Q_{m-1}$,

$$Q_x \subset Q_{m+N_0} \setminus Q_{m-\tilde{N}_0}.$$

This further implies that, for all $m \ge M$ and $x \in Q_m \setminus Q_{m-1}$,

$$\sum_{m\in\mathbb{Z}_+, m\ge M, i\in\Lambda_m} \chi_{Q_{m_i}}(x) \le (N_0 + \tilde{N}_0 + 1)N_D,$$

where N_D is as in Theorem 1.1.1. On the other hand, by Theorem 1.1.1, we know that, for all $m \le M - 1$ and $x \in Q_m \setminus Q_{m-1}$,

$$\sum_{m\in\mathbb{Z}_+, m\le M-1, i\in\Lambda_m} \chi_{Q_{m_i}}(x) \le MN_D.$$

Thus, by these two facts, we conclude that the sequence $\{Q_{m_i}\}_{m\in\mathbb{Z}_+, i\in\Lambda_m}$ has the finite overlapping property.

To prove (b), assume first that the family of cubes, $\{Q_i\}_i$, is finite. We may further suppose that this family of cubes is ordered in such a way that the sizes of the cubes $\{R_i\}_i$ are non decreasing (namely $\ell(R_{i+1}) \ge \ell(R_i)$ for all i). The functions φ_i that we now construct are of the form $\varphi_i = \alpha_i \chi_{A_i}$ with $\alpha_i \in \mathbb{R}$ and $A_i \subset R_i$ such that $\mu(A_i) \ge \mu(R_i)/2$. We let $A_1 := R_1$ and $\varphi_1 := \alpha_1 \chi_{R_1}$, where the constant α_1 is chosen such that

$$\int_{Q_1} f(z)w_1(z)\,d\mu(z) = \int_{\mathbb{R}^D} \varphi_1(z)\,d\mu(z).$$

Suppose that $\varphi_1, \ldots, \varphi_{k-1}$ have been constructed, satisfying (1.4.4) and

$$\sum_{i=1}^{k-1} |\varphi_i| \le B\lambda,$$

where B is some constant which is fixed below.

Let $\{R_{s_1}, \ldots, R_{s_m}\}$ be the subfamily of $\{R_1, \ldots, R_{k-1}\}$ such that $R_{s_j} \cap R_k \ne \emptyset$ and $\{\varphi_{s_j}\}_{j=1}^m$ the corresponding functions. We claim that there exists some positive constant C_1 such that

$$\mu\left(\left\{x \in \mathbb{R}^D : \sum_j \left|\varphi_{s_j}(x)\right| > 2C_1\lambda\right\}\right) \leq \frac{\mu(R_k)}{2}.$$

Indeed, if all $\{R_1, \ldots, R_{k-1}\}$ are disjoint with R_k, then the claim holds true automatically. Otherwise, since $\ell(R_{s_j}) \leq \ell(R_k)$ (because of the non decreasing sizes of $\{R_i\}_i$), it follows that $R_{s_j} \subset 3R_k$. Taking into account that, for $i \in \{1, \ldots, k-1\}$,

$$\int_{\mathbb{R}^D} |\varphi_i(x)|\, d\mu(x) \leq \int_{Q_i} |f(x)|\, d\mu(x),$$

using that R_k is $(6, 6^{D+1})$-doubling, together with the finite overlapping property of $\{Q_i\}_i$ and (1.4.2), we conclude that there exists a positive constant C_1 such that

$$\begin{aligned}
\sum_j \int_{\mathbb{R}^D} |\varphi_{s_j}(x)|\, d\mu(x) &\leq \sum_j \int_{Q_{s_j}} |f(x)|\, d\mu(x) \\
&\lesssim \int_{3R_k} |f(x)|\, d\mu(x) \\
&\lesssim \left[\int_{3R_k} |f(x)|^p\, d\mu(x)\right]^{1/p} [\mu(3R_k)]^{1/p'} \\
&\lesssim \lambda[\mu(6R_k)]^{1/p}[\mu(3R_k)]^{1/p'} \\
&\leq C_1\lambda\mu(R_k).
\end{aligned}$$

This implies the claim.

Let

$$A_k := R_k \bigcap \left\{x \in \mathbb{R}^D : \sum_j \left|\varphi_{s_j}(x)\right| \leq 2C_1\lambda\right\}$$

and $\varphi_k := \alpha_k \chi_{A_k}$, where the constant α_k satisfies that

$$\int_{\mathbb{R}^D} \varphi_k(z)\, d\mu(z) = \int_{Q_k} f(z) w_k(z)\, d\mu(z).$$

Notice that $\mu(A_k) \geq \mu(R_k)/2$. By this fact, together with (1.4.2), we then see that there exists a positive constant C_2 such that

$$|\alpha_k| \leq \frac{1}{\mu(A_k)} \int_{Q_k} |f(x)|\, d\mu(x) \leq \frac{2}{\mu(R_k)} \int_{\frac{1}{2}R_k} |f(x)|\, d\mu(x) \leq C_2\lambda$$

(this calculation also applies to $k = 1$). Thus, we find that, for all $x \in \mathbb{R}^D$,

$$|\varphi_k(x)| + \sum_{j=1}^{k-1} |\varphi_j(x)| \leq (2C_1 + C_2)\lambda.$$

Therefore, (1.4.5) holds true for all k, if we take $B := 2C_1 + C_2$. Also, if $p = 1$, then, by the choices of A_i and φ_i, we have

$$\|\varphi_i\|_{L^\infty(\mu)}\mu(R_i) \lesssim |\alpha_i|\mu(A_i) \sim \left|\int_{\mathbb{R}^D} f(x)w_i(x)\,d\mu(x)\right| \lesssim \int_{Q_i} |f(x)|\,d\mu(x).$$

This implies (1.4.6). If $p \in (1, \infty)$, then we conclude that

$$\begin{aligned}\left[\int_{R_i} |\varphi_i(x)|^p\,d\mu(x)\right]^{1/p} [\mu(R_i)]^{1/p'} &= |\alpha_i|[\mu(A_i)]^{1/p}[\mu(R_i)]^{1/p'} \\ &\lesssim |\alpha_i|\mu(A_i) \\ &\sim \left|\int_{Q_i} f(x)w_i(x)\,d\mu(x)\right| \\ &\lesssim \left[\int_{Q_i} |f(x)|^p\,d\mu(x)\right]^{1/p} [\mu(Q_i)]^{1/p'}.\end{aligned}$$

On the other hand, from (1.4.1), it follows that

$$\left[\int_{Q_i} |f(x)|^p\,d\mu(x)\right]^{1/p} [\mu(2Q_i)]^{1/p'} \lesssim \frac{1}{\lambda^{p-1}} \int_{Q_i} |f(x)|^p\,d\mu(x). \qquad (1.4.10)$$

By these two facts, we obtain (1.4.7).

Suppose now that the collection $\{Q_i\}_i$ of cubes is not finite. For each fixed N, we consider the family $\{Q_i\}_{1 \leq i \leq N}$ of cubes. Then, by the argument as above, we construct functions, $\varphi_1^N, \ldots, \varphi_N^N$, with $\operatorname{supp} \varphi_i^N \subset R_i$ satisfying

$$\int_{\mathbb{R}^D} \varphi_i^N(x)\,d\mu(x) = \int_{Q_i} f(x)w_i(x)\,d\mu(x),$$

$$\sum_{i=1}^{N} |\varphi_i^N| \leq B\lambda \qquad (1.4.11)$$

and, when $p = 1$, it holds true that

$$\|\varphi_i^N\|_{L^\infty(\mu)}\,\mu(R_i) \lesssim \int_{Q_i} |f(x)|\,d\mu(x)$$

or, when $p \in (1, \infty)$, it holds true that

$$\left[\int_{R_i} |\varphi_i^N(x)|^p \, d\mu(x)\right]^{1/p} [\mu(R_i)]^{1/p'} \lesssim \frac{1}{\lambda^{p-1}} \int_{Q_i} |f(x)|^p \, d\mu(x).$$

Notice that the sign of φ_i^N equals the sign of $\int_{Q_i} f(x) w_i(x) \, d\mu(x)$ and hence it is independent of N.

Assume that $p = 1$. Notice that $\{\varphi_1^N\}_{N \in \mathbb{N}} \subset L^\infty(\mu)$ with uniform bound. By [110, Theorem 3.17], we know that there exists a subsequence $\{\varphi_1^k\}_{k \in I_1}$ which is convergent in the weak-$*$ topology of $L^\infty(\mu)$ to some function $\varphi_1 \in L^\infty(\mu)$. Now we consider a subsequence $\{\varphi_2^k\}_{k \in I_2}$, with $I_2 \subset I_1$, which is also convergent in the weak-$*$ topology of $L^\infty(\mu)$ to some function $\varphi_2 \in L^\infty(\mu)$. In general, for each j, we consider a subsequence $\{\varphi_j^k\}_{k \in I_j}$, with $I_j \subset I_{j-1}$, that converges in the weak-$*$ topology of $L^\infty(\mu)$ to some function $\varphi_j \in L^\infty(\mu)$. Observe that the functions $\{\varphi_j\}_j$ satisfy the required properties. Indeed, it follows that[5]

$$\|\varphi_j\|_{L^\infty(\mu)} \le \liminf_{k \to \infty} \left\|\varphi_j^k\right\|_{L^\infty(\mu)} \lesssim \frac{1}{\mu(R_j)} \int_{Q_j} |f(x)| \, d\mu(x),$$

which implies (1.4.6). Similarly, if $p \in (1, \infty)$, then we have (1.4.7).

Fix j. By the argument as above, we may assume that $\{\varphi_j^k\}_k$ are all nonnegative on R_j. The facts that $\{\varphi_j^k\}_k$ converges to φ_j in the weak-$*$ topology of $L^\infty(\mu)$ and $\operatorname{supp} \varphi_j^k \subset R_j$ lead to that, for any $\lambda \in (1, \infty)$,

$$\varphi_j(\chi_{\lambda R_j \setminus R_j} \operatorname{sgn}(\varphi_j)) = 0,$$

where above and in what follows, $\operatorname{sgn}(g)$ denotes the *sign function of the function* g. This implies that $\varphi_j(x) = 0$ for μ-almost every $x \in \mathbb{R}^D \setminus R_j$. Moreover, it is easy to see that φ_j satisfies (1.4.4) and, for μ-almost every $x \in R_j$, $\varphi_j(x) \ge 0$. It remains to show that $\{\varphi_j\}_j$ satisfies (1.4.5). Observe that $\{\varphi_j\}_j \subset L^1_{\rm loc}(\mu)$. By Theorem 1.3.2, we conclude that, for any $m \in \mathbb{N}$ and μ-almost every $x \in \cup_{j=1}^m R_j$,

$$\begin{aligned}
\sum_{j=1}^m |\varphi_j(x)| &= \lim_{r \to 0} \frac{1}{\mu(Q(x,r))} \int_{Q(x,r)} \sum_{j=1}^m |\varphi_j(y)| \, d\mu(y) \\
&= \sum_{j=1}^m \lim_{r \to 0} \frac{1}{\mu(Q(x,r))} \int_{Q(x,r)} \varphi_j(y) \operatorname{sgn}(\varphi_j)(y) \chi_{R_j}(y) \, d\mu(y) \\
&= \sum_{j=1}^m \lim_{r \to 0} \lim_{k \to \infty} \frac{1}{\mu(Q(x,r))} \int_{Q(x,r)} \varphi_j^k(y) \operatorname{sgn}(\varphi_j^k)(y) \chi_{R_j}(y) \, d\mu(y)
\end{aligned}$$

[5]See [157, p. 125].

$$\leq \sum_{j=1}^{m} \lim_{k\to\infty} \lim_{r\to 0} \frac{1}{\mu(Q(x,r))} \int_{Q(x,r)} \left|\varphi_j^k(y)\right| d\mu(y)$$
$$\leq B\lambda,$$

where, in the third-to-last inequality, we used the fact that

$$\operatorname{sgn}(\varphi_j^k)(x) = \operatorname{sgn}(\varphi_j)(x).$$

This finishes the proof of Theorem 1.4.1. □

We now establish another version of the Calderón–Zygmund decomposition. To this end, let $\rho \in (1, \infty)$. We introduce the *maximal operator* $\mathcal{M}_{(\rho)}$ by setting, for any $f \in L^1_{\mathrm{loc}}(\mu)$ and $x \in \mathbb{R}^D$,

$$\mathcal{M}_{(\rho)} f(x) := \sup_{Q \ni x} \frac{1}{\mu(\rho Q)} \int_Q |f(y)|\, d\mu(y). \tag{1.4.12}$$

Theorem 1.4.2. *Let $f \in L^1(\mu)$. For $\lambda \in (0,\infty)$ (with $\lambda \in (2^{D+1}\|f\|_{L^1(\mu)}/\|\mu\|, \infty)$ if $\|\mu\| < \infty$), let*

$$\Omega_\lambda := \{x \in \mathbb{R}^D : \mathcal{M}_{(2)} f(x) > \lambda\}.$$

Then Ω_λ is open and $|f| \leq 2^{D+1}\lambda$ μ-almost everywhere in $\mathbb{R}^D \setminus \Omega_\lambda$. Moreover, if letting the cubes $\{Q_i\}_i$ be the Whitney decomposition of Ω_λ, then

(a) *for each i, there exists a function $\omega_i \in C^\infty(\mathbb{R}^D)$ with $\operatorname{supp}\omega_i \subset \frac{3}{2}Q_i$,*

$$0 \leq \omega_i \leq 1 \text{ and } \|\nabla\omega_i\|_{L^\infty(\mu)} \leq C\ell(Q_i)^{-1}$$

such that $\sum_i \omega_i \equiv 1$ if $x \in \Omega_\lambda$;

(b) *for each i, let R_i be the smallest $(6, 6^{D+1})$-doubling cube of the form $6^k Q_i$, $k \in \mathbb{N}$, with $R_i \setminus \Omega_\lambda \neq \emptyset$. Then there exists a sequence $\{\alpha_i\}_i$ of functions such that, for each i and μ-almost every $x \in \mathbb{R}^D$, $\alpha_i(x) = 0$ if $x \notin R_i$,*

$$\int_{\mathbb{R}^D} \alpha_i(x)\, d\mu(x) = \int_{Q_i} f(x)\omega_i(x)\, d\mu(x), \tag{1.4.13}$$

$$\|\alpha_i\|_{L^\infty(\mu)}\mu(R_i) \leq C \int_{Q_i} |\alpha_i(x)|\, d\mu(x) \tag{1.4.14}$$

and

$$\sum_i |\alpha_i(x)| \leq \tilde{B}\lambda \text{ for } \mu\text{-almost every } x \in \mathbb{R}^D, \tag{1.4.15}$$

where C and $\tilde{B}$ are some positive constants;

(c) *f can be written as $f := g + b$, where*

$$g := f\left[1-\sum_i \omega_i\right]+\sum_i \alpha_i, \quad b := \sum_i (f\omega_i - \alpha_i)$$

and $\|g\|_{L^\infty(\mu)} \lesssim \lambda$.

Proof. The set Ω_λ is open, because $\mathcal{M}_{(2)}$ is lower semi-continuous. Since, for μ-almost every $x \in \mathbb{R}^D$, there exists a sequence of $(2, 2^{D+1})$-doubling cubes centered at x with side length tending to zero, it follows that, for μ-almost every $x \in \mathbb{R}^D$ such that $|f(x)| > 2^{D+1}\lambda$, there exists some $(2, 2^{D+1})$-doubling cube Q centered at x with

$$\int_Q |f|\,d\mu/\mu(Q) > 2^{D+1}\lambda$$

and hence $\mathcal{M}_{(2)} f(x) > \lambda$. Therefore, for μ-almost every $x \in \mathbb{R}^D \setminus \Omega_\lambda$, we find that $|f(x)| \le 2^{D+1}\lambda$.

The existence of the function ω_i of (a) is a standard known fact. Moreover, since $R_i \setminus \Omega_\lambda \neq \emptyset$ for each i, we see that

$$\int_{R_i} |f(x)|\,d\mu(x) \le \lambda\mu(2R_i).$$

By an argument used in the proofs for (1.4.4), (1.4.5) and (1.4.6), together with this observation, we further obtain (b).

Finally, from (a), we deduce that

$$\operatorname{supp}\left(f\left(1-\sum_i w_i\right)\right) \subset \mathbb{R}^D \setminus \Omega_\lambda.$$

Observe that $\sum_i w_i \lesssim 1$. Then we have

$$\left\|f\left(1-\sum_i w_i\right)\right\|_{L^\infty(\mu)} \lesssim \lambda.$$

On the other hand, if (b) holds true, then we see that $\|\sum_i \alpha_i\| \lesssim \lambda$ and hence (c) holds true. This finishes the proof of Theorem 1.4.2. □

1.5 Notes

- The original theorem of Besicovitch deals with Euclidean balls in $\mathbb{R}^D$ by Besicovitch [5] and with more abstract sets by Morse [98]. Theorem 1.1.1 was given by M. de Guzmán [23, pp. 2–5].
- The maximal functions, $\mathcal{M}^{(\rho)}$ and $\mathcal{M}_{(\rho)}$, were introduced by Tolsa [131]. Tolsa also showed that $\mathcal{M}^{(\rho)}$ and $\mathcal{M}_{(\rho)}$ are both bounded on $L^p(\mu)$ for all $p \in (1, \infty)$ and from $L^1(\mu)$ to $L^{1,\infty}(\mu)$. When $\rho = 1$, Journé [75, p. 10] proved that $\mathcal{M}^{(1)}$ is not bounded from $L^1(\mu)$ to $L^{1,\infty}(\mu)$. Thus, the assumption that $\rho \in (1, \infty)$ plays a key role here. Sawano [112] also showed that the *non-centered maximal operator* $M_\rho(f)$, with $\rho \in (1, \infty)$, is bounded from $L^1(\mu)$ to $L^{1,\infty}(\mu)$ by establishing a new covering lemma, where, for all $f \in L^1_{\rm loc}(\mu)$ and $x \in \mathbb{R}^D$,

$$M_\rho(f)(x) := \sup_{B \ni x} \frac{1}{\mu(\rho B)} \int_B |f(y)|\, d\mu(y)$$

and the supremum is taken over all the balls B of $\mathbb{R}^D$ such that $B \ni x$.

Let $(\mathcal{X}, d, \mu)$ be a metric measure space such that μ only satisfies the *polynomial growth condition* as in (0.0.1) with $B(x, r)$ replaced by

$$B(x, r) := \{y \in \mathcal{X} : d(y, x) < r\}.$$

For all $f \in L^1_{\rm loc}(\mathcal{X}, \mu)$ and $x \in \mathcal{X}$, the *centered Hardy–Littlewood maximal operator* $\tilde{M}_\rho(f)$, with $\rho \in [2, \infty)$, is defined by setting

$$\tilde{M}_\rho(f)(x) := \sup_{r \in (0,\infty)} \frac{1}{\mu(B(x, \rho r))} \int_{B(x,r)} |f(y)|\, d\mu(y). \tag{1.5.1}$$

In [103], Nazarov et al. showed that, when $\rho = 3$ in (1.5.1), $\tilde{M}_\rho$ is bounded on $L^p(\mathcal{X}, \mu)$ for all $p \in (1, \infty]$ and from $L^1(\mathcal{X}, \mu)$ to $L^{1,\infty}(\mathcal{X}, \mu)$. Later, using an outer measure, Terasawa in [128] extended the aforementioned result in [103] to any $\rho \in [2, \infty)$. In [112], Sawano further showed that $\rho = 2$ is sharp for the boundedness of $\tilde{M}_\rho$ by giving a counterexample.
- Example 1.2.3 was given by Verdera in [145].
- Example 1.2.4 was given by Tolsa in [137]; see also [37].
- The notion of doubling cubes was introduced by Tolsa in [131].
- Theorem 1.3.2 was established by Tolsa in [131].
- Theorem 1.4.1 was established by Tolsa in [131] (see also [133]), and Theorem 1.4.2 proved by Tolsa in [135]. Another Calderón–Zygmund type decomposition was established by Mateu et al. in [94].

Chapter 2
Approximations of the Identity

In this chapter, we study approximations of the identity on $\mathbb{R}^D$ with the measures μ satisfying (0.0.1). To this end, we first introduce an important notion of coefficients $\delta(Q, R)$ for cubes Q and R in $\mathbb{R}^D$. It turns out that $\delta(Q, R)$ characterizes the geometric relationship between Q and R. Using this notion, we further study cubes of different generations in terms of $\delta(Q, R)$, which are versions of dyadic cubes in the setting $(\mathbb{R}^D, |\cdot|, \mu)$. Then we construct the functions, $\{f_{y,k}\}_{y\in \operatorname{supp}\mu,\, k\in\mathbb{Z}}$, which originate the kernels of approximations of the identity. Via these functions, we introduce and establish some important properties of the approximations of the identity on $\mathbb{R}^D$.

2.1 The Coefficient $\delta(Q, R)$

This section is devoted to the study of the coefficients $\delta(Q, R)$. To begin with, let Q, $R \subset \mathbb{R}^D$ be two cubes and Q_R the *smallest cube* concentric with Q containing Q and R. The following coefficients well describe the geometric properties of cubes in $\mathbb{R}^D$.

Definition 2.1.1. For any given two cubes Q, $R \subset \mathbb{R}^D$, the *coefficient* $\delta(Q, R)$ is defined by

$$\delta(Q, R) := \max\left\{\int_{Q_R\setminus Q} \frac{1}{|x - z_Q|^n}\, d\mu(x),\ \int_{R_Q\setminus R} \frac{1}{|x - z_R|^n}\, d\mu(x)\right\}.$$

Remark 2.1.2. When μ is the D-dimensional Lebesgue measure on $\mathbb{R}^D$, it is easy to see that, for any cubes $Q \subset R$,

$$\log_2 \frac{\ell(Q) + \ell(R)}{\ell(Q)} - 1 \lesssim \delta(Q, R) \lesssim \log_2 \frac{\ell(Q) + \ell(R)}{\ell(Q)} + 1.$$

D. Yang et al., *The Hardy Space H^1 with Non-doubling Measures and Their Applications*, Lecture Notes in Mathematics 2084, DOI 10.1007/978-3-319-00825-7_2,

We may treat points $x \in \mathbb{R}^D$ as if they were cubes (with side length $\ell(\{x\}) = 0$). Thus, for all $x, y \in \mathbb{R}^D$ and cubes Q, the *notation* $\delta(x, Q)$ and $\delta(x, y)$ make sense. From the definition of $\delta(Q, R)$, it is easy to see that

$$\begin{aligned}&\frac{1}{\sqrt{D}}[\ell(Q) + \ell(R) + \operatorname{dist}(Q, R)]\\&\leq \ell(Q_R)\\&\leq 2[\ell(Q) + \ell(R) + \operatorname{dist}(Q, R)],\end{aligned} \tag{2.1.1}$$

where above and in what follows, $\operatorname{dist}(Q, R)$ denotes the *distance between* Q *and* R, namely,

$$\operatorname{dist}(Q, R) := \inf_{x\in Q,\, y\in R} |x - y|.$$

Moreover, for any two cubes $Q \subset R$, we let

$$\delta_{Q,R} := 1 + \sum_{k=1}^{N_{Q,R}} \frac{\mu(2^k Q)}{[\ell(2^k Q)]^n}, \tag{2.1.2}$$

where $N_{Q,R}$ is the *first positive integer* k such that $\ell(2^k Q) \geq \ell(R)$. Notice that, in this case,

$$\ell(R) \leq \ell(Q_R) \leq 2\ell(R) \text{ and } R_Q = R,$$

which implies that

$$\delta(Q, R) = \int_{Q_R \setminus Q} \frac{1}{|x - z_Q|^n}\, d\mu(x).$$

An easy computation shows that

$$1 + \delta(Q, R) \sim \delta_{Q,R}. \tag{2.1.3}$$

The following useful properties of $\delta(\cdot, \cdot)$ play important roles in our context.

Lemma 2.1.3. *The following hold true:*

(a) *If there exist constants* $c \in (1, \infty)$ *and* $C \in (0, \infty)$ *such that*

$$\ell(Q)/c \leq \ell(R) \leq c\ell(Q)$$

and $\operatorname{dist}(Q, R) \leq C\ell(Q)$, *then there exists* $\tilde{C} \in (0, \infty)$, *depending on* C_0, n, D, c *and* C, *such that* $\delta(Q, R) \leq \tilde{C}$. *Moreover,* $\delta(Q, \rho Q) \leq C_0 2^n \rho^n$ *for any* $\rho \in (1, \infty)$;

(b) *Let* $\rho \in (1, \infty)$ *and* $Q \subset R$ *be concentric cubes such that there does not exist any* (ρ, β_ρ)*-doubling cube of the form* $\rho^k Q$, $k \in \mathbb{Z}_+$, *with* $Q \subset \rho^k Q \subset R$.

Then there exists a positive constant C_3, depending on C_0, n, ρ and D, such that $\delta(Q, R) \le C_3$;

(c) *If $Q \subset R$, then there exists a positive constant C, depending on C_0, n and D, such that*

$$\delta(Q, R) \le C \left[1 + \log_2 \frac{\ell(R)}{\ell(Q)}\right];$$

(d) *There exists an $\epsilon_0 \in (0, \infty)$ such that, if $P \subset Q \subset R$, then*

$$\delta(P, R) = \delta(P, Q) + \delta(Q, R) \pm \epsilon_0, \tag{2.1.4}$$

where, for $\epsilon \in (0, \infty)$ and $a, b \in \mathbb{R}$, the notation *$a = b \pm \epsilon$ means that the estimate $|a - b| \le \epsilon$. In particular, $\delta(P, Q) \le \delta(P, R)$ and*

$$\delta(Q, R) \le \delta(P, R) + \epsilon_0.$$

Moreover, if P and Q are concentric, then $\epsilon_0 = 0$;

(e) *There exists a positive constant C, depending on C_0, n and D, such that, for all cubes P, Q, $R \subset \mathbb{R}^D$,*

$$\delta(P, R) \le C + \delta(P, Q) + \delta(Q, R).$$

Proof. To prove (a), let Q and R be two cubes such that $\ell(Q) \sim \ell(R)$ and

$$\operatorname{dist}(Q, R) \le C\ell(Q).$$

Then, from (2.1.1), we deduce that $\ell(Q_R) \lesssim \ell(Q)$. By this and (0.0.1), we see that

$$\int_{Q_R \setminus Q} \frac{1}{|x - z_Q|^n} \, d\mu(x) \lesssim \frac{\mu(Q_R)}{[\ell(Q)]^n} \lesssim \frac{[\ell(Q_R)]^n}{[\ell(Q)]^n} \lesssim 1.$$

Similarly, we have

$$\int_{R_Q \setminus R} \frac{1}{|x - z_R|^n} \, d\mu(x) \lesssim 1.$$

These two estimates imply that $\delta(Q, R) \lesssim 1$.

In particular, for any $\rho \in (1, \infty)$ and cube Q, it follows, from (0.0.1), that

$$\delta(Q, \rho Q) = \int_{\rho Q \setminus Q} \frac{1}{|x - z_Q|^n} \, d\mu(x) \le \frac{\mu(\rho Q)}{[\frac{1}{2}\ell(Q)]^n} \le C_0 2^n \rho^n,$$

which completes the proof of (a).

We now show (b). Since Q and R are concentric, there exists $m \in \mathbb{N}$ such that $\rho^{m-1}Q \subset R \subset \rho^m Q$. From the assumption, it follows that, for any $k \in \{0,\ldots,m-1\}$, $\rho^k Q$ is not (ρ, β_ρ)-doubling, that is, $\mu(\rho^{k+1}Q) > \beta_\rho \mu(\rho^k Q)$. Then, by iteration, we see that, for all $k \in \{0,\ldots,m-1\}$,

$$\mu(\rho^k Q) < \beta_\rho^{-(m-k)} \mu(\rho^m Q).$$

Observe that, in this case, $Q_R = R = R_Q$. This, together with (0.0.1), implies that

$$\begin{aligned}
\delta(Q,R) &= \int_{R\setminus Q} \frac{1}{|x-z_Q|^n}\,d\mu(x) \\
&\le \int_{\rho^m Q\setminus Q} \frac{1}{|x-z_Q|^n}\,d\mu(x) \\
&= \sum_{k=1}^{m} \int_{\rho^k Q\setminus \rho^{k-1}Q} \frac{1}{|x-z_Q|^n}\,d\mu(x) \\
&\lesssim \sum_{k=1}^{m} \frac{\mu(\rho^k Q)}{[\ell(\rho^k Q)]^n} \\
&\lesssim \left[\sum_{k=1}^{m-1} \frac{\beta_\rho^{-(m-k)}}{\rho^{(k-m)n}} + 1\right] \frac{\mu(\rho^m Q)}{[\ell(\rho^m Q)]^n} \\
&\lesssim \sum_{k=1}^{\infty} \rho^{-(D+1-n)k} + 1 \\
&\lesssim 1,
\end{aligned}$$

which completes the proof of (b).

(c) By the definition of $\delta_{Q,R}$ and the polynomial growth condition (0.0.1), we see that

$$\delta_{Q,R} \le C_0 N_{Q,R} + 1.$$

Since $\ell(2^{N_{Q,R}-1}Q) < \ell(R) \le \ell(2^{N_{Q,R}}Q)$, we have

$$\log_2 \frac{\ell(R)}{\ell(Q)} \le N_{Q,R} < 1 + \log_2 \frac{\ell(R)}{\ell(Q)}.$$

This implies that

$$\delta_{Q,R} \lesssim 1 + \log_2 \frac{\ell(R)}{\ell(Q)}.$$

Then, (c) follows from this fact and (2.1.3).

(d) We first assume that P and Q are concentric. Since $P \subset Q \subset R$, in this case, we have $P_R = Q_R$ and $P_Q = Q$ and hence we write

$$\begin{aligned}\delta(P, R) &= \int_{P_R \setminus P} \frac{1}{|x - z_P|^n}\, d\mu(x) \\ &= \int_{P_Q \setminus P} \frac{1}{|x - z_P|^n}\, d\mu(x) + \int_{Q_R \setminus Q} \frac{1}{|x - z_Q|^n}\, d\mu(x) \\ &= \delta(P, Q) + \delta(Q, R).\end{aligned}$$

If P and Q are not concentric, since $Q_P = Q$, it follows that

$$\begin{aligned}\delta(P, R) &= \int_{P_Q \setminus P} \frac{1}{|x - z_P|^n}\, d\mu(x) + \int_{P_R \setminus P_Q} \frac{1}{|x - z_P|^n}\, d\mu(x) \\ &= \delta(P, Q) + \int_{P_R \setminus P_Q} \frac{1}{|x - z_P|^n}\, d\mu(x).\end{aligned}$$

Thus, to show (2.1.4), it suffices to show that

$$S := \left| \int_{P_R \setminus P_Q} \frac{1}{|x - z_P|^n}\, d\mu(x) - \delta(Q, R) \right| \lesssim 1.$$

If $\ell(Q) \geq \frac{1}{2}\ell(R)$, from the fact that $\ell(P_R) \sim \ell(R)$, we deduce that $\ell(P_R) \lesssim \ell(Q)$. This, together with (0.0.1) and (a) of this lemma, implies that

$$\mathrm{S} \lesssim \frac{\mu(P_R)}{[\ell(Q)]^n} + \delta(Q, R) \lesssim 1.$$

Assume that $\ell(Q) < \frac{1}{2}\ell(R)$. We first have the following two observations:

(i) for any sets P, Q and R, if $Q \subset P$ and $R \subset P$, then

$$P \setminus R = ((P \setminus R) \setminus Q) \bigcup (Q \setminus R);$$

(ii) for any sets P, Q and R, if $R \subset P$, then

$$\left(Q \bigcap P\right) \setminus \left(Q \bigcap R\right) = \left(Q \bigcap P\right) \setminus R.$$

Notice that $R_Q = R$. By the above observations, we further write

$$\begin{aligned}
\mathrm{S} &= \left| \int_{P_R \setminus P_Q} \frac{1}{|x - z_P|^n} \, d\mu(x) - \int_{Q_R \setminus Q} \frac{1}{|x - z_Q|^n} \, d\mu(x) \right| \\
&= \left| \left[\int_{(P_R \setminus P_Q) \setminus (Q_R \cap P_R)} \frac{1}{|x - z_P|^n} \, d\mu(x) + \int_{(Q_R \cap P_R) \setminus P_Q} \cdots \right] \right. \\
&\quad - \left[\int_{Q_R \setminus (Q_R \cap P_R)} \frac{1}{|x - z_Q|^n} \, d\mu(x) + \int_{(Q_R \cap P_R) \setminus (Q_R \cap P_Q)} \cdots \right. \\
&\quad \left. \left. + \int_{(Q_R \cap P_Q) \setminus Q} \cdots \right] \right| \\
&\le \int_{P_R \setminus (Q_R \cap P_R)} \frac{1}{|x - z_P|^n} \, d\mu(x) + \int_{Q_R \setminus (Q_R \cap P_R)} \frac{1}{|x - z_Q|^n} \, d\mu(x) \\
&\quad + \int_{(Q_R \cap P_R) \setminus (Q_R \cap P_Q)} \left| \frac{1}{|x - z_P|^n} - \frac{1}{|x - z_Q|^n} \right| d\mu(x) \\
&\quad + \int_{P_Q \setminus Q} \frac{1}{|x - z_Q|^n} \, d\mu(x) \\
&=: \mathrm{S}_1 + \mathrm{S}_2 + \mathrm{S}_3 + \mathrm{S}_4.
\end{aligned}$$

Since $P \subset Q$, we have $\ell(P_Q) \sim \ell(Q)$. For $x \in P_Q \setminus Q$, we see that $|x - z_Q| \ge \frac{1}{2}\ell(Q)$, which implies that

$$\mathrm{S}_4 \lesssim \frac{\mu(P_Q)}{[\ell(Q)]^n} \lesssim 1.$$

We claim that, for any $x \in P_R \triangle Q_R$, $|x - z_P| \gtrsim \ell(R)$ and $|x - z_Q| \gtrsim \ell(R)$, where

$$P_R \triangle Q_R := (P_R \setminus Q_R) \bigcup (Q_R \setminus P_R).$$

Form this claim, it follows that

$$\mathrm{S}_1 + \mathrm{S}_2 \lesssim \frac{\mu(P_R) + \mu(Q_R)}{[\ell(R)]^n} \lesssim 1.$$

To prove the claim, we consider the following four cases.

Case (i) $z_P \in \frac{1}{2}R$ and $z_Q \in \frac{1}{2}R$. In this case, since $R \subset P_R \cap Q_R$, we see that $x \notin R$ if $x \in P_R \triangle Q_R$. This implies that $|x - z_P| \ge \frac{1}{4}\ell(R)$ and $|x - z_P| \ge \frac{1}{4}\ell(R)$.

Case (ii) $z_P \notin \frac{1}{2}R$ and $z_Q \notin \frac{1}{2}R$. In this case, we prove that $z_P \in \frac{1}{3}Q_R$ and $z_Q \in \frac{1}{3}P_R$. Indeed, by the fact that $z_P, z_Q \notin \frac{1}{2}R$, we find that $\ell(P_R) \ge \frac{3}{2}\ell(R)$ and

$\ell(Q_R) \geq \frac{3}{2}\ell(R)$, which, together with the assumption that $\ell(Q) < \frac{1}{2}\ell(R)$, implies that $\ell(Q) \leq \frac{1}{3}\ell(Q_R)$ and $\ell(Q) \leq \frac{1}{3}\ell(P_R)$. Recall that $Q(z_P, \ell(Q))$ is the cube centered at z_P with side length $\ell(Q)$. By these facts and the fact that $P \subset Q$, we see that $z_P \in Q \subset \frac{1}{3}Q_R$ and

$$z_Q \in Q(z_P, \ell(Q)) \subset \frac{1}{3}P_R.$$

The fact that $\ell(P_R) \geq \frac{3}{2}\ell(R)$ further implies that, if $x \notin P_R$, then

$$|x - z_P| \geq \frac{3}{4}\ell(R) \text{ and } |x - z_Q| \geq \frac{1}{2}\ell(R)$$

and, if $x \notin Q_R$, then

$$|x - z_Q| \geq \frac{3}{4}\ell(R) \text{ and } |x - z_P| \geq \frac{1}{2}\ell(R).$$

Case (iii) $z_P \in \frac{1}{2}R$ and $z_Q \notin \frac{1}{2}R$. In this case, we have an analogous estimate that

$$\ell(Q_R) \geq \frac{3}{2}\ell(R) \geq 3\ell(Q).$$

Since $\ell(P_R) \geq \ell(R)$, we conclude that $\ell(P_R) \geq 2\ell(Q)$. It then follows, from these facts, that

$$z_P \in Q \subset \frac{1}{3}Q_R \text{ and } z_Q \in Q(z_P, \ell(Q)) \subset \frac{1}{2}P_R,$$

which in turn implies that, if $x \notin P_R$, then

$$|x - z_P| \geq \frac{1}{2}\ell(R) \text{ and } |x - z_Q| \geq \frac{1}{4}\ell(R)$$

and, if $x \notin Q_R$, then

$$|x - z_Q| \geq \frac{3}{4}\ell(R) \text{ and } |x - z_P| \geq \frac{1}{2}\ell(R).$$

Case (iv) $z_P \notin \frac{1}{2}R$ and $z_Q \in \frac{1}{2}R$. In this case, we find that $\ell(P_R) \geq \frac{3}{2}\ell(R)$, $\ell(Q_R) \geq \ell(R)$, $z_P \in \frac{1}{2}Q_R$ and $z_Q \in \frac{1}{3}P_R$. It then follows, from these facts, that

$$z_P \in Q \subset \frac{1}{3}Q_R \text{ and } z_Q \in Q(z_P, \ell(Q)) \subset \frac{1}{2}P_R,$$

which in turn implies that, if $x \notin P_R$, then

$$|x - z_P| \geq \frac{3}{4}\ell(R) \text{ and } |x - z_Q| \geq \frac{1}{2}\ell(R)$$

and, if $x \notin Q_R$, then

$$|x - z_Q| \geq \frac{1}{2}\ell(R) \text{ and } |x - z_P| \geq \frac{1}{4}\ell(R).$$

The claim follows immediately from the combination of the four estimates above.

Now we estimate S_3. For $x \notin P_Q$, we have

$$\left| \frac{1}{|x - z_P|^n} - \frac{1}{|x - z_Q|^n} \right| \leq n \frac{|z_P - z_Q|}{|x - \xi|^{n+1}},$$

where $\xi := \theta z_P + (1 - \theta) z_Q$ and $\theta \in (0, 1)$. The fact that $z_P, z_Q \in Q$ and the convexity of Q lead to that $\xi \in Q$. Moreover, an easy computation shows that $|x - \xi| \geq \frac{1}{2}\ell(Q)$. Then we see that

$$|x - z_Q| \leq |x - \xi| + |\xi - z_Q| \lesssim |x - \xi|.$$

This implies that

$$S_3 \lesssim \int_{\mathbb{R}^D \setminus Q} \frac{\ell(Q)}{|x - z_Q|^{n+1}} \, d\mu(x) \lesssim 1.$$

Combining the estimates for S_1 through S_4, we find that $S \lesssim 1$. Thus, we conclude that

$$\delta(P, R) = \delta(P, Q) + \delta(Q, R) \pm \epsilon_0.$$

As a consequence of this fact, we further have

$$\delta(Q, R) \leq \delta(P, R) - \delta(P, Q) + \epsilon_0 \leq \delta(P, R) + \epsilon_0.$$

On the other hand, since $P \subset Q \subset R$, it follows that

$$\delta(P, Q) = \int_{P_Q \setminus P} \frac{1}{|x - z_P|^n} \, d\mu(x) \leq \int_{P_R \setminus P} \frac{1}{|x - z_P|^n} \, d\mu(x) = \delta(P, R),$$

which completes the proof of (d).

(e) By similarity, we only prove that, for any cubes P, Q and R,

$$\int_{P_R \setminus P} \frac{1}{|x - z_P|^n} \, d\mu(x) \leq \delta(P, Q) + \delta(Q, R) + C. \tag{2.1.5}$$

Observe that P_Q and P_R are both centered at z_P. If $\ell(P_R) \leq \ell(P_Q)$, then $P_R \subset P_Q$ and

$$\int_{P_R \setminus P} \frac{1}{|x - z_P|^n} \, d\mu(x) \leq \int_{P_Q \setminus P} \frac{1}{|x - z_P|^n} \, d\mu(x) \leq \delta(P, Q),$$

which implies (2.1.5). Now we assume that $\ell(P_Q) < \ell(P_R)$ and let $M \geq 32D\sqrt{D}$ be a fixed constant. Write

$$\int_{P_R \setminus P} \frac{1}{|x - z_P|^n} \, d\mu(x) = \int_{P_Q \setminus P} \frac{1}{|x - z_P|^n} \, d\mu(x) + \int_{P_R \setminus P_Q} \cdots .$$

We further consider the following two cases.

Case (i) $\ell(P_R) \leq M\ell(P_Q)$. In this case, by (0.0.1), we see that

$$\int_{P_R \setminus P_Q} \frac{1}{|x - z_P|^n} \, d\mu(x) \lesssim \frac{\mu(P_R)}{[\ell(P_Q)]^n} \lesssim 1,$$

which in turn implies (2.1.5).

Casc (ii) $\ell(P_R) > M\ell(P_Q)$. In this case, we also have

$$\begin{aligned}
&\int_{P_R \setminus P_Q} \frac{1}{|x - z_P|^n} \, d\mu(x) - \delta(Q, R) \\
&\quad \leq \int_{P_R \setminus P_Q} \frac{1}{|x - z_P|^n} \, d\mu(x) - \int_{Q_R \setminus Q} \frac{1}{|x - z_Q|^n} \, d\mu(x) \\
&\quad = \int_{(P_R \setminus Q_R) \setminus (2\sqrt{D} P_Q)} \frac{1}{|x - z_P|^n} \, d\mu(x) + \int_{(2\sqrt{D} P_Q \setminus P_Q) \cap P_R} \cdots \\
&\qquad + \int_{(P_R \cap Q_R) \setminus (2\sqrt{D} P_Q)} \cdots - \int_{(Q_R \setminus P_R) \setminus (2\sqrt{D} P_Q)} \frac{1}{|x - z_Q|^n} \, d\mu(x) \\
&\qquad - \int_{(2\sqrt{D} P_Q \setminus P_Q) \cap Q_R} \cdots - \int_{(Q_R \cap P_R) \setminus (2\sqrt{D} P_Q)} \cdots \\
&\quad \leq \int_{(P_R \setminus Q_R) \setminus (2\sqrt{D} P_Q)} \frac{1}{|x - z_P|^n} \, d\mu(x) + \int_{2\sqrt{D} P_Q \setminus P_Q} \cdots \\
&\qquad + \int_{(P_R \cap Q_R) \setminus (2\sqrt{D} P_Q)} \left| \frac{1}{|x - z_P|^n} - \frac{1}{|x - z_Q|^n} \right| d\mu(x) \\
&\quad =: \mathrm{J}_1 + \mathrm{J}_2 + \mathrm{J}_3.
\end{aligned}$$

By (a) of this lemma, we see that $J_2 \lesssim 1$. To estimate J_3, we apply the mean value theorem to find that

$$\begin{aligned} J_3 &\lesssim \int_{(P_R \cap Q_R) \setminus (2\sqrt{D}P_Q)} \frac{\ell(P_Q)}{|x-\xi|^{n+1}}\,d\mu(x) \\ &\lesssim \int_{(P_R \cap Q_R) \setminus (2\sqrt{D}P_Q)} \frac{\ell(P_Q)}{|x-z_P|^{n+1}}\,d\mu(x) \\ &\lesssim 1, \end{aligned}$$

where $\xi := (1-\theta)z_P + \theta z_Q$, $\theta \in (0,1)$ and we used the fact that, for any $x \notin 2\sqrt{D}P_Q$,

$$|x-\xi| \ge |x-z_P| - |z_P - z_Q| \ge |x-z_P| - \frac{\sqrt{D}}{2}\ell(P_Q) \ge \frac{1}{2}|x-z_P|.$$

It remains to estimate J_1. To this end, we first claim that

$$\ell(Q_R) \ge \frac{1}{16D}\ell(P_R). \tag{2.1.6}$$

Indeed, applying $\ell(P_R) > M\ell(P_Q)$, the fact that

$$\mathrm{dist}\,(P,R) \le \mathrm{dist}\,(P,Q) + \mathrm{dist}\,(Q,R) + \sqrt{D}[\ell(P)+\ell(Q)+\ell(R)]$$

and (2.1.1), we see that

$$\begin{aligned} \ell(P_R) &\le 2\,[\ell(P)+\ell(R)+\mathrm{dist}\,(P,R)] \\ &\le 2\Big\{\ell(P)+\ell(R)+\mathrm{dist}\,(P,Q)+\mathrm{dist}\,(Q,R) \\ &\quad +\sqrt{D}[\ell(P)+\ell(Q)+\ell(R)]\Big\} \\ &\le 2\Big[(1+2\sqrt{D})\ell(P_Q)+(3\sqrt{D}+1)\ell(Q_R)\Big] \\ &\le \frac{2(1+2\sqrt{D})}{M}\ell(P_R)+2(3\sqrt{D}+1)\ell(Q_R). \end{aligned}$$

Since $M \ge 32D\sqrt{D}$, it follows that $\frac{2(1+2\sqrt{D})}{M} \le \frac{1}{2}$ and hence

$$\left[1-\frac{2(1+2\sqrt{D})}{M}\right]\ell(P_R) \le 2(3\sqrt{D}+1)\ell(Q_R) \le 8D\ell(Q_R),$$

which implies (2.1.6).

By (2.1.6), we see that, for any $x \in (P_R \setminus Q_R) \setminus (2\sqrt{D}P_Q)$,

$$|x - z_Q| \geq \ell(P_R)/(32D),$$

which, together with $M\ell(P_Q) < \ell(P_R)$ and $M > 32D\sqrt{D}$, implies that

$$|x - z_P| \geq |x - z_Q| - |z_Q - z_P| \geq |x - z_Q| - \frac{1}{64D}\ell(P_R) \geq \frac{1}{64D}\ell(P_R).$$

Consequently,

$$\mathrm{J}_1 \lesssim \frac{\mu(P_R)}{[\ell(P_R)]^n} \lesssim 1.$$

By combining the estimates of J_1, J_2 and J_3, we obtain (2.1.5) in this case, which further completes the proof of (e), and hence Lemma 2.1.3. □

The constant in (d) is denoted as ϵ_0. This is because ϵ_0 is much small compared to other constants below (such as A). On the other hand, when μ is the D-dimensional Lebesgue measure, it is not difficult to see that the conclusion of (c) can be improved by that

$$1 + \delta(Q, R) \sim 1 + \log_2 \frac{\ell(R)}{\ell(Q)}.$$

Therefore, $\delta(Q, R)$ characterizes the geometric relationship between Q and R.

We have seen that there exist many big and small doubling cubes. The following two lemmas imply the existence of these doubling cubes in terms of $\delta(\cdot, \cdot)$.

Lemma 2.1.4. *There exists some (big) positive constant γ_0, depending only on C_0, n and D, such that, if R_0 is a cube centered at some point of* $\operatorname{supp}\mu$ *and $\alpha > \gamma_0$, then, for each $x \in R_0 \cap \operatorname{supp}\mu$ such that $\delta(x, 2R_0) > \alpha$, there exists a doubling cube $Q \subset 2R_0$ centered at x satisfying that*

$$\delta(Q, 2R_0) = \alpha \pm \epsilon_1,$$

where ϵ_1 depends only on C_0, n and D (but not on α).

Proof. Take

$$\gamma_0 > \max\{C_3 + 6^n C_0, 12^n C_0\},$$

where C_0 is as in (0.0.1) and C_3 as in Lemma 2.1.3(b). Let Q_1 be the biggest cube centered at x with side length $2^{-k}\ell(R_0)$, $k \in \mathbb{N}$, such that $\delta(Q_1, 2R_0) \geq \alpha$. Indeed, let P_0 be the cube centered at x with side length $\frac{1}{2}\ell(R_0)$ and $P_k := 2^{-k}P_0$, $k \in \mathbb{N}$.

Then it follows, from this fact, together with (a) and (d) of Lemma 2.1.3, that $P_k \subset 2R_0 \subset 6P_0$ and

$$\delta(P_0, 2R_0) \le \delta(P_0, 6P_0) \le 12^n C_0 < \gamma_0 < \alpha.$$

On the other hand, since $x_{2R_0} = (P_k)_{2R_0}$, we see that

$$\begin{aligned} \delta(x, 2R_0) &= \int_{x_{2R_0} \setminus x} \frac{1}{|x-z|^n}\, d\mu(z) \\ &= \lim_{k\to\infty} \int_{x_{2R_0} \setminus P_k} \frac{1}{|x-z|^n}\, d\mu(z) \\ &= \lim_{k\to\infty} \delta(P_k, 2R_0). \end{aligned}$$

Observe that $\{\delta(P_k, 2R_0)\}_k$ is increasing in k. The existence of Q_1 then follows from this combined with the fact that $\delta(x, 2R_0) > \alpha$.

By the choice of Q_1, $\delta(2Q_1, 2R_0) < \alpha$ and $2Q_1 \subset 2R_0$. From this, together with (a) and (d) of Lemma 2.1.3, we deduce that

$$\delta(Q_1, 2R_0) = \delta(Q_1, 2Q_1) + \delta(2Q_1, 2R_0) < 4^n C_0 + \alpha,$$

which, together with $\delta(Q_1, 2R_0) \ge \alpha$, implies that

$$|\delta(Q_1, 2R_0) - \alpha| \le 4^n C_0. \tag{2.1.7}$$

Let Q be the smallest doubling cube of the form $2^i Q_1$, $i \in \mathbb{Z}_+$ (recall that such cube exists). Then Lemma 2.1.3(b) implies that $\delta(Q_1, Q) \le C_3$. We further claim that $\ell(Q) \le \ell(R_0)$. Indeed, if $\ell(Q) > \ell(R_0)$, then $2R_0 \subset 3Q$. This, together with (a) and (d) of Lemma 2.1.3, implies that

$$\delta(Q_1, 2R_0) \le \delta(Q_1, 3Q) = \delta(Q_1, Q) + \delta(Q, 3Q) \le C_3 + 6^n C_0,$$

which contradicts with the fact that $\delta(Q_1, 2R_0) \ge \alpha > \gamma_0$. Therefore, $\ell(Q) \le \ell(R_0)$ and the claim holds true. By this, we conclude that $Q \subset 2R_0$. Moreover, from Lemma 2.1.3(d) and (2.1.7), it follows that

$$\begin{aligned} |\delta(Q, 2R_0) - \alpha| &\le |\delta(Q, 2R_0) - \delta(Q_1, 2R_0)| + |\delta(Q_1, 2R_0) - \alpha| \\ &\le \delta(Q, Q_1) + 4^n C_0 \\ &\le C_3 + 4^n C_0. \end{aligned}$$

This finishes the proof of Lemma 2.1.4. □

Lemma 2.1.5. *Let $\gamma \in (0, \infty)$ and R_0 be a cube centered at some point of $\operatorname{supp} \mu$ and $\alpha > \gamma$. Then, for each $x \in R_0 \cap \operatorname{supp} \mu$ such that $\delta(R_0, \mathbb{R}^D) > \alpha$, there exists a doubling cube $S \supset R_0$ concentric with R_0 satisfying that $\ell(S) \geq 2\ell(R_0)$ and*

$$\delta(R_0, S) = \alpha \pm \epsilon_1,$$

where ϵ_1 depends only on C_0, n and D (but not on α).

Proof. Let $\gamma \in (0, \infty)$ and $\mathcal{Q}_{R_0}$ be the set of all cubes containing R_0 and sharing the same center with R_0. Since

$$0 = \delta(R_0, R_0) = \lim_{S \in \mathcal{Q}_{R_0}, \ell(S) \to \ell(R_0)} \delta(R_0, S),$$

it follows that there exists $\eta_0 \in (0, \infty)$ such that, for any $S \in \mathcal{Q}_{R_0}$ with

$$\ell(R_0) < \ell(S) < (1 + \eta_0)\ell(R_0),$$

it holds true that $\delta(R_0, S) < \gamma$. Let $\rho \in (1, 1 + \eta_0)$ and $S^\rho \in \mathcal{Q}_{R_0}$ be the smallest cube of the form $\rho^k R_0$, $k \in \mathbb{N}$, such that $\delta(R_0, S^\rho) > \alpha$. Indeed, let $R^k := \rho^k R_0$, $k \in \mathbb{N}$. Then

$$\alpha < \delta(R_0, \mathbb{R}^D) = \lim_{k \to \infty} \delta(R_0, R_k).$$

Because $\delta(R_0, R_1) < \gamma < \alpha$ and $\delta(R_0, R_k)$ is increasing in k, S^ρ exists. Moreover, by the choice of S^ρ, we find that $\delta(R_0, \frac{1}{\rho} S^\rho) \leq \alpha$. Therefore, by (d) and (a) of Lemma 2.1.3, we see that

$$\alpha < \delta(R_0, S^\rho) = \delta\left(R_0, \frac{1}{\rho} S^\rho\right) + \delta\left(\frac{1}{\rho} S^\rho, S^\rho\right) \leq \alpha + 2^n \rho^n C_0.$$

This implies that

$$|\delta(R_0, S^\rho) - \alpha| \leq 2^n \rho^n C_0.$$

Let S be the smallest doubling cube of the form $2^i S^\rho$, $i \in \mathbb{N}$. It then follows, from Lemma 2.1.3(b), that $\delta(S, S^\rho) \leq C_3$, which, together with Lemma 2.1.3(d), leads to

$$|\delta(R_0, S) - \alpha| \leq |\delta(R_0, S) - \delta(R_0, S^\rho)| + |\delta(R_0, S^\rho) - \alpha| \leq C_3 + 2^n \rho^n C_0.$$

This finishes the proof of Lemma 2.1.5. □

For convenience, in what follows, we always *assume that the constant ϵ_1 of Lemmas 2.1.4 and 2.1.5 has been chosen such that $\epsilon_1 \geq \epsilon_0$.*

2.2 Cubes of Different Generations

In this section, in terms of the coefficients $\delta(Q, R)$ in Sect. 2.1, we construct cubes of different generations. We start with the following definition.

Definition 2.2.1. A point $x \in \mathbb{R}^D$ is called a *stopping point* (or *stopping cube*) if $\delta(x, Q) < \infty$ for some cube $Q \ni x$ with $\ell(Q) \in (0, \infty)$. On the other hand, $\mathbb{R}^D$ is called an *initial cube* if $\delta(Q, \mathbb{R}^D) < \infty$ for some cube Q with $\ell(Q) \in (0, \infty)$. The cubes Q such that $\ell(Q) \in (0, \infty)$ are called *transit cubes*.

Obviously, if $\delta(x, Q) < \infty$ for some transit cube Q containing x, then $\delta(x, \tilde{Q}) < \infty$ for any transit cube $\tilde{Q}$ containing x; similarly, if $\delta(Q, \mathbb{R}^D) < \infty$ for some transit cube Q, then $\delta(\tilde{Q}, \mathbb{R}^D) < \infty$ for any transit cube $\tilde{Q}$.

Let A_0 be some *big positive constant*. In particular, we *always assume that A_0 is much bigger than the constants ϵ_0, ϵ_1 and γ_0 in Lemmas 2.1.3, 2.1.4 and 2.1.5.* In the following, for any a, $b \in (0, \infty)$, $a \ll b$ means that a is far smaller than b. Moreover, the constants A_0, ϵ_0, ϵ_1 and γ_0 depend only on C_0, n and D.

Now we are ready to introduce the definition of generations of cubes in the case that $\mathbb{R}^D$ is not an initial cube. We point that, if $\mu(\mathbb{R}^D) < \infty$, we also *regard $\mathbb{R}^D$ as a cube.*

Definition 2.2.2. Assume that $\mathbb{R}^D$ is not an initial cube. We fix some doubling cube $R_0 \subset \mathbb{R}^D$. This is the *"reference" cube*. For each $j \in \mathbb{N}$, let R_{-j} be *some doubling cube* concentric with R_0, containing R_0, such that

$$\delta(R_0, R_{-j}) = jA_0 \pm \epsilon_1$$

(which exists because of Lemma 2.1.5). If Q is a transit cube, then Q is called a *cube of generation $k \in \mathbb{Z}$*, if it is a doubling cube and, for some cube R_{-j} containing Q,

$$\delta(Q, R_{-j}) = (j + k)A_0 \pm \epsilon_1.$$

If $Q := \{x\}$ is a stopping cube, then Q is called a *cube of generation $k \in \mathbb{Z}$*, if, for some cube R_{-j} containing x,

$$\delta(Q, R_{-j}) \leq (j + k)A_0 + \epsilon_1.$$

We remark that the cubes of generations are independent of the chosen reference R_{-j} in the sense modulo some small errors. To be precise, since, for each $j \in \mathbb{N}$, $\delta(R_0, \mathbb{R}^D) > jA_0$, by Lemma 2.1.5, we know that R_{-j} exists. Moreover, for $j_1 < j_2$, by the fact $\epsilon_1 \ll A_0$, together with Lemma 2.1.3, we find that $R_{-j_1} \subset R_{-j_2}$ and

$$\delta(R_{-j_1}, R_{-j_2}) = \delta(R_0, R_{-j_2}) - \delta(R_0, R_{-j_1}) = (j_2 - j_1)A_0 \pm 2\epsilon_1. \qquad (2.2.1)$$

Suppose that Q is a transit cube of generation k, then let R_{-j_0} be the doubling cube in Definition 2.2.2 such that $Q \subset R_{-j_0}$ and

$$\delta(Q, R_{-j_0}) = (k + j_0)A_0 \pm \epsilon_1.$$

From Lemma 2.1.3(d), if $j > j_0$, then it follows that $R_{-j_0} \subset R_{-j}$ and

$$\delta(Q, R_{-j}) = \delta(Q, R_{-j_0}) + \delta(R_{-j_0}, R_{-j}) \pm \epsilon_0 = (k + j)A_0 \pm 4\epsilon_1;$$

also, if $j < j_0$, then

$$\delta(Q, R_{-j}) = \delta(Q, R_{-j_0}) - \delta(R_{-j_0}, R_{-j}) \pm \epsilon_0 = (k + j)A_0 \pm 4\epsilon_1.$$

These facts imply that

$$\delta(Q, R_{-j}) = (k + j)A_0 \pm 4\epsilon_1. \tag{2.2.2}$$

Similarly, if Q is a stopping cube of generation k, we have

$$\delta(Q, R_{-j}) \le (k + j)A_0 + 4\epsilon_1.$$

For any $x \in \operatorname{supp}\mu$ and $k \in \mathbb{Z}$, we denote by $Q_{x,k}$ a fixed doubling cube centered at x of generation k. Observe that, if $\mathbb{R}^D$ is not an initial cube, then, for any $x \in \operatorname{supp}\mu$, there are cubes of all generations $k \in \mathbb{Z}$ centered at x. Indeed, Lemma 2.1.3(c) implies that $\ell(R_{-j}) \to \infty$ as $j \to \infty$. Thus, for any $x \in \operatorname{supp}\mu$, we can choose a cube R_{-j} such that $x \in \frac{1}{2}R_{-j}$ and then apply Lemma 2.1.4 to $\frac{1}{2}R_{-j}$. More precisely, for any $k \ge -j + 1$, if $\delta(x, R_{-j}) \le (j + k)A_0$, then we see that $Q_{x,k} = \{x\}$ by Definition 2.2.2; if $\delta(x, R_{-j}) > (j + k)A_0$, then, by Lemma 2.1.4, there exists a doubling cube $Q_{x,k} \subset R_{-j}$ such that

$$\delta(Q_{x,k}, R_{-j}) = (k + j)A_0 \pm \epsilon_1.$$

Similarly, for $k = -j$, if

$$\delta(x, R_{-j-1}) \le (j + k + 1)A,$$

then we have $Q_{x,k} = \{x\}$ by Definition 2.2.2, otherwise the existence of $Q_{x,k}$ comes from Lemma 2.1.4. Going on in this way, we obtain the existence of $Q_{x,k}$ for $k < -j$.

By Definition 2.2.2, if x is not a stopping point and $\mathbb{R}^D$ not an initial cube, then all the cubes $\{Q_{x,k}\}_k$ are transit cubes. On the other hand, we claim that, if x is a stopping point, then there exists some $k_x \in \mathbb{Z}$ such that all the cubes of generations $k < k_x$ centered at x are transit cubes and all the cubes of generation $k \ge k_x$ centered at x coincide with the point x. To see the latter, we choose R_{-j} with $\frac{1}{2}R_{-j}$ containing x as before and consider the following two cases.

Case (i) $\delta(x, R_{-j}) > A_0$. In this case, there exists $k_x \geq -j + 1$ such that

$$(k_x + j)A_0 < \delta(x, R_{-j}) \leq (k_x + j + 1)A_0.$$

Thus, it follows, from Definition 2.2.2, that, for any $k \geq k_x$, $Q_{x,k} = \{x\}$. If there exists $k < k_x$ such that $Q_{x,k} = \{x\}$, then

$$(k_x + j)A_0 < \delta(x, R_{-j}) \leq (k + j)A_0 + 4\epsilon_1 < (k_x + j)A_0.$$

This contradiction shows that $Q_{x,k} \neq \{x\}$ for any $k < k_x$.

Case (ii) $\delta(x, R_{-j}) \leq A_0$. In this case, by Definition 2.2.2, we see that, for any $k \geq -j + 1$, $Q_{x,k} = \{x\}$. If $Q_{x,-j-1} = \{x\}$, then, by Lemma 2.1.3(d) and (2.2.1), we have the following contradiction that

$$A_0 + 5\epsilon_1 < \delta(R_{-j}, R_{-j-2}) \leq \delta(x, R_{-j-2}) + \epsilon_0 < A_0 + 5\epsilon_1.$$

Therefore, $Q_{x,-j-1} \neq \{x\}$, which, together with the fact that $\{Q_{x,k}\}_k$ is decreasing in k, implies that $Q_{x,k} \neq \{x\}$ for $k \leq -j-1$. Thus, $k_x = -j+1$ or $-j$. Combining the two cases, we see that the claim holds true.

Moreover, from Definition 2.2.2, we deduce that, for any $x \in \operatorname{supp}\mu$ and k, $\tilde{k} \in \mathbb{Z}$, if $Q_{x,k}$ and $Q_{x,\tilde{k}}$ are both transit cubes, then

$$\delta(Q_{x,k}, Q_{x,\tilde{k}}) = |k - \tilde{k}|A_0 \pm 5\epsilon_1. \tag{2.2.3}$$

Indeed, choose R_{-j} and $R_{-\tilde{j}}$ as in Definition 2.2.2 such that $Q_{x,k} \subset R_{-j}$ with

$$\delta(Q_{x,k}, R_{-j}) = (k + j)A_0 \pm \epsilon_1,$$

and $Q_{x,k} \subset R_{-\tilde{j}}$ with

$$\delta(Q_{x,\tilde{k}}, R_{-\tilde{j}}) = (\tilde{k} + \tilde{j})A_0 \pm \epsilon_1.$$

This, together with Lemma 2.1.3(d) and (2.2.2), implies that, if $R_{-j} \subset R_{-\tilde{j}}$, then

$$\delta(Q_{x,k}, Q_{x,\tilde{k}}) = \left|\delta(Q_{x,k}, R_{-\tilde{j}}) - \delta(Q_{x,\tilde{k}}, R_{-\tilde{j}})\right| \leq |k - \tilde{k}|A_0 + 5\epsilon_1$$

and, if $R_{-\tilde{j}} \subset R_{-j}$, then

$$\delta(Q_{x,k}, Q_{x,\tilde{k}}) = \left|\delta(Q_{x,k}, R_{-j}) - \delta(Q_{x,\tilde{k}}, R_{-j})\right| \leq |k - \tilde{k}|A_0 + 5\epsilon_1.$$

This shows (2.2.3).

As we have pointed out, for any $x \in \operatorname{supp}\mu$, $\{Q_{x,k}\}_k$ is decreasing in k. Furthermore, from the proposition below, it follows that, when $\mathbb{R}^D$ is not an initial cube, $\ell(Q_{x,k}) \to \infty$ as $k \to -\infty$.

Proposition 2.2.3. *Suppose that $\mathbb{R}^D$ is not an initial cube. Then, for any $x \in \operatorname{supp}\mu$, $\ell(Q_{x,k}) \to \infty$ as $k \to -\infty$.*

Proof. For any given $x \in \operatorname{supp}\mu$, we first assume that $\{x\}$ is not a stopping cube. Then an application of (2.2.3) implies that, for any $N \in \mathbb{N}$, $Q_{x,0}$ and $Q_{x,-N}$ are transit cubes satisfying that $Q_{x,0} \subset Q_{x,-N}$ and

$$\delta(Q_{x,0}, Q_{x,-N}) = NA_0 \pm 5\epsilon_1.$$

Since $\{\ell(Q_{x,k})\}_{k\in\mathbb{Z}}$ is decreasing, if the conclusion of Proposition 2.2.3 is not true, then there exists $M \in \mathbb{N}$ such that, for any $N \in \mathbb{N}$, $\ell(Q_{x,-N}) \le M\ell(Q_{x,0})$. From this and Lemma 2.1.3(c), we deduce that, there exists a positive constant $C_{(D)}$, depending only on D, such that

$$\delta(Q_{x,0}, Q_{x,-N}) \le C_{(D)}\left[1 + \log_2 \frac{\ell(Q_{x,-N})}{\ell(Q_{x,0})}\right] \le C_{(D)}(1 + \log_2 M).$$

On the other hand, since $\epsilon_1 \ll A_0$, then

$$\delta(Q_{x,0}, Q_{x,-N}) \ge NA_0 - 5\epsilon_1 > NA_0/2.$$

Therefore, if we take

$$N > 2C_{(D)}(1 + \log_2 M)/A_0,$$

we then have a contradiction that

$$C_{(D)}(1 + \log_2 M) < \frac{1}{2}NA_0 < \delta(Q_{x,0}, Q_{x,-N}) \le C_{(D)}(1 + \log_2 M),$$

which implies that the conclusion of Proposition 2.2.3 is true in the case that $\{x\}$ is not a stopping cube.

If $\{x\}$ is a stopping cube, then there exists some $k_x \in \mathbb{Z}$ such that all the cubes of generation $k < k_x$ are transit cubes, we conclude that, for $N \in \mathbb{N}$ large enough, $Q_{x,k_x-1} \subset Q_{x,-N}$ and

$$\delta(Q_{x,k_x-1}, Q_{x,-N}) = (N + k_x - 1)A_0 \pm 5\epsilon_1.$$

Furthermore, if there exists $M \in (0, \infty)$ such that, for any $N \in \mathbb{N}$,

$$\ell(Q_{x,-N}) \le M\ell(Q_{x,k_x-1}),$$

then, by taking

$$N > 2\max\left\{|k_x - 1|,\, C_{(D)}(1 + \log_2 M)/A_0\right\},$$

we also have a contradiction, which implies that $\ell(Q_{x,k}) \to \infty$ as $k \to -\infty$. This finishes the proof of Proposition 2.2.3. □

When $\mathbb{R}^D$ is an initial cube, we have to modify a little the definition of cubes of generation k because not all the cubes R_{-j} in Definition 2.2.2 exist.

Definition 2.2.4. Assume that $\mathbb{R}^D$ is an initial cube. Then choose $\mathbb{R}^D$ as the *"reference" cube*: if Q is a transit cube, then Q is called a *cube of generation* $k \in \mathbb{N}$, if Q is doubling and

$$\delta(Q, \mathbb{R}^D) = kA_0 \pm \epsilon_1;$$

if $Q := \{x\}$ is a stopping cube, then Q is called a *cube of generation* $k \in \mathbb{N}$, if

$$\delta(x, \mathbb{R}^D) \le kA_0 + \epsilon_1.$$

Moreover, for all $k \le 0$, then $\mathbb{R}^D$ is called a *cube of generation* k.

Similar to the case that $\mathbb{R}^D$ is not an initial cube, we claim that, when $\mathbb{R}^D$ is an initial cube, for any $x \in \operatorname{supp}\mu$, cubes of all generations centered at x exist. Indeed, we consider the following cases:

Case (i) there exists $k_0 \in \mathbb{N}$ (we may further assume that $k_0 \ge 2$) such that

$$(k_0 - 1)A_0 < \delta(x, \mathbb{R}^D) \le k_0 A_0.$$

In this case, by Definition 2.2.4, we have $Q_{x,k} = \{x\}$ for all $k \ge k_0$. For $k \in \{1, \ldots, k_0 - 1\}$, we find that

$$kA_0 < \delta(x, \mathbb{R}^D) = \lim_{r\to 0}\int_{\mathbb{R}^D\setminus Q(x,r)} \frac{1}{|z-x|^n}\, d\mu(z) = \lim_{r\to 0}\delta(Q(x,r), \mathbb{R}^D).$$

Thus, if r is small enough, we see that

$$\delta(Q(x,r), \mathbb{R}^D) > kA_0$$

for all $k \in \{1, \ldots, k_0 - 1\}$. Then the existence of doubling cube $Q_{x,k}$ such that

$$\delta(Q(x,r), \mathbb{R}^D) = kA_0 \pm \epsilon_1$$

follows from Lemma 2.1.5.

Case (ii) $\delta(x, \mathbb{R}^D) = \infty$. In this case, we have that, for any fixed $k \in \mathbb{N}$, there exists a cube Q_k centered at x such that $\delta(Q_k, \mathbb{R}^D) > kA_0$. By applying Lemma 2.1.5 again, we obtain the existence of doubling cube $Q_{x,k}$ with

$$\delta(Q_{x,k}, \mathbb{R}^D) = kA_0 \pm \epsilon_1.$$

Thus, the claim follows from the combination of the two cases above.

In what follows, for any $x \in \operatorname{supp}\mu$ and $k \in \mathbb{Z}$, we denote by $Q_{x,k}$ a *fixed doubling cube centered at x of generation k*. From Proposition 2.2.3, together with Definition 2.2.4, it follows that, for any $x \in \operatorname{supp}\mu$, $\ell(Q_{x,k}) \to \infty$ as $k \to -\infty$. On the other hand, if we choose A_0 large enough, we have $\ell(Q_{x,k+1}) \le \frac{1}{10}\ell(Q_{x,k})$, which means that $\ell(Q_{x,k}) \to 0$ as $k \to \infty$. Precisely, we have the following more precise result.

Lemma 2.2.5. *There exists some $\eta \in (0, \infty)$ such that, for any $k \in \mathbb{Z}$ and $m \in \mathbb{N}$, if $x, y \in \operatorname{supp}\mu$ are such that $2Q_{x,k} \cap 2Q_{y,k+m} \neq \emptyset$, then*

$$\ell(Q_{y,k+m}) \le 2^{-\eta m}\ell(Q_{x,k}).$$

Proof. It suffices to prove that, for any $k \in \mathbb{Z}$, if $2Q_{x,k} \cap 2Q_{y,k+1} \neq \emptyset$, then

$$\ell(Q_{y,k+1}) \le 2^{-\eta}\ell(Q_{x,k}). \tag{2.2.4}$$

Assume that $Q_{y,k+1} \neq \{y\}$ and let $B \in (1, \infty)$ be fixed later. If

$$\ell(Q_{x,k}) < B\ell(Q_{y,k+1}),$$

then the assumption that $2Q_{x,k} \cap 2Q_{y,k+1} \neq \emptyset$ implies that $2Q_{x,k} \subset 6BQ_{y,k+1}$. Let

$$R_x := Q(x, 12B\ell(Q_{y,k+1})).$$

Then $Q_{x,k} \subset R_x$ and $Q_{y,k+1} \subset R_x$. By Lemma 2.1.3(c), there exists a positive constant $\tilde{C}$ such that

$$\delta(Q_{y,k+1}, R_x) \le \tilde{C}(1 + \log_2 B).$$

By similarity, we only consider that $\mathbb{R}^D$ is not an initial cube. Then we take R_{-j} as in Definition 2.2.2 such that $R_x \subset R_{-j}$. It follows, from Lemma 2.1.3(d), that

$$\delta(Q_{y,k+1}, R_{-j}) = \delta(Q_{y,k+1}, R_x) + \delta(R_x, R_{-j}) \pm \epsilon_0.$$

Let $B := 2^{\gamma A}$, where γ is a small positive constant such that $\tilde{C}\gamma < \frac{1}{2}$. Then we have

$$\delta(R_x, R_{-j}) > (k + 1 + j)A_0 - 4\epsilon_1 - \epsilon_0 - \tilde{C}(1 + \gamma A) > (k + j)A_0 + 4\epsilon_1,$$

which, via Lemma 2.1.3(d), implies that

$$\delta(Q_{x,k}, R_{-j}) \ge \delta(R_x, R_{-j}) > (k + j)A_0 + 4\epsilon_1.$$

This contradicts to the choice of $Q_{x,k}$, which implies (2.2.4) and hence completes the proof of Lemma 2.2.5. □

Remark 2.2.6. Recall that the dyadic cubes, when μ is the D-dimensional Lebesgue measure, are defined as follows[1]: A *dyadic interval* in $\mathbb{R}$ is an interval of the form

$$\left[m2^{-k}, (m+1)2^{-k}\right),$$

where m, k are integers. A *dyadic cube* in $\mathbb{R}^D$ is a product of dyadic intervals of the same length. That is, a dyadic cube is a set of the form

$$\prod_{j=1}^{D} \left[m_j 2^{-k}, (m_j+1)2^{-k}\right)$$

for some integers $k, m_1, \ldots, m_D$.

2.3 The Functions $\varphi_{y,k}$

In this section we construct the functions, $\varphi_{y,k}$, which originate the kernels of approximations of the identity. We define

$$\sigma := 100\epsilon_0 + 100\epsilon_1 + 12^{n+1}C_0$$

and introduce two new positive constants α_1, α_2 such that

$$\epsilon_0,\ \epsilon_1,\ C_0 \ll \sigma \ll \alpha_1 \ll \alpha_2 \ll A_0.$$

Definition 2.3.1. Let $y \in \operatorname{supp} \mu$. If $Q_{y,k}$ is a transit cube, denote by $Q^1_{y,k}$, $\hat{Q}^1_{y,k}$, $Q^2_{y,k}$, $\hat{Q}^2_{y,k}$ and $Q^3_{y,k}$ some *doubling cubes centered at* y *containing* $Q_{y,k}$ *such that*

$$\delta\left(Q_{y,k}, Q^1_{y,k}\right) = \alpha_1 \pm \epsilon_1,$$

$$\delta\left(Q_{y,k}, \hat{Q}^1_{y,k}\right) = \alpha_1 + \sigma \pm \epsilon_1,$$

$$\delta\left(Q_{y,k}, Q^2_{y,k}\right) = \alpha_1 + \alpha_2 \pm \epsilon_1,$$

$$\delta\left(Q_{y,k}, \hat{Q}^2_{y,k}\right) = \alpha_1 + \alpha_2 + \sigma \pm \epsilon_1$$

[1] See, for example, [40].

and

$$\delta\left(Q_{y,k}, Q_{y,k}^3\right) = \alpha_1 + \alpha_2 + 2\sigma \pm \epsilon_1.$$

If $Q_{y,k} = \{y\}$ is a stopping cube and $Q_{y,k-1} = \{y\}$ is also a stopping cube, let

$$Q_{y,k}^1 := \hat{Q}_{y,k}^1 := Q_{y,k}^2 := \hat{Q}_{y,k}^2 := Q_{y,k}^3 := \{y\}.$$

If $Q_{y,k} = \{y\}$ is a stopping cube but $Q_{y,k-1}$ is not, then choose $Q_{y,k}^1$, $\hat{Q}_{y,k}^1$, $Q_{y,k}^2$, $\hat{Q}_{y,k}^2$ and $Q_{y,k}^3$ such that they are doubling and contained in $Q_{y,k-1}$, centered at y and

$$\delta\left(Q_{y,k}^1, Q_{y,k-1}\right) = A_0 - \alpha_1 \pm \epsilon_1,$$

$$\delta\left(\hat{Q}_{y,k}^1, Q_{y,k-1}\right) = A_0 - \alpha_1 - \sigma \pm \epsilon_1$$

$$\delta\left(Q_{y,k}^2, Q_{y,k-1}\right) = A_0 - \alpha_1 - \alpha_2 \pm \epsilon_1,$$

$$\delta\left(\hat{Q}_{y,k}^2, Q_{y,k-1}\right) = A_0 - \alpha_1 - \alpha_2 - \sigma \pm \epsilon_1$$

and

$$\delta\left(Q_{y,k}^3, Q_{y,k-1}\right) = A_0 - \alpha_1 - \alpha_2 - 2\sigma \pm \epsilon_1.$$

If any of these cubes does not exist because $\delta(y, Q_{y,k-1})$ is not big enough, then let this cube be the point $\{y\}$.

Also, if $Q_{y,k} \neq \mathbb{R}^D$, denote by $\hat{Q}_{y,k}^3$, $\hat{\hat{Q}}_{y,k}^3$, $\check{Q}_{y,k}^1$ and $\check{\check{Q}}_{y,k}^1$ some doubling cubes centered at y and contained in $Q_{y,k-1}$ satisfying

$$\delta\left(\hat{Q}_{y,k}^3, Q_{y,k-1}\right) = A_0 - \alpha_1 - \alpha_2 - 3\sigma \pm \epsilon_1,$$

$$\delta\left(\hat{\hat{Q}}_{y,k}^3, Q_{y,k-1}\right) = A_0 - \alpha_1 - \alpha_2 - 4\sigma \pm \epsilon_1,$$

$$\delta\left(\check{Q}_{y,k}^1, Q_{y,k-1}\right) = A_0 - \alpha_1 + \sigma \pm \epsilon_1$$

and

$$\delta\left(\check{\check{Q}}_{y,k}^1, Q_{y,k-1}\right) = A_0 - \alpha_1 + 2\sigma \pm \epsilon_1.$$

If any of the cubes $\hat{Q}_{y,k}^3$, $\hat{\hat{Q}}_{y,k}^3$, $\check{Q}_{y,k}^1$ and $\check{\check{Q}}_{y,k}^1$ does not exist because $\delta(y, Q_{y,k-1})$ is not big enough, then let it be the point $\{y\}$.

If $Q_{y,k} = \mathbb{R}^D$, let

$$Q^1_{y,k} := \hat{Q}^1_{y,k} := Q^2_{y,k} := \hat{Q}^2_{y,k} := Q^3_{y,k} := \hat{Q}^3_{y,k} := \hat{\hat{Q}}^3_{y,k} := \check{Q}^1_{y,k} := \check{\check{Q}}^1_{y,k} := \mathbb{R}^D.$$

Lemma 2.3.2. *Let $y \in \operatorname{supp}\mu$. If choose the constants α_1, α_2 and A_0 big enough, then*

$$Q_{y,k} \subset \check{\check{Q}}^1_{y,k} \subset \check{Q}^1_{y,k} \subset Q^1_{y,k} \subset \hat{Q}^1_{y,k} \subset Q^2_{y,k} \subset \hat{Q}^2_{y,k}$$
$$\subset Q^3_{y,k} \subset \hat{Q}^3_{y,k} \subset \hat{\hat{Q}}^3_{y,k} \subset Q_{y,k-1}.$$

Proof. By the similarity, we only prove the inclusion $\hat{Q}^1_{y,k} \subset Q^2_{y,k}$. Assume first that $Q_{y,k}$ is a transit cube. We then see that

$$\delta\left(Q_{y,k}, \hat{Q}^1_{y,k}\right) < \delta(Q_{y,k}, Q^2_{y,k}),$$

which implies that $\hat{Q}^1_{y,k} \subset Q^2_{y,k}$. If $Q_{y,k} = \{y\} = Q_{y,k-1}$ or $Q_{y,k} = \mathbb{R}^D$, then $\hat{Q}^1_{y,k} = Q^2_{y,k}$ trivially. Assume that $Q_{y,k} = \{y\} \subsetneqq Q_{y,k-1}$ now. We then find that

$$\delta\left(\hat{Q}^1_{y,k}, Q_{y,k-1}\right) > \delta(Q^2_{y,k}, Q_{y,k-1}).$$

This means that $\hat{Q}^1_{y,k} \subset Q^2_{y,k}$, which completes the proof of Lemma 2.3.2. □

For a fixed k, cubes of the k-th generation may have very different sizes for different y. Nevertheless, we still have some kind of regularity as follows.

Lemma 2.3.3. *Given $x, y \in \operatorname{supp}\mu$, let Q_x and Q_y be the cubes centered at x and y, respectively, and assume that $Q_x \cap Q_y \neq \emptyset$ and that there exists some cube R containing $Q_x \cup Q_y$ with*

$$|\delta(Q_x, R) - \delta(Q_y, R)| \le 10\epsilon_1.$$

If R_y is some cube centered at y containing Q_y with

$$\delta(Q_y, R_y) \ge \sigma - 10\epsilon_1,$$

then $Q_x \subset R_y$. As a consequence, for $k \in \mathbb{Z}$,

(a) *if $Q^1_{x,k} \cap Q^1_{y,k} \neq \emptyset$, then $Q^1_{x,k} \subset \hat{Q}^1_{y,k}$ and, in particular, $x \in \hat{Q}^1_{y,k}$;*
(b) *if $Q^2_{x,k} \cap Q^2_{y,k} \neq \emptyset$, then $Q^2_{x,k} \subset \hat{Q}^2_{y,k}$ and, in particular, $x \in \hat{Q}^2_{y,k}$;*
(c) *if $Q_{x,k} \cap Q_{y,k} \neq \emptyset$, then $Q_{x,k} \subset Q_{y,k-1}$.*

Proof. Let x, y, Q_x, Q_y, R and R_y be as in Lemma 2.3.3. We may assume that $\ell(Q_x) > \ell(Q_y)$, for otherwise the fact that $Q_x \cap Q_y \neq \emptyset$ implies that $Q_x \subset 3Q_y \subset R_y$ and Lemma 2.3.3 holds true. Furthermore, if $Q_x = \{x\}$, then $Q_y = \{x\}$ and Lemma 2.3.3 holds true trivially.

Now we assume that $Q_x \neq \{x\}$. We first prove that

$$Q_x \subset Q(y, 3\ell(Q_x)) \subset R_y. \tag{2.3.1}$$

Indeed, from Lemma 2.1.3(a), we deduce that

$$\delta(Q_y, 3Q_y) \le C_0 6^n.$$

By this and the assumption on R_y, we have $3Q_y \subset R_y$. Then we have

$$Q_x \subset Q(y, 3\ell(Q_x)) \subset 6Q_x.$$

Since $Q_x \cup Q_y \subset R$, we have $Q(y, 3\ell(Q_x)) \subset 4R$. It then follows, from the fact that

$$|\delta(Q_x, R) - \delta(Q_y, R)| \le 10\epsilon_1$$

and (d) and (a) of Lemma 2.1.3, that

$$\begin{aligned}\delta(Q_y, R) &\le \delta(Q_x, R) + 10\epsilon_1\\ &\le \delta(Q_x, 4R) + 10\epsilon_1\\ &\le \delta(Q_x, Q(y, 3\ell(Q_x))) + \delta(Q(y, 3\ell(Q_x)), 4R) + 10\epsilon_1 + \epsilon_0\end{aligned}$$

and

$$\begin{aligned}\delta(Q_y, R) &\ge \delta(Q_y, 4R) - \delta(R, 4R) - \epsilon_0\\ &\ge \delta(Q_y, Q(y, 3\ell(Q_x))) + \delta(Q(y, 3\ell(Q_x)), 4R) - 8^n C_0 - \epsilon_0.\end{aligned}$$

Using these two inequalities, together with the definition of σ, and applying Lemma 2.1.3(d), we see that

$$\begin{aligned}\delta(Q_y, Q(y, 3\ell(Q_x))) &\le \delta(Q_x, Q(y, 3\ell(Q_x))) + 12\epsilon_1 + 8^n C_0\\ &\le \delta(Q_x, 6Q_x) + 12\epsilon_1 + 8^n C_0\\ &< \sigma - 10\epsilon_1,\end{aligned}$$

which, together with the facts that $Q(y, 3\ell(Q_x))$ and R_y are both centered at y and that R_y contains Q_y with $\delta(Q_y, R_y) \ge \sigma - 10\epsilon_1$, implies that $Q(y, 3\ell(Q_x)) \subset R_y$ and hence $Q_x \subset R_y$. This implies (2.3.1).

It remains to prove (a), (b) and (c). We only show (a) by similarity. We consider the following two cases:

(i) $\ell(Q^1_{x,k}) \le \ell(Q^1_{y,k})$. In this case, if $Q^1_{y,k} = \{y\}$, then, by $Q^1_{x,k} \cap Q^1_{y,k} \neq \emptyset$, we have $Q^1_{x,k} = \{y\}$ and hence $Q^1_{x,k} \subset \hat{Q}^1_{y,k}$. If $Q^1_{y,k}$ is not a stopping cube, then neither is $\hat{Q}^1_{y,k}$, and hence $\delta(Q^1_{y,k}, \hat{Q}^1_{y,k}) \geq \sigma - 10\epsilon_1$. By $\ell(Q^1_{x,k}) \leq \ell(Q^1_{y,k})$ and $Q^1_{x,k} \cap Q^1_{y,k} \neq \emptyset$, we see that $Q^1_{x,k} \subset 3Q^1_{y,k}$. This, together with the fact that $3Q^1_{y,k} \subset \hat{Q}^1_{y,k}$, finishes the proof of (a).

(ii) $\ell(Q^1_{x,k}) > \ell(Q^1_{y,k})$. In this case, we see that $Q^1_{x,k}$ is not a stopping cube. If $Q^1_{y,k}$ is not a stopping cube, we take R_{-j} containing $Q_{y,k-1}$ and $Q_{x,k-1}$. Then, by (2.2.2) and Definition 2.3.1, we conclude that

$$|\delta(Q^1_{x,k}, R_{-j}) - \delta(Q^1_{y,k}, R_{-j})| \leq 10\epsilon_1.$$

Notice that

$$\delta(Q^1_{y,k}, \hat{Q}^1_{y,k}) \geq \sigma - 2\epsilon_1.$$

By this and (2.3.1), we then have $Q^1_{x,k} \subset \hat{Q}^1_{y,k}$. This shows (a) in the case that $Q^1_{y,k}$ is not a stopping cube.

Assume that $Q^1_{y,k} = \{y\}$. We first see that $Q_{y,k-1} \neq \{y\}$. Indeed, let R_{-j} be the cube such that $Q_{x,k-1} \subset R_{-j}$ and

$$\delta(Q_{x,k-1}, R_{-j}) = (j + k - 1)A_0 \pm \epsilon_1.$$

Observe that Lemma 2.1.3(d) still holds true, if P therein is a stopping cube. Then an application of Lemma 2.1.3(d) implies that

$$\begin{aligned}\delta(y, R_{-j}) &= \delta(y, Q^1_{x,k}) + \delta(Q^1_{x,k}, R_{-j}) \pm \epsilon_0\\ &\geq \delta(Q^1_{x,k}, R_{-j}) - \epsilon_0\\ &> (j+k)A_0 - \sigma - 10\epsilon_1\\ &> (j+k-1)A_0 + \epsilon_1.\end{aligned}$$

By Definition 2.2.2, we see that $Q_{y,k-1} \neq \{y\}$. Moreover, using Lemma 2.1.3(d) again, we have

$$\begin{aligned}\delta(y, Q_{y,k-1}) &= \delta(y, R_{-j}) - \delta(Q_{y,k-1}, R_{-j})\\ &\geq (j+k)A_0 - \sigma - 10\epsilon_1 - (j+k-1)A_0 - \epsilon_1\\ &> A_0 - \alpha_1 - \sigma.\end{aligned}$$

Therefore, by Definition 2.3.1, $\hat{Q}^1_{y,k} \neq \{y\}$.

Finally, we find that

$$\begin{aligned}\delta(y, \hat{Q}^1_{y,k}) &= \delta(y, Q_{y,k-1}) - \delta(\hat{Q}^1_{y,k}, Q_{y,k-1})\\ &\geq A_0 - \sigma - 11\epsilon_1 - (A_0 - \alpha_1 - \sigma + \epsilon_1)\\ &> \sigma - 10\epsilon_1.\end{aligned}$$

By this and the conclusion we obtained in this lemma, we complete the proof of (a) in the case that $Q^1_{y,k} = \{y\}$ and hence Lemma 2.3.3. □

Definition 2.3.4. For any $y \in \operatorname{supp}\mu$ and $k \in \mathbb{Z}$, $\psi_{y,k}$ is a *function* on $\mathbb{R}^D$ such that

(a) for all $x \in \mathbb{R}^D$,

$$0 \le \psi_{y,k}(x) \le \min\left\{\frac{4^n}{[\ell(Q^1_{y,k})]^n}, \frac{1}{|y-x|^n}\right\};$$

(b) if $x \in \hat{Q}^2_{y,k} \setminus Q^1_{y,k}$, then

$$\psi_{y,k}(x) = \frac{1}{|x-y|^n};$$

(c) $\operatorname{supp}\psi_{y,k} \subset Q^3_{y,k}$;
(d) for all $x \in \mathbb{R}^D$,

$$|\nabla_x \psi_{y,k}(x)| \le C \min\left\{\frac{1}{[\ell(Q^1_{y,k})]^{n+1}}, \frac{1}{|y-x|^{n+1}}\right\},$$

where C is a positive constant independent of x, y and k.

It is not difficult to show that such a function exists, if we choose the positive constant C in Definition 2.3.4(d) big enough. Indeed, we can take a function $\omega \in C^1(\mathbb{R}^D)$ such that

$$\chi_{[-1/2,\,1/2]^D} \le \omega \le \chi_{[-1,\,1]^D}.$$

Then there exists a positive constant $\tilde{C}$ such that, for all $x \in \mathbb{R}^D$,

$$|\nabla_x \omega(x)| \le \tilde{C}/|x|.$$

Now, for each $k \in \mathbb{Z}$, $y \in \operatorname{supp}\mu$ and all $x \in \mathbb{R}^D$, define

$$\xi_{y,k}(x) := \omega\left(\frac{2(x-y)}{\ell(Q^3_{y,k})}\right) \quad \text{and} \quad \eta_{y,k}(x) := 1 - \omega\left(\frac{2(x-y)}{\ell(Q^1_{y,k})}\right).$$

Then we have

(i) $\operatorname{supp}\xi_{y,k} \subset Q^3_{y,k}$ and $\operatorname{supp}\eta_{y,k} \subset \mathbb{R}^D \setminus (\frac{1}{2}Q^1_{y,k})$;
(ii) $\xi_{y,k}(x) = 1$ if $x \in \frac{1}{2}Q^3_{y,k}$ and $\eta_{y,k}(x) = 1$ if $x \notin Q^1_{y,k}$.

For all $x \in \mathbb{R}^D$, let

$$\psi_{y,k}(x) := \frac{\xi_{y,k}(x)\eta_{y,k}(x)}{|x-y|^n}.$$

Then we have $\operatorname{supp}\psi_{y,k} \subset Q^3_{y,k} \setminus \frac{1}{2}Q^1_{y,k}$ and (c) holds true. Moreover, since

$$(\hat{Q}^2_{y,k} \setminus Q^1_{y,k}) \subset \left(\frac{1}{2}Q^3_{y,k}\right) \setminus Q^1_{y,k},$$

then $\psi_{y,k}(x) = \frac{1}{|x-y|^n}$ when $x \in \hat{Q}^2_{y,k} \setminus Q^1_{y,k}$, which implies (b).

It remains to prove (a) and (d). To this end, we only need to assume that $x \in Q^3_{y,k} \setminus \frac{1}{2}Q^1_{y,k}$. This implies that $|y-x| \geq \ell(Q^1_{y,k})/4$ and

$$0 \leq \psi_{y,k}(x) \leq \frac{1}{|y-x|^n} \leq \frac{4^n}{[\ell(Q^1_{y,k})]^n}.$$

On the other hand, observe that

$$|\nabla_x \xi_{y,k}(x)| \lesssim \frac{1}{\ell(Q^3_{y,k})}\frac{\ell(Q^3_{y,k})}{|x-y|} \lesssim \frac{1}{|x-y|}$$

and

$$|\nabla_x \eta_{y,k}(x)| \lesssim \frac{1}{\ell(Q^1_{y,k})}\frac{\ell(Q^1_{y,k})}{|x-y|} \lesssim \frac{1}{|x-y|}.$$

It follows, from this, that

$$|\nabla_x \psi_{y,k}(x)| \leq \frac{|\nabla_x \xi_{y,k}(x)|}{|x-y|^n} + \frac{1}{|x-y|^{n+1}} + \frac{|\nabla_x \eta_{y,k}(x)|}{|x-y|^n} \lesssim \frac{1}{|x-y|^{n+1}}.$$

This, together with the support condition of $\psi_{y,k}$, further implies that

$$|\nabla_x \psi_{y,k}(x)| \lesssim \frac{1}{[\ell(Q^1_{y,k})]^{n+1}},$$

which is the desired conclusion.

In the definition of $\psi_{y,k}$, if $Q^1_{y,k} = \{y\}$, then we take $1/\ell(Q^1_{y,k}) = \infty$. If $\hat{Q}^2_{y,k} = \{y\}$ or $Q^1_{y,k} = \mathbb{R}^D$, we let $\psi_{y,k} := 0$. These choices satisfy the conditions in the definition of $\psi_{y,k}$ stated above.

Let $\alpha_2 \in (0,\infty)$ big enough. For all $y \in \operatorname{supp}\mu$, $k \in \mathbb{Z}$ and $x \in \mathbb{R}^D$, we then define $\varphi_{y,k}(x) := \alpha_2^{-1}\psi_{y,k}(x)$.

Lemma 2.3.5. *There exists some constant ϵ_2, depending on n, D, C_0, ϵ_0, ϵ_1 and σ (but not on α_1, α_2 nor A_0), such that, if $Q^1_{y,k} \neq \{y\}$ or $\mathbb{R}^D$, then*

$$\left|\|\psi_{y,k}\|_{L^1(\mu)} - \alpha_2\right| \leq \epsilon_2$$

and

$$\left|\|\psi_{y,k}\|_{L^1(\mu)} - \int_{Q^2_{y,k}\setminus\hat{Q}^1_{y,k}} \frac{1}{|y-x|^n}\,d\mu(x)\right| \leq \epsilon_2.$$

Proof. Let $\epsilon_2 := 3\sigma + 4\epsilon_1 + 4^n C_0$. By Definitions 2.3.1 and 2.3.4, we know that

$$\begin{aligned}\|\psi_{y,k}\|_{L^1(\mu)} &\leq \int_{Q^3_{y,k}\setminus Q^1_{y,k}} \frac{1}{|y-x|^n}\,d\mu(x) + 4^n \frac{\mu(Q^1_{y,k})}{[\ell(Q^1_{y,k})]^n}\\ &\leq \alpha_2 + 2\sigma + 2\epsilon_1 + 4^n C_0.\end{aligned}$$

On the other hand, we find that

$$\|\psi_{y,k}\|_{L^1(\mu)} \geq \int_{\hat{Q}^2_{y,k}\setminus Q^1_{y,k}} \frac{1}{|y-x|^n}\,d\mu(x) \geq \alpha_2 + \sigma - 2\epsilon_1.$$

Then we see that

$$\left|\|\psi_{y,k}\|_{L^1(\mu)} - \alpha_2\right| \leq \epsilon_2.$$

Moreover, another application of Definition 2.3.4 implies that

$$\begin{aligned}&\left|\|\psi_{y,k}\|_{L^1(\mu)} - \int_{Q^2_{y,k}\setminus\hat{Q}^1_{y,k}} \frac{1}{|y-x|^n}\,d\mu(x)\right|\\ &\leq \int_{Q^3_{y,k}\setminus Q^2_{y,k}} \varphi_{y,k}(x)\,d\mu(x) + \int_{\hat{Q}^1_{y,k}\setminus Q^1_{y,k}} \cdots + \int_{Q^1_{y,k}\setminus\frac{1}{2}Q^1_{y,k}} \cdots\\ &\leq \delta(Q^2_{y,k}, Q^3_{y,k}) + \delta(Q^1_{y,k}, \hat{Q}^1_{y,k}) + 4^n C_0\\ &\leq \epsilon_2.\end{aligned}$$

This finishes the proof of Lemma 2.3.5. □

A direct consequence of Lemma 2.3.5 is

$$\lim_{\alpha_2\to\infty} \frac{1}{\alpha_2} \int_{Q^2_{y,k}\setminus\hat{Q}^1_{y,k}} \frac{1}{|y-x|^n}\,d\mu(x) = 1.$$

Lemma 2.3.6. *Let $x, y \in \operatorname{supp}\mu$ and $k \in \mathbb{Z}$. For α_1 and α_2 big enough, the following statements hold true:*

(a) *If $x \in 2Q_{x_0,k}$ and $y \notin \hat{Q}^3_{x_0,k}$ for some $x_0 \in \operatorname{supp}\mu$, then $\varphi_{y,k}(x) = 0$. In particular, $\varphi_{y,k}(x) = 0$ if $y \notin \hat{Q}^3_{x,k}$;*

(b) *There exists a positive constant C, independent of x, y and k, such that, if $y \in \check{Q}^1_{x,k}$, then, for all $x \in \mathbb{R}^D$,*

$$\varphi_{y,k}(x) \leq \frac{C\alpha_2^{-1}}{[\ell(\check{Q}^1_{x,k})]^n};$$

(c) *For all $y \in \mathbb{R}^D$ and $x \in \mathbb{R}^D$, it holds true that*

$$\varphi_{y,k}(x) \leq \frac{\alpha_2^{-1}}{|y-x|^n}$$

and, if $y \in Q^2_{x,k} \setminus \hat{Q}^1_{x,k}$, then

$$\varphi_{y,k}(x) = \frac{\alpha_2^{-1}}{|y-x|^n};$$

(d) *There exists a positive constant $\tilde{C}$, independent of x, y and k, such that, if $x \in Q_{x_0,k}$, then*

$$|\nabla_x \varphi_{y,k}(x)| \leq \tilde{C}\alpha_2^{-1} \min\left\{\frac{1}{[\ell(\check{Q}^1_{x_0,k})]^{n+1}}, \frac{1}{|y-x|^{n+1}}\right\}.$$

Proof. (a) Let $x_0 \in \operatorname{supp}\mu$ and $x \in 2Q_{x_0,k}$. By the definition of $\varphi_{y,k}$ and Definition 2.3.1, we conclude that, if $\varphi_{y,k}(x) \neq 0$, then $x \in Q^3_{y,k}$. This means that

$$Q^3_{x_0,k} \cap Q^3_{y,k} \neq \emptyset,$$

which, together with Lemma 2.3.3, implies that $y \in Q^3_{y,k} \subset \hat{Q}^3_{x_0,k}$.

(b) Let $y \in \check{Q}^1_{x,k}$. From the definitions of $\psi_{y,k}$ and $\varphi_{y,k}$, it follows that, for all $x \in \mathbb{R}^D$,

$$\varphi_{y,k}(x) \lesssim \frac{\alpha_2^{-1}}{[\ell(Q^1_{y,k})]^n}.$$

Therefore, to show (b), it suffices to prove that $\ell(Q^1_{y,k}) \geq \ell(\check{Q}^1_{x,k})$. This can be seen by using the fact that $y \in \check{Q}^1_{x,k} \cap \check{Q}^1_{y,k}$ and applying Lemma 2.3.3.

(c) The first inequality follows from the definitions of $\psi_{y,k}$ and $\varphi_{y,k}$. Thus, it suffices to prove the second. We first observe that, if $y \in Q^2_{x,k} \setminus \hat{Q}^1_{x,k}$, then $x \in \hat{Q}^2_{y,k} \setminus Q^1_{y,k}$. Indeed, if $y \in Q^2_{x,k} \setminus \hat{Q}^1_{x,k}$, then $x \notin Q^1_{y,k}$; for otherwise it follows, from Lemma 2.3.3, that $y \in \hat{Q}^1_{x,k}$, which is impossible. On the other hand, another application of Lemma 2.3.3, via the fact that $y \in Q^2_{x,k}$, implies that $x \in \hat{Q}^2_{y,k}$. From these two facts, the observation follows. We further claim that $\hat{Q}^2_{y,k} \neq \{y\}$. Indeed, if $\hat{Q}^2_{y,k} = \{y\}$, then $Q^1_{y,k} = \{y\}$, which in turn implies that $\hat{Q}^2_{y,k} \setminus Q^1_{y,k} = \emptyset$ and contradicts to the fact that $x \in \hat{Q}^2_{y,k} \setminus Q^1_{y,k}$. This shows the claim. Also, we see that $\hat{Q}^2_{y,k} \neq \mathbb{R}^D$. These facts, together with the definitions of $\psi_{y,k}$ and $\varphi_{y,k}$, imply the second inequality of (c).

(d) From (a), we deduce that, if $\varphi_{y,k}(x) \neq 0$, then $y \in \hat{Q}^3_{x,k}$. Assume that $y \in \check{Q}^1_{x_0,k}$ first. In this case, by the definitions of $\psi_{y,k}$ and $\varphi_{y,k}$, we only need to show that, for all $x \in \mathbb{R}^D$,

$$|\nabla_x \varphi_{y,k}(x)| \lesssim \frac{\alpha_2^{-1}}{[\ell(\check{Q}^1_{x_0,k})]^{n+1}}.$$

To this end, it suffices to show that $\ell(Q^1_{y,k}) \geq \ell(\check{Q}^1_{x_0,k})$. Observe that

$$y \in \check{Q}^1_{y,k} \cap \check{Q}^1_{x_0,k}.$$

Then our desired conclusion that $\check{Q}^1_{x_0,k} \subset Q^1_{y,k}$ comes from another application of Lemma 2.3.3.

If $y \notin \check{Q}^1_{x_0,k}$, by the definitions of $\psi_{y,k}$ and $\varphi_{y,k}$, we see that

$$|\nabla_x \varphi_{y,k}(x)| \lesssim \frac{\alpha_2^{-1}}{|y-x|^{n+1}}.$$

Combining the two cases above then finishes the proof of Lemma 2.3.6. □

Lemma 2.3.7. *For any $\epsilon_3 \in (0,\infty)$, if α_1 and α_2 are big enough, then, for all $z_0 \in \operatorname{supp} \mu$, it holds true that*

$$\int_{\mathbb{R}^D} \varphi_{z_0,k}(x)\, d\mu(x) \leq 1 + \epsilon_3 \tag{2.3.2}$$

and

$$\int_{\mathbb{R}^D} \varphi_{x,k}(z_0)\, d\mu(x) \leq 1 + \epsilon_3. \tag{2.3.3}$$

Moreover, if $z_0 \in \operatorname{supp} \mu$ is such that there exists some transit cube Q_k of the k-th generation with $Q_k \ni z_0$, then

$$1 - \epsilon_3 \le \int_{\mathbb{R}^D} \varphi_{z_0,k}(x)\, d\mu(x) \tag{2.3.4}$$

and

$$1 - \epsilon_3 \le \int_{\mathbb{R}^D} \varphi_{x,k}(z_0)\, d\mu(x). \tag{2.3.5}$$

Proof. Let us see (2.3.4) and (2.3.5) first. Assume that there exists some transit cube Q_k of the k-th generation containing z_0. Since $z_0 \in Q_k \subset \check{Q}_k^1$, we have $\check{Q}_k^1 \subset Q_{z_0,k}^1$ by Lemma 2.3.3. In particular, $\ell(Q_{z_0,k}^1) \in (0,\infty)$. Since $z_0 \in Q_k$ and Q_k is a transit cube, we find that $Q_{z_0,k}^1 \ne \mathbb{R}^D$. Thus, (2.3.4) follows from this and Lemma 2.3.5. On the other hand, if we choose α_2 large enough such that $(\sigma + 2\epsilon_1)/\epsilon_3 \le \alpha_2$, then, by Lemmas 2.3.5 and 2.3.6(c), we know that

$$\int_{\mathbb{R}^D} \varphi_{x,k}(z_0)\, d\mu(x) \ge \int_{Q_{z_0,k}^2 \setminus \hat{Q}_{z_0,k}^1} \varphi_{x,k}(z_0)\, d\mu(x) = \alpha_2^{-1}\delta(\hat{Q}_{z_0,k}^1, Q_{z_0,k}^2) \ge 1 - \epsilon_3,$$

which implies (2.3.5).

Observe that (2.3.2) follows from the definitions of $\psi_{y,k}$ and $\varphi_{y,k}$. Thus, it remains to show (2.3.3). If $\check{Q}_{z_0,k}^1 = \mathbb{R}^D$, then $Q_{x,k}^1 = \mathbb{R}^D$ for all $x \in \check{Q}_{z_0,k}^1$ by Lemma 2.3.3. Thus, by the definition of $\varphi_{x,k}$, we further have $\varphi_{x,k} = 0$ and (2.3.3) holds true. Now we assume that $\check{Q}_{z_0,k}^1 \ne \mathbb{R}^D$ and write

$$\int_{\mathbb{R}^D} \varphi_{x,k}(z_0)\, d\mu(x) = \int_{\hat{Q}_{z_0,k}^3 \setminus \check{Q}_{z_0,k}^1} \varphi_{x,k}(z_0)\, d\mu(x) + \int_{\check{Q}_{z_0,k}^1} \varphi_{x,k}(z_0)\, d\mu(x).$$

Furthermore, by (c) of Lemma 2.3.6, we have

$$\begin{aligned} \int_{\hat{Q}_{z_0,k}^3 \setminus \check{Q}_{z_0,k}^1} \varphi_{x,k}(z_0)\, d\mu(x) &\le \int_{\hat{Q}_{z_0,k}^3 \setminus \check{Q}_{z_0,k}^1} \frac{\alpha_2^{-1}}{|x - z_0|^n}\, d\mu(x) \\ &\le \alpha_2^{-1}(\alpha_2 + 4\sigma + 2\epsilon_1). \end{aligned} \tag{2.3.6}$$

If $\check{Q}_{z_0,k}^1 \ne \{z_0\}$, then it follows, from Lemma 2.3.6(b), that

$$\int_{\check{Q}_{z_0,k}^1} \varphi_{x,k}(z_0)\, d\mu(x) \lesssim \frac{\alpha_2^{-1}\mu(\check{Q}_{z_0,k}^1)}{[\ell(\check{Q}_{z_0,k}^1)]^n} \lesssim \alpha_2^{-1}.$$

This, together with (2.3.6), implies (2.3.3). If $\check{Q}^1_{z_0,k} = \{z_0\}$, then the left hand side of the above inequality is 0 and (2.3.3) still holds true. This finishes the proof of Lemma 2.3.7. □

2.4 Approximations of the Identity

In this section we introduce a class of approximations of the identity. We assume that we have chosen $\epsilon_3 = 1/2$ in Lemma 2.3.7. Recall that

$$1/2 \le \int_{\mathbb{R}^D} \varphi_{x_0,k}(x)\, d\mu(x) \le 3/2$$

and

$$1/2 \le \int_{\mathbb{R}^D} \varphi_{x,k}(x_0)\, d\mu(x) \le 3/2,$$

if x_0 belongs to some transit cube of the k-th generation.

Definition 2.4.1. Let $f \in L^1_{\text{loc}}(\mu)$, $x \in \operatorname{supp}\mu$ and $k \in \mathbb{Z}$. If $Q_{x,k} \neq \mathbb{R}^D$, then let

$$\tilde{S}_k f(x) := \int_{\mathbb{R}^D} \varphi_{y,k}(x) f(y)\, d\mu(y) + \max\left\{0,\ \frac{1}{4} - \int_{\mathbb{R}^D} \varphi_{y,k}(x)\, d\mu(y)\right\} f(x).$$

Observe that, formally, $\tilde{S}_k$ with $k \in \mathbb{Z}$ is an integral operator with the following positive kernel: for all $x, y \in \mathbb{R}^D$

$$\tilde{S}_k(x, y) = \varphi_{y,k}(x) + \max\left\{0,\ \frac{1}{4} - \int_{\mathbb{R}^D} \varphi_{y,k}(x)\, d\mu(y)\right\} \delta_x(y),$$

where δ_x is the *Dirac delta* at x.

Now we can define the operators S_k as follows.

Definition 2.4.2. Let $k \in \mathbb{Z}$. Assume that $Q_{x,k} \neq \mathbb{R}^D$ for some $x \in \operatorname{supp}\mu$. Let M_k be the *operator of multiplication* by $m_k(x) := \frac{1}{\tilde{S}_k 1(x)}$ and W_k the *operator of multiplication* by

$$w_k(x) := \frac{1}{\tilde{S}_k^*(1/\tilde{S}_k 1)(x)} \text{ for all } x \in \mathbb{R}^D.$$

Let

$$S_k := M_k \tilde{S}_k W_k \tilde{S}_k^* M_k.$$

If $Q_{x,k} = \mathbb{R}^D$ for some $x \in \operatorname{supp}\mu$, then let $S_k := 0$.

Notice that, if $Q_{x,k}$ and $Q_{y,k}$ are transit cubes, then S_k is also an integral operator with the following positive kernel: for all $x, y \in \mathbb{R}^D$,

$$S_k(x,y) = \int_{\mathbb{R}^D} M_k(x)\tilde{S}_k(x,z)W_k(z)\tilde{S}_k(y,z)M_k(y)\,d\mu(z). \tag{2.4.1}$$

From definitions of M_k and W_k, we immediately deduce the following estimates.

Lemma 2.4.3. *Let $k \in \mathbb{Z}$. If $x \in \operatorname{supp}\mu$ is such that $Q_{x,k} \neq \mathbb{R}^D$, then*

$$2/3 \le m_k(x) \le 4 \text{ and } 0 \le w_k(x) \le 6.$$

Proof. On the one hand, for all $x \in \operatorname{supp}\mu$, we have $1/4 \le \tilde{S}_k 1(x) \le 3/2$ and hence $2/3 \le m_k(x) \le 4$. On the other hand, we also see that $\tilde{S}_k^* 1(x) \ge 1/4$ and hence

$$\tilde{S}_k^*(1/(\tilde{S}_k 1))(x) \ge \frac{2}{3}\tilde{S}_k^* 1(x) \ge 1/6,$$

which implies that $w_k(x) \le 6$ for all $x \in \operatorname{supp}\mu$ and hence completes the proof of Lemma 2.4.3. □

In the following lemma, we obtain some basic properties of $\{S_k(x,y)\}_{k\in\mathbb{Z}}$ in (2.4.1).

Theorem 2.4.4. *For each $k \in \mathbb{Z}$, the following hold true:*

(a) *Let $x, y \in \operatorname{supp}\mu$. If $Q_{x,k}$ and $Q_{y,k}$ are both transit cubes, then*

$$S_k(x,y) = S_k(y,x);$$

(b) *Let $x \in \operatorname{supp}\mu$. If $Q_{x,k} \neq \mathbb{R}^D$, then $\int_{\mathbb{R}^D} S_k(x,y)\,d\mu(y) = 1$;*

(c) *Let $x \in \operatorname{supp}\mu$. If $Q_{x,k}$ is a transit cube, then $\operatorname{supp}(S_k(x,\cdot)) \subset Q_{x,k-1}$;*

(d) *Let $x, y \in \operatorname{supp}\mu$. If $Q_{x,k}$ and $Q_{y,k}$ are transit cubes, then*

$$0 \le S_k(x,y) \le \frac{C}{(\ell(Q_{x,k}) + \ell(Q_{y,k}) + |x-y|)^n};$$

(e) *Let $x, \tilde{x}, y \in \operatorname{supp}\mu$. If $Q_{x,k}$, $Q_{\tilde{x},k}$, $Q_{y,k}$ are transit cubes, and $x, \tilde{x} \in Q_{x_0,k}$ for some $x_0 \in \operatorname{supp}\mu$, then*

$$|S_k(x,y) - S_k(\tilde{x},y)| \le C\frac{|x-\tilde{x}|}{\ell(Q_{x_0,k})}\frac{1}{(\ell(Q_{x,k}) + \ell(Q_{y,k}) + |x-y|)^n}. \tag{2.4.2}$$

Proof. (a) and (b) are obvious by (2.4.1).

(c) Let x be fixed and assume that $Q_{x,k}$ is a transit cube. From Definition 2.4.1, it follows that

$$\tilde{S}_k(x,z) = \varphi_{z,k}(x).$$

By this and (2.4.1), we see that, if $S_k(x,y) \neq 0$, then there exists some $z \in \operatorname{supp} \mu$ such that $\varphi_{z,k}(x) \neq 0$ and $\varphi_{z,k}(y) \neq 0$. This, combined with Lemma 2.3.6(a), implies that $z \in \hat{Q}^3_{x,k} \cap \hat{Q}^3_{y,k}$. Using Lemma 2.3.3, we obtain

$$y \in \hat{\hat{Q}}{}^3_{x,k} \subset Q_{x,k-1},$$

which completes the proof of (c).

(d) By (2.4.1) and Lemma 2.4.3, we see that, for all $x,\, y \in \operatorname{supp} \mu$,

$$S_k(x,y) \lesssim \int_{\mathbb{R}^D} \tilde{S}_k(x,z)\tilde{S}_k(y,z)\, d\mu(z).$$

Then (d) is reduced to showing that, for all $x,\, y \in \operatorname{supp} \mu$,

$$\int_{\mathbb{R}^D} \tilde{S}_k(x,z)\tilde{S}_k(y,z)\, d\mu(z) \lesssim \frac{1}{[\ell(Q_{x,k}) + \ell(Q_{y,k}) + |x-y|]^n}. \tag{2.4.3}$$

Since $\tilde{S}_k(x,z) = \varphi_{z,k}(x) \lesssim 1/[\ell(Q_{x,k})]^n$ for all $x,\, z \in \operatorname{supp} \mu$, it follows that, for all $x,\, y \in \operatorname{supp} \mu$,

$$S_k(x,y) \lesssim \frac{1}{[\ell(Q_{x,k})]^n} \int_{\mathbb{R}^D} \varphi_{z,k}(y)\, d\mu(z) \lesssim \frac{1}{[\ell(Q_{x,k})]^n}.$$

Similarly, it can be shown that $S_k(x,y) \lesssim 1/[\ell(Q_{y,k})]^n$ for all $x,\, y \in \operatorname{supp} \mu$. Thus, to show (2.4.3), it only remains to prove that $S_k(x,y) \lesssim 1/|x-y|^n$ for all x, $y \in \operatorname{supp} \mu$ with $x \neq y$.

Lemma 2.3.6(c) implies that $\tilde{S}_k(x,z) \lesssim 1/|x-z|^n$ for all $x,\, z \in \operatorname{supp} \mu$ and $x \neq z$ and $\tilde{S}_k(y,z) \lesssim 1/|y-z|^n$ for all $y,\, z \in \operatorname{supp} \mu$ and $y \neq z$. Observe that, if $|x-z| < |x-y|/2$, then $|y-z| > |x-y|/2$. Using these facts, we find that, for all $x,\, y \in \operatorname{supp} \mu$ and $x \neq y$,

$$\begin{aligned}
S_k(x,y) &\lesssim \int_{|x-z|\ge|x-y|/2} \tilde{S}_k(x,z)\tilde{S}_k(y,z)\, d\mu(z) + \int_{|x-z|<|x-y|/2} \cdots \\
&\lesssim \frac{1}{|x-y|^n}\left[\int_{\mathbb{R}^D} \tilde{S}_k(y,z)\, d\mu(z) + \int_{\mathbb{R}^D} \tilde{S}_k(x,z)\, d\mu(z)\right] \\
&\lesssim \frac{1}{|x-y|^n},
\end{aligned}$$

which completes the proof of (d).

(e) Using Lemma 2.4.3, we conclude that, for all $x, \tilde{x}, y \in \operatorname{supp} \mu$,

$$\begin{aligned}
&|S_k(x,y) - S_k(\tilde{x},y)| \\
&\lesssim |M_k(x) - M_k(\tilde{x})| \int_{\mathbb{R}^D} \tilde{S}_k(x,z)\tilde{S}_k(y,z)\,d\mu(z) \\
&\quad + \int_{\mathbb{R}^D} \left|\tilde{S}_k(x,z) - \tilde{S}_k(\tilde{x},z)\right| \tilde{S}_k(y,z)\,d\mu(z) \\
&=: \mathrm{E} + \mathrm{F}.
\end{aligned}$$

We fist estimate E. Since $x, \tilde{x} \in Q_{x_0,k}$, by the facts that $\tilde{S}_k 1 \sim 1$ and that

$$|\nabla \varphi_{z,k}(w)| \lesssim \frac{1}{[\ell(\check{Q}^1_{x_0,k}) + |w - z|]^{n+1}} \tag{2.4.4}$$

for all $w \in Q_{x_0,k}$ and $z \in \mathbb{R}^D$, we conclude that, for all $x, \tilde{x} \in \operatorname{supp} \mu$,

$$\begin{aligned}
|M_k(x) - M_k(\tilde{x})| &\lesssim \left|\tilde{S}_k 1(x) - \tilde{S}_k 1(\tilde{x})\right| \\
&\lesssim \left| \int_{\mathbb{R}^D} [\varphi_{z,k}(x) - \varphi_{z,k}(\tilde{x})]\,d\mu(z) \right| \\
&\lesssim \int_{\mathbb{R}^D} \frac{|x - \tilde{x}|}{[\ell(\check{Q}^1_{x_0,k}) + |x - z|]^{n+1}}\,d\mu(z) \\
&\lesssim \frac{|x - \tilde{x}|}{\ell(\check{Q}^1_{x_0,k})},
\end{aligned} \tag{2.4.5}$$

where, in the second-to-last inequality, we used the fact that

$$\ell(\check{Q}^1_{x_0,k}) + |x - z| \sim \ell(\check{Q}^1_{x_0,k}) + |w - z|$$

for $w \in Q_{x_0,k}$ and all $z \in \mathbb{R}^D$. From this and (2.4.3), it follows that, for all $x, \tilde{x}$, $y \in \operatorname{supp} \mu$,

$$\mathrm{E} \lesssim \frac{|x - \tilde{x}|}{\ell(\check{Q}^1_{x_0,k})} \frac{1}{[\ell(Q_{x,k}) + \ell(Q_{y,k}) + |x - y|]^n}.$$

Let us consider the term F now. Using (2.4.4), we see that, for all $x, \tilde{x}, y \in \operatorname{supp} \mu$,

$$\begin{aligned}
\mathrm{F} &\lesssim \int_{\mathbb{R}^D} \frac{|x - \tilde{x}|}{[\ell(\check{Q}^1_{x_0,k}) + |x - z|]^{n+1}} \tilde{S}_k(y,z)\,d\mu(z) \\
&\sim \int_{|z-y| \ge |x-y|/2} \frac{|x - \tilde{x}|}{[\ell(\check{Q}^1_{x_0,k}) + |x - z|]^{n+1}} \tilde{S}_k(y,z)\,d\mu(z) + \int_{|z-y| < |x-y|/2} \cdots \\
&=: \mathrm{F}_1 + \mathrm{F}_2.
\end{aligned}$$

From (b) and (c) of Lemma 2.3.6, we deduce that, for all $y \in \operatorname{supp}\mu$ and $z \in \mathbb{R}^D$,

$$\varphi_{z,k}(y) \lesssim 1/[\ell(\check{Q}^1_{y,k}) + |y-z|]^n,$$

which in turn implies that, for all $x, \tilde{x}, y \in \operatorname{supp}\mu$,

$$\begin{aligned}
\mathrm{F}_1 &\lesssim \frac{|x-\tilde{x}|}{[\ell(\check{Q}^1_{y,k}) + |x-y|]^n} \int_{\mathbb{R}^D} \frac{1}{[\ell(Q_{x_0,k}) + |x-z|]^{n+1}}\, d\mu(z) \\
&\lesssim \frac{|x-\tilde{x}|}{\ell(Q_{x_0,k})} \frac{1}{[\ell(\check{Q}^1_{y,k}) + |x-y|]^n}.
\end{aligned}$$

It is easy to show that, for all $x, y \in \operatorname{supp}\mu$,

$$\ell(Q_{x,k}) \le 2[\ell(\check{Q}^1_{y,k}) + |x-y|].$$

Indeed, this inequality holds true trivially if $|x-y| > \ell(Q_{x,k})/2$; while if

$$|x-y| \le \ell(Q_{x,k})/2,$$

then $y \in Q_{x,k}$, and hence, $Q_{x,k} \subset \check{Q}^1_{y,k}$ and $\ell(Q_{x,k}) \le \ell(\check{Q}^1_{y,k})$. Thus,

$$\ell(\check{Q}^1_{y,k}) + |x-y| \sim \ell(Q_{x,k}) + \ell(\check{Q}^1_{y,k}) + |x-y|$$

and the term F_1 satisfies (2.4.2).

Let us turn our attention to F_2. In this case, since $|z-y| < |x-y|/2$, we see that $|x-z| > |x-y|/2$ and, for all $x, \tilde{x}, y \in \operatorname{supp}\mu$,

$$\begin{aligned}
\mathrm{F}_2 &\lesssim \frac{|x-\tilde{x}|}{[\ell(\check{Q}^1_{x_0,k}) + |x-y|]^{n+1}} \int_{\mathbb{R}^D} \tilde{S}_k(y,z)\, d\mu(z) \\
&\lesssim \frac{|x-\tilde{x}|}{\ell(\check{Q}^1_{x_0,k})} \frac{1}{[\ell(\check{Q}^1_{x_0,k}) + |x-y|]^n}.
\end{aligned}$$

Thus, to finish the proof of (e), it suffices to show that

$$\ell(Q_{x,k}) + \ell(Q_{y,k}) \lesssim \ell(\check{Q}^1_{x_0,k}) + |x-y|.$$

Because $x \in Q_{x_0,k}$, we have $Q_{x,k} \subset \check{Q}^1_{x_0,k}$ and $\ell(Q_{x,k}) \le \ell(\check{Q}^1_{x_0,k})$. Therefore, (e) is reduced to showing that

$$\ell(Q_{y,k}) \lesssim \ell(\check{Q}^1_{x_0,k}) + |x-y|. \tag{2.4.6}$$

Indeed, if $|x_0 - y| \geq \ell(Q_{y,k})/2$, then

$$\frac{1}{2}\ell(Q_{y,k}) \leq |x_0 - y| \leq |x - x_0| + |x - y| \lesssim \ell(Q_{x_0,k}) + |x - y|.$$

If $|x_0 - y| < \ell(Q_{y,k})/2$, then $x_0 \in Q_{y,k}$ and hence $Q_{y,k} \subset \check{Q}^1_{x_0,k}$, which implies that

$$\ell(Q_{y,k}) \lesssim \ell(\check{Q}^1_{x_0,k}).$$

Thus, (2.4.6) holds true. This finishes the proof of (e) and hence Theorem 2.4.4. □

Remark 2.4.5. Taking the (formal) Definition 2.4.1 of the kernels $\tilde{S}_k(x, y)$, it is easily seen that the properties of the kernels $S_k(x, y)$ in (a), (b), (c), (d) and (e) of Theorem 2.4.4 also hold true without assuming that $Q_{x,k}$, $Q_{\tilde{x},k}$ and $Q_{y,k}$ are transit cubes. Properties (a) through (e) of Theorem 2.4.4 also hold true if any of $Q_{x,k}$, $Q_{\tilde{x},k}$ and $Q_{y,k}$ is a stopping cube, and that (a), (c) through (e) of Theorem 2.4.4 also hold true if any of $Q_{x,k}$, $Q_{\tilde{x},k}$ and $Q_{y,k}$ coincides with $\mathbb{R}^D$, except that (b) of Theorem 2.4.4 is replaced by (b)': if $Q_{x,k} = \mathbb{R}^D$ for some $x \in \operatorname{supp}\mu$, then $S_k \equiv 0$.

2.5 Notes

- The coefficient $\delta(Q, R)$ was first introduced by Tolsa [132, 134, 135] and $\delta_{Q,R}$ in [131].
- Cubes of generations were introduced by Tolsa [132, 134, 135]. Nazarov, Treil, and Volberg in [105] used dyadic martingales associated with random dyadic lattices to establish the $T(b)$ theorem.
- The functions $\psi_{x,k}$ and $\varphi_{x,k}$ were constructed by Tolsa [132, 134, 135].
- The functions S_k were introduced and Theorem 2.4.4 was established by Tolsa in [132]. In [25], it was showed that S_k for each $k \in \mathbb{Z}$ further has the following property: There exists a positive constant C such that, if $Q_{x,k}$, $Q_{\tilde{x},k}$, $Q_{y,k}$ and $Q_{\tilde{y},k}$ are transit cubes, $x,\ \tilde{x} \in Q_{x_0,k}$ and $y,\ \tilde{y} \in Q_{y_0,k}$ for some $x_0,\ y_0 \in \operatorname{supp}\mu$, then

$$|[S_k(x, y) - S_k(\tilde{x}, y)] - [S_k(x, \tilde{y}) - S_k(\tilde{x}, \tilde{y})]|$$
$$\leq C \frac{|x - \tilde{x}|}{\ell(Q_{x_0,k})} \frac{|y - \tilde{y}|}{\ell(Q_{y_0,k})} \frac{1}{[\ell(Q_{x,k}) + \ell(Q_{y,k}) + |x - y|]^n}.$$

Chapter 3
The Hardy Space $H^1(\mu)$

The main purpose of this chapter is to study the Hardy space $H^1(\mu)$. To this end, we introduce the BMO-type space RBMO (μ), establish the John–Nirenberg inequality for functions in RBMO (μ) and some equivalent characterizations of RBMO (μ). We then introduce the atomic Hardy space $H^1(\mu)$ and obtain its basic properties, including that the dual space of $H^1(\mu)$ is RBMO (μ). We also characterize $H^1(\mu)$ in terms of a class of the maximal functions.

3.1 The Space RBMO (μ)

In this section, we introduce the space RBMO (μ) in this context, establish several equivalent characterizations and the corresponding John–Nirenberg inequality for functions in RBMO (μ).

Recall that, for any cube Q, $\tilde{Q}$ denotes the smallest $(2, \beta_2)$-doubling cube which has the form $2^k Q$ with $k \in \mathbb{Z}_+$ (see Sect. 1.2).

Definition 3.1.1. Let $\eta \in (1, \infty)$. A function $f \in L^1_{\rm loc}(\mu)$ is said to be in the *space* RBMO (μ), if there exists some nonnegative constant C such that, for any cube Q,

$$\frac{1}{\mu(\eta Q)} \int_Q \left|f(y) - m_{\tilde{Q}}(f)\right| d\mu(y) \le C \tag{3.1.1}$$

and, for any two doubling cubes $Q \subset R$,

$$|m_Q(f) - m_R(f)| \le C[1 + \delta(Q, R)], \tag{3.1.2}$$

D. Yang et al., *The Hardy Space H^1 with Non-doubling Measures and Their Applications*, Lecture Notes in Mathematics 2084, DOI 10.1007/978-3-319-00825-7_3,

where above and in what follows, for any cube Q, $m_Q(f)$ denotes the *mean* of f over cube Q, that is,

$$m_Q(f) := \frac{1}{\mu(Q)} \int_Q f(x)\, d\mu(x).$$

Moreover, the RBMO (μ) *norm* of f is defined to be the minimal constant C as above and denoted by $\|f\|_{\mathrm{RBMO}\,(\mu)}$.

The space RBMO (μ) is independent of the choice of $\eta \in (1,\infty)$; see Proposition 3.1.6 below.

Remark 3.1.2. (i) When μ is the D-dimensional Lebesgue measure on $\mathbb{R}^D$, a function $f \in L^1_{\mathrm{loc}}(\mathbb{R}^D)$ is said to be in the *space* BMO($\mathbb{R}^D$), if there exists some nonnegative constant C such that

$$\sup_Q \frac{1}{|Q|} \int_Q \left|f(y) - m_Q(f)\right| dy \le C,$$

where the supremum is taken over all cubes in $\mathbb{R}^D$. Moreover, the BMO($\mathbb{R}^D$) *norm* of f is defined to be the minimal constant C as above.[1]

Observe that, when μ is the D-dimensional Lebesgue measure, for any two cubes $Q \subset R$,

$$1 + \delta(Q, R) \sim 1 + \log_2 \frac{\ell(R)}{\ell(Q)}.$$

In this case, (3.1.2) holds true automatically for functions in BMO($\mathbb{R}^D$).

(ii) It can be seen that we obtain equivalent norms of the space RBMO (μ) if we take balls instead of cubes.

For the space RBMO (μ), we have the following properties.

Proposition 3.1.3. *Let $\eta \in (1,\infty)$. The following hold true:*

(i) RBMO (μ) *is a Banach space;*

(ii) $L^\infty(\mu) \subset$ RBMO (μ). *Moreover, for all $f \in L^\infty(\mu)$,*

$$\|f\|_{\mathrm{RBMO}\,(\mu)} \le 2\|f\|_{L^\infty(\mu)};$$

(iii) *If $f \in$ RBMO (μ), then $|f| \in$ RBMO (μ) and there exists a positive constant C such that, for all $f \in$ RBMO (μ),*

$$\||f|\|_{\mathrm{RBMO}\,(\mu)} \le C\|f\|_{\mathrm{RBMO}\,(\mu)};$$

[1] See [74].

(iv) *If* $f, g \in \mathrm{RBMO}(\mu)$ *are real-valued, then* $\min\{f, g\}$, $\max\{f, g\} \in \mathrm{RBMO}(\mu)$, *and there exists a positive constant* C *such that, for all* $f, g \in \mathrm{RBMO}(\mu)$,

$$\|\min\{f, g\}\|_{\mathrm{RBMO}(\mu)} \leq C\{\|f\|_{\mathrm{RBMO}(\mu)} + \|g\|_{\mathrm{RBMO}(\mu)}\}$$

and

$$\|\max\{f, g\}\|_{\mathrm{RBMO}(\mu)} \leq C\{\|f\|_{\mathrm{RBMO}(\mu)} + \|g\|_{\mathrm{RBMO}(\mu)}\}.$$

Proof. The properties (i) and (ii) are easy to show. The property (iii) is a direct consequence of Proposition 3.1.9 below, and (iv) follows from (iii) immediately, which completes the proof of Proposition 3.1.3. □

By Definition 3.3.8, we immediately obtain the following conclusion.

Proposition 3.1.4. *Let* $f \in \mathrm{RBMO}(\mu)$. *Then there exists a positive constant* C *such that, for all doubling cubes* Q *and* R,

$$|m_Q(f) - m_R(f)| \leq C[1 + \delta(Q, R)]\|f\|_{\mathrm{RBMO}(\mu)}.$$

Proof. Assume that $\ell(R_Q) \geq \ell(Q_R)$. Then $Q_R \subset 3R_Q$. Let $\widetilde{3R_Q}$ be the smallest doubling cube of the form $2^k 3R_Q$, $k \in \mathbb{Z}_+$. By Lemma 2.1.3, we have

$$\delta(R, \widetilde{3R_Q}) = \delta(R, R_Q) + \delta(R_Q, \widetilde{3R_Q}) \lesssim 1 + \delta(R, Q).$$

This, together with (3.1.2), implies that

$$\left|m_R(f) - m_{\widetilde{3R_Q}}(f)\right| \lesssim [1 + \delta(R, Q)]\|f\|_{\mathrm{RBMO}(\mu)}. \tag{3.1.3}$$

On the other hand, it holds true that

$$\delta(Q, \widetilde{3R_Q}) \lesssim 1 + \delta(Q, 3R_Q) + \delta(3R_Q, \widetilde{3R_Q}) \lesssim 1 + \delta(Q, Q_R) + \delta(Q_R, 3R_Q).$$

Since Q_R and R_Q have comparable sizes, it follows that $\delta(Q_R, 3R_Q) \lesssim 1$ and hence

$$\delta(Q, \widetilde{3R_Q}) \lesssim 1 + \delta(Q, R).$$

Therefore,

$$\left|m_Q(f) - m_{\widetilde{3R_Q}}(f)\right| \lesssim [1 + \delta(Q, R)]\|f\|_{\mathrm{RBMO}(\mu)}. \tag{3.1.4}$$

By (3.1.3) and (3.1.4), we then obtain the desired conclusion, which completes the proof of Proposition 3.1.4. □

The space RBMO (μ) has a more generalized form as follows.

Definition 3.1.5. Let η, $\rho \in (1,\infty)$, $\gamma \in [1,\infty)$ and $\beta_\rho := \rho^{D+1}$. A function $f \in L^1_{\rm loc}(\mu)$ is said to be in the *space* $\mathrm{RBMO}^\rho_\gamma(\mu)$, if there exists a nonnegative constant $\tilde{C}$ such that, for any cube Q,

$$\frac{1}{\mu(\eta Q)}\int_Q \left|f(y)-m_{\tilde{Q}^\rho}(f)\right|\,d\mu(y)\le \tilde{C} \tag{3.1.5}$$

and, for any two (ρ,β_ρ)-doubling cubes $Q\subset R$,

$$|m_Q(f)-m_R(f)|\le \tilde{C}[1+\delta(Q,R)]^\gamma. \tag{3.1.6}$$

Moreover, the $\mathrm{RBMO}^\rho_\gamma(\mu)$ *norm* of f is defined to be the minimal constant $\tilde{C}$ as above and denoted by $\|f\|_{\mathrm{RBMO}^\rho_\gamma(\mu)}$.

By Definition 3.1.5, we see that, when $\rho=2$ and $\gamma=1$, the space $\mathrm{RBMO}^2_1(\mu)$ is just RBMO (μ). In the following, we show that, for any $\gamma\in(1,\infty)$, the spaces $\mathrm{RBMO}^\rho_\gamma(\mu)$ and RBMO (μ) coincide with equivalent norms, and RBMO (μ) is independent of the choice of $\eta\in(1,\infty)$. To this end, we now introduce another equivalent norm for the space RBMO (μ). Let $\eta\in(1,\infty)$. Suppose that, for a given $f\in L^1_{\rm loc}(\mu)$, there exist a nonnegative constant $\tilde{C}$ and a collection of numbers, $\{f_Q\}_Q$, such that

$$\sup_Q \frac{1}{\mu(\eta Q)}\int_Q |f(y)-f_Q|\,d\mu(y)\le \tilde{C} \tag{3.1.7}$$

and, for any two cubes $Q\subset R$,

$$|f_Q-f_R|\le \tilde{C}[1+\delta(Q,R)]^\gamma. \tag{3.1.8}$$

We then define the *norm* $\|f\|_* := \inf\{\tilde{C}\}$, where the infimum is taken over all the constants $\tilde{C}$ as above and all the numbers $\{f_Q\}_Q$ satisfying (3.1.7) and (3.1.8).

The definition of the norm $\|\cdot\|_*$ depends on the constants η and γ chosen in (3.1.7) and (3.1.8). However, if we write $\|\cdot\|^{(\gamma)}_{*,\eta}$ instead of $\|\cdot\|_*$, we have the following conclusion.

Proposition 3.1.6. *Let η, $\eta_1\in(1,\infty)$, $\eta_2\in(\eta_1,\infty)$ and $\gamma\in(1,\infty)$. Then*

(i) *the norm $\|\cdot\|^{(\gamma)}_{*,\eta_1}$ is equivalent to $\|\cdot\|^{(\gamma)}_{*,\eta_2}$;*
(ii) *the norm $\|\cdot\|^{(\gamma)}_{*,\eta}$ is equivalent to $\|\cdot\|^{(1)}_{*,\eta}$.*

Proof. We first prove (i). Let $\eta_2>\eta_1>1$ be fixed. Obviously, $\|f\|^{(\gamma)}_{*,\eta_2}\le\|f\|^{(\gamma)}_{*,\eta_1}$. To prove the converse, we need to show that, for a fixed collection of numbers, $\{f_Q\}_Q$, satisfying (3.1.7) and (3.1.8) with η and $\tilde{C}$ respectively replaced by η_2 and $\|f\|^{(\gamma)}_{*,\eta_2}$,

$$\sup_Q \frac{1}{\mu(\eta_1 Q)} \int_Q |f(y) - f_Q| \, d\mu(y) \lesssim \|f\|_{*,\eta_2}^{(\gamma)},$$

where the supremum is taken over all cubes Q.

For any $x \in \operatorname{supp} \mu \cap Q$, let Q_x be the cube centered at x with side length $\frac{\eta_1 - 1}{10\eta_2} \ell(Q)$. Then

$$\ell(\eta_2 Q_x) = \frac{\eta_1 - 1}{10} \ell(Q)$$

and hence $\eta_2 Q_x \subset \eta_1 Q$. By Theorem 1.1.1, there exists a sequence $\{x_i\}_i \subset (Q \cap \operatorname{supp} \mu)$ of points such that the family $\{Q_{x_i}\}_i$ of cubes covers $Q \cap \operatorname{supp} \mu$ with bounded overlap. Moreover, since for all i, $\ell(Q_{x_i}) \sim \ell(Q)$, we see that the number of $\{Q_{x_i}\}_i$ is bounded by some constant depending only on η_1, η_2 and D.

Observe that, for any i,

$$|f_{Q_{x_i}} - f_Q| \lesssim \|f\|_{*,\eta_2}^{(\gamma)}.$$

Then we have

$$\begin{aligned} \int_{Q_{x_i}} |f(x) - f_Q| \, d\mu(x) &\le \int_{Q_{x_i}} |f(x) - f_{Q_{x_i}}| \, d\mu(x) + \mu(Q_{x_i}) |f_{Q_{x_i}} - f_Q| \\ &\lesssim \mu(\eta_2 Q_{x_i}) \|f\|_{*,\eta_2}^{(\gamma)}. \end{aligned}$$

Therefore, from the facts that $\{Q_{x_i}\}_i$ are almost disjoint and $\eta_2 Q_{x_i} \subset \eta_1 Q$ for all i, it follows that

$$\begin{aligned} \int_Q |f(x) - f_Q| \, d\mu(x) &\le \sum_i \int_{Q_{x_i}} |f(x) - f_Q| \, d\mu(x) \\ &\lesssim \|f\|_{*,\eta_2}^{(\gamma)} \sum_i \mu(\eta_2 Q_{x_i}) \\ &\lesssim \mu(\eta_1 Q) \|f\|_{*,\eta_2}^{(\gamma)}, \end{aligned}$$

which completes the proof of Proposition 3.1.6(i). □

To show Proposition 3.1.6(ii), we need the following two lemmas.

Lemma 3.1.7. *Let $\zeta \in [2, \infty)$. Then, whenever $Q_1 \subset Q_2 \subset \cdots \subset Q_m$ are concentric cubes with $\delta(Q_i, Q_{i+1}) \ge \zeta$ for all $i \in \{1, \ldots, m-1\}$, it holds true that*

$$\sum_{i=1}^{m-1} [1 + \delta(Q_i, Q_{i+1})] \le 2\delta(Q_1, Q_m).$$

Proof. Let z be the common center of the cubes $\{Q_i\}_{i=1}^m$. Then, since

$$1 \le \zeta/2 \le \frac{1}{2}\delta(Q_i, Q_{i+1}) < \frac{1}{2}[1 + \delta(Q_i, Q_{i+1})],$$

by the assumption, we see that

$$\begin{aligned} 1 + \delta(Q_i, Q_{i+1}) &\le \int_{Q_{i+1}\setminus Q_i} \frac{1}{|x-z|^n} d\mu(x) + \zeta/2 \\ &\le \int_{Q_{i+1}\setminus Q_i} \frac{1}{|x-z|^n} d\mu(x) + \frac{1}{2}[1 + \delta(Q_i, Q_{i+1})] \end{aligned}$$

and finally

$$\begin{aligned} \sum_{i=1}^{m-1}[1 + \delta(Q_i, Q_{i+1})] &\le 2\sum_{i=1}^{m-1} \int_{Q_{i+1}\setminus Q_i} \frac{d\mu(x)}{|x-z|^n} \\ &= 2\int_{Q_m\setminus Q_1} \frac{d\mu(x)}{|x-z|^n} \\ &\le 2\delta(Q_1, Q_m), \end{aligned}$$

which completes the proof of Lemma 3.1.7. □

Lemma 3.1.8. *For a positive constant C_4 large enough, the following statement holds true: let $x \in \mathbb{R}^D$ be a fixed point and $\{f_Q\}_{Q \ni x}$ a collection of numbers. If, for some constant $C_{(x)}$, it holds true that*

$$|f_Q - f_R| \le C_{(x)}[1 + \delta(Q, R)] \tag{3.1.9}$$

for all cubes Q, R with $x \in Q \subset R$ and $\delta(Q, R) \le C_4$, then there exists a positive constant C, independent of x, such that

$$|f_Q - f_R| \le CC_x[1 + \delta(Q, R)] \tag{3.1.10}$$

for all cubes Q, R with $x \in Q \subset R$.

Proof. Consider some cubes $R \supset Q \ni x$, and let $Q_0 := Q$ and $\zeta \in [2, \infty)$. Assuming that a cube Q_i has been chosen, we choose Q_{i+1} to be the smallest cube of the form $2^k Q_i$, $k \in \mathbb{N}$, which satisfies $\delta(Q_i, Q_{i+1}) \ge \zeta$ and $Q_{i+1} \not\supset R$, if one exists. Since $\delta(Q_i, \frac{1}{2}Q_{i+1}) < \zeta$, we also have $\delta(Q_i, Q_{i+1}) \le C$ by (b) and (d) of Lemma 2.1.3.

We continue as long as this is possible; clearly at least the condition $Q_{i+1} \not\supset R$ is violated after finitely many steps, and the process terminates. Let Q_m be the last cube chosen by this algorithm, and let Q_{m+1} be the first cube of the form $2^k Q_m$,

$k \in \mathbb{N}$, such that $Q_{m+1} \supset R$. From $\frac{1}{2}Q_{m+1} \not\supset R$ and $\frac{1}{2}Q_{m+1} \cap R \supset Q$, it follows that $\ell(Q_{m+1}) < 4\ell(R)$; hence $\delta(R, Q_{m+1}) \leq C$ by Lemma 2.1.3 (ii). Also, since it was not possible to find any $2^k Q_m$ with both $\delta(Q_m, 2^k Q_m) \geq \zeta$ and $2^k Q_m \not\supset R$, the cube $\frac{1}{2}Q_{m+1}$ must satisfy $\delta(Q_m, \frac{1}{2}Q_{m+1}) < \zeta$ and hence $\delta(Q_m, Q_{m+1}) \leq C$ by (b) and (d) of Lemma 2.1.3.

Summarizing, we have cubes

$$Q = Q_0 \subset Q_1 \subset \cdots \subset Q_m \subset Q_{m+1},$$

where also $R \subset Q_{m+1}$, $\delta(Q_i, Q_{i+1}) \leq C$ for all $i \in \{0, \ldots, m\}$, and

$$\delta(R, Q_{m+1}) \leq C.$$

This constant qualifies for the large constant in the claim of the lemma: by the assumption,

$$|f_{Q_i} - f_{Q_{i+1}}| \leq C_{(x)}[1 + \delta(Q_i, Q_{i+1})]$$

and also

$$|f_R - f_{Q_{m+1}}| \leq C_{(x)}[1 + \delta(R, Q_{m+1})].$$

Thus,

$$\begin{aligned}
|f_Q - f_R| &\leq \sum_{i=0}^{m} |f_{Q_i} - f_{Q_{i+1}}| + |f_{Q_{m+1}} - f_R| \\
&\leq \sum_{i=0}^{m} C_{(x)}[1 + \delta(Q_i, Q_{i+1})] + C_{(x)}[1 + \delta(R, Q_{m+1})] \\
&\lesssim C_{(x)}[1 + \delta(Q_0, Q_m)],
\end{aligned}$$

where we used Lemma 3.1.7 to estimate the sum over $i \in \{0, \ldots, m-1\}$, here $\delta(Q_i, Q_{i+1}) \geq \zeta$ as required, as well as the bounds $\delta(Q_m, Q_{m+1}), \delta(R, Q_{m+1}) \leq C$. Finally, using (a) and (d) of Lemma 2.1.3 and recalling that $Q_0 = Q$, we have

$$\delta(Q_0, Q_m) \leq \delta(Q_0, Q_{m+1}) \lesssim \delta(Q_0, R) + \delta(R, Q_{m+1}) \lesssim 1 + \delta(Q, R).$$

This finishes the proof of Lemma 3.1.8. □

Proof of Proposition 3.1.6 (ii). By (i) of this proposition, we only need to show (ii) with $\eta = 2$ in (3.1.7). Obviously, $\|f\|_{*,2}^{(\gamma)} \leq \|f\|_{*,2}^{(1)}$.

To see the converse, assume that (3.1.7) and (3.1.8) hold true with $\eta = 2$ and $\tilde{C}$ replaced by $\|f\|^{(\gamma)}_{*,2}$. To show that f satisfies (3.1.8) with $\gamma = 1$, let $x \in \mathbb{R}^D$ and $Q \subset R$ with $x \in Q$ such that $\delta(Q, R) \le C_4$, where C_4 is as in the statement of Lemma 3.1.8. Then we see that

$$|f_Q - f_R| \le [1 + \delta(Q, R)]^\gamma \|f\|^{(\gamma)}_{*,2} \le [1 + \delta(Q, R)](1 + C_4)^{\gamma-1} \|f\|^{(\gamma)}_{*,2},$$

which is like the assumption of Lemma 3.1.8 with $C_{(x)} := (1 + C_4)^{\gamma-1}$. By Lemma 3.1.8, we find that, for all balls $Q \subset R$ with $x \in Q$,

$$|f_Q - f_R| \lesssim [1 + \delta(Q, R)](1 + C_4)^{\gamma-1} \|f\|^{(\gamma)}_{*,2}.$$

This, together with (2.1.3), implies that $\|f\|^{(1)}_{*,2} \lesssim \|f\|^{(\gamma)}_{*,2}$ and hence finishes the proof of Proposition 3.1.6(ii) and hence Proposition 3.1.6. □

By Proposition 3.1.6, we see that, for all $\eta \in (1, \infty)$ and $\gamma \in [1, \infty)$, the norms $\|\cdot\|_*$ are equivalent. With the aid of this fact, we now show that

$$\mathrm{RBMO}^\rho_\gamma(\mu) = \mathrm{RBMO}(\mu).$$

Proposition 3.1.9. *Let η, $\rho \in (1, \infty)$ and $\gamma \in [1, \infty)$. Then the following conclusions hold true:*

(i) *The norms $\|\cdot\|_*$ and $\|\cdot\|_{\mathrm{RBMO}^\rho_\gamma(\mu)}$ are equivalent;*
(ii) *Let $\gamma \in (1, \infty)$. Then the spaces $\mathrm{RBMO}^\rho_\gamma(\mu)$ and $\mathrm{RBMO}(\mu)$ coincide with equivalent norms.*

Proof. Observe that (ii) follows from (i) and Proposition 3.1.6. Therefore, we only need to prove (i). To this end, suppose that $f \in L^1_{\mathrm{loc}}(\mu)$. We first show that

$$\|f\|_* \lesssim \|f\|_{\mathrm{RBMO}^\rho_\gamma(\mu)}. \tag{3.1.11}$$

For any cube Q, let $f_Q := m_{\tilde{Q}^\rho}(f)$. By Definition 3.1.5, we have

$$\frac{1}{\mu(\eta Q)} \int_Q |f(y) - f_Q|\, d\mu(y) \le \|f\|_{\mathrm{RBMO}^\rho_\gamma(\mu)}.$$

Therefore, (3.1.11) is reduced to showing that, for any two cubes $Q \subset R$,

$$|f_Q - f_R| \lesssim [1 + \delta(Q, R)]^\gamma \|f\|_{\mathrm{RBMO}^\rho_\gamma(\mu)}. \tag{3.1.12}$$

To show (3.1.12), we consider two cases.

Case (i) $\ell(\tilde{R}^\rho) \ge \ell(\tilde{Q}^\rho)$. In this case, $\tilde{Q}^\rho \subset 2\tilde{R}^\rho$. Let $R_0 := \widetilde{2\tilde{R}^\rho}^\rho$. It follows, from (a), (b) and (d) of Lemma 2.1.3, that $\delta(\tilde{R}^\rho, R_0) \lesssim 1$ and

$$\delta(\tilde{Q}^{\rho}, R_0) \lesssim 1 + \delta(Q, R).$$

Thus,

$$\begin{aligned}\left|m_{\tilde{Q}^{\rho}}(f) - m_{\tilde{R}^{\rho}}(f)\right| &\le \left|m_{\tilde{Q}^{\rho}}(f) - m_{R_0}(f)\right| + \left|m_{R_0}(f) - m_{\tilde{R}^{\rho}}(f)\right| \\ &\lesssim [1 + \delta(Q, R)]^{\gamma} \|f\|_{\mathrm{RBMO}_{\gamma}^{\rho}(\mu)}.\end{aligned}$$

Case (ii) $\ell(\tilde{R}^{\rho}) < \ell(\tilde{Q}^{\rho})$. In this case, $\tilde{R}^{\rho} \subset 2\rho\tilde{Q}^{\rho}$. Notice that $\ell(\tilde{R}^{\rho}) \ge \ell(Q)$. Thus, there exists a unique $m \in \mathbb{N}$ such that

$$\ell(\rho^{m-1}Q) \le \ell(\tilde{R}^{\rho}) < \ell(\rho^m Q).$$

Therefore,

$$\rho^m Q \subset \widetilde{\rho 2\rho\tilde{Q}^{\rho}}^{\rho} =: Q_0.$$

Then another application of Lemma 2.1.3 implies that $\delta(\tilde{Q}^{\rho}, Q_0) \lesssim 1$ and

$$\delta(\tilde{R}^{\rho}, Q_0) \lesssim 1 + \delta(\tilde{R}^{\rho}, \rho^m Q) + \delta(\rho^m Q, Q_0) \lesssim 1.$$

Thus, an argument similar to Case (i) also establishes (3.1.12) in this case. Therefore, (3.1.12) always holds true.

Now let us establish the converse of (3.1.11). For $f \in L^1_{\mathrm{loc}}(\mu)$, assume that there exists a sequence of numbers, $\{f_Q\}_Q$, satisfying (3.1.7) and (3.1.8) with $\tilde{C}$ replaced by $\|f\|_*$. Since, by Proposition 3.1.6, (3.1.7) holds true with $\rho = \eta$, we find that, if Q is (ρ, β_ρ)-doubling, then

$$\begin{aligned}|f_Q - m_Q(f)| &= \left|\frac{1}{\mu(Q)}\int_Q [f(x) - f_Q]\,d\mu(x)\right| \\ &\le \frac{\mu(\eta Q)}{\mu(Q)}\|f\|_* \\ &\lesssim \|f\|_*. \qquad (3.1.13)\end{aligned}$$

Therefore, for any cube Q, (3.1.8) and (3.1.13) imply that

$$\left|f_Q - m_{\tilde{Q}^{\rho}}(f)\right| \le \left|f_Q - f_{\tilde{Q}^{\rho}}\right| + \left|f_{\tilde{Q}^{\rho}} - m_{\tilde{Q}^{\rho}}(f)\right| \lesssim \|f\|_*.$$

From these estimates and (3.1.7), we deduce that, for any cube Q,

$$\begin{aligned}\int_Q \left|f(x) - m_{\tilde{Q}^{\rho}}(f)\right| d\mu(x) &\le \int_Q |f(x) - f_Q|\,d\mu(x) + \left|f_Q - m_{\tilde{Q}^{\rho}}(f)\right| \mu(Q) \\ &\lesssim \|f\|_* \mu(\eta Q).\end{aligned}$$

Finally, for any two (ρ, β_ρ)-doubling cubes $Q \subset R$, by (3.1.13), together with (3.1.8), we see that

$$|m_Q(f) - m_R(f)| \le |m_Q(f) - f_Q| + |f_Q - f_R| + |f_R - m_R(f)|$$
$$\lesssim [1 + \delta(Q, R)]^\gamma \|f\|_*.$$

Thus, $f \in \mathrm{RBMO}_\gamma^\rho(\mu)$ and

$$\|f\|_{\mathrm{RBMO}_\gamma^\rho(\mu)} \lesssim \|f\|_*,$$

which completes the proof of Proposition 3.1.9(i) and hence Proposition 3.1.9. □

Proposition 3.1.10. *Let $\eta, \rho \in (1, \infty)$. For any $f \in L^1_{\mathrm{loc}}(\mu)$, the following are equivalent:*

(i) $f \in \mathrm{RBMO}(\mu)$;
(ii) *There exists a nonnegative constant C_b such that, for any cube Q,*

$$\int_Q |f(x) - m_Q(f)|\, d\mu(x) \le C_b \mu(\eta Q)$$

and, for any cubes $Q \subset R$,

$$|m_Q(f) - m_R(f)| \le C_b[1 + \delta(Q, R)] \left[\frac{\mu(\eta Q)}{\mu(Q)} + \frac{\mu(\eta R)}{\mu(R)}\right]; \qquad (3.1.14)$$

(iii) *There exists a nonnegative constant C_c such that, for any (ρ, β_ρ)-doubling cube Q,*

$$\int_Q |f(x) - m_Q(f)|\, d\mu(x) \le C_c \mu(Q) \qquad (3.1.15)$$

and, for any (ρ, β_ρ)-doubling cubes $Q \subset R$,

$$|m_Q(f) - m_R(f)| \le C_c[1 + \delta(Q, R)].$$

Moreover, the minimal constants C_b and C_c are equivalent to $\|f\|_{\mathrm{RBMO}(\mu)}$.

Proof. By Propositions 3.1.6 and 3.1.9, it suffices to establish Proposition 3.1.10 with $\eta = \rho = 2$. Assuming that $f \in \mathrm{RBMO}(\mu)$, we now show that (ii) holds true. For any cube Q,

$$\left|m_Q(f) - m_{\tilde{Q}}(f)\right| \le m_Q\left(\left|f - m_{\tilde{Q}}(f)\right|\right) \le \frac{\mu(2Q)}{\mu(Q)} \|f\|_{\mathrm{RBMO}(\mu)}, \qquad (3.1.16)$$

which implies that

$$\begin{aligned}\int_Q |f(x) - m_Q(f)|\,d\mu(x) \\ &\le \int_Q \left|f(x) - m_{\tilde{Q}}(f)\right|\,d\mu(x) + \left|m_Q(f) - m_{\tilde{Q}}(f)\right|\mu(Q) \\ &\le 2\|f\|_{\mathrm{RBMO}\,(\mu)}\mu(2Q).\end{aligned}$$

Moreover, by (3.1.16) and (3.1.12), we see that, for any $Q \subset R$,

$$\begin{aligned}|m_Q(f) - m_R(f)| \\ &\le \left|m_Q(f) - m_{\tilde{Q}}(f)\right| + \left|m_{\tilde{Q}}(f) - m_{\tilde{R}}(f)\right| + \left|m_{\tilde{R}}(f) - m_R(f)\right| \\ &\lesssim [1 + \delta(Q, R)]\left[\frac{\mu(2Q)}{\mu(Q)} + \frac{\mu(2R)}{\mu(R)}\right]\|f\|_{\mathrm{RBMO}\,(\mu)}.\end{aligned}$$

This shows (3.1.14) and hence (ii).

Since (ii) obviously implies (iii), to finish the proof of Proposition 3.1.10, we only need to prove that, if $f \in L^1_{\mathrm{loc}}(\mu)$ satisfies the assumptions in (iii), then $f \in \mathrm{RBMO}\,(\mu)$.

For any cube Q and $x \in Q \cap \operatorname{supp}\mu$, let Q_x be the *biggest doubling cube* with side length $2^{-k}\ell(Q)$ for some $k \in \mathbb{N}$. Then, by Theorem 1.1.1, there exists a sequence $\{Q_{x_i}\}_i$ of cubes with bounded overlap. Observe that, for each i, $\delta(Q_{x_i}, \tilde{Q}) \lesssim 1$. We then know that

$$\left|m_{Q_{x_i}}(f) - m_{\tilde{Q}}(f)\right| \le \left|m_{Q_{x_i}}(f) - m_{\widetilde{2\tilde{Q}}}(f)\right| + \left|m_{\tilde{Q}}(f) - m_{\widetilde{2\tilde{Q}}}(f)\right| \lesssim C_c.$$

Then, from (3.1.15), the facts that, for each i, $Q_{x_i} \subset 2Q$ and $\{Q_{x_i}\}_i$ are almost disjoint, it follows that

$$\begin{aligned}\int_Q \left|f(x) - m_{\tilde{Q}}(f)\right|\,d\mu(x) \\ &\le \sum_i \int_{Q_{x_i}} |f(x) - m_{Q_{x_i}}(f)|\,d\mu(x) + \sum_i \left|m_{Q_{x_i}}(f) - m_{\tilde{Q}}(f)\right|\mu(Q_{x_i}) \\ &\lesssim C_c\mu(2Q).\end{aligned}$$

Thus, $f \in \mathrm{RBMO}\,(\mu)$, which completes the proof of Proposition 3.1.10. □

We have another characterization for RBMO (μ) which is useful in applications. To be precise, let $f \in L^1_{\mathrm{loc}}(\mu)$. If f is real-valued and, for any cube Q, let $m_f(Q)$ be the *real number* such that $\inf_{\alpha\in\mathbb{R}} m_Q(|f - \alpha|)$ is attained when $\mu(Q) \neq 0$, and $m_f(Q) := 0$ when $\mu(Q) = 0$, then $m_f(Q)$ satisfies that

$$\mu(\{x \in Q : f(x) > m_f(Q)\}) \le \mu(Q)/2$$

and

$$\mu(\{x \in Q : f(x) < m_f(Q)\}) \le \mu(Q)/2.$$

If f is complex-valued, we take

$$m_f(Q) := m_{\Re f}(Q) + i m_{\Im f}(Q), \tag{3.1.17}$$

where $i^2 = -1$, and $\Re f$ and $\Im f$ denote, respectively, the *real part* and the *imaginary part* of f. Furthermore, for any $\eta, \rho \in (1,\infty)$, $\beta_\rho := \rho^{D+1}$, $\gamma \in [1,\infty)$ and $f \in L^1_{\rm loc}(\mu)$, we denote by $\|f\|_\circ$ the *minimal nonnegative constant* $\tilde{C}$ such that, for any cube Q,

$$\frac{1}{\mu(\eta Q)} \int_Q |f(x) - m_f(\tilde{Q}^\rho)|\, d\mu(x) \le \tilde{C}$$

and, for any two (ρ, β_ρ)-doubling cubes $Q \subset R$,

$$|m_f(Q) - m_f(R)| \le \tilde{C}[1 + \delta(Q, R)]^\gamma.$$

Proposition 3.1.11. *For any $\eta, \rho \in (1,\infty)$ and $\gamma \in [1,\infty)$, $\|\cdot\|_\circ$ is equivalent to $\|\cdot\|_*$.*

Proof. First, we prove $\|\cdot\|_* \lesssim \|\cdot\|_\circ$. For any $Q \subset \mathbb{R}^D$, let $f_Q := m_f(\tilde{Q}^\rho)$. To show that $\|\cdot\|_* \lesssim \|\cdot\|_\circ$, it suffices to show that, for any $Q \subset R$,

$$\left|m_f(\tilde{Q}^\rho) - m_f(\tilde{R}^\rho)\right| \lesssim [1 + \delta(Q, R)]^\gamma \|f\|_\circ,$$

which can be proved by repeating the proof of (3.1.12).

Now we prove the converse. For any (ρ, β_ρ)-doubling cube Q, by the definition of $m_f(Q)$, we know that

$$\begin{aligned} \left|m_f(Q) - f_Q\right| &\le \frac{1}{\mu(Q)} \int_Q [|f(x) - f_Q| + |f(x) - m_f(Q)|]\, d\mu(x) \\ &\lesssim \|f\|_*. \end{aligned} \tag{3.1.18}$$

This fact, together with (3.1.8), implies that, for any two (ρ, β_ρ)-doubling cubes $Q \subset R$,

$$\begin{aligned} |m_f(Q) - m_f(R)| &\le |m_f(Q) - f_Q| + |f_Q - f_R| + |f_R - m_f(R)| \\ &\lesssim [1 + \delta(Q, R)]^\gamma \|f\|_*. \end{aligned}$$

Finally, for any cube Q, we find that

$$\frac{1}{\mu(\eta Q)}\int_Q |f(x)-m_f(\tilde{Q}^\rho)|\,d\mu(x)$$
$$\leq \frac{1}{\mu(\eta Q)}\int_Q |f(x)-f_Q|\,d\mu(x)+\frac{\mu(Q)}{\mu(\eta Q)}\left[\left|f_Q-f_{\tilde{Q}^\rho}\right|+\left|f_{\tilde{Q}^\rho}-m_f(\tilde{Q}^\rho)\right|\right]$$
$$\lesssim \|f\|_*,$$

where in the last inequality, we used (3.1.7), (3.1.8) and (3.1.18). This finishes the proof of Proposition 3.1.11. □

Remark 3.1.12. By Propositions 3.1.11 and 3.1.6, in the remainder of Part I, unless otherwise stated, we *always assume that both constants* ρ *and* η *in the definition of* $\|\cdot\|_\circ$ *as well as that of* $\|\cdot\|_{\mathrm{RBMO}\,(\mu)}$ *are equal to 2.*

We now present some examples of RBMO (μ). We first recall that, for $\eta \in [1,\infty)$, a function $f \in L^1_{\mathrm{loc}}(\mu)$ is said to belong to the space $\mathrm{BMO}_\eta(\mu)$, if there exists a positive constant C such that, for all cubes Q,

$$\frac{1}{\mu(\eta Q)}\int_Q |f(x)-m_Q(f)|\,d\mu(x)\leq C.$$

Moreover, the $\mathrm{BMO}_\eta(\mu)$ *norm* of f is defined to be the minimal constant C as above and denoted by $\|f\|_{\mathrm{BMO}_\eta(\mu)}$. If $\rho = 1$, we *write* $\mathrm{BMO}_\eta(\mu)$ *simply by* $\mathrm{BMO}(\mu)$.

Example 3.1.13. Assume $D := 2$ and $n := 1$. Thus, we can think that we are in the complex plane. Let $E \subset \mathbb{C}$ be a *one-dimensional Ahlfors–Daivd (AD) regular set.* That is, for all $x \in E$ and $r \in (0, \mathrm{diam}\,(E)]$, it holds true that

$$C^{-1}r \leq \mathcal{H}^1(E\cap B(x,r)) \leq Cr,$$

where $\mathcal{H}^1$ stands for the *one-dimensional Hausdorff measure*. We let $\mu := \mathcal{H}^1_{|E}$. Notice that μ is a doubling measure.

For any Q centered at some point of $\mathrm{supp}\,\mu$, we have

$$\mu(2^kQ)\sim \ell(2^kQ)$$

if $\ell(Q) \leq \mathrm{diam}\,(E)$. Then, given $Q \subset R$, it is easy to show that, if $\ell(R)\leq \mathrm{diam}\,(E)$, then

$$1+\delta(Q,\,R)\sim 1+\log_2\frac{\ell(R)}{\ell(Q)} \tag{3.1.19}$$

and, if $\ell(R) > \mathrm{diam}\,(E)$, then

$$1+\delta(Q,R) \sim 1+\frac{\operatorname{diam}(E)}{\ell(Q)}. \tag{3.1.20}$$

Thus, in this case, we have

$$\mathrm{RBMO}(\mu)=\mathrm{BMO}(\mu),$$

since any function $f \in \mathrm{BMO}(\mu)$ satisfies (3.1.7) and (3.1.8) with $f_Q := m_Q(f)$ for all cubes Q. Notice that (3.1.8) holds true by (3.1.19) and (3.1.20).

Example 3.1.14. Let $D := 2$, $n := 1$ and μ be the *planar Lebesgue measure restricted to the unit square* $[0,1]\times[0,1]$. This measure is doubling, but not AD-regular (for $n = 1$). Now we can show that the coefficients $1+\delta(Q,R)$ are uniformly bounded, namely, for any two squares $Q \subset R$, it holds true that

$$1+\delta(Q,R) \sim 1.$$

Let us take $R_0 := [0,1]^2$ and $Q \subset R_0$. Then, if $f \in \mathrm{RBMO}(\mu)$, we have

$$|m_Q(f-m_{R_0}(f))| = |m_Q(f)-m_{R_0}(f)| \le \delta_{Q,R_0}\|f\|_{\mathrm{RBMO}(\mu)} \lesssim \|f\|_{\mathrm{RBMO}(\mu)},$$

where δ_{Q,R_0} is as in (2.1.2) with R replaced by R_0. Since this holds true for any square $Q \subset R_0$, by the Lebesgue differentiation theorem, $f-m_{R_0}(f)$ is a bounded function and

$$\|f-m_{R_0}(f)\|_{L^\infty(\mu)} \lesssim \|f\|_{\mathrm{RBMO}(\mu)}.$$

Therefore, RBMO (μ) coincides with $L^\infty(\mu)$ modulo the space of constants functions, which is strictly smaller than BMO(μ).

Example 3.1.15. Suppose $D := 2$ and $n := 1$. Let μ be a *measure on the real axis* such that, on the intervals $[-2,-1]$ and $[1,2]$, μ is the linear Lebesgue measure, on the interval $[-1/2,1/2]$ μ is the linear Lebesgue measure times ϵ with $\epsilon \in (0,\infty)$ very small, and $\mu := 0$ elsewhere. We consider the function

$$f := \epsilon^{-1}(\chi_{[1/4,1/2]} - \chi_{[-1/2,-1/4]}).$$

It is easy to show that, for all $\eta \in (1,2]$,

$$\|f\|_{\mathrm{BMO}_\eta(\mu)} \sim \epsilon^{-1}$$

with the implicit equivalent positive constants independent of ϵ. On the other hand, the RBMO (μ) norm of f is $\|f\|_{\mathrm{RBMO}(\mu)} \sim \epsilon^{-1}$ with the implicit equivalent positive constants independent of ϵ, since

$$\|f\|_{\mathrm{RBMO}(\mu)} \gtrsim \left|m_{[-2,2]}(f) - m_{[1/4,1/2]}(f)\right| \sim \epsilon^{-1}$$

and

$$\|f\|_{\mathrm{RBMO}(\mu)} \lesssim \|f\|_{L^\infty(\mu)} \lesssim \epsilon^{-1}.$$

The following theorem is a version of the John–Nirenberg inequality for the space RBMO (μ) .

Theorem 3.1.16. *Let $\eta \in (1,\infty)$ and $f \in \mathrm{RBMO}(\mu)$. If there exists a sequence of numbers, $\{f_Q\}_Q$, such that (3.1.7) and (3.1.8) hold true with $\tilde{C}$ replaced by $C\|f\|_{\mathrm{RBMO}(\mu)}$. Then there exist nonnegative constants C_5 and C_6 such that, for any cube Q and $\lambda \in (0,\infty)$,*

$$\mu(\{x \in Q : |f(x) - f_Q| > \lambda\}) \leq C_5 \mu(\eta Q) \exp\left(\frac{-C_6\lambda}{\|f\|_{\mathrm{RBMO}(\mu)}}\right). \quad (3.1.21)$$

To prove Theorem 3.1.16, we need the following two technical lemmas.

Lemma 3.1.17. *Under the assumption of Theorem 3.1.16, there exists a positive constant C such that, if Q and R are cubes such that, $\ell(Q) \sim \ell(R)$ with the implicit equivalent positive constants independent of Q and R, and*

$$\mathrm{dist}(Q, R) \lesssim \ell(Q),$$

then

$$|f_Q - f_R| \leq C\|f\|_{\mathrm{RBMO}(\mu)}.$$

Proof. As in Definition 2.1.1, let R_Q be the smallest cube concentric with R containing Q and R, then $l(R_Q) \lesssim l(Q)$. By Lemma 2.1.3(a), we see that

$$\delta(Q, R_Q) \lesssim 1 \text{ and } \delta(R, R_Q) \lesssim 1.$$

An application of (3.1.8) implies that

$$|f_Q - f_R| \leq |f_Q - f_{R_Q}| + |f_{R_Q} - f_R| \lesssim \|f\|_{\mathrm{RBMO}(\mu)},$$

which completes the proof of Lemma 3.1.17. □

Lemma 3.1.18. *Let $f \in \mathrm{RBMO}(\mu)$ be a real-valued function. Given $q \in (0,\infty)$, let $f_q(x) := f(x)$ when $|f(x)| \leq q$, and let $f_q(x) := q\frac{f(x)}{|f(x)|}$ when $|f(x)| > q$. Then $f_q \in \mathrm{RBMO}(\mu)$ and*

$$\|f_q\|_{\mathrm{RBMO}(\mu)} \leq C\|f\|_{\mathrm{RBMO}(\mu)},$$

where C is a positive constant independent of q and f.

Proof. For any function g, let $g := g_+ - g_-$, where $g_+ := \max\{g, 0\}$ and

$$g_- := -\min\{g, 0\}.$$

Then, by Proposition 3.1.3, f_+, $f_- \in \mathrm{RBMO}\,(\mu)$,

$$\|f_+\|_{\mathrm{RBMO}\,(\mu)} \lesssim \|f\|_{\mathrm{RBMO}\,(\mu)} \quad \text{and} \quad \|f_-\|_{\mathrm{RBMO}\,(\mu)} \lesssim \|f\|_{\mathrm{RBMO}\,(\mu)}.$$

Observe that $f_{q,+} = \min\{f_+, q\}$ and $f_{q,-} = \min\{f_-, q\}$. We further conclude that

$$\|f_{q,+}\|_{\mathrm{RBMO}\,(\mu)} \lesssim \|f\|_{\mathrm{RBMO}\,(\mu)} \quad \text{and} \quad \|f_{q,-}\|_{\mathrm{RBMO}\,(\mu)} \lesssim \|f\|_{\mathrm{RBMO}\,(\mu)}.$$

Thus, it follows that

$$\|f_q\|_{\mathrm{RBMO}\,(\mu)} \leq \|f_{q,+}\|_{\mathrm{RBMO}\,(\mu)} + \|f_{q,-}\|_{\mathrm{RBMO}\,(\mu)} \lesssim \|f\|_{\mathrm{RBMO}\,(\mu)},$$

which completes the proof of Lemma 3.1.18. □

Remark 3.1.19. Let $f \in \mathrm{RBMO}\,(\mu)$ and $\{f_Q\}_Q$ satisfy the conditions of Theorem 3.1.16. Assume that f and f_Q are real-valued; otherwise we consider their real and imaginary parts, respectively. For any given $q \in (0, \infty)$, let $f_{Q,+} := \max\{f_Q, 0\}$, $f_{Q,-} := -\min\{f_Q, 0\}$ and

$$f_{q,Q} := \min\{f_{Q,+}, q\} - \min\{f_{Q,-}, q\}.$$

Then, by (3.1.7) and (3.1.8), for any given $\eta \in (1, \infty)$, there exists a positive constant $C_{(\eta)}$, depending on η, such that

$$\sup_Q \frac{1}{\mu(\eta Q)} \int_Q |f_q(x) - f_{q,Q}|\, d\mu(x) \leq C_{(\eta)} \|f\|_{\mathrm{RBMO}\,(\mu)}$$

and, for all cubes $Q \subset R$,

$$|f_{q,Q} - f_{q,R}| \leq C_{(\eta)}[1 + \delta(Q, R)]\|f\|_{\mathrm{RBMO}\,(\mu)}.$$

We omit the details.

Proof of Theorem 3.1.16. By similarity, it suffices to establish (3.1.21) for $\eta = 2$. Assume that f and f_Q are real-valued; otherwise we consider their real and imaginary parts, respectively. Let $f \in L^\infty(\mu)$ first and Q be some fixed cube. Without loss of generality, we may assume that $\|f\|_{\mathrm{RBMO}\,(\mu)} = 1$. For $t \in (0, \infty)$, we define

$$X(t) := \sup_Q \frac{1}{\mu(2Q)} \int_Q \exp\left(|f(x) - f_Q|t\right) d\mu(x).$$

If we can prove that, for $t_0 \in (0, \infty)$ small enough,

$$X(t_0) \lesssim 1, \tag{3.1.22}$$

then, for any cube Q,

$$\begin{aligned}\mu\left(\{x \in Q : |f(x) - f_Q| > \lambda/t_0\}\right) &\le \int_Q \exp\left(|f(x) - f_Q|t_0\right) \exp(-\lambda)\, d\mu(x)\\ &\lesssim \mu(2Q)\exp(-\lambda),\end{aligned}$$

and hence (3.1.21) holds true in the case $f \in L^\infty(\mu)$.

To show (3.1.22), we fix a cube Q and let $\tilde{Q} := \frac{3}{2}Q$ and B be a positive constant which is determined later. By the Lebesgue differentiation theorem, for μ-almost every $x \in Q \cap \operatorname{supp}\mu$ such that $|f(x) - f_Q| > B$, there exists a doubling cube Q_x centered at x such that

$$m_{Q_x}(|f - f_Q|) > B. \tag{3.1.23}$$

Moreover, we assume that Q_x is the biggest doubling cube satisfying (3.1.23) with side length $2^k\ell(Q)$ for some $k < 0$ and $\ell(Q_x) \le \frac{1}{20}\ell(Q)$. By Theorem 1.1.1, there exists an almost disjoint subfamily $\{Q_i\}_i$ of the cubes $\{Q_x\}_x$ such that

$$\{x \in Q : |f(x) - f_Q| > B\} \subset \bigcup_i Q_i. \tag{3.1.24}$$

Since (3.1.7) holds true if we replace $\mu(\eta Q)$ by $\mu(\frac{4}{3}Q)$. Thus, if we choose B big enough, by (3.1.23) and the facts that $\{Q_i\}_i$ are almost disjoint and $Q_i \subset \tilde{Q}$ for all i, we conclude that

$$\begin{aligned}\sum_i \mu(Q_i) &\le \sum_i \frac{1}{B}\int_{Q_i} |f(x) - f_Q|\, d\mu(x)\\ &\le \frac{N_D}{B}\int_{\tilde{Q}} |f(x) - f_Q|\, d\mu(x)\\ &\le \frac{\mu(2Q)}{2^{D+3}},\end{aligned} \tag{3.1.25}$$

where N_D is as in Theorem 1.1.1.

Now we claim that there exists a positive constant c such that, for each i,

$$|f_{Q_i} - f_Q| \le c. \tag{3.1.26}$$

We consider in the following three cases.

Case (i) $\ell(\widetilde{2Q_i}) > 10\ell(Q)$. Then there exists some cube $2^m Q_i$, $m \in \mathbb{N}$, containing Q such that $\ell(Q) \sim \ell(2^m Q_i) \le \ell(\widetilde{2Q_i})$. Thus, by (a), (b) and (d) of Lemma 2.1.3, we see that

$$|f_{Q_i} - f_Q| \le |f_{Q_i} - f_{2Q_i}| + |f_{2Q_i} - f_{2^m Q_i}| + |f_{2^m Q_i} - f_Q| \lesssim 1.$$

Case (ii) $\frac{1}{20}\ell(Q) \le \ell(\widetilde{2Q_i}) \le 10\ell(Q)$. In this case, by the fact that $\widetilde{2Q_i} \subset 30Q$ and Lemma 2.1.3, we obtain

$$|f_{Q_i} - f_Q| \le \left|f_{Q_i} - f_{\widetilde{2Q_i}}\right| + \left|f_{\widetilde{2Q_i}} - f_{30Q}\right| + |f_{30Q} - f_Q| \lesssim 1.$$

Case (iii) $\ell(\widetilde{2Q_i}) < \frac{1}{20}\ell(Q)$. By the choice of Q_i, $m_{\widetilde{2Q_i}}(|f - f_Q|) \le B$, which implies that $|m_{\widetilde{2Q_i}}(f - f_Q)| \le B$. It then follows, from this, together with (3.1.7), (3.1.8) and Lemma 2.1.3, that

$$|f_{Q_i} - f_Q| \le \left|f_{Q_i} - f_{\widetilde{2Q_i}}\right| + \left|f_{\widetilde{2Q_i}} - m_{\widetilde{2Q_i}}(f)\right| + \left|m_{\widetilde{2Q_i}}(f) - f_Q\right| \lesssim 1.$$

Combining these cases above, we see that (3.1.26) holds true.

Now we finish the proof of (3.1.22). Indeed, from (3.1.24) through (3.1.26) and the doubling property of Q_i, it follows that

$$\begin{aligned}
&\frac{1}{\mu(2Q)}\int_Q \exp\left(|f(x) - f_Q|t\right)\,d\mu(x)\\
&\le \frac{1}{\mu(2Q)}\int_{Q\setminus\bigcup_i Q_i} \exp(Bt)\,d\mu(x)\\
&\quad + \frac{1}{\mu(2Q)}\sum_i \int_{Q_i} \exp\left(|f(x) - f_{Q_i}|t\right)\,d\mu(x)\exp(ct)\\
&\le \exp(Bt) + \frac{1}{4}X(t)\exp(ct). \qquad (3.1.27)
\end{aligned}$$

Since $f \in L^\infty(\mu)$, $X(t) < \infty$, which implies that

$$X(t)\left[1 - \frac{1}{4}\exp(ct)\right] \le \exp(Bt).$$

We then take t_0 small enough to see that $X(t_0) \lesssim 1$. Thus, (3.1.22) holds true.

When f is not bounded, consider the function f_q of Lemma 3.1.18. From Lemma 3.1.18 and Remark 3.1.19, we deduce that

$$\mu\left(\{x \in Q : |f_q(x) - f_{q,Q}| > \lambda\}\right) \lesssim \mu(2Q)\exp\left(-C_6\lambda\right).$$

A limiting argument then completes the proof of Theorem 3.1.16. □

From Theorem 3.1.16, we can easily deduce that the following spaces $\mathrm{RBMO}^p(\mu)$ coincide for all $p \in [1, \infty)$.

Let $\eta, \rho \in (1, \infty)$, $\beta_\rho := \rho^{D+1}$, $\gamma \in [1, \infty)$ and $p \in [1, \infty)$. A function $f \in L^1_{\mathrm{loc}}(\mu)$ is said to belong to the *space* $\mathrm{RBMO}^p(\mu)$ if there exists a nonnegative constant $\tilde{C}$ such that, for all cubes Q,

$$\left\{\frac{1}{\mu(\eta Q)}\int_Q \left|f(x) - m_{\tilde{Q}^\rho}(f)\right|^p d\mu(x)\right\}^{1/p} \le \tilde{C} \tag{3.1.28}$$

and, for any two (ρ, β_ρ)-doubling cubes $Q \subset R$,

$$|m_Q(f) - m_R(f)| \le \tilde{C}[1 + \delta(Q, R)]^\gamma. \tag{3.1.29}$$

Moreover, the minimal constant $\tilde{C}$ as above is defined to be the $\mathrm{RBMO}^p(\mu)$ *norm* of f and denoted by $\|f\|_{\mathrm{RBMO}^p(\mu)}$.

Arguing as for $p = 1$, we can show that another equivalent definition for the space $\mathrm{RBMO}^p(\mu)$ can be given in terms of the numbers, $\{f_Q\}_Q$, as in (3.1.7) and (3.1.8) without depending on the constants η, $\rho \in (1, \infty)$ and $\gamma \in [1, \infty)$.

Using Theorem 3.1.16, we have the following conclusion.

Corollary 3.1.20. *Let* $p \in [1, \infty)$, η, $\rho \in (1, \infty)$ *and* $\gamma \in [1, \infty)$. *Then the spaces* $\mathrm{RBMO}^p(\mu)$ *coincide with equivalent norms.*

Proof. For any function $f \in L^1_{\mathrm{loc}}(\mu)$, the inequality

$$\|f\|_{\mathrm{RBMO}(\mu)} \le \|f\|_{\mathrm{RBMO}^p(\mu)}$$

follows from the Hölder inequality immediately. The conditions (3.1.6) and (3.1.29) coincide. Thus, it suffices to compare (3.1.5) with (3.1.28). To see the converse, we assume that $f \in \mathrm{RBMO}(\mu)$. Then

$$\begin{aligned}
&\frac{1}{\mu(\eta Q)}\int_Q \left|f - m_{\tilde{Q}^\rho}(f)\right|^p d\mu(x)\\
&\quad= \frac{1}{\mu(\eta Q)}\int_0^\infty p\lambda^{p-1}\mu\left(\left\{x \in Q : |f(x) - m_{\tilde{Q}^\rho}(f)| > \lambda\right\}\right) d\lambda\\
&\quad\lesssim p\int_0^\infty \lambda^{p-1}\exp\left(\frac{-C_6\lambda}{\|f\|_{\mathrm{RBMO}(\mu)}}\right) d\lambda\\
&\quad\lesssim \|f\|^p_{\mathrm{RBMO}(\mu)},
\end{aligned}$$

which implies that

$$\|f\|_{\mathrm{RBMO}^p(\mu)} \lesssim \|f\|_{\mathrm{RBMO}(\mu)},$$

and hence completes the proof of Corollary 3.1.20. □

We have another corollary of the John–Nirenberg inequality which is useful in applications.

Definition 3.1.21. Given a doubling cube Q, denote by $Z(Q,\lambda)$ the *set* of all points $x \in Q$ such that any doubling cube P, with $x \in P$ and $\ell(P) \le \ell(Q)/4$, satisfies that $|m_P(f) - m_Q(f)| \le \lambda$.

In other words, $Q \setminus Z(Q,\lambda)$ is the subset of Q such that, for some doubling cube P, with $x \in P$ and $\ell(P) \le \ell(Q)/4$, we have $|m_P(f) - m_Q(f)| > \lambda$.

Corollary 3.1.22. *Let $Q \subset \mathbb{R}^D$ be a doubling cube. If $f \in \mathrm{RBMO}\,(\mu)$, then there exist positive constant c and C such that, for any $\lambda \in (0,\infty)$,*

$$\mu(Q \setminus Z(Q,\lambda)) \le c\mu(Q) \exp\left(\frac{-C\lambda}{\|f\|_{\mathrm{RBMO}\,(\mu)}}\right).$$

Proof. For any $x \in Q \setminus Z(Q,\lambda)$, there exists some doubling cube P_x which contains x and satisfies that $\ell(P_x) \le \ell(Q)/4$ and $|m_{P_x}(f) - m_Q(f)| > \lambda$. Then, by Theorem 1.1.1 and Remark 1.1.2, there are points $\{x_i\}_i \subset Q \setminus Z(Q,\lambda)$ such that

$$Q \setminus Z(Q,\lambda) \subset \bigcup_i (2P_i)$$

and the cubes $\{2P_i\}_i$ form an almost disjoint family.

Since, for each i, we have $\ell(P_i) \le \ell(Q)/4$ and $P_i \cap Q \ne \emptyset$, it is easy to see that $2P_i \subset \frac{7}{4}Q$. On the other hand, it follows, from the Jensen inequality, that

$$\begin{aligned}
&\frac{1}{\mu(P_i)} \int_{P_i} \exp(|f(x) - m_Q(f)|k)\, d\mu(x) \\
&\quad \ge \exp\left(\frac{1}{\mu(P_i)} \int_{P_i} |f(x) - m_Q(f)|k\, d\mu(x)\right) \\
&\quad \ge \exp(|m_{P_i} f - m_{Q_i} f|k) \\
&\quad \ge \exp(\lambda k),
\end{aligned}$$

where k is some positive constant that is fixed below. Then, by these facts and the doubling property of P_i, we have

$$\begin{aligned}
\mu(Q \setminus Z(Q,\lambda)) &\le \sum_i \mu(2P_i) \\
&\lesssim \sum_i \int_{P_i} \exp(|f(x) - m_Q(f)k|) \exp(-\lambda k)\, d\mu(x) \\
&\lesssim \int_{\frac{7}{4}Q} \exp(|f(x) - m_Q(f)|k) \exp(-\lambda k)\, d\mu(x). \quad (3.1.30)
\end{aligned}$$

Since $f \in \mathrm{RBMO}\,(\mu)$, we see that there exists a positive constant $\tilde{c}$ such that

$$\begin{aligned}
&\exp(|f(x)-m_Q(f)|k)\\
&\quad\le \exp\left(|f(x)-m_{\frac74 Q}(f)|k\right)\exp\left(\left|m_{\frac74 Q}(f)-m_Q(f)\right|k\right)\\
&\quad\le \exp\left(|f(x)-m_{\frac74 Q}(f)|k\right)\exp(\tilde{c}\|f\|_{{\rm RBMO}\,(\mu)}k). \qquad (3.1.31)
\end{aligned}$$

Notice that $\frac74 Q$ is a $(\frac87, 2^{D+1})$-doubling cube. Then Theorem 3.1.16 implies that, for any $\lambda\in(0,\infty)$,

$$\mu\left(\left\{x\in\frac74 Q:\ \left|f(x)-m_{\frac74 Q}(f)\right|k>\lambda\right\}\right)\lesssim\mu\left(\frac74 Q\right)\exp\left(\frac{-C_6\lambda}{\|f\|_{{\rm RBMO}\,(\mu)}}\right).$$

By this inequality, together with (3.1.30) and (3.1.31), we find that

$$\begin{aligned}
&\mu(Q\setminus Z(Q,\lambda))\\
&\quad\lesssim \exp(-\lambda k)\exp(\tilde{c}\|f\|_{{\rm RBMO}\,(\mu)}k)\int_{\frac74 Q}\exp\left(\left|f(x)-m_{\frac74 Q}(f)\right|k\right)d\mu(x)\\
&\quad\sim \exp(-\lambda k)\exp(\tilde{c}\|f\|_{{\rm RBMO}\,(\mu)}k)\\
&\qquad\times\int_0^\infty\mu\left(\left\{x\in\frac74 Q:\ \exp\left(\left|f(x)-m_{\frac74 Q}(f)\right|k\right)>t\right\}\right)dt\\
&\quad\lesssim\mu\left(\frac74 Q\right)\exp(-\lambda k)\exp(\tilde{c}\|f\|_{{\rm RBMO}\,(\mu)}k)\int_1^\infty\exp\left(\frac{-C_6\ln t}{k\|f\|_{{\rm RBMO}\,(\mu)}}\right)dt.
\end{aligned}$$

Thus, if we choose $k:=C_6/(2\|f\|_{{\rm RBMO}\,(\mu)})$, by the doubling property of Q, we conclude that

$$\begin{aligned}
\mu(Q\setminus Z(Q,\lambda))&\lesssim\mu\left(\frac74 Q\right)\exp\left(\frac{-C_6\lambda}{2\|f\|_{{\rm RBMO}\,(\mu)}}\right)\\
&\lesssim\mu(Q)\exp\left(\frac{-C_6\lambda}{2\|f\|_{{\rm RBMO}\,(\mu)}}\right).
\end{aligned}$$

This finishes the proof of Corollary 3.1.22. □

3.2 The Atomic Hardy Space $H^{1,p}_{\rm atb}(\mu)$

In this section, we focus our attention to the atomic Hardy space. To this end, we first introduce the notion of (p,γ)-atomic blocks as follows.

Definition 3.2.1. Let $p\in(1,\infty]$, $\eta\in(1,\infty)$ and $\gamma\in[1,\infty)$. A function $b\in L^1_{\rm loc}(\mu)$ is called a (p,γ)-*atomic block* if

(i) there exists some cube R such that $\operatorname{supp} b \subset R$;
(ii)

$$\int_{\mathbb{R}^D} b(x)\,d\mu(x) = 0;$$

(iii) for any $j \in \{1, 2\}$, there exist a function a_j supported on a cube $Q_j \subset R$ and $\lambda_j \in \mathbb{C}$ such that $b = \lambda_1 a_1 + \lambda_2 a_2$ and

$$\|a_j\|_{L^p(\mu)} \le \left[\mu(\eta Q_j)\right]^{1/p-1} [1 + \delta(Q_j, R)]^{-\gamma}. \tag{3.2.1}$$

Then let

$$|b|_{H^{1,p}_{\mathrm{atb},\gamma}(\mu)} := |\lambda_1| + |\lambda_2|.$$

A function f is said to belong to the *atomic Hardy space* $H^{1,p}_{\mathrm{atb},\gamma}(\mu)$ if there exist (p, γ)-atomic blocks $\{b_i\}_{i\in\mathbb{N}}$ such that

$$f = \sum_{i=1}^{\infty} b_i \text{ and } \sum_{i=1}^{\infty} |b_i|_{H^{1,p}_{\mathrm{atb},\gamma}(\mu)} < \infty.$$

The $H^{1,p}_{\mathrm{atb},\gamma}(\mu)$ *norm* of f is defined by

$$\|f\|_{H^{1,p}_{\mathrm{atb},\gamma}(\mu)} := \inf\left\{\sum_i |b_i|_{H^{1,p}_{\mathrm{atb},\gamma}(\mu)}\right\},$$

where the infimum is taken over all the possible decompositions of f in (p, γ)-atomic blocks.

Remark 3.2.2. (i) Let $p \in (1, \infty]$ and μ be the D-dimensional Lebesgue measure. A function $b \in L^1_{\mathrm{loc}}(\mathbb{R}^D)$ is called a *p-atom* if

(a) there exists some cube Q such that $\operatorname{supp} b \subset Q$,
(b)

$$\int_{\mathbb{R}^D} b(x)\,dx = 0,$$

(c) the function b satisfies that

$$\|b\|_{L^p(\mathbb{R}^D)} \le |Q|^{1/p-1}.$$

A function f is said to belong to the *atomic Hardy space* $H^{1,p}(\mathbb{R}^D)$ if there exist p-atoms $\{b_i\}_{i\in\mathbb{N}}$ such that

$$f = \sum_{i=1}^{\infty} \lambda_i b_i \quad \text{and} \quad \sum_{i=1}^{\infty} |\lambda_i| < \infty.$$

The $H^{1,p}(\mathbb{R}^D)$ *norm* of f is defined by

$$\|f\|_{H^{1,p}(\mathbb{R}^D)} := \inf\left\{\sum_i |\lambda_i|\right\},$$

where the infimum is taken over all the possible decompositions of f in p-atoms.[2]

(ii) Let $\rho \in (1,\infty)$ and $\beta_\rho := \rho^{D+1}$. As in Remark 3.2.2, due to the fact that, for any cubes $Q \subset R$,

$$1 + \delta(Q, \tilde{R}^\rho) \sim 1 + \delta(Q, R), \tag{3.2.2}$$

if necessary, we may assume that the cube R in Definition 3.2.1 is (ρ, β_ρ)-doubling.

When $\gamma = 1$, we *denote the atomic Hardy space* $H^{1,p}_{\rm atb,1}(\mu)$ *simply by* $H^{1,p}_{\rm atb}(\mu)$, and we *call the* $(p,1)$*-atomic block simply by* the p*-atomic block*. We will see that

$$H^{1,p}_{\rm atb,\gamma}(\mu) = H^{1,p}_{\rm atb}(\mu)$$

for any $\gamma \in (1,\infty)$ and $p \in (1,\infty]$.

We first establish an equivalent characterization of $H^{1,p}_{\rm atb,\gamma}(\mu)$. Let $p \in (1,\infty]$. Instead of Definition 3.2.1, we can also define the atomic block b in the following way: b satisfies (i) and (ii) of Definition 3.2.1 and

$$b = \sum_{j=1}^{\infty} \lambda_j a_j, \tag{3.2.3}$$

where $\{a_j\}_{j=1}^{\infty}$ and $\{\lambda_j\}_{j=1}^{\infty}$ satisfy (3.2.1) and

$$|b|_{\tilde{H}^{1,p}_{\rm atb,\gamma}(\mu)} := \sum_{j=1}^{\infty} |\lambda_j| < \infty, \tag{3.2.4}$$

respectively. Correspondingly, we obtain an *atomic Hardy space*, temporarily denoted by $\tilde{H}^{1,p}_{\rm atb,\gamma}(\mu)$. Then we have the following conclusion.

---

[2] See [15, 78].

Proposition 3.2.3. *Let $p \in (1, \infty]$ and $\gamma \in (1, \infty)$. The spaces $\tilde{H}^{1,p}_{\mathrm{atb},\gamma}(\mu)$ and $H^{1,p}_{\mathrm{atb},\gamma}(\mu)$ coincide with equivalent norms.*

Proof. Indeed, obviously,

$$H^{1,p}_{\mathrm{atb},\gamma}(\mu) \subset \tilde{H}^{1,p}_{\mathrm{atb},\gamma}(\mu).$$

To prove the converse, assume that b satisfies (i) and (ii) of Definition 3.2.1 and (3.2.3) with $\{a_j\}_{j=1}^{\infty}$ and $\{\lambda_j\}_{j=1}^{\infty}$, respectively, satisfying (3.2.1) and (3.2.4). To prove

$$\tilde{H}^{1,p}_{\mathrm{atb},\gamma}(\mu) \subset H^{1,p}_{\mathrm{atb},\gamma}(\mu),$$

we only need to show that $b \in H^{1,p}_{\mathrm{atb},\gamma}(\mu)$ and

$$\|b\|_{H^{1,p}_{\mathrm{atb},\gamma}(\mu)} \lesssim \sum_{j=1}^{\infty} |\lambda_j|.$$

To this end, for each function a_j with $\operatorname{supp} a_j \subset Q_j$, let $A_j := \lambda_j a_j + c_j \chi_{\tilde{R}^\rho}$, where

$$c_j := -\lambda_j \int_{Q_j} a_j(x)\, d\mu(x)/\mu(\tilde{R}^\rho).$$

Then

$$\int_{\mathbb{R}^D} A_j(x)\, d\mu(x) = 0.$$

Clearly, we have $\sum_{j=1}^{\infty} A_j = b$. Furthermore, we write

$$A_j := \lambda_j a_j + c_j \chi_{\tilde{R}^\rho} = \lambda_j a_j + \left[c_j \mu\left(\rho \tilde{R}^\rho\right)\right]\left[\frac{\chi_{\tilde{R}^\rho}}{\mu(\rho \tilde{R}^\rho)}\right] =: \lambda_j a_j + \tilde{\lambda}_j \tilde{a}_j.$$

By (3.2.2), we conclude that A_j is a harmless constant multiple of a (p, γ)-atomic block supported in $\tilde{R}^\rho$. Moreover, for $p \in (1, \infty]$, by the Hölder inequality and the size condition of a_j, we obtain

$$\begin{aligned}
|c_j| &\le |\lambda_j| \frac{1}{\mu(\tilde{R}^\rho)} \|a_j\|_{L^p(\mu)} [\mu(Q_j)]^{1-1/p} \\
&\lesssim |\lambda_j| \frac{1}{\mu(\tilde{R}^\rho)} [\mu(Q_j)]^{1-1/p} [\mu(\rho Q_j)]^{1/p-1} [1 + \delta(Q_j, R)]^{-\gamma} \\
&\lesssim \frac{|\lambda_j|}{\mu(\tilde{R}^\rho)}
\end{aligned}$$

and

$$|\tilde{\lambda}_j| = |c_j|\mu(\rho\tilde{R}^\rho) \lesssim \frac{|\lambda_j|}{\mu(\tilde{R}^\rho)}\mu(\rho\tilde{R}^\rho) \lesssim |\lambda_j|,$$

which shows that $b \in H^{1,p}_{\rm atb,\gamma}(\mu)$ and

$$\|b\|_{H^{1,p}_{\rm atb,\gamma}(\mu)} \lesssim \sum_{j=1}^{\infty} |\lambda_j|.$$

Thus, $\tilde{H}^{1,p}_{\rm atb,\gamma}(\mu)$ and $H^{1,p}_{\rm atb,\gamma}(\mu)$ coincide with equivalent norms. This finishes the proof of Proposition 3.2.3. □

Proposition 3.2.4. *Let $p \in (1,\infty]$, $\gamma \in [1,\infty)$ and $\eta \in (1,\infty)$. Then the following conclusions hold true:*

(i) *The space $H^{1,p}_{\rm atb,\gamma}(\mu)$ is a Banach space;*
(ii) *For $1 < p_1 \le p_2 \le \infty$, $H^{1,p_2}_{\rm atb,\gamma}(\mu) \subset H^{1,p_1}_{\rm atb,\gamma}(\mu) \subset L^1(\mu)$;*
(iii) *The space $H^{1,p}_{\rm atb,\gamma}(\mu)$ is independent of the choice of the constant η;*
(iv) *$H^{1,\infty}_{\rm atb,\gamma}(\mu)$ is dense in $H^{1,p}_{\rm atb,\gamma}(\mu)$.*

Proof. The proofs for (i) and (ii) are standard and the details omitted.

To prove (iii), we only consider the case $p \in (1,\infty)$ by similarity. Assume that $\eta_1 > \eta_2 > 1$. For $i \in \{1,2\}$, write the atomic Hardy spaces corresponding to η_i as $H^{1,p}_{\rm atb,\gamma,\eta_i}(\mu)$ for the moment. Clearly, $H^{1,p}_{\rm atb,\gamma,\eta_1}(\mu) \subset H^{1,p}_{\rm atb,\gamma,\eta_2}(\mu)$.

Conversely, let

$$b := \sum_{j=1}^{2} \lambda_j a_j \in H^{1,p}_{\rm atb,\gamma,\eta_2}(\mu)$$

be a (p,γ)-atomic block, where, for any $j \in \{1,2\}$, $\operatorname{supp} a_j \subset Q_j \subset R$ for some cubes Q_j and R as in Definition 3.2.1. By Remark 3.2.2, we assume that R is (ρ,β_ρ)-doubling with $\rho \ge \eta_1$. Then, for each j, we find that

$$\|a_j\|_{L^p(\mu)} \le [\mu(\eta_2 Q_j)]^{1/p-1}[1+\delta(Q_j,R)]^{-\gamma}. \qquad (3.2.5)$$

From Theorem 1.1.1, it follows that there exists a sequence $\{Q_{j,k}\}_{k=1}^{N}$ of cubes such that

$$Q_j \subset \bigcup_{k=1}^{N} Q_{j,k} \quad \text{and} \quad \ell(Q_{j,k}) = \frac{\eta_2 - 1}{10\eta_1}\ell(Q_j)$$

and the center of $Q_{j,k}$ belongs to Q_j for every j and k, where N is bounded by some positive constant depending only on η_1, η_2 and D. Observe that $\eta_1 Q_{j,k} \subset \eta_2 Q_j$. For each $k \in \{1, \ldots, N\}$, define

$$a_{j,k} := a_j \frac{\chi_{Q_{j,k}}}{\sum_{k=1}^N \chi_{Q_{j,k}}}$$

and $\lambda_{j,k} := \lambda_j$. Then we have

$$b = \sum_{j=1}^2 \lambda_j a_j = \sum_{j=1}^2 \sum_{k=1}^N \lambda_{j,k} a_{j,k}.$$

Moreover, by (3.2.5), together with (a), (b) and (d) of Lemma 2.1.3, we see that

$$\begin{aligned}
\|a_{j,k}\|_{L^p(\mu)} &\le [\mu(\eta_2 Q_j)]^{1/p-1} [1 + \delta(Q_j, R)]^{-\gamma} \\
&\le [\mu(\eta_1 Q_{j,k})]^{1/p-1} [1 + \delta(Q_j, R)]^{-\gamma} \\
&\lesssim [\mu(\eta_1 Q_{j,k})]^{1/p-1} [1 + \delta(Q_{j,k}, \rho R)]^{-\gamma}.
\end{aligned} \tag{3.2.6}$$

Let

$$c_{j,k} := \lambda_{j,k}(a_{j,k} + \gamma_{j,k}\chi_R),$$

where

$$\gamma_{j,k} := -\frac{1}{\mu(R)} \int_{\mathbb{R}^D} a_{j,k}(x)\, d\mu(x).$$

We claim that $c_{j,k}$ is a (p, γ)-atomic block. Indeed, $\operatorname{supp} c_{j,k} \subset \rho R$ and

$$\int_{\mathbb{R}^D} c_{j,k}(x)\, d\mu(x) = 0.$$

Moreover, since R is (ρ, β_ρ)-doubling and $Q_{j,k} \subset \rho R$, by the Hölder inequality and (3.2.6), we know that

$$\begin{aligned}
\left\|\gamma_{j,k}\chi_R\right\|_{L^p(\mu)} &\lesssim [\mu(R)]^{1/p-1} [\mu(Q_{j,k})]^{1-1/p} [\mu(\eta_1 Q_{j,k})]^{1/p-1} \\
&\quad \times [1 + \delta(Q_{j,k}, \rho R)]^{-\gamma} \\
&\lesssim [\mu(\eta_1 R)]^{1/p-1}.
\end{aligned}$$

This, together with (3.2.6), implies that

$$|c_{j,k}|_{H^{1,p}_{\mathrm{atb},\gamma,\eta_1}(\mu)} \lesssim |\lambda_{j,k}|.$$

Thus, the claim is true.

By the claim, we see that

$$b = \sum_{j=1}^{2}\sum_{k=1}^{N} c_{j,k} \in H^{1,p}_{\mathrm{atb},\gamma,\eta_1}(\mu)$$

and

$$\|b\|_{H^{1,p}_{\mathrm{atb},\gamma,\eta_1}(\mu)} \lesssim \sum_{j=1}^{2} |\lambda_j|.$$

Thus, we conclude that

$$H^{1,p}_{\mathrm{atb},\gamma,\eta_2}(\mu) \subset H^{1,p}_{\mathrm{atb},\gamma,\eta_1}(\mu),$$

which shows (iii).

To prove (iv), by Definition 3.2.1, we know that, for every $f \in H^{1,p}_{\mathrm{atb},\gamma}(\mu)$ and $\varepsilon \in (0,\infty)$, there exist $M \in \mathbb{N}$ and

$$g := \sum_{j=1}^{M} b_j$$

such that

$$\|f - g\|_{H^{1,p}_{\mathrm{atb},\gamma}(\mu)} < \varepsilon/2,$$

where, for $j \in \{1,\dots,M\}$, b_j is a (p,γ)-atomic block and $\operatorname{supp} b_j \subset R_j$, some cube. Moreover, for any fixed $\eta \in (1,\infty)$ and all $j \in \{1,\dots,M\}$, there exists $h_j \in L^\infty(\mu)$ such that $\operatorname{supp} h_j \subset R_j$ and

$$\|b_j - h_j\|_{L^p(\mu)} < \frac{\varepsilon}{2^{M+1}\mu(\eta R_j)^{1-1/p}}.$$

Let

$$d_j := \int_{\mathbb{R}^D} h_j(x)\,d\mu(x).$$

From the Hölder inequality and the fact that

$$\int_{\mathbb{R}^D} b_j(x)\,d\mu(x) = 0,$$

we deduce that

$$\begin{aligned}|d_j| &= \left|\int_{\mathbb{R}^D} [h_j(x) - b_j(x)]\, d\mu(x)\right| \\ &\le \mu(R_j)^{1-1/p}\|h_j - b_j\|_{L^p(\mu)} \\ &\le \frac{\varepsilon\mu(R_j)^{1-1/p}}{2^{M+1}\mu(\eta R_j)^{1-1/p}}\end{aligned}$$

for all $j \in \{1, \ldots, M\}$. Define

$$\tilde{h}_j := \frac{d_j}{\mu(R_j)}\chi_{R_j},$$

where χ_{R_j} is the characteristic function of R_j, and

$$\tilde{b}_j := h_j - \tilde{h}_j$$

for all $j \in \{1, \ldots, M\}$. Then, it holds true that $\tilde{b}_j \in L^\infty(\mu)$, $\operatorname{supp}(\tilde{b}_j) \subset R_j$ and

$$\int_{\mathbb{R}^D} \tilde{b}_j(x)\, d\mu(x) = 0.$$

From this, it is easy to see that $\tilde{b}_j \in H^{1,\infty}_{\mathrm{atb},\gamma}(\mu)$. Moreover, from the fact that

$$\left\|\tilde{h}_j\right\|_{L^p(\mu)} = |d_j|\left[\mu(R_j)\right]^{1/p-1} \le \frac{\varepsilon}{2^{M+1}\left[\mu(\eta R_j)\right]^{1-1/p}},$$

it follows that

$$\left\|b_j - \tilde{b}_j\right\|_{L^p(\mu)} \le \|b_j - h_j\|_{L^p(\mu)} + \left\|\tilde{h}_j\right\|_{L^p(\mu)} \le \frac{\varepsilon}{2^M\left[\mu(\eta R_j)\right]^{1-1/p}},$$

which in turn implies that $b_j - \tilde{b}_j \in H^{1,p}_{\mathrm{atb},\gamma}(\mu)$ and

$$\left\|b_j - \tilde{b}_j\right\|_{H^{1,p}_{\mathrm{atb},\gamma}(\mu)} < \frac{\varepsilon}{2^M}.$$

Let

$$\tilde{g} := \sum_{j=1}^{M} \tilde{b}_j.$$

Then $\tilde{g} \in H_{\mathrm{atb},\gamma}^{1,\infty}(\mu)$ and

$$\|f - \tilde{g}\|_{H_{\mathrm{atb},\gamma}^{1,p}(\mu)} < \frac{\varepsilon}{2} + \frac{\varepsilon}{2} = \varepsilon,$$

which completes the proof of Proposition 3.2.4(iv) and hence Proposition 3.2.4. □

Remark 3.2.5. By Proposition 3.2.4, unless otherwise stated, we *always assume that the constants η in Definition 3.2.1 and ρ in Remark 3.2.2 are equal to 2.*

We now prove that the dual space of $H_{\mathrm{atb},\gamma}^{1,p}(\mu)$ is RBMO (μ). We begin with one of the inclusions for $p = \infty$.

Lemma 3.2.6. *Let $\gamma \in [1, \infty)$. Then*

$$\mathrm{RBMO}_\gamma^2(\mu) \subset \left(H_{\mathrm{atb},\gamma}^{1,\infty}(\mu)\right)^*.$$

That is, for any $g \in \mathrm{RBMO}_\gamma^2(\mu)$, the linear functional

$$L_g(f) := \int_{\mathbb{R}^D} f(x)g(x)\,d\mu(x),$$

defined on bounded functions f with compact support, can be extended to a continuous linear functional L_g over $H_{\mathrm{atb},\gamma}^{1,\infty}(\mu)$ and

$$\|L_g\|_{(H_{\mathrm{atb},\gamma}^{1,\infty}(\mu))^*} \le C\|g\|_{\mathrm{RBMO}_\gamma^2(\mu)},$$

where C is a positive constant independent of g.

Proof. It suffices to show that, if

$$b := \sum_{i=1}^{2} \lambda_i a_i$$

is an (∞, γ)-atomic block with $\mathrm{supp}\, b \subset R$ as in Definition 3.2.1, then, for any $g \in \mathrm{RBMO}_\gamma^2(\mu)$,

$$\left|\int_{\mathbb{R}^D} b(x)g(x)\,d\mu(x)\right| \lesssim |b|_{H_{\mathrm{atb},\gamma}^{1,\infty}(\mu)} \|g\|_{\mathrm{RBMO}_\gamma^2(\mu)}.$$

By

$$\int_{\mathbb{R}^D} b(x)\,d\mu(x) = 0,$$

we have

$$\left|\int_{\mathbb{R}^D} b(x)g(x)\,d\mu(x)\right| = \left|\int_R b(x)\left[g(x) - m_{\tilde{R}}(g)\right]\,d\mu(x)\right|$$
$$\leq \sum_{i=1}^{2} |\lambda_i|\|a_j\|_{L^\infty(\mu)} \int_{Q_i} \left|g(x) - m_{\tilde{R}}(g)\right|\,d\mu(x).$$

Now, for $i \in \{1, 2\}$, it follows, from (3.1.5) and (3.1.6), that

$$\int_{Q_i} \left|g(x) - m_{\tilde{R}}(g)\right|\,d\mu(x)$$
$$\leq \int_{Q_i} \left|g(x) - m_{\widetilde{Q_i}}(g)\right|\,d\mu(x) + \left|m_{\widetilde{Q_i}}(g) - m_{\tilde{R}}(g)\right|\mu(Q_i)$$
$$\lesssim \mu(2Q_i)[1 + \delta(Q_i, R)]\|g\|_{\mathrm{RBMO}_\gamma^2(\mu)}.$$

By this and Definition 3.2.1, we see that

$$\left|\int_{\mathbb{R}^D} b(x)g(x)\,d\mu(x)\right| \lesssim \sum_{i=1}^{2} \lambda_i \|g\|_{\mathrm{RBMO}_\gamma^2(\mu)},$$

which completes the proof of Lemma 3.2.6. □

In the following lemma we prove the converse inequality to the one in Lemma 3.2.6.

Lemma 3.2.7. *Let $\gamma \in [1, \infty)$. If $g \in \mathrm{RBMO}_\gamma^2(\mu)$ and L_g is as in Lemma 3.2.6, then*

$$\|L_g\|_{(H_{\mathrm{atb},\gamma}^{1,\infty}(\mu))^*} \sim \|g\|_{\mathrm{RBMO}_\gamma^2(\mu)}$$

with the implicit equivalent positive constants independent of g.

Proof. By Lemma 3.2.6, it suffices to show

$$\|L_g\|_{(H_{\mathrm{atb},\gamma}^{1,\infty}(\mu))^*} \gtrsim \|g\|_{\mathrm{RBMO}_\gamma^\rho(\mu)}.$$

Without loss of generality, we may assume that g is real-valued. With the aid of Lemma 3.2.6, we only need to prove that there exists some function $f \in H_{\mathrm{atb},\gamma}^{1,\infty}(\mu)$ such that

$$|L_g(f)| \gtrsim \|g\|_\circ \|f\|_{H_{\mathrm{atb},\gamma}^{1,\infty}(\mu)}. \tag{3.2.7}$$

Let $\epsilon \in (0, 1/8]$. There exist two possibilities.

Case (a) There exists some doubling cube $Q \subset \mathbb{R}^D$ such that

$$\int_Q |g(x) - m_g(Q)|\, d\mu(x) \geq \epsilon \|g\|_\circ \mu(Q). \tag{3.2.8}$$

In this case, if Q is doubling and satisfies (3.2.8), then we take f satisfying that $f(x) := 1$ if $g(x) > m_g(Q)$, $f(x) := -1$ if $g(x) < m_g(Q)$, and $f(x) := \pm 1$ if $g(x) = m_g(Q)$, such that

$$\int_{\mathbb{R}^D} f(x)\, d\mu(x) = 0.$$

Then we see that

$$\begin{aligned}\left|\int_{\mathbb{R}^D} g(x) f(x)\, d\mu(x)\right| &= \left|\int_{\mathbb{R}^D} [g(x) - m_g(Q)] f(x)\, d\mu(x)\right| \\ &= \int_{\mathbb{R}^D} |g(x) - m_g(Q)|\, d\mu(x) \\ &\geq \epsilon \|g\|_\circ \mu(Q).\end{aligned}$$

Since f is an (∞, γ)-atomic block and Q is doubling, it follows that

$$\|f\|_{H_{\rm atb,\gamma}^{1,\infty}(\mu)} \leq |f|_{H_{\rm atb,\gamma}^{1,\infty}(\mu)} \lesssim \mu(Q).$$

Therefore, we find that

$$|L_g(f)| = \left|\int_{\mathbb{R}^D} g(x) f(x)\, d\mu(x)\right| \gtrsim \epsilon \|g\|_\circ \|f\|_{H_{\rm atb,\gamma}^{1,\infty}(\mu)},$$

which shows (3.2.7) in this case.

Case (b) For any doubling $Q \subset \mathbb{R}^D$, (3.2.8) fails. In this case, we further consider the following two subcases.

Subcase (i) For any two doubling cubes $Q \subset R$,

$$|m_g(Q) - m_g(R)| \leq \frac{1}{2}[1 + \delta(Q, R)]^\gamma \|g\|_\circ.$$

In this subcase, by the definition of $\|g\|_\circ$, there exists some cube Q such that

$$\int_Q \left|g(x) - m_g(\tilde{Q})\right|\, d\mu(x) \geq \frac{1}{2} \|g\|_\circ \mu(2Q).$$

Let $f := a_1 + a_2$, where

$$a_1 := \chi_{Q \cap \{g > m_g(\tilde{Q})\}} - \chi_{Q \cap \{g < m_g(\tilde{Q})\}},$$

a_2 is supported on $\tilde{Q}$ and constant on $\tilde{Q}$ such that

$$\int_{\mathbb{R}^D} [a_1(x) + a_2(x)]\, d\mu(x) = 0.$$

We then have

$$\|a_2\|_{L^\infty(\mu)}\mu(\tilde{Q}) = \left|\int_{\mathbb{R}^D} a_2(x)\, d\mu(x)\right| = \left|\int_{\mathbb{R}^D} a_1(x)\, d\mu(x)\right| \le \mu(Q).$$

Since Q is doubling and $\delta(Q, 2\tilde{Q}) \lesssim 1$, we see that

$$\|f\|_{H^{1,\infty}_{\mathrm{atb},\gamma}(\mu)} \lesssim \|a_1\|_{L^\infty(\mu)}\mu(2Q) + \|a_2\|_{L^\infty(\mu)}\mu(2\tilde{Q}) \lesssim \mu(2Q).$$

Now we find that

$$\begin{aligned} L_g(f) &= \int_{\tilde{Q}} [g - m_g(\tilde{Q})]\, f\, d\mu \\ &= \int_{\tilde{Q}} [g - m_g(\tilde{Q})]\, a_1\, d\mu + \int_{\tilde{Q}} [g - m_g(\tilde{Q})]\, a_2\, d\mu. \end{aligned}$$

From the definition of a_1 and the choice of Q, it follows that

$$\left|\int_{\tilde{Q}} [g(x) - m_g(\tilde{Q})]\, a_1(x) d\mu(x)\right| = \int_Q |g(x) - m_g(\tilde{Q})|\, d\mu(x) \ge \frac{1}{2}\|g\|_\circ \mu(2Q).$$

On the other hand, since (3.2.8) fails, we see that

$$\left|\int_{\tilde{Q}} [g - m_g(\tilde{Q})]\, a_2 d\mu\right| \le \frac{\mu(Q)}{\mu(\tilde{Q})}\int_{\tilde{Q}} |g - m_g(\tilde{Q})|\, d\mu \le \epsilon\|g\|_\circ \mu(2Q).$$

From the choice of ϵ, we then deduce that

$$|L_g(f)| \ge \frac{1}{4}\|g\|_\circ \mu(2Q) \gtrsim \|g\|_\circ \|f\|_{H^{1,\infty}_{\mathrm{atb},\gamma}(\mu)},$$

which shows (3.2.7) in this subcase.

Subcase (ii) There exist some doubling cubes $Q \subset R$ such that

$$|m_g(Q) - m_g(R)| > \frac{1}{2}[1 + \delta(Q, R)]^\gamma \|g\|_\circ. \tag{3.2.9}$$

Let Q and R be such cubes. We take

$$f := \frac{1}{\mu(R)}\chi_R - \frac{1}{\mu(Q)}\chi_Q.$$

Then

$$\int_{\mathbb{R}^D} f(x)\,d\mu(x) = 0,$$

and hence f is an (∞, γ)-atomic block satisfying

$$\|f\|_{H^{1,\infty}_{\mathrm{atb},\gamma}(\mu)} \lesssim [1 + \delta(Q, R)]^{\gamma}.$$

Since (3.2.8) fails for Q and R, it follows, from (3.2.9) and the fact that $\epsilon \le \frac{1}{8}$, that

$$\begin{aligned}
|L_g(f)| &= \left|\int_R [g(x) - m_g(R)]\, f(x)\,d\mu(x)\right| \\
&= \left|\frac{1}{\mu(R)}\int_R [g(x) - m_g(R)]\;d\mu(x)\right. \\
&\quad \left. -\frac{1}{\mu(Q)}\int_Q [g(x) - m_g(R)]\;d\mu(x)\right| \\
&\ge |m_g(Q) - m_g(R)| - \left|\frac{1}{\mu(R)}\int_R [g(x) - m_g(R)]\;d\mu(x)\right| \\
&\quad - \left|\frac{1}{\mu(Q)}\int_Q [g(x) - m_g(Q)]\;d\mu(x)\right| \\
&\ge \frac{1}{4}[1 + \delta(Q, R)]^{\gamma}\,\|g\|_{\circ} \\
&\gtrsim \|g\|_{\circ}\|f\|_{H^{1,\infty}_{\mathrm{atb},\gamma}(\mu)}.
\end{aligned}$$

Thus, (3.2.7) also holds true in this subcase, which completes the proof of Lemma 3.2.7. □

For $H^{1,p}_{\mathrm{atb},\gamma}(\mu)$ with $p \in (1, \infty)$, we also show the easier one of the inclusions first.

Lemma 3.2.8. *For any $p \in (1, \infty)$ and $\gamma \in [1, \infty)$,*

$$\mathrm{RBMO}^2_{\gamma}(\mu) \subset \left(H^{1,p}_{\mathrm{atb},\gamma}(\mu)\right)^{*}.$$

That is, for any $g \in \mathrm{RBMO}^2_{\gamma}(\mu)$, the linear functional

$$L_g(f) := \int_{\mathbb{R}^D} f(x) g(x)\,d\mu(x),$$

defined over $f \in L^{\infty}(\mu)$ with compact support, can be extended to a unique continuous linear functional L_g over $H^{1,p}_{\mathrm{atb},\gamma}(\mu)$ and

$$\|L_g\|_{(H^{1,p}_{\mathrm{atb},\gamma}(\mu))^*} \le C\|g\|_{\mathrm{RBMO}^2_\gamma(\mu)},$$

where C is a positive constant independent of g.

Proof. It suffices to show that, if $b := \sum_{j=1}^2 \lambda_j a_j$ is a (p,γ)-atomic block with $\operatorname{supp} b \subset R$ as in Definition 3.2.1, then, for any $g \in \mathrm{RBMO}^2_\gamma(\mu)$,

$$\left|\int_{\mathbb{R}^D} b(x)g(x)\,d\mu(x)\right| \lesssim |b|_{H^{1,p}_{\mathrm{atb},\gamma}(\mu)}\|g\|_{\mathrm{RBMO}^2_\gamma(\mu)}.$$

Since

$$\int_{\mathbb{R}^D} b(x)\,d\mu(x) = 0,$$

by the Hölder inequality, we see that

$$\begin{aligned}\left|\int_{\mathbb{R}^D} b(x)g(x)\,d\mu(x)\right| &= \left|\int_R b(x)\left[g(x) - m_{\tilde{R}}(g)\right]\,d\mu(x)\right| \\ &\le \sum_{i=1}^2 |\lambda_i|\|a_j\|_{L^p(\mu)}\left[\int_{Q_i} |g(x) - m_{\tilde{R}}(g)|^{p'}\,d\mu(x)\right]^{1/p'}.\end{aligned}$$

Applying Corollary 3.1.20, we have

$$\begin{aligned}&\left[\int_{Q_i} |g(x) - m_{\tilde{R}}(g)|^{p'}\,d\mu(x)\right]^{1/p'} \\ &\le \left[\int_{Q_i} \left|g(x) - m_{\widetilde{Q_i}}(g)\right|^{p'}\,d\mu(x)\right]^{1/p'} + \left|m_{\widetilde{Q_i}}(g) - m_{\widetilde{R}}(g)\right|[\mu(Q_j)]^{1/p'} \\ &\lesssim \|g\|_{\mathrm{RBMO}^2_\gamma(\mu)}[\mu(2Q_j)]^{1/p'} + [1+\delta(Q_j,R)]^\gamma\|g\|_{\mathrm{RBMO}^2_\gamma(\mu)}[\mu(Q_j)]^{1/p'} \\ &\lesssim [1+\delta(Q_j,R)]^\gamma\|g\|_{\mathrm{RBMO}^2_\gamma(\mu)}[\mu(2Q_j)]^{1/p'}.\end{aligned}$$

This, together with (3.2.1), implies that

$$\left|\int_{\mathbb{R}^D} b(x)g(x)\,d\mu(x)\right| \lesssim \sum_j |\lambda_j|\|g\|_{\mathrm{RBMO}^2_\gamma(\mu)} \sim |b|_{H^{1,p}_{\mathrm{atb},\gamma}(\mu)}\|g\|_{\mathrm{RBMO}^2_\gamma(\mu)},$$

which completes the proof of Lemma 3.2.8. □

Lemma 3.2.9. *Let $p \in (1,\infty)$ and $\gamma \in [1,\infty)$. Then*

$$\left(H^{1,p}_{\mathrm{atb},\gamma}(\mu)\right)^* \subset L^{p'}_{\mathrm{loc}}(\mu).$$

Proof. This lemma is an easy consequence of the Riesz representation theorem, and it can be proved by a slight modification of the argument in [75, pp. 39–40]. We omit the details, which completes the proof of Lemma 3.2.9. □

Now we have the duality of $H^{1,p}_{\mathrm{atb},\gamma}(\mu)$ and RBMO (μ) as follows.

Theorem 3.2.10. *For any $p \in (1, \infty)$ and $\gamma \in [1, \infty)$,*

$$\left(H^{1,p}_{\mathrm{atb},\gamma}(\mu)\right)^* = \mathrm{RBMO}^2_\gamma(\mu).$$

Proof. By Lemma 3.2.8, to prove Theorem 3.2.10, it suffices to show that, for any $p \in (1, \infty)$ and $\gamma \in [1, \infty)$,

$$\left(H^{1,p}_{\mathrm{atb},\gamma}(\mu)\right)^* \subset \mathrm{RBMO}^2_\gamma(\mu).$$

Based on Lemma 3.2.9, we let $g \in L^{p'}_{\mathrm{loc}}(\mu)$ such that $L_g \in (H^{1,p}_{\mathrm{atb},\gamma}(\mu))^*$. We prove that $g \in \mathrm{RBMO}^2_\gamma(\mu)$ by showing that, for any cube Q,

$$\frac{1}{\mu(2Q)} \int_Q \left|g(x) - m_g(\tilde{Q})\right| d\mu(x) \lesssim \|L_g\|_{(H^{1,p}_{\mathrm{atb},\gamma}(\mu))^*} \tag{3.2.10}$$

and, for any two doubling cubes $Q \subset R$,

$$|m_g(Q) - m_g(R)| \lesssim [1 + \delta(Q, R)]^\gamma \|L_g\|_{(H^{1,p}_{\mathrm{atb},\gamma}(\mu))^*}. \tag{3.2.11}$$

We fist consider the case that Q is doubling. Without loss of generality, we may assume that

$$\begin{aligned} &\int_{Q \cap \{g > m_g(Q)\}} |g(x) - m_g(Q)|^{p'} d\mu(x) \\ &\quad \geq \int_{Q \cap \{g < m_g(Q)\}} |g(x) - m_g(Q)|^{p'} d\mu(x). \end{aligned} \tag{3.2.12}$$

We further define an atomic block as follows:

$$a(x) := \begin{cases} |g(x) - m_g(Q)|^{p'-1}, & \text{if } x \in Q^*; \\ C_Q, & \text{if } x \in Q \setminus Q^*; \\ 0, & \text{if } x \notin Q, \end{cases}$$

where

$$Q^* := Q \bigcap \{x \in \mathbb{R}^D : g(x) > m_g(Q)\}$$

and C_Q is a constant such that $\int_{\mathbb{R}^D} a(x)\, d\mu(x) = 0$.

By the definition of $m_g(Q)$, we have

$$\mu(Q^*) \le \frac{1}{2}\mu(Q) \le \mu(Q \setminus Q^*). \tag{3.2.13}$$

Since Q is doubling, it follows that

$$\begin{aligned}\|a\|_{H^{1,p}_{\mathrm{atb},\gamma}(\mu)} &\lesssim \|a\|_{L^p(\mu)}[\mu(Q)]^{1/p'}\\ &\lesssim \mu(Q)\left[\frac{1}{\mu(Q)}\int_{Q^*}|g(x)-m_g(Q)|^{p'}\,d\mu(x)\right.\\ &\quad\left.+\frac{1}{\mu(Q)}\int_{Q\setminus Q^*}|C_Q|^p\,d\mu(x)\right]^{1/p}.\end{aligned}$$

From (3.2.13), we deduce that

$$\begin{aligned}&\frac{1}{\mu(Q)}\int_{Q\setminus Q^*}|C_Q|^p\,d\mu(x)\\ &\quad\lesssim \frac{1}{\mu(Q\setminus Q^*)}\int_{Q\setminus Q^*}|C_Q|^p\,d\mu(x)\\ &\quad\sim \left|\frac{1}{\mu(Q\setminus Q^*)}\int_{Q\setminus Q^*}C_Q\,d\mu(x)\right|^p\\ &\quad\sim \left[\frac{1}{\mu(Q\setminus Q^*)}\int_{Q^*}|g(x)-m_g(Q)|^{p'-1}\,d\mu(x)\right]^p\\ &\quad\lesssim \frac{1}{\mu(Q\setminus Q^*)}\int_{Q^*}|g(x)-m_g(Q)|^{p'}\,d\mu(x).\end{aligned}$$

Therefore, we obtain

$$\|a\|_{H^{1,p}_{\mathrm{atb},\gamma}(\mu)} \lesssim \mu(Q)\left[\frac{1}{\mu(Q)}\int_{Q^*}|g(x)-m_g(Q)|^{p'}\,d\mu(x)\right]^{1/p}. \tag{3.2.14}$$

Since

$$(g-m_g(Q))a \ge 0$$

on Q, it follows, from (3.2.12), that

$$\int_Q g(x)a(x)\,d\mu(x) = \int_Q [g(x)-m_g(Q)]\,a(x)\,d\mu(x)$$

$$\geq \int_{Q^*} |g(x) - m_g(Q)|^{p'} \, d\mu(x)$$
$$\geq \frac{1}{2} \int_Q |g(x) - m_g(Q)|^{p'} \, d\mu(x).$$

By this and (3.2.14), we see that

$$\left[\frac{1}{\mu(Q)} \int_Q |g(x) - m_g(Q)|^{p'} \, d\mu(x)\right]^{1/p'} \|a\|_{H^{1,p}_{\rm atb,\gamma}(\mu)}$$
$$\lesssim \int_Q |g(x) - m_g(Q)|^{p'} \, d\mu(x)$$
$$\lesssim \int_Q g(x)a(x) \, d\mu(x)$$
$$\lesssim \|L_g\|_{(H^{1,p}_{\rm atb,\gamma}(\mu))^*} \|a\|_{H^{1,p}_{\rm atb,\gamma}(\mu)}.$$

Thus, (3.2.10) holds true in this case.

Assume now that Q is not doubling. Let $b := a_1 + a_2$, where

$$a_1 := \frac{|g - m_g(\tilde{Q})|^{p'}}{g - m_g(\tilde{Q})} \chi_{Q \cap \{g \neq m_g(\tilde{Q})\}},$$

$a_2 := C_{\tilde{Q}} \chi_{\tilde{Q}}$ and $C_{\tilde{Q}}$ is a constant such that

$$\int_{\mathbb{R}^D} b(x) \, d\mu(x) = 0.$$

Now we estimate $\|b\|_{H^{1,p}_{\rm atb,\gamma}(\mu)}$. Since $\tilde{Q}$ is doubling and $\delta(Q, \tilde{Q}) \lesssim 1$, it follows that

$$\|b\|_{H^{1,p}_{\rm atb,\gamma}(\mu)} \lesssim \left[\int_Q |g(x) - m_g(\tilde{Q})|^{p'} \, d\mu(x)\right]^{1/p} [\mu(2Q)]^{1/p'}$$
$$+ |C_{\tilde{Q}}| \mu(\tilde{Q}). \tag{3.2.15}$$

By

$$\int_{\mathbb{R}^D} b(x) \, d\mu(x) = 0,$$

we have

$$\begin{aligned}\mu(\tilde{Q})\left|C_{\tilde{Q}}\right| &= \left|\int_{\mathbb{R}^D} a_1(x)\,d\mu(x)\right| \\ &\le \int_Q |g(x) - m_g(\tilde{Q})|^{p'-1}\,d\mu(x) \\ &\le \left[\int_Q |g(x) - m_g(\tilde{Q})|^{p'}\,d\mu(x)\right]^{1/p} [\mu(Q)]^{1/p'}. \end{aligned} \tag{3.2.16}$$

Thus,

$$\|b\|_{H^{1,p}_{\mathrm{atb},\gamma}(\mu)} \lesssim \left[\int_Q |g(x) - m_g(\tilde{Q})|^{p'}\,d\mu(x)\right]^{1/p} [\mu(2Q)]^{1/p'}. \tag{3.2.17}$$

As

$$\int_{\mathbb{R}^D} b(x)\,d\mu(x) = 0,$$

we also find that

$$\begin{aligned}\int_{\tilde{Q}} g(x)b(x)\,d\mu(x) &= \int_{\tilde{Q}} [g(x) - m_g(\tilde{Q})]\,b(x)\,d\mu(x) \\ &= \int_Q [g(x) - m_g(\tilde{Q})]\,a_1(x)\,d\mu(x) \\ &\quad + C_{\tilde{Q}} \int_{\tilde{Q}} [g(x) - m_g(\tilde{Q})]\,d\mu(x).\end{aligned}$$

From (3.2.16) and the fact that $\tilde{Q}$ satisfies (3.2.10), it follows that

$$\begin{aligned}&\int_Q |g(x) - m_g(\tilde{Q})|^{p'}\,d\mu(x) \\ &\quad = \int_Q [g(x) - m_g(\tilde{Q})]\,a_1(x)\,d\mu(x) \\ &\quad \le \left|\int_{\mathbb{R}^D} g(x)b(x)\,d\mu(x)\right| + \left|C_{\tilde{Q}}\right| \int_{\tilde{Q}} |g(x) - m_g(\tilde{Q})|\,d\mu(x) \\ &\quad \lesssim \|L_g\|_{(H^{1,p}_{\mathrm{atb},\gamma}(\mu))^*} \left[\|b\|_{H^{1,p}_{\mathrm{atb},\gamma}(\mu)} + \left|C_{\tilde{Q}}\right| \mu(\tilde{Q})\right] \\ &\quad \lesssim \|L_g\|_{(H^{1,p}_{\mathrm{atb},\gamma}(\mu))^*} \\ &\qquad \times \left\{\|b\|_{H^{1,p}_{\mathrm{atb},\gamma}(\mu)} + \left[\int_Q |g(x) - m_g(\tilde{Q})|^{p'}\,d\mu(x)\right]^{1/p} [\mu(Q)]^{1/p'}\right\}.\end{aligned}$$

By (3.2.17), we see that

$$\begin{aligned}&\int_Q |g(x)-m_g(\tilde{Q})|^{p'}\,d\mu(x)\\&\quad\lesssim \|L_g\|_{(H_{\mathrm{atb},\gamma}^{1,p}(\mu))^*}\left[\int_Q |g(x)-m_g(\tilde{Q})|^{p'}\,d\mu(x)\right]^{1/p}[\mu(2Q)]^{1/p'}.\end{aligned}$$

That is,

$$\left[\frac{1}{\mu(2Q)}\int_Q |g(x)-m_g(\tilde{Q})|^{p'}\,d\mu(x)\right]^{1/p'}\lesssim \|L_g\|_{(H_{\mathrm{atb},\gamma}^{1,p}(\mu))^*},$$

which implies (3.2.10) in this case and hence completes the proof of (3.2.10).

It remains to show that (3.2.11) holds true for doubling cubes $Q\subset R$. Let $b:=a_1+a_2$ be a (p,γ)-atomic block, where

$$a_1:=\frac{|g-m_g(R)|^{p'}}{g-m_g(R)}\chi_{Q\cap\{g\neq m_g(R)\}},$$

$a_2:=C_R\chi_R$ and C_R is a constant such that

$$\int_{\mathbb{R}^D} b(x)\,d\mu(x)=0.$$

Arguing as in (3.2.15), (3.2.16) and (3.2.17), we obtain

$$\begin{aligned}&\|b\|_{H_{\mathrm{atb},\gamma}^{1,p}(\mu)}\\&\quad\lesssim [1+\delta(Q,R)]^{\gamma}\\&\quad\quad\times\left[\int_Q |g(x)-m_g(R)|^{p'}\,d\mu(x)\right]^{1/p}[\mu(2Q)]^{1/p'}.\end{aligned}\tag{3.2.18}$$

From this, we deduce that

$$\begin{aligned}&\int_Q |g(x)-m_g(R)|^{p'}\,d\mu(x)\\&\quad\lesssim \|L_g\|_{(H_{\mathrm{atb},\gamma}^{1,p}(\mu))^*}\left[\|b\|_{H_{\mathrm{atb},\gamma}^{1,p}(\mu)}+|C_R|\,\mu(R)\right]\\&\quad\lesssim \|L_g\|_{(H_{\mathrm{atb},\gamma}^{1,p}(\mu))^*}\left\{\|b\|_{H_{\mathrm{atb},\gamma}^{1,p}(\mu)}\right.\\&\quad\quad\left.+\left[\int_Q |g(x)-m_g(R)|^{p'}\,d\mu(x)\right]^{1/p}[\mu(Q)]^{1/p'}\right\}.\end{aligned}$$

Using (3.2.18), we further have

$$\int_Q |g(x) - m_g(R)|^{p'} \, d\mu(x)$$

$$\lesssim \|L_g\|_{(H^{1,p}_{\text{atb},\gamma}(\mu))^*} [1 + \delta(Q,R)]^\gamma \left[\int_Q |g(x) - m_g(R)|^{p'} \, d\mu(x) \right]^{1/p} [\mu(Q)]^{1/p'}.$$

Consequently, we see that

$$\left[\frac{1}{\mu(Q)} \int_Q |g(x) - m_g(R)|^{p'} \, d\mu(x) \right]^{1/p'} \lesssim \|L_g\|_{(H^{1,p}_{\text{atb},\gamma}(\mu))^*} [1 + \delta(Q,R)]^\gamma.$$

Recall that Q is doubling and hence satisfies (3.2.10). We then find that

$$\begin{aligned}
&|m_g(Q) - m_g(R)| \\
&\quad = \frac{1}{\mu(Q)} \int_Q |m_g(Q) - m_g(R)| \, d\mu(x) \\
&\quad \le \frac{1}{\mu(Q)} \int_Q |g(x) - m_g(Q)| \, d\mu(x) + \frac{1}{\mu(Q)} \int_Q |g(x) - m_g(R)| \, d\mu(x) \\
&\quad \lesssim \|L_g\|_{(H^{1,p}_{\text{atb},\gamma}(\mu))^*} [1 + \delta(Q,R)]^\gamma,
\end{aligned}$$

which shows (3.2.11) and hence completes the proof of Theorem 3.2.10. □

By Theorem 3.2.10, we now show that

$$H^{1,p}_{\text{atb},\gamma}(\mu) = H^{1,\infty}_{\text{atb},\gamma}(\mu)$$

and the dual space of $H^{1,\infty}_{\text{atb},\gamma}(\mu)$ is $\text{RBMO}^\rho_\gamma(\mu)$ for any $\gamma \in [1,\infty)$ and $p \in (1,\infty)$.

Theorem 3.2.11. *Let $p \in (1,\infty)$ and $\gamma \in [1,\infty)$. Then $H^{1,p}_{\text{atb},\gamma}(\mu) = H^{1,\infty}_{\text{atb},\gamma}(\mu)$ and*

$$\left(H^{1,\infty}_{\text{atb},\gamma}(\mu) \right)^* = \text{RBMO}^2_\gamma(\mu).$$

Proof. By Proposition 3.2.4, we see that, if $f \in (H^{1,p}_{\text{atb},\gamma}(\mu))^*$, then $f \in (H^{1,\infty}_{\text{atb},\gamma}(\mu))^*$. With the aid of Theorem 3.2.10, we consider the maps

$$i : H^{1,\infty}_{\text{atb},\gamma}(\mu) \to H^{1,p}_{\text{atb},\gamma}(\mu)$$

and

$$i^* : \mathrm{RBMO}_\gamma^2(\mu) = \left(H_{\mathrm{atb},\gamma}^{1,p}(\mu)\right)^* \to \left(H_{\mathrm{atb},\gamma}^{1,\infty}(\mu)\right)^*.$$

Notice that the map i is an inclusion and i^* the canonical injection of $\mathrm{RBMO}_\gamma^2(\mu)$ in $(H_{\mathrm{atb},\gamma}^{1,\infty}(\mu))^*$ (with the identification $g \equiv L_g$ for $g \in \mathrm{RBMO}_\gamma^2(\mu)$). By Lemma 3.2.7, $i^*(\mathrm{RBMO}_\gamma^2(\mu))$ is closed in $(H_{\mathrm{atb},\gamma}^{1,\infty}(\mu))^*$. An application of the Banach closed range theorem[3] shows that $H_{\mathrm{atb},\gamma}^{1,\infty}(\mu)$ is closed in $H_{\mathrm{atb},\gamma}^{1,p}(\mu)$, which, together with Proposition 3.2.4, implies that

$$H_{\mathrm{atb},\gamma}^{1,\infty}(\mu) = H_{\mathrm{atb},\gamma}^{1,p}(\mu)$$

as a set. Thus, i maps $H_{\mathrm{atb},\gamma}^{1,\infty}(\mu)$ onto $H_{\mathrm{atb},\gamma}^{1,p}(\mu)$. Observing that both $H_{\mathrm{atb},\gamma}^{1,\infty}(\mu)$ and $H_{\mathrm{atb},\gamma}^{1,p}(\mu)$ are Banach spaces, by the corollary of the open mapping theorem,[4] we conclude that

$$H_{\mathrm{atb},\gamma}^{1,\infty}(\mu) = H_{\mathrm{atb},\gamma}^{1,p}(\mu)$$

with equivalent norms, which completes the proof of Theorem 3.2.11. □

Based on Theorem 3.2.11, we now show that $H_{\mathrm{atb},\gamma}^{1,p}(\mu) = H_{\mathrm{atb}}^{1,p}(\mu)$ for any $p \in (1,\infty]$ and $\gamma \in (1,\infty)$.

Theorem 3.2.12. *For any $p \in (1,\infty]$ and any $\gamma \in (1,\infty)$, the spaces $H_{\mathrm{atb},\gamma}^{1,p}(\mu)$ and $H_{\mathrm{atb}}^{1,p}(\mu)$ coincide with equivalent norms.*

Proof. Using Theorem 3.2.11, it suffices to show Theorem 3.2.12 for $p = \infty$. Obviously,

$$H_{\mathrm{atb},\gamma}^{1,\infty}(\mu) \subseteq H_{\mathrm{atb}}^{1,\infty}(\mu).$$

From another application of Theorem 3.2.11, we deduce that

$$\left(H_{\mathrm{atb},\gamma}^{1,\infty}(\mu)\right)^* = \left(H_{\mathrm{atb}}^{1,\infty}(\mu)\right)^*. \tag{3.2.19}$$

By the Banach closed range theorem, $H_{\mathrm{atb},\gamma}^{1,\infty}(\mu)$ is closed in $H_{\mathrm{atb}}^{1,\infty}(\mu)$. From the corollary of the Hahn–Banach theorem[5] and (3.2.19), it follows that

---

[3] See [157, p. 205].

[4] See [157, p. 77].

[5] See [157, p. 108].

$$H^{1,\infty}_{\mathrm{atb},\gamma}(\mu) = H^{1,\infty}_{\mathrm{atb}}(\mu)$$

as a set. Invoking the open mapping theorem again, we see that

$$H^{1,\infty}_{\mathrm{atb},\gamma}(\mu) = H^{1,\infty}_{\mathrm{atb}}(\mu)$$

with equivalent norms. This finishes the proof of Theorem 3.2.12. □

3.3 An Equivalent Characterization of $H^{1,p}_{\mathrm{atb}}(\mu)$ Via the Maximal Function

It is known that the classical Hardy space $H^1(\mathbb{R}^D)$ on $\mathbb{R}^D$ with the D-dimensional Lebesgue measure can be characterized by various maximal functions.[6] In this section, we establish an equivalent characterization of $H^{1,p}_{\mathrm{atb}}(\mu)$ in terms of a class of maximal functions on $\mathbb{R}^D$ with measures μ satisfying (0.0.1). We start with the definition of the maximal functions.

Definition 3.3.1. Given $f \in L^1_{\mathrm{loc}}(\mu)$, let

$$\mathcal{M}_\Phi(f)(x) := \sup_{\varphi \sim x} \left| \int_{\mathbb{R}^D} f(y)\varphi(y)\, d\mu(y) \right|,$$

where the notation $\varphi \sim x$ means that $\varphi \in L^1(\mu) \cap C^1(\mathbb{R}^D)$ satisfying

(i) $\|\varphi\|_{L^1(\mu)} \le 1$,
(ii) $0 \le \varphi(y) \le \frac{1}{|y-x|^n}$ for all $y \in \mathbb{R}^D$,
(iii) $|\nabla\varphi(y)| \le \frac{1}{|y-x|^{n+1}}$ for all $y \in \mathbb{R}^D$.

Using $\mathcal{M}_\Phi$, we now introduce a Hardy space $H^1_\Phi(\mu)$ as follows.

Definition 3.3.2. The *Hardy space* $H^1_\Phi(\mu)$ is defined to be the set of all functions $f \in L^1(\mu)$ satisfying that

$$\int_{\mathbb{R}^D} f(x)\, d\mu(x) = 0 \quad \text{and} \quad \mathcal{M}_\Phi(f) \in L^1(\mu).$$

Moreover, the *norm* of $f \in H^1_\Phi(\mu)$ is defined by

$$\|f\|_{H^1_\Phi(\mu)} := \|f\|_{L^1(\mu)} + \|\mathcal{M}_\Phi(f)\|_{L^1(\mu)}.$$

The main result of this section is the following theorem.

---

[6] See [26, 121].

Theorem 3.3.3. *A function f belongs to $H^{1,\infty}_{\rm atb}(\mu)$ if and only if $f \in H^1_\Phi(\mu)$. Moreover, in this case, there exists a constant $C \in (1,\infty)$, independent of f, such that*

$$\|f\|_{H^{1,\infty}_{\rm atb}(\mu)}/C \le \|f\|_{H^1_\Phi(\mu)} \le C\|f\|_{H^{1,\infty}_{\rm atb}(\mu)}.$$

We first show the "only if" part of Theorem 3.3.3 as follows.

Lemma 3.3.4. *There exists a positive constant $\tilde{C}$ such that, for all $f \in H^{1,\infty}_{\rm atb}(\mu)$,*

$$\|\mathcal{M}_\Phi f\|_{L^1(\mu)} \le C\|f\|_{H^{1,\infty}_{\rm atb}(\mu)}. \tag{3.3.1}$$

Proof. Let $b := \lambda_1 a_1 + \lambda_2 a_2$ be an atomic block supported on some cube R, where a_i, $i \in \{1,2\}$, is a function supported on a cube $Q_i \subset R$ such that

$$\|a_i\|_{L^\infty(\mu)} \le [1+\delta(Q,R)][\mu(2Q_i)]^{-1}.$$

We show that

$$\|\mathcal{M}_\Phi b\|_{L^1(\mu)} \lesssim |\lambda_1| + |\lambda_2|. \tag{3.3.2}$$

First we estimate the integral

$$\int_{\mathbb{R}^D\setminus(2R)} \mathcal{M}_\Phi b(x)\,d\mu(x).$$

For $x \in \mathbb{R}^D \setminus (2R)$ and $\varphi \sim x$, since

$$\int_{\mathbb{R}^D} b(x)\,d\mu(x) = 0,$$

we have

$$\begin{aligned}\left|\int_R b(y)\varphi(y)\,d\mu(y)\right| &= \left|\int_R b(y)[\varphi(y)-\varphi(z_R)]\,d\mu(y)\right|\\ &\lesssim \int_R |b(y)|\frac{\ell(R)}{|x-z_R|^{n+1}}\,d\mu(y).\end{aligned}$$

Thus,

$$\begin{aligned}\int_{\mathbb{R}^D\setminus(2R)} \mathcal{M}_\Phi b(x)\,d\mu(x) &\lesssim \|b\|_{L^1(\mu)}\int_{\mathbb{R}^D\setminus(2R)} \frac{\ell(R)}{|x-z_R|^{n+1}}\,d\mu(x)\\ &\lesssim \|b\|_{L^1(\mu)}\\ &\lesssim |\lambda_1|+|\lambda_2|.\end{aligned}$$

Now we show that, for each i,

$$\int_{2R} M_\Phi a_i(x)\,d\mu(x) \lesssim 1.$$

If $x \in 2Q$ and $\varphi \sim x$, then

$$\left|\int_{Q_i} a_i(x)\varphi(x)\,d\mu(x)\right| \le \|a_i\|_{L^\infty(\mu)}\|\varphi\|_{L^1(\mu)} \le \|a_i\|_{L^\infty(\mu)}.$$

Thus,

$$\int_{2Q_i} \mathcal{M}_\Phi a_i(x)\,d\mu(x) \le \|a_i\|_{L^\infty(\mu)}\mu(2Q_i) \le 1.$$

For $x \in 2R \setminus (2Q_i)$ and $\varphi \sim x$, we see that

$$\left|\int_{Q_i} a_i(y)\varphi(y)\,d\mu(y)\right| \lesssim \|a_i\|_{L^1(\mu)}\frac{1}{|x - z_{Q_i}|^n}.$$

Therefore,

$$\begin{aligned}\int_{2R\setminus(2Q_i)} \mathcal{M}_\Phi a_i(x)\,d\mu(x) &\lesssim \|a_i\|_{L^1(\mu)}\int_{2R\setminus(2Q_i)} \frac{1}{|x - z_{Q_i}|^n}\,d\mu(x)\\ &\lesssim \|a_i\|_{L^1(\mu)}[1 + \delta(Q, R)]\\ &\lesssim 1. \end{aligned} \tag{3.3.3}$$

This finishes the proof of Lemma 3.3.4. □

To show the "if" part of Theorem 3.3.3, we first have the following result.

Lemma 3.3.5. *The subspace $H^1_\Phi(\mu) \cap L^\infty(\mu)$ is dense in $H^1_\Phi(\mu)$.*

Proof. Given $f \in H^1_\Phi(\mu)$, for each integer $k \in \mathbb{Z}_+$, we consider the generalized Calderón–Zygmund decomposition of f in Theorem 1.4.2 with $\lambda := 2^k$. We adopt the convention that all the elements of that decomposition carry the subscript k. Thus, we write $f := g_k + b_k$ as in Theorem 1.4.2. We know that g_k is bounded and satisfies that

$$\int_{\mathbb{R}^D} g_k(x)\,d\mu(x) = 0.$$

We show that $g_k \to f$ in $L^1(\mu)$ and

$$\|\mathcal{M}_\Phi(g_k - f)\|_{L^1(\mu)} \to 0 \text{ as } k \to \infty.$$

It is not difficult to show that b_k tends to 0 in $L^1(\mu)$ as $k \to \infty$. Indeed, if we let

$$\Omega_k := \{x \in \mathbb{R}^D : \mathcal{M}_{(2)} f(x) > 2^k\},$$

then $\mu(\Omega_k) \to 0$ as $k \to \infty$, since $f \in L^1(\mu)$. Thus,

$$\int_{\mathbb{R}^D} |b_k(x)|\, d\mu(x) \lesssim \sum_i \int_{Q_{i,k}} |f(x)w_{i,k}(x)|\, d\mu(x) \lesssim \int_{\Omega_k} |f(x)|\, d\mu(x) \to 0,$$

as $k \to \infty$, and hence $g_k \to f$ in $L^1(\mu)$.

We now show that $\|\mathcal{M}_\Phi b_k\|_{L^1(\mu)} \to 0$ as $k \to \infty$. Let

$$b_{i,k} := f w_{i,k} - \alpha_{i,k}.$$

Then we have

$$\|\mathcal{M}_\Phi b_k\|_{L^1(\mu)} \le \sum_i \|\mathcal{M}_\Phi b_{i,k}\|_{L^1(\mu)}.$$

We write

$$\begin{aligned}\|\mathcal{M}_\Phi b_{i,k}\|_{L^1(\mu)} \le& \int_{\mathbb{R}^D \setminus (2R_{i,k})} \mathcal{M}_\Phi b_{i,k}(x)\, d\mu(x)\\ &+ \int_{2R_{i,k}} \mathcal{M}_\Phi (f w_{i,k})(x)\, d\mu(x) + \int_{2R_{i,k}} \mathcal{M}_\Phi \alpha_{i,k}(x)\, d\mu(x).\end{aligned}$$

Taking into account

$$\int_{\mathbb{R}^D} b_{i,k}(x)\, d\mu(x) = 0,$$

we see that

$$\int_{\mathbb{R}^D \setminus (2R_{i,k})} \mathcal{M}_\Phi b_{i,k}(x)\, d\mu(x) \lesssim \|b_{i,k}\|_{L^1(\mu)} \lesssim \|f w_{i,k}\|_{L^1(\mu)}.$$

By Theorem 1.4.2, we conclude that

$$\int_{2R_{i,k}} \mathcal{M}_\Phi \alpha_{i,k}(x)\, d\mu(x) \le \|\alpha_{i,k}\|_{L^\infty(\mu)} \mu(2R_{i,k}) \lesssim \|f w_{i,k}\|_{L^1(\mu)}.$$

Write

$$\int_{2R_{i,k}} \mathcal{M}_\Phi (f w_{i,k})(x)\, d\mu(x) = \int_{(2R_{i,k}) \setminus (2Q_{i,k})} \mathcal{M}_\Phi (f w_{i,k})(x)\, d\mu(x) + \int_{2Q_{i,k}} \cdots.$$

By Definition 3.3.1, we then have

$$\begin{aligned}
&\int_{(2R_{i,k})\setminus(2Q_{i,k})} \mathcal{M}_\Phi(fw_{i,k})(x)\,d\mu(x)\\
&\quad\lesssim \|fw_{i,k}\|_{L^1(\mu)} \int_{(2R_{i,k})\setminus(2Q_{i,k})} \frac{1}{|x - z_{Q_{i,k}}|^n}\,d\mu(x)\\
&\quad\lesssim \|fw_{i,k}\|_{L^1(\mu)}[1 + \delta(Q_{i,k}, R_{i,k})]\\
&\quad\lesssim \|fw_{i,k}\|_{L^1(\mu)}.
\end{aligned}$$

It remains to estimate

$$\int_{2Q_{i,k}} \mathcal{M}_\Phi(fw_{i,k})(x)\,d\mu(x).$$

Consider $x \in 2Q_{i,,k}$ and $\varphi \sim x$. We claim that $C\varphi w_{i,k} \sim x$ for some positive constant C. Indeed, for $y \in \mathbb{R}^D$, we have

$$0 \le \varphi(y)w_{i,k}(y) \le \frac{1}{|y - x|^n}$$

and

$$\begin{aligned}
|\nabla(\varphi w_{i,k})(y)| &\le |\nabla\varphi(y)w_{i,,k}(y)| + |\varphi(y)\nabla w_{i,k}(y)|\\
&\lesssim \frac{1}{|y - x|^{n+1}} + \frac{1}{|y - x|^n}|\nabla w_{i,k}(y)|.
\end{aligned}$$

Recall that

$$|\nabla w_{i,k}(y)| \lesssim [\ell(Q_{i,k})]^{-1} \quad \text{and} \quad \operatorname{supp} w_{i,k} \subset 2Q_{i,k}.$$

Then we see that, for all $y \in \mathbb{R}^D$,

$$|\nabla w_{i,k}(y)| \lesssim |y - x|^{-1}.$$

Thus,

$$|\nabla(\varphi w_{i,k})(y)| \lesssim |y - x|^{-n-1}$$

and hence the claim holds true.

Using the claim, we further see that

$$\left|\int_{\mathbb{R}^D} \varphi(y)[f(y)w_{i,k}(y)]\,d\mu(y)\right| = \left|\int_{\mathbb{R}^D} [\varphi(y)w_{i,k}(y)]f(y)\,d\mu(y)\right| \lesssim \mathcal{M}_\Phi f(x),$$

which implies that

$$\int_{2Q_{i,k}} \mathcal{M}_\Phi(f w_{i,k})(x)\,d\mu(x) \lesssim \int_{2Q_{i,k}} \mathcal{M}_\Phi f(x)\,d\mu(x).$$

Combining the estimates above, we have

$$\|\mathcal{M}_\Phi b_{i,k}\|_{L^1(\mu)} \lesssim \|f w_{i,k}\|_{L^1(\mu)} + \int_{2Q_{i,k}} \mathcal{M}_\Phi f(x)\,d\mu(x).$$

Taking into account the finite overlap of the cubes $2Q_{i,k}$, we further find that

$$\|\mathcal{M}_\Phi b_k\|_{L^1(\mu)} \lesssim \int_{\Omega_k} [|f(x)| + \mathcal{M}_\Phi f(x)]\,d\mu(x) \to 0,$$

as $k \to \infty$, which completes the proof of Lemma 3.3.5. □

Lemma 3.3.6. *Let $f \in H^1_\Phi(\mu)$. Then there exists a sequence $\{f_k\}_{k\in\mathbb{N}}$ of functions, bounded with compact support, such that*

$$\int_{\mathbb{R}^D} f_k(x)\,d\mu(x) = 0, \quad f_k \to f \quad \text{in} \quad L^1(\mu) \quad \text{and} \quad \|\mathcal{M}_\Phi(f - f_k)\|_{L^1(\mu)} \to 0.$$

Proof. By Lemma 3.3.5, we may assume that $f \in H^1_\Phi(\mu) \cap L^\infty(\mu)$. Consider the infinite increasing sequence of the cubes $Q_k := 4^{N_k}[-1,1]^D$ that are $(4, 4^{D+1})$-doubling, where $\{N_k\}_{k\in\mathbb{N}} \subset \mathbb{N}$ is increasing. Let w be a C^∞ function such that, for all x,

$$\chi_{[-1,1]^D}(x) \le w(x) \le \chi_{[-2,2]^D}(x).$$

Define $w_k(x) := w(4^{-N_k}x)$ for all $x \in \mathbb{R}^D$. Then we see that

$$\chi_{Q_k} \le w_k \le \chi_{2Q_k}.$$

Let

$$f_k := w_k f - \frac{\chi_{Q_k}}{\mu(Q_k)} \int_{\mathbb{R}^D} w_k(x) f(x)\,d\mu(x).$$

It is clear that f_k is bounded, has compact support and converges to f in $L^1(\mu)$ as $k \to \infty$. We prove that

$$\begin{aligned}\|\mathcal{M}_\Phi(f - f_k)\|_{L^1(\mu)} \lesssim{}& \left|\int_{\mathbb{R}^D} w_k(x) f(x)\,d\mu(x)\right| + \int_{\mathbb{R}^D\setminus(4Q_k)} \mathcal{M}_\Phi f(x)\,d\mu(x)\\ &+ \int_{4Q_k} \mathcal{M}_\Phi([1-w_k]f)(x)\,d\mu(x).\end{aligned} \tag{3.3.4}$$

Finally we show that the terms on the right hand side of (3.3.4) tend to 0 as $k \to \infty$.

Now we consider the integral of $\mathcal{M}_\Phi(f - f_k)$ over $\mathbb{R}^D \setminus (4Q_k)$. Write

$$\int_{\mathbb{R}^D\setminus(4Q_k)} \mathcal{M}_\Phi(f - f_k)(x)\, d\mu(x) \le \int_{\mathbb{R}^D\setminus(4Q_k)} \mathcal{M}_\Phi(f)(x)\, d\mu(x) + \int_{\mathbb{R}^D\setminus(4Q_k)} \mathcal{M}_\Phi(f_k)(x)\, d\mu(x).$$

It suffices to estimate the second term on the right hand side. Take $x \in \mathbb{R}^D \setminus (4Q_k)$, $\varphi \sim x$ and let $y_0 \in 2Q_k$ be the point where φ attains its minimum over $2Q_k$. Let

$$c_k := \int_{\mathbb{R}^D} w_k(x) f(x)\, d\mu(x)/\mu(Q_k)$$

and then we write

$$\begin{aligned}\int_{\mathbb{R}^D} f_k(y)\varphi(y)\, d\mu(y) &= \int_{\mathbb{R}^D} f_k(y)[\varphi(y) - \varphi(y_0)]\, d\mu(y)\\ &= \int_{\mathbb{R}^D} w_k(y) f(y)[\varphi(y) - \varphi(y_0)]\, d\mu(y)\\ &\quad -c_k \int_{Q_k} [\varphi(y) - \varphi(y_0)]\, d\mu(y)\\ &=: \mathrm{I}_1 - \mathrm{I}_2.\end{aligned}$$

Define $\psi(y) := w_k(y)[\varphi(y) - \varphi(y_0)]$ for all $y \in \mathbb{R}^D$. Then ψ satisfies that

$$0 \le \psi(y) \le \varphi(y)$$

and

$$\begin{aligned}|\nabla\psi(y)| &\le |w_k(y)\nabla\varphi(y)| + |\nabla w_k(y)||\varphi(y) - \varphi(y_0)|\\ &\lesssim \frac{1}{|x - y|^{n+1}} + [\ell(Q_k)]^{-1}\frac{\ell(Q_k)}{|x - y|^{n+1}}\\ &\sim \frac{1}{|x - y|^{n+1}}.\end{aligned}$$

Therefore, $C\psi \sim x$ for some positive constant C in the sense of Definition 3.3.1, and hence

$$|\mathrm{I}_1| \lesssim \mathcal{M}_\Phi f(x).$$

For I_2, we have

$$|\mathrm{I}_2| \lesssim |c_k|\mu(Q_k)\frac{\ell(Q_k)}{|x-y_0|}.$$

Thus,

$$\mathcal{M}_\Phi f_k(x) \lesssim \mathcal{M}_\Phi f(x) + |c_k|\mu(Q_k)\frac{\ell(Q_k)}{|x-y_0|^{n+1}}.$$

Since

$$\int_{\mathbb{R}^D\setminus(4Q_k)} \frac{1}{|x-y_0|^{n+1}}\,d\mu(x) \lesssim [\ell(Q_k)]^{-1},$$

it follows that

$$\begin{aligned}
&\int_{\mathbb{R}^D\setminus(4Q_k)} \mathcal{M}_\Phi f_k(x)\,d\mu(x)\\
&\quad\lesssim \int_{\mathbb{R}^D\setminus(4Q_k)} \mathcal{M}_\Phi f(x)\,d\mu(x) + |c_k|\mu(Q_k)\\
&\quad\lesssim \int_{\mathbb{R}^D\setminus(4Q_k)} \mathcal{M}_\Phi f(x)\,d\mu(x) + \left|\int_{\mathbb{R}^D} w_k(x)f(x)\,d\mu(x)\right|. \qquad (3.3.5)
\end{aligned}$$

To deal with

$$\int_{4Q_k} \mathcal{M}_\Phi(f-f_k)(x)\,d\mu(x) \text{ for } x\in 4Q_k,$$

we write

$$\mathcal{M}_\Phi(f-f_k)(x) \le \mathcal{M}_\Phi([1-w_k]f)(x) + \mathcal{M}_\Phi\left(c_k\chi_{Q_k}\right)(x). \qquad (3.3.6)$$

Since $\mathcal{M}_\Phi\chi_{Q_k}(x)\le 1$ and Q_k is $(4, 4^{n+1})$-doubling, it follows that

$$\int_{4Q_k} \mathcal{M}_\Phi\left(c_k\chi_{Q_k}\right)(x)\,d\mu(x) \lesssim |c_k|\mu(4Q_k) \sim \left|\int_{\mathbb{R}^D} w_k f\,d\mu\right|. \qquad (3.3.7)$$

Combining (3.3.5), (3.3.6) and (3.3.7), we then obtain (3.3.4).

It remains to show that the terms on the right hand side of (3.3.4) tend to 0 as $k\to\infty$. Since $f, \mathcal{M}_\Phi f \in L^1(\mu)$, by the dominated convergence theorem, we see that

$$\lim_{k\to\infty}\left[\left|\int_{\mathbb{R}^D} w_k(x)f(x)\,d\mu(x)\right| + \int_{\mathbb{R}^D\setminus(4Q_k)} \mathcal{M}_\Phi f(x)\,d\mu(x)\right] = 0.$$

It suffices to estimate the third term on the right hand side of (3.3.4). Take $x\in 4Q_k$ and $\varphi\sim x$. By an argument similar to that used in the proof of Lemma 3.3.5, we easily see that $C w_k \varphi \sim x$ for some positive constant C. Thus, we have

$$\mathcal{M}_\Phi(w_k f)(x) \lesssim \mathcal{M}_\Phi f(x)$$

and, for any $x \in \mathbb{R}^D$,

$$\begin{aligned}\chi_{4Q_k}(x)\mathcal{M}_\Phi([1-w_k]f)(x) &\le \chi_{4Q_k}(x)[\mathcal{M}_\Phi f(x) + \mathcal{M}_\Phi(w_k f)(x)] \\ &\lesssim \mathcal{M}_\Phi f(x).\end{aligned}$$

Therefore, if we show that $\chi_{4Q_k}(x)\mathcal{M}_\Phi([1-w_k]f)(x)$ tends to 0 pointwise as $k \to \infty$, then an application of the dominated convergence theorem implies the lemma.

For a fixed $x \in \mathbb{R}^D$, let k_0 be such that $x \in \frac{1}{2}Q_k$ for $k \ge k_0$. Notice that, if $\varphi \sim x$ and $y \notin Q_k$, then $|\varphi(y)| \lesssim 1/[\ell(Q_k)]^n$. Thus,

$$\left|\int_{\mathbb{R}^D} \varphi(y)[1-w_k(y)]f(y)\,d\mu(y)\right| \le \|f\|_{L^1(\mu)}\|(1-w_k)\varphi\|_{L^\infty(\mu)} \lesssim \frac{\|f\|_{L^1(\mu)}}{[\ell(Q_k)]^n}.$$

Then we conclude that

$$\chi_{4Q_k}(x)\mathcal{M}_\Phi([1-w_k]f)(x) \lesssim \frac{\|f\|_{L^1(\mu)}}{[\ell(Q_k)]^n} \to 0$$

as $k \to \infty$. This finishes the proof of Lemma 3.3.6. □

To complete the proof of Theorem 3.3.3, we also need the following important lemma.

Lemma 3.3.7. *Let $f \in \mathrm{RBMO}(\mu)$ with compact support and*

$$\int_{\mathbb{R}^D} f(x)\,d\mu(x) = 0.$$

Then there exist functions $h_m \in L^\infty(\mu)$, $m \in \mathbb{Z}_+$, such that

$$f(x) = h_0(x) + \sum_{m=1}^{\infty} \int_{\mathbb{R}^D} \varphi_{y,m}(x) h_m(y)\,d\mu(y) \tag{3.3.8}$$

with convergence in $L^1(\mu)$, where, for each $m \in \mathbb{N}$, $\varphi_{y,m} \sim y$, and there exists a positive constant C, independent of f, such that

$$\sum_{m=0}^{\infty} |h_m| \le C\|f\|_{\mathrm{RBMO}(\mu)}. \tag{3.3.9}$$

Assume that Lemma 3.3.7 holds true for the moment. Then we can complete the proof of Theorem 3.3.3.

Proof of Theorem 3.3.3. By Lemma 3.3.4, we only need to show the "if" part of Theorem 3.3.3. Let $f \in H^1_\Phi(\mu)$ such that $f \in L^\infty(\mu)$ and has compact support. In this case, $f \in H^{1,\infty}_{\mathrm{atb}}(\mu)$ and hence we only have to estimate the norm of f.

Since

$$\left(H^{1,\infty}_{\mathrm{atb}}(\mu)\right)^* = \mathrm{RBMO}\,(\mu),$$

given $f \in H^{1,\infty}_{\mathrm{atb}}(\mu)$, by the Hahn–Banach theorem, we see that

$$\|f\|_{H^{1,\infty}_{\mathrm{atb}}(\mu)} = \sup_{\|g\|_{\mathrm{RBMO}\,(\mu)} \le 1} |\langle f, g\rangle|.$$

Since

$$\int_{\mathbb{R}^D} f(x)\, d\mu(x) = 0,$$

we may assume that g has compact support and

$$\int_{\mathbb{R}^D} g(x)\, d\mu(x) = 0.$$

Then, by applying Lemma 3.3.7 to g, we find that

$$\begin{aligned}|\langle f, g\rangle| \le &\left|\int_{\mathbb{R}^D} f(x) h_0(x)\, d\mu(x)\right| \\ &+ \left|\sum_{m=1}^{\infty} \int_{\mathbb{R}^D}\int_{\mathbb{R}^D} \varphi_{y,m}(x) h_m(y) f(x)\, d\mu(y)\, d\mu(x)\right|.\end{aligned}$$

By the fact that

$$\left|\int_{\mathbb{R}^D} \varphi_{y,m}(x) f(x)\, d\mu(x)\right| \le \mathcal{M}_\Phi f(y),$$

we have

$$\begin{aligned}|\langle f, g\rangle| &\le \|f\|_{L^1(\mu)} \|h_0\|_{L^\infty(\mu)} + \sum_{m=1}^{\infty} \int_{\mathbb{R}^D} \mathcal{M}_\Phi f(y) |h_m(y)|\, d\mu(y) \\ &\le \|f\|_{L^1(\mu)} \|h_0\|_{L^\infty(\mu)} + \|\mathcal{M}_\Phi f\|_{L^1(\mu)} \left\|\sum_{m=1}^{\infty} |h_m|\right\|_{L^\infty(\mu)} \\ &\lesssim [\|f\|_{L^1(\mu)} + \|\mathcal{M}_\Phi f\|_{L^1(\mu)}] \|g\|_{\mathrm{RBMO}\,(\mu)}.\end{aligned}$$

Therefore,

$$\|f\|_{H^{1,\infty}_{\rm atb}(\mu)} \lesssim \|f\|_{L^1(\mu)} + \|\mathcal{M}_\Phi f\|_{L^1(\mu)}.$$

In the general case where we do not know a priori that $f \in H^{1,\infty}_{\rm atb}(\mu)$. However, by Lemma 3.3.6, we can consider a sequence $\{f_k\}_k \subset H^1_\Phi(\mu)$ of functions which are bounded with compact support and satisfy that $f_k \to f$ in $L^1(\mu)$ and

$$\|\mathcal{M}_\Phi(f - f_k)\|_{L^1(\mu)} \to 0, \text{ as } k \to \infty.$$

Then, by the standard arguments, we finish the proof of Theorem 3.3.3. □

We assume that the support of f in Lemma 3.3.7 is in a doubling cube R_0. Let $A_0, \alpha_1, \alpha_2, \epsilon_1, \epsilon_2$ and σ be as in Chap. 2. Then we introduce cubes of generations with respect to R_0.

Definition 3.3.8. Suppose that the support of the function f in Lemma 3.3.7 is contained in a doubling cube R_0. Let $m \in \mathbb{N}$ and $x \in \operatorname{supp}\mu \cap R_0$. If

$$\delta(x, 2R_0) > mA_0,$$

denote by $Q_{x,m}$ *a doubling cube such that*

$$|\delta(Q_{x,m}, 2R_0)| \leq \epsilon_1.$$

Also, $\tilde{\mathcal{D}}_m := \{Q_{i,m}\}_{i \in \tilde{I}_m}$ is a *subfamily, with finite overlap, of the cubes*

$$\left\{Q_{x,m} : x \in \operatorname{supp}\mu \bigcap R_0\right\}$$

such that each cube $Q_{i,m} := Q_{y_i,m}$ is centered at some point $y_i \in \operatorname{supp}\mu \cap R_0$ with $\delta(y_i, 2R_0) > mA_0$ and

$$\left\{x \in \operatorname{supp}\mu \bigcap R_0 : \delta(x, 2R_0) > mA_0\right\} \subset \bigcup_{i \in \tilde{I}_m} Q_{i,m}.$$

If $\delta(x, 2R_0) \leq mA_0$, let $Q_{x,m} := \{x\}$. Denote by $\hat{\mathcal{D}}_m$ the *family of cubes* $Q_{x,m} := \{x\}$ *such that* $\delta(x, 2R_0) \leq mA_0$ *and* $x \notin \cup_{i \in \tilde{I}_m} Q_{i,m}$. Let

$$\mathcal{D}_m := \tilde{\mathcal{D}}_m \bigcup \hat{\mathcal{D}}_m.$$

The cubes

$$\left\{Q_{x,m} : x \in \operatorname{supp}\mu \bigcap R_0\right\}$$

are called *cubes of the m-th generation with respect to* R_0.

As in Sect. 2.3, for each $m \in \mathbb{N}$ and $y \in \operatorname{supp}\mu \cap R_0$, we also introduce the cubes

$$Q^1_{y,m},\ \hat{Q}^1_{y,m},\ Q^2_{y,m},\ \hat{Q}^2_{y,m},\ Q^3_{y,m},\ \hat{Q}^3_{y,m},\ \hat{\hat{Q}}^3_{y,m},\ \check{Q}^1_{y,m} \text{ and } \check{\check{Q}}^1_{y,m}.$$

Furthermore, as in Definition 2.3.4, we define the functions $\{\psi_{y,m}\}$ for $y \in \operatorname{supp}\mu \cap 2R_0$ and $m \in \mathbb{N}$ with respect to R_0. We see that $\{\psi_{y,m}\}$ also satisfy Lemma 2.3.5.

Definition 3.3.9. Let $\psi_{y,m}$, with $m \in \mathbb{N}$ and $y \in \operatorname{supp}\mu \cap (2R_0)$, be as in Definition 2.3.4 and $w_{i,m}$ the weight function defined for $y \in \bigcup_{i\in\tilde{I}_m} Q_{i,m}$ by

$$w_{i,m}(y) := \frac{\chi_{Q_{i,m}}(y)}{\sum_{j\in\tilde{I}_m}\chi_{Q_{j,m}}(y)}.$$

If $y \in \operatorname{supp}\mu \cap (2R_0)$ belongs to some cube $Q_{i,m}$ centered at some point y_i, with $\ell(Q_{i,m}) \in (0,\infty)$, then let

$$\varphi_{y,m}(x) := \alpha_2^{-1}\sum_i w_{i,m}(y)\psi_{y_i,m}(x) \quad \text{for all} \quad x \in \mathbb{R}^D.$$

If y does not belong to any cube $Q_{i,m}$ with $\ell(Q_{i,m}) \in (0,\infty)$ (this implies $\delta(y,2R_0) \le mA_0$ and $Q_{y,m} = \{y\}$), then let

$$\varphi_{y,m}(x) := \alpha_2^{-1}\psi_{y,m}(x) \quad \text{for all} \quad x \in \mathbb{R}^D.$$

Define $w_{i,m}(y) := \chi_{Q_{i,m}}(y)$ for all $y \in \mathbb{R}^D$ if $\ell(Q_{i,m}) = 0$. Then, for all $m \in \mathbb{N}$ and $x, y \in \mathbb{R}^D$, it holds true that

$$\varphi_{y,m}(x) = \alpha_2^{-1}\sum_i w_{i,m}(y)\psi_{y_i,m}(x).$$

Let us remark that the functions

$$\left\{\varphi_{y,m} : y \in \operatorname{supp}\mu \bigcap 2R_0,\ m \in \mathbb{N}\right\}$$

satisfy Lemmas 2.3.6 and (2.3.7). On the other hand, a more natural definition for $\varphi_{y,m}$ would have been the choice $\varphi_{y,m}(x) := \alpha^{-1}\psi_{y,m}(x)$ for all y and $\alpha \in (0,\infty)$. However, as we shall see, for some of the argument in the proof of Lemma 3.3.7 below, the choice of Definition 3.3.9 is better.

Proof of Lemma 3.3.7. At the level of generation m, we construct a function h_m yielding the 'potential'

$$U_m(x) := \int_{\mathbb{R}^D} \varphi_{y,m}(x)h_m(y)\,d\mu(y),\ \forall x \in \mathbb{R}^D.$$

We assume that the support of f is contained in some doubling cube R_0 and, for each integer $m \in \mathbb{N}$, we consider the family $\mathcal{D}_m$ of "dyadic" cubes $Q_{i,m}$ introduced in Definition 3.3.8, and we let

$$\mathcal{D} := \bigcup_{m \in \mathbb{N}} \mathcal{D}_m.$$

Recall that the elements of $\mathcal{D}$ may be cubes with side length 0, namely, points.

For each m we will construct functions g_m and b_m. The function g_m is supported on a subfamily $\mathcal{D}_m^G$ of the cubes in $\mathcal{D}_m$. On the other hand, b_m is supported on a subfamily $\mathcal{D}_m^B$ of the cubes in $\mathcal{D}_m$. We let

$$\mathcal{D}^G := \bigcup_{m \in \mathbb{N}} \mathcal{D}_m^G \quad \text{and} \quad \mathcal{D}^B := \bigcup_{m \in \mathbb{N}} \mathcal{D}_m^B.$$

The cubes in $\mathcal{D}^G$ are called *good cubes* and the ones in $\mathcal{D}^B$ *bad cubes* which are determined later (In the family $\mathcal{D}_m$, there exist also cubes which are neither good nor bad, in general).

From g_m and b_m, we will obtain the following potentials: for $x \in \mathbb{R}^D$,

$$U_m^G(x) := \int_{\mathbb{R}^D} \varphi_{y,m}(x) g_m(y)\, d\mu(y),$$

$$U_m^B(x) := \int_{\mathbb{R}^D} \varphi_{y,m}(x) b_m(y)\, d\mu(y)$$

and

$$U_m(x) := U_m^G(x) + U_m^B(x).$$

These potential is successively subtracted from f. For $m \in \mathbb{N}$, we let $f_1 := f$,

$$f_{m+1} := f - \sum_{j=1}^{m} U_j = f_m - U_m,$$

and

$$h_0 := f - \sum_{m=1}^{\infty} U_m = \lim_{m \to \infty} f_m \tag{3.3.10}$$

in $L^1(\mu)$. The supports of the functions, g_m, b_m, U_m^G and U_m^B, are contained in $2R_0$.

By induction we show that there exist positive constants C_7 and C_8 such that the functions, g_m, b_m, U_m and f_m, possess the following properties:

(a) $|g_m|, |b_m| \le C_7 A_0 \|f\|_{\mathrm{RBMO}(\mu)}$;
(b) $|m_Q(f_{m+1})| \le A_0 \|f\|_{\mathrm{RBMO}(\mu)}$ if $Q \in \mathcal{D}_m$ and $\ell(Q) \in (0, \infty)$;
(c) If $Q \in \mathcal{D}_m$, $\ell(Q) \in (0, \infty)$ and $g_m \not\equiv 0$ on Q, then

$$|m_Q(f_{m+1})| \le \frac{7}{20} A_0 \|f\|_{\mathrm{RBMO}(\mu)};$$

(d) If $Q \in \mathcal{D}_m$ and

$$|m_Q(f_m)| \le \frac{8}{20} A_0 \|f\|_{\mathrm{RBMO}(\mu)},$$

then $U_m \equiv 0$ and $g_m \equiv b_m \equiv 0$ on Q;
(e) If $Q \in \mathcal{D}_m$ and

$$\delta(Q, 2R_0) \le \left(m - \frac{1}{10}\right) A_0 \quad (\text{hence } \ell(Q) = 0),$$

then $U_m \equiv 0$ and $g_m \equiv b_m \equiv 0$ on Q;
(f) If $\delta(x, 2R_0) < \infty$, then

$$|h_0(x)| \le C_8 A_0 \|f\|_{\mathrm{RBMO}(\mu)}$$

and, if $Q \in \mathcal{D}_m$ and $\ell(Q) = 0$, then

$$|m_Q(f_{m+1})| = |f_{m+1}(z_Q)| \le C_8 A_0 \|f\|_{\mathrm{RBMO}(\mu)};$$

(g) For each m, there exist functions, $g_m^1, \ldots, g_m^{B_D}$, where B_D is as in Theorem 1.1.1, such that

(g.1) for all $x \in \mathbb{R}^D$,

$$U_m^G(x) = \sum_{p=1}^{B_D} \int_{\mathbb{R}^D} \varphi_{y,m}^p(x) g_m^p(y)\, d\mu(y),$$

where $\varphi_{y,m}^p$ is defined below;
(g.2) $|g_m^p| \le 2C_7 A_0 \|f\|_{\mathrm{RBMO}(\mu)}$ for $p \in \{1, \ldots, B_D\}$;
(g.3) the functions $\{\sum_{p=1}^{B_D} |g_m^p|\}_m$ have disjoint supports.

(h) The family of cubes, $\mathcal{D}^B$, that support the functions b_m, $m \in \mathbb{N}$, satisfies the following *Carleson packing condition* that, for each cube $R \in \mathcal{D}_m$ with $\ell(R) \in (0, \infty)$,

$$\sum_{\{Q:\, Q \cap R \neq \emptyset,\, Q \in \mathcal{D}_k^B,\, k > m\}} \mu(Q) \lesssim \mu(R). \tag{3.3.11}$$

If some cube Q coincides with a point $\{x\}$, then we let $m_Q(f_m) := f_m(x)$. Also, the notation for the summation in (3.3.11) is an abuse of the notation. This summation has to be understood as

$$\sum_{\{Q:\ Q\subset 2R,\ Q\in\mathcal{D}_k^B,\ k>m\}} \mu(Q)$$

$$:= \sum_{\substack{\{Q:\ \ell(Q)\in(0,\infty),\ Q\subset 2R\\ Q\in\mathcal{D}_k^B,\ k>m\}}} \mu(Q) + \sum_{k>m} \mu(\{x\in 2R:\ \{x\}\in D_k^B\}).$$

On the other hand, the number B_D that appears in (g) is the number of disjoint families of cubes given in Theorem 1.1.1 and depends only on D.

The function $\varphi_{y,m}^p$ of (g) is defined as follows. We let

$$\mathcal{D}_m := \mathcal{D}_m^1 \bigcup \cdots \bigcup \mathcal{D}_m^{B_D},$$

where each subfamily $\mathcal{D}_m^p$ is disjoint. Then we let

$$\varphi_{y,m}^p(x) := \varphi_{y_i,m}(x) \quad \text{for} \quad x\in\mathbb{R}^D,$$

if $y\in Q_{i,m}$ with $Q_{i,m}\in\mathcal{D}_m^p$, and $\varphi_m^p(x) := 0$ for $x\in\mathbb{R}^D$, if there does not exist any cube of the subfamily $\mathcal{D}_m^p$ containing y.

First we show that, if there exist functions g_m and b_m satisfying (a)–(h), then Lemma 3.3.7 holds true, and we show the existence of these functions.

It is not difficult to show that, if (3.3.8) and (3.3.9) hold true, then the summation of (3.3.10) converges in $L^1_{\mathrm{loc}}(\mu)$. Since the support of all the functions involved is contained in $2R_0$, the convergence is in $L^1(\mu)$.

We then show that, if (b) and (f) hold true, then

$$\|h_0\|_{L^\infty(\mu)} \lesssim A_0\|f\|_{\mathrm{RBMO}(\mu)}.$$

Taking (f) into account, we only have to see that, for μ-almost every $x\in\operatorname{supp}\mu\cap R_0$ such that $\delta(x,2R_0)=\infty$,

$$|h_0(x)| \lesssim A_0\|f\|_{\mathrm{RBMO}(\mu)}. \tag{3.3.12}$$

In this case, if $Q\in\mathcal{D}_k$ is such that $x\in Q$, then $\ell(Q)\in(0,\infty)$. We are going to prove that, for $Q\in\mathcal{D}_k$, $k\in\{1,\ldots,m-1\}$,

$$|m_Q(f_m)| \lesssim A_0\|f\|_{\mathrm{RBMO}(\mu)}. \tag{3.3.13}$$

By (b), we only need to consider $Q\in\mathcal{D}_k$, $k\in\{1,\ldots,m-2\}$. This cube is covered with finite overlap by the family of the cubes $\mathcal{D}_{m-1}$. Moreover, if $P\in\mathcal{D}_{m-1}$

and $P \cap Q \neq \emptyset$, then $\ell(P) \leq \ell(Q)/10$ and hence $P \subset 2Q$. Thus, by this fact and (b), we see that

$$\int_Q |f_m(x)|\,d\mu(x) \leq \sum_i \int_{Q\cap Q_{i,m-1}} |f_m(x)|\,d\mu(x)$$
$$\lesssim A_0\|f\|_{\mathrm{RBMO}\,(\mu)}\mu(2Q)$$
$$\lesssim A_0\|f\|_{\mathrm{RBMO}\,(\mu)}\mu(Q).$$

Therefore, (3.3.13) holds true.

Then h_0 satisfies that

$$|m_Q(h_0)| \lesssim A_0\|f\|_{\mathrm{RBMO}\,(\mu)}$$

for all $Q \in \mathcal{D}$ containing x, because the sequence $\{f_m\}_m$ converges to h_0 in $L^1(\mu)$. By the Lebesgue differentiation theorem, we conclude that

$$|h_0(x)| \lesssim A_0\|f\|_{\mathrm{RBMO}\,(\mu)}$$

for μ-almost every $x \in \operatorname{supp}\mu$ with $\delta(x, 2R_0) = \infty$. Therefore,

$$\|h_0\|_{L^\infty(\mu)} \lesssim A_0\|f\|_{\mathrm{RBMO}\,(\mu)}.$$

Observe that the function g_m^p in (g.1) originates with the same potential as g_m. Indeed, it is constructed by modifying the function g_m slightly such that it is supported in disjoint sets for different m. By (g.2) we have

$$\sum_{m=1}^{\infty}\sum_{p=1}^{B_D}|g_m^p| \lesssim B_D A_0\|f\|_{\mathrm{RBMO}\,(\mu)}.$$

The supports of the functions $\{b_m\}_{m\in\mathbb{N}}$ may be not disjoint. To solve this problem, we construct "corrected" versions of $w_{i,m}b_m$. Moreover, as in the case of g_m, the modifications are made in such a way that the potentials U_m^B are not changed.

We assume that the functions $\{b_m\}_{m\in\mathbb{N}}$, have been obtained and they satisfy (a)–(h). We start the construction of some new functions in the small cubes, and then we go over the cubes from previous generations. However, since there exists an infinite number of generations, we need to use a limiting argument.

By Definition 3.3.9, for each j, we write the potential originated by b_j as

$$U_j^B(x) := \sum_{i\in I_j}\alpha_2^{-1}\psi_{y_i,j}(x)\int_{\mathbb{R}^D} w_{i,j}(y)b_j(y)\,d\mu(y)$$

for all $x \in \mathbb{R}^D$. For a fixed $m \in \mathbb{N}$, we are going to define functions $v^m_{i,j}$ for all $j \in \{1, \ldots, m\}$ and $i \in I_j$. The functions $v^m_{i,j}$ satisfy that

$$\operatorname{supp} v^m_{i,j} \subset Q_{i,j}, \tag{3.3.14}$$

where $Q_{i,j} \in \mathcal{D}^B_j$, the sign of $v^m_{i,j}$ is constant on $Q_{i,j}$, and

$$\int_{\mathbb{R}^D} v^m_{i,j}(y)\,d\mu(y) = \int_{\mathbb{R}^D} w_{i,j} b_j(y)\,d\mu(y). \tag{3.3.15}$$

Moreover, we also find that

$$\sum_{j=1}^m \sum_{i \in I_j} |v^m_{i,j}| \lesssim A_0 \|f\|_{\mathrm{RBMO}(\mu)}. \tag{3.3.16}$$

We let $v^m_{i,m}(y) := w_{i,m}(y) b_m(y)$ for all $y \in \mathbb{R}^D$ and $i \in I_m$. Assume that we have obtained functions, $v^m_{i,m}, v^m_{i,m-1}, \ldots, v^m_{i,k+1}$, for all $i \in I_j$ and $j \in \{k+1, \ldots, m\}$, fulfilling (3.3.14), (3.3.15) and

$$\sum_{j=k+1}^m \sum_{i \in I_j} |v^m_{i,j}| \le B A_0 \|f\|_{\mathrm{RBMO}(\mu)},$$

where B is some positive constant that is fixed below.

We now construct $v^m_{i,k}$. Let $Q_{i_0,k} \in \mathcal{D}_k$ be some fixed cube of the k-th generation. Assume that $Q_{i_0,k}$ is not a single point. Since the cubes in the family $\mathcal{D}^B$ satisfy the packing condition (3.3.11), by (3.3.15), we conclude that, for any $t \in (0, \infty)$,

$$\begin{aligned}
&\mu\left(\left\{y \in Q_{i_0,k} : \sum_{j=k+1}^m \sum_{i \in I_j} |v^m_{i,j}(y)| > t\right\}\right)\\
&\quad \le \frac{1}{t} \sum_{j=k+1}^m \sum_{i \in I_j} \int_{Q_{i_0,k}} |v^m_{i,j}(y)|\,d\mu(y)\\
&\quad \le \frac{1}{t} \sum_{j=k+1}^m \sum_{i \in I_j} \int_{Q_{i_0,k}} |w_{i,k}(y) b_j(y)|\,d\mu(y)\\
&\quad \le \frac{C_7 A_0 \|f\|_{\mathrm{RBMO}(\mu)}}{t} \sum_{\{Q:\, Q \cap Q_{i_0,k} \ne \emptyset,\, Q \in \mathcal{D}^B_j,\, j>k\}} \mu(Q)\\
&\quad \le \frac{C_9 A_0 \|f\|_{\mathrm{RBMO}(\mu)}}{t} \mu(Q_{i_0,k}).
\end{aligned}$$

Therefore, if we choose $t := 2C_9A_0\|f\|_{\mathrm{RBMO}(\mu)}$ and we let

$$V^m_{i_0,k} := \left\{ y \in Q_{i_0,k} : \sum_{j=k+1}^{m} \sum_{i \in I_j} |v^m_{i,j}(y)| \le t \right\},$$

we have $\mu(V^m_{i_0,k}) \ge \mu(Q_{i_0,k})/2$. If we let $v^m_{i_0,k} := c^m_{i_0,k}\chi_{V^m_{i_0,k}}$, where $c^m_{i_0,k} \in \mathbb{R}$ is such that (3.3.15) holds true for $i = i_0$, then it follows, from Definition 3.3.9, that

$$|c^m_{i_0,k}| \le \frac{1}{\mu(V^m_{i_0,k})} \int_{\mathbb{R}^D} |w_{i_0,k}(y)b_k(y)|\,d\mu(y) \le 2C_7A_0\|f\|_{\mathrm{RBMO}(\mu)}.$$

By the finite overlap of the cubes in $\mathcal{D}_k$, we obtain

$$\sum_{\{i_0:\, Q_{i_0,k} \in \mathcal{D}^B_k,\, \ell(Q_{i_0,k}) \ne 0\}} |v^m_{i_0,k}| \le 2C_7B_DA_0\|f\|_{\mathrm{RBMO}(\mu)},$$

where B_D is the constant in Theorem 1.1.1. Now, if we take $B := 2C_7B_D + 2C_9$, we conclude that

$$\sum_{\{i_0:\, Q_{i_0,k} \in \mathcal{D}_k,\, \ell(Q_{i_0,k}) \ne 0\}} |v^m_{i_0,k}| + \sum_{j=k+1}^{m} \sum_{i \in I_j} |v^m_{i,j}| \le BA_0\|f\|_{\mathrm{RBMO}(\mu)}. \quad (3.3.17)$$

If $Q_{i_0,k}$ is a single point $\{y_0\}$, then we let $v^m_{i_0,k} := w_{i_0,k}b_k$. By (a), we see that, for all $y \in \mathbb{R}^D$,

$$|v^m_{i_0,k}(y)| \le |b_k(y_0)| \le C_7A_0\|f\|_{\mathrm{RBMO}(\mu)}.$$

Thus, we know that, for all $y \in \mathbb{R}^D$,

$$\sum_{j=k}^{m} \sum_{i \in I_j} |v^m_{i,j}(y)| = |b_k(y)| \le C_7A_0\|f\|_{\mathrm{RBMO}(\mu)} \le BA_0\|f\|_{\mathrm{RBMO}(\mu)}. \quad (3.3.18)$$

From the fact that $\operatorname{supp} v^m_{i,j} \subset Q_{i,j}$ for $Q_{i,j} \in \mathcal{D}^B_j$, (3.3.17) and (3.3.18), we deduce that

$$\sum_{j=k}^{m} \sum_{i \in I_j} |v^m_{i,j}| \le BA_0\|f\|_{\mathrm{RBMO}(\mu)}.$$

Operating in this way, the functions $v^m_{i,j}$, $j \in \{1, \dots, m\}$ and $i \in I_j$, satisfy the conditions (3.3.14) through (3.3.16).

For each $m \in \mathbb{N}$, we let

$$\mathcal{D}_m^{p,B} := \mathcal{D}_m^p \bigcap \mathcal{D}_m^B.$$

Now we take a subsequence $\{m_k\}_k$ such that, for all $i \in I_1$, the functions $\{v_{i,1}^{m_k}\}_k$ converge weakly in $L^\infty(\mu)$ to some function $v_{i,1} \in L^\infty(\mu)$. We remark that the sequence $\{m_k\}_k$ can be chosen independent of i, since, by Theorem 1.1.1, there exists a bounded number B_D of subfamilies $\{\mathcal{D}_1^1, \ldots, \mathcal{D}_1^{B_D}\}$ of $\mathcal{D}_1$ such that each subfamily $\mathcal{D}_1^p$ is disjoint, where B_D is as in Theorem 1.1.1. We write

$$\sum_{i \in I_1} v_{i,1}^m = \sum_{p=1}^{B_D} \sum_{\{i:\, Q_{i,1} \in \mathcal{D}_1^{p,B}\}} v_{i,1}^m,$$

and we choose $\{m_k\}_k$ such that, for each p,

$$\sum_{\{i:\, Q_{i,1} \in \mathcal{D}_1^{p,B}\}} v_{i,1}^{m_k} \quad \text{converges weakly to} \quad \sum_{\{i:\, Q_{i,1} \in \mathcal{D}_1^{p,B}\}} v_{i,1}.$$

In a similar way, we consider another subsequence $\{m_{k_j}\}_j$ of $\{m_k\}_k$ such that for all $i \in I_2$, the functions $\{v_{i,2}^{m_{k_j}}\}_j$ converge weakly in $L^\infty(\mu)$ to some function $v_{i,2} \in L^\infty(\mu)$. Going on with this process, we obtain functions $\{v_{i,j}\}_{j \in \mathbb{N}}$, which satisfy (3.3.14), (3.3.15) and

$$\sum_{j=1}^{\infty} \sum_{i \in I_j} |v_{i,j}| \lesssim A_0 \|f\|_{\mathrm{RBMO}(\mu)}. \tag{3.3.19}$$

Also, for all $x \in \mathbb{R}^D$, we have

$$U_j^B(x) = \sum_{i \in I_j} \alpha_2^{-1} \psi_{y_i,j} \int_{\mathbb{R}^D} v_{i,j}(y)\, d\mu(y).$$

We define

$$b_m^p(y) := \sum_{\{i:\, Q_{i,m} \in \mathcal{D}_m^{p,B}\}} v_{i,m}(y) \quad \text{for all} \quad y \in \mathbb{R}^D.$$

Recall that $\varphi_{y,m}^p(x) = \varphi_{y_i,m}(x)$ for all $x \in \mathbb{R}^D$ if $y \in Q_{i,m}$ with $Q_{i,m} \in \mathcal{D}_m^p$, and $\varphi_{y,m}^p(x) = 0$ for all $x \in \mathbb{R}^D$ if there does not exist any cube of the subfamily $\mathcal{D}_m^p$ containing y. Then we have

$$U_m^B(x) = \sum_{p=1}^{B_D} \int_{\mathbb{R}^D} \varphi_{y,m}^p(x) b_m^p(y)\, d\mu(y) \quad \text{for all} \quad x \in \mathbb{R}^D.$$

Now we let $h^p_m := g^p_m + b^p_m$. Then, by (3.3.19), (3.3.12), (f) and (g), we conclude that, for all $x \in \mathbb{R}^D$,

$$f(x) = h_0(x) + \sum_{p=1}^{B_D} \sum_{m=1}^{\infty} \int_{\mathbb{R}^D} \varphi^p_{y,m}(x) h^p_m(y)\, d\mu(y),$$

with $C\varphi^p_{y,m} \sim y$ for some positive constant C, and

$$|h_0| + \sum_{p=1}^{B_D} \sum_{m=1}^{\infty} |h^p_m| \lesssim A_0 \|f\|_{\mathrm{RBMO}(\mu)},$$

Thus, Lemma 3.3.7 follows.

The Constructions of g_m and b_m. In this subsection we construct inductively functions g_m and b_m satisfying the properties (a)–(e). We show in next step that these functions satisfy (f)–(h) too.

Assume that $g_1, \ldots, g_{m-1}$ and $b_1, \ldots, b_{m-1}$ have been constructed and they satisfy (a)–(e). Let Ω_m be the set of points $x \in \operatorname{supp} \mu$ with $\delta(x, 2R_0) > mA_0$ such that there exists some cube $Q \in \mathcal{D}_m$, $\ell(Q) \in (0, \infty)$, with $Q \ni x$ and

$$|m_Q(f_m)| \geq \frac{3}{4} A_0 \|f\|_{\mathrm{RBMO}(\mu)}.$$

If $\Omega_m = \emptyset$, we let $b_m := 0$ and $g_m := 0$. Then (a)–(h) follow. Thus, we only consider $\Omega_m \neq \emptyset$. For each $x \in \Omega_m$, we consider a doubling cube $S_{x,m}$ centered at x such that

$$\delta(S_{x,m}, 2R_0) = mA_0 - \alpha_1 - \alpha_2 - \alpha_3 \pm \epsilon_1,$$

where α_3 is some big constant with $10\alpha_2 < \alpha_3 \ll A_0$, whose precise value is fixed below. One has to think that $S_{x,m}$ is much bigger than $Q^3_{x,m}$ but much smaller than $Q_{x,m-1}$.

Now we take a Besicovitch covering of Ω_m with cubes of type $S_{x,m}$, $x \in \Omega_m$:

$$\Omega_m \subset \bigcup_j S_{j,m},$$

where $S_{j,m}$ stands for $S_{x_j,m}$ with $x_j \in \Omega_m$. We say that a cube $Q \in \mathcal{D}_m$ is *good* if

$$Q \subset \bigcup_j \frac{3}{2} S_{j,m},$$

and we say that it is *bad* if it is not good and

$$Q \subset \bigcup_j 2S_{j,m}.$$

Both good and bad cubes are contained in $\cup_j 2S_{j,m}$.

Now we define g_m and b_m by

$$g_m := \sum_{\{i:\, Q_{i,m} \in \mathcal{D}_m^G\}} w_{i,m} m_{Q_{i,m}}(f_m)$$

and

$$b_m := \sum_{\{i:\, Q_{i,m} \in \mathcal{D}_m^B\}} w_{i,m} m_{Q_{i,m}}(f_m).$$

Because there exists some overlapping among the cubes in $\mathcal{D}_m$, we have used the weights $w_{i,m}$ in the definition of these functions. However, one should think that g_m and b_m are approximations of the mean of f over the cubes of $\mathcal{D}_m^G$ and $\mathcal{D}_m^B$, respectively.

The following claims are useful.

Claim 1. Let $Q_{h,m} \in \mathcal{D}_m$ be such that either $g_m \not\equiv 0$, $b_m \not\equiv 0$ or $U_m \not\equiv 0$ on $2Q_{h,m}$. Then there exists some j such that $\hat{Q}_{h,m}^3 \subset 4S_{j,m}$ and hence $Q_{h,m} \subset 4S_{j,m}$.

Proof. In the first two cases, it holds true that

$$2Q_{h,m} \bigcap 2S_{j,m} \neq \emptyset \quad \text{for some} \quad j.$$

In the latter case, by (a) of Lemma 2.3.6 and our construction, there exists some j such that

$$\hat{Q}_{h,m}^3 \bigcap 2S_{j,m} \neq \emptyset.$$

Thus, in any case, we have

$$\hat{Q}_{h,m}^3 \bigcap 2S_{j,m} \neq \emptyset \text{ for some } j.$$

Arguing as in Lemma 2.2.5, it is easy to show that

$$\ell(\hat{Q}_{h,m}^3) \leq \ell(S_{j,m})/4 \quad \text{and hence} \quad \hat{Q}_{h,m}^3 \subset 4S_{j,m}.$$

This finishes the proof of Claim 1. □

We now show that (e) is satisfied.

Claim 2. If $Q \in \mathcal{D}_m$ and $\delta(Q, 2R_0) \le (m - \frac{1}{10})A_0$ (hence $\ell(Q) = 0$), then

$$U_m \equiv g_m \equiv b_m \equiv 0$$

on Q and $Q \notin (\mathcal{D}_m^G \cup \mathcal{D}_m^B)$.

Proof. Assume that $Q \equiv \{x\}$ and that either $g_m \not\equiv 0$, $b_m \not\equiv 0$ or $U_m \not\equiv 0$ on Q, or $Q \in (\mathcal{D}_m^G \cup \mathcal{D}_m^B)$. By the preceding claim, $Q \subset 4S_{j,m}$ for some j. Then, by Lemma 2.1.3, we see that

$$\begin{aligned}\delta(x, 2R_0) &= \delta(x, 4S_{j,m}) + \delta(4S_{j,m}, 2R_0) \pm \epsilon_0\\ &\ge \delta(4S_{j,m}, 2R_0) - \epsilon_0\\ &\ge \delta(S_{j,m}, 2R_0) - 8^n C_0 - \epsilon_0\\ &> \left(m - \frac{1}{10}\right) A_0,\end{aligned}$$

which completes the proof of Claim 2.

The following estimate is necessary in our construction.

Claim 3. Let Q be some cube of the m-th generation with respect to R_0, and let x, $y \in 2Q$. If $g_1, \dots, g_m$ and $b_1, \dots, b_m$ satisfy (a), then, for all x, $y \in \mathbb{R}^D$,

$$\sum_{k=1}^{m} |U_k(x) - U_k(y)| \le \frac{A_0}{100} \|f\|_{\mathrm{RBMO}(\mu)}.$$

Proof. By (a), we only need to show that

$$\sum_{k=1}^{m} C_7 A_0 \int_{\mathbb{R}^D} |\varphi_{z,k}(x) - \varphi_{z,k}(y)|\, d\mu(z) \le \frac{A_0}{100}.$$

Let $x_0 \in \operatorname{supp}\mu$ be such that x, $y \in 2Q_{x_0,m}$. Obviously, we can assume $\ell(Q_{x_0,m}) \in (0, \infty)$. For each $k \le m$, we let

$$\begin{aligned}\int_{\mathbb{R}^D} |\varphi_{z,k}(x) - \varphi_{z,k}(y)|\, d\mu(z) &= \int_{\mathbb{R}^D \setminus \check{Q}^1_{x_0,k}} |\varphi_{z,k}(x) - \varphi_{z,k}(y)|\, d\mu(z) + \int_{\check{Q}^1_{x_0,k}} \cdots\\ &=: \mathrm{I}_{1,k} + \mathrm{I}_{2,k}.\end{aligned}$$

We now estimate $\mathrm{I}_{1,k}$. Notice that, if x, $y \in 2Q_{x_0,m}$, then

$$x,\ y \in 2Q_{x_0,k} \subset \frac{1}{2}\check{Q}^1_{x_0,k}.$$

Thus,

$$|x - z| \sim |y - z| \sim |x_0 - z|$$

for $z \in \mathbb{R}^D \setminus \check{Q}^1_{x_0,k}$. Thus, by (d) of Lemma 2.3.6, we see that

$$\mathrm{I}_{1,k} \lesssim \alpha_2^{-1} \int_{\mathbb{R}^D \setminus \check{Q}^1_{x_0,k}} \frac{|x - y|}{|x - z|^{n+1}}\, d\mu(z) \le \tilde{c}\alpha_2^{-1} \frac{\ell(Q_{x_0,m})}{\ell(\check{Q}^1_{x_0,k})}. \tag{3.3.20}$$

In case $k < m$, by Lemma 2.2.5, we find that

$$\mathrm{I}_{1,k} \lesssim \alpha_2^{-1} \frac{\ell(Q_{x_0,m})}{\ell(Q_{x_0,k})} \le c\alpha_2^{-1} 2^{-\gamma(m-k)A_0}.$$

Therefore,

$$C_7 A_0 \sum_{k=1}^{m} \mathrm{I}_{1,k} \le C_7 c\alpha_2^{-1} A \sum_{k=1}^{m-1} 2^{-\gamma(m-k)A_0} + C_7 \tilde{c}\alpha_2^{-1} A_0 \frac{\ell(Q_{x_0,m})}{\ell(\check{Q}^1_{x_0,m})}.$$

The first sum on the right-hand side is less than $A_0/400$ for A_0 big enough and $\alpha_2 \in (1, \infty)$. The second term on the right-hand side is also less than $A_0/400$ if we choose α_2 big enough. Thus,

$$C_7 A_0 \sum_{k=1}^{m} \mathrm{I}_{1,k} \le \frac{A_0}{200}.$$

We consider the integrals $\mathrm{I}_{2,k}$. By Lemma 2.3.6(d), we know that, for all $u \in Q_{x_0,k}$,

$$|\nabla \varphi_{z,k}(u)| \lesssim \frac{\alpha_2^{-1}}{[\ell(\check{Q}^1_{x_0,k})]^{n+1}}.$$

Therefore,

$$\mathrm{I}_{2,k} \lesssim \alpha_2^{-1} \int_{\check{Q}^1_{x_0,k}} \frac{|x - y|}{[\ell(\check{Q}^1_{x_0,k})]^{n+1}}\, d\mu(z) \lesssim \alpha_2^{-1} \frac{\ell(Q_{x_0,m})}{\ell(\check{Q}^1_{x_0,k})}.$$

This is the same estimate that we have obtained for $\mathrm{I}_{1,k}$ in (3.3.20) and hence we also have

$$C_7A_0\sum_{k=1}^m \mathrm{I}_{2,k}\le \frac{A_0}{200},$$

if we choose A_0 and α_2 big enough. Combining the estimates of $\mathrm{I}_{1,k}$ and $\mathrm{I}_{2,k}$ finishes the proof of Claim 3.

We next see that (a) holds true.

Claim 4. If $Q\in(\mathcal{D}_m^G\cup\mathcal{D}_m^B)$, then

$$|m_Q(f_m)|\le C_7A_0\|f\|_{\mathrm{RBMO}(\mu)}.$$

Also, we have

$$|g_m|,\ |b_m|\le C_7A_0\|f\|_{\mathrm{RBMO}(\mu)}.$$

Proof. We first prove the first statement. By Claim 2, we know that

$$\delta(Q,2R_0)>\left(m-\frac{1}{10}\right)A_0.$$

Let $R\in\mathcal{D}_{m-1}$ be such that $Q\cap R\ne\emptyset$. We must have $\ell(R)\in(0,\infty)$. Otherwise, $R=\{x\}$ for some $x\in Q$ and hence

$$\delta(R,2R_0)>\left(m-\frac{1}{10}\right)A_0>(m-1)A_0+\epsilon_1,$$

which is not possible because $R\in\mathcal{D}_{m-1}$.

Since $\ell(Q)\le\ell(R)/10$, we have $Q\subset 2R$. We know that

$$|m_R(f_m)|\le A_0\|f\|_{\mathrm{RBMO}(\mu)},$$

because (b) hods true for $m-1$. By Claim 3 for $m-1$ and R, we find that

$$\begin{aligned}|m_Q(f_m)|&\le|m_R(f_m)|+|m_Q(f_m)-m_R(f_m)|\\&\le|m_R(f_m)|+|m_Q(f)-m_R(f)|+\left|m_Q\left(\sum_{k=1}^{m-1}U_k\right)-m_R\left(\sum_{k=1}^{m-1}U_k\right)\right|\\&\lesssim A_0\|f\|_{\mathrm{RBMO}(\mu)}+|m_Q(f)-m_R(f)|.\end{aligned}$$

The term $|m_Q(f)-m_R(f)|$ is also bounded above by $CA_0\|f\|_{\mathrm{RBMO}(\mu)}$, because Q and R are doubling, $f\in\mathrm{RBMO}(\mu)$, and it is easy to show that $\delta(Q,R)\le CA_0$.

The estimates on g_m and b_m follow from the definitions of these functions and the estimate

$$|m_Q(f_m)| \lesssim A_0\|f\|_{\mathrm{RBMO}(\mu)} \quad \text{for} \quad Q \in \left(\mathcal{D}_m^G \bigcup \mathcal{D}_m^B\right).$$

This finishes the proof of Claim 4. □

Now we turn our attention to (d).

Claim 5. If $Q \in \mathcal{D}_m$ and

$$|m_Q(f_m)| \leq \frac{8}{20} A_0\|f\|_{\mathrm{RBMO}(\mu)},$$

then $U_m \equiv 0$ and $g_m \equiv b_m \equiv 0$ on $2Q$.

Proof. Suppose that $Q \equiv Q_{h,m} \in \mathcal{D}_m$ is such that either $g_m \not\equiv 0$, $b_m \not\equiv 0$ or $U_m \not\equiv 0$ on $2Q_{h,m}$. By Claim 1, we have $Q_{h,m} \subset 4S_{j,m}$ for some j. By construction, the center of $S_{j,m}$ belongs to some cube $Q_{i,m}$ with

$$|m_{Q_{i,m}}(f_m)| \geq \frac{3}{4} A_0\|f\|_{\mathrm{RBMO}(\mu)}.$$

It is easy to see that

$$\delta(Q_{h,m}, 4S_{j,m}) \leq C + \alpha_1 + \alpha_2 + \alpha_3$$

for some positive constant C. Thus,

$$|m_{Q_{i,m}}(f) - m_{Q_{h,m}}(f)| \leq (C + 2\alpha_1 + 2\alpha_2 + 2\alpha_3)\|f\|_{\mathrm{RBMO}(\mu)}.$$

Since $Q_{i,m}$ and $Q_{h,m}$ are contained in a common cube of the generation $m-1$, by Claim 3, we see that

$$\begin{aligned}
|m_{Q_{i,m}}(f_m) - m_{Q_{h,m}}(f_m)| &\leq |m_{Q_{i,m}}(f) - m_{Q_{h,m}}(f)| \\
&\quad + \left| m_{Q_{i,m}}\left(\sum_{k=1}^{m-1} U_k\right) - m_{Q_{h,m}}\left(\sum_{k=1}^{m-1} U_k\right)\right| \\
&\leq (C + 2\alpha_1 + 2\alpha_2 + 2\alpha_3 + A_0/100)\|f\|_{\mathrm{RBMO}(\mu)} \\
&\leq \frac{1}{10} A_0\|f\|_{\mathrm{RBMO}(\mu)}.
\end{aligned}$$

Therefore,

$$|m_{Q_{h,m}}(f_m)| \geq \left(\frac{3}{4} - \frac{1}{10}\right) A_0\|f\|_{\mathrm{RBMO}(\mu)} > \frac{8}{20} A_0\|f\|_{\mathrm{RBMO}(\mu)},$$

which completes the proof of Claim 5. □

Claim 6. If $g_m \not\equiv 0$ on Q and $Q \in \mathcal{D}_m$ with $\ell(Q) \in (0, \infty)$, then

$$|m_Q(f_{m+1})| \le \frac{7}{20} A_0 \|f\|_{\mathrm{RBMO}(\mu)}.$$

Proof. Assume that $Q := Q_{i,m}$. By Lemma 2.3.6, we have to deal with the cube $\hat{Q}_{i,m}^3$.

We now show that, if $P \in \mathcal{D}_m$ is such that $P \cap \hat{Q}_{i,m}^3 \neq \emptyset$, then $P \in (\mathcal{D}_m^G \cup \mathcal{D}_m^B)$. Indeed, notice that $P \subset \hat{\hat{Q}}_{i,m}^3$. By the definition of g_m and our assumption in Claim 6, there exists some j such that $Q_{i,m} \cap \frac{3}{2} S_{j,m} \neq \emptyset$, which implies that

$$\hat{\hat{Q}}_{i,m}^3 \bigcap \frac{3}{2} S_{j,m} \neq \emptyset.$$

For α_3 big enough, we have

$$\ell(\hat{\hat{Q}}_{i,m}^3) \ll \ell(S_{j,m}) \quad \text{and hence} \quad \hat{\hat{Q}}_{i,m}^3 \subset 2S_{j,m}.$$

Thus, $P \in (\mathcal{D}_m^G \cup \mathcal{D}_m^B)$.

We now estimate the term

$$\sup_{y \in \hat{Q}_{i,m}^3} |g_m(y) + b_m(y) - m_{Q_{i,m}}(f_m)|.$$

Recall that

$$g_m(y) + b_m(y) = \sum_{\{h:\ Q_{h,m} \in \mathcal{D}_m^G \bigcup \mathcal{D}_m^B\}} w_{h,m}(y) m_{Q_{h,m}}(f_m).$$

By the arguments above, if $y \in \hat{Q}_{i,m}^3$ and $w_{h,m}(y) \neq 0$, then $Q_{h,m} \cap \hat{Q}_{i,m}^3 \neq \emptyset$ and hence $Q_{h,m} \in \mathcal{D}_m^G \cup \mathcal{D}_m^B$. Thus, for all $y \in \mathbb{R}^D$,

$$g_m(y) + b_m(y) - m_{Q_{i,m}}(f_m) = \sum_{\{h:\ Q_{h,m} \in (\mathcal{D}_m^G \bigcup \mathcal{D}_m^B)\}} w_{h,m}(y)[m_{Q_{h,m}}(f_m) - m_{Q_{i,m}}(f_m)].$$

By Claim 3 and Lemma 2.1.3, we know that

$$\begin{aligned}
|m_{Q_{h,m}}(f_m) - m_{Q_{i,m}}(f_m)| &\le \frac{1}{100} A_0 \|f\|_{\mathrm{RBMO}(\mu)} + |m_{Q_{h,m}}(f) - m_{Q_{i,m}}(f)| \\
&\le \left[\frac{1}{100} A_0 + C + 2\delta(Q_{h,m}, Q_{i,m})\right] \|f\|_{\mathrm{RBMO}(\mu)} \\
&\le \frac{1}{50} A_0 \|f\|_{\mathrm{RBMO}(\mu)}.
\end{aligned}$$

Thus, for all $y \in \mathbb{R}^D$, we see that

$$|g_m(y) + b_m(y) - m_{Q_{i,m}}(f_m)| \le \frac{1}{50} A_0 \|f\|_{\mathrm{RBMO}(\mu)}. \tag{3.3.21}$$

For $x \in Q_{i,m}$, we have

$$\begin{aligned} |U_m(x) - m_{Q_{i,m}}(f_m)| &\le \left| U_m(x) - m_{Q_{i,m}}(f_m) \int_{\mathbb{R}^D} \varphi_{y,m}(x)\, d\mu(y) \right| \\ &\quad + |m_{Q_{i,m}}(f_m)| \left| 1 - \int_{\mathbb{R}^D} \varphi_{y,m}(x)\, d\mu(y) \right|. \end{aligned} \tag{3.3.22}$$

By (3.3.21), (2.3.3) and Lemma 2.3.6, we conclude that

$$\begin{aligned} &\left| U_m(x) - m_{Q_{i,m}}(f_m) \int_{\mathbb{R}^D} \varphi_{y,m}(x)\, d\mu(y) \right| \\ &= \left| \int_{\hat{Q}^3_{i,m}} \varphi_{y,m}(x)[g_m(y) + b_m(y) - m_{Q_{i,m}}(f_m)]\, d\mu(y) \right| \\ &\le (1 + \epsilon_3) \frac{1}{50} A_0 \|f\|_{\mathrm{RBMO}(\mu)}. \end{aligned}$$

On the other hand, from Lemma 2.3.6 and Claim 4, it follows that

$$|m_{Q_{i,m}}(f_m)| \left| 1 - \int_{\mathbb{R}^D} \varphi_{y,m}(x)\, d\mu(y) \right| \le C_7 \epsilon_3 A_0 \|f\|_{\mathrm{RBMO}(\mu)}.$$

Thus, if we choose ϵ_3 small enough, we have

$$\begin{aligned} |m_{Q_{i,m}}(f_{m+1})| &= |m_{Q_{i,m}}(f_m - U_m)| \\ &= |m_{Q_{i,m}}(U_m - m_{Q_{i,m}}(f_m))| \\ &\le \left[\frac{1 + \epsilon_3}{50} + C_7 \epsilon_3 \right] A_0 \|f\|_{\mathrm{RBMO}(\mu)} \\ &\le \frac{7}{20} A_0 \|f\|_{\mathrm{RBMO}(\mu)}. \end{aligned}$$

This finishes the proof of Claim 6.

Now we show that (b) holds true.

Claim 7. If $Q \in \mathcal{D}_m$ and $\ell(Q) \in (0, \infty)$, then

$$|m_Q(f_{m+1})| \le A_0 \|f\|_{\mathrm{RBMO}(\mu)}.$$

Proof. If $Q \in \mathcal{D}_m^G$, by Claim 6, we know that

$$|m_Q(f_{m+1})| \le \frac{7}{20} A_0 \|f\|_{\mathrm{RBMO}(\mu)}.$$

If $Q \in \mathcal{D}_m \setminus \mathcal{D}_m^G$, then from $\ell(Q) \ll \ell(S_{j,m})$ and $Q \not\subset \cup_j \frac{3}{2} S_{j,m}$, we deduce that $Q \cap (\cup_j S_{j,m}) = \emptyset$. This, together with the construction of $S_{j,m}$, implies that $Q \cap \Omega_m = \emptyset$. On the other hand, by $\ell(Q) \in (0, \infty)$, we know that $Q \in \tilde{\mathcal{D}}_m$, which in turn implies that z_Q, the center of Q, satisfies that $\delta(z_Q, 2R_0) > mA_0$. By these two facts and the definition of Ω_m, we conclude that

$$|m_Q(f_m)| < \frac{3}{4} A_0 \|f\|_{\mathrm{RBMO}(\mu)}. \tag{3.3.23}$$

Assume that $Q := Q_{h,m}$. If $U_m \equiv 0$ on Q, then (3.3.23) implies that

$$|m_Q(f_{m+1})| = |m_Q(f_m)| < \frac{3}{4} A_0 \|f\|_{\mathrm{RBMO}(\mu)}.$$

Now we consider the case that

$$Q_{h,m} \bigcap \left(\bigcup_j S_{j,m} \right) = \emptyset \quad \text{and} \quad U_m \not\equiv 0 \text{ on } Q_{h,m}.$$

By Claim 1, there exists some j with $\hat{Q}_{h,m}^3 \subset 4S_{j,m}$. By Lemma 2.3.6(a), if $x \in Q_{h,m}$, we then have

$$U_m(x) = \int_{\hat{Q}_{h,m}^3} \varphi_{y,m}(x)[g_m(y) + b_m(y)]\, d\mu(y).$$

Thus, if $y \in Q_{i,m}$ and $\varphi_{y,m}(x) \neq 0$, we find that $Q_{i,m} \cap \hat{Q}_{h,m}^3 \neq \emptyset$. Therefore, $Q_{i,m} \subset \hat{\hat{Q}}_{h,m}^3$. Then, from this and Lemma 2.1.3(d), it follows that

$$\delta(Q_{i,m}, Q_{h,m}) \le \epsilon_0 + \delta\left(Q_{i,m}, \hat{\hat{Q}}_{h,m}^3\right) + \delta\left(Q_{h,m}, \hat{\hat{Q}}_{h,m}^3\right) \le \frac{A_0}{400}.$$

This, together with $f \in \mathrm{RBMO}(\mu)$, implies that

$$|m_{Q_{i,m}}(f) - m_{Q_{h,m}}(f)| \le \frac{A_0}{100} \|f\|_{\mathrm{RBMO}(\mu)}.$$

Combining this and Claim 3, we conclude that

$$\begin{aligned}&\left|m_{Q_{i,m}}(f_m) - m_{Q_{h,m}}(f_m)\right|\\&\quad\le |m_{Q_{i,m}}(f) - m_{Q_{h,m}}(f)| + \left|m_{Q_{i,m}}\left(\sum_{k=1}^{m-1} U_k\right) - m_{Q_{h,m}}\left(\sum_{k=1}^{m-1} U_k\right)\right|\\&\quad\le \frac{1}{10} A_0 \|f\|_{\mathrm{RBMO}\,(\mu)}.\end{aligned} \tag{3.3.24}$$

By (d) and the assumption that $U_m \not\equiv 0$ on $Q_{h,m}$, we see that

$$|m_{Q_{h,m}}(f_m)| > \frac{8}{20} A_0 \|f\|_{\mathrm{RBMO}\,(\mu)}. \tag{3.3.25}$$

From the definitions of g_m and b_m, (3.3.24) and (3.3.25), we deduce that $m_{Q_{h,m}}(f_m)$ and $U_m(x)$ have the same sign.

On the other hand, by (3.3.23) and (3.3.24), we know that

$$|m_{Q_{i,m}}(f_m)| \le \frac{34}{40} A_0 \|f\|_{\mathrm{RBMO}\,(\mu)}.$$

Thus, from the definitions of g_m and b_m, we deduce that

$$\|g_m + b_m\|_{L^\infty(\mu)} \le \frac{34}{40} A_0 \|f\|_{\mathrm{RBMO}\,(\mu)},$$

which, together with (2.3.3), implies that

$$\begin{aligned}|U_m(x)| &\le \frac{34}{40} A_0 \|f\|_{\mathrm{RBMO}\,(\mu)} \int_{\mathbb{R}^D} \varphi_{y,m}(x)\, d\mu(y)\\&\le (1+\epsilon_3)\frac{34}{40} A_0 \|f\|_{\mathrm{RBMO}\,(\mu)}\\&\le A_0 \|f\|_{\mathrm{RBMO}\,(\mu)}.\end{aligned} \tag{3.3.26}$$

Since $m_{Q_{h,m}}(f_m)$ and $U_m(x)$ have the same sign, combining (3.3.23) and (3.3.26), we conclude that

$$|m_Q(f_{m+1})| = |m_Q(f_m - U_m)| \le m_Q(|m_Q(f_m) - U_m|) \le A_0 \|f\|_{\mathrm{RBMO}\,(\mu)}$$

and hence (b) also holds true in this case. □

Therefore, (a) through (e) are satisfied.

The statement (f) is a direct consequence of the following assertion.

Claim 8. If $\delta(x, 2R_0) < \infty$ and $Q = \{x\} \in \mathcal{D}_m$, then

$$h_0(x) = f_{m+1}(x) \text{ and } |h_0(x)| \le C_8 A_0 \|f\|_{\mathrm{RBMO}\,(\mu)}.$$

Proof. Take $m \in \mathbb{N}$ such that

$$(m-1)A_0 < \delta(x, 2R_0) \le mA_0.$$

By (e), we see that $U_{m+k}(x) = 0$ for $k \in \mathbb{N}$. Therefore, for all $x \in \mathbb{R}^D$,

$$f_{m+1}(x) = f_{m+2}(x) = \cdots = h_0(x).$$

By (a) and (2.3.3), for all $x \in \mathbb{R}^D$, we have

$$\begin{aligned}|f_{m+1}(x)| &\le |f_m(x)| + |U_m(x)|\\ &\le |f_m(x)| + 2C_7(1+\epsilon_3)A_0\|f\|_{\mathrm{RBMO}(\mu)}.\end{aligned}$$

Thus, we only have to estimate $|f_m(x)|$ for all $x \in \mathbb{R}^D$.

Take $Q_{i,m-1} \in \mathcal{D}_{m-1}$ with $x \in Q_{i,m-1}$. Since $\ell(Q_{i,m-1}) \in (0,\infty)$, by (b), we know that

$$|m_{Q_{i,m-1}}(f_m)| \le A_0\|f\|_{\mathrm{RBMO}(\mu)}.$$

Applying Claim 3, we conclude that

$$\begin{aligned}|m_{Q_{i,m-1}}(f_m) - f_m(x)| &\le |m_{Q_{i,m-1}}(f) - f(x)| + \frac{A_0}{100}\|f\|_{\mathrm{RBMO}(\mu)}\\ &\lesssim \left[1 + \delta(x, Q_{i,m-1}) + \frac{A_0}{100}\right]\|f\|_{\mathrm{RBMO}(\mu)}.\end{aligned}$$

It is easy to show that

$$\delta(x, Q_{i,m-1}) \le A_0 + \epsilon_0 + \epsilon_1.$$

Then we obtain

$$|f_m(x)| \lesssim A_0\|f\|_{\mathrm{RBMO}(\mu)},$$

which completes the proof of Claim 8.

Now we turn our attention to (g). Given some good cube $Q_{i,m} \in \mathcal{D}_m^G$ with $\ell(Q_{i,m}) \in (0,\infty)$, we let

$$Z_{i,m} := Z(Q_{i,m}, A_0\|f\|_{\mathrm{RBMO}(\mu)}/30),$$

where, for a cube Q and $\lambda \in (0,\infty)$, $Z(Q,\lambda)$ is as in Definition 3.1.21. If $Q_{i,m} \in \mathcal{D}_m^G$ and $\ell(Q_{i,m}) = 0$, we let $Z_{i,m} := Q_{i,m}$. The set $Z_{i,m}$ has a very nice property:

Claim 9. Let $k > m$ and $Q_{i,m} \in \mathcal{D}_m^G$. If $P \in \mathcal{D}_k$ is such that $P \cap Z_{i,m} \ne \emptyset$, then $g_k \equiv b_k \equiv 0$ on $2P$ and $P \notin (\mathcal{D}_k^G \cup \mathcal{D}_k^B)$.

Proof. Consider first the case $\ell(Q_{i,m}) = 0$. If $P \in \mathcal{D}_k$ is such that $P \cap Q_{i,m} \neq \emptyset$, then $\ell(P) \leq \ell(Q_{i,m})/10 = 0$ and hence $P = Q_{i,m}$. Therefore,

$$\delta(P, 2R_0) \leq mA_0 < \left(k - \frac{1}{10}\right) A_0.$$

By (e), we obtain $b_k \equiv g_k \equiv 0$ on P. Thus, $U_k \equiv 0$ and $P \notin (\mathcal{D}_k^G \cup \mathcal{D}_k^B)$.

Assume that $\ell(Q_{i,m}) \in (0, \infty)$. Let $x \in P \cap Z_{i,m}$. From the definition of $Z_{i,m}$, we deduce that

$$|m_{Q_{i,m}}(f) - m_S(f)| \leq \frac{A_0}{30} \|f\|_{\mathrm{RBMO}(\mu)} \tag{3.3.27}$$

for any $S \in \mathcal{D}_{m+j}$, $j \geq 1$, with $x \in S$. Also, Claim 6 implies that

$$|m_{Q_{i,m}}(f_{m+1})| \leq \frac{7}{20} A_0 \|f\|_{\mathrm{RBMO}(\mu)}.$$

Consider now $P_{m+1} \in \mathcal{D}_{m+1}$ with $x \in P_{m+1}$. Observe that

$$\ell(P) \leq \ell(Q_{i,m})/10 \text{ and } P_{m+1} \subset 2Q_{i,m}.$$

From this, it then follows that

$$\begin{aligned}
|m_{P_{m+1}}(f_{m+1})| &\leq |m_{Q_{i,m}}(f_{m+1})| + |m_{Q_{i,m}}(f_{m+1}) - m_{P_{m+1}}(f_{m+1})| \\
&\leq \frac{7}{20} A_0 \|f\|_{\mathrm{RBMO}(\mu)} + \left|m_{Q_{i,m}}(f) - m_{P_{m+1}}(f)\right| \\
&\quad + \left|m_{Q_{i,m}}\left(\sum_{k=1}^{m} U_k\right) - m_{P_{m+1}}\left(\sum_{k=1}^{m} U_k\right)\right|.
\end{aligned}$$

By (3.3.27) and Claim 3, we see that

$$|m_{P_{m+1}}(f_{m+1})| \leq \frac{8}{20} A_0 \|f\|_{\mathrm{RBMO}(\mu)}.$$

By (d), on $2P_{m+1}$ we have $g_m \equiv b_m \equiv 0$ and $U_{m+1} \equiv 0$. Thus, $f_{m+2} \equiv f_{m+1}$ on any cube $2P_{m+1}$ with $P_{m+1} \in \mathcal{D}_{m+1}$ containing x. Moreover, if $P_{m+1} \in (\mathcal{D}_{m+1}^G \cup \mathcal{D}_{m+1}^B)$, then there exists $j \in \mathbb{N}$ such that $P_{m+1} \subset 2S_{j,m+1}$. By the definition of $S_{j,m+1}$, we see that there exists a cube $Q_{i,m+1}$ with

$$|m_{Q_{i,m+1}}(f_{m+1})| \geq \frac{3}{4} A_0 \|f\|_{\mathrm{RBMO}(\mu)}.$$

Notice that

$$\left|m_{Q_{i,m+1}}(f_{m+1}) - m_{P_{m+1}}(f_{m+1})\right| \le \frac{1}{10} A_0 \|f\|_{\mathrm{RBMO}\,(\mu)}.$$

From this, we deduce that

$$\left|m_{P_{m+1}}(f_{m+1})\right| \ge \left(\frac{3}{4} - \frac{1}{10}\right) A_0 \|f\|_{\mathrm{RBMO}\,(\mu)} > \frac{8}{20} A_0 \|f\|_{\mathrm{RBMO}\,(\mu)},$$

which contradicts to the fact that

$$|m_{P_{m+1}}(f_{m+1})| \le \frac{8}{20} A_0 \|f\|_{\mathrm{RBMO}\,(\mu)}.$$

Thus, we have

$$P_{m+1} \notin \left(\mathcal{D}_{m+1}^G \bigcup \mathcal{D}_{m+1}^B\right).$$

Take $P_{m+2} \in \mathcal{D}_{m+2}$ such that $x \in P_{m+2}$. Notice that $P_{m+2} \subset 2P_{m+1}$ for some $P_{m+1} \in \mathcal{D}_{m+1}$ containing x. Thus, on P_{m+2}, $f_{m+2} \equiv f_{m+1}$. Then, from Claim 3, we deduce that

$$\begin{aligned}|m_{P_{m+2}}(f_{m+2})| &\le |m_{Q_{i,m}}(f_{m+1})| + |m_{Q_{i,m}}(f_{m+1}) - m_{P_{m+2}}(f_{m+1})| \\ &\le \frac{7}{20} A_0 \|f\|_{\mathrm{RBMO}\,(\mu)} + \left|m_{Q_{i,m}}(f) - m_{P_{m+2}}(f)\right| \\ &\quad + \left|m_{Q_{i,m}}\left(\sum_{k=1}^{m} U_k\right) - m_{P_{m+2}}\left(\sum_{k=1}^{m} U_k\right)\right| \\ &\le \frac{8}{20} A_0 \|f\|_{\mathrm{RBMO}\,(\mu)}.\end{aligned}$$

Again by (d), we see that

$$g_{m+2} \equiv b_{m+2} \equiv U_{m+2} \equiv 0$$

on $2P_{m+2}$. Thus,

$$f_{m+3} = f_{m+1} \text{ on } 2P_{m+2} \quad \text{and} \quad P_{m+2} \notin \left(\mathcal{D}_{m+2}^G \bigcup \mathcal{D}_{m+2}^B\right).$$

Going on, for all $j \in \mathbb{N}$ and $P_{m+j} \in \mathcal{D}_{m+j}$ containing x, we conclude that

$$g_{m+j} \equiv b_{m+j} \equiv U_{m+j} \equiv 0$$

on $2P_{m+j}$ and $P_{m+j} \notin (\mathcal{D}_{m+j}^G \cup \mathcal{D}_{m+j}^B)$. This finishes the proof of Claim 9.

As a consequence of Claim 9, $Z_{i,m}$ is a good place for supporting g_m. If, for each m, g_m was supported on $\cup_i Z_{i,m}$, then the supports of $\{g_m\}_{m\in\mathbb{N}}$ would be disjoint.

Thus, we are going to make some "corrections" according to this argument. For all $x \in \mathbb{R}^D$, we have

$$U_m^G(x) = \sum_{i\in I_m} \varphi_{y_i,m} \int_{\mathbb{R}^D} w_{i,m}(y) g_m(y)\, d\mu(y).$$

For each $Q_{i,m}$ with $\ell(Q_{i,m}) \in (0,\infty)$ and all $y \in \mathbb{R}^D$, we let

$$u_{i,m}(y) := \int_{\mathbb{R}^D} w_{i,m}(z) g_m(z)\, d\mu(z) \frac{\chi_{Z_{i,m}}(y)}{\mu(Z_{i,m})}.$$

If $\ell(Q_{i,m}) = 0$, for all $y \in \mathbb{R}^D$, we let

$$u_{i,m}(y) := w_{i,m}(y) g_m(y) = g_m(y).$$

Then U_m^G can be written as

$$U_m^G = \sum_{i\in I_m} \varphi_{y_i,m} \int_{\mathbb{R}^D} u_{i,m}(y)\, d\mu(y).$$

As in the case of U_m^B, if we let

$$\mathcal{D}_m^G := \mathcal{D}_m^{1,G} \bigcup \cdots \bigcup \mathcal{D}_m^{B_D,G},$$

where each subfamily $\mathcal{D}_m^{p,G}$ is disjoint, we can write U_m^G in the following way: for all $x \in \mathbb{R}^D$,

$$U_m^G(x) = \sum_{p=1}^{B_D} \int_{\mathbb{R}^D} \varphi_{y,m}^p(x) g_m^p(y)\, d\mu(y)$$

with

$$g_m^p(y) := \sum_{\{i:\ Q_{i,m}\in\mathcal{D}_m^{p,G}\}} u_{i,m}(y)$$

and $\varphi_{y,m}^p(x) := \varphi_{y,m}(x)$ if $y \in Q_{i,m}$ and $Q_{i,m} \in \mathcal{D}_m^p$.

By Corollary 3.1.22, if A_0 is big enough, we have $\mu(Z_{i,m}) \geq \mu(Q_{i,m})/2$. Then, it easy to show that

$$\|u_{i,m}\|_{L^\infty(\mu)} \leq 2\|g_m\|_{L^\infty(\mu)}$$

for all i. Thus, from (a), (g.2) follows. Moreover, because of Claim 9, (g.3) also holds true. □

Claim 10. For any $R \in \mathcal{D}_m$ with $\ell(R) \in (0, \infty)$, the bad cubes satisfy the packing condition

$$\sum_{\{Q:\ Q\cap R\neq\emptyset,\ Q\in\mathcal{D}_k^B,\ k>m\}} \mu(Q) \le C\mu(R).$$

Proof. Let $k > m$ be fixed. We now estimate the term

$$\sum_{\{Q:\ Q\cap R\neq\emptyset,\ Q\in\mathcal{D}_k^B\}} \mu(Q).$$

Let $Q \in \mathcal{D}_k^B$ be such that $Q \cap R \neq \emptyset$. Since Q is a bad cube, there exists some j such that $2S_{j,k} \cap Q \neq \emptyset$. Then we have $Q \subset 4S_{j,k}$. Since

$$A_0 \gg \alpha_1 + \alpha_2 + \alpha_3 \quad \text{and} \quad 4S_{j,k} \bigcap R \neq \emptyset,$$

we obtain $\ell(S_{j,k}) \le \ell(R)/20$ and hence $4S_{j,k} \subset 2R$.

By the finite overlapping of the cubes Q in $\mathcal{D}_k$ and the doubling property of $\{S_{j,k}\}_j$, we see that

$$\begin{aligned}\sum_{\{Q:\ Q\cap R\neq\emptyset,\ Q\in\mathcal{D}_k^B\}} \mu(Q) &\lesssim \mu\left(\bigcup_{\{j:\ S_{j,k}\subset 2R\}} 2S_{j,k}\right)\\ &\lesssim \sum_{\{j:\ S_{j,k}\subset 2R\}} \mu(2S_{j,k})\\ &\lesssim \sum_{\{j:\ S_{j,k}\subset 2R\}} \mu(S_{j,k}).\end{aligned} \tag{3.3.28}$$

On the other hand, from the construction of g_k^p, it is easy to show that

$$\mu(S_{j,k}) \lesssim \mu\left(S_{j,k} \bigcap \left\{x \in \mathbb{R}^D:\ \sum_{p=1}^{B_D} |g_k^p(x)| \neq 0\right\}\right). \tag{3.3.29}$$

Indeed, we first see that, if $Q \in \mathcal{D}_k$ such that $Q \cap S_{j,k} \neq \emptyset$, then $\ell(Q) \le \ell(S_{j,k})/4$, which implies that $Q \subset \frac{3}{2}S_{j,k}$ and hence $Q \in \mathcal{D}_k^G$. Now we let

$$Z_S := Z\left(S_{j,k}, \frac{A_0}{60}\|f\|_{\mathrm{RBMO}(\mu)}\right).$$

Then, by applying Corollary 3.1.22, we have

$$\mu(S_{j,k}) = \mu(S_{j,k} \setminus Z_S) + \mu(Z_S) \le c\mu(S_{j,k}) + \mu(Z_S)$$

for some $c \in (0,1)$. This implies that $\mu(S_{j,k}) \lesssim \mu(Z_S)$. Thus, (3.3.29) is reduced to the fact that

$$\begin{aligned}\mu(Z_S) &\le \mu\left(S_{j,k} \bigcap \bigcup_{Q_{i,k}\in\mathcal{D}_k^G} Z_{i,k}\right)\\ &= \mu\left(S_{j,k} \bigcap \left\{x \in \mathbb{R}^D : \sum_{p=1}^{B_D} |g_k^p(x)| \ne 0\right\}\right).\end{aligned} \tag{3.3.30}$$

The second inclusion is obvious. To see the first one, for μ-almost every $x \in Z_S$, there exists i such that $Q_{i,k} \ni x$. Since $Q_{i,k} \cap S_{j,k} \ne \emptyset$, we have $Q_{i,k} \in \mathcal{D}_k^G$. By the definition of Z_S, for any doubling cube $P \ni x$ with $\ell(P) \le \ell(Q_{i,k})/4$, we know that

$$\begin{aligned}|m_P(f) - m_{Q_{i,k}}(f)| &\le |m_P(f) - m_{S_{j,k}}(f)| + |m_{S_{j,k}}(f) - m_{Q_{i,k}}(f)|\\ &\le \frac{A_0}{30}\|f\|_{\mathrm{RBMO}(\mu)}.\end{aligned}$$

Thus, $x \in S_{j,k} \cap Z_{i,k}$. Therefore, the first inclusion in (3.3.30) holds true and hence (3.3.29) is true.

By (3.3.28), (3.3.29) and the bounded overlapping of the cubes $S_{j,k}$, we have

$$\sum_{\{Q:\ Q\cap R\ne\emptyset,\ Q\in\mathcal{D}_k^B\}} \mu(Q) \lesssim \mu\left(2R \bigcap \left\{x \in \mathbb{R}^D : \sum_{p=1}^{B_D} |g_k^p(x)| \ne 0\right\}\right).$$

Summing over $k > m$, as the supports of the functions $\{g_k^p\}_k$ are disjoint, we obtain

$$\begin{aligned}\sum_{\{Q:\ Q\cap R\ne\emptyset,\ Q\in\mathcal{D}_k^B,\, k>m\}} \mu(Q) &\lesssim \sum_{k>m} \mu\left(2R \cap \left\{x \in \mathbb{R}^D : \sum_{p=1}^{B_D} |g_k^p(x)| \ne 0\right\}\right)\\ &\lesssim \mu(2R)\\ &\lesssim \mu(R).\end{aligned}$$

This finishes the proof of Claim 10 and hence Lemma 3.3.7. □

Remark 3.3.10. By Theorems 3.2.11, 3.2.12 and 3.3.3, in what follows, we *identify the atomic Hardy space* $H^{1,p}_{\mathrm{atb},\gamma}(\mu)$ *with* $H^1_\Phi(\mu)$ *and write* $H^{1,p}_{\mathrm{atb},\gamma}(\mu)$ *simply by* $H^1(\mu)$. Moreover, when we use the atomic characterization of $H^1(\mu)$, unless explicitly pointed out, we *always assume that* $\gamma = 1$, $\eta = 2$ *and* $p = \infty$.

3.4 Notes

- The space $\mathrm{RBMO}^\rho_\gamma(\mu)$ when $\gamma = 1$ was introduced by Tolsa in [131] and when $\gamma \in (1, \infty)$ by Hu et al. in [55]. In [94], Mateu et al. introduced the spaces $\mathrm{BMO}(\mu)$ and $H^1_{\mathrm{at}}(\mu)$ with definitions similar to the classical ones. They showed that the dual space of $H^1_{\mathrm{at}}(\mu)$ is $\mathrm{BMO}(\mu)$ and $\mathrm{BMO}(\mu)$ satisfies a version of the John–Nirenberg inequality. However, unlike in the classical case, Calderón–Zygmund operators may be bounded on $L^2(\mu)$ but not from $L^\infty(\mu)$ to $\mathrm{BMO}(\mu)$ or from $H^1_{\mathrm{at}}(\mu)$ to $L^1(\mu)$, as it was showed by Verdera [144]. On the other hand, Nazarov et al. [105] introduced another version space of BMO type and proved that Calderón–Zygmund operators bounded on $L^2(\mu)$ are also bounded from $L^\infty(\mu)$ to their BMO space. Nevertheless, the BMO space in [105] does not satisfy the John–Nirenberg inequality.
- Examples 3.1.13, 3.1.14 and 3.1.15 were given by Tolsa [131]. For Example 3.1.15, see also [105].
- Theorem 3.1.16 and Corollary 3.1.20 were proved in [131]; see also [94] for another version of the John–Nirenberg inequality. Corollary 3.1.20, for $p \in (0, 1)$ and $\gamma = 1$, is also true, which was proved by Hu et al. in [63]. See also Sawano and Tanaka [118] for a localized and weighted version of the John–Nirenberg inequality.
- A characterization of $\mathrm{RBMO}(\mu)$ in terms of the John–Strömberg sharp maximal function was established by Hu et al. in [67]. Let μ be an absolutely continuous measure on $\mathbb{R}^D$, namely, there exists a weight ω such that $d\mu = \omega\,dx$. Lerner [79] also established the John–Strömberg characterization of $\mathrm{BMO}(\omega)$ in [94].
- Let $(\mathcal{X}, d, \mu)$ be a metric measure space with μ satisfying the polynomial growth condition. In [33], García-Cuerva and Gatto obtained a necessary and sufficient condition for the boundedness of Calderón–Zygmund operators associated to the measure μ, on Lipschitz spaces $\mathrm{Lip}_\alpha(\mathcal{X}, \mu)$ with $\alpha \in (0, 1)$. Also, when $(\mathcal{X}, d, \mu) := (\mathbb{R}^D, |\cdot|, \mu)$ with μ satisfying (0.0.1), they established several characterizations of $\mathrm{Lip}_\alpha(\mu)$, in terms of mean oscillations. This allows us to view the space $\mathrm{RBMO}(\mu)$ as a limit case for $\alpha \to 0$ of the spaces $\mathrm{Lip}_\alpha(\mu)$.
- Definition 3.2.1 and Theorems 3.2.10 and 3.2.11 were given by Tolsa in [131] when $\gamma = 1$ and in [55] when $\gamma \in (1, \infty)$. Theorem 3.2.12 was established in [55]. A version of the atomic Hardy space was introduced in [94].
- A version of the atomic decomposition of $H^{1,q}_{\mathrm{atb}}(\mu)$ was established by Hu and Liang in [51].
- Definition 3.3.1 and Theorem 3.3.3 were given by Tolsa in [134].

- In [28], a molecular characterization of the Hardy space $H^1(\mu)$ was established by Fu, Da. Yang and Do. Yang.
- Let $p \in (0, 1)$. To the best of our knowledge, there does not exist *any result* for the Hardy space $H^p(\mu)$ for μ being a non-negative Radon measure on $\mathbb{R}^D$ which only satisfies the polynomial growth condition (0.0.1).

Chapter 4
The Local Atomic Hardy Space $h^1(\mu)$

This chapter is mainly devoted to the study of the local version of $H^1(\mu)$ and its dual space. First, we introduce a local atomic Hardy space $h^1(\mu)$ and a local BMO-type space $\mathrm{rbmo}\,(\mu)$. After presenting some basic properties of these spaces, we then prove that the space $\mathrm{rbmo}\,(\mu)$ satisfies the John–Nirenberg inequality and its predual space is $h^1(\mu)$. Moreover, we also establish the relations between $H^1(\mu)$ and $h^1(\mu)$ as well as between $\mathrm{RBMO}\,(\mu)$ and $\mathrm{rbmo}\,(\mu)$. In addition, we also introduce a BLO-type space $\mathrm{RBLO}\,(\mu)$ and its local version $\mathrm{rblo}\,(\mu)$ on $(\mathbb{R}^D, |\cdot|, \mu)$ with μ as in (0.0.1) and establish some characterizations of both $\mathrm{RBLO}\,(\mu)$ and $\mathrm{rblo}\,(\mu)$.

4.1 The Local Atomic Hardy Space $h^1(\mu)$

To introduce our local spaces, we first introduce a special set of cubes via the coefficients $\delta(Q, R)$. To be precise, let A_0, ϵ_0, ϵ_1 and σ_0 be the positive constants as in Sect. 2.2. In the case that $\mathbb{R}^D$ is not an initial cube, letting $\{R_{-j}\}_{j\in\mathbb{Z}_+}$ be the cubes as in Definition 2.2.2, we then define the *set*

$$\mathcal{Q} := \big\{Q \subset \mathbb{R}^D : \text{ there exists a cube } P \subset Q \text{ and } j \in \mathbb{Z}_+ \text{ such that } P \subset R_{-j} \text{ with } \delta(P, R_{-j}) \le (j+1)A_0 + \epsilon_1\big\}.$$

If $\mathbb{R}^D$ is an initial cube, we define the *set*

$$\mathcal{Q} := \big\{Q \subset \mathbb{R}^D : \text{ there exists a cube } P \subset Q \text{ such that } \delta(P, \mathbb{R}^D) \le A_0 + \epsilon_1\big\}.$$

It is easy to see that, if $Q \in \mathcal{Q}$, then any R containing Q is also in $\mathcal{Q}$ and the set $\mathcal{Q}$ is independent of the chosen reference cubes $\{R_{-j}\}_{j\in\mathbb{Z}_+}$ in the sense modulo some small error (the error is no more than $2\epsilon_1 + \epsilon_0$). Moreover, the following observation implies that, in the case that μ is the D-dimensional Lebesgue measure on $\mathbb{R}^D$, then,

D. Yang et al., *The Hardy Space H^1 with Non-doubling Measures and Their Applications*, Lecture Notes in Mathematics 2084, DOI 10.1007/978-3-319-00825-7_4,

for any cube $Q \subset \mathbb{R}^D$, $Q \in \mathcal{Q}$ if and only if $\ell(Q) \gtrsim 1$. Based on this observation, we can think that our local spaces are the local spaces in the spirit of Goldberg.[1]

Proposition 4.1.1. *Let μ be the D-dimensional Lebesgue measure on $\mathbb{R}^D$. Then, for any cube $Q \subset \mathbb{R}^D$, $Q \in \mathcal{Q}$ if and only if $\ell(Q) \geq a_0$, where a_0 is a positive constant independent of Q.*

Proof. In this case, we choose $\{R_{-j}\}_{j\in\mathbb{Z}_+}$ as the cubes centered at the origin with side length 2^j. We first see the sufficiency. For any cube Q with $\ell(Q) \geq a_0$, it is easy to see that there exists a nonnegative constant $\tilde{C}$, which depends only on D and $j \in \mathbb{Z}_+$, such that $Q \subset R_{-j}$ and

$$\delta(Q, R_{-j}) \leq \int_{B(x_Q,\sqrt{D}\ell(R_{-j}))\setminus B(x_Q,\frac{a_0}{2})} \frac{1}{|x - x_Q|^D}\,dx \leq (j+1)A_0,$$

where

$$A_0 \geq \tilde{C} \log_2(2\sqrt{D}\max(1/a_0, 1)).$$

Thus, $Q \in \mathcal{Q}$.

Conversely, if $Q \in \mathcal{Q}$, then there exist a cube $\tilde{Q} \subset Q$ and $j \in \mathbb{Z}_+$ such that

$$\tilde{Q} \subset R_{-j} \quad \text{and} \quad \delta(\tilde{Q}, R_{-j}) \leq (j+1)A_0 + \epsilon_1.$$

To finish the proof, it suffices to show that $\ell(\tilde{Q}) \gtrsim 1$. Moreover, we only need to consider the case that $\ell(\tilde{Q}) < (\sqrt{D})^{-1}$. Let ω_{D-1} be the $(D-1)$*-dimensional Lebesgue measure of the unit sphere in* $\mathbb{R}^D$. By Definition 2.1.1, we have

$$\delta(\tilde{Q},\, R_{-j}) = \int_{\tilde{Q}_{R_{-j}}\setminus\tilde{Q}} \frac{1}{|x - x_{\tilde{Q}}|^D}dx.$$

From this and the fact that

$$\{x \in \mathbb{R}^D : \sqrt{D}\ell(\tilde{Q})/2 \leq |x - x_{\tilde{Q}}| \leq \ell(R_{-j})/2\} \subset (\tilde{Q}_{R_{-j}} \setminus \tilde{Q}),$$

it follows that

$$\delta(\tilde{Q},\, R_{-j}) \geq \omega_{D-1} \int_{\frac{\sqrt{D}\ell(\tilde{Q})}{2}}^{\frac{\ell(R_{-j})}{2}} \frac{1}{r}\,dr = \omega_{D-1} \log_2\left(\frac{\ell(R_{-j})}{\sqrt{D}\ell(\tilde{Q})}\right),$$

[1] See [38].

which implies that

$$\ell(R_{-j}) \lesssim 2^{\delta(\tilde{Q}, R_{-j})/\omega_{D-1}} \ell(\tilde{Q}) \lesssim 2^{(jA_0)/\omega_{D-1}} \ell(\tilde{Q}). \tag{4.1.1}$$

On the other hand, since $\ell(R_0) = 1$, it follows that

$$\begin{aligned}\delta(R_0, R_{-j}) &= \int_{R_{-j} \setminus R_0} \frac{1}{|x|^D}\, dx \\ &\leq \omega_{D-1} \int_{\frac{1}{2}}^{\sqrt{D}\ell(R_{-j})} \frac{1}{r}\, dr \\ &= \omega_{D-1} \log_2 \left(2\sqrt{D}\ell(R_{-j})\right),\end{aligned}$$

which, together with Definition 2.2.2, implies that

$$\ell(R_{-j}) \gtrsim 2^{\delta(R_0, R_{-j})/\omega_{D-1}} \gtrsim 2^{(jA_0)/\omega_{D-1}}. \tag{4.1.2}$$

Combining (4.1.1) and (4.1.2), we then see that $\ell(\tilde{Q}) \gtrsim 1$, which completes the proof of Proposition 4.1.1. □

In what follows, for any cube R and $x \in R \cap \operatorname{supp} \mu$, let $\{Q_{x,k}\}_k$ be as in Sect. 2.2 and H_R^x the *largest integer* k such that $R \subset Q_{x,k}$. The following properties on H_R^x are useful in applications.

Lemma 4.1.2. *The following properties hold true:*

(a) *For any cube R and $x \in R \cap \operatorname{supp} \mu$, $Q_{x, H_R^x+1} \subset 3R$ and $5R \subset Q_{x, H_R^x-1}$;*
(b) *For any cube R, $x \in R \cap \operatorname{supp} \mu$ and $k \in \mathbb{Z}$ with $k \geq H_R^x + 2$, $Q_{x,k} \subset \frac{7}{5}R$;*
(c) *For any cube $R \subset \mathbb{R}^D$ and $x, y \in R \cap \operatorname{supp} \mu$, $|H_R^x - H_R^y| \leq 1$;*
(d) *For any cube R and $x \in R \cap \operatorname{supp} \mu$, $H_R^x \in \mathbb{Z}_+$ when $R \notin \mathcal{Q}$. Moreover, if $R \in \mathcal{Q}$, then $H_R^x \leq 1$ when $\mathbb{R}^D$ is not an initial cube, and $0 \leq H_R^x \leq 1$ when $\mathbb{R}^D$ is an initial cube;*
(e) *When $k \geq 2$, for any $x \in \operatorname{supp} \mu$, $Q_{x,k} \notin \mathcal{Q}$;*
(f) *For any cube $R \notin \mathcal{Q}$ and $x \in R \cap \operatorname{supp} \mu$, if any cube $\tilde{R} \subset Q_{x, H_R^x+2}$, then $\tilde{R} \notin \mathcal{Q}$;*
(g) *For any cube R and $x \in R \cap \operatorname{supp} \mu$, there exists a positive constant C such that $\delta(R, Q_{x,H_R^x}) \leq C$ and $\delta(Q_{x,H_R^x+1}, R) \leq C$.*

Proof. We first show (a). For any $x \in R \cap \operatorname{supp} \mu$, by the definition of H_R^x, together with the decreasing property of $\{Q_{x,k}\}_{k\in\mathbb{Z}}$ in k, we know that $R \subset Q_{x, H_R^x}$ and $R \not\subset Q_{x, H_R^x+1}$, which imply that

$$\ell(R) \leq \ell(Q_{x, H_R^x}) \quad \text{and} \quad \ell(Q_{x, H_R^x+1}) \leq 2\ell(R).$$

These facts, together with the fact that $\ell(Q_{x,H_R^x}) \le \frac{1}{10}\ell(Q_{x,H_R^x-1})$ (see Lemma 2.2.5), imply (a).

To see (b), for any $x \in R \cap \operatorname{supp} \mu$, by the fact that

$$\ell(Q_{x,H_R^x+2}) \le \frac{1}{10}\ell(Q_{x,H_R^x+1}),$$

together with the fact that $\ell(Q_{x,H_R^x+1}) \le 2\ell(R)$, we have $\ell(Q_{x,H_R^x+2}) \le \frac{1}{5}\ell(R)$. Thus, $Q_{x,H_R^x+2} \subset \frac{7}{5}R$, which, together with the decreasing property of $Q_{x,k}$ in k again, shows (b).

For any $R \subset \mathbb{R}^D$ and $x, y \in R \cap \operatorname{supp}\mu$, it is clear that $y \in Q_{x,H_R^x} \cap Q_{y,H_R^x}$. Then Lemma 2.3.3, together with the definition of H_R^x, implies that

$$R \subset Q_{x,H_R^x} \subset Q_{y,H_R^x-1}.$$

This shows that $H_R^y \ge H_R^x - 1$. Symmetrically, we have $H_R^y \le H_R^x + 1$, which shows (c).

We now prove (d). Assume that $R \notin \mathcal{Q}$. By similarity, we only consider the case that $\mathbb{R}^D$ is not an initial cube. By the definitions of σ_0 and A_0, together with Lemma 2.1.3(a), we see that, for any cube Q,

$$\delta(Q,3Q) \le 6^n C_0 < \sigma_0 \ll A_0.$$

Now assume that $3R \subset R_{-j}$ for some $j \in \mathbb{Z}_+$. If $R \notin \mathcal{Q}$, by the conclusion of (a) and Lemma 2.1.3(d), we conclude that

$$\begin{aligned}(j+1)A_0 + \epsilon_1 &< \delta(R,R_{-j})\\ &= \delta(R,3R) + \delta(3R,R_{-j})\\ &< \sigma_0 + \delta(Q_{x,H_R^x+1},R_{-j}) + \epsilon_0\\ &= \sigma_0 + (H_R^x + 1 + j)A_0 \pm 4\epsilon_1 + \epsilon_0.\end{aligned}$$

This estimate, together with the fact that

$$\epsilon_0 \le \epsilon_1 \ll \sigma \ll A_0,$$

implies that $H_R^x \in \mathbb{Z}_+$.

If $\mathbb{R}^D$ is not an initial cube and $R \in \mathcal{Q}$, by the definitions of $\mathcal{Q}$ and cubes of generations, there exist a cube $\tilde{Q} \subset R$ and $j_1, j_2 \in \mathbb{Z}_+$ such that $\tilde{Q} \subset R_{-j_1}$ with

$$\delta(\tilde{Q},R_{-j_1}) \le (j_1+1)A_0 + \epsilon_1,$$

and $Q_{x,H_R^x} \subset R_{-j_2}$ with

$$\delta(Q_{x,H_R^x},R_{-j_2}) = (j_2 + H_R^x)A_0 \pm \epsilon_1.$$

Let $j := \max(j_1, j_2)$. By Lemma 2.1.3, we have

$$\delta(Q_{x,H_R^x}, R_{-j}) = (j + H_R^x)A_0 \pm 4\epsilon_1.$$

Lemma 2.1.3(d), together with the definition of $Q \in \mathcal{Q}$ and the fact that $\epsilon_0 \le \epsilon_1$, implies that

$$\delta(Q_{x,H_R^x}, R_{-j}) \le (j+1)A_0 + 4\epsilon_1.$$

On the other hand, if $H_R^x \ge 2$, then, by the fact that $\epsilon_1 \ll A_0$,

$$\delta(Q_{x,H_R^x}, R_{-j}) = (j + H_R^x)A_0 \pm 4\epsilon_1 > (j+1)A_0 + 4\epsilon_1.$$

This is a contradiction, which shows that $H_R^x \le 1$ when $R \in \mathcal{Q}$.

Similarly, if $\mathbb{R}^D$ is an initial cube, then, for any cube $R \in \mathcal{Q}$ and $x \in R \cap \operatorname{supp}\mu$, we also find that $H_R^x \le 1$. On the other hand, recall that, if $\mathbb{R}^D$ is an initial cube, then, for any cube R, $x \in \operatorname{supp}\mu$ and $k \in \mathbb{Z}$ with $k \le 0$, $Q_{x,k} = \mathbb{R}^D$. Therefore, obviously, $H_R^x \in \mathbb{Z}_+$, which proves (d).

To see (e), by similarity, we only consider the case that $\mathbb{R}^D$ is an initial cube. Assume that $Q_{x,k} \in \mathcal{Q}$. By the definition, there exists a cube $Q \subset Q_{x,k}$ such that $\delta(Q, \mathbb{R}^D) \le A_0 + \epsilon_1$. By Lemma 2.1.3(d), we then have

$$A_0 + \epsilon_1 \ge \delta(Q, \mathbb{R}^D) = \delta(Q, Q_{x,k}) + \delta(Q_{x,k}, \mathbb{R}^D) \pm \epsilon_0 \ge kA_0 - \epsilon_1 - \epsilon_0,$$

which is impossible when $k \ge 2$, since $A_0 \gg \epsilon_1 \ge \epsilon_0$. Thus, $Q_{x,k} \notin \mathcal{Q}$, which completes the proof of (e).

To prove (f), we only consider the case that $\mathbb{R}^D$ is not an initial cube, since the argument for the case that $\mathbb{R}^D$ is an initial cube is similar. If any cube $\tilde{R} \in \mathcal{Q}$, by the definition of $\mathcal{Q}$, there exist a cube $\hat{R} \subset \tilde{R}$ and $j_1 \in \mathbb{Z}_+$ such that

$$\hat{R} \subset R_{-j_1} \quad \text{and} \quad \delta(\hat{R}, R_{-j_1}) \le (j_1 + 1)A_0 + \epsilon_1.$$

By Definition 2.2.2, there exists a $j_2 \in \mathbb{Z}_+$ such that

$$Q_{x,H_R^x+2} \subset R_{-j_2} \quad \text{and} \quad \delta(Q_{x,H_R^x+2}, R_{-j_2}) \le (j_2+1)A_0 + \epsilon_1.$$

Let $j := \max\{j_1, j_2\}$. By the assumption that $\tilde{R} \subset Q_{x,H_R^x+2}$, we also know that $\hat{R} \subset Q_{x,H_R^x+2}$, which, combined with Lemma 2.1.3(d), implies that

$$\begin{aligned}(j+1)A_0 + 3\epsilon_1 + 2\epsilon_0 &\ge \delta(\hat{R}, R_{-j}) + \epsilon_0 \\ &\ge \delta(Q_{x,H_R^x+2}, R_{-j}) \\ &= (H_R^x + 2 + j)A_0 \pm 2\epsilon_1 \pm \epsilon_0,\end{aligned}$$

where we used the fact that

$$\delta(R_{-j}, R_{-j_i}) = (j - j_i)A_0 \pm 2\epsilon_1.$$

This, together with the choice of the constant A_0, shows that $H_R^x < 0$, which contradicts to (d). Thus, $\tilde{R} \notin \mathcal{Q}$, which completes the proof of (f).

Finally, by the properties (a) and (b) above and Lemma 2.1.3, we see that

$$\delta(2R, Q_{x, H_R^x - 1}) \leq \epsilon_0 + \delta(Q_{x, H_R^x + 2}, Q_{x, H_R^x - 1}) \lesssim 1$$

and hence

$$\delta(R, Q_{x, H_R^x}) \leq \epsilon_0 + \delta(R, Q_{x, H_R^x - 1}) \lesssim 1 + \delta(R, 2R) + \delta(2R, Q_{x, H_R^x - 1}) \lesssim 1.$$

Also, the above property (a) and Lemma 2.1.3 imply that

$$\begin{aligned}
\delta(Q_{x, H_R^x + 1}, R) &\lesssim 1 + \delta(Q_{x, H_R^x + 1}, 3R) + \delta(3R, R) \\
&\lesssim 1 + \delta(Q_{x, H_R^x + 1}, Q_{x, H_R^x - 1}) \\
&\lesssim 1,
\end{aligned}$$

which proves (g) and hence completes the proof of Lemma 4.1.2. □

We now introduce local Hardy spaces.

Definition 4.1.3. Let $p \in (1, \infty]$ and $\eta \in (1, \infty)$. A function $b \in L^1_{\mathrm{loc}}(\mu)$ is called a *p-block* if only (i) and (iii) of Definition 3.2.1 hold true. Moreover, let

$$|b|_{h^{1,p}_{\mathrm{atb}}(\mu)} := \sum_{j=1}^{2} |\lambda_j|.$$

A function $f \in L^1(\mu)$ is said to belong to the *space* $h^{1,p}_{\mathrm{atb}}(\mu)$ if there exist p-atomic blocks or p-blocks $\{b_i\}_i$ such that $f = \sum_i b_i$ and $\sum_i |b_i|_{h^{1,p}_{\mathrm{atb}}(\mu)} < \infty$, where b_i is a p-atomic block as in Definition 3.2.1 if $\operatorname{supp} b_i \subset R_i$ and $R_i \notin \mathcal{Q}$, while b_i is a p-block if $\operatorname{supp} b_i \subset R_i$ and $R_i \in \mathcal{Q}$. Moreover, the $h^{1,p}_{\mathrm{atb}}(\mu)$ *norm* of f is defined by

$$\|f\|_{h^{1,p}_{\mathrm{atb}}(\mu)} := \inf\left\{\sum_i |b_i|_{h^{1,p}_{\mathrm{atb}}(\mu)}\right\},$$

where the infimum is taken over all decompositions of f in p-atomic blocks or p-blocks as above.

Remark 4.1.4. (i) Let $\rho \in (1,\infty)$ and $\beta_\rho := \rho^{D+1}$. Due to (3.2.2), if necessary, we may assume that the cube R in Definition 4.1.3 is (ρ, β_ρ)-doubling.
(ii) Let $p \in (1,\infty]$ and μ be the D-dimensional Lebesgue measure. A function $f \in L^1(\mathbb{R}^D)$ is said to belong to the *space* $h^{1,p}(\mathbb{R}^D)$ if $f = \sum_i \lambda_i b_i$ and $\sum_i |\lambda_i| < \infty$, where, for each i, b_i is a p-atom as in Remark 3.2.2(i) if $\operatorname{supp} b_i \subset Q_i$ and $\ell(Q_i) \le 1$, while b_i only satisfies (a) and (c) of Remark 3.2.2(i) if $\operatorname{supp} b_i \subset Q_i$ and $\ell(Q_i) \in (1,\infty)$. Moreover, the $h^{1,p}(\mathbb{R}^D)$ *norm* of f is defined by

$$\|f\|_{h^{1,p}(\mathbb{R}^D)} := \inf\left\{\sum_i |b_i|_{h^{1,p}(\mathbb{R}^D)}\right\},$$

where the infimum is taken over all decompositions of f as above.[2]

It is easy to see that $H^{1,p}_{\rm atb}(\mu) \subsetneq h^{1,p}_{\rm atb}(\mu) \subsetneq L^1(\mu)$. Moreover, we have the following basic properties on the space $h^{1,p}_{\rm atb}(\mu)$.

Proposition 4.1.5. *Let $p \in (1,\infty]$. The following four properties hold true:*

(i) *The space $h^{1,p}_{\rm atb}(\mu)$ is a Banach space;*
(ii) *For $1 < p_1 \le p_2 \le \infty$, $h^{1,p_2}_{\rm atb}(\mu) \subset h^{1,p_1}_{\rm atb}(\mu) \subset L^1(\mu)$;*
(iii) *The local atomic Hardy space $h^{1,p}_{\rm atb}(\mu)$ is independent of the choice of the constant $\eta \in (1,\infty)$;*
(iv) *Let $p \in (1,\infty)$. The local atomic Hardy space $h^{1,\infty}_{\rm atb}(\mu)$ is dense in the local Hardy space $h^{1,p}_{\rm atb}(\mu)$.*

Proof. The proofs of the first two properties are similar to the usual proofs for the classical atomic Hardy spaces with μ being the D-dimensional Lebesgue measure. Thus, we omit the details.

Moreover, we only show Property (iii) for the case $p = \infty$ by similarity. Let $\eta_1 > \eta_2 > 1$. It is obvious that, for any $b \in h^{1,\infty}_{{\rm atb},\,\eta_1}(\mu)$, we have

$$b \in h^{1,\infty}_{{\rm atb},\,\eta_2}(\mu) \quad \text{and} \quad \|b\|_{h^{1,\infty}_{{\rm atb},\,\eta_2}(\mu)} \le \|b\|_{h^{1,\infty}_{{\rm atb},\,\eta_1}(\mu)}.$$

To prove the converse, let

$$b := \sum_{j=1}^{2} \lambda_j a_j \in h^{1,\infty}_{{\rm atb},\,\eta_2}(\mu)$$

[2]See [38].

be an ∞-atomic block with $\operatorname{supp} b \subset R \notin \mathcal{Q}$ or an ∞-block with $\operatorname{supp} b \subset R \in \mathcal{Q}$ as in Definition 4.1.3. By Remark 4.1.4, we may assume that R is (ρ, β_ρ)-doubling with $\rho \geq \eta_1$. Then, for each $j \in \{1, 2\}$,

$$\|a_j\|_{L^\infty(\mu)} \leq \{\mu(\eta_2 Q_j)[1 + \delta(Q_j, R)]\}^{-1}.$$

For any $x \in Q_j \cap \operatorname{supp} \mu$, let Q_x be the cube centered at x with side length $\frac{\eta_2 - 1}{10\eta_1}\ell(Q_j)$. We then see that $\eta_1 Q_x \subset \eta_2 Q_j$. By Theorem 1.1.1, there exists an almost disjoint subfamily $\{Q_{j,k}\}_k$ of the cubes $\{Q_x\}_x$ covering $Q_j \cap \operatorname{supp} \mu$. Moreover, for each $j \in \{1, 2\}$, the number of cubes $\{Q_{j,k}\}_k$ of the Besicovitch covering is bounded by some constant $N \in \mathbb{N}$ depending only on η_1, η_2 and D; see Theorem 1.1.1. Since $\ell(Q_{j,k}) \sim \ell(Q_j)$ for all k, by Lemma 2.1.3, we have $\delta(Q_{j,k}, Q_j) \lesssim 1$. Moreover, it follows, from Lemma 2.1.3 again, that

$$\delta(Q_{j,k}, \eta_2 R) \lesssim 1 + \delta(Q_j, R).$$

Therefore, by letting

$$a_{j,k} := a_j \frac{\chi_{Q_{j,k}}}{\sum_{k=1}^N \chi_{Q_{j,k}}}$$

and $\lambda_{j,k} := \lambda_j$ for $k \in \{1, \ldots, N\}$, we see that

$$b = \sum_{j=1}^2 \lambda_j a_j = \sum_{j=1}^2 \sum_{k=1}^N \lambda_{j,k} a_{j,k}$$

and

$$\|a_{j,k}\|_{L^\infty(\mu)} \leq \|a_j\|_{L^\infty(\mu)} \lesssim \{\mu(\eta_1 Q_{j,k})[1 + \delta(Q_{j,k}, \eta_2 R)]\}^{-1}.$$

If b is an ∞-atomic block and $R \notin \mathcal{Q}$, then arguing as in (iii) of Proposition 3.2.4, we see that

$$b \in h^{1,\infty}_{\mathrm{atb},\,\eta_1}(\mu) \quad \text{and} \quad \|b\|_{h^{1,\infty}_{\mathrm{atb},\,\eta_1}(\mu)} \lesssim \|b\|_{h^{1,\infty}_{\mathrm{atb},\,\eta_2}(\mu)}.$$

If b is an ∞-block and $R \in \mathcal{Q}$, then, for each j, k, let $c_{j,k} := \lambda_{j,k} a_{j,k}$. It is obvious that, for each j and k, $c_{j,k}$ is an ∞-block with

$$|c_{j,k}|_{h^{1,\infty}_{\mathrm{atb},\,\eta_1}(\mu)} \lesssim |\lambda_{j,k}| \quad \text{and} \quad \operatorname{supp} c_{j,k} \subset \eta_2 R.$$

Moreover,

$$b = \sum_{j=1}^2 \sum_{k=1}^N c_{j,k} \quad \text{and} \quad \|b\|_{h^{1,\infty}_{\mathrm{atb},\,\eta_1}(\mu)} \lesssim \sum_{j=1}^2 \sum_{k=1}^N |\lambda_{j,k}| \lesssim \sum_{j=1}^2 |\lambda_j|,$$

which implies that

$$\|b\|_{h^{1,\infty}_{\mathrm{atb},\eta_1}(\mu)} \lesssim \|b\|_{h^{1,\infty}_{\mathrm{atb},\eta_2}(\mu)}.$$

Thus, (iii) holds true.

Now we prove (iv). By Definition 4.1.3, for every $f \in h^{1,p}_{\mathrm{atb}}(\mu)$ and $\epsilon \in (0,\infty)$, there exist $m \in \mathbb{N}$ and $g := \sum_{j=1}^m b_j$ such that $\|f - g\|_{h^{1,p}_{\mathrm{atb}}(\mu)} < \frac{\epsilon}{2}$, where, for $j \in \{1,\ldots,m\}$, b_j is a p-atomic block if $\operatorname{supp} b_j \subset R_j \notin \mathcal{Q}$ or a p-block if $\operatorname{supp} b_j \subset R_j \in \mathcal{Q}$. Moreover, for any $j \in \{1,\ldots,m\}$, it is easy to scc that there exists an $h_j \in L^\infty(\mu)$ such that $\operatorname{supp} h_j \subset R_j$ and

$$\|b_j - h_j\|_{L^p(\mu)} < \frac{\epsilon}{2^{m+1}[\mu(2R_j)]^{1-1/p}}.$$

For each j, if b_j is a p-atomic block with $\operatorname{supp} b_j \subset R_j \notin \mathcal{Q}$, then take

$$\widetilde{b_j} := h_j - \frac{\chi_{R_j}}{\mu(R_j)} \int_{\mathbb{R}^D} h_j \, d\mu.$$

Observe that $\widetilde{b_j}$ is an ∞-atomic block with

$$\operatorname{supp} b_j \subset R_j \quad \text{and} \quad \|b_j - \widetilde{b_j}\|_{h^{1,p}_{\mathrm{atb}}(\mu)} < \frac{\epsilon}{2m}.$$

If b_j is a p-block with $\operatorname{supp} b_j \subset R_j \in \mathcal{Q}$, then take $\widetilde{b_j} := h_j$. It is easy to see that $\widetilde{b_j}$ is an ∞-block with

$$\operatorname{supp} \widetilde{b_j} \subset R_j \in \mathcal{Q} \quad \text{and} \quad \|b_j - \widetilde{b_j}\|_{h^{1,p}_{\mathrm{atb}}(\mu)} < \frac{\epsilon}{2m}.$$

Now let $\tilde{g} := \sum_{j=1}^m \widetilde{b_j}$. From Definition 4.1.3, it further follows that $\tilde{g} \in h^{1,\infty}_{\mathrm{atb}}(\mu)$ and

$$\|f - \tilde{g}\|_{h^{1,p}_{\mathrm{atb}}(\mu)} \le \|f - g\|_{h^{1,p}_{\mathrm{atb}}(\mu)} + \|g - \tilde{g}\|_{h^{1,p}_{\mathrm{atb}}(\mu)} < \epsilon,$$

which completes the proof of Proposition 4.1.5. □

Remark 4.1.6. By Proposition 4.1.5, unless otherwise stated, we *always assume that the constant η in Definition 4.1.3 is equal to 2.*

4.2 The Space rbmo (μ)

In this section, we further consider the local space rbmo (μ), which turns out to be the dual space of $h_{\mathrm{atb}}^{1,p}(\mu)$ for $p \in (1, \infty]$. We begin with the definition of rbmo (μ).

Definition 4.2.1. Let $\eta \in (1, \infty)$, $\rho \in [\eta, \infty)$ and $\beta_\rho := \rho^{D+1}$. A function $f \in L^1_{\mathrm{loc}}(\mu)$ is said to be in the *space* $\mathrm{rbmo}_{\eta,\rho}(\mu)$, if there exists a nonnegative constant $\tilde{C}$ such that, for any cube $Q \notin \mathcal{Q}$,

$$\frac{1}{\mu(\eta Q)} \int_Q \left| f(y) - m_{\tilde{Q}^\rho}(f) \right| d\mu(y) \le \tilde{C},$$

that, for any two (ρ, β_ρ)-doubling cubes $Q \subset R$ with $Q \notin \mathcal{Q}$,

$$|m_Q(f) - m_R(f)| \le \tilde{C}[1 + \delta(Q, R)]$$

and that, for any cube $Q \in \mathcal{Q}$,

$$\frac{1}{\mu(\eta Q)} \int_Q |f(y)| \, d\mu(y) \le \tilde{C}. \tag{4.2.1}$$

Moreover, the $\mathrm{rbmo}_{\eta,\rho}(\mu)$ *norm* of f is defined to be the minimal constant $\tilde{C}$ as above and denoted by $\|f\|_{\mathrm{rbmo}_{\eta,\rho}(\mu)}$.

Remark 4.2.2. Let μ be the D-dimensional Lebesgue measure. A function $f \in L^1_{\mathrm{loc}}(\mathbb{R}^D)$ is said to be in the *space* bmo $(\mathbb{R}^D)$, if there exists a nonnegative constant C such that, for any cube Q with $\ell(Q) \le 1$,

$$\frac{1}{|Q|} \int_Q |f(y) - m_Q(f)| \, dy \le C$$

and that, for any cube Q with $\ell(Q) \in (1, \infty)$,

$$\frac{1}{|Q|} \int_Q |f(y)| \, dy \le C.$$

Moreover, the bmo $(\mathbb{R}^D)$ *norm* of f is defined to be the minimal constant C as above and denoted by $\|f\|_{\mathrm{bmo}(\mathbb{R}^D)}$.[3]

It follows, from Definition 3.1.5, that, for any fixed $\eta \in (1, \infty)$ and $\rho \in [\eta, \infty)$,

$$\mathrm{rbmo}_{\eta,\rho}(\mu) \subset \mathrm{RBMO}(\mu).$$

[3]See [38].

Moreover, from the propositions below, we see that the space $\mathrm{rbmo}_{\eta,\rho}(\mu)$ enjoys properties similar to the space RBMO (μ). First of all, we have the following basic properties, the details being omitted.

Proposition 4.2.3. *Let $\eta \in (1,\infty)$ and $\rho \in [\eta,\infty)$. The following properties hold true:*

(i) $\mathrm{rbmo}_{\eta,\rho}(\mu)$ *is a Banach space;*

(ii) $L^\infty(\mu) \subset \mathrm{rbmo}_{\eta,\rho}(\mu) \subset \mathrm{RBMO}(\mu)$. *Moreover, for all $f \in L^\infty(\mu)$,*

$$\|f\|_{\mathrm{rbmo}_{\eta,\rho}(\mu)} \le 2\|f\|_{L^\infty(\mu)}$$

and there exists a positive constant C such that, for all $f \in \mathrm{rbmo}_{\eta,\rho}(\mu)$,

$$\|f\|_{\mathrm{RBMO}(\mu)} \le C\|f\|_{\mathrm{rbmo}_{\eta,\rho}(\mu)};$$

(iii) *If $f \in \mathrm{rbmo}_{\eta,\rho}(\mu)$, then $|f| \in \mathrm{rbmo}_{\eta,\rho}(\mu)$ and there exists a positive constant C such that, for all $f \in \mathrm{rbmo}_{\eta,\rho}(\mu)$,*

$$\||f|\|_{\mathrm{rbmo}_{\eta,\rho}(\mu)} \le C\|f\|_{\mathrm{rbmo}_{\eta,\rho}(\mu)};$$

(iv) *If $f, g \in \mathrm{rbmo}_{\eta,\rho}(\mu)$ are real-valued, then $\min(f,g)$, $\max(f,g) \in \mathrm{rbmo}_{\eta,\rho}(\mu)$, and there exists a positive constant C such that, for all $f, g \in \mathrm{rbmo}_{\eta,\rho}(\mu)$,*

$$\|\min(f,g)\|_{\mathrm{rbmo}_{\eta,\rho}(\mu)} \le C(\|f\|_{\mathrm{rbmo}_{\eta,\rho}(\mu)} + \|g\|_{\mathrm{rbmo}_{\eta,\rho}(\mu)})$$

and

$$\|\max(f,g)\|_{\mathrm{rbmo}_{\eta,\rho}(\mu)} \le C(\|f\|_{\mathrm{rbmo}_{\eta,\rho}(\mu)} + \|g\|_{\mathrm{rbmo}_{\eta,\rho}(\mu)}).$$

We now introduce another equivalent norm for $\mathrm{rbmo}_{\eta,\rho}(\mu)$. Let $\eta \in (1,\infty)$. Suppose that, for a given $f \in L^1_{\mathrm{loc}}(\mu)$, there exist a nonnegative constant $\tilde{C}$ and a collection of numbers, $\{f_Q\}_Q$, such that

$$\sup_{Q \notin \mathcal{Q}} \frac{1}{\mu(\eta Q)} \int_Q |f(y) - f_Q|\, d\mu(y) \le \tilde{C}, \tag{4.2.2}$$

that, for any two cubes $Q \subset R$ with $Q \notin \mathcal{Q}$,

$$|f_Q - f_R| \le \tilde{C}[1 + \delta(Q,R)] \tag{4.2.3}$$

and that, for any cube $Q \in \mathcal{Q}$,

$$|f_Q| \le \tilde{C}. \tag{4.2.4}$$

We then define the *norm* $\|f\|_{\star,\eta} := \inf\{\tilde{C}\}$, where the infimum is taken over all the constants $\tilde{C}$ as above and all the numbers $\{f_Q\}_Q$ satisfying (4.2.2) through (4.2.4).

With a minor modification of the proof for Proposition 3.1.6, we have the following conclusion.

Proposition 4.2.4. *The norms* $\|\cdot\|_{\star,\eta}$ *for* $\eta \in (1,\infty)$ *are equivalent.*

Proof. Let $\eta_1 > \eta_2 > 1$ be fixed. Obviously, $\|f\|_{\star,\eta_1} \le \|f\|_{\star,\eta_2}$.

To prove the converse, we need to show that, for a fixed collection of numbers, $\{f_Q\}_Q$, satisfying (4.2.2) through (4.2.4) with η and $\tilde{C}$ respectively replaced by η_1 and $\|f\|_{\star,\eta_1}$, it holds true that

$$\sup_{Q \notin \mathcal{Q}} \frac{1}{\mu(\eta_2 Q)} \int_Q |f(y) - f_Q|\, d\mu(y) \lesssim \|f\|_{\star,\eta_1}.$$

Fix $\rho \in [\eta_1, \infty)$ and $\beta_\rho = \rho^{D+1}$. For any cube $Q \notin \mathcal{Q}$ and $x \in \operatorname{supp}\mu \cap Q$, we choose $\tilde{Q}_{x,2}$ as follows. If

$$\ell(Q_{x,H_Q^x+2}) \le \frac{\eta_2 - 1}{10\eta_1}\ell(Q),$$

we then let $\tilde{Q}_{x,2} := Q_{x,H_Q^x+2}$. Otherwise, let k_0 be the *maximal negative integer* such that

$$\rho^{k_0}\ell(Q_{x,H_Q^x+2}) \le \frac{\eta_2 - 1}{10\eta_1}\ell(Q)$$

and we then let $\tilde{Q}_{x,2}$ be the biggest (ρ,β_ρ)-doubling cube centered at x with side length $\rho^k\ell(Q_{x,H_Q^x+2})$ with $k \le k_0$. By Lemma 2.1.3, we have $\delta(\tilde{Q}_{x,2}, Q_{x,H_Q^x+2}) \lesssim 1$. From (d), (e) and (f) of Lemma 4.1.2, it follows that $\tilde{Q}_{x,2} \notin \mathcal{Q}$. By Theorem 1.1.1, there exists a subsequence of cubes, $\{\tilde{Q}_{x_i,2}\}_i$, which still covers $Q \cap \operatorname{supp}\mu$ and has a bounded overlap. For any i, by Lemma 4.1.2(g) and $Q \notin \mathcal{Q}$, we have

$$\begin{aligned}
|f_{\tilde{Q}_{x_i,2}} - f_Q| &\le |f_{\tilde{Q}_{x_i,2}} - f_{Q_{x_i,H_Q^{x_i}}}| + |f_Q - f_{Q_{x_i,H_Q^{x_i}}}| \\
&\lesssim \left[1 + \delta(\tilde{Q}_{x_i,2}, Q_{x_i,H_Q^{x_i}}) + \delta(Q, Q_{x_i,H_Q^{x_i}})\right]\|f\|_{\star,\eta_1} \\
&\lesssim \|f\|_{\star,\eta_1}.
\end{aligned}$$

From this estimate, together with the facts that, for each i, $\tilde{Q}_{x_i,2} \notin \mathcal{Q}$ and $Q_{x_i,2}$ is (ρ,β_ρ)-doubling and that $\rho \ge \eta_1$, we see that

$$\begin{aligned}
&\int_{\tilde{Q}_{x_i,2}} |f(x) - f_Q|\, d\mu(x) \\
&\quad\le \int_{\tilde{Q}_{x_i,2}} |f(x) - f_{\tilde{Q}_{x_i,2}}|\, d\mu(x) + \mu(\tilde{Q}_{x_i,2})|f_{\tilde{Q}_{x_i,2}} - f_Q| \\
&\quad\lesssim \mu(\tilde{Q}_{x_i,2})\|f\|_{\star,\eta_1}.
\end{aligned}$$

Therefore, by the facts that $\{\tilde{Q}_{x_i,2}\}_i$ are almost disjoint and that $\tilde{Q}_{x_i,2} \subset \eta_2 Q$ for all i, we find that

$$\int_Q |f(x) - f_Q|\, d\mu(x) \leq \sum_i \int_{\tilde{Q}_{x_i,2}} |f(x) - f_Q|\, d\mu(x) \lesssim \mu(\eta_2 Q)\|f\|_{\star,\eta_1},$$

which completes the proof of Proposition 4.2.4. □

Based on Proposition 4.2.4, from now on, we write $\|\cdot\|_\star$ *instead of* $\|\cdot\|_{\star,\eta}$.

Proposition 4.2.5. *Let* $\eta \in (1,\infty)$, $\rho \in [\eta,\infty)$ *and* $\beta_\rho := \rho^{D+1}$. *Then the norms* $\|\cdot\|_\star$ *and* $\|\cdot\|_{\mathrm{rbmo}_{\eta,\rho}(\mu)}$ *are equivalent.*

Proof. Suppose that $f \in L^1_{\mathrm{loc}}(\mu)$. We first show that

$$\|f\|_\star \lesssim \|f\|_{\mathrm{rbmo}_{\eta,\rho}(\mu)}. \tag{4.2.5}$$

For any cube Q, let $f_Q := m_{\tilde{Q}^\rho}(f)$ if $\tilde{Q}^\rho \notin \mathcal{Q}$, and otherwise, let $f_Q := 0$. For any $Q \notin \mathcal{Q}$, if $\tilde{Q}^\rho \notin \mathcal{Q}$, by Definition 4.2.1, we have

$$\frac{1}{\mu(\eta Q)}\int_Q |f(y) - f_Q|\, d\mu(y) \leq \|f\|_{\mathrm{rbmo}_{\eta,\rho}(\mu)}.$$

If $\tilde{Q}^\rho \in \mathcal{Q}$, then $f_Q = 0$. The (ρ,β_ρ)-doubling property of $\tilde{Q}^\rho$, together with Definition 4.2.1 and the assumption that $\rho \geq \eta$, further implies that

$$\begin{aligned}
&\frac{1}{\mu(\eta Q)}\int_Q |f(y) - f_Q|\, d\mu(y)\\
&\quad \leq \frac{1}{\mu(\eta Q)}\int_Q \left|f(y) - m_{\tilde{Q}^\rho}(f)\right| d\mu(y) + \frac{\mu(Q)}{\mu(\eta Q)}\left|m_{\tilde{Q}^\rho}(f)\right|\\
&\quad \lesssim \|f\|_{\mathrm{rbmo}_{\eta,\rho}(\mu)}.
\end{aligned}$$

Notice that $Q \subset \tilde{Q}^\rho$. If $Q \in \mathcal{Q}$, then $\tilde{Q}^\rho \in \mathcal{Q}$ and $f_Q = 0$. Obviously,

$$|f_Q| \lesssim \|f\|_{\mathrm{rbmo}_{\eta,\rho}(\mu)}.$$

Therefore (4.2.5) is reduced to showing that, for any two cubes $Q \subset R$ with $Q \notin \mathcal{Q}$,

$$|f_Q - f_R| \lesssim [1 + \delta(Q,R)]\|f\|_{\mathrm{rbmo}_{\eta,\rho}(\mu)}. \tag{4.2.6}$$

To show (4.2.6), we first observe that, for any $f \in \mathrm{rbmo}_{\eta,\rho}(\mu)$ and cubes $Q \subset R$,

$$\left|m_{\tilde{Q}^\rho}(f) - m_{\tilde{R}^\rho}(f)\right| \lesssim [1 + \delta(Q,R)]\|f\|_{\mathrm{rbmo}_{\eta,\rho}(\mu)}. \tag{4.2.7}$$

If $Q \in \mathcal{Q}$, then $\tilde{Q}^\rho \in \mathcal{Q}$ and $\tilde{R}^\rho \in \mathcal{Q}$. In this case, (4.2.7) follows directly from Definition 4.2.1. If $Q \notin \mathcal{Q}$, to show (4.2.7), we consider two cases.

Case (i) $\ell(\tilde{R}^\rho) \geq \ell(\tilde{Q}^\rho)$. In this case, $\tilde{Q}^\rho \subset 2\tilde{R}^\rho$. Let $R_0 = \widetilde{2\tilde{R}^\rho}^\rho$. It follows, from Lemma 2.1.3, that

$$\delta(\tilde{R}^\rho, R_0) \lesssim 1 \quad \text{and} \quad \delta(\tilde{Q}^\rho, R_0) \lesssim 1 + \delta(Q, R).$$

Therefore, if neither $\tilde{Q}^\rho$ nor $\tilde{R}^\rho$ are in $\mathcal{Q}$, then

$$\begin{aligned}\left|m_{\tilde{Q}^\rho}(f) - m_{\tilde{R}^\rho}(f)\right| &\leq \left|m_{\tilde{Q}^\rho}(f) - m_{R_0}(f)\right| + \left|m_{R_0}(f) - m_{\tilde{R}^\rho}(f)\right| \\ &\lesssim [1 + \delta(Q, R)]\|f\|_{\mathrm{rbmo}_{\eta,\rho}(\mu)}.\end{aligned}$$

If both $\tilde{Q}^\rho$ and $\tilde{R}^\rho$ are in $\mathcal{Q}$, then, by $\rho \geq \eta$ and the (ρ, β_ρ)-doubling property of $\tilde{Q}^\rho$ and $\tilde{R}^\rho$, we know that

$$\left|m_{\tilde{Q}^\rho}(f) - m_{\tilde{R}^\rho}(f)\right| \leq \left|m_{\tilde{Q}^\rho}(f)\right| + \left|m_{\tilde{R}^\rho}(f)\right| \lesssim \|f\|_{\mathrm{rbmo}_{\eta,\rho}(\mu)}.$$

Thus, we only need to consider the case that only one of $\tilde{Q}^\rho$ and $\tilde{R}^\rho$ is in $\mathcal{Q}$. By similarity, we may assume that $\tilde{Q}^\rho \in \mathcal{Q}$ while $\tilde{R}^\rho \notin \mathcal{Q}$. Since $\tilde{Q}^\rho \subset R_0$, we then have $R_0 \in \mathcal{Q}$ and

$$\begin{aligned}\left|m_{\tilde{Q}^\rho}(f) - m_{\tilde{R}^\rho}(f)\right| &\leq \left|m_{\tilde{Q}^\rho}(f)\right| + |m_{R_0}(f)| + \left|m_{R_0}(f) - m_{\tilde{R}^\rho}(f)\right| \\ &\lesssim \|f\|_{\mathrm{rbmo}_{\eta,\rho}(\mu)}.\end{aligned}$$

Case (ii) $\ell(\tilde{R}^\rho) < \ell(\tilde{Q}^\rho)$. In this case, $\tilde{R}^\rho \subset 2\rho\tilde{Q}^\rho$. Notice that $\ell(\tilde{R}^\rho) \geq \ell(Q)$. Thus, there exists a unique $m \in \mathbb{N}$ such that

$$\ell(\rho^{m-1}Q) \leq \ell(\tilde{R}^\rho) < \ell(\rho^m Q).$$

Therefore, $\rho^m Q \subset \widetilde{2\rho\tilde{Q}^\rho}^\rho$. Let $Q_0 := \widetilde{2\rho\tilde{Q}^\rho}^\rho$. Then another application of Lemma 2.1.3 implies that $\delta(\tilde{Q}^\rho, Q_0) \lesssim 1$ and

$$\delta(\tilde{R}^\rho, Q_0) \lesssim 1 + \delta(\tilde{R}^\rho, \rho^m Q) + \delta(\rho^m Q, Q_0) \lesssim 1.$$

Therefore, an argument similar to Case (i) also establishes (4.2.7) in this case. Thus, (4.2.7) always holds true.

We now establish (4.2.6) by using (4.2.7) and considering the following three cases.

Case (i) $\tilde{Q}^\rho, \tilde{R}^\rho \in \mathcal{Q}$ or $\tilde{Q}^\rho, \tilde{R}^\rho \notin \mathcal{Q}$. In this case, (4.2.6) follows directly from (4.2.7).

Case (ii) $\tilde{Q}^\rho \notin \mathcal{Q}$ and $\tilde{R}^\rho \in \mathcal{Q}$. In this case, the estimate (4.2.7), together with $\rho \ge \eta$ and the (ρ, β_ρ)-doubling property of $\tilde{R}^\rho$, implies that

$$\begin{aligned}|f_Q - f_R| &\le \left|m_{\tilde{Q}^\rho}(f) - m_{\tilde{R}^\rho}(f)\right| + \left|m_{\tilde{R}^\rho}(f)\right| \\ &\lesssim [1 + \delta(Q, R)]\|f\|_{\mathrm{rbmo}_{\eta,\rho}(\mu)}.\end{aligned}$$

Case (iii) $\tilde{Q}^\rho \in \mathcal{Q}$ and $\tilde{R}^\rho \notin \mathcal{Q}$. In this case, an argument similar to that used in Case (ii) also leads to that

$$\begin{aligned}|f_Q - f_R| &\le \left|m_{\tilde{Q}^\rho}(f) - m_{\tilde{R}^\rho}(f)\right| + \left|m_{\tilde{Q}^\rho}(f)\right| \\ &\lesssim [1 + \delta(Q, R)]\|f\|_{\mathrm{rbmo}_{\eta,\rho}(\mu)}.\end{aligned}$$

Thus, (4.2.6) holds true and hence (4.2.5) is also true.

Now let us establish the converse of (4.2.5). For $f \in L^1_{\mathrm{loc}}(\mu)$, assume that there exists a sequence of numbers, $\{f_Q\}_Q$, satisfying (4.2.2), (4.2.3) and (4.2.4) with $\tilde{C}$ replaced by $\|f\|_\star$. First we claim that, for any cube $Q \in \mathcal{Q}$,

$$\frac{1}{\mu(\eta Q)}\int_Q |f(x)|\,d\mu(x) \lesssim \|f\|_\star. \tag{4.2.8}$$

For any cube Q and $x \in \operatorname{supp}\mu \cap Q$, let $\tilde{Q}_{x,2}$ be the *biggest* (ρ, β_ρ)*-doubling cube* centered at x with side length $\rho^k \ell(Q_{x,2})$, $k \le 0$ and $\ell(\tilde{Q}_{x,2}) \le \frac{\eta-1}{10\eta}\ell(Q)$. From Lemma 2.1.3, it is easy to see that $\delta(\tilde{Q}_{x,2}, Q_{x,2}) \lesssim 1$. By Lemma 4.1.2(e), we then have $\tilde{Q}_{x,2} \notin \mathcal{Q}$. Applying Theorem 1.1.1, we obtain a subsequence of cubes, $\{\tilde{Q}_{x_i,2}\}_i$, covering $Q \cap \operatorname{supp}\mu$ with a bounded overlap. From the bounded overlap and the (ρ, β_ρ)-doubling property of $\{\tilde{Q}_{x_i,2}\}_i$, (4.2.2), (4.2.4) and the facts that $\tilde{Q}_{x_i,2} \subset \eta Q$, $\tilde{Q}_{x_i,2} \notin \mathcal{Q}$ and $\rho \ge \eta$, it follows that

$$\begin{aligned}&\frac{1}{\mu(\eta Q)}\int_Q |f(x)|\,d\mu(x) \\ &\le \sum_i \frac{1}{\mu(\eta Q)}\int_{\tilde{Q}_{x_i,2}} |f(x) - f_{\tilde{Q}_{x_i,2}}|\,d\mu(x) \\ &\quad + \sum_i \frac{\mu(\tilde{Q}_{x_i,2})}{\mu(\eta Q)}\left[|f_{\tilde{Q}_{x_i,2}} - f_{Q_{x_i,1}}| + |f_{Q_{x_i,1}}|\right] \\ &\lesssim \sum_i \frac{\mu(\tilde{Q}_{x_i,2})}{\mu(\eta Q)}\left[1 + \delta(\tilde{Q}_{x_i,2}, Q_{x_i,1})\right]\|f\|_\star \\ &\lesssim \|f\|_\star.\end{aligned}$$

We now claim that, for any cube $Q \notin \mathcal{Q}$,

$$\frac{1}{\mu(\eta Q)} \int_Q \left| f(x) - m_{\tilde{Q}^\rho}(f) \right| d\mu(x) \lesssim \|f\|_\star. \tag{4.2.9}$$

Notice that, if $Q \notin \mathcal{Q}$ and Q is (ρ, β_ρ)-doubling, then using the fact $\rho \geq \eta$, we have

$$\begin{aligned}|f_Q - m_Q(f)| &= \left| \frac{1}{\mu(Q)} \int_Q [f(x) - f_Q]\, d\mu(x) \right| \\ &\leq \frac{\mu(\eta Q)}{\mu(Q)} \|f\|_\star \\ &\lesssim \|f\|_\star. \end{aligned} \tag{4.2.10}$$

Therefore, for any cube $Q \notin \mathcal{Q}$, if $\tilde{Q}^\rho \notin \mathcal{Q}$, then by (4.2.3) and (4.2.10), we see that

$$\left| f_Q - m_{\tilde{Q}^\rho}(f) \right| \leq \left| f_Q - f_{\tilde{Q}^\rho} \right| + \left| f_{\tilde{Q}^\rho} - m_{\tilde{Q}^\rho}(f) \right| \lesssim \|f\|_\star;$$

if $\tilde{Q}^\rho \in \mathcal{Q}$, then from (4.2.3), (4.2.4), (4.2.8) and $\rho \geq \eta$, it follows that

$$\left| f_Q - m_{\tilde{Q}^\rho}(f) \right| \leq \left| f_Q - f_{\tilde{Q}^\rho} \right| + \left| f_{\tilde{Q}^\rho} \right| + \left| m_{\tilde{Q}^\rho}(f) \right| \lesssim \|f\|_\star.$$

From these estimates and (4.2.2), we deduce that, for any cube $Q \notin \mathcal{Q}$,

$$\begin{aligned}\int_Q \left| f(x) - m_{\tilde{Q}^\rho}(f) \right| d\mu(x) &\leq \int_Q |f(x) - f_Q|\, d\mu(x) + \left| f_Q - m_{\tilde{Q}^\rho}(f) \right| \mu(Q) \\ &\lesssim \|f\|_\star \mu(\eta Q),\end{aligned}$$

which shows (4.2.9).

Finally, for any two (ρ, β_ρ)-doubling cubes $Q \subset R$ with $Q \notin \mathcal{Q}$, if $R \notin \mathcal{Q}$, (4.2.10), together with (4.2.3), implies that

$$\begin{aligned}|m_Q(f) - m_R(f)| &\leq |m_Q(f) - f_Q| + |f_Q - f_R| + |f_R - m_R(f)| \\ &\lesssim [1 + \delta(Q, R)]\|f\|_\star.\end{aligned}$$

If $R \in \mathcal{Q}$, (4.2.10), together with (4.2.3), (4.2.4), (4.2.8) and the (ρ, β_ρ)-doubling property of R, leads to that

$$\begin{aligned}|m_Q(f) - m_R(f)| &\leq |m_Q(f) - f_Q| + |f_Q - f_R| + |f_R| + |m_R(f)| \\ &\lesssim [1 + \delta(Q, R)]\|f\|_\star.\end{aligned}$$

Thus,

$$f \in \text{rbmo}_{\eta,\rho}(\mu) \quad \text{and} \quad \|f\|_{\text{rbmo}_{\eta,\rho}(\mu)} \lesssim \|f\|_{\star},$$

which completes the proof of Proposition 4.2.5. □

Remark 4.2.6. Let $\eta \in (1,\infty)$ and $\rho \in [\eta,\infty)$. From Propositions 4.2.4 and 4.2.5, it follows that the space $\text{rbmo}_{\eta,\rho}(\mu)$ is independent of the choices of η and ρ. From now on, we simply write rbmo (μ) *instead of* $\text{rbmo}_{\eta,\rho}(\mu)$ for any η and ρ as above.

Proposition 4.2.7. *Let $\eta \in (1,\infty)$, $\rho \in [\eta,\infty)$ and $\beta_\rho := \rho^{D+1}$. For any $f \in L^1_{\text{loc}}(\mu)$, the following are equivalent:*

(i) $f \in \text{rbmo}_{\eta,\rho}(\mu)$;

(ii) *There exists a nonnegative constant C_b such that, for any cube $Q \notin \mathcal{Q}$,*

$$\int_Q |f(x) - m_Q(f)|\,d\mu(x) \le C_b\mu(\eta Q), \tag{4.2.11}$$

that, for any cubes $Q \subset R$ with $Q \notin \mathcal{Q}$,

$$|m_Q(f) - m_R(f)| \le C_b[1 + \delta(Q,R)]\left[\frac{\mu(\eta Q)}{\mu(Q)} + \frac{\mu(\eta R)}{\mu(R)}\right] \tag{4.2.12}$$

and that, for any cube $Q \in \mathcal{Q}$,

$$\int_Q |f(x)|\,d\mu(x) \le C_b\mu(\eta Q); \tag{4.2.13}$$

(iii) *There exists a nonnegative constant C_c such that, for any (ρ,β_ρ)-doubling cube $Q \notin \mathcal{Q}$,*

$$\int_Q |f(x) - m_Q(f)|\,d\mu(x) \le C_c\mu(Q), \tag{4.2.14}$$

that, for any (ρ,β_ρ)-doubling cubes $Q \subset R$ with $Q \notin \mathcal{Q}$,

$$|m_Q(f) - m_R(f)| \le C_c[1 + \delta(Q,R)] \tag{4.2.15}$$

and that, for any (ρ,β_ρ)-doubling cube $Q \in \mathcal{Q}$,

$$\int_Q |f(x)|\,d\mu(x) \le C_c\mu(Q). \tag{4.2.16}$$

Moreover, the minimal constants C_b and C_c are equivalent to $\|f\|_{\text{rbmo}_{\eta,\rho}(\mu)}$.

Proof. By Propositions 4.2.4 and 4.2.5, it suffices to establish Proposition 4.2.7 with $\eta = \rho = 2$. We write rbmo (μ) instead of $\mathrm{rbmo}_{\eta,\rho}(\mu)$ for simplicity. Assuming that $f \in \mathrm{rbmo}(\mu)$, we now show that (ii) holds true. First, (4.2.13) follows from Definition 4.2.1 directly. On the other hand, by

$$\mathrm{rbmo}(\mu) \subset \mathrm{RBMO}(\mu)$$

and Proposition 3.1.10(ii), we see that (4.2.11) and (4.2.12) hold true.

Since (ii) obviously implies (iii), to finish the proof of Proposition 4.2.7, we only need to prove that, if $f \in L^1_{\mathrm{loc}}(\mu)$ satisfies the assumptions in (iii), then $f \in \mathrm{rbmo}(\mu)$.

For any $Q \notin \mathcal{Q}$, let $\{\tilde{Q}_{x_i,2}\}_i$ be the sequence of cubes as in the proof of Proposition 4.2.4 with $\eta_1 = \eta_2 = 2$, which covers $Q \cap \operatorname{supp}\mu$ with a bounded overlap. We then find that, for each i,

$$\tilde{Q}_{x_i,2} \notin \mathcal{Q} \quad \text{and} \quad \delta(\tilde{Q}_{x_i,2}, Q_{x_i, H_Q^{x_i}+2}) \lesssim 1.$$

By the last assertion, together with Lemmas 4.1.2 and 2.1.3, we further see that

$$\delta(\tilde{Q}_{x_i,2}, \widetilde{2Q}) \lesssim 1.$$

Obviously, by the choice of $\{\tilde{Q}_{x_i,2}\}_i$, we have $\tilde{Q}_{x_i,2} \subset 2Q$. These facts, together with (4.2.15) and Lemma 2.1.3, imply that

$$\left|m_{\tilde{Q}_{x_i,2}}(f) - m_{\tilde{Q}}(f)\right| \le \left|m_{\tilde{Q}_{x_i,2}}(f) - m_{\widetilde{2Q}}(f)\right| + \left|m_{\tilde{Q}}(f) - m_{\widetilde{2Q}}(f)\right| \lesssim C_c.$$

Then, from (4.2.14), the facts that, for each i, $\tilde{Q}_{x_i,2} \subset 2Q$ and that $\tilde{Q}_{x_i,2}$ are almost disjoint, it follows that

$$\begin{aligned}
&\int_Q \left|f(x) - m_{\tilde{Q}}(f)\right| d\mu(x) \\
&\quad \le \sum_i \int_{\tilde{Q}_{x_i,2}} |f(x) - m_{\tilde{Q}_{x_i,2}}(f)|\, d\mu(x) + \sum_i \left|m_{\tilde{Q}_{x_i,2}}(f) - m_{\tilde{Q}}(f)\right| \mu(\tilde{Q}_{x_i,2}) \\
&\quad \lesssim C_c \mu(2Q).
\end{aligned}$$

On the other hand, if $Q \in \mathcal{Q}$, let $\{\tilde{Q}_{x_i,2}\}_i$ be the sequence of cubes as in the proof of Proposition 4.2.5 with $\eta = 2$, which covers $Q \cap \operatorname{supp}\mu$ with a bounded overlap. We then see that, for each i,

$$\tilde{Q}_{x_i,2} \notin \mathcal{Q} \quad \text{and} \quad \delta(\tilde{Q}_{x_i,2}, Q_{x_i,2}) \lesssim 1.$$

The last assertion further implies that, for all i,

$$\delta(\tilde{Q}_{x_i,2}, Q_{x_i,1}) \lesssim 1,$$

which, together with (4.2.15), $Q_{x_i,1} \in \mathcal{Q}$ and (4.2.16), leads to that

$$\left|m_{\tilde{Q}_{x_i,2}}(f)\right| \le \left|m_{\tilde{Q}_{x_i,2}}(f) - m_{Q_{x_i,1}}(f)\right| + \left|m_{Q_{x_i,1}}(f)\right| \lesssim C_c.$$

Using the almost disjoint property and the doubling property of $\{\tilde{Q}_{x_i,2}\}_i$, $\tilde{Q}_{x_i,2} \notin \mathcal{Q}$, (4.2.14), (4.2.16) and $\tilde{Q}_{x_i,2} \subset 2Q$, we obtain

$$\begin{aligned}
&\int_Q |f(x)|\,d\mu(x)\\
&\quad \le \sum_i \left\{\int_{\tilde{Q}_{x_i,2}} \left|f(x) - m_{\tilde{Q}_{x_i,2}}(f)\right| d\mu(x) + \mu(\tilde{Q}_{x_i,2}) \left|m_{\tilde{Q}_{x_i,2}}(f)\right|\right\}\\
&\quad \lesssim C_c \mu(2Q),
\end{aligned}$$

which implies (4.2.1). Thus, $f \in \mathrm{rbmo}\,(\mu)$, which completes the proof of Proposition 4.2.7. □

The following theorem is a local version of the John–Nirenberg inequality for the space rbmo (μ).

Theorem 4.2.8. *Let $\eta \in (1,\infty)$ and $f \in \mathrm{rbmo}\,(\mu)$. If there exists a sequence of numbers, $\{f_Q\}_Q$, such that* (4.2.2), (4.2.3) *and* (4.2.4) *hold true with $\tilde{C}$ replaced by $c_1\|f\|_{\mathrm{rbmo}\,(\mu)}$ for some positive constant c_1. Then there exist nonnegative constants C and c_2 such that, for any cube $Q \in \mathcal{Q}$ and $\lambda \in (0,\infty)$,*

$$\mu\left(\{x \in Q : |f(x)| > \lambda\}\right) \le C\mu(\eta Q) \exp\left(\frac{-c_2\lambda}{\|f\|_{\mathrm{rbmo}\,(\mu)}}\right) \tag{4.2.17}$$

and, for any $Q \notin \mathcal{Q}$ and $\lambda \in (0,\infty)$,

$$\mu\left(\{x \in Q : |f(x) - f_Q| > \lambda\}\right) \le C\mu(\eta Q) \exp\left(\frac{-c_2\lambda}{\|f\|_{\mathrm{rbmo}\,(\mu)}}\right). \tag{4.2.18}$$

Proof. Since $f \in \mathrm{rbmo}\,(\mu)$, it follows that, if $\{f_Q\}$ satisfies (4.2.2), (4.2.3) and (4.2.4), then $\{f_Q\}$ also satisfies (3.1.7) and (3.1.8). Then (4.2.18) follows from Theorem 3.1.16 immediately.

To show (4.2.17), let $Q \in \mathcal{Q}$. If $\lambda > 2c_1\|f\|_{\mathrm{rbmo}\,(\mu)}$, then $|f_Q| \le \lambda/2$, which, together with Theorem 3.1.16, implies that

$$\begin{aligned}
\mu(\{x \in Q :\ |f(x)| > \lambda\}) &\le \mu(\{x \in Q :\ |f(x) - f_Q| > \lambda/2\})\\
&\lesssim \mu(\eta Q) \exp\left(\frac{-c_2\lambda}{\|f\|_{\mathrm{rbmo}\,(\mu)}}\right).
\end{aligned}$$

If $\lambda \in (0, 2c_1\|f\|_{\mathrm{rbmo}\,(\mu)}]$, then

$$\mu(\{x \in Q : |f(x)| > \lambda\}) \le \mu(Q) \lesssim \mu(\eta Q) \exp\left(\frac{-c_2\lambda}{\|f\|_{\mathrm{rbmo}\,(\mu)}}\right).$$

This finishes the proof of Theorem 4.2.8. □

From Theorem 4.2.8, we can easily deduce that the following spaces, $\mathrm{rbmo}^p_{\eta,\rho}(\mu)$, coincide for all $p \in [1, \infty)$.

For any $\eta \in (1, \infty)$, $\rho \in [\eta, \infty)$, $\beta_\rho := \rho^{D+1}$ and $p \in [1, \infty)$, a function $f \in L^1_{\mathrm{loc}}(\mu)$ is said to belong to the *space* $\mathrm{rbmo}^p_{\eta,\rho}(\mu)$ if there exists a nonnegative constant $\tilde{C}$ such that, for all $Q \notin \mathcal{Q}$,

$$\left\{\frac{1}{\mu(\eta Q)} \int_Q \left|f(x) - m_{\tilde{Q}^\rho}(f)\right|^p d\mu(x)\right\}^{1/p} \le \tilde{C},$$

that, for any two (ρ, β_ρ)-doubling cubes $Q \subset R$ with $Q \notin \mathcal{Q}$,

$$|m_Q(f) - m_R(f)| \le \tilde{C}[1 + \delta(Q, R)]$$

and that, for any $Q \in \mathcal{Q}$,

$$\left\{\frac{1}{\mu(\eta Q)} \int_Q |f(x)|^p \, d\mu(x)\right\}^{1/p} \le \tilde{C}.$$

Moreover, the minimal constant $\tilde{C}$ as above is defined to be the $\mathrm{rbmo}^p_{\eta,\rho}(\mu)$ *norm* of f and denoted by $\|f\|_{\mathrm{rbmo}^p_{\eta,\rho}(\mu)}$.

Arguing as for $p = 1$, we show that another equivalent definition for the space $\mathrm{rbmo}^p_{\eta,\rho}(\mu)$ can be given in terms of the numbers $\{f_Q\}_Q$ as in (4.2.2) through (4.2.4) without depending on the constants $\eta \in (1, \infty)$ and $\rho \in [\eta, \infty)$.

Using Theorem 4.2.8, by following an argument as in Corollary 3.1.20, we have the following conclusion, the details being omitted.

Corollary 4.2.9. *For any* $p \in [1, \infty)$, $\eta \in (1, \infty)$ *and* $\rho \in [\eta, \infty)$, *the spaces* $\mathrm{rbmo}^p_{\eta,\rho}(\mu)$ *coincide with equivalent norms.*

We have another characterization for $\mathrm{rbmo}\,(\mu)$ which is useful in applications. To be precise, for $f \in L^1_{\mathrm{loc}}(\mu)$ and cube Q, let $m_f(Q)$ be as in (3.1.17). Furthermore, for any $p \in [1, \infty)$, $\eta \in (1, \infty)$, $\rho \in [\eta, \infty)$, $\beta_\rho := \rho^{D+1}$ and $f \in L^1_{\mathrm{loc}}(\mu)$, we denote by $\|f\|_{**}$ the *minimal nonnegative constant* $\tilde{C}$ such that, for any $Q \notin \mathcal{Q}$,

$$\frac{1}{\mu(\eta Q)} \int_Q |f(x) - m_f(\tilde{Q}^\rho)| \, d\mu(x) \le \tilde{C},$$

that, for any two (ρ, β_ρ)-doubling cubes $Q \subset R$ with $Q \notin \mathcal{Q}$,

$$|m_f(Q) - m_f(R)| \le \tilde{C}[1 + \delta(Q, R)]$$

and that, for any $Q \in \mathcal{Q}$,

$$|m_f(Q)| \le \tilde{C}\frac{\mu(\eta Q)}{\mu(Q)}.$$

Lemma 4.2.10. *For any $\eta \in (1, \infty)$ and $\rho \in [\eta, \infty)$, $\|\cdot\|_{**}$ is equivalent to $\|\cdot\|_{\mathrm{rbmo}\,(\mu)}$.*

Proof. By Proposition 4.2.5, it suffices to show $\|\cdot\|_{**} \sim \|\cdot\|_{\star}$. First, we prove $\|\cdot\|_{\star} \lesssim \|\cdot\|_{**}$. For any $Q \subset \mathbb{R}^D$, let $f_Q := m_f(\tilde{Q}^\rho)$ if $\tilde{Q}^\rho \notin \mathcal{Q}$ and, otherwise, let $f_Q := 0$. Arguing as in (4.2.5), to show that $\|\cdot\|_{\star} \lesssim \|\cdot\|_{**}$, it suffices to prove that, for any $Q \subset R$ with $Q \notin \mathcal{Q}$,

$$|f_Q - f_R| \lesssim [1 + \delta(Q, R)]\|f\|_{**}.$$

Moreover, as in the proof of (4.2.5), this can be deduced from the fact

$$|m_f(\tilde{Q}^\rho) - m_f(\tilde{R}^\rho)| \lesssim [1 + \delta(Q, R)]\|f\|_{**},$$

which can be proved by repeating the proof of (4.2.7). Thus, $\|\cdot\|_{\star} \lesssim \|\cdot\|_{**}$.

Now we prove the converse. For any cube $Q \in \mathcal{Q}$, by the definition of $m_f(Q)$, we have

$$\begin{aligned}\left|m_f(Q)\mu(Q) - \int_Q f(x)\,d\mu(x)\right| &\le \int_Q |f(x) - m_f(Q)|\,d\mu(x)\\ &\le \int_Q |f(x)|\,d\mu(x),\end{aligned}$$

which implies in turn that

$$|m_f(Q)|\mu(Q) \lesssim \int_Q |f(x)|\,d\mu(x). \tag{4.2.19}$$

Therefore, by Proposition 4.2.5, we see that

$$|m_f(Q)| \lesssim \frac{\mu(\eta Q)}{\mu(Q)}\|f\|_{\mathrm{rbmo}\,(\mu)} \lesssim \frac{\mu(\eta Q)}{\mu(Q)}\|f\|_{\star}. \tag{4.2.20}$$

On the other hand, for any (ρ, β_ρ)-doubling cube $Q \notin \mathcal{Q}$, by the definition of $m_f(Q)$ again, we know that

$$\left|m_f(Q) - f_Q\right| \le \frac{1}{\mu(Q)} \int_Q [|f(x) - f_Q| + |f(x) - m_f(Q)|] \, d\mu(x)$$
$$\lesssim \|f\|_\star. \tag{4.2.21}$$

This fact, together with (4.2.3), implies that, for any two (ρ, β_ρ)-doubling cubes $Q \subset R$ with $Q \notin \mathcal{Q}$ and $R \notin \mathcal{Q}$,

$$|m_f(Q) - m_f(R)| \le |m_f(Q) - f_Q| + |f_Q - f_R| + |f_R - m_f(R)|$$
$$\lesssim [1 + \delta(Q, R)]\|f\|_\star.$$

Moreover, by (4.2.21), together with (4.2.3), (4.2.4) and (4.2.20), we know that, for any two (ρ, β_ρ)-doubling cubes $Q \subset R$ with $Q \notin \mathcal{Q}$ and $R \in \mathcal{Q}$,

$$|m_f(Q) - m_f(R)| \le |m_f(Q) - f_Q| + |f_Q - f_R| + |f_R| + |m_f(R)|$$
$$\lesssim [1 + \delta(Q, R)]\|f\|_\star.$$

Combining these estimates above, we see that, for any two (ρ, β_ρ)-doubling cubes $Q \subset R$ with $Q \notin \mathcal{Q}$,

$$|m_f(Q) - m_f(R)| \lesssim [1 + \delta(Q, R)]\|f\|_\star.$$

Finally, for any cube $Q \notin \mathcal{Q}$, we find that

$$\frac{1}{\mu(\eta Q)} \int_Q \left|f(x) - m_f(\tilde{Q}^\rho)\right| d\mu(x)$$
$$\le \frac{1}{\mu(\eta Q)} \int_Q |f(x) - f_Q| \, d\mu(x) + \frac{\mu(Q)}{\mu(\eta Q)} \left[\left|f_Q - f_{\tilde{Q}^\rho}\right| + \left|f_{\tilde{Q}^\rho} - m_f(\tilde{Q}^\rho)\right|\right]$$
$$\lesssim \|f\|_\star,$$

where in the last inequality, we used (4.2.2) through (4.2.4) and (4.2.20) in the case $\tilde{Q}^\rho \in \mathcal{Q}$, and (4.2.2), (4.2.3) and (4.2.21) in the case $\tilde{Q}^\rho \notin \mathcal{Q}$. This finishes the proof of Lemma 4.2.10. □

Remark 4.2.11. By Lemma 4.2.10, Propositions 4.2.4 and 4.2.5, in the remainder of Part I of this book, unless otherwise stated, we *always assume that both constants* ρ *and* η *in the definition of* $\|\cdot\|_{**}$ *as well as that of* $\|\cdot\|_{\mathrm{rbmo}\,(\mu)}$ *are equal to 2.*

Inspired by the duality between $H^1(\mu)$ and RBMO (μ), we show that the space $h^{1,\infty}_{\mathrm{atb}}(\mu)$ is the predual space of the space rbmo (μ).

Lemma 4.2.12. rbmo $(\mu) \subset (h^{1,\infty}_{\mathrm{atb}}(\mu))^*$. *That is, for any* $g \in$ rbmo (μ), *the linear functional*

$$L_g(f) := \int_{\mathbb{R}^D} f(x)g(x)\,d\mu(x),$$

defined on bounded functions f *with compact support, can be extended to a continuous linear functional* L_g *over* $h^{1,\infty}_{\mathrm{atb}}(\mu)$ *and*

$$\|L_g\|_{(h^{1,\infty}_{\mathrm{atb}}(\mu))^*} \le C\|g\|_{\mathrm{rbmo}\,(\mu)},$$

where C *is a positive constant independent of* g.

Proof. By Remarks 4.2.11 and 4.1.6, we take $\rho = \eta = 2$ in Definitions 4.2.1 and 4.1.3. Following some standard arguments,[4] we only need to show that, if $b := \sum_{j=1}^2 \lambda_j a_j$ is an ∞-atomic block with supp $b \subset R \notin \mathcal{Q}$ as in Definition 3.2.1 or an ∞-block with supp $b \subset R \in \mathcal{Q}$ as in Definition 4.1.3, then, for any $g \in$ rbmo (μ),

$$\left|\int_{\mathbb{R}^D} b(x)g(x)\,d\mu(x)\right| \lesssim |b|_{h^{1,\infty}_{\mathrm{atb}}(\mu)}\|g\|_{\mathrm{rbmo}\,(\mu)}.$$

If b is an ∞-atomic block with supp $b \subset R \notin \mathcal{Q}$, by an argument similar to that used in Lemma 3.2.6, we conclude that

$$\left|\int_{\mathbb{R}^D} b(x)g(x)\,d\mu(x)\right| \lesssim \sum_{i=1}^2 |\lambda_i|\|g\|_{\mathrm{rbmo}\,(\mu)}.$$

If b is an ∞-block with supp $b \subset R \in \mathcal{Q}$, we have

$$\left|\int_{\mathbb{R}^D} b(x)g(x)\,d\mu(x)\right| \le \sum_{i=1}^2 |\lambda_i| \int_{Q_i} |a_i(x)||g(x)|\,d\mu(x).$$

Now, for $i \in \{1, 2\}$, if $Q_i \in \mathcal{Q}$, it follows, from (3.2.1), that

$$\begin{aligned}\int_{Q_i} |a_i(x)||g(x)|\,d\mu(x) &\le \{\mu(2Q_i)[1 + \delta(Q_i, R)]\}^{-1} \int_{Q_i} |g(x)|\,d\mu(x)\\ &\le \|g\|_{\mathrm{rbmo}\,(\mu)}.\end{aligned}$$

[4] See [39, pp. 294–296].

If $Q_i \notin \mathcal{Q}$, then, by (3.2.1), together with Definition 4.2.1 and (4.2.7), we see that

$$\begin{aligned}
&\int_{Q_i} |a_i(x)||g(x)|\,d\mu(x)\\
&\leq \int_{Q_i} |a_i(x)|\left|g(x) - m_{\widetilde{Q_i}}(g)\right|\,d\mu(x) + \left|m_{\widetilde{Q_i}}(g)\right|\int_{Q_i} |a_i(x)|\,d\mu(x)\\
&\leq \|a_i\|_{L^\infty(\mu)}\left[\int_{Q_i}\left|g(x) - m_{\widetilde{Q_i}}(g)\right|\,d\mu(x)\right.\\
&\qquad\left.+\mu(Q_i)\left|m_{\widetilde{Q_i}}(g) - m_{\widetilde{R}}(g)\right| + \mu(Q_i)\left|m_{\widetilde{R}}(g)\right|\right]\\
&\lesssim \|g\|_{\mathrm{rbmo}\,(\mu)}.
\end{aligned}$$

Therefore, we have

$$\left|\int_{\mathbb{R}^D} b(x)g(x)\,d\mu(x)\right| \lesssim \sum_{i=1}^{2} |\lambda_i|\|g\|_{\mathrm{rbmo}\,(\mu)} \sim |b|_{h^{1,\infty}_{\mathrm{atb}}(\mu)}\|g\|_{\mathrm{rbmo}\,(\mu)},$$

which completes the proof of Lemma 4.2.12. □

Lemma 4.2.13. *If* $g \in \mathrm{rbmo}\,(\mu)$ *and* L_g *is as in Lemma 4.2.12, then*

$$\|L_g\|_{(h^{1,\infty}_{\mathrm{atb}}(\mu))^*} \sim \|g\|_{\mathrm{rbmo}\,(\mu)}$$

with the implicit equivalent positive constants independent of g.

Proof. By Lemma 4.2.12, it suffices to show

$$\|L_g\|_{(h^{1,\infty}_{\mathrm{atb}}(\mu))^*} \gtrsim \|g\|_{\mathrm{rbmo}\,(\mu)}.$$

Without loss of generality, we may assume that g is real-valued. With the aid of Lemma 4.2.10, we only need to prove that there exists some function $f \in h^{1,\infty}_{\mathrm{atb}}(\mu)$ such that

$$|L_g(f)| \gtrsim \|g\|_{**}\|f\|_{h^{1,\infty}_{\mathrm{atb}}(\mu)}. \tag{4.2.22}$$

By Remarks 4.2.11 and 4.1.6, we take $\rho = \eta = 2$ in the definition of $\|\cdot\|_{**}$ and Definition 4.1.3. Let $\epsilon \in (0, 1/8]$. There exist two possibilities.

Case (a) There exists some doubling cube $Q \subset \mathbb{R}^D$ with $Q \notin \mathcal{Q}$ such that

$$\int_Q |g(x) - m_g(Q)|\,d\mu(x) \geq \epsilon\|g\|_{\circ}\mu(Q), \tag{4.2.23}$$

or there exists some doubling cube $Q \subset \mathbb{R}^D$ with $Q \in \mathcal{Q}$ such that

$$|m_g(Q)| \geq \epsilon \|g\|_\circ. \tag{4.2.24}$$

If (4.2.23) holds true, then, for such a cube $Q \notin \mathcal{Q}$ satisfying (4.2.23), by an argument similar to that used in the proof of Case (a) in Lemma 3.2.7, we find an $f \in h^{1,\infty}_{\rm atb}(\mu)$ such that (4.2.22) holds true.

If (4.2.24) holds true, for such a cube $Q \in \mathcal{Q}$ satisfying (4.2.24), we take

$$f := \mathrm{sgn}\,(g)\chi_Q.$$

It follows immediately that f is an ∞-block with

$$\mathrm{supp}\, f \subset Q \quad \text{and} \quad |f|_{h^{1,\infty}_{\rm atb}(\mu)} \lesssim \mu(Q).$$

By this fact, (4.2.19) and (4.2.24), we see that

$$|L_g(f)| = \left|\int_Q g(x)f(x)\,d\mu(x)\right| = \int_Q |g(x)|\,d\mu(x) \gtrsim \epsilon\|g\|_{**}\|f\|_{h^{1,\infty}_{\rm atb}(\mu)}.$$

Thus, in Case (a), (4.2.22) holds true.

Case (b) For any doubling $Q \subset \mathbb{R}^D$ with $Q \notin \mathcal{Q}$, (4.2.23) fails and, for any doubling cube $Q \subset \mathbb{R}^D$ with $Q \in \mathcal{Q}$, (4.2.24) fails. In this case, we further consider the following two subcases.

Subcase (i) For any two doubling cubes $Q \subset R$ with $Q \notin \mathcal{Q}$,

$$|m_g(Q) - m_g(R)| \leq \frac{1}{2}[1 + \delta(Q, R)]\|g\|_{**}.$$

In this subcase, from the definition of $\|g\|_{**}$, it follows that there exists some cube $Q \notin \mathcal{Q}$ such that

$$\int_Q \left|g(x) - m_g(\tilde{Q})\right| d\mu(x) \geq \frac{1}{2}\|g\|_{**}\mu(2Q). \tag{4.2.25}$$

If $\tilde{Q} \notin \mathcal{Q}$, then, by the argument used in the proof of Subcase (i) of Lemma 3.2.7, we conclude that (4.2.22) holds true. If $\tilde{Q} \in \mathcal{Q}$, we then let

$$f := \chi_{Q\cap\{g>m_g(\tilde{Q})\}} - \chi_{Q\cap\{g<m_g(\tilde{Q})\}}.$$

It is easy to see that f is an ∞-block with

$$\mathrm{supp}\, f \subset \tilde{Q} \quad \text{and} \quad \|f\|_{h^{1,\infty}_{\rm atb}(\mu)} \lesssim \mu(2Q).$$

Moreover, since (4.2.24) fails for $\tilde{Q}$, using (4.2.25), we have

$$\begin{aligned}|L_g(f)| &\geq \left|\int_Q [g(x) - m_g(\tilde{Q})]\, f(x)\, d\mu(x)\right| - \left|m_g(\tilde{Q})\right|\left|\int_Q f(x)\, d\mu(x)\right| \\ &\geq \frac{1}{2}\|g\|_{**}\mu(2Q) - \epsilon\|g\|_{**}\mu(Q) \\ &\gtrsim \|g\|_{**}\|f\|_{h^{1,\infty}_{\rm atb}(\mu)}.\end{aligned}$$

Subcase (ii) There exist some doubling cubes $Q \subset R$ with $Q \notin \mathcal{Q}$ such that

$$|m_g(Q) - m_g(R)| > \frac{1}{2}[1 + \delta(Q, R)]\|g\|_{**}.$$

In this subcase, we also only need to consider the case that $R \in \mathcal{Q}$, because, if $R \notin \mathcal{Q}$, the argument used in the proof of Subcase (ii) of Lemma 3.2.7 works here as well. Assume that $R \in \mathcal{Q}$ and take $f := \chi_Q$. Then f is an ∞-block with $\operatorname{supp} f \subset R$ and

$$\|f\|_{h^{1,\infty}_{\rm atb}(\mu)} \lesssim [1 + \delta(Q, R)]\mu(Q).$$

Since (4.2.23) fails for Q and (4.2.24) fails for R, it follows, from the assumption of this subcase and the fact that $\epsilon \leq \frac{1}{8}$, that

$$\begin{aligned}|L_g(f)| &= \left|\int_Q [g(x) - m_g(Q)]f(x)\, d\mu(x) + m_g(Q)\mu(Q)\right| \\ &\geq |m_g(Q) - m_g(R)|\mu(Q) - \int_Q |g(x) - m_g(Q)|\, d\mu(x) \\ &\quad - |m_g(R)|\mu(Q) \\ &> \frac{1}{2}[1 + \delta(Q, R)]\|g\|_{**}\mu(Q) - 2\epsilon\|g\|_{**}\mu(Q) \\ &\geq \frac{1}{4}[1 + \delta(Q, R)]\|g\|_{**}\mu(Q).\end{aligned}$$

Therefore (4.2.22) also holds true in this case, which completes the proof of Lemma 4.2.13. □

Lemma 4.2.14. *For any $p \in (1, \infty)$, $\operatorname{rbmo}(\mu) \subset (h^{1,p}_{\rm atb}(\mu))^*$. That is, for any $g \in \operatorname{rbmo}(\mu)$, the linear functional*

$$L_g(f) := \int_{\mathbb{R}^D} f(x)g(x)\, d\mu(x),$$

defined over $f \in L^\infty(\mu)$ with compact support, can be extended to a unique continuous linear functional L_g over $h^{1,p}_{\rm atb}(\mu)$ and

$$\|L_g\|_{(h_{\mathrm{atb}}^{1,\,p}(\mu))^*} \le C\|g\|_{\mathrm{rbmo}\,(\mu)},$$

where C is a positive constant independent of g.

Proof. By Remarks 4.2.11 and 4.1.6, we take $\rho = \eta = 2$ in Definitions 4.2.1 and 4.1.3. Similar to the proof of Lemma 4.2.12, it suffices to show that, if $b := \sum_{i=1}^{2} \lambda_i a_i$ is a p-atomic block with $\operatorname{supp} b \subset R \notin \mathcal{Q}$ as in Definition 3.2.1 or a p-block with $\operatorname{supp} b \subset R \in \mathcal{Q}$ as in Definition 4.1.3, then, for any $g \in \mathrm{rbmo}\,(\mu)$,

$$\left|\int_{\mathbb{R}^D} b(x)g(x)\,d\mu(x)\right| \lesssim |b|_{h_{\mathrm{atb}}^{1,\,p}(\mu)}\|g\|_{\mathrm{rbmo}\,(\mu)}.$$

If b is a p-atomic block with $\operatorname{supp} b \subset R \notin \mathcal{Q}$, then an argument similar to that used in the proof of Lemma 3.2.8 implies the desired estimate.

Now suppose b is a p-block with $\operatorname{supp} b \subset R \in \mathcal{Q}$. In this case, we also see that

$$\left|\int_{\mathbb{R}^D} b(x)g(x)\,d\mu(x)\right| \le \sum_{i=1}^{2} |\lambda_i| \int_{Q_i} |a_i(x)||g(x)|\,d\mu(x).$$

For each i, if $Q_i \in \mathcal{Q}$, then it follows, from the Hölder inequality, (3.2.1) and Corollary 4.2.9, that

$$\begin{aligned}\int_{Q_i} |a_i(x)||g(x)|\,d\mu(x) &\le \left[\int_{Q_i} |a_i(x)|^p\,d\mu(x)\right]^{1/p}\left[\int_{Q_i} |g(x)|^{p'}\,d\mu(x)\right]^{1/p'}\\ &\lesssim \|g\|_{\mathrm{rbmo}\,(\mu)}.\end{aligned}$$

If $Q_i \notin \mathcal{Q}$, then using the Hölder inequality, (4.2.7), (3.2.1), Definition 4.2.1 and Corollary 4.2.9 again, we have

$$\begin{aligned}&\int_{Q_i} |a_i(x)||g(x)|\,d\mu(x)\\ &\quad\le \int_{Q_i} |a_i(x)|\left|g(x) - m_{\widetilde{Q_i}}(g)\right|\,d\mu(x) + \left|m_{\widetilde{Q_i}}(g)\right|\int_{Q_i} |a_i(x)|\,d\mu(x)\\ &\quad\le \|a_i\|_{L^p(\mu)}\left\{\left\|[g - m_{\widetilde{Q_i}}(g)]\chi_{Q_i}\right\|_{L^{p'}(\mu)} + \left|m_{\widetilde{Q_i}}(g) - m_{\tilde{R}}(g)\right|[\mu(Q_i)]^{1/p'}\right.\\ &\qquad\left. + \left|m_{\tilde{R}}(g)\right|[\mu(Q_i)]^{1/p'}\right\}\\ &\quad\lesssim \|g\|_{\mathrm{rbmo}\,(\mu)}.\end{aligned}$$

Therefore,

$$\left|\int_{\mathbb{R}^D} b(x)g(x)\,d\mu(x)\right| \lesssim \sum_{i=1}^{2} |\lambda_i| \|g\|_{\mathrm{rbmo}\,(\mu)} \sim |b|_{h^{1,\,p}_{\mathrm{atb}}(\mu)} \|g\|_{\mathrm{rbmo}\,(\mu)}.$$

This finishes the proof of Lemma 4.2.14. □

The proof of the following lemma is similar to that of Lemma 3.2.9, the details being omitted.

Lemma 4.2.15. *Let* $p \in (1,\infty)$ *and* $1/p + 1/p' = 1$. *Then* $(h^{1,\,p}_{\mathrm{atb}}(\mu))^* \subset L^{p'}_{\mathrm{loc}}(\mu)$.

Lemma 4.2.16. *For any* $p \in (1,\infty)$, *it holds true that*

$$\left(h^{1,\,p}_{\mathrm{atb}}(\mu)\right)^* = \mathrm{rbmo}\,(\mu).$$

Proof. By Lemma 4.2.14, to prove the lemma, it suffices to show that, for any $p \in (1,\infty)$,

$$\left(h^{1,\,p}_{\mathrm{atb}}(\mu)\right)^* \subset \mathrm{rbmo}\,(\mu).$$

Based on Lemma 4.2.15, we let $g \in L^{p'}_{\mathrm{loc}}(\mu)$ such that $L_g \in (h^{1,\,p}_{\mathrm{atb}}(\mu))^*$. We prove that $g \in \mathrm{rbmo}\,(\mu)$ by showing that, for any $Q \notin \mathcal{Q}$,

$$\frac{1}{\mu(2Q)} \int_Q \left|g(x) - m_g(\tilde{Q})\right| d\mu(x) \lesssim \|L_g\|_{(h^{1,\,p}_{\mathrm{atb}}(\mu))^*}, \tag{4.2.26}$$

that, for any two doubling cubes $Q \subset R$ with $Q \notin \mathcal{Q}$,

$$|m_g(Q) - m_g(R)| \lesssim [1 + \delta(Q,R)] \|L_g\|_{(h^{1,\,p}_{\mathrm{atb}}(\mu))^*} \tag{4.2.27}$$

and that, for any $Q \in \mathcal{Q}$,

$$|m_g(Q)| \lesssim \frac{\mu(2Q)}{\mu(Q)} \|L_g\|_{(h^{1,\,p}_{\mathrm{atb}}(\mu))^*}. \tag{4.2.28}$$

We first show (4.2.28). Let $Q \in \mathcal{Q}$ and $f := \mathrm{sgn}\,(g)\chi_Q$. Then f is a p-block with

$$\mathrm{supp}\, f \subset Q \quad \text{and} \quad |f|_{h^{1,\,p}_{\mathrm{atb}}(\mu)} \lesssim \mu(2Q).$$

By the definition of $m_g(Q)$, we see that

$$\begin{aligned}|m_g(Q)\mu(Q)| &\le \int_Q |g(x) - m_g(Q)|\, d\mu(x) + \left|\int_Q g(x)\, d\mu(x)\right|\\ &\lesssim \int_Q |g(x)|\, d\mu(x)\\ &\sim |L_g(f)|\\ &\lesssim \|L_g\|_{(h_{\mathrm{atb}}^{1,p}(\mu))^*}\mu(2Q),\end{aligned}$$

which implies (4.2.28).

If Q is doubling and $Q \notin \mathcal{Q}$, then (4.2.26) is true by following the argument used for (3.2.10), therefore we only need to show that (4.2.26) holds true when Q is not doubling and $Q \notin \mathcal{Q}$. Moreover, we may assume that $\tilde{Q} \in \mathcal{Q}$, since the proof of (3.2.10) also works here for $\tilde{Q} \notin \mathcal{Q}$. Let

$$f := \frac{|g - m_g(\tilde{Q})|^{p'}}{g - m_g(\tilde{Q})}\chi_{Q\cap\{g\neq m_g(\tilde{Q})\}}.$$

Then f is a p-block with $\operatorname{supp} f \subset \tilde{Q}$ and

$$|f|_{h_{\mathrm{atb}}^{1,p}(\mu)} \lesssim \left[\int_Q |g(x) - m_g(\tilde{Q})|^{p'}\, d\mu(x)\right]^{1/p} [\mu(2Q)]^{1/p'}. \tag{4.2.29}$$

On the other hand, by (4.2.28), together with the doubling property of $\tilde{Q}$ and Proposition 4.1.5(ii), we have

$$\begin{aligned}&\int_Q |g(x) - m_g(\tilde{Q})|^{p'}\, d\mu(x)\\ &\quad= \int_Q [g(x) - m_g(\tilde{Q})]\, f(x)\, d\mu(x)\\ &\quad\le \left|\int_Q g(x) f(x)\, d\mu(x)\right| + |m_g(\tilde{Q})| \int_Q |f(x)|\, d\mu(x)\\ &\quad\lesssim \|L_g\|_{(h_{\mathrm{atb}}^{1,p}(\mu))^*}\|f\|_{h_{\mathrm{atb}}^{1,p}(\mu)},\end{aligned}$$

which, together with (4.2.29) and the Hölder inequality, implies (4.2.26) in this case.

To prove (4.2.27), let $Q \subset R$ with $Q \notin \mathcal{Q}$ be any two doubling cubes. If $R \notin \mathcal{Q}$, then, by (3.2.11), we obtain (4.2.27). Now suppose that $R \in \mathcal{Q}$. We choose

$$f := \frac{|g - m_g(R)|^{p'}}{g - m_g(R)}\chi_{Q\cap\{g\neq m_g(R)\}}.$$

Then f is a p-block with $\operatorname{supp} f \subset R$ and

$$|f|_{h^{1,p}_{\rm atb}(\mu)} \lesssim [1+\delta(Q,R)]\left[\int_Q |g(x)-m_g(R)|^{p'}\,d\mu(x)\right]^{1/p} [\mu(2Q)]^{1/p'}.$$

Consequently, by applying (4.2.28), Proposition 4.1.5(ii) and the doubling property of R, we see that

$$\begin{aligned}
&\int_Q |g(x)-m_g(R)|^{p'}\,d\mu(x)\\
&\quad = \int_Q [g(x)-m_g(R)]f(x)\,d\mu(x)\\
&\quad \le \left|\int_Q g(x)f(x)\,d\mu(x)\right| + |m_g(R)|\int_Q |f(x)|\,d\mu(x)\\
&\quad \lesssim \|L_g\|_{(h^{1,p}_{\rm atb}(\mu))^*}\|f\|_{h^{1,p}_{\rm atb}(\mu)}\\
&\quad \lesssim \|L_g\|_{(h^{1,p}_{\rm atb}(\mu))^*}[1+\delta(Q,R)]\left[\int_Q |g(x)-m_g(R)|^{p'}\,d\mu(x)\right]^{1/p} [\mu(2Q)]^{1/p'}.
\end{aligned}$$

Therefore,

$$\left[\frac{1}{\mu(2Q)}\int_Q |g(x)-m_g(R)|^{p'}\,d\mu(x)\right]^{1/p'} \lesssim [1+\delta(Q,R)]\|L_g\|_{(h^{1,p}_{\rm atb}(\mu))^*}.$$

Recall that Q is doubling. From this fact, the last estimate as above, (4.2.26) and the Hölder inequality, it follows that

$$\begin{aligned}
&|m_g(Q)-m_g(R)|\\
&\quad \le \frac{1}{\mu(Q)}\int_Q |g(x)-m_g(Q)|\,d\mu(x) + \frac{1}{\mu(Q)}\int_Q |g(x)-m_g(R)|\,d\mu(x)\\
&\quad \lesssim [1+\delta(Q,R)]\|L_g\|_{(h^{1,p}_{\rm atb}(\mu))^*},
\end{aligned}$$

which shows (4.2.27) and hence completes the proof of Lemma 4.2.16. □

Now we can show that the dual space of $h^{1,p}_{\rm atb}(\mu)$ is ${\rm rbmo}\,(\mu)$.

Theorem 4.2.17. *For any fixed $p \in (1,\infty)$,*

$$h^{1,p}_{\rm atb}(\mu) = h^{1,\infty}_{\rm atb}(\mu) \quad \text{and} \quad \left(h^{1,\infty}_{\rm atb}(\mu)\right)^* = {\rm rbmo}\,(\mu).$$

Proof. By Proposition 4.1.2(iv), we see that, if $f \in (h^{1,p}_{\rm atb}(\mu))^*$, then $f \in (h^{1,\infty}_{\rm atb}(\mu))^*$. With the aid of Lemma 4.2.16, we consider the maps

$$i : h^{1,\infty}_{\rm atb}(\mu) \to h^{1,p}_{\rm atb}(\mu)$$

and

$$i^* : \mathrm{rbmo}\,(\mu) = \left(h^{1,p}_{\mathrm{atb}}(\mu)\right)^* \to \left(h^{1,\infty}_{\mathrm{atb}}(\mu)\right)^*.$$

Notice that the map i is an inclusion and i^* the canonical injection of rbmo (μ) in $(h^{1,\infty}_{\mathrm{atb}}(\mu))^*$ (with the identification $g \equiv L_g$ for $g \in \mathrm{rbmo}\,(\mu)$). By Lemma 4.2.13, $i^*(\mathrm{rbmo}\,(\mu))$ is closed in $(h^{1,\infty}_{\mathrm{atb}}(\mu))^*$. An application of the Banach closed range theorem shows that $h^{1,\infty}_{\mathrm{atb}}(\mu)$ is closed in $h^{1,p}_{\mathrm{atb}}(\mu)$, which, together with Proposition 4.1.5(iv), implies that

$$h^{1,\infty}_{\mathrm{atb}}(\mu) = h^{1,p}_{\mathrm{atb}}(\mu)$$

as a set. Thus i maps $h^{1,\infty}_{\mathrm{atb}}(\mu)$ onto $h^{1,p}_{\mathrm{atb}}(\mu)$. Observing that both $h^{1,\infty}_{\mathrm{atb}}(\mu)$ and $h^{1,p}_{\mathrm{atb}}(\mu)$ are Banach spaces, by the corollary of the open mapping theorem, we conclude that

$$h^{1,\infty}_{\mathrm{atb}}(\mu) = h^{1,p}_{\mathrm{atb}}(\mu)$$

with equivalent norms, which completes the proof of Theorem 4.2.17. □

Remark 4.2.18. By Theorem 4.2.17, in what follows, for all $p \in (1,\infty]$, we *denote $h^{1,p}_{\mathrm{atb}}(\mu)$ simply by $h^1(\mu)$.*

4.3 Relations Between $H^1(\mu)$ and $h^1(\mu)$ or Between RBMO (μ) and rbmo (μ)

We next come to establish relations between spaces $H^1(\mu)$ and $h^1(\mu)$, or between spaces RBMO (μ) and rbmo (μ), respectively.

Theorem 4.3.1. *For any $k \in \mathbb{Z}$, let S_k be as in Sect. 2.4. Then there exists a positive constant C, independent of k, such that, for all $f \in H^1(\mu)$,*

$$\|S_k(f)\|_{H^1(\mu)} \le C\|f\|_{H^1(\mu)}.$$

Proof. By the Fatou lemma, to show Theorem 4.3.1, it suffices to prove that, for any ∞-atomic block $b = \sum_{j=1}^2 \lambda_j a_j$ as in Definition 3.2.1,

$$\mathcal{M}_\Phi(S_k(b)) \in L^1(\mu) \quad \text{and} \quad \|\mathcal{M}_\Phi(S_k(b))\|_{L^1(\mu)} \lesssim \sum_{j=1}^2 |\lambda_j|.$$

Moreover, if $k \le 0$ and $\mathbb{R}^D$ is an initial cube, then $S_k = 0$ and Theorem 4.3.1 holds true automatically in this case. Therefore, we may assume that $\mathbb{R}^D$ is not an initial cube when $k \le 0$. Using the notation as in Definition 3.2.1 and choosing any $x_0 \in \operatorname{supp}\mu \cap R$, we now consider the following two cases: (i) $k \le H_R^{x_0}$; (ii) $k \ge H_R^{x_0} + 1$.

In Case (i), write

$$\|\mathcal{M}_\Phi(S_k(b))\|_{L^1(\mu)} = \int_{8R} \mathcal{M}_\Phi(S_k(b))(x)\,d\mu(x) + \int_{\mathbb{R}^D\setminus 8R} \cdots =: \mathrm{I} + \mathrm{II}.$$

Since $\mathcal{M}_\Phi$ is sublinear, it follows that

$$\begin{aligned}
\mathrm{I} &\le \sum_{j=1}^{2} |\lambda_j| \int_{8R} \mathcal{M}_\Phi(S_k(a_j))(x)\,d\mu(x) \\
&= \sum_{j=1}^{2} |\lambda_j| \int_{2Q_j} \mathcal{M}_\Phi(S_k(a_j))(x)\,d\mu(x) + \sum_{j=1}^{2} |\lambda_j| \int_{8R\setminus 2Q_j} \cdots \\
&=: \mathrm{I}_1 + \mathrm{I}_2.
\end{aligned}$$

By (b) and (d) of Theorem 2.4.4, we see that, for any $x \in 2Q_j$, $j \in \{1,2\}$, and $\varphi \sim x$,

$$\begin{aligned}
\left| \int_{\mathbb{R}^D} \varphi(y) S_k(a_j)(y)\,d\mu(y) \right| &\le \int_{\mathbb{R}^D} \int_{\mathbb{R}^D} \varphi(y) S_k(y,z) |a_j(z)|\,d\mu(z)\,d\mu(y) \\
&\le \|a_j\|_{L^\infty(\mu)},
\end{aligned}$$

which implies that

$$\mathcal{M}_\Phi(S_k(a_j))(x) \le \|a_j\|_{L^\infty(\mu)}.$$

From this, together with (3.2.1), we further deduce that

$$\mathrm{I}_1 \le \sum_{j=1}^{2} |\lambda_j| \|a_j\|_{L^\infty(\mu)} \mu(2Q_j) \lesssim \sum_{j=1}^{2} |\lambda_j|.$$

On the other hand, for any $x \in 8R \setminus 2Q_j$ and $z \in Q_j$, $j \in \{1,2\}$, it holds true that $|x-z| \sim |x-x_j|$, where x_j denotes the center of Q_j. This observation, together with the fact that, for any $x, y, z \in \mathbb{R}^D$, if $|y-z| < \frac{1}{2}|x-z|$, then $|x-z| < 2|x-y|$, the properties (b) and (e) of Theorem 2.4.4, implies that, for any $x \in 8R \setminus 2Q_j$, $\varphi \sim x$ and $z \in Q_j$,

$$\begin{aligned}
&\int_{\mathbb{R}^D} \varphi(y) S_k(y,z)\, d\mu(y) \\
&\lesssim \int_{|y-z|\geq \frac{1}{2}|x-z|} \frac{\varphi(y)}{|y-z|^n}\, d\mu(y) + \int_{|y-z|<\frac{1}{2}|x-z|} \frac{S_k(y,z)}{|x-y|^n}\, d\mu(y) \\
&\lesssim \int_{|y-z|\geq \frac{1}{2}|x-z|} \frac{\varphi(y)}{|x-z|^n}\, d\mu(y) + \int_{|y-z|<\frac{1}{2}|x-z|} \frac{S_k(y,z)}{|x-z|^n}\, d\mu(y) \\
&\lesssim \frac{1}{|x-x_j|^n}.
\end{aligned}$$

From this fact and (3.2.1), it then follows that

$$\begin{aligned}
\left| \int_{\mathbb{R}^D} \varphi(y) S_k(a_j)(y)\, d\mu(y) \right| &\leq \int_{Q_j} |a_j(z)| \int_{\mathbb{R}^D} \varphi(y) S_k(y,z)\, d\mu(y)\, d\mu(z) \\
&\lesssim \frac{1}{|x-x_j|^n} \|a_j\|_{L^\infty(\mu)} \mu(Q_j) \\
&\lesssim \frac{1}{|x-x_j|^n} \frac{1}{1+\delta(Q_j,R)}.
\end{aligned}$$

Thus, for any $x \in 8R \setminus 2Q_j$,

$$\mathcal{M}_\Phi(S_k(a_j))(x) \lesssim \frac{1}{|x-x_j|^n} \frac{1}{1+\delta(Q_j,R)}.$$

Moreover, by Lemma 2.1.3, we obtain

$$\delta(2Q_j, 8R) \leq \delta(Q_j, 8R) \lesssim 1+\delta(Q_j,R)+\delta(R,8R) \lesssim 1+\delta(Q_j,R). \tag{4.3.1}$$

Therefore, we conclude that

$$\mathrm{I}_2 \lesssim \sum_{j=1}^{2} |\lambda_j| \frac{\delta(2Q_j, 8R)}{1+\delta(Q_j,R)} \lesssim \sum_{j=1}^{2} |\lambda_j|.$$

To estimate II, by the observation that

$$\int_{\mathbb{R}^D} S_k(b)(x)\, d\mu(x) = 0,$$

we write

$$\mathrm{II} \leq \int_{\mathbb{R}^D \setminus 8R} \left| \sup_{\varphi \sim x} \int_{\mathbb{R}^D} S_k(b)(y)[\varphi(y) - \varphi(x_0)]\, d\mu(y) \right| d\mu(x)$$

$$\leq \int_{\mathbb{R}^D\setminus 8R} \sup_{\varphi\sim x} \int_{2R} |S_k(b)(y)||\varphi(y)-\varphi(x_0)|\,d\mu(y)\,d\mu(x)$$

$$+ \int_{\mathbb{R}^D\setminus 8R} \left| \sup_{\varphi\sim x} \int_{\mathbb{R}^D\setminus 2R} S_k(b)(y)[\varphi(y)-\varphi(x_0)]\,d\mu(y) \right| d\mu(x)$$

$$=: \mathrm{II}_1 + \mathrm{II}_2.$$

Notice that, for any $y \in 2R$ and $x \in 2^{m+1}R \setminus 2^m R$ with $m \geq 3$,

$$|x - x_0| \geq \ell(2^{m-2}R) \quad \text{and} \quad |x_0 - y| \leq 2\sqrt{D}\ell(R),$$

which implies that $|y - x| \sim |x_0 - x|$. This fact, together with the mean value theorem, implies that, for any $\varphi \sim x$,

$$|\varphi(y) - \varphi(x_0)| \lesssim \frac{|y - x_0|}{|x_0 - x|^{n+1}}. \tag{4.3.2}$$

Moreover, let N_j be the smallest integer k such that $2R \subset 2^k Q_j$. Observe that $\{S_k\}_k$ are bounded on $L^2(\mu)$ uniformly. Then Theorem 2.4.4(d), together with the Hölder inequality, Lemma 2.1.3, (4.3.2) and (3.2.1), leads to that

$$\mathrm{II}_1 \leq \sum_{j=1}^{2} |\lambda_j| \sum_{m=3}^{\infty} \int_{2^{m+1}R\setminus 2^m R} \left\{ \sup_{\varphi\sim x} \int_{2R\setminus 2Q_j} |S_k(a_j)(y)||\varphi(y)-\varphi(x_0)|\,d\mu(y) \right.$$

$$\left. + \sup_{\varphi\sim x} \int_{2Q_j} \cdots \right\} d\mu(x)$$

$$\lesssim \sum_{j=1}^{2} |\lambda_j| \sum_{m=3}^{\infty} \int_{2^{m+1}R\setminus 2^m R} \frac{\ell(R)}{\left[\ell(2^m R)\right]^{n+1}} \left\{ \int_{2R\setminus 2Q_j} \int_{Q_j} \frac{|a_j(z)|}{|y-z|^n}\,d\mu(z)\,d\mu(y) \right.$$

$$\left. + [\mu(2Q_j)]^{\frac{1}{2}} \left[\int_{2Q_j} |S_k(a_j)(y)|^2\,d\mu(y) \right]^{\frac{1}{2}} \right\} d\mu(x)$$

$$\lesssim \ell(R) \sum_{j=1}^{2} |\lambda_j| \sum_{m=3}^{\infty} \frac{\mu(2^{m+1}R)}{\left[\ell(2^m R)\right]^{n+1}}$$

$$\times \left\{ \sum_{i=1}^{N_j-1} \int_{2^{i+1}Q_j\setminus 2^i Q_j} \int_{Q_j} \frac{\|a_j\|_{L^\infty(\mu)}}{|y-z|^n}\,d\mu(z)\,d\mu(y) \right.$$

$$+ [\mu(2Q_j)]^{\frac{1}{2}} \left[\int_{Q_j} |a_j(y)|^2\, d\mu(y)\right]^{\frac{1}{2}} \Bigg\}$$

$$\lesssim \sum_{j=1}^{2} |\lambda_j| \|a_j\|_{L^\infty(\mu)} \left\{ \sum_{i=1}^{N_j-1} \frac{\mu(2^{i+1}Q_j)}{[\ell(2^i Q_j)]^n} \mu(Q_j) + \mu(2Q_j) \right\}$$

$$\lesssim \sum_{j=1}^{2} |\lambda_j| \left[\frac{1+\delta(2Q_j, 2R)}{1+\delta(Q_j, R)} + 1 \right]$$

$$\lesssim \sum_{j=1}^{2} |\lambda_j|.$$

To estimate II_2, we write

$$\begin{aligned}
\mathrm{II}_2 \le & \sum_{m=3}^{\infty} \int_{2^{m+1}R \setminus 2^m R} \mathcal{M}_\Phi(S_k(b)\chi_{2^{m+2}R\setminus 2^{m-1}R})(x)\, d\mu(x) \\
& + \sum_{m=3}^{\infty} \int_{2^{m+1}R \setminus 2^m R} \sup_{\varphi \sim x} \int_{2^{m+2}R \setminus 2^{m-1}R} |S_k(b)(y)| \varphi(x_0)\, d\mu(y)\, d\mu(x) \\
& + \sum_{m=3}^{\infty} \int_{2^{m+1}R \setminus 2^m R} \sup_{\varphi \sim x} \int_{\mathbb{R}^D \setminus 2^{m+2}R} |S_k(b)(y)| |\varphi(y) - \varphi(x_0)|\, d\mu(y)\, d\mu(x) \\
& + \sum_{m=3}^{\infty} \int_{2^{m+1}R \setminus 2^m R} \sup_{\varphi \sim x} \int_{2^{m-1}R \setminus 2R} \cdots \\
=: & \ \mathrm{E}_1 + \mathrm{E}_2 + \mathrm{E}_3 + \mathrm{E}_4.
\end{aligned}$$

Recall that $\mathcal{M}_\Phi$ is bounded on $L^p(\mu)$ for any $p \in (1, \infty)$. On the other hand, by (c) and (a) of Theorem 2.4.4, we have

$$\operatorname{supp}(S_k(b)) \subset \bigcup_{y \in R} Q_{y,k-1},$$

which, together with $k \le H_R^{x_0}$ and Lemma 2.3.3, further implies that

$$\operatorname{supp}(S_k(b)) \subset Q_{x_0,k-2}.$$

These facts, together with the Hölder inequality, lead to that

$$\mathrm{E}_1 \le \sum_{m=3}^{\infty} \left\{ \int_{2^{m+1}R\setminus 2^m R} \left[\mathcal{M}_\Phi(S_k(b)\chi_{2^{m+2}R\setminus 2^{m-1}R})(x)\right]^2 d\mu(x) \right\}^{\frac{1}{2}} [\mu(2^{m+1}R)]^{\frac{1}{2}}$$

$$\lesssim \sum_{m=3}^{\infty} \left\{ \int_{(2^{m+2}R\setminus 2^{m-1}R)\cap(Q_{x_0,k-2})} [S_k(b)(x)]^2 \, d\mu(x) \right\}^{\frac{1}{2}} [\mu(2^{m+1}R)]^{\frac{1}{2}}.$$

Let m_0 be the largest integer and m_1 the smallest integer satisfying

$$2^{m_0}R \subset 2Q_{x_0,k} \subset Q_{x_0,k-2} \subset 2^{m_1}R.$$

Then Lemma 2.1.3, along with the facts that $\ell(2^{m_0}R) \sim \ell(2Q_{x_0,k})$ and

$$\ell(2^{m_1}R) \sim \ell(Q_{x_0,k-2}),$$

implies that

$$\delta(2^{m_0}R, 2^{m_1}R) \lesssim 1 + \delta(2Q_{x_0,k}, Q_{x_0,k-2}) \lesssim 1. \tag{4.3.3}$$

If $m \ge m_1 + 1$, then

$$Q_{x_0,k-2} \bigcap (2^{m+2}R \setminus 2^{m-1}R) = \emptyset$$

and, if $m \le m_0 - 2$, then

$$(Q_{x_0,k-2} \setminus 2Q_{x_0,k}) \bigcap (2^{m+2}R \setminus 2^{m-1}R) = \emptyset.$$

From this, it then follows that

$$\mathrm{E}_1 \lesssim \sum_{m=3}^{m_1} \left\{ \int_{(2^{m+2}R\setminus 2^{m-1}R)\cap(2Q_{x_0,k})} [S_k(b)(x)]^2 \, d\mu(x) \right\}^{\frac{1}{2}} [\mu(2^{m+1}R)]^{\frac{1}{2}}$$
$$+ \sum_{m=m_0-1}^{m_1} \left\{ \int_{(2^{m+2}R\setminus 2^{m-1}R)\cap(Q_{x_0,k-2}\setminus 2Q_{x_0,k})} \cdots \right\}^{\frac{1}{2}} [\mu(2^{m+1}R)]^{\frac{1}{2}}.$$

Let us estimate the first term. By

$$\int_{\mathbb{R}^D} b(x)\, d\mu(x) = 0,$$

together with (e), (a) of Theorem 2.4.4 and $R \subset Q_{x_0,k}$ for $k \le H_R^{x_0}$, we see that

$$|S_k(b)(x)| \le \int_R |S_k(x,z) - S_k(x,x_0)||b(z)|\, d\mu(z)$$

$$
\begin{aligned}
&\lesssim \int_R \frac{|x_0 - z||b(z)|}{\ell(Q_{x_0,k})[\ell(Q_{x_0,k}) + |x_0 - x|]^n} \, d\mu(z) \\
&\lesssim \frac{\ell(R)\|b\|_{L^1(\mu)}}{\ell(Q_{x_0,k})[\ell(Q_{x_0,k}) + |x_0 - x|]^n}. \qquad (4.3.4)
\end{aligned}
$$

For any $x \in 2^{m+2}R \setminus 2^{m-1}R$ with $m \geq 3$, if $x \in 2Q_{x_0,k}$, then

$$|x - x_0| \lesssim \ell(Q_{x_0,k}).$$

This observation, together with (4.3.4), implies that

$$
\begin{aligned}
&\left\{\int_{(2^{m+2}R\setminus 2^{m-1}R)\cap 2Q_{x_0,k}} [S_k(b)(x)]^2 \, d\mu(x)\right\}^{\frac{1}{2}} \\
&\quad\lesssim \ell(R)\|b\|_{L^1(\mu)} \left\{\int_{2^{m+2}R\setminus 2^{m-1}R} \frac{1}{|x_0 - x|^{2(n+1)}} \, d\mu(x)\right\}^{\frac{1}{2}} \\
&\quad\lesssim \ell(R)\|b\|_{L^1(\mu)} \frac{[\mu(2^{m+2}R)]^{\frac{1}{2}}}{[\ell(2^m R)]^{n+1}}.
\end{aligned}
$$

Moreover, another application of (4.3.4) leads to that

$$
\begin{aligned}
&\left\{\int_{(2^{m+2}R\setminus 2^{m-1}R)\cap (Q_{x_0,k-2}\setminus 2Q_{x_0,k})} [S_k(b)(x)]^2 \, d\mu(x)\right\}^{\frac{1}{2}} \\
&\quad\lesssim \|b\|_{L^1(\mu)} \left\{\int_{2^{m+2}R\setminus 2^{m-1}R} \frac{1}{|x_0 - x|^{2n}} \, d\mu(x)\right\}^{\frac{1}{2}} \\
&\quad\lesssim \|b\|_{L^1(\mu)} \frac{[\mu(2^{m+2}R)]^{\frac{1}{2}}}{[\ell(2^m R)]^{n}}.
\end{aligned}
$$

Combining these estimates above, by (0.0.1), we conclude that

$$
\begin{aligned}
\mathrm{E}_1 &\lesssim \|b\|_{L^1(\mu)} \left\{\sum_{m=3}^{m_1} \frac{\ell(R)\mu(2^{m+2}R)}{[\ell(2^m R)]^{n+1}} + \sum_{m=m_0-1}^{m_1} \frac{\mu(2^{m+2}R)}{[\ell(2^m R)]^n}\right\} \\
&\lesssim \|b\|_{L^1(\mu)}[1 + \delta(2Q_{x_0,k}, Q_{x_0,k-2})] \\
&\lesssim \sum_{j=1}^{2} |\lambda_j|,
\end{aligned}
$$

where, in the last-to-second inequality, we used the following fact that, for any cube R,

$$\sum_{m=m_0-1}^{m_1} \frac{\mu(2^{m+1}R)}{[\ell(2^m R)]^n} \sim 1 + \delta(2^{m_0}R, 2^{m_1}R). \tag{4.3.5}$$

Similarly, it follows, from (4.3.3) through (4.3.5), (0.0.1) and

$$\sup_{\varphi \sim x} \varphi(x_0) \leq \frac{1}{|x - x_0|^n},$$

that

$$\begin{aligned}
\mathrm{E}_2 &\lesssim \sum_{m=3}^{m_1} \int_{2^{m+1}R \setminus 2^m R} \sup_{\varphi \sim x} \varphi(x_0) \int_{2^{m+2}R \setminus 2^{m-1}R} \frac{\ell(R)\|b\|_{L^1(\mu)}}{\ell(Q_{x_0,k})|x_0 - y|^n} \, d\mu(y)\, d\mu(x) \\
&\lesssim \|b\|_{L^1(\mu)} \left\{ \sum_{m=3}^{m_1} \int_{2^{m+1}R \setminus 2^m R} \frac{\ell(R)}{|x_0 - x|^n} \right. \\
&\quad \times \int_{(2^{m+2}R \setminus 2^{m-1}R) \cap 2Q_{x_0,k}} \frac{1}{|x_0 - y|^{n+1}} \, d\mu(y)\, d\mu(x) \\
&\quad + \sum_{m=m_0-1}^{m_1} \int_{2^{m+1}R \setminus 2^m R} \frac{1}{|x_0 - x|^n} \\
&\quad \left. \times \int_{(2^{m+2}R \setminus 2^{m-1}R) \cap (Q_{x_0,k-2} \setminus 2Q_{x_0,k})} \frac{1}{|x_0 - y|^n} \, d\mu(y)\, d\mu(x) \right\} \\
&\lesssim \|b\|_{L^1(\mu)} \left\{ \sum_{m=3}^{m_1} \frac{\ell(R)\mu(2^{m+2}R)}{[\ell(2^m R)]^{n+1}} + \sum_{m=m_0-1}^{m_1} \frac{\mu(2^{m+1}R)}{[\ell(2^m R)]^n} \delta(2Q_{x_0,k}, Q_{x_0,k-2}) \right\} \\
&\lesssim \sum_{j=1}^{2} |\lambda_j|.
\end{aligned}$$

Now we estimate E_3. Recalling that

$$\mathrm{supp}\,(S_k(b)) \subset Q_{x_0,k-2} \subset 2^{m_1}R,$$

we see that

$$\mathrm{E}_3 = \sum_{m=3}^{m_1-3} \int_{2^{m+1}R \setminus 2^m R} \sup_{\varphi \sim x} \int_{\mathbb{R}^D \setminus 2^{m+2}R} |S_k(b)(y)||\varphi(y) - \varphi(x_0)| \, d\mu(y)\, d\mu(x).$$

For any $m \le m_1 - 3$, any $x \in 2^{m+1}R \setminus 2^m R$ and $y \in 2^{i+1}R \setminus 2^i R$ with $i \ge m+2$, it is easy to see that

$$|x_0 - x| \gtrsim 2^m \ell(R) \quad \text{and} \quad |y - x| \gtrsim 2^m \ell(R). \tag{4.3.6}$$

Using (4.3.4) again, we have

$$\begin{aligned}
&\sup_{\varphi \sim x} \int_{\mathbb{R}^D \setminus 2^{m+2}R} |S_k(b)(y)||\varphi(y) - \varphi(x_0)| \, d\mu(y) \\
&\lesssim \sum_{i=m+2}^{\infty} \int_{(2^{i+1}R \setminus 2^i R) \cap Q_{x_0,k-2}} \frac{\ell(R)\|b\|_{L^1(\mu)}}{\ell(Q_{x_0,k})|x_0 - y|^n} \\
&\quad \times \left[\frac{1}{|y - x|^n} + \frac{1}{|x_0 - x|^n}\right] d\mu(y) \\
&\lesssim \frac{\|b\|_{L^1(\mu)}}{[\ell(2^m R)]^n} \sum_{i=m+2}^{m_1-3} \int_{(2^{i+1}R \setminus 2^i R) \cap Q_{x_0,k-2}} \frac{\ell(R)}{\ell(Q_{x_0,k})|x_0 - y|^n} \, d\mu(y) \\
&\lesssim \frac{\|b\|_{L^1(\mu)}}{[\ell(2^m R)]^n} \sum_{i=m+2}^{m_1-3} \left\{ \int_{(2^{i+1}R \setminus 2^i R) \cap 2Q_{x_0,k}} \frac{\ell(R)}{|x_0 - y|^{n+1}} \, d\mu(y) \right. \\
&\quad \left. + \int_{(2^{i+1}R \setminus 2^i R) \cap (Q_{x_0,k-2} \setminus 2Q_{x_0,k})} \frac{\ell(R)}{\ell(Q_{x_0,k})|x_0 - y|^n} \, d\mu(y) \right\}.
\end{aligned}$$

Therefore, from (4.3.3) through (4.3.5) and (0.0.1), it follows that

$$\begin{aligned}
\mathrm{E}_3 &\lesssim \|b\|_{L^1(\mu)} \left\{ \sum_{m=3}^{m_1-3} \frac{\mu(2^{m+1}R)}{[\ell(2^m R)]^n} \sum_{i=m+2}^{m_1-3} \int_{(2^{i+1}R \setminus 2^i R) \cap 2Q_{x_0,k}} \frac{\ell(R)}{|x_0 - y|^{n+1}} \, d\mu(y) \right. \\
&\quad + \sum_{m=m_0-1}^{m_1-3} \frac{\mu(2^{m+1}R)}{[\ell(2^m R)]^n} \sum_{i=m+2}^{m_1-3} \int_{(2^{i+1}R \setminus 2^i R) \cap (Q_{x_0,k-2} \setminus 2Q_{x_0,k})} \frac{1}{|x_0 - y|^n} \, d\mu(y) \\
&\quad + \sum_{m=3}^{m_0-2} \frac{\mu(2^{m+1}R)}{[\ell(2^m R)]^n} \\
&\quad \left. \times \sum_{i=m+2}^{m_1-3} \int_{(2^{i+1}R \setminus 2^i R) \cap (Q_{x_0,k-2} \setminus 2Q_{x_0,k})} \frac{\ell(R)}{\ell(Q_{x_0,k})|x_0 - y|^n} \, d\mu(y) \right\} \\
&\lesssim \|b\|_{L^1(\mu)} \left\{ \sum_{m=3}^{m_1-3} \sum_{i=m+2}^{m_1-3} \frac{\mu(2^{i+1}R)\ell(R)}{[\ell(2^i R)]^{n+1}} \right.
\end{aligned}$$

$$+\sum_{m=m_0-1}^{m_1-3}\frac{\mu(2^{m+1}R)}{[\ell(2^mR)]^n}\sum_{i=m_0+1}^{m_1-3}\frac{\mu(2^{i+1}R)}{[\ell(2^iR)]^n}$$

$$\left.+\sum_{m=3}^{m_0-2}\sum_{i=m+2}^{m_0}\frac{\mu(2^{i+1}R)\ell(R)}{[\ell(2^iR)]^{n+1}}+\sum_{m=3}^{m_0-2}\sum_{i=m_0}^{m_1-3}\frac{\mu(2^{i+1}R)}{[\ell(2^iR)]^n}\frac{\ell(R)}{\ell(2^mR)}\right\}$$

$$\lesssim \|b\|_{L^1(\mu)}[1+\delta(2Q_{x_0,k},Q_{x_0,k-2})]^2$$

$$\lesssim \sum_{j=1}^{2}|\lambda_j|,$$

where, in the third-to-last inequality, we used the facts that, if $i\le m_0$, then

$$\ell(2^iR)\le \ell(Q_{x_0,k})$$

and that, if $m\le m_0-2$, then

$$\ell(2^mR)\le \ell(Q_{x_0,k}).$$

Now we estimate E_4. Notice that, if $m\le m_0+1$, then

$$(2^{m-1}R\setminus 2R)\bigcap(Q_{x_0,k-2}\setminus 2Q_{x_0,k})=\emptyset.$$

Therefore, by $\mathrm{supp}\,(S_k(b))\subset Q_{x_0,k-2}$, we have

$$\mathrm{E}_4\le \sum_{m=3}^{\infty}\int_{2^{m+1}R\setminus 2^mR}\sup_{\varphi\sim x}\int_{(2^{m-1}R\setminus 2R)\cap 2Q_{x_0,k}}|S_k(b)(y)|$$
$$\times|\varphi(y)-\varphi(x_0)|\,d\mu(y)\,d\mu(x)$$
$$+\sum_{m=m_0+2}^{m_1-1}\int_{2^{m+1}R\setminus 2^mR}\sup_{\varphi\sim x}\int_{(2^{m-1}R\setminus 2R)\cap(Q_{x_0,k-2}\setminus 2Q_{x_0,k})}\cdots$$
$$+\sum_{m=m_1}^{\infty}\int_{2^{m+1}R\setminus 2^mR}\sup_{\varphi\sim x}\int_{(2^{m-1}R\setminus 2R)\cap(Q_{x_0,k-2}\setminus 2Q_{x_0,k})}\cdots$$
$$=:\mathrm{J}_1+\mathrm{J}_2+\mathrm{J}_3.$$

Observing that (4.3.2) holds true for any $y\in 2^{m-1}R\setminus 2R$ and $x\in 2^{m+1}R\setminus 2^mR$ with $m\ge 3$, by (4.3.2), (4.3.4) and (0.0.1), we see that

$$\sup_{\varphi\sim x}\int_{(2^{m-1}R\setminus 2R)\cap 2Q_{x_0,k}}|S_k(b)(y)||\varphi(y)-\varphi(x_0)|\,d\mu(y)$$

$$\begin{aligned}
&\lesssim \int_{(2^{m-1}R\setminus 2R)\cap 2Q_{x_0,k}} |S_k(b)(y)| \frac{\ell(Q_{x_0,k})}{|x_0-x|^{n+1}}\,d\mu(y)\\
&\lesssim \frac{\ell(R)\|b\|_{L^1(\mu)}}{|x_0-x|^{n+1}} \int_{(2^{m-1}R\setminus 2R)\cap 2Q_{x_0,k}} \frac{1}{[\ell(Q_{x_0,k})+|x_0-y|]^n}\,d\mu(y)\\
&\lesssim \frac{\ell(R)\|b\|_{L^1(\mu)}}{|x_0-x|^{n+1}}.
\end{aligned}$$

From this fact and (0.0.1), it follows that

$$\mathrm{J}_1 \lesssim \|b\|_{L^1(\mu)}\ell(R)\sum_{m=3}^{\infty}\int_{2^{m+1}R\setminus 2^m R}\frac{1}{|x_0-x|^{n+1}}\,d\mu(x) \lesssim \sum_{j=1}^{2}|\lambda_j|.$$

On the other hand, since (4.3.6) holds true for any

$$x \in 2^{m+1}R\setminus 2^m R \quad \text{and} \quad y \in 2^{m-1}R\setminus 2R$$

with $m \geq 3$, by (4.3.3) through (4.3.5), together with Definition 3.3.1(ii), we conclude that

$$\begin{aligned}
\mathrm{J}_2 &\lesssim \sum_{m=m_0+2}^{m_1-1}\int_{2^{m+1}R\setminus 2^m R}\int_{(2^{m-1}R\setminus 2R)\cap(Q_{x_0,k-2}\setminus 2Q_{x_0,k})}\frac{\|b\|_{L^1(\mu)}\ell(R)}{\ell(Q_{x_0,k})|x_0-y|^n}\\
&\quad\times\left[\frac{1}{|y-x|^n}+\frac{1}{|x_0-x|^n}\right]d\mu(y)\,d\mu(x)\\
&\lesssim \|b\|_{L^1(\mu)}\sum_{m=m_0+2}^{m_1-1}\frac{\mu(2^{m+1}R)}{[\ell(2^m R)]^n}\int_{Q_{x_0,k-2}\setminus 2Q_{x_0,k}}\frac{1}{|x_0-y|^n}\,d\mu(y)\\
&\lesssim \sum_{j=1}^{2}|\lambda_j|.
\end{aligned}$$

Finally, using (4.3.6), (4.3.2) through (4.3.4), (0.0.1) and the fact that, for any $y \in Q_{x_0,k-2}$, $|x_0-y| \lesssim \ell(2^{m_1}R)$, we have

$$\begin{aligned}
\mathrm{J}_3 &\lesssim \sum_{m=m_1}^{\infty}\int_{2^{m+1}R\setminus 2^m R}\int_{Q_{x_0,k-2}\setminus 2Q_{x_0,k}}\frac{\|b\|_{L^1(\mu)}}{|x_0-y|^n}\frac{\ell(2^{m_1}R)}{|x_0-x|^{n+1}}\,d\mu(y)\,d\mu(x)\\
&\lesssim \|b\|_{L^1(\mu)}\sum_{m=m_1}^{\infty}\frac{\ell(2^{m_1}R)\mu(2^{m+1}R)}{[\ell(2^m R)]^{n+1}}\\
&\lesssim \sum_{j=1}^{2}|\lambda_j|.
\end{aligned}$$

Combining the estimates for J_1, J_2 and J_3, we then complete the proof of Theorem 4.3.1 in Case (i).

In Case (ii), we consider the following two subcases.

Subcase (i) $k \geq H_R^{x_0} + 1$ and, for all $y \in R \cap \operatorname{supp} \mu$, $R \not\subset Q_{y,k-1}$. In this subcase, it is easy to see that, for any $y \in R$, $Q_{y,k-1} \subset 4R$, which, together with

$$\operatorname{supp}(S_k(b)) \subset \bigcup_{y \in R} Q_{y,k-1},$$

implies that $\operatorname{supp}(S_k(b)) \subset 4R$. Let I and II be as in Case (i). We also have

$$\|\mathcal{M}_\Phi(S_k(b))\|_{L^1(\mu)} \leq \mathrm{I} + \mathrm{II} \quad \text{and} \quad \mathrm{I} \lesssim \sum_{j=1}^{2} |\lambda_j|.$$

On the other hand, since $\operatorname{supp}(S_k(b)) \subset 4R$, similar to the estimate for II_1 in Case (i) with $2R$ replaced by $4R$, we obtain

$$\mathrm{II} \leq \int_{\mathbb{R}^D \setminus 8R} \sup_{\varphi \sim x} \int_{4R} |S_k(b)(y)||\varphi(y) - \varphi(x_0)| \, d\mu(y) \, d\mu(x) \lesssim \sum_{j=1}^{2} |\lambda_j|.$$

Subcase (ii) $k \geq H_R^{x_0} + 1$ and there exists some $y_0 \in R \cap \operatorname{supp} \mu$ such that $R \subset Q_{y_0,k-1}$. In this subcase, by applying Lemma 2.3.3, we see that

$$\operatorname{supp}(S_k(b)) \subset \bigcup_{y \in R} Q_{y,k-1} \subset Q_{y_0,k-2} \subset Q_{x_0,k-3}.$$

Then

$$\begin{aligned} \|\mathcal{M}_\Phi(S_k(b))\|_{L^1(\mu)} &= \int_{4Q_{x_0,k-3}} \mathcal{M}_\Phi(S_k(b))(x) \, d\mu(x) + \int_{\mathbb{R}^D \setminus 4Q_{x_0,k-3}} \cdots \\ &=: \mathrm{F}_1 + \mathrm{F}_2. \end{aligned}$$

Arguing as in the estimate for II_1 in Case (i) with $2R$ replaced by $Q_{x_0,k-3}$ again, we have

$$\mathrm{F}_2 \lesssim \sum_{j=1}^{2} |\lambda_j|.$$

On the other hand, by the fact that $\mathcal{M}_\Phi$ is sublinear, we obtain

$$\begin{aligned} \mathrm{F}_1 &\leq \sum_{j=1}^{2} |\lambda_j| \int_{2Q_j} \mathcal{M}_\Phi(S_k(a_j))(x) \, d\mu(x) + \sum_{j=1}^{2} |\lambda_j| \int_{4Q_{x_0,k-3} \setminus 2Q_j} \cdots \\ &=: \mathrm{L}_1 + \mathrm{L}_2. \end{aligned}$$

Since the argument of I_1 in Case (i) still works for L_1, it suffices to show

$$\mathrm{L}_2 \lesssim \sum_{j=1}^{2} |\lambda_j|.$$

However, because $R \subset Q_{y_0,k-1}$, we conclude that $k \le H_R^{y_0} + 1$. This fact, together with Lemma 4.1.2(c), leads to that $k \le H_R^{x_0} + 2$. Then, by the assumption that $H_R^{x_0} + 1 \le k$, together with Lemma 2.1.3 and Lemma 4.1.2(e), we see that

$$\delta(R, Q_{x_0,k-2}) \lesssim 1 + \delta(R, Q_{x_0, H_R^{x_0}}) + \delta(Q_{x_0, H_R^{x_0}}, Q_{x_0,k-2}) \lesssim 1.$$

Moreover, another application of Lemma 2.1.3 implies that

$$\begin{aligned}\delta(2Q_j, 4Q_{x_0,k-2}) &\le \delta(Q_j, 4Q_{x_0,k-2}) \\ &\lesssim 1 + \delta(Q_j, R) + \delta(R, Q_{x_0,k-2}) + \delta(Q_{x_0,k-2}, 4Q_{x_0,k-2}) \\ &\lesssim 1 + \delta(Q_j, R).\end{aligned}$$

Therefore, arguing as in Case (i), we have

$$\mathrm{L}_2 \lesssim \sum_{j=1}^{2} |\lambda_j| \frac{\delta(2Q_j, 4Q_{x_0,k-2})}{1 + \delta(Q_j, R)} \lesssim \sum_{j=1}^{2} |\lambda_j|,$$

which completes the proof of Theorem 4.3.1. □

For any $k \in \mathbb{Z}$, from Theorem 4.3.1, the linearity of S_k, the fact that

$$(H^1(\mu))^* = \mathrm{RBMO}\,(\mu)$$

and a dual argument, we deduce the uniform boundedness of S_k in RBMO (μ). We omit the details.

Corollary 4.3.2. *For any $k \in \mathbb{Z}$, let S_k be as in Sect. 2.4. Then there exists a positive constant C, independent of k, such that, for all $f \in$ RBMO (μ),*

$$\|S_k(f)\|_{\mathrm{RBMO}\,(\mu)} \le C \|f\|_{\mathrm{RBMO}\,(\mu)}.$$

As a consequence of Theorem 4.3.1, we also have the following result.

Proposition 4.3.3. *Let $k \in \mathbb{N}$ and S_k be as in Sect. 2.4. If $f \in h_{\mathrm{atb}}^{1,\infty}(\mu)$, then*

$$f - S_k(f) \in H^1(\mu) \quad \text{and} \quad \|f - S_k(f)\|_{H^1(\mu)} \le C \|f\|_{h_{\mathrm{atb}}^{1,\infty}(\mu)},$$

where C is a positive constant independent of f and k.

Proof. Notice that, for any ∞-block or ∞-atomic block

$$b := \sum_{j=1}^{2} \lambda_j a_j$$

as in Definitions 4.1.3 or 3.2.1, and $k \in \mathbb{N}$, by (b) of Theorem 2.4.4 and the Tonelli theorem, we have

$$\|S_k(b)\|_{L^1(\mu)} \leq \|b\|_{L^1(\mu)} \lesssim \sum_{j=1}^{2} |\lambda_j|. \tag{4.3.7}$$

From this and Definition 3.3.2, to prove Proposition 4.3.3, it suffices to show that

$$\|\mathcal{M}_\Phi(b - S_k(b))\|_{L^1(\mu)} \lesssim \sum_{j=1}^{2} |\lambda_j|. \tag{4.3.8}$$

Let b be an ∞-atomic block with $\operatorname{supp} b \subset R \notin \mathcal{Q}$. By Theorem 4.3.1, we know that $\{S_k\}_{k\in\mathbb{N}}$ are uniformly bounded on $H^1(\mu)$. Then we have

$$\|\mathcal{M}_\Phi(S_k(b))\|_{L^1(\mu)} \lesssim \sum_{j=1}^{2} |\lambda_j|.$$

This, together with the sublinear property of $\mathcal{M}_\Phi$ and Definition 3.3.2, implies (4.3.8) in this case.

Now assume that b is an ∞-block with $\operatorname{supp} b \subset R \in \mathcal{Q}$. Fix any $x_0 \in R \cap \operatorname{supp} \mu$. We consider the following two cases: (i) $k \leq H_R^{x_0}$; (ii) $k \geq H_R^{x_0} + 1$.

In Case (i), write

$$\begin{aligned}\|\mathcal{M}_\Phi(b - S_k(b))\|_{L^1(\mu)} &= \int_{8R} \mathcal{M}_\Phi(b - S_k(b))(x)\,d\mu(x) + \int_{\mathbb{R}^D\setminus 8R} \cdots \\ &=: \mathrm{I}_1 + \mathrm{I}_2.\end{aligned}$$

Since $\mathcal{M}_\Phi$ is sublinear, it follows that

$$\begin{aligned}\mathrm{I}_1 &\leq \sum_{j=1}^{2} |\lambda_j| \int_{2Q_j} \mathcal{M}_\Phi(a_j)(x)\,d\mu(x) + \sum_{j=1}^{2} |\lambda_j| \int_{8R\setminus 2Q_j} \cdots \\ &\quad + \int_{8R} \mathcal{M}_\Phi(S_k(b))(x)\,d\mu(x) \\ &=: \mathrm{J}_1 + \mathrm{J}_2 + \mathrm{J}_3.\end{aligned}$$

By an argument similar to that used in the estimate for the term I in the proof of Theorem 4.3.1, we have

$$\mathrm{J}_3 \lesssim \sum_{j=1}^{2} |\lambda_j|.$$

Thus, I_1 is reduced to showing that

$$\mathrm{J}_1 + \mathrm{J}_2 \lesssim \sum_{j=1}^{2} |\lambda_j|.$$

For each $j \in \{1, 2\}$, by Definition 4.1.3, we see that, for any $x \in 2Q_j$ and $\varphi \sim x$,

$$\left| \int_{\mathbb{R}^D} \varphi(y) a_j(y)\, d\mu(y) \right| \le \|a_j\|_{L^\infty(\mu)} \|\varphi\|_{L^1(\mu)} \le \|a_j\|_{L^\infty(\mu)},$$

which implies that

$$\mathcal{M}_\Phi(a_j)(x) \le \|a_j\|_{L^\infty(\mu)} \text{ for any } x \in 2Q_j.$$

From this fact, together with Definition 4.1.3, we further deduce that

$$\mathrm{J}_1 \le \sum_{j=1}^{2} |\lambda_j| \|a_j\|_{L^\infty(\mu)} \mu(2Q_j) \lesssim \sum_{j=1}^{2} |\lambda_j|.$$

On the other hand, for any $j \in \{1, 2\}$ and $x \in 8R \setminus 2Q_j$, we see that, for any $y \in Q_j$, $|x - y| \sim |x - x_j|$, where x_j is the center of Q_j. From this, it follows that, for any $x \in 8R \setminus 2Q_j$,

$$\mathcal{M}_\Phi(a_j)(x) \le \sup_{\varphi \sim x} \left| \int_{\mathbb{R}^D} \varphi(y) a_j(y)\, d\mu(y) \right| \lesssim \frac{\|a_j\|_{L^1(\mu)}}{|x - x_j|^n},$$

which, together with Lemma 2.1.3 and Definition 4.1.3, implies that

$$\begin{aligned}
\int_{8R \setminus 2Q_j} \mathcal{M}_\Phi(a_j)(x)\, d\mu(x) &\lesssim \|a_j\|_{L^1(\mu)} \delta(2Q_j, 8R) \\
&\lesssim \|a\|_{L^\infty(\mu)} \mu(Q_j)[1 + \delta(Q_j, R)] \\
&\lesssim 1.
\end{aligned}$$

Therefore, it holds true that

$$\mathrm{J}_2 \lesssim \sum_{j=1}^{2} |\lambda_j|,$$

which, together with estimates for J_1 and J_3, implies that

$$\mathrm{I}_1 \lesssim \sum_{j=1}^{2} |\lambda_j|.$$

Now we estimate I_2. By the facts that

$$\int_{\mathbb{R}^D} [b(x) - S_k(b)(x)]\, d\mu(x) = 0$$

and that $\operatorname{supp} b \subset R$, we write

$$\begin{aligned}
\mathrm{I}_2 \le& \int_{\mathbb{R}^D \setminus 8R} \sup_{\varphi \sim x} \int_R |b(y)||\varphi(y) - \varphi(x_0)|\, d\mu(y)\, d\mu(x) \\
&+ \int_{\mathbb{R}^D \setminus 8R} \sup_{\varphi \sim x} \int_{2R} |S_k(b)(y)||\varphi(y) - \varphi(x_0)|\, d\mu(y)\, d\mu(x) \\
&+ \int_{\mathbb{R}^D \setminus 8R} \sup_{\varphi \sim x} \left| \int_{\mathbb{R}^D \setminus 2R} S_k(b)(y)[\varphi(y) - \varphi(x_0)]\, d\mu(y) \right| d\mu(x) \\
=:& \mathrm{L}_1 + \mathrm{L}_2 + \mathrm{L}_3.
\end{aligned}$$

By arguing as in the proofs for II_1 and II_2 in Theorem 4.3.1, we see that

$$\mathrm{L}_2 + \mathrm{L}_3 \lesssim \sum_{j=1}^{2} |\lambda_j|.$$

Thus, we only need to show that

$$\mathrm{L}_1 \lesssim \sum_{j=1}^{2} |\lambda_j|.$$

From (4.3.2), Definition 4.1.3 and (0.0.1), it follows that

$$\begin{aligned}
\mathrm{L}_1 &\le \sum_{j=1}^{2} |\lambda_j| \sum_{m=3}^{\infty} \int_{2^{m+1}R \setminus 2^m R} \sup_{\varphi \sim x} \int_{Q_j} |a_j(y)||\varphi(y) - \varphi(x_0)|\, d\mu(y)\, d\mu(x) \\
&\lesssim \sum_{j=1}^{2} |\lambda_j| \sum_{m=3}^{\infty} \int_{2^{m+1}R \setminus 2^m R} \frac{\ell(R)}{[\ell(2^m R)]^{n+1}} \|a_j\|_{L^\infty(\mu)} \mu(Q_j)\, d\mu(x) \\
&\lesssim \sum_{j=1}^{2} |\lambda_j|.
\end{aligned}$$

Therefore, we obtain (4.3.8) in Case (i).

In Case (ii), we further consider the following two subcases.

Subcase (i) $k \geq H_R^{x_0} + 1$ and for all $y \in R \cap \operatorname{supp} \mu$, $R \not\subset Q_{y,k-1}$. In this subcase, it is easy to see that, for any $y \in R$, $Q_{y,k-1} \subset 4R$, which, together with

$$\operatorname{supp}(S_k(b)) \subset \bigcup_{y \in R} Q_{y,k-1},$$

implies that $\operatorname{supp}(S_k(b)) \subset 4R$. Let I_1 and I_2 be as in Case (i). We also have

$$\|\mathcal{M}_\Phi(b - S_k(b))\|_{L^1(\mu)} =: \mathrm{I}_1 + \mathrm{I}_2 \quad \text{and} \quad \mathrm{I}_1 \lesssim \sum_{j=1}^{2} |\lambda_j|.$$

On the other hand, since $\operatorname{supp}(S_k(b)) \subset 4R$, by

$$\int_{\mathbb{R}^D} [b(x) - S_k(b)(x)]\, d\mu(x) = 0,$$

we have

$$\begin{aligned}
\mathrm{I}_2 &\leq \int_{\mathbb{R}^D \setminus 8R} \sup_{\varphi \sim x} \left| \int_{\mathbb{R}^D} [b(y) - S_k(b)(y)][\varphi(y) - \varphi(x_0)]\, d\mu(y) \right| d\mu(x) \\
&\leq \int_{\mathbb{R}^D \setminus 8R} \sup_{\varphi \sim x} \int_R |b(y)||\varphi(y) - \varphi(x_0)|\, d\mu(y)\, d\mu(x) \\
&\quad + \int_{\mathbb{R}^D \setminus 8R} \sup_{\varphi \sim x} \int_{4R} |S_k(b)(y)||\varphi(y) - \varphi(x_0)|\, d\mu(y)\, d\mu(x).
\end{aligned}$$

Moreover, using estimates similar to those for L_1 and L_2 in Case (i) with $2R$ in L_2 replaced by $4R$, we obtain

$$\mathrm{I}_2 \lesssim \sum_{j=1}^{2} |\lambda_j|.$$

Subcase (ii) $k \geq H_R^{x_0} + 1$ and there exists some $y_0 \in R \cap \operatorname{supp} \mu$ such that $R \subset Q_{y_0,k-1}$. In this subcase, by applying Lemma 2.3.3, we see that

$$\operatorname{supp}(S_k(b)) \subset \bigcup_{y \in R} Q_{y,k-1} \subset Q_{y_0,k-2} \subset Q_{x_0,k-3}.$$

Then

$$\begin{aligned}
&\|\mathcal{M}_\Phi(b - S_k(b))\|_{L^1(\mu)}\\
&= \int_{4Q_{x_0,k-3}} \mathcal{M}_\Phi(b - S_k(b))(x)\,d\mu(x) + \int_{\mathbb{R}^D \setminus 4Q_{x_0,k-3}} \cdots\\
&=: \mathrm{E}_1 + \mathrm{E}_2.
\end{aligned}$$

Arguing as in estimates for L_1 and L_2 in Case (i) with $2R$ in L_2 replaced by $Q_{x_0,k-3}$, we have

$$\mathrm{E}_2 \lesssim \sum_{j=1}^{2} |\lambda_j|.$$

On the other hand, by the fact that $\mathcal{M}_\Phi$ is sublinear, we obtain

$$\begin{aligned}
\mathrm{E}_1 &\le \sum_{j=1}^{2} |\lambda_j| \int_{2Q_j} \mathcal{M}_\Phi(a_j)(x)\,d\mu(x) + \sum_{j=1}^{2} |\lambda_j| \int_{4Q_{x_0,k-3} \setminus 2Q_j} \cdots\\
&\quad + \int_{4Q_{x_0,k-3}} \mathcal{M}_\Phi(S_k(b))(x)\,d\mu(x)\\
&=: \mathrm{F}_1 + \mathrm{F}_2 + \mathrm{F}_3.
\end{aligned}$$

By using an argument similar to that used in the proof of J_1 in Case (i), we see that

$$\mathrm{F}_1 \lesssim \sum_{j=1}^{2} |\lambda_j|.$$

On the other hand, because $R \subset Q_{y_0,k-1}$, we conclude that

$$k \le H_R^{y_0} + 1.$$

This fact, together with Lemma 4.1.2(c), implies that $k \le H_R^{x_0} + 2$. Then the assumption that $H_R^{x_0} + 1 \le k$, together with Lemmas 2.1.3 and 4.1.2(g), implies that $\delta(R, Q_{x_0,k-3}) \lesssim 1$. Moreover, by another application of Lemma 2.1.3, we find that

$$\delta(2Q_j, 4Q_{x_0,k-3}) \lesssim 1 + \delta(Q_j, R).$$

Therefore, arguing as in Case (i), we see that, for any $x \in 4Q_{x_0,k-3} \setminus 2Q_j$,

$$\mathcal{M}_\Phi(a_j)(x) \lesssim \frac{\|a_j\|_{L^\infty(\mu)}\mu(Q_j)}{|x - x_j|^n}.$$

This, together with Definition 4.1.3, implies that

$$\mathrm{F}_2 \lesssim \sum_{j=1}^{2} |\lambda_j| \|a_j\|_{L^\infty(\mu)} \mu(Q_j) \delta(2Q_j, 4Q_{x_0,k-3}) \lesssim \sum_{j=1}^{2} |\lambda_j|.$$

Similarly, by (b) and (d) of Theorem 2.4.4, we have

$$\begin{aligned}
\mathrm{F}_3 &\leq \sum_{j=1}^{2} |\lambda_j| \left[\int_{2Q_j} \mathcal{M}_\Phi(S_k(a_j))(x)\, d\mu(x) + \int_{4Q_{x_0,k-3}\setminus 2Q_j} \cdots \right] \\
&\lesssim \sum_{j=1}^{2} |\lambda_j| \|a_j\|_{L^\infty(\mu)} \mu(2Q_j) \left[1 + \delta(2Q_j, 4Q_{x_0,k-3})\right] \\
&\lesssim \sum_{j=1}^{2} |\lambda_j|,
\end{aligned}$$

which completes the proof of Proposition 4.3.3. □

Remark 4.3.4. Indeed, from Theorem 4.3.1, we see that Proposition 4.3.3 and Corollary 4.3.6 below also hold true for S_k with $k \leq 0$ when $\mathbb{R}^D$ is not an initial cube, the details being omitted.

To establish the relation between RBMO (μ) and rbmo (μ), we need the following estimate.

Lemma 4.3.5. *There exists a positive constant C such that, for any cubes $Q \subset R$ and $f \in$ RBMO (μ),*

$$\int_R \frac{|f(y) - m_{\tilde{Q}}(f)|}{[|y - x_Q| + \ell(Q)]^n}\, d\mu(y) \leq C[1 + \delta(Q, R)]^2 \|f\|_{\mathrm{RBMO}(\mu)}.$$

Proof. Without loss of generality, we may assume that $\|f\|_{\mathrm{RBMO}(\mu)} = 1$. Notice that, from (0.0.1) and Definition 3.1.1, it follows that

$$\int_Q \frac{|f(y) - m_{\tilde{Q}}(f)|}{[|y - x_Q| + \ell(Q)]^n}\, d\mu(y) \leq \frac{1}{[\ell(Q)]^n} \int_Q \left| f(y) - m_{\tilde{Q}}(f) \right| d\mu(y) \lesssim 1.$$

Therefore, to show Lemma 4.3.5, it suffices to show that

$$\int_{R\setminus Q} \frac{|f(y) - m_{\tilde{Q}}(f)|}{|y - x_Q|^n}\, d\mu(y) \lesssim [1 + \delta(Q, R)]^2. \tag{4.3.9}$$

By (0.0.1), (2.1.3) and Lemma 2.1.3, together with Definition 3.1.1, we see that

$$\begin{aligned}
&\int_{R\setminus Q} \frac{|f(y)-m_{\tilde{Q}}(f)|}{|y-x_Q|^n}\,d\mu(y)\\
&\lesssim \sum_{k=0}^{N_{Q,R}} \frac{1}{[\ell(2^{k+1}Q)]^n} \int_{2^{k+1}Q\setminus 2^k Q} \left|f(y)-m_{\tilde{Q}}(f)\right| d\mu(y)\\
&\le \sum_{k=0}^{N_{Q,R}} \frac{1}{[\ell(2^{k+1}Q)]^n} \int_{2^{k+1}Q\setminus 2^k Q} \left|f(y)-m_{\widetilde{2^{k+1}Q}}(f)\right| d\mu(y)\\
&\quad+\sum_{k=0}^{N_{Q,R}} \frac{\mu\left(2^{k+1}Q\right)}{[\ell(2^{k+1}Q)]^n} \left|m_{\widetilde{2^{k+1}Q}}(f)-m_{\tilde{Q}}(f)\right|\\
&\lesssim \sum_{k=1}^{N_{Q,R}+1} \frac{\mu(2^kQ)}{\left[\ell\left(2^kQ\right)\right]^n} + \sum_{k=1}^{N_{Q,R}+1} \frac{\mu(2^kQ)}{\left[\ell\left(2^kQ\right)\right]^n}\left[1+\delta\left(Q,2^kQ\right)\right]\\
&\lesssim \delta_{Q,R}+\delta_{Q,R}[1+\delta(Q,\,R)]\\
&\lesssim [1+\delta(Q,\,R)]^2,
\end{aligned}$$

which completes the proof of Lemma 4.3.5. □

Corollary 4.3.6. *Let $k \in \mathbb{N}$ and S_k be as in Sect. 2.4. Then*

(i)

$$\mathrm{rbmo}\,(\mu) = \{b \in \mathrm{RBMO}\,(\mu) : S_k(b) \in L^\infty(\mu)\};$$

moreover, for any $b \in \mathrm{rbmo}\,(\mu)$,

$$k\|b\|_{\mathrm{rbmo}\,(\mu)} \sim \|S_k(b)\|_{L^\infty(\mu)} + \|b\|_{\mathrm{RBMO}\,(\mu)}$$

with the implicit equivalent positive constants independent of k and b;

(ii) *if $f \in \mathrm{RBMO}\,(\mu)$, then*

$$f - S_k(f) \in \mathrm{rbmo}\,(\mu);$$

moreover, there exists a positive constant C, independent of k and f, such that

$$\|f - S_k(f)\|_{\mathrm{rbmo}\,(\mu)} \le C\|f\|_{\mathrm{RBMO}\,(\mu)}.$$

Proof. To prove (i), assume that $b \in \mathrm{RBMO}\,(\mu)$ with $S_k(b) \in L^\infty(\mu)$ first. For any $f \in h^{1,\infty}_{\mathrm{atb}}(\mu)$, by Proposition 4.3.3, together with Theorems 2.4.4(a) and 3.2.11, we conclude that

$$\begin{aligned}
&\left|\int_{\mathbb{R}^D} b(x) f(x)\, d\mu(x)\right| \\
&\leq \left|\int_{\mathbb{R}^D} b(x)\, [f(x) - S_k(f)(x)]\, d\mu(x)\right| + \left|\int_{\mathbb{R}^D} b(x) S_k(f)(x)\, d\mu(x)\right| \\
&\lesssim \|f\|_{h^{1,\infty}_{\rm atb}(\mu)} [\|b\|_{\mathrm{RBMO}\,(\mu)} + \|S_k(b)\|_{L^\infty(\mu)}].
\end{aligned}$$

Thus, by Theorem 4.2.17, we see that $b \in \mathrm{rbmo}\,(\mu)$ and

$$\|b\|_{\mathrm{rbmo}\,(\mu)} \lesssim \|b\|_{\mathrm{RBMO}\,(\mu)} + \|S_k(b)\|_{L^\infty(\mu)}.$$

Conversely, assume that $b \in \mathrm{rbmo}\,(\mu)$. If $k \geq 2$, by (e) of Lemma 4.1.2, we know that, for any $x \in \operatorname{supp}\mu$, $Q_{x,k} \notin \mathcal{Q}$. Therefore, from (b) through (d) of Theorem 2.4.4, the facts that $Q_{x,1} \in \mathcal{Q}$ and

$$\mathrm{rbmo}\,(\mu) \subset \mathrm{RBMO}\,(\mu),$$

Definition 4.2.1 and Lemma 4.3.5, it follows that, for any $x \in \operatorname{supp}\mu$,

$$\begin{aligned}
|S_k(b)(x)| &\leq \int_{Q_{x,k-1}} S_k(x, y) \left|b(y) - m_{Q_{x,k}}(b)\right| d\mu(y) \\
&\quad + \left|m_{Q_{x,k}}(b) - m_{Q_{x,1}}(b)\right| + \left|m_{Q_{x,1}}(b)\right| \\
&\lesssim k\|b\|_{\mathrm{rbmo}\,(\mu)}.
\end{aligned}$$

Let $k = 1$. If $\mathbb{R}^D$ is an initial cube, we first claim that, for any $x \in \operatorname{supp}\mu$,

$$\int_{\mathbb{R}^D\setminus Q_{x,1}} \frac{|b(y)|}{|x - y|^n}\, d\mu(y) \lesssim \|b\|_{\mathrm{rbmo}\,(\mu)}.$$

Indeed, by the fact that $2^{j+1}Q_{x,1} \in \mathcal{Q}$ for all $j \geq 0$, together with Definition 4.2.1 and the fact that $\delta(Q_{x,1}, \mathbb{R}^D) \lesssim 1$, we see that, for any $j_0 \in \mathbb{N}$,

$$\begin{aligned}
\int_{2^{j_0} Q_{x,1}\setminus Q_{x,1}} \frac{|b(y)|}{|y-x|^n}\, d\mu(y) &\lesssim \sum_{j=0}^{j_0-1} \frac{1}{[\ell(2^{j+2}Q_{x,1})]^n} \int_{2^{j+1}Q_{x,1}\setminus 2^j Q_{x,1}} |b(y)|\, d\mu(y) \\
&\lesssim [1 + \delta(Q_{x,1}, \mathbb{R}^D)]\|b\|_{\mathrm{rbmo}\,(\mu)} \\
&\lesssim \|b\|_{\mathrm{rbmo}\,(\mu)}.
\end{aligned}$$

By letting $j_0 \to \infty$, we know that the above claim holds true.

By this claim, (d) and Definition 4.2.1, together with $Q_{x,1} \in \mathcal{Q}$, we conclude that, for any $x \in \operatorname{supp}\mu$,

$$\begin{aligned}
|S_1(b)(x)| &\lesssim \int_{\mathbb{R}^D} \frac{|b(z)|}{[|x-z| + \ell(Q_{x,1})]^n}\,d\mu(z) \\
&\lesssim \int_{\mathbb{R}^D\setminus Q_{x,1}} \frac{|b(z)|}{|x-z|^n}\,d\mu(z) + \int_{Q_{x,1}} \frac{|b(z)|}{[\ell(Q_{x,1})]^n}\,d\mu(z) \\
&\lesssim \|b\|_{\mathrm{rbmo}\,(\mu)}. \qquad (4.3.10)
\end{aligned}$$

If $\mathbb{R}^D$ is not an initial cube, then, by (c), (d) and Definition 4.2.1, together with $Q_{x,0}$, $Q_{x,1} \in \mathcal{Q}$, we see that, for any $x \in \operatorname{supp}\mu$,

$$\begin{aligned}
|S_1(b)(x)| &\lesssim \int_{Q_{x,0}} \frac{|b(z)|}{[|x-z| + \ell(Q_{x,1})]^n}\,d\mu(z) \\
&\lesssim \int_{Q_{x,0}\setminus Q_{x,1}} \frac{|b(z)|}{|x-z|^n}\,d\mu(z) + \int_{Q_{x,1}} \frac{|b(z)|}{[\ell(Q_{x,1})]^n}\,d\mu(z) \\
&\lesssim \|b\|_{\mathrm{rbmo}\,(\mu)}.
\end{aligned}$$

Combining these estimates above, we know that, for each $k \in \mathbb{N}$,

$$S_k(b) \in L^\infty(\mu) \quad \text{and} \quad \|S_k(b)\|_{L^\infty(\mu)} \lesssim k\|b\|_{\mathrm{rbmo}\,(\mu)},$$

which imply that

$$\|S_k(b)\|_{L^\infty(\mu)} + \|b\|_{\mathrm{RBMO}\,(\mu)} \lesssim k\|b\|_{\mathrm{rbmo}\,(\mu)}.$$

This establishes (i).

For any $b \in h^{1,\infty}_{\mathrm{atb}}(\mu)$, it follows, from Proposition 4.3.3, that

$$\|b - S_k(b)\|_{H^1(\mu)} \lesssim \|b\|_{h^{1,\infty}_{\mathrm{atb}}(\mu)}.$$

By this fact and Theorem 3.2.11, we find that, for any $f \in \mathrm{RBMO}\,(\mu)$,

$$\begin{aligned}
\left|\int_{\mathbb{R}^D} [f(x) - S_k(f)(x)]b(x)\,d\mu(x)\right| &= \left|\int_{\mathbb{R}^D} f(x)[b(x) - S_k(b)(x)]\,d\mu(x)\right| \\
&\lesssim \|f\|_{\mathrm{RBMO}\,(\mu)}\|b\|_{h^{1,\infty}_{\mathrm{atb}}(\mu)},
\end{aligned}$$

which, via Theorem 4.2.17, implies that

$$f - S_k(f) \in \mathrm{rbmo}\,(\mu).$$

This establishes (ii) and hence finishes the proof of Corollary 4.3.6. □

4.4 The Spaces RBLO (μ) and rblo (μ)

In this section, we introduce the space RBLO (μ) and its local version rblo (μ).

Definition 4.4.1. Let $\eta,\ \rho \in (1,\infty)$ and $\beta_\rho := \rho^{D+1}$. A function $f \in L^1_{\rm loc}(\mu)$ is said to belong to the *space* RBLO (μ), if there exists some constant $C \in \mathbb{Z}_+$ such that, for any cube Q,

$$\frac{1}{\mu(\eta Q)}\int_Q \left[f(x) - \operatorname*{ess\,inf}_{\tilde{Q}^\rho} f\right] d\mu(x) \le C \tag{4.4.1}$$

and, for any two (ρ,β_ρ)-doubling cubes $Q \subset R$,

$$\operatorname*{ess\,inf}_{Q} f - \operatorname*{ess\,inf}_{R} f \le C[1 + \delta(Q,R)]. \tag{4.4.2}$$

The minimal constant C as above is defined to be the *norm* of f in the space RBLO (μ) and denoted by $\|f\|_{{\rm RBLO}\,(\mu)}$.

Remark 4.4.2. Let μ be the D-dimensional Lebesgue measure. A function $f \in L^1_{\rm loc}(\mathbb{R}^D)$ is said to be in the *space* BLO ($\mathbb{R}^D$), if there exists some nonnegative constant C such that

$$\sup_Q \frac{1}{|Q|}\int_Q \left[f(x) - \operatorname*{ess\,inf}_{Q} f\right] dx \le C,$$

where the supremum is taken over all cubes in $\mathbb{R}^D$. Moreover, the BLO ($\mathbb{R}^{\rm D}$) *norm* of f is defined to be the minimal constant C as above.[5]

To begin with, we prove that the space RBLO (μ) is independent of the chosen constants $\eta \in (1,\infty)$ and $\rho \in (1,\infty)$.

Let $\eta \in (1,\infty)$. Suppose that, for a given $f \in L^1_{\rm loc}(\mu)$, there exist a nonnegative constant $\tilde{C}$ and a collection of numbers, $\{f_Q\}_Q$, such that

$$\sup_Q \frac{1}{\mu(\eta Q)}\int_Q [f(y) - f_Q]\,d\mu(y) \le \tilde{C}, \tag{4.4.3}$$

that, for any two cubes $Q \subset R$,

$$|f_Q - f_R| \le \tilde{C}[1 + \delta(Q,R)] \tag{4.4.4}$$

and that, for any cube Q,

---

[5] See [17].

$$f_Q \leq \operatorname*{ess\,inf}_Q f. \tag{4.4.5}$$

We then define the *norm* $\|f\|_{\star\star,\eta} := \inf\{\tilde{C}\}$, where the infimum is taken over all the constants $\tilde{C}$ as above and all the numbers $\{f_Q\}_Q$ satisfying (4.4.3) through (4.4.5).

With a minor modification of the proof for Proposition 3.1.6, we have the following proposition, the details being omitted.

Proposition 4.4.3. *Let $\eta_1, \eta_2 \in (1,\infty)$. Then there exists a constant $C \in [1,\infty)$ such that, for any $f \in L^1_{\rm loc}(\mu)$,*

$$\|f\|_{\star\star,\eta_1}/C \leq \|f\|_{\star\star,\eta_2} \leq C\|f\|_{\star\star,\eta_1}.$$

Based on Proposition 4.4.3, from now on, we write $\|\cdot\|_{\star\star}$ *instead of* $\|\cdot\|_{\star\star,\eta}$.

Proposition 4.4.4. *Let $\eta \in (1,\infty)$, $\rho \in (1,\infty)$ and $\beta_\rho := \rho^{D+1}$. Then the norms $\|\cdot\|_{\star\star}$ and $\|\cdot\|_{{\rm RBLO}\,(\mu)}$ are equivalent.*

Proof. Suppose that $f \in L^1_{\rm loc}(\mu)$. We first show that

$$\|f\|_{\star\star} \lesssim \|f\|_{{\rm RBLO}\,(\mu)}. \tag{4.4.6}$$

For any cube Q, let

$$f_Q := \operatorname*{ess\,inf}_{\tilde{Q}^\rho} f.$$

Then (4.4.3) and (4.4.5) hold true with

$$\tilde{C} := \|f\|_{{\rm RBLO}\,(\mu)}.$$

To show that (4.4.4) also holds true, let $R_0 := \widetilde{2\tilde{R}^\rho}^\rho$ if $\ell(\tilde{R}^\rho) \geq \ell(\tilde{Q}^\rho)$ and $R_0 := \widetilde{2\rho\tilde{Q}^\rho}^\rho$ if $\ell(\tilde{R}^\rho) < \ell(\tilde{Q}^\rho)$. Then, arguing as in the proof of (4.2.7), we conclude that, for any two cubes $Q \subset R$,

$$\left|\operatorname*{ess\,inf}_{\tilde{Q}^\rho} f - \operatorname*{ess\,inf}_{\tilde{R}^\rho} f\right| \lesssim [1+\delta(Q,R)]\|f\|_{{\rm RBLO}\,(\mu)}. \tag{4.4.7}$$

Now let us establish the converse of (4.4.6). For $f \in L^1_{\rm loc}(\mu)$, assume that there exists a sequence of numbers, $\{f_Q\}_Q$, satisfying (4.4.3) through (4.4.5) with $\tilde{C}$ replaced by $\|f\|_{\star\star}$. For any cube Q, by (4.4.4), (4.4.5) and Lemma 2.1.3, we see that

$$\begin{aligned} f_Q - \operatorname*{ess\,inf}_{\tilde{Q}^\rho} f &= f_Q - f_{\tilde{Q}^\rho} + f_{\tilde{Q}^\rho} - \operatorname*{ess\,inf}_{\tilde{Q}^\rho} f \\ &\leq \left[1+\delta(Q,\tilde{Q}^\rho)\right]\|f\|_{\star\star} \\ &\lesssim \|f\|_{\star\star}. \end{aligned}$$

This fact, together with (4.4.3), implies that, for any cube Q,

$$\frac{1}{\mu(\eta Q)} \int_Q \left[f(y) - \operatorname*{ess\,inf}_{\tilde{Q}^\rho} f \right] d\mu(y)$$

$$= \frac{1}{\mu(\eta Q)} \int_Q [f(y) - f_Q] \, d\mu(y) + \frac{\mu(Q)}{\mu(\eta Q)} \left[f_Q - \operatorname*{ess\,inf}_{\tilde{Q}^\rho} f \right]$$

$$\lesssim \|f\|_{\star\star}.$$

On the other hand, for any (ρ, β_ρ)-doubling cube Q, since (4.4.3) holds true with $\eta = \rho$ by Proposition 4.4.3, we see that (4.4.5) implies that

$$m_Q(f) - f_Q = \frac{1}{\mu(Q)} \int_Q [f(x) - f_Q] \, d\mu(x) \leq \frac{\mu(\rho Q)}{\mu(Q} \|f\|_{\star\star} \lesssim \|f\|_{\star\star}.$$

Then, from (4.4.4) and (4.4.5), it follows that, for any two (ρ, β_ρ)-doubling cubes $Q \subset R$,

$$\begin{aligned} \operatorname*{ess\,inf}_Q f - \operatorname*{ess\,inf}_R f &\leq \operatorname*{ess\,inf}_Q f - f_Q + f_Q - f_R \\ &\leq m_Q(f) - f_Q + [1 + \delta(Q, R)] \|f\|_{\star\star} \\ &\lesssim [1 + \delta(Q, R)] \|f\|_{\star\star}. \end{aligned}$$

This establishes the converse of (4.4.6) and hence finishes the proof of Proposition 4.4.4. □

From Definition 4.4.1 and Proposition 4.4.4, we deduce that

$$L^\infty(\mu) \subset \text{RBLO}\,(\mu) \subset \text{RBMO}\,(\mu).$$

Moreover, we also have the following characterization of RBLO (μ).

Proposition 4.4.5. *Let $\eta \in (1, \infty)$, $\rho \in (1, \infty)$ and $\beta_\rho := \rho^{D+1}$. For $f \in L^1_{\text{loc}}(\mu)$, the following statements are equivalent:*

(i) $f \in \text{RBLO}\,(\mu)$;
(ii) *There exists a nonnegative constant C, satisfying* (4.4.2), *such that, for any (ρ, β_ρ)-doubling cube Q,*

$$\frac{1}{\mu(Q)} \int_Q \left[f(x) - \operatorname*{ess\,inf}_Q f \right] d\mu(x) \leq C; \tag{4.4.8}$$

(iii) *There exists a nonnegative constant $\tilde{C}$, satisfying* (4.4.8), *such that, for any (ρ, β_ρ)-doubling cubes $Q \subset R$,*

$$m_Q(f) - m_R(f) \le \tilde{C}[1 + \delta(Q, R)]. \tag{4.4.9}$$

Moreover, the minimal constants C and $\tilde{C}$ are equivalent to $\|f\|_{\mathrm{RBLO}(\mu)}$.

Proof. By Propositions 4.4.3 and 4.4.4, it suffices to establish Proposition 4.4.5 with $\eta = \rho = 2$. Notice that (i) automatically implies (ii). We now prove that (ii) implies (iii). From (4.4.2), together with (4.4.8), it follows that, for any doubling cubes $Q \subset R$,

$$\begin{aligned} m_Q(f) - m_R(f) &\le m_Q(f) - \operatorname*{ess\,inf}_Q f + \operatorname*{ess\,inf}_Q f - \operatorname*{ess\,inf}_R f \\ &\lesssim C[1 + \delta(Q, R)], \end{aligned}$$

which implies (iii).

Finally, assume that (iii) holds true. For any cube $Q \subset \mathbb{R}^D$ and $x \in Q \cap \operatorname{supp} \mu$, let Q_x be the *biggest doubling cube* centered at x

$$\text{with side length} \quad 2^k \ell(Q),\ k \le 0, \quad \text{and} \quad \ell(Q_x) \le \frac{1}{20}\ell(Q).$$

Then Lemma 2.1.3 implies that $\delta(Q_x, Q) \lesssim 1$. By Theorem 1.1.1, there exists a subsequence of cubes, $\{Q_{x_i}\}_i$, which covers $Q \cap \operatorname{supp} \mu$ and has a bounded overlap. Moreover, from (4.4.8), (4.4.9) and $Q_{x_i} \subset 2Q$, it follows that

$$\begin{aligned} \operatorname*{ess\,inf}_{Q_{x_i}} f - \operatorname*{ess\,inf}_{\widetilde{2Q}} f &\le m_{Q_{x_i}}(f) - m_{\widetilde{2Q}}(f) + m_{\widetilde{2Q}}(f) - \operatorname*{ess\,inf}_{\widetilde{2Q}} f \\ &\lesssim \tilde{C}\left[1 + \delta(Q_{x_i}, \widetilde{2Q})\right] \\ &\lesssim \tilde{C}. \end{aligned}$$

This fact, together with the facts that $\{Q_{x_i}\}_i$ covers $Q \cap \operatorname{supp} \mu$ with a bounded overlap, $Q_{x_i} \subset 2Q$ and Q_{x_i} is doubling, and (4.4.8), implies that

$$\begin{aligned} &\int_Q \left[f(x) - \operatorname*{ess\,inf}_{\tilde{Q}} f\right] d\mu(x) \\ &\le \sum_i \int_{Q_{x_i}} \left[f(x) - \operatorname*{ess\,inf}_{Q_{x_i}} f\right] d\mu(x) + \sum_i \mu(Q_{x_i}) \left[\operatorname*{ess\,inf}_{Q_{x_i}} f - \operatorname*{ess\,inf}_{\widetilde{2Q}} f\right] \\ &\lesssim \tilde{C} \sum_i \mu(Q_{x_i}) \\ &\lesssim \tilde{C} \mu(2Q). \end{aligned}$$

On the other hand, from (4.4.9) and (4.4.8), it follows that, for any doubling cubes $Q \subset R$,

$$\operatorname*{ess\,inf}_{Q} f - \operatorname*{ess\,inf}_{R} f \le m_Q(f) - m_R(f) + m_R(f) - \operatorname*{ess\,inf}_{R} f$$
$$\lesssim \tilde{C}[1 + \delta(Q, R)].$$

Therefore, we see that $f \in \mathrm{RBLO}\,(\mu)$, which implies (i). This finishes the proof of Proposition 4.4.5. □

Remark 4.4.6. Let $\eta \in (1, \infty)$ and $\rho \in (1, \infty)$. From Propositions 4.4.3 and 4.4.4, it follows that the space RBLO (μ) is independent of the choices of η and ρ. From now on, we *always assume* $\eta = \rho = 2$ when we consider RBLO (μ).

We next recall the notion of the *natural maximal operator*. For any locally integrable function f and $x \in \mathbb{R}^D$, define

$$\mathcal{M}(f)(x) := \sup_{\substack{Q \ni x \\ Q \text{ doubling}}} \frac{1}{\mu(Q)} \int_Q f(y)\, d\mu(y).$$

Also, the *non-centered doubling Hardy–Littlewood maximal operator* $\mathcal{M}^d(f)$ is defined by $\mathcal{M}^d(f) := \mathcal{M}(|f|)$.

Lemma 4.4.7. *If* $f \in \mathrm{RBMO}\,(\mu)$, *then there exists a positive constant* C *such that, for any doubling cube* Q,

$$\frac{1}{\mu(Q)} \int_Q \mathcal{M}(f)(x)\, d\mu(x) \le C \|f\|_{\mathrm{RBMO}\,(\mu)} + \operatorname*{ess\,inf}_{x \in Q} \mathcal{M}(f)(x). \tag{4.4.10}$$

Moreover, if $\mathcal{M}(f)$ *is finite* μ*-almost everywhere, then*

$$\frac{1}{\mu(Q)} \int_Q \mathcal{M}(f)(x)\, d\mu(x) - \operatorname*{ess\,inf}_{x \in Q} \mathcal{M}(f)(x) \le C \|f\|_{\mathrm{RBMO}\,(\mu)}. \tag{4.4.11}$$

Proof. Assume that $f \in \mathrm{RBMO}\,(\mu)$ and Q is a doubling cube. We write

$$f = (f - m_Q(f))\chi_{\frac{4}{3}Q} + [m_Q(f)\chi_{\frac{4}{3}Q} + f\chi_{\mathbb{R}^D \setminus \frac{4}{3}Q}].$$

Observe that $\mathcal{M}$ is bounded on $L^2(\mu)$. Then, by the Hölder inequality and Corollary 3.1.20, we have

$$\int_Q \mathcal{M}[(f - m_Q(f))\chi_{\frac{4}{3}Q}](x)\, d\mu(x)$$
$$\le [\mu(Q)]^{1/2} \left\{ \int_Q \left| \mathcal{M}[(f - m_Q(f))\chi_{\frac{4}{3}Q}](x) \right|^2 d\mu(x) \right\}^{1/2}$$

$$\begin{aligned}&\lesssim [\mu(Q)]^{1/2}\left\{\int_{\frac{4}{3}Q}\left|f(x)-m_{\widetilde{\frac{4}{3}Q}}(f)\right|^2\,d\mu(x)\right\}^{1/2}\\&\quad+\mu\left(\frac{4}{3}Q\right)\left|m_{\widetilde{\frac{4}{3}Q}}(f)-m_Q(f)\right|\\&\lesssim \mu(Q)\|f\|_{\mathrm{RBMO}\,(\mu)}.\end{aligned}$$

Next, we show that there exists a positive constant $\tilde{C}$, independent of Q and f, such that

$$\begin{aligned}&\frac{1}{\mu(Q)}\int_Q \mathcal{M}[m_Q(f)\chi_{\frac{4}{3}Q}+f\chi_{\mathbb{R}^D\setminus\frac{4}{3}Q}](x)\,d\mu(x)\\&\quad\le \tilde{C}\|f\|_{\mathrm{RBMO}\,(\mu)}+\operatorname*{ess\,inf}_{x\in Q}\mathcal{M}(f)(x).\end{aligned}$$

To this end, it suffices to show that, for μ-almost every $x\in Q$,

$$\mathcal{M}[m_Q(f)\chi_{3Q}+f\chi_{\mathbb{R}^D\setminus\frac{4}{3}Q}](x)\le \tilde{C}\|f\|_{\mathrm{RBMO}\,(\mu)}+\operatorname*{ess\,inf}_{x\in Q}\mathcal{M}(f)(x),$$

which is reduced to showing that, for any doubling cube R containing x,

$$\begin{aligned}\mathrm{E}:&=\frac{1}{\mu(R)}\int_R [m_Q(f)\chi_{\frac{4}{3}Q}+f\chi_{\mathbb{R}^D\setminus\frac{4}{3}Q}](x)\,d\mu(x)\\&\le \tilde{C}\|f\|_{\mathrm{RBMO}\,(\mu)}+\operatorname*{ess\,inf}_{x\in Q}\mathcal{M}(f)(x).\end{aligned}$$

If $R\subset\frac{4}{3}Q$, then we conclude that

$$\mathrm{E}=m_Q(f)\le \operatorname*{ess\,inf}_{x\in Q}\mathcal{M}(f)(x).$$

Assume that $R\not\subset\frac{4}{3}Q$. We then see that $\ell(R)\ge\frac{1}{6}\ell(Q)$. There exist two cases.

Case (i) $\ell(R)\le 4\ell(Q)$. In this case, it is easy to see that $Q\subset 13R\subset 57Q$ with the aid of the fact that $\ell(R)\ge\frac{1}{6}\ell(Q)$. From this fact, Propositions 3.1.10 and 3.1.4, and Lemma 2.1.3, it follows that

$$\begin{aligned}\mathrm{E}-\operatorname*{ess\,inf}_{x\in Q}\mathcal{M}(f)(x)&\le\frac{1}{\mu(R)}\int_R [f(z)-m_Q(f)]\,\chi_{\mathbb{R}^D\setminus\frac{4}{3}Q}(z)\,d\mu(z)\\&\le\frac{1}{\mu(R)}\int_R |f(z)-m_R(f)|\,d\mu(z)+|m_R(f)-m_Q(f)|\\&\lesssim 1+\delta(Q,13R)\\&\lesssim 1.\end{aligned}$$

Case (ii) $\ell(R)>4\ell(Q)$. In this case, $Q\subset\frac{3}{2}R$. We also have

$$
\begin{aligned}
&\mathrm{E} - \operatorname*{ess\,inf}_{x\in Q} \mathcal{M}(f)(x) \\
&\le \frac{1}{\mu(R)} \int_R \left\{ \left[m_Q(f) - m_{\widetilde{\frac{3}{2}R}}(f) \right] \chi_{\frac{4}{3}Q} \right. \\
&\quad \left. + \left[f(z) - m_{\widetilde{\frac{3}{2}R}}(f) \right] \chi_{\mathbb{R}^D \setminus \frac{4}{3}Q}(z) \right\} d\mu(z) \\
&\le \frac{\mu(\frac{4}{3}Q)}{\mu(R)} \left| m_Q(f) - m_{\widetilde{\frac{3}{2}R}}(f) \right| + \frac{1}{\mu(R)} \int_R \left| f(z) - m_{\widetilde{\frac{3}{2}R}}(f) \right| d\mu(z) \\
&\lesssim \frac{1}{\mu(R)} \int_R \left| f(z) - m_{\widetilde{\frac{3}{2}R}}(f) \right| d\mu(z) \\
&\lesssim 1.
\end{aligned}
$$

This finishes the proof of Lemma 4.4.7. □

Theorem 4.4.8. *Let $f \in$ RBMO (μ). Then $\mathcal{M}(f)$ is either infinite everywhere or finite almost everywhere and, in the latter case, there exists a positive constant C, independent of f, such that*

$$\|\mathcal{M}(f)\|_{\mathrm{RBLO}\,(\mu)} \le C\|f\|_{\mathrm{RBMO}\,(\mu)}.$$

Proof. Suppose that $f \in$ RBMO (μ) and there exists some point $x_0 \in \mathbb{R}^D$ such that $\mathcal{M}(f)(x_0) < \infty$. It then follows, from Lemma 4.4.7, that (4.4.10) holds true. By (4.4.10) and Proposition 4.4.5, Theorem 4.4.8 is reduced to proving that, for any doubling cubes $Q \subset R$,

$$m_Q[\mathcal{M}(f)] - m_R[\mathcal{M}(f)] \lesssim [1 + \delta(Q, R)]\|f\|_{\mathrm{RBMO}\,(\mu)}. \tag{4.4.12}$$

To prove (4.4.12), for any point $x \in Q$, we further let

$$\mathcal{M}_1(f)(x) := \sup_{\substack{P\ni x,\, P \text{ doubling}\\ \ell(P)\le 4\ell(R)}} \frac{1}{\mu(P)} \int_P f(y)\, d\mu(y),$$

$$\mathcal{M}_2(f)(x) := \sup_{\substack{P\ni x,\, P \text{ doubling}\\ \ell(P)> 4\ell(R)}} \frac{1}{\mu(P)} \int_P f(y)\, d\mu(y),$$

$$\mathcal{U}_{1,Q} := \{x \in Q : \ \mathcal{M}_1(f)(x) \ge \mathcal{M}_2(f)(x)\} \quad \text{and} \quad \mathcal{U}_{2,Q} := Q \setminus \mathcal{U}_{1,Q}.$$

Then, for any $x \in Q$,

$$\mathcal{M}(f)(x) = \max\{\mathcal{M}_1(f)(x), \mathcal{M}_2(f)(x)\}.$$

By writing

$$f = [f - m_R(f)]\chi_{\frac{4}{3}Q} + [f - m_R(f)]\chi_{\mathbb{R}^D\setminus\frac{4}{3}Q} + m_R(f)$$

and using the fact that $m_R(f) \le m_R[\mathcal{M}(f)]$, we see that

$$\begin{aligned}
&m_Q(\mathcal{M}(f)) - m_R(\mathcal{M}(f))\\
&\quad\le \frac{1}{\mu(Q)}\int_{\mathcal{U}_{1,Q}} \mathcal{M}_1([f - m_R(f)]\chi_{\frac{4}{3}Q})(x)\,d\mu(x)\\
&\qquad + \frac{1}{\mu(Q)}\int_{\mathcal{U}_{1,Q}} \mathcal{M}_1([f - m_R(f)]\chi_{\mathbb{R}^D\setminus\frac{4}{3}Q})(x)\,d\mu(x)\\
&\qquad + \frac{1}{\mu(Q)}\int_{\mathcal{U}_{2,Q}} \{\mathcal{M}_2(f)(x) - m_R(\mathcal{M}(f))\}\,d\mu(x)\\
&\quad =: \mathrm{I}_1 + \mathrm{I}_2 + \mathrm{I}_3.
\end{aligned}$$

From the boundedness of $\mathcal{M}(f)$ on $L^2(\mu)$, the Hölder inequality, Corollary 3.1.20, the doubling property of Q and Lemma 2.1.3, it follows that

$$\begin{aligned}
\mathrm{I}_1 &\le \left\{\frac{1}{\mu(Q)}\int_Q \left\{\mathcal{M}\left([f - m_R(f)]\chi_{\frac{4}{3}Q}\right)(x)\right\}^2 d\mu(x)\right\}^{1/2}\\
&\lesssim \left\{\frac{1}{\mu(Q)}\int_{\frac{4}{3}Q} |f(x) - m_R(f)|^2\,d\mu(x)\right\}^{1/2}\\
&\lesssim \left\{\frac{1}{\mu(Q)}\int_{\frac{4}{3}Q} \left|f(x) - m_{\widetilde{\frac{4}{3}Q}}(f)\right|^2 d\mu(x)\right\}^{1/2} + \left|m_{\widetilde{\frac{4}{3}Q}}(f) - m_Q(f)\right|\\
&\quad + |m_Q(f) - m_R(f)|\\
&\lesssim [1 + \delta(Q,R)]\|f\|_{\mathrm{RBMO}(\mu)}.
\end{aligned}$$

To estimate I_2, we prove that, for any point $x \in Q$ and doubling cube $P \ni x$ with $\ell(P) \le 4\ell(R)$,

$$\begin{aligned}
\mathrm{J} &:= \frac{1}{\mu(P)}\int_P |f(y) - m_R(f)|\chi_{\mathbb{R}^D\setminus\frac{4}{3}Q}(y)\,d\mu(y)\\
&\lesssim [1 + \delta(Q,R)]\|f\|_{\mathrm{RBMO}(\mu)}. \qquad (4.4.13)
\end{aligned}$$

If $P \subset \frac{4}{3}Q$, then $\mathrm{J} \equiv 0$ and (4.4.13) holds true automatically. Assume that $P \not\subset \frac{4}{3}Q$. We then see that $\ell(P) \ge \frac{1}{6}\ell(Q)$, which, together with the fact that $\ell(P) \le 4\ell(R)$, implies that $Q \subset 13P \subset 57R$. Thus, by Lemma 2.1.3, together with (3.1.12), we conclude that

$$\begin{aligned}
\mathrm{J} &\le \frac{1}{\mu(P)}\int_P |f(y)-m_P(f)|\,d\mu(y) + \left|m_P(f)-m_{\widetilde{13P}}(f)\right| \\
&\quad + \left|m_{\widetilde{13P}}(f)-m_Q(f)\right| + \left|m_Q(f)-m_R(f)\right| \\
&\lesssim [1+\delta(Q,R)]\|f\|_{\mathrm{RBMO}\,(\mu)}.
\end{aligned}$$

Now we estimate I_3. Notice that, for any $x \in Q$ and doubling cube P containing x with $\ell(P) > 4\ell(R)$, $R \subset \frac{3}{2}P$. Then, from the fact that

$$m_{\widetilde{\frac{3}{2}P}}(f) \le m_R(\mathcal{M}(f)),$$

it follows that

$$\begin{aligned}
m_P(f) - m_R(\mathcal{M}(f)) &\le \left|m_P(f) - m_{\widetilde{\frac{3}{2}P}}(f)\right| + m_{\widetilde{\frac{3}{2}P}}(f) - m_R(\mathcal{M}(f)) \\
&\lesssim \|f\|_{\mathrm{RBMO}\,(\mu)}.
\end{aligned}$$

Taking the supremum over all doubling cubes P containing x with $\ell(P) > 4\ell(R)$, we know that, for any $x \in Q$,

$$\mathcal{M}_2(f)(x) - m_R(\mathcal{M}(f)) \lesssim \|f\|_{\mathrm{RBMO}\,(\mu)}.$$

This implies that

$$\mathrm{I}_3 \lesssim \|f\|_{\mathrm{RBMO}\,(\mu)}.$$

Combining the estimates for I_1 through I_3, we obtain (4.4.12), which, together with (4.4.10), implies that $\mathcal{M}$ is bounded from RBMO (μ) to RBLO (μ) and hence completes the proof of Theorem 4.4.8. □

Lemma 4.4.9. *Let f be a locally integrable function. Then, $f \in$ RBLO (μ) if and only if $\mathcal{M}(f) - f \in L^\infty(\mu)$ and f satisfies* (4.4.9) *with $\rho = 2$. Furthermore,*

$$\|\mathcal{M}(f) - f\|_{L^\infty(\mu)} \sim \|f\|_{\mathrm{RBLO}\,(\mu)}$$

with the implicit equivalent positive constants independent of f.

Proof. Assuming that $f \in$ RBLO (μ), we then see that (4.4.9) holds true. For μ-almost every $x \in \mathbb{R}^D$, there exists a sequence of doubling cubes, $\{Q_k\}_k$, centered at x with $\ell(Q_k) \to 0$ such that

$$\lim_{k\to\infty} \frac{1}{\mu(Q_k)}\int_{Q_k} f(y)\,d\mu(y) = f(x). \tag{4.4.14}$$

Let x be any point satisfying (4.4.14) and Q a doubling cube containing x. Then there exists some Q_k as in (4.4.14) such that $Q_k \subset \widetilde{2Q}$. We conclude that

$$f(x) \geq \operatorname*{ess\,inf}_{Q_k} f \geq \operatorname*{ess\,inf}_{\widetilde{2Q}} f$$

and hence

$$\begin{aligned} m_Q(f) - f(x) &\leq m_Q(f) - m_{\widetilde{2Q}}(f) + m_{\widetilde{2Q}}(f) - \operatorname*{ess\,inf}_{\widetilde{2Q}} f \\ &\lesssim \|f\|_{\mathrm{RBLO}\,(\mu)}. \end{aligned}$$

Taking the supremum over all doubling cubes containing x, we have

$$\mathcal{M}(f)(x) - f(x) \lesssim \|f\|_{\mathrm{RBLO}\,(\mu)}.$$

Conversely, assume that f satisfies (4.4.9) and $\mathcal{M}(f) - f \in L^\infty(\mu)$. Then, it is easy to see that, for any doubling cube Q and μ-almost every $x \in Q$,

$$f(x) \geq m_Q(f) - \|\mathcal{M}(f) - f\|_{L^\infty(\mu)},$$

which implies that

$$\operatorname*{ess\,inf}_{Q} f \geq m_Q(f) - \|\mathcal{M}(f) - f\|_{L^\infty(\mu)}.$$

From this, together with (4.4.8), (4.4.9) and Proposition 4.4.15, we conclude that $f \in \mathrm{RBLO}\,(\mu)$ and

$$\|f\|_{\mathrm{RBLO}\,(\mu)} \lesssim \|\mathcal{M}(f) - f\|_{L^\infty(\mu)}.$$

Therefore, the proof of Lemma 4.4.9 is completed. □

As a corollary of Lemma 4.4.9 and Theorem 4.4.8, we immediately have the following result, the details being omitted.

Theorem 4.4.10. *A locally integrable function f belongs to* RBLO (μ) *if and only if there exist $h \in L^\infty(\mu)$ and $g \in$* RBMO (μ) *with $\mathcal{M}(g)$ finite μ-almost everywhere such that*

$$f = \mathcal{M}(g) + h. \tag{4.4.15}$$

Furthermore,

$$\|f\|_{\mathrm{RBLO}\,(\mu)} \sim \inf\{\|g\|_{\mathrm{RBMO}\,(\mu)} + \|h\|_{L^\infty(\mu)}\}$$

with the implicit equivalent positive constants independent of f, g and h, where the infimum is taken over all representations of f as in (4.4.15).

Proof. If there exist g and h satisfying (4.4.15), then from Theorem 4.4.8, it follows that $\mathcal{M}(g) \in \mathrm{RBLO}(\mu)$. This implies that $f \in \mathrm{RBLO}(\mu)$ and

$$\begin{aligned}\|f\|_{\mathrm{RBLO}(\mu)} &\lesssim \|\mathcal{M}(g)\|_{\mathrm{RBLO}(\mu)} + \|h\|_{L^\infty(\mu)}\\ &\lesssim \|g\|_{\mathrm{RBMO}(\mu)} + \|h\|_{L^\infty(\mu)}.\end{aligned}$$

To see the converse, suppose that $f \in \mathrm{RBLO}(\mu)$. By Theorem 4.4.8 again, we see $\mathcal{M}(f) \in \mathrm{RBLO}(\mu)$. Let $h := f - \mathcal{M}(f)$ and $g := f$. Then Theorem 4.4.10 follows from Lemma 4.4.9. □

Now we introduce the definition of the space rblo (μ).

Definition 4.4.11. Let $\eta \in (1, \infty)$, $\rho \in [\eta, \infty)$ and $\beta_\rho := \rho^{D+1}$. A function $f \in L^1_{\mathrm{loc}}(\mu)$ is said to belong to the *space* rblo (μ) if there exists a nonnegative constant $\tilde{C}$ such that, for any cube $Q \notin \mathcal{Q}$,

$$\frac{1}{\mu(\eta Q)} \int_Q \left[f(x) - \operatorname*{ess\,inf}_{\tilde{Q}^\rho} f\right] d\mu(x) \le \tilde{C},$$

that, for any two (ρ, β_ρ)-doubling cubes $Q \subset R$ with $Q \notin \mathcal{Q}$,

$$\operatorname*{ess\,inf}_{Q} f - \operatorname*{ess\,inf}_{R} f \le \tilde{C}[1 + \delta(Q, R)], \tag{4.4.16}$$

that, for any cube $Q \in \mathcal{Q}$,

$$\frac{1}{\mu(\eta Q)} \int_Q |f(y)|\, d\mu(y) \le \tilde{C} \tag{4.4.17}$$

and that, for any cube $Q \in \mathcal{Q}$,

$$\left|\operatorname*{ess\,inf}_{\tilde{Q}^\rho} f\right| \le \tilde{C}. \tag{4.4.18}$$

Moreover, the rblo (μ) *norm* of f is defined to be the minimal constant $\tilde{C}$ as above and denoted by $\|f\|_{\mathrm{rblo}(\mu)}$.

Remark 4.4.12. Let μ be the D-dimensional Lebesgue measure on $\mathbb{R}^D$. A function $f \in L^1_{\mathrm{loc}}(\mathbb{R}^D)$ is said to belong to the *space* blo $(\mathbb{R}^D)$ if there exists a nonnegative constant C such that, for any cube Q with $\ell(Q) \le 1$,

$$\frac{1}{|Q|} \int_Q \left[f(x) - \operatorname*{ess\,inf}_{Q} f\right] dx \le C$$

and that, for any cube Q with $\ell(Q) \in (1, \infty)$,

$$\frac{1}{|Q|} \int_Q |f(y)|\, dy \le C.$$

Moreover, the blo $(\mathbb{R}^D)$ *norm* of f is defined to be the minimal constant C as above and denoted by $\|f\|_{\mathrm{blo}\,(\mathbb{R}^D)}$.

We now prove that the space rblo (μ) is independent of the chosen constants η and ρ. To this end, let $\eta \in (1, \infty)$. Suppose that, for a given $f \in L^1_{\mathrm{loc}}(\mu)$, there exist a nonnegative constant $\tilde{C}$ and a collection of numbers, $\{f_Q\}_Q$, such that

$$\sup_{Q \notin \mathcal{Q}} \frac{1}{\mu(\eta Q)} \int_Q [f(y) - f_Q]\, d\mu(y) \le \tilde{C}, \tag{4.4.19}$$

that, for any two cubes $Q \subset R$ with $Q \notin \mathcal{Q}$,

$$|f_Q - f_R| \le \tilde{C}[1 + \delta(Q, R)], \tag{4.4.20}$$

that, for any cube $Q \in \mathcal{Q}$,

$$\frac{1}{\mu(\eta Q)} \int_Q |f(y)|\, d\mu(y) + |f_Q| \le \tilde{C} \tag{4.4.21}$$

and that, for any cube Q,

$$f_Q \le \operatorname*{ess\,inf}_Q f. \tag{4.4.22}$$

We then define the *norm* $\|f\|_{**,\eta} := \inf\{\tilde{C}\}$, where the infimum is taken over all the constants $\tilde{C}$ as above and all the numbers $\{f_Q\}_Q$ satisfying (4.4.19) through (4.4.22).

Similar to Proposition 4.2.4, we have the following property on the norm $\|\cdot\|_{**,\eta}$, the details being omitted.

Proposition 4.4.13. *The norm $\|\cdot\|_{**,\eta}$ is independent of the choice of the constant $\eta \in (1, \infty)$.*

Based on Proposition 4.4.13, from now on, we write $\|\cdot\|_{**}$ *instead of* $\|\cdot\|_{**,\eta}$. The proofs of the following two propositions are slight modifications of the proofs for Propositions 4.2.5 and 4.2.7, the details being omitted.

Proposition 4.4.14. *Let η, ρ and β_ρ be as in Definition 4.4.11. Then the norms $\|\cdot\|_{**}$ and $\|\cdot\|_{\mathrm{rblo}\,(\mu)}$ are equivalent.*

Proposition 4.4.15. *Let $\eta \in (1, \infty)$, $\rho \in [\eta, \infty)$ and $\beta_\rho := \rho^{D+1}$. For $f \in L^1_{\mathrm{loc}}(\mu)$, the following statements are equivalent:*

(i) $f \in \mathrm{rblo}\,(\mu)$;
(ii) *There exists a nonnegative constant* $\tilde{C}$, *satisfying* (4.4.16) *through* (4.4.18), *such that, for any* (ρ, β_ρ)*-doubling cube* $Q \notin \mathcal{Q}$,

$$\frac{1}{\mu(Q)} \int_Q \left[f(x) - \operatorname*{ess\,inf}_{Q} f \right] d\mu(x) \leq \tilde{C}; \qquad (4.4.23)$$

(iii) *There exists a nonnegative constant* $\tilde{c}$, *satisfying* (4.4.17), (4.4.18) *and* (4.4.23), *such that, for any* (ρ, β_ρ)*-doubling cubes* $Q \subset R$ *with* $Q \notin \mathcal{Q}$,

$$m_Q(f) - m_R(f) \leq \tilde{c}[1 + \delta(Q, R)].$$

Moreover, the minimal constants $\tilde{C}$ *and* $\tilde{c}$ *as above are equivalent to* $\|f\|_{\mathrm{rblo}\,(\mu)}$ *with the equivalent positive constants independent of* f.

Remark 4.4.16. Let $\eta \in (1, \infty)$ and $\rho \in [\eta, \infty)$. From Propositions 4.4.13 and 4.4.14, it follows that the definition of the space rblo (μ) is independent of the choices of η and ρ. From now on, we *always assume* $\eta = \rho = 2$ when we consider rblo (μ).

From Definitions 4.4.1 and 4.4.11, together with Propositions 4.2.4 and 4.4.13, it is easy to see that

$$\mathrm{rblo}\,(\mu) \subset \left\{ \mathrm{RBLO}\,(\mu) \bigcap \mathrm{rbmo}\,(\mu) \right\}.$$

Therefore, as a consequence of Corollary 4.3.6, we have the following result, the details being omitted.

Corollary 4.4.17. *Let* $k \in \mathbb{N}$ *and* S_k *be as in Sect. 2.4. Then*

$$\mathrm{rblo}\,(\mu) \subset \{b \in \mathrm{RBLO}\,(\mu) : \ S_k(b) \in L^\infty(\mu)\}.$$

We now establish the relation between the space RBLO (μ) and the space rblo (μ) and some characterizations of the space rblo (μ) by some maximal function.

Proposition 4.4.18. *Let* $k \in \mathbb{N}$ *and* S_k *be as in Sect. 2.4. If* $f \in \mathrm{RBLO}\,(\mu)$, *then* $f - S_k f \in \mathrm{rblo}\,(\mu)$ *and*

$$\|f - S_k(f)\|_{\mathrm{rblo}\,(\mu)} \leq C \|f\|_{\mathrm{RBLO}\,(\mu)},$$

where C *is a positive constant independent of* k *and* f.

Proof. Without loss of generality, we may assume that $\|f\|_{\mathrm{RBLO}\,(\mu)} = 1$. We first show that, for any cube $Q \in \mathcal{Q}$,

$$\frac{1}{\mu(2Q)}\int_Q |f(x) - S_k(f)(x)|\,d\mu(x) \lesssim 1. \tag{4.4.24}$$

To this end, let us consider the following two cases:

Case (i) There exists some $x_0 \in Q \cap \operatorname{supp}\mu$ such that $Q \subset Q_{x_0,k-2}$. In this case, we have $H_Q^{x_0} \geq k-2$. On the other hand, by the fact that $Q \in \mathcal{Q}$ and Lemma 4.1.2(d), we see that $H_Q^{x_0} \leq 1$, which in turn implies that $1 \leq k \leq 3$. Moreover, from the facts that

$$-2 \leq H_Q^{x_0} - k \leq 0 \quad \text{and} \quad Q \subset Q_{x_0,k-2},$$

Lemmas 4.1.2(c) and 2.3.3, it follows that, for any $x \in \operatorname{supp}\mu \cap Q$,

$$-1 \leq H_Q^x - k + 2 \leq 3 \quad \text{and} \quad Q \subset Q_{x,k-3}.$$

By this fact, (4.4.7), Lemmas 2.1.3 and 4.1.2(g), we find that, for any $x \in \operatorname{supp}\mu \cap Q$,

$$\begin{aligned}
&\left|\operatorname*{ess\,inf}_{\tilde{Q}} f - \operatorname*{ess\,inf}_{Q_{x,k}} f\right| \\
&\quad\leq \left|\operatorname*{ess\,inf}_{\tilde{Q}} f - \operatorname*{ess\,inf}_{Q_{x,k-3}} f\right| + \left|\operatorname*{ess\,inf}_{Q_{x,k-3}} f - \operatorname*{ess\,inf}_{Q_{x,k}} f\right| \\
&\quad\lesssim 1.
\end{aligned} \tag{4.4.25}$$

For each $x \in Q \cap \operatorname{supp}\mu$, write

$$\begin{aligned}
&|f(x) - S_k(f)(x)| \\
&\quad\leq f(x) - \operatorname*{ess\,inf}_{\tilde{Q}} f + \left|\operatorname*{ess\,inf}_{\tilde{Q}} f - \operatorname*{ess\,inf}_{Q_{x,k}} f\right| + \left|\operatorname*{ess\,inf}_{Q_{x,k}} f - S_k(f)(x)\right|.
\end{aligned}$$

Notice that an easy argument, involving (b) through (d) of Theorem 2.4.4, implies that, for any $x \in \operatorname{supp}\mu$,

$$\left|\operatorname*{ess\,inf}_{Q_{x,k}} f - S_k(f)(x)\right| \lesssim 1. \tag{4.4.26}$$

Then (4.4.24) follows from the combination of (4.4.25), (4.4.26) and the following trivial fact that

$$\frac{1}{\mu(2Q)}\int_Q \left[f(x) - \operatorname*{ess\,inf}_{\tilde{Q}} f\right] d\mu(x) \leq 1.$$

Case (ii) For any $x \in Q \cap \operatorname{supp}\mu$, $Q \not\subset Q_{x,k-2}$. In this case, notice that, by Lemma 4.1.2(b), for any $x \in Q$, $Q_{x,k-1} \subset \frac{7}{5}Q$. Then, from (a) and (b) of Theorem 2.4.4, the Tonelli theorem and Proposition 4.4.3, it follows that

$$\begin{aligned}
&\frac{1}{\mu(2Q)}\int_Q |f(x) - S_k(f)(x)|\,d\mu(x)\\
&\le \frac{1}{\mu(2Q)}\int_Q \left|f(x) - \operatorname*{ess\,inf}_{\frac{7}{5}Q} f\right| d\mu(x)\\
&\quad+\frac{1}{\mu(2Q)}\int_Q \left|\operatorname*{ess\,inf}_{\frac{7}{5}Q} f - S_k(f)(x)\right| d\mu(x)\\
&\le \frac{2}{\mu(2Q)}\int_{\frac{7}{5}Q}\left[f(y) - \operatorname*{ess\,inf}_{\frac{7}{5}Q} f\right] d\mu(y)\\
&\lesssim 1.
\end{aligned}$$

Now we prove that, for any doubling cube Q,

$$m_Q(f - S_k(f)) - \operatorname*{ess\,inf}_Q [f - S_k(f)] \lesssim 1. \tag{4.4.27}$$

From Proposition 4.4.5 and (4.4.26), it follows that

$$\begin{aligned}
&m_Q(f - S_k(f)) - \operatorname*{ess\,inf}_Q [f - S_k(f)]\\
&\le \frac{1}{\mu(Q)}\int_Q \left\{[f(x) - S_k(f)(x)] - \operatorname*{ess\,inf}_Q f - \operatorname*{ess\,inf}_Q[-S_k(f)]\right\} d\mu(x)\\
&\lesssim 1 + \frac{1}{\mu(Q)}\int_Q \left\{\left[-S_k(f)(x) + \operatorname*{ess\,inf}_{Q_{x,k}} f\right]\right.\\
&\quad\left.+\left(-\operatorname*{ess\,inf}_{Q_{x,k}} f - \operatorname*{ess\,inf}_Q[-S_k(f)]\right)\right\} d\mu(x)\\
&\lesssim 1.
\end{aligned}$$

Thus, (4.4.27) holds true.

By (4.4.24) and (4.4.27), for any cube $Q \in \mathcal{Q}$, we see that

$$\begin{aligned}
&\left|\operatorname*{ess\,inf}_{\tilde{Q}}[f - S_k f]\right|\\
&\le \left|\operatorname*{ess\,inf}_{\tilde{Q}}[f - S_k f] - m_{\tilde{Q}}(f - S_k(f))\right| + \left|m_{\tilde{Q}}(f - S_k(f))\right|\\
&\lesssim 1.
\end{aligned}$$

By this, together with (4.4.24), (4.4.27) and Proposition 4.4.15, to complete the proof of Proposition 4.4.18, it remains to prove that, for any two doubling cubes $Q \subset R$ with $Q \notin \mathcal{Q}$,

$$m_Q(f - S_k(f)) - m_R(f - S_k(f)) \lesssim 1 + \delta(Q, R).$$

By Proposition 4.4.5, we first write

$$\begin{aligned} & m_Q(f - S_k(f)) - m_R(f - S_k(f)) \\ & = m_Q(f) - m_R(f) - m_Q(S_k(f)) + m_R(S_k(f)) \\ & \lesssim [1 + \delta(Q, R)] - m_Q(S_k(f)) + m_R(S_k(f)). \end{aligned}$$

As in the proof of (4.4.24), we consider the following three cases.

Case (i) There exists some $x_0 \in R \cap \operatorname{supp} \mu$ such that $R \subset Q_{x_0, k-2}$. In this case, Lemmas 2.3.3 and 2.1.3 imply that, for any $x \in Q \cap \operatorname{supp} \mu$ and $y \in R \cap \operatorname{supp} \mu$,

$$R \subset Q_{x_0, k-2} \subset Q_{x, k-3} \subset Q_{y, k-4} \subset Q_{x, k-5} \text{ and } \delta(Q_{y, k-4}, Q_{x, k-5}) \lesssim 1.$$

By this, we further conclude that

$$\begin{aligned} & \left| \operatorname*{ess\,inf}_{Q_{x,k}} f - \operatorname*{ess\,inf}_{Q_{y,k}} f \right| \\ & \le \left| \operatorname*{ess\,inf}_{Q_{x,k}} f - \operatorname*{ess\,inf}_{Q_{x,k-5}} f \right| + \left| \operatorname*{ess\,inf}_{Q_{x,k-5}} f - \operatorname*{ess\,inf}_{Q_{y,k-4}} f \right| + \left| \operatorname*{ess\,inf}_{Q_{y,k-4}} f - \operatorname*{ess\,inf}_{Q_{y,k}} f \right| \\ & \lesssim 1, \end{aligned}$$

which, together with (4.4.26), implies that

$$\begin{aligned} & -m_Q(S_k(f)) + m_R(S_k(f)) \\ & \le \frac{1}{\mu(Q)} \frac{1}{\mu(R)} \int_Q \int_R \left\{ \left| S_k(f)(x) - \operatorname*{ess\,inf}_{Q_{x,k}} f \right| + \left| \operatorname*{ess\,inf}_{Q_{x,k}} f - \operatorname*{ess\,inf}_{Q_{y,k}} f \right| \right. \\ & \quad \left. + \left| \operatorname*{ess\,inf}_{Q_{y,k}} f - S_k(f)(y) \right| \right\} d\mu(x)\, d\mu(y) \\ & \lesssim 1. \end{aligned}$$

Case (ii) For any $x \in R \cap \operatorname{supp} \mu$, $Q \not\subset Q_{x, k-2}$. In this case, for any $x \in R \cap \operatorname{supp} \mu$, it holds true that $R \not\subset Q_{x, k-2}$. From Lemma 4.1.2(b), we deduce that, for any $x \in Q \cap \operatorname{supp} \mu$ and $y \in R \cap \operatorname{supp} \mu$,

$$Q_{x,k} \subset Q_{x,k-1} \subset \frac{7}{5} Q \quad \text{and} \quad Q_{y,k} \subset \frac{7}{5} R.$$

By the Tonelli theorem and Lemma 2.1.3, we see that

$$\begin{aligned}
&-m_Q(S_k(f)) + m_R(S_k(f)) \\
&\quad \le \frac{1}{\mu(Q)}\frac{1}{\mu(R)}\int_Q\int_R \left\{ \left| S_k(f)(x) - \operatorname*{ess\,\widetilde{inf}}_{\frac{7}{5}Q} f \right| + \left| \operatorname*{ess\,\widetilde{inf}}_{\frac{7}{5}Q} f - \operatorname*{ess\,\widetilde{inf}}_{\frac{7}{5}R} f \right| \right. \\
&\qquad \left. + \left| \operatorname*{ess\,\widetilde{inf}}_{\frac{7}{5}R} f - S_k(f)(y) \right| \right\} d\mu(x)\, d\mu(y) \\
&\quad \lesssim 1 + \delta(Q, R).
\end{aligned}$$

Case (iii) For any $x \in R \cap \operatorname{supp}\mu$, $R \not\subset Q_{x,k-2}$ and there exists some point $x_0 \in R \cap \operatorname{supp}\mu$ such that $Q \subset Q_{x_0,k-2}$. In this case, Lemma 2.3.3 implies that, for any $x \in Q \cap \operatorname{supp}\mu$,

$$Q \subset Q_{x_0,k-2} \subset Q_{x,k-3}$$

and, moreover, Lemma 4.1.2(b) implies that, for any $x \in Q \cap \operatorname{supp}\mu$,

$$Q_{x,k} \subset Q_{x,k-1} \subset \frac{7}{5}R.$$

By these facts, Lemma 2.1.3(e) and (4.4.7), we see that, for any $x \in Q \cap \operatorname{supp}\mu$,

$$\left| \operatorname*{ess\,inf}_{Q_{x,k}} f - \operatorname*{ess\,\widetilde{inf}}_{\frac{7}{5}R} f \right| \le 1 + \delta\left(Q_{x,k}, \frac{7}{5}R\right) \lesssim 1 + \delta(Q, R).$$

From this, (4.4.26) and the Tonelli theorem, we deduce that

$$\begin{aligned}
&-m_Q(S_k(f)) + m_R(S_k(f)) \\
&\quad \le \frac{1}{\mu(Q)}\frac{1}{\mu(R)}\int_Q\int_R \left\{ \left| S_k(f)(x) - \operatorname*{ess\,inf}_{Q_{x,k}} f \right| + \left| \operatorname*{ess\,inf}_{Q_{x,k}} f - \operatorname*{ess\,\widetilde{inf}}_{\frac{7}{5}R} f \right| \right. \\
&\qquad \left. + \left| \operatorname*{ess\,\widetilde{inf}}_{\frac{7}{5}R} f - S_k(f)(y) \right| \right\} d\mu(x)\, d\mu(y) \\
&\quad \lesssim 1 + \delta(Q, R),
\end{aligned}$$

which completes the proof of Proposition 4.4.18. □

We next introduce the *local natural maximal operator*, which is a local variant of $\mathcal{M}$. For any locally integrable function f and $x \in \mathbb{R}^D$, let

$$\mathcal{M}_l(f)(x) := \sup_{\substack{Q \ni x,\, Q \notin \mathcal{Q} \\ Q \text{ doubling}}} \frac{1}{\mu(Q)} \int_Q f(y)\, d\mu(y).$$

The following lemma is a local version of Lemma 4.4.9 and its proof omitted.

Lemma 4.4.19. *Let f be a locally integrable function. Then $f \in \mathrm{rblo}\,(\mu)$ if and only if*

$$\mathcal{M}_l(f) - f \in L^\infty(\mu)$$

and f satisfies (4.4.16), (4.4.17) *and* (4.4.18) *with $\eta = 2 = \rho$. Furthermore,*

$$\|\mathcal{M}_l(f) - f\|_{L^\infty(\mu)} \sim \|f\|_{\mathrm{rblo}\,(\mu)}$$

with the implicit equivalent positive constants independent of f.

Lemma 4.4.20. *If $f \in \mathrm{rbmo}\,(\mu)$, then there exists a nonnegative constant $\tilde{C}$, independent of f, such that, for any doubling cube $Q \notin \mathcal{Q}$,*

$$\frac{1}{\mu(Q)} \int_Q \mathcal{M}_l(f)(x)\, d\mu(x) \le \tilde{C} \|f\|_{\mathrm{rbmo}\,(\mu)} + \operatorname*{ess\,inf}_Q \mathcal{M}_l(f).$$

Moreover, if $\mathcal{M}_l(f)$ is μ-almost everywhere finite, then

$$\frac{1}{\mu(Q)} \int_Q \mathcal{M}_l(f)(x)\, d\mu(x) - \operatorname*{ess\,inf}_Q \mathcal{M}_l(f) \le \tilde{C} \|f\|_{\mathrm{rbmo}\,(\mu)}.$$

Proof. Fix $f \in \mathrm{rbmo}\,(\mu)$. Without loss of generality, we may assume that

$$\|f\|_{\mathrm{rbmo}\,(\mu)} = 1.$$

For any doubling cube $Q \notin \mathcal{Q}$, write

$$f = [f - m_Q(f)]\chi_{\frac{4}{3}Q} + m_Q(f)\chi_{\frac{4}{3}Q} + f\chi_{\mathbb{R}^D \setminus \frac{4}{3}Q}.$$

Obviously, $\mathcal{M}_l$ is bounded on $L^2(\mu)$ since $\mathcal{M}_l(f) \le \mathcal{M}^d(f)$. Then, by this fact, the Hölder inequality, the doubling property of Q, Corollary 4.2.9 and Lemma 2.1.3, we conclude that

$$\begin{aligned}
&\int_Q \mathcal{M}_l\left[(f - m_Q(f))\chi_{\frac{4}{3}Q}\right](x)\, d\mu(x) \\
&\quad \lesssim \left\{\int_{\frac{4}{3}Q} |f(x) - m_Q(f)|^2\, d\mu(x)\right\}^{1/2} [\mu(Q)]^{1/2}
\end{aligned}$$

$$\lesssim \left\{\int_{\frac{4}{3}Q} \left|f(x) - m_{\widetilde{\frac{4}{3}Q}}(f)\right|^2 d\mu(x)\right\}^{1/2} [\mu(Q)]^{1/2}$$
$$+\mu(Q)\left|m_Q(f) - m_{\widetilde{\frac{4}{3}Q}}(f)\right|$$
$$\lesssim \mu(Q).$$

From this, it follows that

$$\frac{1}{\mu(Q)}\int_Q \mathcal{M}_l\left[(f - m_Q(f))\chi_{\frac{4}{3}Q}\right](x)\,d\mu(x) \lesssim 1.$$

Therefore, with the aid of Proposition 4.4.15, Lemma 4.4.20 is reduced to proving that there exists a positive constant C such that, for μ-almost every $x \in Q$,

$$\mathcal{M}_l\left[m_Q(f)\chi_{\frac{4}{3}Q} + f\chi_{\mathbb{R}^D\setminus\frac{4}{3}Q}\right](x) \le C + \operatorname*{ess\,inf}_Q \mathcal{M}_l(f). \tag{4.4.28}$$

For any doubling cube R containing x with $R \notin Q$ and any $y \in Q$, if $R \subset \frac{4}{3}Q$, then

$$\mathrm{E} := \frac{1}{\mu(R)}\int_R \left[m_Q(f)\chi_{\frac{4}{3}Q}(z) + f(z)\chi_{\mathbb{R}^D\setminus\frac{4}{3}Q}(z)\right] d\mu(z) - \mathcal{M}_l(f)(y)$$
$$\le m_Q(f) - \mathcal{M}_l(f)(y)$$
$$\le 0.$$

Assume that $R \not\subset \frac{4}{3}Q$ now. We then see that $\ell(R) \ge \frac{1}{6}\ell(Q)$. There exist two cases.

Case (i) $\ell(R) \le 4\ell(Q)$. In this case, it is easy to see that

$$Q \subset 13R \subset 57Q$$

with the aid of the fact that $\ell(R) \ge \frac{1}{6}\ell(Q)$. From this fact, Proposition 4.2.7, (4.2.7) and Lemma 2.1.3, it follows that

$$\mathrm{E} = \frac{1}{\mu(R)}\int_R [f(z) - m_Q(f)]\,\chi_{\mathbb{R}^D\setminus\frac{4}{3}Q}(z)\,d\mu(z) + m_Q(f) - \mathcal{M}_l(f)(y)$$
$$\le \frac{1}{\mu(R)}\int_R [|f(z) - m_R(f)| + |m_R(f) - m_{\widetilde{13R}}(f)|$$
$$+ |m_{\widetilde{13R}}(f) - m_Q(f)|]\,d\mu(z)$$
$$\lesssim 1 + \delta(Q, 13R)$$
$$\lesssim 1.$$

Case (ii) $\ell(R) > 4\ell(Q)$. In this case, Lemma 4.1.2(a) and (d) imply that

$$Q \subset \frac{3}{2}R \subset Q_{z, H_R^z - 1}$$

and $H_R^z \in \mathbb{Z}_+$ for any $z \in R \cap \operatorname{supp} \mu$. Let

$$R_1 := \{z \in R : Q_{z, H_R^z - 1} \notin \mathcal{Q}\}$$

and $R_2 := R \setminus R_1$. Then we write

$$\begin{aligned}
\mathrm{E} &= \frac{1}{\mu(R)} \int_{R_1} \left[m_Q(f)\chi_{\frac{4}{3}Q}(z) + f(z)\chi_{\mathbb{R}^D \setminus \frac{4}{3}Q}(z) - \mathcal{M}_l(f)(y) \right] d\mu(z) \\
&\quad + \frac{1}{\mu(R)} \int_{R_2} \cdots \\
&=: \mathrm{E}_1 + \mathrm{E}_2.
\end{aligned}$$

If $z \in R_1$, then

$$m_{Q_{z, H_R^z - 1}}(f) \le \mathcal{M}_l(f)(y) \quad \text{since} \quad y \in Q_{z, H_R^z - 1}.$$

Therefore, by (4.2.7), Proposition 4.2.7, Lemmas 2.1.3 and 4.1.2(g), and the doubling property of Q and R, we conclude that

$$\begin{aligned}
\mathrm{E}_1 &\le \frac{1}{\mu(R)} \int_{R_1} \left[\left| m_Q(f) - m_{\widetilde{\frac{3}{2}R}}(f) \right| \chi_{\frac{4}{3}Q}(z) + \left| f(z) - m_{\widetilde{\frac{3}{2}R}}(f) \right| \chi_{\mathbb{R}^D \setminus \frac{4}{3}Q}(z) \right. \\
&\quad \left. + \left| m_{\widetilde{\frac{3}{2}R}}(f) - m_{Q_{z, H_R^z - 1}}(f) \right| + m_{Q_{z, H_R^z - 1}}(f) - \mathcal{M}_l(f)(y) \right] d\mu(z) \\
&\lesssim \frac{\mu(\frac{4}{3}Q)}{\mu(R)} \left| m_Q(f) - m_{\widetilde{\frac{3}{2}R}}(f) \right| + \frac{1}{\mu(R)} \int_R \left[1 + \delta\left(\frac{3}{2}R, Q_{z, H_R^z - 1}\right) \right] d\mu(z) \\
&\lesssim \frac{1}{\mu(R)} \int_{\frac{3}{2}R} \left| f(z) - m_{\widetilde{\frac{3}{2}R}}(f) \right| d\mu(z) + 1 \\
&\lesssim 1.
\end{aligned}$$

On the other hand, for $z \in R_2$, Lemma 4.1.2(d) implies that $H_R^z \le 2$, from which it follows that $Q_{y,2} \subset Q_{y, H_R^z - 1}$. The fact that $y \in (Q_{y, H_R^z - 1} \cap Q_{z, H_R^z - 1})$, together with Lemma 2.3.3, implies that

$$Q_{y, H_R^z - 1} \subset Q_{z, H_R^z - 2} \subset Q_{y, H_R^z - 3}.$$

Moreover, by the fact that $H_R^z \in \mathbb{Z}_+$, we find that $Q_{y, H_R^z - 3} \subset Q_{y, -3}$ (Recall that by Definition 2.2.4, if $\mathbb{R}^D$ is an initial cube, then, for any $y \in \operatorname{supp} \mu$ and $k \le 0$,

$Q(y, k) = \mathbb{R}^D$). Thus, from the fact that

$$m_{Q_{y,2}}(f) \le \mathcal{M}_l(f)(y),$$

(4.2.7), Proposition 4.2.7, Lemmas 2.1.3 and 4.1.2(g), and the doubling property of Q and R, it follows that

$$\begin{aligned} \mathrm{E}_2 &\le \frac{1}{\mu(R)} \int_{R_2} \Big[\Big| m_Q(f) - m_{\widetilde{\frac{3}{2}R}}(f) \Big| \chi_{\frac{4}{3}Q}(z) + \Big| f(z) - m_{\widetilde{\frac{3}{2}R}}(f) \Big| \chi_{\mathbb{R}^D \setminus \frac{4}{3}Q}(z) \\ &\quad + \Big| m_{\widetilde{\frac{3}{2}R}}(f) - m_{Q_{z, H_R^z - 2}}(f) \Big| + \Big| m_{Q_{z, H_R^z - 2}}(f) - m_{Q_{y,2}}(f) \Big| \Big] \, d\mu(z) \\ &\lesssim \frac{1}{\mu(R)} \int_R [1 + \delta(Q_{y,2}, Q_{z, H_R^z - 2})] \, d\mu(z) \\ &\lesssim 1. \end{aligned}$$

Then (4.4.28) holds true, which completes the proof of Lemma 4.4.20. □

The following Theorems 4.4.21 and 4.4.22 are local variants of Theorems 4.4.8 and 4.4.10. We remark that, unlike the case RBLO (μ), if $f \in$ rbmo (μ), then $\mathcal{M}_l(f)$ is finite almost everywhere.

Theorem 4.4.21. *$\mathcal{M}_l$ is bounded from* rbmo (μ) *to* rblo (μ), *namely, there exists a positive constant C such that, for all $f \in$* rbmo (μ),

$$\|\mathcal{M}_l(f)\|_{\mathrm{rblo}\,(\mu)} \le C \|f\|_{\mathrm{rbmo}\,(\mu)}.$$

Proof. Fix $f \in$ rbmo (μ). By the homogeneity of $\mathcal{M}_l$, we only need to prove the conclusion of Theorem 4.4.21 for $\|f\|_{\mathrm{rbmo}\,(\mu)} = 1$. We first prove that, for any cube $Q \in \mathcal{Q}$,

$$\frac{1}{\mu(2Q)} \int_Q |\mathcal{M}_l(f)(x)| \, d\mu(x) \lesssim 1. \tag{4.4.29}$$

Write

$$f = f\chi_{\frac{4}{3}Q} + f\chi_{\mathbb{R}^D \setminus \frac{4}{3}Q}.$$

Then, from the Hölder inequality, the boundedness of $\mathcal{M}_l$ in $L^2(\mu)$ and Corollary 4.2.9, we deduce that

$$\begin{aligned} \frac{1}{\mu(2Q)} \int_Q \Big| \mathcal{M}_l \left(f\chi_{\frac{4}{3}Q} \right)(x) \Big| \, d\mu(x) &\lesssim \left\{ \frac{1}{\mu(2Q)} \int_{\frac{4}{3}Q} |f(x)|^2 \, d\mu(x) \right\}^{1/2} \\ &\lesssim 1. \end{aligned}$$

On the other hand, for any $x \in Q$ and doubling cube $P \ni x$ with

$$P \notin \mathcal{Q} \quad \text{and} \quad P \bigcap \left(\mathbb{R}^D \setminus \frac{4}{3}Q\right) \neq \emptyset,$$

it is easy to see that $\ell(P) \geq \frac{1}{6}\ell(Q)$. This implies that $Q \subset 13P$ and hence $\widetilde{13P} \in \mathcal{Q}$. Therefore, from Proposition 4.2.7 and Lemma 2.1.3, we deduce that

$$\begin{aligned}
&\frac{1}{\mu(P)} \int_P |f(z)|\, d\mu(z) \\
&\leq \frac{1}{\mu(P)} \int_P |f(z) - m_P(f)|\, d\mu(z) + \left|m_P(f) - m_{\widetilde{13P}}(f)\right| + \left|m_{\widetilde{13P}}(f)\right| \\
&\lesssim 1.
\end{aligned}$$

This further implies that (4.4.29) holds true and hence $\mathcal{M}_l(f)$ is finite almost everywhere.

Now we prove that, for any doubling cubes $Q \subset R$ with $Q \notin \mathcal{Q}$,

$$m_Q(\mathcal{M}_l(f)) - m_R(\mathcal{M}_l(f)) \lesssim 1 + \delta(Q, R). \tag{4.4.30}$$

Let

$$\begin{aligned}
Q_1 := \{x \in Q :\ &\text{for any doubling cube } P \text{ containing } x, \\
&\text{if } \ell(P) > 4\ell(R), \text{ then } P \in \mathcal{Q}\}
\end{aligned}$$

and $Q_2 := Q \setminus Q_1$. Moreover, for any $x \in Q$, let

$$\mathcal{M}_l^1(f)(x) := \sup_{\substack{P \ni x,\ P \text{ doubling} \\ P \notin \mathcal{Q} \text{ and } \ell(P) \leq 4\ell(R)}} \frac{1}{\mu(P)} \int_P f(y)\, d\mu(y)$$

and, for any $x \in Q_2$, let

$$\mathcal{M}_l^2(f)(x) := \sup_{\substack{P \ni x,\ P \text{ doubling} \\ P \notin \mathcal{Q} \text{ and } \ell(P) > 4\ell(R)}} \frac{1}{\mu(P)} \int_P f(y)\, d\mu(y).$$

Let

$$\mathcal{U}_{1,Q} := \{x \in Q_2 : \mathcal{M}_l^1(f)(x) \geq \mathcal{M}_l^2(f)(x)\} \quad \text{and} \quad \mathcal{U}_{2,Q} := Q_2 \setminus \mathcal{U}_{1,Q}.$$

Then, for any $x \in (Q_1 \cup \mathcal{U}_{1,Q})$,

$$\mathcal{M}_l(f)(x) = \mathcal{M}_l^1(f)(x)$$

and, for any $x \in \mathcal{U}_{2,Q}$,

$$\mathcal{M}_l(f)(x) = \mathcal{M}_l^2(f)(x).$$

By writing

$$f = [f - m_R(f)]\chi_{\frac{4}{3}Q} + [f - m_R(f)]\chi_{\mathbb{R}^D \setminus \frac{4}{3}Q} + m_R(f)$$

and using the fact that $m_R(f) \le m_R(\mathcal{M}_l(f))$, we see that

$$\begin{aligned}
&m_Q(\mathcal{M}_l(f)) - m_R(\mathcal{M}_l(f)) \\
&\le \frac{1}{\mu(Q)} \int_{(Q_1 \bigcup \mathcal{U}_{1,Q})} \mathcal{M}_l^1([f - m_R(f)]\chi_{\frac{4}{3}Q})(x)\, d\mu(x) \\
&\quad + \frac{1}{\mu(Q)} \int_{(Q_1 \bigcup \mathcal{U}_{1,Q})} \mathcal{M}_l^1([f - m_R(f)]\chi_{\mathbb{R}^D \setminus \frac{4}{3}Q})(x)\, d\mu(x) \\
&\quad + \frac{1}{\mu(Q)} \int_{\mathcal{U}_{2,Q}} \{\mathcal{M}_l^2(f)(x) - m_R[\mathcal{M}_l(f)]\}\, d\mu(x) \\
&=: \mathrm{F}_1 + \mathrm{F}_2 + \mathrm{F}_3.
\end{aligned}$$

Using the estimate for I_1 in the proof of Theorem 4.4.8, we know that

$$\mathrm{F}_1 \lesssim 1 + \delta(Q, R).$$

On the other hand, using an argument similar to that used in the estimate (4.4.13), we conclude that, for any point $x \in (Q_1 \cup \mathcal{U}_{1,Q})$ and doubling cube $P \ni x$ with $P \notin Q$ and $\ell(P) \le 4\ell(R)$,

$$\frac{1}{\mu(P)} \int_P |f(y) - m_R(f)| \chi_{\mathbb{R}^D \setminus \frac{4}{3}Q}(y)\, d\mu(y) \lesssim 1 + \delta(Q, R).$$

This implies that

$$\mathrm{F}_2 \lesssim 1 + \delta(Q, R).$$

To estimate F_3, it suffices to prove that, for any doubling cube P containing x with $P \notin Q$ and $\ell(P) > 4\ell(R)$,

$$m_P(f) - m_R(\mathcal{M}_l(f)) \lesssim 1.$$

For any $z \in P \cap \operatorname{supp} \mu$, (a) and (d) of Lemma 4.1.2 imply that

$$R \subset \frac{3}{2}P \subset Q_{z, H_P^z - 1} \quad \text{and} \quad H_P^z \in \mathbb{Z}_+.$$

Let

$$R_1 := \{y \in R : \text{ there exists a point } z_y \in P \text{ such that } Q_{z_y, H_P^{z_y}-1} \notin \mathcal{Q}\}$$

and $R_2 := R \setminus R_1$. Then, we have

$$\begin{aligned}
&m_P(f) - m_R(\mathcal{M}_l(f)) \\
&= \frac{1}{\mu(R)} \int_{R_1} [m_P(f) - \mathcal{M}_l(f)(y)]\, d\mu(y) + \frac{1}{\mu(R)} \int_{R_2} \cdots .
\end{aligned}$$

Observe that, for any $y \in R_1$,

$$m_{Q_{z_y, H_P^{z_y}-1}}(f) \le \mathcal{M}_l(f)(y).$$

This, together with (4.2.7), Lemmas 2.1.3 and 4.1.2(g), implies that

$$\begin{aligned}
m_P(f) - \mathcal{M}_l(f)(y) &\le \left| m_P(f) - m_{Q_{z_y, H_P^{z_y}-1}}(f) \right| \\
&\quad + m_{Q_{z_y, H_P^{z_y}-1}}(f) - \mathcal{M}_l(f)(y) \\
&\le \left| m_P(f) - m_{\widetilde{\frac{3}{2}P}}(f) \right| + \left| m_{\widetilde{\frac{3}{2}P}}(f) - m_{Q_{z_y, H_P^{z_y}-1}}(f) \right| \\
&\lesssim 1.
\end{aligned}$$

On the other hand, by Lemma 4.1.2(d), we see that $H_P^z \le 2$ for any $z \in (P \cap \operatorname{supp} \mu)$. Moreover, for any $y \in (R_2 \cap \operatorname{supp} \mu)$, an easy argument, involving the facts that $y \in (Q_{y, H_P^z-1} \cap Q_{z, H_P^z-1})$ and $H_P^z \ge 0$, and Lemma 2.3.3, implies that

$$Q_{y,2} \subset Q_{y, H_P^z-1} \subset Q_{z, H_P^z-2} \subset Q_{y, H_P^z-3} \subset Q_{y,-3}.$$

Thus, from Lemma 4.1.2(g) and the fact that

$$\mathcal{M}_l(f)(y) \ge m_{Q_{y,2}}(f),$$

it follows that

$$\begin{aligned}
m_P(f) - \mathcal{M}_l(f)(y) &\le \left| m_P(f) - m_{Q_{z, H_P^z-2}}(f) \right| + \left| m_{Q_{z, H_P^z-2}}(f) - m_{Q_{y,2}}(f) \right| \\
&\lesssim 1 + \delta(P, Q_{z, H_P^z-2}) + \delta(Q_{y,2}, Q_{z, H_P^z-2}) \\
&\lesssim 1.
\end{aligned}$$

Therefore, combining these estimates above, we conclude that, for any doubling cube $P \ni x$ with $P \notin \mathcal{Q}$ and $\ell(P) > 4\ell(R)$,

$$m_P(f) - m_R[\mathcal{M}_l(f)] \lesssim 1,$$

which implies that $F_3 \lesssim 1$. The combination of estimates for F_1 through F_3 implies (4.4.30).

By Lemma 4.4.20, to finish the proof of Theorem 4.4.21, we need to show that, for any cube $Q \in \mathcal{Q}$,

$$\left| \operatorname*{ess\,inf}_{\tilde{Q}} \mathcal{M}_l(f) \right| \lesssim 1. \tag{4.4.31}$$

If $\operatorname{ess\,inf}_{\tilde{Q}} \mathcal{M}_l(f) \geq 0$, then (4.4.29) implies (4.4.31). Assume that

$$\operatorname*{ess\,inf}_{\tilde{Q}} \mathcal{M}_l(f) < 0.$$

Then we see that

$$\left| \operatorname*{ess\,inf}_{\tilde{Q}} \mathcal{M}_l(f) \right| = \operatorname*{ess\,sup}_{\tilde{Q}} \{-\mathcal{M}_l(f)\}.$$

Recall that, for any $x \in \operatorname{supp} \mu$, $Q_{x,2} \notin \mathcal{Q}$ and $Q_{x,1} \in \mathcal{Q}$ (see (d) and (e) of Lemma 4.1.2). By these facts and Proposition 4.2.3, for all $x \in \operatorname{supp} \mu$, we know that

$$\begin{aligned} -\mathcal{M}_l(f)(x) &\leq \inf_{\substack{P \ni x,\, P \notin \mathcal{Q} \\ P \text{ doubling}}} m_P(|f|) \\ &\leq \left| m_{Q_{x,2}}(|f|) - m_{Q_{x,1}}(|f|) \right| + m_{Q_{x,1}}(|f|) \\ &\lesssim 1, \end{aligned}$$

which completes the proof of (4.4.31) and hence Theorem 4.4.21. □

The following theorem is similar to Theorem 4.4.10 and its proof omitted.

Theorem 4.4.22. *A locally integrable function f belongs to* rblo (μ) *if and only if there exist $h \in L^\infty(\mu)$ and $g \in$* rbmo (μ) *such that*

$$f = \mathcal{M}_l(g) + h. \tag{4.4.32}$$

Furthermore,

$$\|f\|_{\mathrm{rblo}\,(\mu)} \sim \inf\{\|g\|_{\mathrm{rbmo}\,(\mu)} + \|h\|_{L^\infty(\mu)}\}$$

with the implicit equivalent positive constants independent of f, g and h, where the infimum is taken over all representations of f as in (4.4.32).

4.5 Notes

- The local atomic Hardy space $h^{1,p}_{\mathrm{atb}}(\mu)$ and rbmo(μ) were introduced by Goldberg in [38] when μ is the D-dimensional Lebesgue measure and by Hu et al. in [66] when μ only satisfies the polynomial growth condition (0.0.1). Versions of the local Hardy space and the local BMO-type space were also introduced by Yang in [151].
- Theorem 4.2.8 was proved in [66].
- Theorem 4.2.17 was established in [66].
- Theorem 4.3.1 was given by Da. Yang and Do. Yang in [152].
- The space RBLO(μ) was firstly introduced by Jiang in [72] in the following way: A function $f \in L^1_{\mathrm{loc}}(\mu)$ is said to belong to the *space* RBLO(μ), if it satisfies (4.4.8) and (4.4.9) with (ρ, β_ρ)-doubling cube replaced by

$$(4\sqrt{D}, (4\sqrt{D})^{n+1})\text{–doubling cube.}$$

  Definition 4.4.1 was given in [66]. Due to the existence of small (ρ, ρ^{D+1})-doubling cubes, where $\rho \in (1,\infty)$, it seems that Definition 4.4.1 is more convenient in applications. Moreover, Definition 4.4.1 is convenient in proving that the space RBLO(μ) is independent of the choices of the constants η, $\rho \in (1,\infty)$; see Propositions 4.4.3 and 4.4.4.
- The natural maximal operator $\mathcal{M}$ is a variant in the non-doubling context of the so-called natural maximal operator on $\mathbb{R}^D$ in [2, 107], which was introduced by Jiang in [72].
- If μ is the D-dimensional Lebesgue measure, Theorem 4.4.8 was obtained by Bennett in [2]. When μ only satisfies (0.0.1), Theorems 4.4.8 and 4.4.10 were first established by Jiang [72] and improved in [66] by proving that, under the assumption that $f \in$ RBMO(μ), (4.4.12) holds true automatically and $\mathcal{M}(f)$ is finite almost everywhere.
- From Theorem 4.4.8, Proposition 3.1.3(iii) and the fact that $\mathcal{M}^d(f) = \mathcal{M}(|f|)$, it follows that, if $f \in$ RBMO(μ), then $\mathcal{M}^d(f)$ is either infinite everywhere or finite almost everywhere and, in the latter case, $\mathcal{M}^d(f) \in$ RBLO(μ) and there exists a positive constant C, independent of f, such that

$$\|\mathcal{M}^d(f)\|_{\mathrm{RBLO}(\mu)} \le C\|f\|_{\mathrm{RBMO}(\mu)}.$$

  If μ is the D-dimensional Lebesgue measure and RBLO(μ) is replaced by RBMO(μ), this conclusion was obtained by Bennett, DeVore and Sharpley in [3].
- Theorems 4.4.21 and 4.4.22 were established in [66].

Chapter 5
Boundedness of Operators over $(\mathbb{R}^D, \mu)$

In this chapter, we focus our attention on the boundedness of singular integral operators in the Lebesgue space and the Hardy space over $(\mathbb{R}^D, \mu)$. We first establish some weighted estimates for local sharp maximal operator as well as several interpolation results which are useful in applications. Then we investigate the boundedness of singular integral operators on $L^p(\mu)$ and $H^1(\mu)$, and the boundedness of maximal singular integral operators and commutators on $L^p(\mu)$ as well as their endpoint estimates.

5.1 The Local Sharp Maximal Operator

As is well known, in the Euclidean space, the sharp maximal operator of Fefferman and Stein and its local version, which link Calderón–Zygmund operators and the Hardy–Littlewood maximal operator, play important roles in the study of boundedness of operators on function spaces. By the good-λ inequality, we know that, for suitable functions f and $p \in (0, \infty)$, the $L^p(\mathbb{R}^D)$ norm of the Hardy–Littlewood maximal function of f, in some sense, is equivalent to that of the sharp maximal function of f. Our goal in this section is to show that the John–Strömberg sharp maximal operator, defined by (5.1.6) below, enjoys the same properties.

For a measurable function f and any cube Q, $m_f(Q)$, the *median value of f on the cube Q*, is defined as in Sect. 3.1. Let

$$\rho \in [1, \infty), \quad \beta_{2\rho} := (2\rho)^{D+1} \quad \text{and} \quad s \in (0, \beta_{2\rho}^{-1}/4).$$

In what follows, we call a $(2\rho, \beta_{2\rho})$-doubling cube simply by a $(2\rho, \beta_{2\rho})$ *-doubling cube*. For any fixed cube Q and μ-locally integrable function f, define $m_{0,s;Q}^{\rho}(f)$ by

D. Yang et al., *The Hardy Space H^1 with Non-doubling Measures and Their Applications*, Lecture Notes in Mathematics 2084, DOI 10.1007/978-3-319-00825-7_5,

$$m^{\rho}_{0,s;Q}(f) := \inf\left\{t \in (0,\infty) : \mu\left(\{y \in Q : |f(y)| > t\}\right) < s\mu\left(\frac{3}{2}\rho Q\right)\right\} \quad (5.1.1)$$

when $\mu(Q) \neq 0$, and $m^{\rho}_{0,s;Q}(f) := 0$ when $\mu(Q) = 0$.

For the preliminary properties of $m^{\rho}_{0,s,Q}$ and $m_f(Q)$, we have the following two lemmas.

Lemma 5.1.1. *Let $\rho \in [1,\infty)$, $s \in (0, \beta_{2\rho}^{-1}/4)$ and Q be a 2ρ-doubling cube with $\mu(Q) \neq 0$. For any constant $c \in \mathbb{C}$ and μ-locally integrable function f, it holds true that*

$$\left|m^{\rho}_{0,s;Q}(f) - |c|\right| \leq m^{\rho}_{0,s;Q}(f - c).$$

Proof. To prove Lemma 5.1.1, it suffices to prove that

$$m^{\rho}_{0,s;Q}(f) + m^{\rho}_{0,s;Q}(f - c) \geq |c|$$

and

$$m^{\rho}_{0,s;Q}(f) \leq m^{\rho}_{0,s;Q}(f - c) + |c|.$$

To show the first one, we observe that, if $t_1,\, t_2 \in (0,\infty)$ satisfy that

$$\mu\left(\{y \in Q : |f(y) - c| > t_1\}\right) < s\mu\left(\frac{3}{2}\rho Q\right)$$

and

$$\mu\left(\{y \in Q : |f(y)| > t_2\}\right) < s\mu\left(\frac{3}{2}\rho Q\right),$$

then $t_1 + t_2 \geq |c|$. Otherwise, we have

$$\begin{aligned}\mu\left(\{y \in Q : |f(y)| \leq t_2\}\right) &\leq \mu\left(\{y \in Q : |f(y) - c| > t_1\}\right)\\ &< s\mu\left(\frac{3}{2}\rho Q\right)\\ &< \frac{1}{2}\mu(Q),\end{aligned}$$

which is contradictory to the fact that

$$\mu(\{y \in Q : |f(y)| \le t_2\}) > \mu(Q) - s\mu\left(\frac{3}{2}\rho Q\right) \ge \frac{1}{2}\mu(Q).$$

Therefore,

$$m^{\rho}_{0,s;Q}(f) + m^{\rho}_{0,s;Q}(f-c) \ge |c|.$$

To see the second one, for any $\epsilon \in (0,\infty)$, we take $t_0 \in (0,\infty)$ such that

$$t_0 < m^{\rho}_{0,s;Q}(f-c) + \epsilon$$

and

$$\mu(\{y \in Q : |f(y) - c| > t_0\}) < s\mu\left(\frac{3}{2}\rho Q\right).$$

This in turn implies that

$$\mu(\{y \in Q : |f(y)| > t_0 + |c|\}) < s\mu\left(\frac{3}{2}\rho Q\right).$$

Thus,

$$m^{\rho}_{0,s;Q}(f) \le t_0 + |c| < m^{\rho}_{0,s;Q}(f-c) + |c| + \epsilon,$$

which implies the desired estimate by letting $\epsilon \to 0$ and hence completes the proof of Lemma 5.1.1. □

Lemma 5.1.2. *Let $\rho, p \in [1,\infty)$, $s \in (0, \beta_{2\rho,d}^{-1}/4)$ and Q be a 2ρ-doubling cube. Then, for any μ-locally integrable real-valued function f, it holds true that*

$$|m_f(Q)| \le m^{\rho}_{0,s;Q}(f).$$

Proof. Without loss of generality, we may assume that $\mu(Q) > 0$. If f is real-valued and $m_f(Q) \ge 0$, we see that

$$\begin{aligned}&\{y \in Q : |f(y)| \ge |m_f(Q)|\}\\ &\quad = \{y \in Q : f(y) \ge m_f(Q)\} \bigcup \{y \in Q : f(y) \le -m_f(Q)\}\end{aligned}$$

and, if $m_f(Q) < 0$, then

$$\begin{aligned}&\{y \in Q : |f(y)| \ge |m_f(Q)|\}\\ &\quad = \{y \in Q : f(y) \ge -m_f(Q)\} \bigcup \{y \in Q : f(y) \le m_f(Q)\}.\end{aligned}$$

Therefore, by the definition of $m_f(Q)$, we know that

$$\begin{aligned}
&\mu(\{y \in Q : |f(y)| \ge |m_f(Q)|\}) \\
&\quad \ge \min\{\mu(\{y \in Q : f(y) \ge m_f(Q)\}), \mu(\{y \in Q : f(y) \le m_f(Q)\})\} \\
&\quad \ge \frac{1}{2}\mu(Q). \qquad (5.1.2)
\end{aligned}$$

This implies that, for any $t \in (0, \infty)$ satisfying

$$\mu(\{y \in Q : |f(y)| > t\}) < s\mu\left(\frac{3}{2}\rho Q\right),$$

it holds true that $t \ge |m_f(Q)|$. Otherwise, we then obtain

$$\mu(\{y \in Q : |f(y)| \ge |m_f(Q)|\}) < s\mu(2\rho Q) \le s\beta_{2\rho}\mu(Q) < \frac{1}{4}\mu(Q),$$

which is contradictory to (5.1.2). By taking the infimum over t, we then obtain the desired conclusion, which completes the proof of Lemma 5.1.2. □

The *John–Strömberg maximal operators* $\mathcal{M}^{\rho}_{0,s}$ and $\mathcal{M}^{\rho,d}_{0,s}$ are defined by setting, for all $f \in L^1_{\rm loc}(\mu)$ and $x \in \mathbb{R}^D$,

$$\mathcal{M}^{\rho}_{0,s} f(x) := \sup_{Q \ni x} m^{\rho}_{0,s;Q}(f)$$

and

$$\mathcal{M}^{\rho,d}_{0,s} f(x) := \sup_{Q \ni x,\, Q\ 2\rho\text{-doubling}} m^{\rho}_{0,s;Q}(f). \qquad (5.1.3)$$

For any two cubes $Q_1 \subset Q_2$, let

$$\delta^{\rho}_{Q_1,Q_2} := 1 + \sum_{k=1}^{N^{\rho}_{Q_1,Q_2}} \frac{\mu((2\rho)^k Q_1)}{[\ell((2\rho)^k Q_1)]^n}, \qquad (5.1.4)$$

where $N^{\rho}_{Q_1,Q_2}$ is the *smallest integer* $k \in \mathbb{N}$ such that $\ell((2\rho)^k Q_1) \ge \ell(Q_2)$. Arguing as in (2.1.3), we see that there exists a constant $C \in (1, \infty)$, depending on ρ, such that, for any cubes $Q \subset R$,

$$\delta^{\rho}_{Q,R}/C \le 1 + \delta(Q, R) \le C\delta^{\rho}_{Q,R}. \qquad (5.1.5)$$

The *John–Strömberg sharp maximal operator* $\mathcal{M}^{\rho,\natural}_{0,s}$ is defined by setting, for all $f \in L^1_{\rm loc}(\mathbb{R}^n)$ and $x \in \mathbb{R}^D$,

$$\mathcal{M}_{0,s}^{\rho,\natural} f(x) := \sup_{Q \ni x} m_{0,s;Q}^{\rho}\left(f - m_f(\tilde{Q}^{2\rho})\right) + \sup_{\substack{x \in Q \subset R \\ Q,R,\,2\rho\text{—doubling}}} \frac{|m_f(Q) - m_f(R)|}{\delta_{Q,R}^{\rho}}. \tag{5.1.6}$$

In the case that $\rho = 1$, we *denote* $\mathcal{M}_{0,s}^{\rho}$, $\mathcal{M}_{0,s}^{\rho,d}$ *and* $\mathcal{M}_{0,s}^{\rho,\sharp}$ *simply by* $\mathcal{M}_{0,s}$, $\mathcal{M}_{0,s}^{d}$ *and* $\mathcal{M}_{0,s}^{\sharp}$, respectively.

It is easy to show that, for any $f \in L_{\rm loc}^1(\mu)$, cube $Q \ni x$ and $\epsilon \in (0,\infty)$,

$$\mu\left(\left\{y \in Q : \left|f(y) - m_f(\tilde{Q}^{2\rho})\right| > \mathcal{M}_{0,s}^{\rho,\natural} f(x) + \epsilon\right\}\right) < s\mu\left(\frac{3}{2}\rho Q\right). \tag{5.1.7}$$

Let $r \in (0,\infty)$. The *sharp maximal operator* $\mathcal{M}_r^{\rho,\natural}$ is defined by setting, for all $f \in L_{\rm loc}^{\max\{1,r\}}(\mu)$ and $x \in \mathbb{R}^D$,

$$\mathcal{M}_r^{\rho,\natural} f(x) := \sup_{x \in Q}\left[\frac{1}{\mu(\frac{3}{2}\rho Q)}\int_Q \left|f(y) - m_f(\tilde{Q}^{2\rho})\right|^r d\mu(y)\right]^{1/r} + \sup_{\substack{x \in Q \subset R \\ Q,R,\,2\rho\text{—doubling}}} \frac{|m_f(Q) - m_f(R)|}{\delta_{Q,R}^{\rho}}.$$

If $r = 1$, we *write* $\mathcal{M}_r^{\rho,\natural}$ *simply by* $\mathcal{M}^{\rho,\natural}$. It is obvious that, for any $f \in L_{\rm loc}^{\max\{1,r\}}(\mu)$, cube Q and $r \in (0,\infty)$,

$$m_{0,s;Q}^{\rho}\left(f - m_f(\tilde{Q})\right) \le s^{-1/r}\left[\frac{1}{\mu(\frac{3}{2}\rho Q)}\int_Q \left|f(y) - m_f(\tilde{Q}^{2\rho})\right|^r d\mu(y)\right]^{1/r}. \tag{5.1.8}$$

Therefore, for all $f \in L_{\rm loc}^{\max\{1,r\}}(\mu)$ and $x \in \mathbb{R}^D$, we know that

$$\mathcal{M}_{0,s}^{\rho,\natural} f(x) \le s^{-1/r}\mathcal{M}_r^{\rho,\natural} f(x). \tag{5.1.9}$$

Moreover, we have the following several technical lemmas.

Lemma 5.1.3. *Let* $\rho \in [1,\infty)$ *and* $s \in (0, \beta_{2\rho}^{-1}/4)$. *Then, for any* μ*-locally integrable function* f *and* $x \in \mathbb{R}^D$, *it holds true that*

$$\mathcal{M}_{0,s}^{\rho,\natural}(|f|)(x) \le 8\mathcal{M}_{0,s}^{\rho,\natural}(f)(x).$$

Proof. Notice that, for any cube Q with $\mu(Q) > 0$, $c \in \mathbb{C}$ and μ-locally integrable function h, it holds true that

$$m_h(Q) - c = m_{h-c}(Q). \tag{5.1.10}$$

It follows, from Lemmas 5.1.1 and 5.1.2, that

$$\begin{aligned}
&m^{\rho}_{0,s;\,Q}\left(|f| - m_{|f|}(\tilde{Q}^{2\rho})\right)\\
&\quad \le m^{\rho}_{0,s;\,Q}\left(|f| - |m_f(\tilde{Q}^{2\rho})|\right) + \left||m_f(\tilde{Q}^{2\rho})| - m_{|f|}(\tilde{Q}^{2\rho})\right|\\
&\quad \le m^{\rho}_{0,s;\,Q}\left(f - m_f(\tilde{Q}^{2\rho})\right) + \left|m_{|f|-|m_f(\tilde{Q}^{2\rho})|}(\tilde{Q}^{2\rho})\right|\\
&\quad \le m^{\rho}_{0,s;\,Q}\left(f - m_f(\tilde{Q}^{2\rho})\right) + 2m^{\rho}_{0,s;\,\tilde{Q}^{2\rho}}\left(|f| - |m_f(\tilde{Q}^{2\rho})|\right)\\
&\quad \le m^{\rho}_{0,s;\,Q}\left(f - m_f(\tilde{Q}^{2\rho})\right) + 2m^{\rho}_{0,s;\,\tilde{Q}^{2\rho}}\left(f - m_f(\tilde{Q}^{2\rho})\right).
\end{aligned}$$

On the other hand, by Lemma 5.1.2 again, we see that, for any 2ρ-doubling cubes $Q \subset R$,

$$\begin{aligned}
&|m_{|f|}(Q) - m_{|f|}(R)|\\
&\quad \le \left|m_{|f|}(Q) - |m_f(Q)|\right| + \left||m_f(Q)| - |m_f(R)|\right| + \left|m_{|f|}(R) - |m_f(R)|\right|\\
&\quad \le |m_{|f|-|m_f(Q)|}(Q)| + |m_f(Q) - m_f(R)| + |m_{|f|-|m_f(R)|}(R)|\\
&\quad \le 2m^{\rho}_{0,s;\,Q}(f - m_f(Q)) + |m_f(Q) - m_f(R)| + 2m^{\rho}_{0,s;\,R}(f - m_f(R)).
\end{aligned}$$

Combining the last two estimates above, we then obtain the desired estimate, which completes the proof of Lemma 5.1.3. □

Lemma 5.1.4. *Let $\rho \in [1,\infty)$, $s \in (0, \beta_{2\rho}^{-1}/4)$ and $r \in (0,\infty)$. For any cube Q and μ-locally integrable function f, it holds true that*

$$m^{\rho}_{0,s;\,Q}(f - m_f(Q)) \le 3s^{-1/r} \inf_{c\in\mathbb{C}} \left\{\frac{1}{\mu(\frac{3}{2}\rho Q)} \int_Q |f(y) - c|^r \, d\mu(y)\right\}^{1/r}.$$

Proof. For any cube Q and μ-locally integrable function f, let

$$m^{\rho,\natural}_{0,s;\,Q}(f) := \inf_{c\in\mathbb{C}} m^{\rho}_{0,s;\,Q}(f - c).$$

Then, to prove Lemma 5.1.4, it suffices to show that

$$m^{\rho}_{0,s;\,Q}(f - m_f(Q)) \le 3m^{\rho,\natural}_{0,s;\,Q}(f). \tag{5.1.11}$$

The estimate (5.1.11) is trivial if $\mu(Q)=0$; hence we only need to consider the case that $\mu(Q)>0$. For any fixed $\epsilon\in(0,\infty)$, we choose $c_Q=a+ib$ such that

$$\mu(\{y\in Q:\ |f(y)-c_Q|>m_{0,s;\,Q}^{\rho,\natural}(f)+\epsilon\})<s\mu\left(\frac{3}{2}\rho Q\right). \tag{5.1.12}$$

Thus,

$$\mu(\{y\in Q:\ |\Re f(y)-a|>m_{0,s;\,Q}^{\rho,\natural}(f)+\epsilon\})<s\mu\left(\frac{3}{2}\rho Q\right) \tag{5.1.13}$$

and

$$\mu(\{y\in Q:\ |\Im f(y)-b|>m_{0,s;\,Q}^{\rho,\natural}(f)+\epsilon\})<s\mu\left(\frac{3}{2}\rho Q\right). \tag{5.1.14}$$

If

$$\Re[m_f(Q)]>a+m_{0,s;\,Q}^{\rho,\natural}(f)+\epsilon,$$

then, by (5.1.13), we see that

$$\begin{aligned}
&\mu(\{y\in Q:\ \Re f(y)<\Re[m_f(Q)]\})\\
&\quad\geq\mu(\{y\in Q:\ \Re f(y)\leq a+m_{0,s;\,Q}^{\rho,\natural}(f)+\epsilon\})\\
&\quad\geq(1-s\beta_{2\rho})\mu(Q)\\
&\quad>\frac{1}{2}\mu(Q),
\end{aligned}$$

which is contradictory to the definition of $\Re[m_f(Q)]$. Therefore,

$$\Re[m_f(Q)]\leq a+m_{0,s;\,Q}^{\rho,\natural}(f)+\epsilon.$$

We also deduce, from (5.1.13), that

$$\Re[m_f(Q)]\geq a-m_{0,s;\,Q}^{\rho,\natural}(f)-\epsilon$$

and, from (5.1.14), that

$$b-m_{0,s;\,Q}^{\rho,\natural}(f)-\epsilon\leq\Im[m_f(Q)]\leq b+m_{0,s;\,Q}^{\rho,\natural}(f)+\epsilon.$$

Combining these estimates, we find that

$$|m_f(Q)-c_Q|\leq 2m_{0,s;\,Q}^{\rho,\natural}(f)+2\epsilon,$$

which leads to that

$$\begin{aligned}&\{y \in Q : |f(y) - m_f(Q)| > 3m_{0,s;Q}^{\rho,\natural}(f) + 3\epsilon\}\\ &\quad\subset \{y \in Q : |f(y) - c_Q| > m_{0,s;Q}^{\rho,\natural}(f) + \epsilon\}.\end{aligned}$$

By this and (5.1.12), we then conclude that

$$\mu(\{y \in Q : |f(y) - m_f(Q)| > 3m_{0,s;Q}^{\rho,\natural}(f) + 3\epsilon\}) < s\mu\left(\frac{3}{2}\rho Q\right)$$

and hence

$$m_{0,s;Q}^{\rho}(f - m_f(Q)) \le 3m_{0,s;Q}^{\rho,\natural}(f) + 3\epsilon.$$

The inequality (5.1.11) then follows from letting $\epsilon \to 0$, which completes the proof of Lemma 5.1.4. □

Let $r \in (0, \infty)$ and $\rho \in [1, \infty)$. The *sharp maximal operator* $\mathcal{M}_r^{\rho,\sharp}$ is defined by setting, for all $f \in L_{\rm loc}^{\max\{1,r\}}(\mu)$ and $x \in \mathbb{R}^D$,

$$\begin{aligned}\mathcal{M}_r^{\rho,\sharp} f(x) := &\sup_{Q \ni x} \left[\frac{1}{\mu\left(\frac{3}{2}\rho Q\right)} \int_Q |f(y) - m_{\widetilde{Q}^{2\rho}}(f)|^r \, d\mu(y)\right]^{1/r}\\ &+ \sup_{\substack{x \in Q \subset R\\ Q,\,R,\,2\rho\text{—doubling}}} \frac{|m_Q(f) - m_R(f)|}{\delta_{Q,R}^{\rho}}.\end{aligned} \tag{5.1.15}$$

If $r = 1$, we *write* $\mathcal{M}_r^{\rho,\sharp}$ *simply by* $\mathcal{M}^{\rho,\sharp}$. If $r = 1$ and $\rho = 1$, we *write* $\mathcal{M}_r^{\rho,\sharp}$ *simply by* $\mathcal{M}^{\sharp}$.

The following lemma shows that the sharp maximal operator $\mathcal{M}_{0,s}^{\rho,\natural}$ is fairly smaller than the operator $\mathcal{M}_r^{\rho,\sharp}$.

Lemma 5.1.5. *Let $\rho \in [1, \infty)$. Then there exists a positive constant C, depending on s, such that, for all $f \in L_{\rm loc}^1(\mu)$ and $x \in \mathbb{R}^D$,*

$$\mathcal{M}_{0,s}^{\rho,\natural} f(x) \le C\mathcal{M}^{\rho,\sharp} f(x). \tag{5.1.16}$$

Moreover, if f is nonnegative, then, for any $r \in (0, \infty)$, it holds true that

$$\mathcal{M}_{0,s}^{\rho,\natural} f(x) \le C\mathcal{M}_r^{\rho,\sharp} f(x) \tag{5.1.17}$$

with the positive constant C independent of f and x.

Proof. From (5.1.9) with $r = 1$, we first deduce that, for all $f \in L^1_{\mathrm{loc}}(\mu)$ and $x \in \mathbb{R}^D$,

$$\mathcal{M}^{\rho,\natural}_{0,s} f(x) \lesssim \mathcal{M}^{\rho,\natural} f(x).$$

Thus, to show (5.1.16), it suffices to show that, for all $f \in L^1_{\mathrm{loc}}(\mu)$ and $x \in \mathbb{R}^D$,

$$\mathcal{M}^{\rho,\natural} f(x) \lesssim \mathcal{M}^{\rho,\sharp} f(x). \tag{5.1.18}$$

Notice that, for any cube Q,

$$\begin{aligned}
&\frac{1}{\mu(\frac{3}{2}Q)} \int_Q \left|f(y) - m_f(\widetilde{Q}^{2\rho})\right| d\mu(y) \\
&\quad \le \frac{1}{\mu(\frac{3}{2}Q)} \int_Q |f(y) - m_{\widetilde{Q}^{2\rho}}(f)|\, d\mu(y) + \left|m_f(\widetilde{Q}^{2\rho}) - m_{\widetilde{Q}^{2\rho}}(f)\right|
\end{aligned}$$

and

$$\left|m_f(\widetilde{Q}^{2\rho}) - m_{\widetilde{Q}^{2\rho}}(f)\right| \le \frac{1}{\mu(\widetilde{Q}^{2\rho})} \int_{\widetilde{Q}^{2\rho}} \left|f(y) - m_f(\widetilde{Q}^{2\rho})\right| d\mu(y).$$

By the definition of $m_f(Q)$, we see that, for any cube I,

$$\frac{1}{\mu(I)} \int_I |f(y) - m_f(I)|\, d\mu(y) \le \frac{1}{\mu(I)} \int_I |f(y) - m_I(f)|\, d\mu(y).$$

Combining these three inequalities, we see that, for any x and cube Q containing x,

$$\frac{1}{\mu(\frac{3}{2}Q)} \int_Q \left|f(y) - m_f(\widetilde{Q}^{2\rho})\right| d\mu(y) \lesssim \mathcal{M}^{\rho,\sharp} f(x). \tag{5.1.19}$$

On the other hand, for any two 2ρ-doubling cubes Q and R, with $Q \subset R$, we have

$$\begin{aligned}
&|m_f(Q) - m_f(R)| \\
&\quad \le |m_Q(f) - m_f(Q)| + |m_R(f) - m_f(R)| + |m_Q(f) - m_R(f)| \\
&\quad \le \frac{1}{\mu(Q)} \int_Q |f(y) - m_f(Q)|\, d\mu(y) \\
&\qquad + \frac{1}{\mu(R)} \int_R |f(y) - m_f(R)|\, d\mu(y) + |m_Q(f) - m_R(f)| \\
&\quad \le 3\delta^{\rho}_{Q,R} \inf_{x \in Q} \mathcal{M}^{\rho,\sharp} f(x).
\end{aligned}$$

Combining this with (5.1.19), we obtain (5.1.18) and hence complete the proof of (5.1.16).

We now turn our attention to (5.1.17). Assume that f is nonnegative. We know that, for any cube Q, it holds true that $m_f(Q) \geq 0$ and hence by Lemmas 5.1.1 and 5.1.4, we see that

$$\begin{aligned} &\left|m_f(\tilde{Q}^{2\rho}) - m_{\tilde{Q}^{2\rho}}(f)\right| \\ &\leq \left|m^{\rho}_{0,s;\,\tilde{Q}^{2\rho}}(f) - m_f(\tilde{Q}^{2\rho})\right| + \left|m^{\rho}_{0,s;\,\tilde{Q}^{2\rho}}(f) - m_{\tilde{Q}^{2\rho}}(f)\right| \\ &\leq m^{\rho}_{0,s;\,\tilde{Q}^{2\rho}}\left(f - m_f(\tilde{Q}^{2\rho})\right) + m^{\rho}_{0,s;\,\tilde{Q}^{2\rho}}\left(f - m_{\tilde{Q}^{2\rho}}(f)\right) \\ &\lesssim \left[\frac{1}{\mu(\tilde{Q}^{2\rho})}\int_{\tilde{Q}^{2\rho}} |f(y) - m_{\tilde{Q}^{2\rho}}(f)|^r\, d\mu(y)\right]^{1/r}. \end{aligned}$$

Also, for any two 2ρ-doubling cubes $Q \subset R$ with $Q \ni x$, we have

$$\begin{aligned} &|m_f(Q) - m_f(R)| \\ &\leq |m_f(Q) - m_Q(f)| + |m_Q(f) - m_R(f)| + |m_f(R) - m_R(f)| \\ &\lesssim \mathcal{M}^{\rho,\sharp}_r f(x). \end{aligned}$$

Combining the last two estimates, we obtain (5.1.17), which completes the proof of Lemma 5.1.5. □

Let Φ be a function on $[0, \infty)$, which is called a *Young function*, if it is continuous, convex and increasing, $\Phi(0) = 0$ and $\lim_{t\to\infty} \Phi(t) = \infty$. Let $\rho \in (1, \infty)$ and ω be a locally integrable function. For a suitable function f and a cube Q, define

$$\begin{aligned} &\|f\|_{\Phi,\rho,Q,\omega} \\ &:= \inf\left\{\lambda \in (0,\infty) : \frac{1}{\omega(\rho Q)}\int_Q \Phi\left(\frac{|f(x)|}{\lambda}\right)\omega(x)\, d\mu(x) \leq 1\right\}. \end{aligned}$$

Notice that

$$\Psi_r(t) := \exp t^r - 1 \quad \text{for all} \quad r \in (0, \infty)$$

is a Young function. For a suitable function f and a cube Q, define

$$\begin{aligned} &\|f\|_{\exp L^r,\rho,Q,\omega} \\ &:= \inf\left\{\lambda \in (0,\infty) : \frac{1}{\omega(\rho Q)}\int_Q \exp\left(\frac{|f(x)|}{\lambda}\right)^r \omega(x)\, d\mu(x) \leq 2\right\}. \end{aligned}$$

Let $p \in (0, \infty)$. For the special case

$$\Phi(t) := t^p \log^r(e + t) \quad \text{for all} \quad t \in (0, \infty),$$

we *denote* $\|f\|_{\Phi,\rho,Q,\omega}$ *by* $\|f\|_{L^p(\log L)^r,\rho,Q,\omega}$.

For the case $\omega \equiv 1$, we *denote* $\|f\|_{\Phi,\rho,Q,\omega}$ *simply by* $\|f\|_{\Phi,\rho,Q}$ *and* $\|f\|_{\exp L^r,\rho,Q,\omega}$ *by* $\|f\|_{\exp L^r,\rho,Q}$. Moreover, we *denote* $\|f\|_{\Phi,1,Q,1}$ (resp. $\|f\|_{\exp L^r,1,Q,1}$) *simply by* $\|f\|_{\Phi,Q,\mu}$ *(resp.* $\|f\|_{\exp L^r,Q,\mu}$*)*.

Let $\sigma \in \mathbb{R}$, $p \in (0, \infty)$, $\rho \in (1, \infty)$ and ω be a locally integrable function. The *maximal operators* $\mathcal{M}_{L^p(\log L)^\sigma,\rho,w}$ and $\mathcal{M}_{L^p(\log L)^\sigma,\rho}$ are defined, respectively, by setting, for all suitable functions f and $x \in \mathbb{R}^D$,

$$\mathcal{M}_{L^p(\log L)^\sigma,\rho,w} f(x) := \sup_{Q \ni x} \|f\|_{L^p(\log L)^\sigma,\rho,Q,w}$$

and

$$\mathcal{M}_{L^p(\log L)^\sigma,\rho} f(x) := \sup_{Q \ni x} \|f\|_{L^p(\log L)^\sigma,\rho,Q},$$

where the supremums are taken over all cubes $Q \ni x$. We *denote* $\mathcal{M}_{L^1(\log L)^\sigma,\rho}$ *simply by* $\mathcal{M}_{L(\log L)^\sigma,\rho}$.

Definition 5.1.6. Let $\rho \in [1, \infty)$, $\sigma \in [0, \infty)$ and $p \in (1, \infty)$. A weight w is said to belong to the *weight class* $A^\rho_{p,(\log L)^\sigma}(\mu)$, if there exists a positive constant C such that, for any cube Q,

$$\left[\frac{1}{\mu(\rho Q)} \int_Q w(x) d\mu(x)\right]^{1/p} \|w^{-1/p}\|_{L^{p'}(\log L)^{\sigma p'},\rho,Q} \le C.$$

For the case that $\sigma = 0$, *denote* $A^\rho_{p,(\log L)^\sigma}(\mu)$ *simply by* $A^\rho_p(\mu)$. Also, a weight w is said to belong to the *weight class* $A^\rho_1(\mu)$, if there exists a positive constant C such that, for any cube Q,

$$\frac{1}{\mu(\rho Q)} \int_Q w(x)\, d\mu(x) \le C \inf_{x \in Q} w(x).$$

As in the classical setting, let

$$A^\rho_\infty(\mu) := \bigcup_{p=1}^{\infty} A^\rho_p(\mu).$$

In what follows, for any $u \in A^\rho_\infty(\mu)$ and $p \in (0, \infty)$, we use $L^p(u)$ and $L^{p,\infty}(u)$ to denote, respectively, the *weighted* $L^p(\mu)$ *space with weight* u and the *weak weighted* $L^p(\mu)$ *space with weight* u. For any $f \in L^p(u)$, its *norm* $\|f\|_{L^p(u)}$ is defined by

$$\|f\|_{L^p(u)} := \left\{\int_{\mathbb{R}^D} |f(x)|^p u(x)\, d\mu(x)\right\}^{1/p}$$

and, for any $f \in L^{p,\infty}(u)$, its *norm* $\|f\|_{L^{p,\infty}(u)}$ is defined by

$$\|f\|_{L^{p,\infty}(u)} := \inf_{\lambda \in (0,\infty)} \lambda^{1/p} u(\{x \in \mathbb{R}^D : |f(x)| > \lambda\}).$$

Let $\rho \in [1, \infty)$. The *non-centered* 2ρ*-doubling maximal operator* $\mathcal{M}^{\rho,d}$ is defined by setting, for all $f \in L^1_{\text{loc}}(\mu)$ and $x \in \mathbb{R}^D$,

$$\mathcal{M}^{\rho,d} f(x) := \sup_{Q \ni x,\, Q\ 2\rho\text{-doubling}} \frac{1}{\mu(Q)} \int_Q |f(y)|\, d\mu(y), \tag{5.1.20}$$

where the supremum is taken over all 2ρ-doubling cubes $Q \ni x$. If $\rho = 1$, then we *write* $\mathcal{M}^{\rho,d}$ *simply by* $\mathcal{M}^d$.

For the behavior of the maximal operators $\mathcal{M}_{(\eta)}$ and $\mathcal{M}^{\rho,d}$ on weighted L^p spaces with $A^\rho_p(\mu)$ weight, we have the following conclusion.

Lemma 5.1.7. *Let* $\rho \in [1, \infty)$, $\eta \in (\rho, \infty)$ *and* $\mathcal{M}_{(\eta)}$ *and* $\mathcal{M}^{\rho,d}$ *be the maximal operators defined by* (1.4.12) *and* (5.1.20), *respectively. Then, for any* $p \in [1, \infty)$ *and* $u \in A^\rho_p(\mu)$, *both* $\mathcal{M}_{(\eta)}$ *and* $\mathcal{M}^{\rho,d}$ *are bounded from* $L^p(u)$ *to* $L^{p,\infty}(u)$.

Proof. Notice that, for any 2ρ-doubling cube Q,

$$\frac{1}{\mu(Q)} \int_Q |f(y)|\, d\mu(y) \le \beta_{2\rho} \frac{1}{\mu(2\rho Q)} \int_Q |f(y)|\, d\mu(y) \lesssim \inf_{x \in Q} \mathcal{M}_{(2\rho)} f(x).$$

Thus, it suffices to consider the operator $\mathcal{M}_{(\eta)}$ for any $\eta \in (\rho, \infty)$. It is easy to see that, for any $p \in [1, \infty)$, all suitable functions f and $x \in \mathbb{R}^D$,

$$\mathcal{M}_{(\eta)} f(x) \lesssim \mathcal{M}_{p,(\eta)} f(x),$$

where, for any $p \in (0, \infty)$ and $\eta \in (1, \infty)$, the *maximal operator* $\mathcal{M}_{p,(\eta)}$ is defined by setting, for all $f \in L^p_{\text{loc}}(\mu)$ and $x \in \mathbb{R}^D$,

$$\mathcal{M}_{p,(\eta)} f(x) := \sup_{Q \ni x} \left[\frac{1}{\mu(\eta Q)} \int_Q |f(y)|^p\, d\mu(y)\right]^{1/p}, \tag{5.1.21}$$

where the supremum is taken over all cubes $Q \ni x$. For any fixed $R \in (0, \infty)$, the *operator* $\mathcal{M}^R_{p,(\eta)}$ is defined by setting, for all $f \in L^p_{\text{loc}}(\mu)$ and $x \in \mathbb{R}^D$,

$$\mathcal{M}_{p,(\eta)}^{R} f(x) := \sup_{Q \ni x, \ell(Q)<R} \left[\frac{1}{\mu(\eta Q)} \int_Q |f(y)|^p \, d\mu(y) \right]^{1/p},$$

where the supremum is taken over all cubes $Q \ni x$ with side lengths $\ell(Q) < R$. Let

$$E_\lambda^R := \left\{ x \in \mathbb{R}^D : \mathcal{M}_{p,(\eta)}^R f(x) > \lambda \right\}.$$

For any $x \in E_\lambda^R$, there exists a cube Q_x, with side length $\ell(Q_x) \in (0, R)$, such that

$$\mu(\eta Q) < \lambda^{-p} \int_{Q_x} |f(y)|^p \, d\mu(y).$$

Applying Theorem 1.1.1, we know that there exist disjoint subfamilics, $\mathcal{D}_k := \{Q_j^k\}_j$, $k \in \{1, \ldots, B_D\}$, of cubes $\{Q_x\}_x$ such that

$$E_\lambda^R \subset \bigcup_{k=1}^{B_D} \bigcup_j \frac{\eta}{\rho} Q_j^k,$$

where B_D is as in Theorem 1.1.1. Therefore,

$$\begin{aligned} u(E_\lambda^R) &\le \sum_{k=1}^{B_D} \sum_j u\left(\frac{\eta}{\rho} Q_j^k\right) \\ &\lesssim \lambda^{-p} \sum_{k=1}^{B_D} \sum_j \int_{Q_j^k} |f(y)|^p u(y) \, d\mu(y) \\ &\lesssim \lambda^{-p} \int_{\mathbb{R}^D} |f(y)|^p u(y) \, d\mu(y). \end{aligned}$$

Taking $R \to \infty$, we then see that

$$u(\{x \in \mathbb{R}^D : \mathcal{M}_{p,(\eta)} f(x) > \lambda\} \lesssim \lambda^{-p} \int_{\mathbb{R}^D} |f(y)|^p u(y) \, d\mu(y),$$

which completes the proof of Lemma 5.1.7. □

As an easy consequence of Lemma 5.1.7, we obtain the following conclusion.

Lemma 5.1.8. *Let ρ, $p \in [1, \infty)$, $u \in A_p^\rho(\mu)$ and $\eta \in (\rho, \infty)$. Then there exist constants C, $\tilde{C}_1 \in [1, \infty)$ such that,*

(i) *for any cube Q and μ-measurable set $E \subset Q$,*

$$\frac{u(E)}{u(Q)} \ge C^{-1} \left[\frac{\mu(E)}{\mu(\eta Q)} \right]^p;$$

(ii) *for any* 2ρ*-doubling cube* Q *and* μ*-measurable set* $E \subset Q$,

$$\frac{u(E)}{u(Q)} \geq \tilde{C}_1^{-1} \left[\frac{\mu(E)}{\mu(Q)}\right]^p;$$

(iii) *for any* 2ρ*-doubling cube* Q *and* μ*-measurable set* $E \subset Q$,

$$\frac{u(E)}{u(Q)} \leq 1 - \tilde{C}_1^{-1} \left[1 - \frac{\mu(E)}{\mu(Q)}\right]^p.$$

Proof. Obviously, (ii) follows from (i) with $\eta := 2\rho$, and (iii) is an easy consequence of (ii) with E replaced by $Q \setminus E$. Thus, it suffices to prove (i), whose proof is similar to that of the classical case.[1] Indeed, it is easy to see that, for any cube Q and μ-measurable set $E \subset Q$,

$$\inf_{x \in Q} \mathcal{M}_{(\eta)} \chi_E(x) \geq \mu(E)/\mu(\eta Q).$$

On the other hand, Lemma 5.1.7 states that, for any $p \in [1, \infty)$, $u \in A_p^\rho(\mu)$ and $\eta > \rho$, there exists a positive constant C such that, for any $\lambda \in (0, \infty)$,

$$u(\{x \in \mathbb{R}^D : \mathcal{M}_{(\eta)} f(x) > \lambda\}) \leq C\lambda^{-p} \int_{\mathbb{R}^D} |f(x)|^p u(x)\, d\mu(x).$$

Thus, the last two estimates imply that, for any $\lambda \in (0, \mu(E)/\mu(\eta Q))$,

$$u(Q) \leq u(\{x \in \mathbb{R}^D : \mathcal{M}_{(\eta)} \chi_E(x) > \lambda\}) \leq C\lambda^{-p} u(E)$$

and hence, for any $\lambda \in (0, \mu(E)/\mu(\eta Q))$,

$$\frac{u(E)}{u(Q)} \geq C^{-1}\lambda^p.$$

Letting $\lambda \to \mu(E)/\mu(\eta Q)$, we then obtain the conclusion (i), which completes the proof of Lemma 5.1.8. □

Lemma 5.1.9. *Let* $\rho,\ p \in [1, \infty)$ *and* $s \in (0, \beta_{2\rho}^{-1}/4)$. *Then, for all* μ*-locally integrable functions* f *and* $\lambda \in (0, \infty)$,

(i)

$$\{x \in \mathbb{R}^D : |f(x)| > \lambda\} \subset \{x \in \mathbb{R}^D : \mathcal{M}_{0,s}^{\rho,d} f(x) \geq \lambda\} \bigcup \Theta$$

with $\mu(\Theta) = 0$;

[1] See [121].

(ii) *for* $u \in A_p^\rho(\mu)$,

$$u\left(\left\{x \in \mathbb{R}^D : \mathcal{M}_{0,s}^{\rho,d} f(x) > \lambda\right\}\right) \le C s^{-p} u\left(\{x \in \mathbb{R}^D : |f(x)| > \lambda\}\right),$$

where C *is a positive constant depending on* D *and* ρ, *but not on* s *and the weight* u.

Proof. By the Lebesgue differentiation theorem, we know that, for μ-almost every $x \in \mathbb{R}^D$,

$$|f(x)| \le \mathcal{M}^{\rho,d} f(x) \tag{5.1.22}$$

and hence

$$\begin{aligned}
&\{x \in \mathbb{R}^D : |f(x)| > \lambda\} \\
&\quad = \{x \in \mathbb{R}^D : \chi_{\{y \in \mathbb{R}^D : |f(y)| > \lambda\}}(x) = 1\} \\
&\quad \subset \left\{x \in \mathbb{R}^D : \mathcal{M}^{\rho,d}\left(\chi_{\{y \in \mathbb{R}^D : |f(y)| > \lambda\}}\right)(x) > s\beta_{2\rho}\right\} \bigcup \Theta,
\end{aligned}$$

where $\mu(\Theta) = 0$. On the other hand, if

$$\mathcal{M}^{\rho,d}\left(\chi_{\{y \in \mathbb{R}^D : |f(y)| > \lambda\}}\right)(x) > s\beta_{2\rho},$$

then there exists a 2ρ-doubling cube Q containing x such that $\mu(Q) > 0$ and

$$\mu(\{y \in Q : |f(y)| > \lambda\}) > s\beta_{2\rho}\mu(Q) \ge s\mu\left(\frac{3}{2}\rho Q\right).$$

Notice that, for any $t > m_{0,s;\,Q}^\rho(f)$,

$$\mu(\{y \in Q : |f(y)| > t\}) < s\mu\left(\frac{3}{2}\rho Q\right).$$

Thus, $m_{0,s;\,Q}^\rho(f) \ge \lambda$ and hence $\mathcal{M}_{0,s}^{\rho,d} f(x) \ge \lambda$, which implies (i).

We now turn our attention to (ii). For any fixed $\lambda \in (0,\infty)$, $r \in (0,\infty)$ and $x \in \mathbb{R}^D$, let

$$\mathcal{M}_{0,s}^{\rho,d,r} f(x) := \sup_{\substack{Q \ni x,\, \ell(Q) < r,\\ Q\ 2\rho\text{—doubling}}} m_{0,s;\,Q}^\rho(f),$$

where the supremum is taken over all 2ρ-doubling cubes $Q \ni x$ with side lengths $\ell(Q) < r$, and

$$E_{r,\lambda} := \{x \in \mathbb{R}^D : \mathcal{M}_{0,s}^{\rho,d,r} f(x) > \lambda\}.$$

For any $x \in E_{r,\lambda}$, there exists a 2ρ-doubling cube Q_x such that $x \in Q_x$, $\ell(Q_x) < r$, and

$$\mu(\{y \in Q_x : |f(y)| > \lambda\}) \geq s\mu\left(\frac{3}{2}\rho Q_x\right).$$

It now follows, from (ii) of Lemma 5.1.8, that

$$u(\{y \in Q_x : |f(y)| > \lambda\}) \gtrsim s^p u\left(\frac{3}{2}\rho Q_x\right).$$

By Theorem 1.1.1, we obtain a family of cubes,

$$\{Q_\tau\}_\tau \subset \{Q_x\}_{x \in E_{r,\lambda}}, \quad \text{such that} \quad E_{r,\lambda} \subset \bigcup_\tau Q_\tau \quad \text{and} \quad \sum_\tau \chi_{Q_\tau} \leq N_D,$$

where N_D is as in Theorem 1.1.1. Therefore,

$$\begin{aligned} u(E_{r,\lambda}) &\lesssim s^{-p} \sum_\tau u(\{y \in Q_\tau : |f(y)| > \lambda\}) \\ &\lesssim s^{-p} u(\{y \in \mathbb{R}^D : |f(y)| > \lambda\}), \end{aligned}$$

which, together with some basic properties of measures, shows (ii). This finishes the proof of Lemma 5.1.9. □

Remark 5.1.10. Repeating the proof of Lemma 5.1.9, we can prove that, for ρ, $p \in [1, \infty)$ and $s \in (0, \beta_{2\rho}^{-1}/4)$, there exists a positive constant C such that, for any locally integrable function f and $\lambda \in (0, \infty)$,

$$\mu\left(\{x \in \mathbb{R}^D : \mathcal{M}_{0,s}^\rho f(x) > \lambda\}\right) \leq C s^{-p} \mu(\{x \in \mathbb{R}^D : |f(x)| > \lambda\}).$$

We now give a good-λ inequality linking the John–Strömberg sharp maximal operator $\mathcal{M}_{0,s}^{\rho,\natural}$ and the John–Strömberg maximal operator $\mathcal{M}_{0,s}^{\rho,d}$ as follows.

Theorem 5.1.11. *Let $\rho \in [1, \infty)$, $s_1 \in (0, \beta_{2\rho}^{-1}/4)$ and $u \in A_p^\rho(\mu)$ with $p \in [1, \infty)$. Then there exists a constant $\tilde{C}_2 \in (1, \infty)$ such that, for any $s_2 \in (0, \tilde{C}_2^{-1} s_1)$, $\gamma \in (0, \infty)$ and real-valued function $f \in L^{p_0,\infty}(\mu)$ with some $p_0 \in [1, \infty)$,*

$$\begin{aligned} &u\left(\{x \in \mathbb{R}^D : \mathcal{M}_{0,s_1}^{\rho,d} f(x) > (1+\gamma)\lambda, \mathcal{M}_{0,s_2}^{\rho,\natural} f(x) \leq \theta_2 \gamma \lambda\}\right) \\ &\quad \leq \theta_1 u(\{x \in \mathbb{R}^D : \mathcal{M}_{0,s_1}^{\rho,d} f(x) > \lambda\}), \end{aligned}$$

provided that

(i) $\mu(\mathbb{R}^D) = \infty$ *and* $\lambda \in (0, \infty)$, *or*
(ii) $\mu(\mathbb{R}^D) < \infty$ *and*

$$\lambda > \lambda_f := \left(s_1 \mu(\mathbb{R}^D)\right)^{-1/p_0} \|f\|_{L^{p_0,\infty}(\mu)},$$

where $\theta_1, \theta_2 \in (0, 1)$ *are two constants depending only on* D, ρ *and* μ.

Proof. Let $\lambda_f := 0$ if $\mu(\mathbb{R}^D) = \infty$, and

$$\lambda_f := \left(s_1 \mu(\mathbb{R}^D)\right)^{-1/p_0} \|f\|_{L^{p_0,\infty}(\mu)}$$

if $\mu(\mathbb{R}^D) < \infty$. For each fixed $\lambda > \lambda_f$, let

$$\Omega_\lambda := \left\{x \in \mathbb{R}^D : \mathcal{M}^{\rho,d}_{0,s_1} f(x) > \lambda\right\}$$

and

$$E_\lambda := \left\{x \in \mathbb{R}^D : \mathcal{M}^{\rho,d}_{0,s_1} f(x) > (1+\gamma)\lambda,\ \mathcal{M}^{\rho,\natural}_{0,s_2} f(x) \le \theta_2 \gamma \lambda\right\}.$$

Notice that, if $\lambda > \lambda_f$, then

$$\mu(\{y \in \mathbb{R}^D : |f(y)| > \lambda\}) \le \frac{\|f\|^{p_0}_{L^{p_0,\infty}(\mu)}}{\lambda^{p_0}} < s_1 \mu(\mathbb{R}^D)$$

and hence

$$\frac{1}{\mu(\mathbb{R}^D)} \int_{\{y \in \mathbb{R}^D :\ |f(y)| > \lambda\}} d\mu(y) < s_1.$$

On the other hand, for μ-almost every $x \in \mathbb{R}^D$, there exists a sequence of 2ρ-doubling cubes, $\{I_k\}_k$, such that $\ell(I_k) \to \infty$ when $k \to \infty$. Therefore, by the basic property of μ, we may assume that, for any $\lambda > \lambda_f$ and any $x \in \mathbb{R}^D$,

$$\lim_{\substack{I \ni x,\, \ell(I) \to \infty, \\ I\ 2\rho-\text{doubling}}} \frac{1}{\mu(I)} \int_{\{y \in I :\ |f(y)| > \lambda\}} d\mu(y) < s_1,$$

which implies that

$$\lim_{\substack{I \ni x,\, \ell(I) \to \infty, \\ I\ 2\rho-\text{doubling}}} m^{\rho}_{0,s_1;I}(f) < \lambda. \tag{5.1.23}$$

On the other hand, for each fixed $x \in E_\lambda$, there exists a 2ρ-doubling cube Q containing x such that

$$m^{\rho}_{0,s_1;Q}(f) > (1 + \gamma/2)\lambda.$$

The inequality (5.1.23) tells us that among these 2ρ-doubling cubes, there exists one 2ρ-doubling cube, denoted by Q_x, which has almost maximal side length in the sense that, if some 2ρ-doubling cube I contains x and has side length no less than $2\rho\ell(Q_x)$, then

$$m^{\rho}_{0,s_1;I}(f) \le (1+\gamma/2)\lambda.$$

Let R_x be the cube centered at x and having side length $3\rho\ell(Q_x)$, and $S_x := \widetilde{R_x}^{2\rho}$. An application of Lemma 5.1.1 implies that

$$\begin{aligned}
&|m^{\rho}_{0,s_1;Q_x}(f) - m^{\rho}_{0,s_1;S_x}(f)|\\
&\quad\le \left|m^{\rho}_{0,s_1;Q_x}(f) - |m_f(Q_x)|\right|\\
&\qquad + \left|m_f(Q_x) - m_f(S_x)\right| + \left|m^{\rho}_{0,s_1;S_x}(f) - |m_f(S_x)|\right|\\
&\quad\le m^{\rho}_{0,s_1;Q_x}(f - m_f(Q_x))\\
&\qquad + |m_f(Q_x) - m_f(S_x)| + m^{\rho}_{0,s_1;S_x}(f - m_f(S_x))\\
&\quad\le 3\delta^{\rho}_{Q_x,S_x} \inf_{y\in Q_x} \mathcal{M}^{\rho,\natural}_{0,s_1} f(y)\\
&\quad\le 3\delta^{\rho}_{Q_x,S_x} \mathcal{M}^{\rho,\natural}_{0,s_2} f(x)\\
&\quad\le \tilde{C}_3\theta_2\gamma\lambda,
\end{aligned}$$

where $\tilde{C}_3$ is a positive constant, depending only on D, ρ and μ, such that

$$3\delta^{\rho}_{Q_x,S_x} \le \tilde{C}_3.$$

If we choose $\theta_2 \in (0,\infty)$ small enough, then $m^{\rho}_{0,s_1;S_x}(f) > \lambda$ and hence $S_x \subset \Omega_\lambda$.

By Theorem 1.1.1, there exist B_D subfamilies $\mathcal{D}_k := \{S^k_j\}_j$, $k \in \{1,\ldots,B_D\}$, of cubes $\{S_x\}_{x\in E_\lambda}$ such that

(i)

$$E_\lambda \subset \bigcup_{k=1}^{B_D} \bigcup_j S^k_j$$

and, for each j and k,

$$m^{\rho}_{0,s_1;S^k_j}(f) > \lambda;$$

(ii) for each subfamily $\mathcal{D}_k$, $k \in \{1,\ldots,B_D\}$, the cubes in $\mathcal{D}_k$ are pairwise disjoint;
(iii) each cube S^k_j is 2ρ-doubling and centered at some point $x^k_j \in E_\lambda$.

We obtain, at least, one family, which, without loss of generality, may be supposed to be $\mathcal{D}_1$, such that

$$u\left(\bigcup_j S_j^1\right) \geq B_D^{-1} u\left(\bigcup_{j,k} S_j^k\right). \tag{5.1.24}$$

If we can prove that there exists a positive constant $\tilde{C}_4$ such that, for each $S_j^1 \in \mathcal{D}_1$,

$$\mu\left(S_j^1 \bigcap E_\lambda\right) \leq \tilde{C}_4 s_1^{-1} s_2 \mu(S_j^1), \tag{5.1.25}$$

it then follows, from Lemma 5.1.8, that

$$u\left(S_j^1 \bigcap E_\lambda\right) \leq \left[1 - \tilde{C}_1^{-1}(1 - \tilde{C}_4 s_1^{-1} s_2)^p\right] u(S_j^1).$$

Let $\tilde{C}_2 := 1 + \tilde{C}_4$ and

$$\tilde{C}_5 := 1 - \tilde{C}_1^{-1}(1 - \tilde{C}_4 s_1^{-1} s_2)^p.$$

Recall that $\{S_j^1\}_j$ are pairwise disjoint. Thus, for $s_2 \in (0,\ \tilde{C}_2^{-1} s_1)$,

$$u\left(E_\lambda \bigcap \left(\bigcup_j S_j^1\right)\right) \leq \tilde{C}_5 \sum_j u(S_j^1) = \tilde{C}_5 u\left(\bigcup_j S_j^1\right).$$

This, via (5.1.24), in turn implies that

$$\begin{aligned}
u(E_\lambda) &\leq u\left(\left[\bigcup_{k=1}^{B_D} \bigcup_j S_j^k\right] \setminus \left[\bigcup_j S_j^1\right]\right) + u\left(E_\lambda \bigcap \left[\bigcup_j S_j^1\right]\right) \\
&\leq \tilde{C}_5 u\left(\left[\bigcup_{k=1}^{B_D} \bigcup_j S_j^k\right] \setminus \left[\bigcup_j S_j^1\right]\right) + \tilde{C}_5 u\left(\bigcup_j S_j^1\right) \\
&\quad + (1 - \tilde{C}_5) u\left(\left[\bigcup_{k=1}^{B_D} \bigcup_j S_j^k\right] \setminus \left[\bigcup_j S_j^1\right]\right) \\
&\leq \tilde{C}_5 u\left(\bigcup_{k=1}^{B_D} \bigcup_j S_j^k\right) + (1 - \tilde{C}_5)\left(1 - \frac{1}{B_D}\right) u\left(\bigcup_{k=1}^{B_D} \bigcup_j S_j^k\right) \\
&\leq \left(1 - \frac{1 - \tilde{C}_5}{B_D}\right) u(\Omega_\lambda).
\end{aligned}$$

Thus, the desired conclusion holds true with

$$\theta_1 := 1 - \frac{1-\tilde{C}_5}{B_D}.$$

We now prove (5.1.25). For each fixed $y \in (S_j^1 \cap E_\lambda)$, we claim that, if Q is a 2ρ-doubling cube containing y and satisfying

$$m^{\rho}_{0,s_1;Q}(f) > (1+\gamma)\lambda,$$

then $\ell(Q) \le \ell(S_j^1)/8$. Otherwise,

$$Q_{x_j^1} \subset S_j^1 \subset \widetilde{30Q}^{2\rho}$$

and hence

$$\begin{aligned}\left|m^{\rho}_{0,s_1;Q}(f) - m^{\rho}_{0,s_1;\widetilde{30Q}^{2\rho}}(f)\right| &\le 3\delta^{\rho}_{Q,\widetilde{30Q}^{2\rho}}\mathcal{M}^{\rho,\natural}_{0,s_1}f(y)\\ &\le 3\delta^{\rho}_{Q,\widetilde{30Q}^{2\rho}}\mathcal{M}^{\rho,\natural}_{0,s_2}f(y)\\ &\le \tilde{C}_6\theta_2\gamma\lambda,\end{aligned}$$

where $\tilde{C}_6 \in (1,\infty)$ satisfies that $3\delta^{\rho}_{Q,\widetilde{30Q}} \le \tilde{C}_6$. Let

$$\theta_2 := 1/(2\tilde{C}_3 + 2\tilde{C}_6).$$

We then see that

$$m^{\rho}_{0,s_1;\widetilde{30Q}^{2\rho}}(f) > (1+\gamma/2)\lambda,$$

which contradicts the facts that

$$Q_{x_j^1} \subset \widetilde{30Q}^{2\rho},\ \ell(\widetilde{30Q}^{2\rho}) > 2\rho\ell(Q_{x_j^1})$$

and $Q_{x_j^1}$ is the chosen maximal 2ρ-doubling cube.

For each fixed $y \in (S_j^1 \cap E_\lambda)$, we find that there exists a 2ρ-doubling cube I such that

$$y \in I \quad \text{and} \quad m^{\rho}_{0,s_1;I}(f) > (1+\gamma)\lambda.$$

Our claim then tells us that $\ell(I) \le \ell(S_j^1)$ and $I \subset \frac{5}{4}S_j^1$. Thus,

$$m^{\rho}_{0,s_1;I}(f\chi_{\frac{5}{4}S^1_j}) = m^{\rho}_{0,s_1;I}(f) > (1+\gamma)\lambda.$$

On the other hand, by Lemma 5.1.2, we have

$$\left| m_f\left(\widetilde{\frac{5}{4}S^1_j}^{2\rho}\right)\right| \le m^{\rho}_{0,s_1;\,\widetilde{\frac{5}{4}S^1_j}^{2\rho}}(f) \le (1+\gamma/2)\lambda.$$

This, via Lemma 5.1.1, implies that

$$m^{\rho}_{0,s_1;I}\left(\left[f - m_f\left(\widetilde{\frac{5}{4}S^1_j}^{2\rho}\right)\right]\chi_{\frac{5}{4}S^1_j}\right) > \gamma\lambda/2$$

and hence

$$\left(S^1_j\bigcap E_\lambda\right) \subset \left\{y\in\mathbb{R}^D : \mathcal{M}^{\rho,d}_{0,s_1}\left(\left[f - m_f\left(\widetilde{\frac{5}{4}S^1_j}^{2\rho}\right)\right]\chi_{\frac{5}{4}S^1_j}\right)(y) > \gamma\lambda/2\right\}.$$

Invoking (ii) of Lemma 5.1.9 and the inequality (5.1.7) and noticing that $\theta_2 < 1/4$ by our choice, we conclude that, for some $\sigma\in(0,\infty)$ small enough,

$$\begin{aligned}
&\mu\left(S^1_j\bigcap E_\lambda\right)\\
&\lesssim s_1^{-1}\mu\left(\left\{y\in\frac{5}{4}S^1_j : \left|f(y) - m_f\left(\widetilde{\frac{5}{4}S^1_j}^{2\rho}\right)\right| \ge \gamma\lambda/2\right\}\right)\\
&\lesssim s_1^{-1}\mu\left(\left\{y\in\frac{5}{4}S^1_j : \left|f(y) - m_f\left(\widetilde{\frac{5}{4}S^1_j}^{2\rho}\right)\right| > 2\mathcal{M}^{\rho,\natural}_{0,s_2}f(x^1_j)+\sigma\right\}\right)\\
&\lesssim s_1^{-1}s_2\mu(2\rho S^1_j)\\
&\lesssim s_1^{-1}s_2\mu(S^1_j).
\end{aligned}$$

This finishes the proof of Theorem 5.1.11. □

Theorem 5.1.12. *Let* $\rho\in[1,\infty)$, $s_1\in(0,\beta_{2\rho}^{-1}/4)$, $p\in(0,\infty)$ *and* $u\in A^{\rho}_{\infty}(\mu)$. *Let* Θ *be a nonnegative increasing function on* $[0,\infty)$ *which satisfies the condition that*

$$\Theta(0)=0 \quad\text{and}\quad \Theta(t_1t_2)\le\tilde{\Theta}(t_1)\Theta(t_2) \quad \textit{for any } t_1\in[0,\infty) \quad\textit{and}\quad t_2\in[0,\infty),$$

where $\tilde{\Theta}$ is a continuous function on $[0, \infty)$ such that $\tilde{\Theta}(1) \le 1$. Then there exist a constant $c \in (0,1)$, depending on s_1 and u, and a positive constant C such that, for any $s_2 \in (0, cs_1)$,

(i) *if $\mu(\mathbb{R}^D) = \infty$, $f \in L^{p_0,\infty}(\mu)$ with $p_0 \in [1,\infty)$, and, for any $R \in (0,\infty)$,*

$$\sup_{\lambda\in(0,R)} \Theta(\lambda)u(\{x \in \mathbb{R}^D : |f(x)| > \lambda\}) < \infty,$$

then

$$\begin{aligned}&\sup_{\lambda\in(0,\infty)} \Theta(\lambda)u\left(\left\{x \in \mathbb{R}^D : \mathcal{M}^{\rho,d}_{0,s_1} f(x) > \lambda\right\}\right)\\&\quad\le C\sup_{\lambda\in(0,\infty)} \Theta(\lambda)u\left(\left\{x \in \mathbb{R}^D : \mathcal{M}^{\rho,\natural}_{0,s_2} f(x) > \lambda\right\}\right);\end{aligned}$$

(ii) *if $\mu(\mathbb{R}^D) < \infty$ and $f \in L^{p_0,\infty}(\mu)$ with $p_0 \in [1,\infty)$, then*

$$\begin{aligned}&\sup_{\lambda\in(0,\infty)} \Theta(\lambda)u\left(\left\{x \in \mathbb{R}^D : \mathcal{M}^{\rho,d}_{0,s_1} f(x) > \lambda\right\}\right)\\&\quad\le C\sup_{\lambda\in(0,\infty)} \Theta(\lambda)u\left(\left\{x \in \mathbb{R}^D : \mathcal{M}^{\rho,\natural}_{0,s_2} f(x) > \lambda\right\}\right)\\&\quad\quad+Cu(\mathbb{R}^D)\Theta(\lambda_f),\end{aligned}$$

where

$$\lambda_f := \left[s_1\mu(\mathbb{R}^D)\right]^{-1/p_0} \|f\|_{L^{p_0,\infty}(\mu)}.$$

Proof. By Lemma 5.1.3, we may assume that f is real-valued. We first consider the case that $\mu(\mathbb{R}^D) = \infty$. By Theorem 5.1.11, we see that, for any $\gamma \in (0,\infty)$ and $\lambda \in (0,\infty)$,

$$\begin{aligned}&u\left(\left\{x \in \mathbb{R}^D : \mathcal{M}^{\rho,d}_{0,s_1} f(x) > (1+\gamma)\lambda\right\}\right)\\&\quad\le \theta_1 u\left(\left\{x \in \mathbb{R}^D : \mathcal{M}^{\rho,d}_{0,s_1} f(x) > \lambda\right\}\right)\\&\quad\quad+u\left(\left\{x \in \mathbb{R}^D : \mathcal{M}^{\rho,\natural}_{0,s_2} f(x) > \theta_2\gamma\lambda\right\}\right)\end{aligned}$$

and consequently

$$\begin{aligned}&\Theta((1+\gamma)\lambda)\, u\left(\left\{x \in \mathbb{R}^D : \mathcal{M}^{\rho,d}_{0,s_1} f(x) > (1+\gamma)\lambda\right\}\right)\\&\quad\le \theta_1\Theta((1+\gamma)\lambda)\, u\left(\left\{x \in \mathbb{R}^D : \mathcal{M}^{\rho,d}_{0,s_1} f(x) > \lambda\right\}\right)\\&\quad\quad+\Theta((1+\gamma)\lambda)\, u\left(\left\{x \in \mathbb{R}^D : \mathcal{M}^{\rho,\natural}_{0,s_2} f(x) > \theta_2\gamma\lambda\right\}\right),\end{aligned}$$

where Θ_1, $\Theta_2 \in (0, 1)$ are as in Theorem 5.1.11. Taking the supremum in the last inequality implies that, for any $R \in (0, \infty)$,

$$\begin{aligned}
&\sup_{\lambda\in(0,(1+\gamma)R)} \Theta(\lambda)u\left(\left\{x \in \mathbb{R}^D : \mathcal{M}_{0,s_1}^{\rho,d} f(x) > \lambda\right\}\right) \\
&\quad \le \theta_1\tilde{\Theta}(1+\gamma) \sup_{\lambda\in(0,R)} \Theta(\lambda)u\left(\left\{x \in \mathbb{R}^D : \mathcal{M}_{0,s_1}^{\rho,d} f(x) > \lambda\right\}\right) \\
&\qquad + \tilde{\Theta}(1+\gamma) \sup_{\lambda\in(0,\infty)} \Theta(\lambda)u\left(\left\{x \in \mathbb{R}^D : \mathcal{M}_{0,s_2}^{\rho,\natural} f(x) > \theta_2\gamma\lambda\right\}\right).
\end{aligned}$$

By (ii) of Lemma 5.1.9, we see that

$$\begin{aligned}
&\sup_{\lambda\in(0,R)} \Theta(\lambda)u\left(\left\{x \in \mathbb{R}^D : \mathcal{M}_{0,s_1}^{\rho,d} f(x) > \lambda\right\}\right) \\
&\quad \lesssim \sup_{\lambda\in(0,R)} \Theta(\lambda)u\left(\left\{x \in \mathbb{R}^D : |f(x)| > \lambda\right\}\right).
\end{aligned}$$

Our hypotheses guarantee that, in this case,

$$\sup_{\lambda\in(0,R)} \Theta(\lambda)u\left(\left\{x \in \mathbb{R}^D : \mathcal{M}_{0,s_1}^{\rho,d} f(x) > \lambda\right\}\right) < \infty.$$

When $\mu(\mathbb{R}^D) = \infty$, choosing γ small enough such that $\tilde{\Theta}(1+\gamma)\theta_1 < 1$, we then conclude that

$$\begin{aligned}
&\sup_{\lambda\in(0,R)} \Theta(\lambda)u\left(\left\{x \in \mathbb{R}^D : \mathcal{M}_{0,s_1}^{\rho,d} f(x) > \lambda\right\}\right) \\
&\quad \lesssim \sup_{\lambda\in(0,R)} \Theta(\lambda)u\left(\left\{x \in \mathbb{R}^D : \mathcal{M}_{0,s_2}^{\rho,\natural} f(x) > \lambda\right\}\right).
\end{aligned}$$

We turn our attention to the case of $\mu(\mathbb{R}^D) < \infty$. Another application of Theorem 5.1.11 gives that, for any $R > \lambda_f$ and $\gamma \in (0, 1)$,

$$\begin{aligned}
&\sup_{\lambda\in(0,(1+\gamma)R)} \Theta(\lambda)u\left(\left\{x \in \mathbb{R}^D : \mathcal{M}_{0,s_1}^{\rho,d} f(x) > \lambda\right\}\right) \\
&\le \sup_{\lambda\in[(1+\gamma)\lambda_f,(1+\gamma)R)} \Theta(\lambda)u\left(\left\{x \in \mathbb{R}^D : \mathcal{M}_{0,s_1}^{\rho,d} f(x) > \lambda\right\}\right) + \sup_{\lambda\in(0,(1+\gamma)\lambda_f)} \cdots \\
&\le \tilde{\Theta}(1+\gamma) \sup_{\lambda\in[\lambda_f,R)} \Theta(\lambda)u\left(\left\{x \in \mathbb{R}^D : \mathcal{M}_{0,s_1}^{\rho,d} f(x) > (1+\gamma)\lambda\right\}\right) \\
&\quad + \tilde{\Theta}(1+\gamma)\Theta(\lambda_f)u(\mathbb{R}^D)
\end{aligned}$$

$$\leq \tilde{\Theta}(1+\gamma)\theta_1 \sup_{\lambda\in[\lambda_f, R)} \Theta(\lambda)u\left(\left\{x \in \mathbb{R}^D : \mathcal{M}_{0,s_1}^{\rho,d} f(x) > \lambda\right\}\right)$$
$$+\tilde{\Theta}(1+\gamma) \sup_{\lambda\in(0,\infty)} \Theta(\lambda)u\left(\left\{x \in \mathbb{R}^D : \mathcal{M}_{0,s_2}^{\rho,\natural} f(x) > \theta_2\gamma\lambda\right\}\right)$$
$$+\tilde{\Theta}(1+\gamma)\Theta(\lambda_f)u(\mathbb{R}^D),$$

where $\Theta_1,\ \Theta_2 \in (0,1)$ are as in Theorem 5.1.11. By the implicity that $\mu(\mathbb{R}^D) < \infty$ implies $u(\mathbb{R}^D) < \infty$, we then conclude that, when $\mu(\mathbb{R}^D) < \infty$,

$$\sup_{\lambda\in(0,R)} \Theta(\lambda)u\left(\left\{x \in \mathbb{R}^D : \mathcal{M}_{0,s_1}^{\rho,d} f(x) > \lambda\right\}\right)$$
$$\lesssim \sup_{\lambda\in(0,R)} \Theta(\lambda)u\left(\left\{x \in \mathbb{R}^D : \mathcal{M}_{0,s_2}^{\rho,\natural} f(x) > \lambda\right\}\right) + \Theta(\lambda_f)u(\mathbb{R}^D).$$

Taking $R \to \infty$ we then obtain the desired conclusion (ii), which completes the proof of Theorem 5.1.12. □

Similar to Theorem 5.1.12, we have the following conclusion.

Theorem 5.1.13. *Let $\rho \in [1,\infty)$, $s_1 \in (0, \beta_{2\rho}^{-1}/4)$, $p \in (0,\infty)$ and $u \in A_\infty^\rho(\mu)$. Let Θ be a nonnegative increasing function on $[0, \infty)$ which satisfies the condition that*

$$\Theta(0)=0 \quad \text{and} \quad \tilde{\Theta}(t_1t_2)\leq\tilde{\Theta}(t_1)\tilde{\Theta}(t_2) \quad \textit{for any} \quad t_1\in[0,\infty) \quad \textit{and} \quad t_2\in[0,\infty),$$

where $\tilde{\Theta}$ is a continuous function on $[0, \infty)$ such that $\tilde{\Theta}(1) \leq 1$. Then there exist a constant $\tilde{c} \in (0, 1)$, depending on s_1 and u, and a positive constant C such that, for any $s_2 \in (0, \tilde{c}s_1)$,

(i) *if $\mu(\mathbb{R}^D) = \infty$, $f \in L^{p_0,\infty}(\mu)$ with $p_0 \in [1, \infty)$, and, for any $R \in (0,\infty)$,*

$$\int_0^R u(\{x \in \mathbb{R}^D : |f(x)| > \lambda\})\, d\Theta(\lambda) < \infty,$$

then

$$\int_{\mathbb{R}^D} \Theta\left(\mathcal{M}_{0,s_1}^{\rho,d} f(x)\right) u(x)\, d\mu(x) \leq C \int_{\mathbb{R}^D} \Theta\left(\mathcal{M}_{0,s_2}^{\rho,\natural} f(x)\right) u(x)\, d\mu(x);$$

(ii) *if $\mu(\mathbb{R}^D) < \infty$ and $f \in L^{p_0,\infty}(\mu)$ with $p_0 \in [1, \infty)$, then*

$$\int_{\mathbb{R}^D} \Theta\left(\mathcal{M}_{0,s_1}^{\rho,d} f(x)\right) u(x)\, d\mu(x) \leq C \int_{\mathbb{R}^D} \Theta\left(\mathcal{M}_{0,s_2}^{\rho,\natural} f(x)\right) u(x)\, d\mu(x)$$
$$+Cu(\mathbb{R}^D)\Theta\left(\lambda_f\right),$$

where

$$\lambda_f := \left(s_1\mu(\mathbb{R}^D)\right)^{-1/p_0} \|f\|_{L^{p_0,\infty}(\mu)}.$$

Proof. Notice that

$$\int_{\mathbb{R}^D} \Theta\left(\mathcal{M}_{0,s}^{\rho,d} f(x)\right) u(x)\,d\mu(x)$$
$$= \lim_{R\to\infty} \int_0^R u(\{x \in \mathbb{R}^D : \mathcal{M}_{0,s}^{\rho,d} f(x) > \lambda\})\,d\Theta(\lambda).$$

On the other hand, by Theorem 5.1.11, we see that

$$u\left(\left\{x \in \mathbb{R}^D : \mathcal{M}_{0,s_1}^{\rho,d} f(x) > (1+\gamma)\lambda\right\}\right)$$
$$\leq \theta_1 u\left(\left\{x \in \mathbb{R}^D : \mathcal{M}_{0,s_1}^{\rho,d} f(x) > \lambda\right\}\right)$$
$$+u(\{x \in \mathbb{R}^D : \mathcal{M}_{0,s_2}^{\rho,\natural} f(x) > \theta_2\gamma\lambda\})$$

and hence, for any $R \in (0,\infty)$,

$$\int_0^R u(\{x \in \mathbb{R}^D : \mathcal{M}_{0,s_1}^{\rho,d} f(x) > \lambda\})\,d\Theta(\lambda)$$
$$\leq \tilde{\Theta}(1+\gamma) \int_0^R u(\{x \in \mathbb{R}^D : \mathcal{M}_{0,s_1}^{\rho,d} f(x) > (1+\gamma)\lambda\})\,d\Theta(\lambda)$$
$$\leq \theta_1\tilde{\Theta}(1+\gamma) \int_0^R u\left(\left\{x \in \mathbb{R}^D : \mathcal{M}_{0,s_1}^{\rho,d} f(x) > \lambda\right\}\right) d\Theta(\lambda)$$
$$+\tilde{\Theta}(1+\gamma) \int_0^R u(\{x \in \mathbb{R}^D : \mathcal{M}_{0,s_2}^{\rho,\natural} f(x) > \theta_2\gamma\lambda\})\,d\Theta(\lambda)$$
$$\leq \theta_1\tilde{\Theta}(1+\gamma) \int_0^R u\left(\left\{x \in \mathbb{R}^D : \mathcal{M}_{0,s_1}^{\rho,d} f(x) > \lambda\right\}\right) d\Theta(\lambda)$$
$$+C \int_{\mathbb{R}^D} \Theta(\mathcal{M}_{0,s_2}^{\rho,\natural}(f)(x))\,u(x)\,d\mu(x),$$

where C is a positive constant and $\Theta_1,\ \Theta_2 \in (0,1)$ are as in Theorem 5.1.11. Our assumption, along with (ii) of Lemma 5.1.9, implies that

$$\int_0^R u\left(\left\{x \in \mathbb{R}^D : \mathcal{M}_{0,s_1}^{\rho,d} f(x) > \lambda\right\}\right) d\Theta(\lambda) < \infty.$$

If we choose γ sufficiently small such that $\theta_1\tilde{\Theta}(1+\gamma) < 1$, we then obtain the conclusion (i).

To prove (ii), we notice that

$$\int_{\mathbb{R}^D} \Theta\left(\mathcal{M}_{0,s}^{\rho,d} f(x)\right) u(x)\, d\mu(x)$$
$$\leq \int_{\lambda_f}^{\infty} u\left(\{\mathcal{M}_{0,s}^{\rho,d} f(x) > \lambda\}\right) d\Theta(\lambda) + \Theta(\lambda_f)u(\mathbb{R}^D).$$

An argument, invoking Theorem 5.1.12 which is now familiar for us, leads to that, when $s_2 \in (0, \tilde{C}_2^{-1}s_1)$,

$$\int_{\lambda_f}^{\infty} u\left(\{\mathcal{M}_{0,s}^{\rho,d} f(x) > \lambda\}\right) d\Theta(\lambda) \lesssim \int_{\mathbb{R}^D} \Theta(\mathcal{M}_{0,s_2}^{\rho,\natural}(f)(x))u(x)\, d\mu(x).$$

The conclusion (ii) now follows directly, which completes the proof of Theorem 5.1.13. □

5.2 Interpolation Theorems Related to $H^1(\mu)$

This section is devoted to some interpolation theorems related to $H^1(\mu)$. We see that, in the interpolation theory, the space $H^1(\mu)$ with the measure μ as in (0.0.1), is a good substitution of the classical Hardy space $H^1(\mathbb{R}^D)$. We begin with some preliminary lemmas.

Lemma 5.2.1. *Let T, T_1 and T_2 be three operators such that, for any $f_1, f_2 \in L_c^\infty(\mu)$, and μ-almost every $x \in \mathbb{R}^D$,*

$$|T(f_1 + f_2)(x)| \leq |T_1 f_1(x)| + |T_2 f_2(x)|.$$

Suppose that

(i) *for some p_0, q_0 with $p_0 \leq q_0$ and $p_0, q_0 \in (1, \infty]$, T_1 is bounded from $L^{p_0}(\mu)$ to $L^{q_0,\infty}(\mu)$;*
(ii) *for some $q_1 \in [1, q_0)$, T_2 is bounded from $H^1(\mu)$ to $L^{q_1,\infty}(\mu)$, that is, there exists a positive constant C such that, for any $\lambda \in (0,\infty)$ and $f \in H^1(\mu)$,*

$$\mu(\{x \in \mathbb{R}^D : |T_2 f(x) > \lambda\}) \leq C\left[\lambda^{-1}\|f\|_{H^1(\mu)}\right]^{q_1}.$$

Then, for any p, q with

$$\frac{1}{p} = t + \frac{1-t}{p_0},\ \frac{1}{q} = \frac{t}{q_1} + \frac{1-t}{q_0},\quad t \in (0, 1),$$

there exists a positive constant C such that, for any bounded function f with compact support,

$$\sup_{\lambda\in(0,\infty)} \lambda^q \mu(\{x \in \mathbb{R}^D : |Tf(x)| > \lambda\}) \le C\|f\|^q_{L^p(\mu)}.$$

Proof. Our goal is to prove that, for any positive constant λ and bounded function f with compact support,

$$\lambda^q \mu(\{x \in \mathbb{R}^D : |Tf(x)| > \lambda\}) \lesssim \left(\int_{\mathbb{R}^D} |f(x)|^p \, d\mu(x)\right)^{\frac{q}{p}}. \tag{5.2.1}$$

For each fixed $\lambda \in (0,\infty)$ and bounded function f with compact support, observe that, if $\mu(\mathbb{R}^D) < \infty$ and

$$\lambda \le \|f\|_{L^p(\mu)}/\left[\mu\left(\mathbb{R}^D\right)\right]^{1/q},$$

the inequality (5.2.1) follows directly, since

$$\lambda^q \mu(\{x \in \mathbb{R}^D : |Tf(x)| > \lambda\}) \lesssim \|f\|^q_{L^p(\mu)}.$$

Thus, we may assume $\mu(\mathbb{R}^D) = \infty$, or

$$\mu\left(\mathbb{R}^D\right) < \infty \quad \text{and} \quad \lambda > \|f\|_{L^p(\mu)}/\left[\mu\left(\mathbb{R}^D\right)\right]^{1/q}.$$

Notice that

$$\frac{\frac{1}{q}-\frac{1}{q_0}}{\frac{1}{q_1}-\frac{1}{q_0}} = \frac{\frac{1}{p}-\frac{1}{p_0}}{1-\frac{1}{p_0}}, \quad \frac{\frac{1}{q_1}-\frac{1}{q}}{\frac{1}{q_1}-\frac{1}{q_0}} = \frac{1-\frac{1}{p}}{1-\frac{1}{p_0}}.$$

We then see that

$$\frac{\frac{1}{q}-\frac{1}{q_0}}{\frac{1}{q_1}-\frac{1}{q}} = \frac{\frac{1}{p}-\frac{1}{p_0}}{1-\frac{1}{p}}$$

and hence

$$\frac{(q_0-q)p_0}{(p_0-p)q_0} = \frac{q-q_1}{(p-1)q_1}.$$

Let

$$\theta(p) := \frac{q-q_1}{(p-1)q_1}.$$

By homogeneity, we may assume that $\|f\|_{L^p(\mu)} = 1$. Applying the Calderón–Zygmund decomposition to $|f|^p$ at level $\lambda^{p\theta(p)}$, we know that there exists a sequence $\{Q_j\}_j$ of cubes such that

(a) the cubes $\{Q_j\}_j$ have bounded overlaps, that is,

$$\sum_j \chi_{Q_j}(x) \lesssim 1;$$

(b)

$$\frac{1}{\mu(2Q_j)} \int_{Q_j} |f(x)|^p \, d\mu(x) > \frac{\lambda^{p\theta(p)}}{2^{D+1}};$$

(c) for any $\eta \in (2, \infty)$,

$$\frac{1}{\mu(2\eta Q_j)} \int_{\eta Q_j} |f(x)|^p \, d\mu(x) \le \frac{\lambda^{p\theta(p)}}{2^{D+1}};$$

(d) $|f(x)| \le \lambda^{\theta(p)}$ for μ-almost every $x \in \mathbb{R}^D \setminus \cup_j Q_j$;
(e) for each fixed j, let R_j be the *smallest* $(6, 6^{D+1})$*-doubling cube of the form* $6^k Q_j$ *with* $k \in \mathbb{N}$. Let

$$w_j := \chi_{Q_j} / \sum_k \chi_{Q_k}.$$

Then there exists a function ϕ_j with $\operatorname{supp} \phi_j \subset R_j$ satisfying that

$$\int_{\mathbb{R}^D} \phi_j(x) \, d\mu(x) = \int_{Q_j} f(x) w_j(x) \, d\mu(x),$$

$$\sum_j |\phi_j(x)| \lesssim \lambda^{\theta(p)} \quad \text{for } \mu - \text{almost every } x \in \mathbb{R}^D,$$

and

$$\left[\int_{R_j} |\phi(x)|^p \, d\mu(x)\right]^{1/p} [\mu(R_j)]^{1/p'} \lesssim \frac{1}{\lambda^{(p-1)\theta(p)}} \int_{Q_j} |f(x)|^p \, d\mu(x).$$

Decompose f as

$$f(x) = g(x) + b(x),$$

where

$$g(x) := f(x)\chi_{\mathbb{R}^D\setminus\bigcup_j Q_j}(x) + \sum_j \phi_j(x)$$

and

$$b(x) := \sum_j \left[f(x)w_j(x) - \phi_j(x)\right].$$

It is easy to show that

$$\|g\|_{L^\infty(\mu)} \lesssim \lambda^{\theta(p)},$$

$b \in H^1(\mu)$ and

$$\|b\|_{H^{1,p}_{\mathrm{atb}}(\mu)} \lesssim \lambda^{-(p-1)\theta(p)}.$$

Moreover,

$$\|g\|^{p_0}_{L^{p_0}(\mu)} \le \|g\|^{p_0-p}_{L^\infty(\mu)}\|g\|^p_{L^p(\mu)} \lesssim \lambda^{(p_0-p)\theta(p)}.$$

This in turn leads to that

$$\begin{aligned}\mu(\{x \in \mathbb{R}^D : |T_1 g(x)| > \lambda/2\}) &\lesssim \lambda^{-q_0}\|g\|^{q_0}_{L^{p_0}(\mu)}\\ &\lesssim \lambda^{-q_0}\lambda^{(p_0-p)q_0\theta(p)/p_0}\\ &\lesssim \lambda^{-q} \end{aligned} \tag{5.2.2}$$

and

$$\mu(\{x \in \mathbb{R}^D : |T_2 b(x)| > \lambda/2\}) \lesssim \lambda^{-q_1}\|b\|^{q_1}_{H^{1,p}_{\mathrm{atb}}(\mu)} \lesssim \lambda^{-q}.$$

Combining these inequalities, we then complete the proof of Lemma 5.2.1. □

For the special case that $q_1 = 1$ and $p_0 = q_0$ in Lemma 5.2.1, we have the following more general conclusion.

Lemma 5.2.2. *Let T, T_1 and T_2 be three operators such that, for all $f_1, f_2 \in L^\infty_c(\mu)$, and μ-almost every $x \in \mathbb{R}^D$,*

$$|T(f_1 + f_2)(x)| \le C[|T_1 f_1(x)| + |T_2 f_2(x)|].$$

Suppose that

(i) *there exists an operator T_3, which is bounded on $L^{p_0}(\mu)$ for some $p_0 \in (1, \infty]$, such that, for any bounded function f with compact support and $x \in \mathbb{R}^D$,*

$$|T_1(f)(x)| \le |T_3(f)(x)| + A\|f\|_{L^\infty(\mu)};$$

(ii) *T_2 is bounded from $H^1(\mu)$ to $L^{1,\infty}(\mu)$.*

Then, for any $p \in (1, p_0)$, there exists a positive constant C such that, for any bounded function f with compact support,

$$\sup_{\lambda\in(0,\infty)} \lambda^p \mu(\{x \in \mathbb{R}^D : |Tf(x)| > \lambda\}) \leq C\|f\|^p_{L^p(\mu)}.$$

Proof. If $q_1 = 1$ and $p_0 = q_0$, then we see that $q = p$ and $\theta(p) = 1$. Decompose $f = g + b$ as in Lemma 5.2.1. Replacing the estimate (5.2.2) by that

$$\begin{aligned}\mu(\{x \in \mathbb{R}^D : |T_1 g(x)| > (A_0 + 1)\lambda\}) &\leq \mu(\{x \in \mathbb{R}^D : |T_3 g(x)| > \lambda\})\\ &\lesssim \lambda^{-p_0}\|g\|^{p_0}_{L^{p_0}(\mu)}\\ &\lesssim \lambda^{-p}\end{aligned}$$

and repeating the argument similar to that used in the proof of Lemma 5.2.1, we conclude that, for any $\lambda \in (0, \infty)$,

$$\mu\left(\left\{x \in \mathbb{R}^D : |Tf(x)| \geq \left(A_0 + \frac{3}{2}\right)\lambda\right\}\right) \lesssim \lambda^p\|f\|^p_{L^p(\mu)}.$$

This finishes the proof of Lemma 5.2.1. □

Lemma 5.2.3. *Let $s \in (0, 2^{-D-2})$ and T be an operator which satisfies that, for suitable functions f_1 and f_2, and μ-almost every $x \in \mathbb{R}^D$,*

$$|Tf_1(x) - Tf_2(x)| \leq |T(f_1 - f_2)(x)|.$$

Then there exists a positive constant C such that, for any f_1 and f_2,

$$\mathcal{M}^\natural_{0,s}[T(f_1 + f_2)](x) \leq C\left[\mathcal{M}^\natural_{0,s/2}(Tf_1)(x) + \mathcal{M}_{0,s/2}(Tf_2)(x)\right].$$

Proof. For any cube Q, via a straightforward computation, we know that

$$\begin{aligned}m_{0,s;Q}\left(T(f_1 + f_2) - m_{T(f_1+f_2)}(\tilde{Q})\right) \leq\ & m_{0,s/2;Q}\left(Tf_1 - m_{Tf_1}(\tilde{Q})\right)\\ &+m_{0,s/2;Q}\left(T(f_1 + f_2) - Tf_1\right)\\ &+\left|m_{T(f_1+f_2)}(\tilde{Q}) - m_{Tf_1}(\tilde{Q})\right|.\end{aligned}$$

By (5.1.10), we see that

$$\begin{aligned}\left|m_{T(f_1+f_2)}(\tilde{Q}) - m_{Tf_1}(\tilde{Q})\right| &= \left|m_{T(f_1+f_2)-m_{Tf_1}(\tilde{Q})}(\tilde{Q})\right|\\ &\leq 2m_{0,s;\tilde{Q}}\left(T(f_1 + f_2) - m_{Tf_1}(\tilde{Q})\right)\end{aligned}$$

$$\begin{aligned}&\leq 2m_{0,s;\tilde{Q}}\left(Tf_1 - m_{Tf_1}(\tilde{Q})\right)\\&\quad+2m_{0,s;\tilde{Q}}\left(T(f_1+f_2) - Tf_1\right),\end{aligned}$$

where the second inequality follows from Lemma 5.1.2. This in turn implies that

$$\begin{aligned}&m_{0,s;Q}\left(T(f_1+f_2) - m_{T(f_1+f_2)}(\tilde{Q})\right)\\&\quad\leq 3\inf_{x\in Q}\mathcal{M}^{\natural}_{0,s/2}(Tf_1)(x) + 3\inf_{x\in Q}\mathcal{M}_{0,s/2}(Tf_2)(x).\end{aligned}$$

On the other hand, for any two doubling cubes $Q \subset R$, we know that

$$\begin{aligned}\left|m_{T(f_1+f_2)}(Q) - m_{T(f_1+f_2)}(R)\right| &\leq \left|m_{T(f_1+f_2)}(Q) - m_{Tf_1}(Q)\right|\\&\quad+\left|m_{T(f_1+f_2)}(R) - m_{Tf_1}(R)\right|\\&\quad+\left|m_{Tf_1}(Q) - m_{Tf_1}(R)\right|\\&\leq 2m_{0,2^{-(d+3)};Q}\left(Tf_1 - m_{Tf_1}(Q)\right)\\&\quad+2m_{0,2^{-(d+3)};Q}\left(T(f_1+f_2) - Tf_1\right)\\&\quad+2m_{0,2^{-(d+3)};R}\left(Tf_1 - m_{Tf_1}(R)\right)\\&\quad+2m_{0,2^{-(d+3)};R}\left(T(f_1+f_2) - Tf_1\right)\\&\quad+|m_{Tf_1}(Q) - m_{Tf_1}(R)|\\&\leq 4\inf_{x\in Q}\mathcal{M}^{\natural}_{0,s/2}(Tf_1) + 4\inf_{x\in Q}\mathcal{M}_{0,s/2}(Tf_2)(x)\\&\quad+|m_{Tf_1}(Q) - m_{Tf_1}(R)|,\end{aligned}$$

which completes the proof of Lemma 5.2.3. □

We now formulate the main results of this section as follows.

Theorem 5.2.4. *Let $\mu(\mathbb{R}^D) = \infty$ and T be an operator which satisfies that*

(i)

$$|Tf_1 - Tf_2| \leq |T(f_1 - f_2)|$$

μ-almost everywhere for all $f_1, f_2 \in L^\infty_c(\mu)$;

(ii) *there exists another operator T_1, which is bounded from $L^{p_0}(\mu)$ to $L^{q_0}(\mu)$ for some p_0, q_0 with $1 < p_0 \leq q_0 \leq \infty$, such that, for any $s \in (0, 2^{-D-2})$ and bounded function f with compact support,*

$$\mathcal{M}^{\natural}_{0,s}(Tf)(x) \leq |T_1 f(x)|;$$

(iii) *for some $q_1 \in [1, q_0)$, T is bounded from $H^1(\mu)$ to $L^{q_1,\infty}(\mu)$.*

Then, for any $p, q \in (1, \infty)$ *with*

$$\frac{1}{p} = t + \frac{1-t}{p_0}, \quad \frac{1}{q} = \frac{t}{q_1} + \frac{1-t}{q_0} \quad \text{and} \quad t \in (0, 1),$$

T is bounded from $L^p(\mu)$ to $L^q(\mu)$.

Proof. Observe that, for any $s \in (0, 2^{-D-2})$, our assumption (ii) tells us that the operator $\mathcal{M}_{0,s}^{\natural} \circ T$ is bounded from $L^{p_0}(\mu)$ to $L^{q_0,\infty}(\mu)$. On the other hand, the assumption (iii) in Theorem 5.2.4, along with Lemma 5.1.9, states that $\mathcal{M}_{0,s} \circ T$ is bounded from $H^1(\mu)$ to $L^{q_1,\infty}(\mu)$. It then follows, from Lemmas 5.2.1 and 5.2.3, that $\mathcal{M}_{0,s}^{\natural} \circ T$ is bounded from $L^p(\mu)$ to $L^{q,\infty}(\mu)$ for any p, q with

$$\frac{1}{p} = t + \frac{1-t}{p_0}, \quad \frac{1}{q} = \frac{t}{q_1} + \frac{1-t}{q_0} \quad \text{and} \quad t \in (0, 1).$$

Let

$$L_{c,0}^{\infty}(\mu) := \left\{ f : \ f \text{ is bounded, has compact support, } \int_{\mathbb{R}^D} f(x) d\mu(x) = 0 \right\}.$$

It is well known that $L_{c,0}^{\infty}(\mu)$ is a density subset of $L^p(\mu)$ for any $p \in [1, \infty)$. For each fixed $f \in L_{c,0}^{\infty}(\mu)$, our hypotheses guarantee that $Tf \in L^{q_1,\infty}(\mu)$ and hence, for any $R \in (0, \infty)$, if $q > q_1$, then

$$\begin{aligned} &\int_0^R \lambda^{q-1} \mu(\{x \in \mathbb{R}^D : |Tf(x)| > \lambda\}) \, d\lambda \\ &\quad \le R^{q-q_1} \sup_{\lambda \in (0,\infty)} \lambda^{q_1} \mu(\{x \in \mathbb{R}^D : |Tf(x)| > \lambda\}). \end{aligned}$$

Thus, from (i) of Lemma 5.1.9 and Theorem 5.1.12, it follows that, for all $f \in L_{c,0}^{\infty}(\mu)$,

$$\|Tf\|_{L^{q,\infty}(\mu)} \le \|\mathcal{M}_{0,s}^{d}(Tf)\|_{L^{q,\infty}(\mu)} \lesssim \|f\|_{L^p(\mu)}.$$

Notice that T is sublinear. Thus, for all $f \in L_{c,0}^{\infty}(\mu)$, it holds true that

$$\|Tf\|_{L^q(\mu)} \lesssim \|f\|_{L^p(\mu)},$$

which completes the proof of Theorem 5.2.4. □

At the end of this section, we list some variants of Theorem 5.2.4, whose proofs are similar to that of Theorem 5.2.4. We omit the details.

Theorem 5.2.5. *Let $\mu(\mathbb{R}^D) = \infty$ and T be an operator which satisfies that*

(i)

$$|Tf_1 - Tf_2| \le |T(f_1 - f_2)|$$

μ-almost everywhere for all bounded functions f_1 and f_2 with compact support;

(ii) *there exists an operator T_1, which is bounded on $L^{p_0}(\mu)$ for some $p_0 \in (1, \infty]$, such that, for any $s \in (0, 2^{-D-2})$, bounded function f with compact support and μ-almost every $x \in \mathbb{R}^D$,*

$$\mathcal{M}^{\natural}_{0,s}(Tf)(x) \le |T_1 f(x)| + A\|f\|_{L^\infty(\mu)};$$

(iii) *T is bounded from $H^1(\mu)$ to $L^{1,\infty}(\mu)$.*

Then, for any $p \in (1, p_0)$, T is bounded on $L^p(\mu)$.

Remark 5.2.6. For the case that $\mu(\mathbb{R}^D) < \infty$, we do not know whether Theorems 5.2.4 and 5.2.5 arc still true or not.

Theorem 5.2.7. *Let T be an operator which satisfies that*

(i)

$$|Tf_1 - Tf_2| \le |T(f_1 - f_2)|$$

μ-almost everywhere for all bounded functions f_1 and f_2 with compact support;

(ii) *there exists another operator T_1, which is bounded from $L^{p_0}(\mu)$ to $L^{q_0}(\mu)$ for some p_0, q_0 with $1 < p_0 \le q_0 \le \infty$, such that, for any $s \in (0, 2^{-D-2})$, bounded function f with compact support and μ-almost every $x \in \mathbb{R}^D$,*

$$\mathcal{M}^{\natural}_{0,s}(Tf)(x) \le |T_1 f(x)|;$$

(iii) *for some $p_1 \in (1, p_0)$ and $q_1 \in [1, q_0)$ with $p_1 \le q_1$, T is bounded from $L^{p_1}(\mu)$ to $L^{q_1,\infty}(\mu)$.*

Then, for any $p, q \in (1, \infty)$ with

$$\frac{1}{p} = \frac{t}{p_1} + \frac{1-t}{p_0}, \quad \frac{1}{q} = \frac{t}{q_1} + \frac{1-t}{q_0} \quad \text{and} \quad t \in (0, 1),$$

T is bounded from $L^p(\mu)$ to $L^q(\mu)$.

Theorem 5.2.8. *Let T be an operator which satisfies that*

(i)

$$|Tf_1 - Tf_2| \le |T(f_1 - f_2)|$$

μ-almost everywhere for all bounded functions f_1 and f_2 with compact support;

(ii) *there exists an operator T_1, which is bounded on $L^{p_0}(\mu)$ for some $p_0 \in (1, \infty]$, such that, for any $s \in (0, 2^{-D-2})$, bounded function f with compact support and μ-almost every $x \in \mathbb{R}^D$,*

$$\mathcal{M}^{\natural}_{0,s}(Tf)(x) \leq |T_1 f(x)| + A\|f\|_{L^\infty(\mu)};$$

(iii) *T is bounded from $L^{p_1}(\mu)$ to $L^{p_1,\infty}(\mu)$.*

Then T is bounded on $L^p(\mu)$ for any $p \in (p_1, p_0)$.

5.3 $L^p(\mu)$ Boundedness for Singular Integral Operators

The definition of singular integral operators with measure as in (0.0.1) is similar to that of the classical singular integral. Let K be a function on

$$(\mathbb{R}^D \times \mathbb{R}^D)\backslash\{(x, y) : x = y\},$$

for which there exists a positive constant C such that, for all $x, y \in \mathbb{R}^D$ and $x \neq y$,

$$|K(x, y)| \leq C|x - y|^{-n} \tag{5.3.1}$$

and, for all $y, \tilde{y} \in \mathbb{R}^D$,

$$\int_{|x-y|\geq 2|y-\tilde{y}|} \left[|K(x, y) - K(x, \tilde{y})| + |K(y, x) - K(\tilde{y}, x)|\right] d\mu(x) \leq C. \tag{5.3.2}$$

Associated with the kernel K, we define the *singular integral operator* T by setting, for any bounded function f with compact support and $x \notin \operatorname{supp} f$,

$$Tf(x) := \int_{\mathbb{R}^D} K(x, y) f(y)\, d\mu(y).$$

However, this integral may not be convergent for many functions f, because the kernel K may have a singularity for $x = y$. Thus, we consider the *truncated operator* T_ϵ for any $\epsilon \in (0, \infty)$, which is defined by setting, for any bounded functions f with compact support and $x \notin \operatorname{supp} f$,

$$T_\epsilon f(x) := \int_{|x-y|\geq\epsilon} K(x, y) f(y)\, d\mu(y). \tag{5.3.3}$$

Proposition 5.3.1. *Let $p \in (1,\infty)$. If $\{T_\epsilon\}_{\epsilon\in(0,\infty)}$ is bounded on $L^p(\mu)$ uniformly on $\epsilon \in (0,\infty)$, then there exists an operator $\tilde{T}$ which is the weak limit as $\epsilon \to 0$ of some subsequence of the uniformly bounded operators $\{T_\epsilon\}_{\epsilon\in(0,\infty)}$. Moreover, the operator $\tilde{T}$ is also bounded on $L^p(\mu)$ and satisfies that, for $f \in L^p(\mu)$ with compact support and μ-almost every $x \notin \operatorname{supp} f$,*

$$\tilde{T} f(x) = \int_{\mathbb{R}^D} K(x,y) f(y)\, d\mu(y).$$

Proof. Let c_0 be a positive constant such that, for any $\epsilon \in (0,\infty)$ and $f \in L^p(\mu)$,

$$\|T_\epsilon f\|_{L^p(\mu)} \le c_0 \|f\|_{L^p(\mu)}.$$

Since $L^p(\mu)$ is separable, there exists a countable subset $\{f_i\}_{i\in\mathbb{N}}$ which is dense in $L^p(\mu)$. Let

$$\{\epsilon_j\}_{j\in\mathbb{N}} \subset (0,\infty)$$

such that $\epsilon_j \to 0$ as $j \to \infty$. Because $\{T_{\epsilon_j} f_1\}_{j\in\mathbb{N}}$ is uniformly bounded on $L^p(\mu)$, by the Alaoglu theorem, we conclude that there exists a subsequence of $\{T_{\epsilon_j} f_1\}$, which is denoted by $\{T_{\epsilon_j^{(1)}} f_1\}_{j\in\mathbb{N}}$, such that $\{T_{\epsilon_j^{(1)}} f_1\}_{j\in\mathbb{N}}$ has a $*$-weak limit. Likewise, for $f_2 \in L^p(\mu)$, for the uniformly bounded sequence $\{T_{\epsilon_j^{(1)}} f_2\}_{j\in\mathbb{N}}$ on $L^p(\mu)$, another application of the Alaoglu theorem implies that there exists a subsequence of $\{T_{\epsilon_j^{(1)}} f_2\}_{j\in\mathbb{N}}$, denoted by $\{T_{\epsilon_j^{(2)}} f_2\}_{j\in\mathbb{N}}$, such that $\{T_{\epsilon_j^{(2)}} f_2\}_{j\in\mathbb{N}}$ has a $*$-weak limit. Observe that $\{T_{\epsilon_j^{(2)}} f_1\}_{j\in\mathbb{N}}$ also has a $*$-weak limit. Repeating the procedure above, we find a subsequence of $\{T_{\epsilon_j}\}_{j\in\mathbb{N}}$, denoted by $\{T_n\}_{n\in\mathbb{N}}$, such that, for any f_i, $i \in \mathbb{N}$, $\{T_n f_i\}_{n\in\mathbb{N}}$ has a $*$-weak limit.

Now we show that, for any $f \in L^p(\mu)$, $\{T_n f\}_{n\in\mathbb{N}}$ has $*$-weak limit. For any $f \in L^p(\mu)$, by the fact that $\{T_n\}_{n\in\mathbb{N}}$ is uniformly bounded on $L^p(\mu)$, there exists a subsequence $\{T_{n_k}\}_{k\in\mathbb{N}}$ such that $n_k \to \infty$ as $k \to \infty$ and $\{T_{n_k} f\}_{k\in\mathbb{N}}$ has $*$-weak limit, denoted by $\tilde{T} f$. That is, for any $g \in L^{p'}(\mu)$,

$$\lim_{k\to\infty} \int_{\mathbb{R}^D} T_{n_k} f(x) g(x)\, d\mu(x) = \int_{\mathbb{R}^D} \tilde{T} f(x) g(x)\, d\mu(x).$$

Therefore, for any $\delta \in (0,\infty)$, there exists $N_1 \in \mathbb{N}$ such that, when $k > N_1$, then

$$\left| \int_{\mathbb{R}^D} T_{n_k} f(x) g(x)\, d\mu(x) - \int_{\mathbb{R}^D} \tilde{T} f(x) g(x)\, d\mu(x) \right| < \delta/2. \tag{5.3.4}$$

We claim that

$$\left\{ \int_{\mathbb{R}^D} T_n f(x) g(x)\, d\mu(x) \right\}_{n\in\mathbb{N}}$$

is a Cauchy sequence. Indeed, for $f \in L^p(\mu)$, by the density of $\{f_i\}_{i\in\mathbb{N}}$ in $L^p(\mu)$, we conclude that, for any $\epsilon \in (0, \infty)$, there exists $i \in \mathbb{N}$ such that

$$\|f - f_i\|_{L^p(\mu)} < \frac{\epsilon}{3c_0\|g\|_{L^{p'}(\mu)} + 1}. \tag{5.3.5}$$

Because $\{T_n f_i\}_{n\in\mathbb{N}}$ has $*$-weak limit, for the fixed $g \in L^{p'}(\mu)$, we see that

$$\lim_{n\to\infty} \int_{\mathbb{R}^D} T_n f_i(x) g(x)\, d\mu(x) = \int_{\mathbb{R}^D} \tilde{T} f_i(x) g(x)\, d\mu(x).$$

Thus, for $\epsilon \in (0, \infty)$ as above, there exists $N \in \mathbb{N}$ such that, when $m, n > N$, we have

$$\left|\int_{\mathbb{R}^D} T_n f_i(x) g(x)\, d\mu(x) - \int_{\mathbb{R}^D} T_m f_i(x) g(x)\, d\mu(x)\right| < \epsilon/3.$$

From this, together with (5.3.5), we deduce that, for given $\epsilon \in (0, \infty)$ and N, when $m, n > N$, by the Hölder inequality, we see that

$$\begin{aligned}
&\left|\int_{\mathbb{R}^D} T_n f(x) g(x)\, d\mu(x) - \int_{\mathbb{R}^D} T_m f(x) g(x)\, d\mu(x)\right| \\
&\quad\le \left|\int_{\mathbb{R}^D} T_n f(x) g(x)\, d\mu(x) - \int_{\mathbb{R}^D} T_n f_i(x) g(x)\, d\mu(x)\right| \\
&\qquad + \left|\int_{\mathbb{R}^D} T_n f_i(x) g(x)\, d\mu(x) - \int_{\mathbb{R}^D} T_m f_i(x) g(x)\, d\mu(x)\right| \\
&\qquad + \left|\int_{\mathbb{R}^D} T_m f_i(x) g(x)\, d\mu(x) - \int_{\mathbb{R}^D} T_m f(x) g(x)\, d\mu(x)\right| \\
&\quad\le \|T_n(f - f_i)\|_{L^p(\mu)} \|g\|_{L^{p'}(\mu)} + \epsilon/3 + \|T_m(f - f_i)\|_{L^p(\mu)} \|g\|_{L^{p'}(\mu)} \\
&\quad< \epsilon,
\end{aligned}$$

which implies that

$$\left\{\int_{\mathbb{R}^D} T_n f(x) g(x)\, d\mu(x)\right\}_{n\in\mathbb{N}}$$

is a Cauchy sequence.

From the claim, for given δ as in (5.3.4), there exists $N_2 \in \mathbb{N}$ such that, when $n_k, n > N_2$, we have

$$\left|\int_{\mathbb{R}^D} T_{n_k} f(x) g(x)\, d\mu(x) - \int_{\mathbb{R}^D} T_n f(x) g(x)\, d\mu(x)\right| < \delta/2.$$

Recall that $n_k \to \infty$ as $k \to \infty$. Then there exists $N_3 \in \mathbb{N}$ such that $n_k > N_2$ when $k > N_3$. Therefore, for given δ, take $k > \max\{N_3, N_1\}$. This implies that $n_k > N_2$. Consequently,

$$\begin{aligned}
&\left|\int_{\mathbb{R}^D} T_n f(x)g(x)\,d\mu(x) - \int_{\mathbb{R}^D} \tilde{T} f(x)g(x)\,d\mu(x)\right| \\
&\quad \le \left|\int_{\mathbb{R}^D} T_n f(x)g(x)\,d\mu(x) - \int_{\mathbb{R}^D} T_{n_k} f(x)g(x)\,d\mu(x)\right| \\
&\qquad + \left|\int_{\mathbb{R}^D} T_{n_k} f(x)g(x)\,d\mu(x) - \int_{\mathbb{R}^D} \tilde{T} f(x)g(x)\,d\mu(x)\right| \\
&\quad < \delta/2 + \delta/2 \\
&\quad = \delta.
\end{aligned}$$

Therefore, for any $f \in L^p(\mu)$, we know that $\{T_n f\}_{n\in\mathbb{N}}$ has a $*$-weak limit. That is, for any $g \in L^{p'}(\mu)$, it holds true that

$$\lim_{n\to\infty} \int_{\mathbb{R}^D} T_n f(x)g(x)\,d\mu(x) = \int_{\mathbb{R}^D} \tilde{T} f(x)g(x)\,d\mu(x).$$

Then the operator $\tilde{T}$ above is well defined on $L^p(\mu)$. We now show that $\tilde{T}$ is bounded on $L^p(\mu)$. Indeed, for any $f \in L^p(\mu)$, we see that

$$\begin{aligned}
\|\tilde{T} f\|_{L^p(\mu)} &= \sup_{\|g\|_{L^{p'}(\mu)}\le 1} \left|\int_{\mathbb{R}^D} \tilde{T} f(x)g(x)\,d\mu(x)\right| \\
&= \sup_{\|g\|_{L^{p'}(\mu)}\le 1} \left|\lim_{n\to\infty}\int_{\mathbb{R}^D} T_n f(x)g(x)\,d\mu(x)\right| \\
&\le \sup_{\|g\|_{L^{p'}(\mu)}\le 1} \lim_{n\to\infty} \|T_n f\|_{L^p(\mu)} \|g\|_{L^{p'}(\mu)} \\
&\le c_0 \|f\|_{L^p(\mu)}.
\end{aligned}$$

We further claim that, for any $f \in L^p(\mu)$ such that $\operatorname{supp} f \neq \mathbb{R}^D$,

$$\tilde{T} f(x) = \int_{\mathbb{R}^D} K(x,y) f(y)\,d\mu(y) \tag{5.3.6}$$

for almost every $x \notin \operatorname{supp} f$.

Indeed, for each $n \in \mathbb{N}$, let $T_n := T_{\epsilon_n}$, where $\epsilon_n \to 0$ as $n \to \infty$. For each $k \in \mathbb{Z}$, let

$$A_k := \left\{x \in \mathbb{R}^D : \ \operatorname{dist}(x, \operatorname{supp} f) > 2^k\right\}.$$

Then, for given k, since $\epsilon_n \to 0$ as $n \to \infty$, there exists $N \in \mathbb{N}$ such that $\epsilon_n < 2^k$ when $n > N$. Thus, for any $\delta \in (0, \infty)$, there exists $N \in \mathbb{N}$ such that, when $n > N$,

$$\begin{aligned}&\left|\int_{\mathbb{R}^D} \tilde{T} f(x)g(x)\chi_{A_k}(x)\,d\mu(x) - \int_{\mathbb{R}^D} T_n f(x)g(x)\chi_{A_k}(x)\,d\mu(x)\right|\\&\quad= \left|\int_{\mathbb{R}^D} \tilde{T} f(x)g(x)\chi_{A_k}(x)\,d\mu(x)\right.\\&\qquad\left.- \int_{\mathbb{R}^D}\left\{\int_{\mathbb{R}^D} K(x,y)f(y)\,d\mu(y)\right\} g(x)\chi_{A_k}(x)\,d\mu(x)\right|\\&\quad< \delta.\end{aligned}$$

This implies that, for any $g \in L^{p'}(\mu)$,

$$\begin{aligned}&\int_{\mathbb{R}^D} \tilde{T} f(x)g(x)\chi_{A_k}(x)\,d\mu(x)\\&\quad= \int_{\mathbb{R}^D}\left\{\int_{\mathbb{R}^D} K(x,y)f(y)\,d\mu(y)\right\} g(x)\chi_{A_k}(x)\,d\mu(x),\end{aligned}$$

from which we further deduce that, for almost every $x \in A_k$,

$$\tilde{T} f(x) = \int_{\mathbb{R}^D} K(x,y)f(y)\,d\mu(y).$$

Combining this, with the fact that

$$(\mathbb{R}^D \setminus \operatorname{supp} f) = \bigcup_{k\in\mathbb{Z}} A_k,$$

we obtain (5.3.6). □

Therefore, we define the operator T associated with kernel K by setting, for any bounded function f with compact support and μ-almost every $x \notin \operatorname{supp} f$,

$$Tf(x) := \int_{\mathbb{R}^D} K(x,\, y)f(y)\,d\mu(y). \tag{5.3.7}$$

We say that T is a *Calderón–Zygmund operator* with kernel K and measure μ as in (0.0.1) if T satisfies (5.3.7), T is bounded on $L^2(\mu)$, K satisfies (5.3.1) and the *regularity condition* that there exist positive constants $\delta \in (0,\, 1]$ and $C \in (0, \infty)$ such that, for any $x, \tilde{x}, y$ with $|x-y| \ge 2|x-\tilde{x}|$,

$$|K(x,\, y) - K(\tilde{x},\, y)| + |K(y,\, x) - K(y,\, \tilde{x})| \le C\frac{|x-\tilde{x}|^\delta}{|x-y|^{n+\delta}}. \tag{5.3.8}$$

The *maximal singular integral operator* associated with T is defined by setting, for any bounded function f with compact support and μ-almost every $x \notin \operatorname{supp} f$,

$$T^{\sharp} f(x) := \sup_{\epsilon \in (0,\infty)} |T_{\epsilon} f(x)|. \tag{5.3.9}$$

Obviously, if the operator $T^{\sharp}$ is bounded on $L^p(\mu)$ with $p \in (1,\infty)$ (or bounded from $L^1(\mu)$ to $L^{1,\infty}(\mu)$), then T is also bounded on $L^p(\mu)$ with $p \in (1,\infty)$ (or bounded from $L^1(\mu)$ to $L^{1,\infty}(\mu)$). It is interesting to see whether the $L^p(\mu)$ mapping properties, with $p \in [1,\infty)$, of $T^{\sharp}$ can be deduced from the $L^2(\mu)$ boundedness of T or not, as in the setting that $d\mu$ is the D-dimensional Lebesgue measure. To this end, we first establish an inequality of Cotlar type, which links the operator T and the maximal operator $T^{\sharp}$.

Theorem 5.3.2. *Let K be a function on*

$$(\mathbb{R}^D \times \mathbb{R}^D) \setminus \{(x, y) : x = y\},$$

which satisfies (5.3.1) *and* (5.3.2), *and $T^{\sharp}$ be the operator defined by* (5.3.9). *If T is bounded on $L^2(\mu)$, then there exists a positive constant $\tilde{C}_7$ such that, for all $f \in L^2(\mu) \cap L^{\infty}(\mu)$ and μ-almost every $x \in \operatorname{supp} \mu$,*

$$T^{\sharp} f(x) \le \tilde{C}_7 \left\{ \mathcal{M}_{(3/2)}(Tf)(x) + \|f\|_{L^{\infty}(\mu)} \right\}.$$

Proof. As is pointed out at the beginning of this section, T is also bounded on $L^2(\mu)$ in the sense of Proposition 5.3.1. Thus, for each fixed function $f \in L^2(\mu) \cap L^{\infty}(\mu)$, we know that Tf is finite almost everywhere. Let $x \in \mathbb{R}^D$ be a point such that $|Tf(x)| < \infty$, $\epsilon \in (0,\infty)$ and Q_x be the biggest doubling cube centered at x and having side length $2^{-k}\epsilon$ for some $k \in \mathbb{N}$ such that $2^k \ge \sqrt{D}$. Decompose f as

$$f(y) = f(y)\chi_{2Q_x}(y) + f(y)\chi_{\mathbb{R}^D \setminus 2Q_x}(y) =: f_1(y) + f_2(y)$$

for all $y \in \mathbb{R}^D$. Notice that $2Q_x \subset B(x, \epsilon)$. From this, it follows that, for each $z \in Q_x$,

$$\begin{aligned}
|T_{\epsilon}(f)(x)| &= |T_{\epsilon}(f_2)(x)| \\
&\le |T(f_2)(x)| + |T(f\chi_{B(x,\epsilon)\setminus 2Q_x})(x)| \\
&\le |T(f_2)(x) - T(f_2)(z)| + |T(f_2)(z)| \\
&\quad + |T(f\chi_{B(x,\epsilon)\setminus 2Q_x})(x)| \\
&\le |T(f_2)(x) - T(f_2)(z)| + |Tf(z)| + |T(f_1)(z)| \\
&\quad + |T(f\chi_{B(x,\epsilon)\setminus 2Q_x})(x)|.
\end{aligned} \tag{5.3.10}$$

By (5.3.1) and (5.3.2), we see that, for μ-almost every $z \in Q_x$,

$$|T(f_2)(x) - T(f_2)(z)| \lesssim \|f\|_{L^\infty(\mu)}. \tag{5.3.11}$$

Let k_0 be the positive integer such that

$$2^{k_0}\ell(Q_x)\sqrt{D} \le \epsilon < 2^{k_0+1}\ell(Q_x)\sqrt{D}.$$

Then, by the fact that Q_x is the biggest doubling cube centered at x and having side length $2^{-k}\epsilon$, (5.3.1) and Lemma 2.1.3(b), we easily obtain

$$\begin{aligned}
|T(f\chi_{B(x,\epsilon)\setminus 2Q_x})(x)| &\le \int_{B(x,\epsilon)\setminus 2^{k_0}Q_x} |K(x, y)f(y)|\,d\mu(y) \\
&\quad + \int_{2^{k_0}Q_x\setminus 2Q_x} |K(x, y)f(y)|\,d\mu(y) \\
&\lesssim \|f\|_{L^\infty(\mu)}\left[\frac{\mu(B(x,\epsilon))}{[\ell(2^{k_0}Q_x)]^n} + 1 + \delta(2Q_x, 2^{k_0}Q_x)\right] \\
&\lesssim \|f\|_{L^\infty(\mu)},
\end{aligned}$$

which, together with (5.3.10), (5.3.11), the boundedness of T on $L^2(\mu)$ and the fact Q_x is doubling, shows that

$$\begin{aligned}
|T_\epsilon(f)(x)| &\lesssim \|f\|_{L^\infty(\mu)} + \frac{1}{\mu(2Q_x)}\int_{Q_x} |Tf(z)|\,d\mu(z) \\
&\quad + \frac{1}{\mu(2Q_x)}\int_{Q_x} |T(f_1)(z)|\,d\mu(z) \\
&\lesssim \|f\|_{L^\infty(\mu)} + \mathcal{M}_{(3/2)}(Tf)(x) + \frac{1}{\mu(2Q_x)}\|T(f_1)\|_{L^2(\mu)}[\mu(Q_x)]^{1/2} \\
&\lesssim \|f\|_{L^\infty(\mu)} + \mathcal{M}_{(3/2)}(Tf)(x) + \frac{1}{\mu(2Q_x)}\|f_1\|_{L^2(\mu)}[\mu(Q_x)]^{1/2} \\
&\lesssim \|f\|_{L^\infty(\mu)} + \mathcal{M}_{(3/2)}(Tf)(x).
\end{aligned}$$

From this, it is easy to deduce the desired conclusion, which completes the proof of Theorem 5.3.2. □

The following theorem states that the $L^p(\mu)$ boundedness, with $p \in (1, \infty)$, and the weak type $(1, 1)$ boundedness of $T^\sharp$ are equivalent.

Theorem 5.3.3. *Let K be a function on*

$$\left(\mathbb{R}^D \times \mathbb{R}^D\right) \setminus \{(x, y) : x = y\},$$

which satisfies (5.3.1) *and* (5.3.2)*, and $T^\sharp$ the maximal singular integral operator defined by* (5.3.9)*. Then the following three statements are equivalent:*

(i) *$T^\sharp$ is bounded on $L^{p_0}(\mu)$ with $p_0 \in (1, \infty)$;*
(ii) *$T^\sharp$ is bounded from $L^1(\mu)$ into $L^{1,\infty}(\mu)$; namely, there exists a positive constant C such that, for all $f \in L^1(\mu)$ and $\lambda \in (0, \infty)$,*

$$\mu\left(\left\{x \in \mathbb{R}^D : T^\sharp f(x) > \lambda\right\}\right) \le C\lambda^{-1}\|f\|_{L^1(\mu)};$$

(iii) *$T^\sharp$ is bounded on $L^p(\mu)$ for any $p \in (1, \infty)$.*

Proof. (i) $\Longrightarrow$ (ii) Our goal is to show that, for any $f \in L^1(\mu)$ and $\lambda \in (0, \infty)$,

$$\mu\left(\left\{x \in \mathbb{R}^D : T^\sharp f(x) > \lambda\right\}\right) \lesssim \lambda^{-1} \int_{\mathbb{R}^D} |f(x)|\, d\mu(x). \tag{5.3.12}$$

Observing that,

$$\text{if} \quad \mu(\mathbb{R}^D) < \infty \quad \text{and} \quad \lambda \in \left(0, 2^{D+1}\|f\|_{L^1(\mu)}/\mu\left(\mathbb{R}^D\right)\right],$$

then the inequality (5.3.12) is trivial, we may assume that

$$\lambda \in \left(2^{D+1}\|f\|_{L^1(\mu)}/\mu\left(\mathbb{R}^D\right), \infty\right) \quad \text{if} \quad \|\mu\| < \infty.$$

For each fixed

$$\lambda \in (2^{D+1}\|f\|_{L^1(\mu)}/\mu(\mathbb{R}^D), \infty) \quad \text{and} \quad f \in L^1(\mu),$$

applying the Calderón–Zygmund decomposition to f at level λ, we then decompose f as $f = g + h$, where

$$g := f\chi_{\mathbb{R}^D \setminus (\bigcup_j Q_j)} + \sum_j \varphi_j$$

and

$$h := f - g = \sum_j \left[w_j f - \varphi_j\right] =: \sum_j h_j.$$

It is easy to see that there exists a positive constant $\tilde{C}_8$ such that, for μ-almost every x,

$$|g(x)| \le \tilde{C}_8 \lambda$$

and

$$\|g\|_{L^1(\mu)} \lesssim \|f\|_{L^1(\mu)}.$$

Thus, from the boundedness of $T^\sharp$ on $L^2(\mu)$, it follows that

$$\mu\left(\{x \in \mathbb{R}^D : |T^\sharp g(x)| > \lambda\}\right) \lesssim \frac{1}{\lambda^2} \int_{\mathbb{R}^D} |T^\sharp g(x)|^2 \, d\mu(x) \lesssim \frac{1}{\lambda^2} \|g\|^2_{L^2(\mu)}$$
$$\lesssim \frac{1}{\lambda} \|f\|_{L^1(\mu)}.$$

Noticing that

$$\mu\left(\bigcup_j 2Q_j\right) \lesssim \lambda^{-1} \sum_j \int_{Q_j} |f(x)| \, d\mu(x) \lesssim \lambda^{-1} \int_{\mathbb{R}^D} |f(x)| \, d\mu(x),$$

we see that the proof of (5.3.12) is reduced to proving that

$$\mu\left(\left\{x \in \mathbb{R}^D \setminus \left(\bigcup_j 2Q_j\right) : |T^\sharp h(x)| > \lambda\right\}\right) \lesssim \lambda^{-1} \|f\|_{L^1(\mu)}. \qquad (5.3.13)$$

To prove (5.3.13), denote by x_j the center of Q_j and R_j the smallest $(6, 6^{D+1})$-doubling cube of the family $\{6^k Q_i\}_{k \in \mathbb{N}}$. For any $x \in \mathbb{R}^D \setminus 2R_j$, by

$$\int_{\mathbb{R}^D} h_j(x) \, d\mu(x) = 0$$

and (5.3.2), we easily obtain the following facts:

(i) if $\epsilon \le \operatorname{dist}\{x, R_j\}$, then

$$|T_\epsilon h_j(x)| \le \int_{\mathbb{R}^D} |K(x, y) - K(x, x_j)||h_j(y)| \, d\mu(y);$$

(ii)

$$\text{if } \epsilon > \operatorname{dist}\{x, R_j\} + \sqrt{D}\ell(R_j), \text{ then } |T_\epsilon h_j(x)| = 0;$$

(iii) if

$$\operatorname{dist}\{x, R_j\} < \epsilon \le \operatorname{dist}\{x, R_j\} + \sqrt{D}\ell(R_j),$$

then

$$\operatorname{dist}\{x, R_j\} \ge \frac{1}{2}\ell(R_j)$$

and there exists a positive constant $\tilde{c}$ such that

$$\epsilon < \operatorname{dist}\{x, R_j\} + \sqrt{D}\ell(R_j) \le \tilde{c} \operatorname{dist}\{x, R_j\}$$

and hence

$$
\begin{aligned}
|T_\epsilon h_j(x)| &\le \left|\int_{|x-y|>\mathrm{dist}\{x,R_j\}} K(x,y)h_j(y)\,d\mu(y)\right| \\
&\quad + \left|\int_{\mathrm{dist}\{x,R_j\}\le|x-y|<\epsilon} K(x,y)h_j(y)\,d\mu(y)\right| \\
&\le \int_{\mathbb{R}^D} |K(x,y)-K(x,x_j)||h_j(y)|\,d\mu(y) \\
&\quad + \int_{\epsilon/\tilde{c}\le|x-y|<\epsilon} |K(x,y)h_j(y)|\,d\mu(y) \\
&\le \int_{\mathbb{R}^D} |K(x,y)-K(x,x_j)||h_j(y)|\,d\mu(y) \\
&\quad + \frac{C}{\epsilon^n}\int_{|x-y|<\epsilon} |h_j(y)|\,d\mu(y)
\end{aligned}
$$

for some positive constant C. Therefore, for $x \in \mathbb{R}^D \setminus 2R_j$ and any $\epsilon \in (0, \infty)$, we have

$$
|T_\epsilon h_j(x)| \le \int_{\mathbb{R}^D} |K(x,y)-K(x,x_j)||h_j(y)|\,d\mu(y) + \frac{C}{\epsilon^n}\int_{|x-y|<\epsilon} |h_j(y)|\,d\mu(y).
$$

This, along with the fact that

$$
|T_\epsilon(w_j f)(x)| \lesssim \frac{1}{|x-x_j|^n}\int_{Q_j} |f(y)|\,d\mu(y), \quad x \in \mathbb{R}^D \setminus 2Q_j,
$$

implies that, for each fixed $x \in \mathbb{R}^D \setminus (\cup_j 2Q_j)$,

$$
\begin{aligned}
T^\sharp h(x) &\le \sup_{\epsilon\in(0,\infty)} \left|\sum_j T_\epsilon h_j(x)\chi_{2R_j\setminus 2Q_j}(x)\right| + \sup_{\epsilon\in(0,\infty)} \left|\sum_j T_\epsilon h_j(x)\chi_{\mathbb{R}^D\setminus 2R_j}(x)\right| \\
&\lesssim \sum_j T^\sharp \varphi_j(x)\chi_{2R_j}(x) + \sum_j \frac{\chi_{2R_j\setminus 2Q_j}(x)}{|x-x_j|^n}\int_{Q_j} |f(y)|\,d\mu(y) \\
&\quad + \sum_j \int_{\mathbb{R}^D} |K(x,y)-K(x,x_j)||h_j(y)|\,d\mu(y)\chi_{\mathbb{R}^D\setminus 2R_j}(x)
\end{aligned}
$$

$$+\mathcal{M}_{(3/2)}\left(\sum_j |h_j|\right)(x)$$
$$=: \mathrm{E}(x) + \mathrm{F}(x) + \mathrm{G}(x) + \mathrm{H}(x).$$

By the boundedness of $T^\sharp$ on $L^2(\mu)$, we find that

$$\begin{aligned}\mu\left(\left\{x \in \mathbb{R}^D \setminus \left(\bigcup_j 2Q_j\right) : \mathrm{E}(x) > \lambda\right\}\right) &\leq \lambda^{-1}\sum_j \int_{2R_j} |T^\sharp\varphi_j(x)|\,d\mu(x)\\ &\leq \lambda^{-1}\sum_j \|T^\sharp\varphi_j\|_{L^2(\mu)}[\mu(2R_j)]^{1/2}\\ &\lesssim \lambda^{-1}\sum_j \|\varphi_j\|_{L^\infty(\mu)}\mu(2R_j)\\ &\lesssim \lambda^{-1}\int_{\mathbb{R}^D} |f(x)|\,d\mu(x).\end{aligned}$$

Recall that R_j is the smallest $(6, 6^{D+1})$-doubling cube of the form $6^k Q_j$ with $k \in \mathbb{N}$. We have

$$\int_{2R_j\setminus 2Q_j} \frac{1}{|x-x_j|^n}\,d\mu(x) \lesssim 1 + \delta(2Q_j, 2R_j) \lesssim 1.$$

Therefore,

$$\begin{aligned}\mu\left(\left\{x \in \mathbb{R}^D \setminus \left(\bigcup_j 2Q_j\right) : \mathrm{F}(x) > \lambda\right\}\right) &\lesssim \lambda^{-1}\sum_j \int_{Q_j} |f(x)|\,d\mu(x)\\ &\lesssim \lambda^{-1}\|f\|_{L^1(\mu)}.\end{aligned}$$

A trivial computation implies that

$$\begin{aligned}&\mu\left(\left\{x \in \mathbb{R}^D \setminus \left(\bigcup_j 2Q_j\right) : \mathrm{G}(x) > \lambda\right\}\right)\\ &\leq \lambda^{-1}\sum_j \int_{\mathbb{R}^D\setminus 2R_j}\int_{\mathbb{R}^D} |K(x,y) - K(x,x_j)||h_j(y)|\,d\mu(y)\,d\mu(x)\\ &\lesssim \lambda^{-1}\|f\|_{L^1(\mu)}.\end{aligned}$$

From the fact that $\mathcal{M}_{(3/2)}$ is bounded from $L^1(\mu)$ into weak $L^1(\mu)$, it follows that

$$\mu\left(\left\{x \in \mathbb{R}^D \setminus \left(\bigcup_j 2Q_j\right) : \mathrm{H}(x) > \lambda\right\}\right) \lesssim \lambda^{-1} \sum_j \int_{\mathbb{R}^D} |h_j(x)|\, d\mu(x)$$
$$\lesssim \lambda^{-1} \|f\|_{L^1(\mu)}.$$

Combining the estimates for E(x), F(x), G(x) and H(x), we obtain (5.3.13) and hence completes the proof of (i) $\Longrightarrow$ (ii).

(ii) $\Longrightarrow$ (iii) Let $r \in (0,1)$. We claim that, for any $f \in L^p(\mu) \cap L^\infty(\mu)$ with $p \in (1,\infty)$,

$$\|\mathcal{M}_r^\sharp(T^\sharp f)\|_{L^\infty(\mu)} \lesssim \|f\|_{L^\infty(\mu)}. \tag{5.3.14}$$

To show this, let

$$h_Q := m_Q\left(\left[T^\sharp\left(f\chi_{\mathbb{R}^D\setminus\frac{4}{3}Q}\right)\right]^r\right).$$

Observe that, for any cube Q,

$$\begin{aligned}
&\frac{1}{\mu(\frac{3}{2}Q)} \int_Q \left||T^\sharp f(x)|^r - m_{\tilde{Q}}[(T^\sharp f)^r]\right| d\mu(x) \\
&\leq \frac{1}{\mu(\frac{3}{2}Q)} \int_Q \left||T^\sharp f(x)|^r - h_Q\right| d\mu(x) + \left|h_Q - h_{\tilde{Q}}\right| \\
&\quad + \frac{1}{\mu(\tilde{Q})} \int_{\tilde{Q}} \left||T^\sharp f(x)|^r - h_{\tilde{Q}}\right| d\mu(x)
\end{aligned}$$

and, for two doubling cubes $Q \subset R$,

$$\begin{aligned}
&\left|m_Q[(T^\sharp f)^r] - m_R[(T^\sharp f)^r]\right| \\
&\leq \left|m_Q[(T^\sharp f)^r] - h_Q\right| + \left|h_Q - h_R\right| + \left|h_R - m_R[(T^\sharp f)^r]\right|.
\end{aligned}$$

If we can prove that, for any cube Q,

$$\frac{1}{\mu(\frac{3}{2}Q)} \int_Q \left||T^\sharp f(x)|^r - h_Q\right| d\mu(x) \lesssim \|f\|^r_{L^\infty(\mu)} \tag{5.3.15}$$

and, for any two cubes Q and R with $Q \subset R$,

$$|h_Q - h_R| \lesssim [1 + \delta(Q,R)]^r \|f\|^r_{L^\infty(\mu)}, \tag{5.3.16}$$

our claim (5.3.14) follows directly.

Now we prove (5.3.15). Write

$$\begin{aligned}
&\frac{1}{\mu(\frac{3}{2}Q)}\int_Q \left||T^\sharp f(x)|^r - h_Q\right| \,d\mu(x) \\
&\quad\le \frac{1}{\mu(\frac{3}{2}Q)}\int_Q \left|T^\sharp(f\chi_{\frac{4}{3}Q})(x)\right|^r \,d\mu(x) \\
&\qquad+\frac{1}{\mu(\frac{3}{2}Q)}\int_Q \left||T^\sharp(f\chi_{\mathbb{R}^D\setminus\frac{4}{3}Q})(x)|^r - h_Q\right| \,d\mu(x) \\
&\quad=: \mathrm{I}+\mathrm{J}.
\end{aligned}$$

The weak type (1, 1) estimate of $T^\sharp$, via the Kolmogorov inequality, tells us that

$$\mathrm{I}\lesssim \frac{[\mu(Q)]^{1-r}}{\mu(\frac{3}{2}Q)}\|f\chi_{\frac{4}{3}Q}\|^r_{L^1(\mu)}\lesssim \|f\|^r_{L^\infty(\mu)}.$$

On the other hand, notice that there exists a constant $C\in(1,\infty)$ such that, for any $x,\,y\in Q$,

$$\begin{aligned}
&\left|T^\sharp\left(f\chi_{\mathbb{R}^D\setminus\frac{4}{3}Q}\right)(x) - T^\sharp\left(f\chi_{\mathbb{R}^D\setminus\frac{4}{3}Q}\right)(y)\right| \\
&\quad\le \sup_{\epsilon\in(0,\infty)}\left|\int_{|x-z|>\epsilon} K(x,z)f(z)\chi_{\mathbb{R}^D\setminus\frac{4}{3}Q}(z)\,d\mu(z)\right. \\
&\qquad\left.-\int_{|x-z|>\epsilon} K(y,z)f(z)\chi_{\mathbb{R}^D\setminus\frac{4}{3}Q}(z)\,d\mu(z)\right| \\
&\qquad+\sup_{\epsilon\in(0,\infty)}\left|\int_{|y-z|>\epsilon} K(y,z)f(z)\chi_{\mathbb{R}^D\setminus\frac{4}{3}Q}(z)\,d\mu(z)\right. \\
&\qquad\left.-\int_{|x-z|>\epsilon} K(y,z)f(z)\chi_{\mathbb{R}^D\setminus\frac{4}{3}Q}(z)\,d\mu(z)\right| \\
&\quad\le \int_{\mathbb{R}^D\setminus\frac{4}{3}Q}|K(x,z)-K(y,z)||f(z)|\,d\mu(z) \\
&\qquad+\sup_{\epsilon\in(0,\infty)}\int_{C^{-1}\epsilon\le|y-z|\le C\epsilon}|K(y,z)||f(z)|\,d\mu(z) \\
&\quad\lesssim \|f\|_{L^\infty(\mu)}.
\end{aligned}\tag{5.3.17}$$

From this, it follows that

$$\mathrm{J}\lesssim \|f\|^r_{L^\infty(\mu)}$$

and hence the estimate (5.3.15) holds true.

We now turn our attention to (5.3.16). Denote $N_{Q,R}+1$ simply by N_2. Write

$$\begin{aligned}|h_Q - h_R| &= \left|m_Q\left[\left(T^\sharp\left(f\chi_{\mathbb{R}^D\setminus\frac{4}{3}Q}\right)\right)^r\right] - m_R\left[\left(T^\sharp\left(f\chi_{\mathbb{R}^D\setminus\frac{4}{3}R}\right)\right)^r\right]\right|\\ &\le m_Q\left[\left(T^\sharp\left(f\chi_{2Q\setminus\frac{4}{3}Q}\right)\right)^r\right] + m_Q\left[\left(T^\sharp\left(f\chi_{2^{N_2}Q\setminus 2Q}\right)\right)^r\right]\\ &\quad + \left|m_Q\left[\left(T^\sharp\left(f\chi_{\mathbb{R}^D\setminus 2^{N_2}Q}\right)\right)^r\right] - m_R\left[\left(T^\sharp\left(f\chi_{\mathbb{R}^D\setminus 2^{N_2}Q}\right)\right)^r\right]\right|\\ &\quad + m_R\left[\left(T^\sharp\left(f\chi_{2^{N_2}Q\setminus\frac{4}{3}R}\right)\right)^r\right]\\ &=: \mathrm{L}_1 + \mathrm{L}_2 + \mathrm{L}_3 + \mathrm{L}_4.\end{aligned}$$

A familiar argument, invoking the regularity condition (5.3.2), implies that

$$\mathrm{L}_3 \lesssim \|f\|^r_{L^\infty(\mu)}.$$

Notice that, by the size condition (5.3.1) and the growth condition, for any $y \in Q$, we have

$$T^\sharp(f\chi_{2Q\setminus\frac{4}{3}Q})(y) \lesssim \int_{2Q\setminus\frac{4}{3}Q} \frac{|f(u)|}{|y-u|^n}\, d\mu(u) \lesssim \|f\|_{L^\infty(\mu)}$$

and, similarly, for $z \in R$, it holds true that

$$T^\sharp(f\chi_{2R\setminus\frac{4}{3}R})(z) \lesssim \int_{2R\setminus\frac{4}{3}R} \frac{|f(u)|}{|z-u|^n}\, d\mu(u) \lesssim \|f\|_{L^\infty(\mu)}.$$

Thus,

$$\mathrm{L}_1 + \mathrm{L}_4 \lesssim \|f\|^r_{L^\infty(\mu)}.$$

On the other hand, observing that, for any $x \in Q$,

$$\begin{aligned}\left|T^\sharp\left(f\chi_{2^{N_2}Q\setminus 2Q}\right)(x)\right| &\lesssim \|f\|_{L^\infty(\mu)} \int_{(2^{N_2}Q)\setminus(2Q)} \frac{1}{|x-z|^n}\, d\mu(z)\\ &\lesssim [1+\delta(Q,R)]\|f\|_{L^\infty(\mu)},\end{aligned}$$

we thus have

$$\mathrm{L}_2 \lesssim [1+\delta(Q,R)]^r \|f\|^r_{L^\infty(\mu)}.$$

The inequality (5.3.16) then follows.

We can now conclude the proof of (ii) $\Longrightarrow$ (iii). Notice that, for any suitable functions f_1, f_2,

$$|T^\sharp(f_1) - T^\sharp(f_2)| \le |T^\sharp(f_1 - f_2)|.$$

Theorem 5.2.8 then implies the desired result.

(iii) $\Longrightarrow$ (i) is obvious. This finishes the proof of Theorem 5.3.3. □

By Theorem 5.3.3, we can obtain the following conclusion.

Theorem 5.3.4. *Let K be a function on*

$$(\mathbb{R}^D \times \mathbb{R}^D) \setminus \{(x, y) : x = y\},$$

which satisfies (5.3.1) *and* (5.3.2)*. If the operator T, defined by* (5.3.7)*, is bounded on $L^2(\mu)$, then*

(i) *the corresponding maximal singular integral operator $T^\sharp$, defined by* (5.3.9)*, is bounded on $L^p(\mu)$ for any $p \in (1, \infty)$;*
(ii) *$T^\sharp$ is bounded from $L^1(\mu)$ to $L^{1,\infty}(\mu)$.*

Proof. By Theorem 5.3.3, it suffices to show that $T^\sharp$ is bounded from $L^1(\mu)$ to $L^{1,\infty}(\mu)$. For each fixed $f \in L^1(\mu)$ and $\lambda \in (0, \infty)$, applying the Calderón–Zygmund decomposition to f at level λ, we conclude that, with the notation same as in the proof of Theorem 5.3.3,

$$f = g + h.$$

We may assume that

$$\lambda \in \left(2^{D+1}\|f\|_{L^1(\mu)}/\mu\left(\mathbb{R}^D\right), \infty\right)$$

if $\mu(\mathbb{R}^D) < \infty$. Notice that,

$$\text{if} \quad \mu(\mathbb{R}^D) < \infty \quad \text{and} \quad \lambda \in \left(0, 2^{D+1}\|f\|_{L^1(\mu)}/\mu\left(\mathbb{R}^D\right)\right],$$

then (ii) obviously holds true. Applying Theorem 5.3.2 and the boundedness of $\mathcal{M}_{(3/2)} \circ T$ on $L^2(\mu)$, we have

$$\begin{aligned}
&\mu\left(\left\{x \in \mathbb{R}^D : |T^\sharp g(x)| > (\tilde{C}_8 + 1)\tilde{C}_7\lambda\right\}\right) \\
&\quad \le \mu\left(\left\{x \in \mathbb{R}^D : |\mathcal{M}_{(3/2)}(Tg)(x)| > \lambda\right\}\right) \\
&\quad \le \lambda^{-2} \int_{\mathbb{R}^D} |\mathcal{M}_{(3/2)}(Tg)(x)|^2 \, d\mu(x) \\
&\quad \lesssim \lambda^{-2}\|g\|_{L^2(\mu)}^2 \\
&\quad \lesssim \lambda^{-1}\|f\|_{L^1(\mu)}.
\end{aligned}$$

As in the proof of (i) $\Longrightarrow$ (ii) in Theorem 5.3.3, we see that the proof of Theorem 5.3.4 is reduced to proving that

$$\mu\left(\left\{x \in \mathbb{R}^D \setminus \left(\bigcup_j 2Q_j\right) : |T^\sharp h(x)| > \lambda\right\}\right) \lesssim \lambda^{-1}\|f\|_{L^1(\mu)}. \tag{5.3.18}$$

We now prove (5.3.18). Notice that, for each fixed $x \in \mathbb{R}^D \setminus (\cup_j 2Q_j)$, we have

$$\begin{aligned}
T^\sharp h(x) &\lesssim \sum_j T^\sharp \varphi_j(x)\chi_{2R_j}(x) + \sum_j \frac{\chi_{2R_j\setminus 2Q_j}(x)}{|x-x_j|^n}\int_{Q_j}|f(y)|\,d\mu(y)\\
&\quad + \sum_j \int_{\mathbb{R}^D}|K(x,y)-K(x,x_j)||h_j(y)|\,d\mu(y)\chi_{\mathbb{R}^D\setminus 2R_j}(x)\\
&\quad + \mathcal{M}_{(3/2)}\left(\sum_j |h_j|\right)(x)\\
&=: \mathrm{E}(x)+\mathrm{F}(x)+\mathrm{G}(x)+\mathrm{H}(x).
\end{aligned}$$

The estimates for $\mathrm{F}(x)$, $\mathrm{G}(x)$ and $\mathrm{H}(x)$ are the same as in the proof of Theorem 5.3.3. We only need to estimate $\mathrm{E}(x)$. Another application of Theorem 5.3.2 implies that

$$\begin{aligned}
&\mu\left(\left\{x \in \mathbb{R}^D \setminus \left(\bigcup_j 2Q_j\right) : \mathrm{E}(x) > \lambda\right\}\right)\\
&\le \lambda^{-1}\sum_j \int_{2R_j}\left|T^\sharp\varphi_j(x)\right|\,d\mu(x)\\
&\lesssim \lambda^{-1}\left\{\sum_j \int_{2R_j}\mathcal{M}_{(3/2)}(T\varphi_j)(x)\,d\mu(x) + \sum_j \|\varphi_j\|_{L^\infty(\mu)}\,\mu(2R_j)\right\}\\
&\lesssim \lambda^{-1}\left\{\sum_j \left\|\mathcal{M}_{(3/2)}(T\varphi_j)\right\|_{L^2(\mu)}[\mu(2R_j)]^{1/2} + \sum_j \|\varphi_j\|_{L^\infty(\mu)}\mu(2R_j)\right\}\\
&\lesssim \lambda^{-1}\sum_j \|\varphi_j\|_{L^\infty(\mu)}\mu(2R_j)\\
&\lesssim \lambda^{-1}\int_{\mathbb{R}^D}|f(x)|\,d\mu(x).
\end{aligned}$$

This finishes the proof of Theorem 5.3.4. $\square$

5.4 Behavior on $H^1(\mu)$ of Singular Integral Operators

As is well known, the building block of the Hardy space plays an important role in the study of the boundedness of operators on Hardy spaces. In this section, we study the behavior of singular integral operators on $H^1(\mu)$ via atomic blocks.

Let $p \in (1, \infty]$. For each cube Q, we denote by $L^p(Q)$ the *subspace of functions in* $L^p(\mu)$ *supported in* Q and

$$L_0^p(Q) := L_{c,0}^p(\mu) \bigcap L^p(Q).$$

Then the unions of $L_0^p(Q)$ and $L^p(Q)$ as Q varies over all cubes coincide with $L_{c,0}^p(\mu)$ and $L_c^p(\mu)$, respectively. Now let $\{Q_j\}_{j\in\mathbb{N}}$ be a sequence of increasing concentric cubes with

$$\mathbb{R}^D = \bigcup_{j\in\mathbb{N}} Q_j.$$

We topologize $L_{c,0}^p(\mu)$ (resp. $L_c^p(\mu)$) as the strict inductive limit of the spaces $L_0^p(Q_j)$ (resp. $L^p(Q_j)$).[2] It is known that the topology of $L_{c,0}^p(\mu)$ (resp. $L_c^p(\mu)$) is independent of the choice of $\{Q_j\}_{j\in\mathbb{N}}$.

Theorem 5.4.1. *Let $\eta \in (1, \infty)$, $\gamma \in [1, \infty)$, $p \in (1, \infty)$, T be a linear operator and $\mathcal{Y}$ a Banach space. If there exists a non-negative constant C such that, for all (p, γ)-atomic blocks b,*

$$\|Tb\|_{\mathcal{Y}} \le C|b|_{H_{\mathrm{atb},\gamma}^{1,p}(\mu)}, \tag{5.4.1}$$

then T can be extended to a bounded linear operator from $H^1(\mu)$ to $\mathcal{Y}$.

Proof. Without loss of generality, we may assume $p = 2$. Moreover, by Theorem 3.2.12, we choose $\eta = 2$ and $\gamma = 1$ in the definition of $H_{\mathrm{atb},\gamma}^{1,p}(\mu)$. Let Q be a fixed cube. If $f \in L_0^2(Q)$, then f is a $(2, 1)$-atomic block and

$$|f|_{H_{\mathrm{atb}}^{1,2}(\mu)} \le \|f\|_{L^2(\mu)}[\mu(2Q)]^{1/2}. \tag{5.4.2}$$

Moreover, from this and (5.4.1), it follows that, for any sequence of increasing concentric cubes, $\{Q_j\}_{j\in\mathbb{N}}$, with

$$\mathbb{R}^D = \bigcup_{j\in\mathbb{N}} Q_j,$$

[2] See [6, II, p. 33].

T is bounded from $L_0^2(Q_j)$ to $\mathcal{Y}$ for each $j \in \mathbb{N}$. Then T is bounded from $L^2_{c,0}(\mu)$ to $\mathcal{Y}$, which implies that the adjoint operator T^* of T is bounded from the dual space $\mathcal{Y}^*$ of $\mathcal{Y}$ to $(L^2_{c,0}(\mu))^*$. Moreover, for all functions $f \in \mathcal{Y}^*$ and $(2,1)$-atomic blocks b, we have

$$\left|\int_{\mathbb{R}^D} b(x)T^*(f)(x)\,d\mu(x)\right| = |\langle Tb, f\rangle| \lesssim \|f\|_{\mathcal{Y}^*}|b|_{H^{1,2}_{\rm atb}(\mu)}. \tag{5.4.3}$$

We claim that, for all $f \in \mathcal{Y}^*$,

$$T^*f \in \mathrm{RBMO}\,(\mu) \quad \text{and} \quad \|T^*f\|_{\mathrm{RBMO}\,(\mu)} \lesssim \|f\|_{\mathcal{Y}^*}.$$

Indeed, observe that, for any doubling cube Q and $\phi \in L^2(Q)$ with $\|\phi\|_{L^2(Q)} = 1$, $[\phi - m_Q(\phi)]\chi_Q$ is a $(2,1)$-atomic block. From this, (5.4.2) and (5.4.3), we deduce that

$$\begin{aligned}
&\left[\int_Q \left|T^*f(x) - m_Q(T^*f)\right|^2\,d\mu(x)\right]^{1/2}\\
&\quad= \sup_{\|\phi\|_{L^2(Q)}=1}\left|\int_Q \phi(x)\left[T^*f(x) - m_Q(T^*f)\right]\,d\mu(x)\right|\\
&\quad= \sup_{\|\phi\|_{L^2(Q)}=1}\left|\int_Q [\phi(x) - m_Q(\phi)]T^*f(x)\,d\mu(x)\right|\\
&\quad\lesssim \|f\|_{\mathcal{Y}^*}[\mu(Q)]^{1/2},
\end{aligned}$$

which implies that

$$\left[\frac{1}{\mu(Q)}\int_Q \left|T^*f(x) - m_Q(T^*f)\right|^2\,d\mu(x)\right]^{1/2} \lesssim \|f\|_{\mathcal{Y}^*}. \tag{5.4.4}$$

By (5.4.4) and Proposition 3.1.10, the proof of the claim is reduced to showing that, for all doubling cubes $Q \subset R$,

$$\left|m_Q(T^*f) - m_R(T^*f)\right| \lesssim [1 + \delta(Q,R)]\|f\|_{\mathcal{Y}^*}. \tag{5.4.5}$$

Let

$$a_1 := \frac{|T^*f - m_R(T^*f)|^2}{T^*f - m_R(T^*f)}\chi_{Q\cap\{T^*f \neq m_R(T^*f)\}},$$

$$a_2 := C_R\chi_R \quad \text{and} \quad b := a_1 + a_2,$$

where C_R is a constant such that b has integral 0. Then b is a $(2,1)$-atomic block and

$$\begin{aligned}|b|_{H^{1,2}_{\rm atb}(\mu)} &\lesssim \|a_1\|_{L^2(\mu)}[\mu(Q)]^{1/2}[1+\delta(Q,R)] + |C_R|\mu(R)\\ &\lesssim \left[\int_Q |T^*f(x) - m_R(T^*f)|^2\,d\mu(x)\right]^{1/2}[\mu(Q)]^{1/2}[1+\delta(Q,R)].\end{aligned}$$

By this, (5.4.3) and (5.4.4) with Q replaced by R, we have

$$\begin{aligned}&\int_Q |T^*f(x) - m_R(T^*f)|^2\,d\mu(x)\\ &\quad= \int_{\mathbb{R}^D} a_1(x)[T^*f(x) - m_R(T^*f)]\,d\mu(x)\\ &\quad\le \left[\left|\int_{\mathbb{R}^D} b(x)T^*f(x)\,d\mu(x)\right| + |C_R|\int_R |T^*f(x) - m_R(T^*f)|\,d\mu(x)\right]\\ &\quad\lesssim \left[\int_Q |T^*f(x) - m_R(T^*f)|^2\,d\mu(x)\right]^{1/2}[\mu(Q)]^{1/2}[1+\delta(Q,R)]\|f\|_{\mathcal{Y}^*},\end{aligned}$$

which implies that

$$\left[\frac{1}{\mu(Q)}\int_Q |T^*f(x) - m_R(T^*f)|^2\,d\mu(x)\right]^{1/2} \lesssim [1+\delta(Q,R)]\|f\|_{\mathcal{Y}^*}.$$

From this, the Hölder inequality and (5.4.4), it then follows that

$$\begin{aligned}&|m_Q(T^*f) - m_R(T^*f)|\\ &\quad\le \frac{1}{\mu(Q)}\int_Q \left[|m_Q(T^*f) - T^*f(x)| + |T^*f(x) - m_R(T^*f)|\right]\,d\mu(x)\\ &\quad\lesssim [1+\delta(Q,R)]\|f\|_{\mathcal{Y}^*},\end{aligned}$$

which is (5.4.5). By this and (5.4.3), we conclude that

$$T^*f \in \mathrm{RBMO}\,(\mu) \quad \text{and} \quad \|T^*f\|_{\mathrm{RBMO}\,(\mu)} \lesssim \|f\|_{\mathcal{Y}^*}.$$

Thus, the claim is true.

Let $H^{1,2}_{\rm fin}(\mu)$ be the *set* of all finite linear combinations of $(2,1)$-atomic blocks. Then $H^{1,2}_{\rm fin}(\mu)$ is dense in $H^1(\mu)$. On the other hand, $H^{1,2}_{\rm fin}(\mu)$ coincides with $L^2_{c,0}(\mu)$ as vector spaces. Recall that RBMO (μ) is the dual space of $H^1(\mu)$. Thus, by the above claim, we know that, for all $g \in H^{1,2}_{\rm fin}(\mu)$ and $f \in \mathcal{Y}^*$ with $\|f\|_{\mathcal{Y}^*} = 1$,

$$|\langle Tg, f\rangle| = |\langle g, T^*f\rangle| \lesssim \|g\|_{H^1(\mu)}\|T^*f\|_{\mathrm{RBMO}\,(\mu)} \lesssim \|g\|_{H^1(\mu)}.$$

From this and (5.4.1), it follows that

$$Tg \in \mathcal{Y} \quad \text{and} \quad \|Tg\|_{\mathcal{Y}} \lesssim \|g\|_{H^1(\mu)},$$

which, via a density argument, then completes the proof of Theorem 5.4.1. □

Remark 5.4.2. Observe that (5.4.1) is also necessary for an operator T to be bounded from $H^1(\mu)$ to $\mathcal{Y}$. From this fact and Theorem 5.4.1, we further deduce that, if T is linear, then T can be extended to a bounded linear operator from $H^1(\mu)$ to $\mathcal{Y}$ if and only if T satisfies (5.4.1).

For sublinear operators bounded from $L^1(\mu)$ to $L^{1,\infty}(\mu)$, we also have the following conclusion.

Theorem 5.4.3. *Let $\eta \in (1,\infty)$, $\gamma \in [1,\infty)$, $p \in (1,\infty)$ and T be a sublinear operator bounded from $L^1(\mu)$ to $L^{1,\infty}(\mu)$. If T satisfies* (5.4.1) *with $\mathcal{Y} = L^1(\mu)$, then T can be extended to a bounded sublinear operator from $H^1(\mu)$ to $L^1(\mu)$.*

Proof. As in the proof of Theorem 5.4.1, we choose $\eta = 2$ and $\gamma = 1$ in the definition of $H^{1,p}_{\mathrm{atb},\gamma}(\mu)$. Let

$$f \in H^1(\mu) \quad \text{and} \quad f = \sum_{i=1}^{\infty} b_i,$$

where, for each $i \in \mathbb{N}$, b_i is a $(p,1)$-atomic block with p as in Theorem 5.4.3. Since $H^1(\mu) \subset L^1(\mu)$ and T is bounded from $L^1(\mu)$ to $L^{1,\infty}(\mu)$, we see that Tf is well defined. Furthermore, by the boundedness of T from $L^1(\mu)$ to $L^{1,\infty}(\mu)$, we know that, for any $\epsilon \in (0,\infty)$,

$$\lim_{N\to\infty} \mu\left(\left\{x \in \mathbb{R}^D : \left|T\left(\sum_{i=N+1}^{\infty} b_i\right)(x)\right| > \epsilon\right\}\right) \lesssim \lim_{N\to\infty} \frac{1}{\epsilon} \sum_{i=N+1}^{\infty} \|b_i\|_{L^1(\mu)} = 0.$$

This, via the Riesz theorem, implies that there exists a subsequence

$$\left\{T\left(\sum_{i=1}^{j_k} b_i\right)\right\}_{j_k} \quad \text{of} \quad \left\{T\left(\sum_{i=1}^{j} b_i\right)\right\}_j$$

such that, for μ-almost every $x \in \mathbb{R}^D$,

$$|Tf(x)| \le \left|T\left(\sum_{i=1}^{j_k-1} b_i\right)(x)\right| + \left|T\left(\sum_{i=j_k}^{\infty} b_i\right)(x)\right|$$

$$\leq \sum_{i=1}^{j_k-1} |Tb_i(x)| + \left| T\left(\sum_{i=j_k}^{\infty} b_i\right)(x) \right|$$

$$\to \sum_{i=1}^{\infty} |Tb_i(x)|, \quad j_k \to \infty.$$

Since T is sublinear, then from this fact, we deduce that, for μ-almost every $x \in \mathbb{R}^D$,

$$|Tf(x)| \lesssim \sum_{i=1}^{\infty} |Tb_i(x)|,$$

which, via (5.4.1), in turn implies that

$$\|Tf\|_{L^1(\mu)} \lesssim \sum_{i=1}^{\infty} \|Tb_i\|_{L^1(\mu)} \lesssim \sum_{i=1}^{\infty} |b_i|_{H^{1,p}_{\rm atb}(\mu)}.$$

By this, we find that

$$Tf \in L^1(\mu) \quad \text{and} \quad \|Tf\|_{L^1(\mu)} \lesssim \|f\|_{H^1(\mu)}.$$

This finishes the proof of Theorem 5.4.3. □

Theorem 5.4.4. *Let K be a function on*

$$(\mathbb{R}^D \times \mathbb{R}^D) \setminus \{(x, y) : x = y\},$$

which satisfies (5.3.1) *and* (5.3.2), *and T the operator defined by* (5.3.7). *Then, for any $\rho \in (1, \infty)$, the following statements are equivalent:*

(i) *There exists a positive constant C such that, for any cube Q and function $a \in L^\infty(\mu)$ supported on Q,*

$$\int_Q |Ta(x)|\, d\mu(x) \leq C \|a\|_{L^\infty(\mu)} \mu(\rho Q); \tag{5.4.6}$$

(ii) *There exists a positive constant C such that, for any bounded function f with compact support,*

$$\|Tf\|_{{\rm RBMO}(\mu)} \leq C \|f\|_{L^\infty(\mu)};$$

(iii) *There exists a positive constant C such that, for any ∞-atomic block b,*

$$\|Tb\|_{L^1(\mu)} \leq C \|b\|_{H^1(\mu)}.$$

Proof. For simplicity, we may assume that $\rho = 2$. The proof of (i) $\Longrightarrow$ (ii) is similar to that of (5.3.14). We omit the details.

(ii) $\Longrightarrow$ (i) Let $a \in L^\infty(\mu)$ be supported on some cube Q. We consider the following two cases.

Case I $\ell(Q) \le \operatorname{diam}(\operatorname{supp}\mu)/(20\sqrt{D})$ (this is always the case if $\mu(\mathbb{R}^D) = \infty$). Our assumption says that

$$\int_Q \left|Ta(x) - m_{\tilde{Q}}(Ta)\right| \, d\mu(x) \lesssim \|a\|_{L^\infty(\mu)}\mu(2Q).$$

Thus, to show (ii), it suffices to show that

$$\left|m_{\tilde{Q}}(Ta)\right| \lesssim \|a\|_{L^\infty(\mu)}. \tag{5.4.7}$$

We first claim that there exists a $j_0 \in \mathbb{N}$ such that

$$\mu(2^{j_0} Q \setminus 2Q) > 0. \tag{5.4.8}$$

Indeed, if for all $j \in \mathbb{N}$, $\mu(2^j Q \setminus 2Q) = 0$, then we see that $\mu(\mathbb{R}^D \setminus 2Q) = 0$, which implies that $\operatorname{supp}\mu \subset 2Q$. This contradicts to the fact that

$$\ell(Q) \le \operatorname{diam}\,(\operatorname{supp}\mu)/\left(20\sqrt{D}\right)$$

and hence the claim holds true.

Now we assume that R is the smallest cube of the form $2^j Q$ such that (5.4.8) holds true. We then see that

$$\mu(2^{-1}R \setminus 2Q) = 0 \quad \text{and} \quad \mu(R \setminus 2Q) > 0.$$

Thus,

$$\mu\left(R \setminus \left(2^{-1}R \bigcup 2Q\right)\right) > 0.$$

By this, we choose $x_0 \in R \setminus (2^{-1}R \cup 2Q)$ such that some cube centered at x_0 with the side length $2^{-k}\ell(R)$ for some $k \ge 3$ is doubling. Let $d_0 :=$ dist (x_0, Q) and Q_0 the biggest cube of this form. Then we know that

$$Q_0 \subset 2R \quad \text{and} \quad \operatorname{dist}(Q_0, Q) \gtrsim \ell(Q).$$

We now claim that

$$\delta(Q, R) \lesssim 1. \tag{5.4.9}$$

Indeed, if $R \subset 2^5 Q$, then, by Lemma 2.1.3, we have (5.4.9). If $R \supset 2^6 Q$, then $\frac{1}{12} R \supset 3Q$. Notice that, in this case,

$$\mu(2^{-1} R \setminus 2Q) = 0 \quad \text{implies that} \quad \delta\left(2Q, \frac{1}{12} R\right) = 0.$$

Thus, by this, together with Lemma 2.1.3, we further have

$$\begin{aligned}
\delta(Q, 2R) &\le \delta(Q, 2Q) + \delta\left(2Q, \frac{1}{12} R\right) + \delta\left(\frac{1}{12} R, 2R\right) \\
&= \delta(Q, 2Q) + \delta\left(\frac{1}{12} R, 2R\right) \\
&\lesssim 1.
\end{aligned}$$

Thus, (5.4.9) also holds true in this case.

On the other hand, by the definition of Q_0, we see that

$$4\ell(\widetilde{2Q_0}) \ge \ell(R) \text{ and } 2R \subset 16(\widetilde{2Q_0}).$$

Therefore, by Lemma 2.1.3, we have

$$\delta(Q_0, 2R) \le \delta(Q_0, 16(\widetilde{2Q_0})) \le \delta(Q_0, \widetilde{2Q_0}) + \delta(\widetilde{2Q_0}, 16(\widetilde{2Q_0})) \lesssim 1.$$

From this, we deduce that

$$\begin{aligned}
\left| m_{Q_0}(Ta) - m_{\tilde{Q}}(Ta) \right| &\le \left| m_{Q_0}(Ta) - m_R(Ta) \right| + \left| m_R(Ta) - m_{\tilde{Q}}(Ta) \right| \\
&\lesssim \|Ta\|_{\mathrm{RBMO}(\mu)} \\
&\lesssim \|a\|_{L^\infty(\mu)}.
\end{aligned} \tag{5.4.10}$$

Moreover, $\mathrm{dist}(Q_0, Q) \sim d_0$ and hence, for any $y \in Q_0$,

$$|Ta(y)| \lesssim \frac{\mu(Q)}{d_0^n} \|a\|_{L^\infty(\mu)}.$$

Because $\ell(Q) \lesssim d_0$, we then see that

$$\left| m_{Q_0}(Ta) \right| \lesssim \|a\|_{L^\infty(\mu)},$$

which, via (5.4.10), implies (5.4.7) in this case.

Case II $\ell(Q) > \mathrm{diam}(\mathrm{supp}\, \mu)/(20\sqrt{D})$. Since Q is centered at some point of $\mathrm{supp}\, \mu$, we may assume that

$$\ell(Q) \le 4\, \mathrm{diam}(\mathrm{supp}\, \mu)/\sqrt{D}.$$

Then $Q \cap \operatorname{supp} \mu$ is covered by a finite number of cubes Q_j centered at points of $\operatorname{supp} \mu$ with side length $\ell(Q)/200$. We further find that the number of cubes Q_j is bounded above by some fixed constant N depending only on D. Let

$$a_j := \frac{\chi_{Q_j}}{\sum_k \chi_{Q_k}} a.$$

Notice that

$$\ell(2Q_j) \le \operatorname{diam}(\operatorname{supp} \mu)/(20\sqrt{D}).$$

We know, by the conclusion of Case I, that

$$\begin{aligned}\int_Q |Ta(x)\, d\mu(x) &\le \sum_j \int_{Q \setminus 2Q_j} |Ta_j(x)|\, d\mu(x) + \sum_j \int_{2Q_j} \cdots \\ &\lesssim \sum_j \|a_j\|_{L^\infty(\mu)} \mu(Q) + \sum_j \|a_j\|_{L^\infty(\mu)} \mu(4Q_j) \\ &\lesssim \|a\|_{L^\infty(\mu)} \mu(2Q),\end{aligned}$$

which completes the proof of (ii) $\Longrightarrow$ (i).

(i) $\Longrightarrow$ (iii). It suffices to prove that, for any ∞-atomic block b,

$$\|Tb\|_{L^1(\mu)} \lesssim |b|_{H^{1,\infty}_{\rm atb}(\mu)}. \tag{5.4.11}$$

Let b be an ∞-atomic block with $\operatorname{supp} b \subset R$ for some cube R. Assume that, for all $x \in \mathbb{R}^D$,

$$b(x) = \sum_{k=1}^2 \lambda_k a_k(x),$$

where $\{a_j\}_{j=1}^2$ are functions as in Definition 3.2.1 with $p = \infty$ and $\gamma = 1$. Write

$$\int_{\mathbb{R}^D} |Tb(x)|\, d\mu(x) = \int_{\mathbb{R}^D \setminus 2R} |Tb(x)|\, d\mu(x) + \int_{2R} \cdots.$$

A standard argument, invoking the regularity (5.3.2) and

$$\int_{\mathbb{R}^D} b(x)\, d\mu(x) = 0,$$

now implies that

$$\int_{\mathbb{R}^D \setminus 2R} |Tb(x)|\, d\mu(x) \lesssim \|b\|_{L^1(\mu)} \lesssim |b|_{H^{1,\infty}_{\rm atb}(\mu)}.$$

On the other hand, we have

$$\int_{2R} |Tb(x)|\,d\mu(x) \le \sum_j |\lambda_j| \int_{2R} |Ta_j(x)|\,d\mu(x)$$
$$= \sum_j |\lambda_j| \int_{2Q_j} |Ta_j(x)|\,d\mu(x) + \sum_j |\lambda_j| \int_{2R\setminus 2Q_j} \cdots .$$

Our assumption states that

$$\int_{2Q_j} |Ta_j(x)|\,d\mu(x) \lesssim \|a_j\|_{L^\infty(\mu)}\mu(4Q_j).$$

Also, notice that, for any $x \notin 2Q_j$,

$$|Ta_j(x)| \lesssim \frac{1}{[\ell(2^k Q_j)]^n}\|a_j\|_{L^1(\mu)}.$$

A trivial computation leads to that

$$\int_{2R\setminus 2Q_j} |Ta_j(x)|\,d\mu(x) \lesssim \int_{2R\setminus 2Q_j} \frac{1}{|x-x_j|^n}\,d\mu(x)\|a_j\|_{L^1(\mu)}$$
$$\lesssim [1+\delta(Q_j, R)]\|a_j\|_{L^\infty(\mu)}\mu(Q_j).$$

By Proposition 3.2.4(iii), we see that

$$\int_{2R} |Ta_j(x)|\,d\mu(x) \lesssim [1+\delta(Q_j, R)]\|a_j\|_{L^\infty(\mu)}\mu(2Q_j) \lesssim |b|_{H^{1,\infty}_{\rm atb}(\mu)},$$

which implies (iii) and hence completes the proof of (i) $\Longrightarrow$ (iii).

(iii) $\Longrightarrow$ (i). Let a be a bounded function supported on some cube Q. We first consider the case that

$$\ell(Q) \le \operatorname{diam}(\operatorname{supp}\mu)/\left(20\sqrt{D}\right).$$

Let Q_0, R be the cubes same as in Case I in the proof for (ii) $\Longrightarrow$ (i). Recall that Q_0 is doubling,

$$Q\bigcup Q_0 \subset 2R,\ \delta(Q, R) \lesssim 1,\ \delta(Q_0, R) \lesssim 1 \quad \text{and} \quad \operatorname{dist}(Q_0, Q) \ge \ell(Q).$$

Let

$$b := a + c_{Q_0}\chi_{Q_0}$$

with C_{Q_0} a constant such that

$$\int_{\mathbb{R}^D} b(x)\,d\mu(x) = 0.$$

Notice that, for each fixed $y \in Q$,

$$|T(c_{Q_0}\chi_{Q_0})(y)| \lesssim \frac{|c_{Q_0}|\mu(Q_0)}{\mathrm{dist}(Q, Q_0)} \lesssim \frac{\|a\|_{L^1(\mu)}}{\mathrm{dist}(Q, Q_0)} \lesssim \|a\|_{L^\infty(\mu)}.$$

We then see that

$$\begin{aligned}\int_Q |Ta(x)\,d\mu(x) &\lesssim \int_Q |Tb(x)|\,d\mu(x) + \|a\|_{L^\infty(\mu)}\mu(Q)\\ &\lesssim |b|_{H^{1,\infty}_{\mathrm{atb}}(\mu)} + \|a\|_{L^\infty(\mu)}\mu(Q)\\ &\lesssim [1+\delta(Q, R)]\left[\|a\|_{L^\infty(\mu)}\mu(2Q) + |c_{Q_0}|\mu(2Q_0)\right]\\ &\quad + \|a\|_{L^\infty(\mu)}\mu(Q).\end{aligned}$$

Since Q_0 is doubling, it follows that

$$|c_{Q_0}|\mu(2Q_0) \lesssim \|a\|_{L^1(\mu)} \lesssim \|a\|_{L^\infty(\mu)}\mu(Q).$$

Therefore,

$$\int_Q |Ta(x)|\,d\mu(x) \lesssim \|a\|_{L^\infty(\mu)}\mu(2Q).$$

If $\ell(Q) > \mathrm{diam}(\mathrm{supp}\mu)/(20\sqrt{D})$, the proof for the implication (iii) $\Longrightarrow$ (i) is similar to Case II in the proof of (ii) $\Longrightarrow$ (i), the details being omitted. This finishes the proof of Theorem 5.4.4. □

Using an argument similar to that used in the proof of the implication (i) $\Longrightarrow$ (iii) in Theorem 5.4.4, we see that, if a kernel K satisfies (5.3.1) and (5.3.2), and the operator defined by (5.3.7) is bounded on $L^p(\mu)$ for some $p \in (1,\infty)$, then T is also bounded from $H^1(\mu)$ to $L^1(\mu)$. Also, notice that Theorem 5.4.4 is true for the truncated operator T_ϵ, with the bound appearing in Theorem 5.4.4 independent of ϵ. Since the boundedness of T in (5.3.7) on $L^2(\mu)$ implies (5.4.6), if T is bounded on $L^2(\mu)$, then from Theorem 5.4.4 and Lemma 5.2.2, we immediately deduce that T is bounded on $L^p(\mu)$ for $p \in (1, 2]$ as follows.

Corollary 5.4.5. *Let K satisfy* (5.3.1) *and* (5.3.2) *and T be as in* (5.3.7). *If T is bounded on $L^2(\mu)$, then T can be extended boundedly on $L^p(\mu)$ for all $p \in (1, 2]$.*

One may ask whether the Calderón–Zygmund operator T is bounded on $H^1(\mu)$ or not. Notice that, by Theorem 5.4.4, for $b \in H^1(\mu)$, we have $Tb \in L^1(\mu)$. Also, if $Tb \in H^1(\mu)$, then

$$\int_{\mathbb{R}^D} Tb(x)\, d\mu(x) = 0 \tag{5.4.12}$$

by the definition of the Hardy space $H^1(\mu)$. Let T^* be the *adjoint operator of* T. We say $T^*1 = 0$ if T satisfies (5.4.12) for any bounded function b with compact support and integral zero. We then have the following boundedness of T on $H^1(\mu)$.

Theorem 5.4.6. *Let T be a Calderón–Zygmund operator as in* (5.3.7) *with kernel K satisfying* (5.3.1) *and* (5.3.8). *If $T^*1 = 0$ as in* (5.4.12), *then T is bounded on $H^1(\mu)$.*

Proof. Let $\mathcal{M}_\Phi$ be as in Definition 3.3.1 and take $\eta = 4$ in the definition of $H^{1,2}_{\mathrm{atb},2}(\mu)$. For any $f \in H^{1,2}_{\mathrm{atb},2}(\mu)$, by Definition 3.2.1, there exist $(2,2)$-atomic blocks $\{b_i\}_{i=1}^\infty$ such that

$$f = \sum_{i=1}^\infty b_i \quad \text{and} \quad \sum_{i=1}^\infty |b_i|_{H^{1,2}_{\mathrm{atb},2}(\mu)} < \infty.$$

We now show that, for any $(2,2)$-atomic block b as in Definition 3.2.1, Tb is in $H^1(\mu)$ with norm $C|b|_{H^{1,2}_{\mathrm{atb},2}(\mu)}$, where C is independent of b. Let all the notation be the same as in Definition 3.2.1. By our choices, a_j now satisfies the following size condition that

$$\|a_j\|_{L^2(\mu)} \le [\mu(4Q_j)]^{-1/2}[1 + \delta(Q_j, R)]^{-2}, \tag{5.4.13}$$

where $j \in \{1, 2\}$.

The assumption that $T^*1 = 0$ tells us that

$$\int_{\mathbb{R}^D} Tb(x)\, d\mu(x) = 0.$$

Recalling that T is bounded from $H^1(\mu)$ into $L^1(\mu)$, we obtain

$$\|Tb\|_{L^1(\mu)} \lesssim |b|_{H^{1,2}_{\mathrm{atb},2}(\mu)}.$$

By this and Theorem 3.3.3, it remains to show that

$$\|\mathcal{M}_\Phi(Tb)\|_{L^1(\mu)} \lesssim |b|_{H^{1,2}_{\mathrm{atb},2}(\mu)}. \tag{5.4.14}$$

Write

$$\begin{aligned}\|\mathcal{M}_\Phi(Tb)\|_{L^1(\mu)} &= \int_{4R} \mathcal{M}_\Phi(Tb)(x)\,d\mu(x) + \int_{\mathbb{R}^D\setminus 4R} \cdots \\ &=: \mathrm{I} + \mathrm{II}.\end{aligned}$$

Noticing that $\mathcal{M}_\Phi$ is sublinear, we control I by

$$\begin{aligned}\mathrm{I} &\le \int_{4R} \mathcal{M}_\Phi([Tb]\chi_{8R})(x)\,d\mu(x) + \int_{4R} \mathcal{M}_\Phi\left([Tb]\chi_{\mathbb{R}^D\setminus 8R}\right)(x)\,d\mu(x) \\ &=: \mathrm{I}_1 + \mathrm{I}_2.\end{aligned}$$

From the fact that, for $j \in \{1, 2\}$, $Q_j \subset R$, it follows that, for any $z \in Q_j$ and $y \in 2^{k+1}R \setminus 2^k R$ with $k \ge 3$, it holds true that $|y - z| \ge \ell(2^{k-2}R)$. By this fact, the Hölder inequality, (ii) of Definition 3.3.1, (5.3.1) and (3.2.1), we obtain

$$\begin{aligned}\mathrm{I}_2 &\le \int_{4R} \sup_{\varphi\sim x}\left[\int_{\mathbb{R}^D\setminus 8R} |Tb(y)|\varphi(y)\,d\mu(y)\right] d\mu(x) \\ &\le \sum_{j=1}^2 |\lambda_j| \int_{4R} \sum_{k=3}^\infty \int_{2^{k+1}R\setminus 2^k R} \left|\int_{Q_j} K(y, z)a_j(z)\,d\mu(z)\right| \frac{1}{|x-y|^n}\,d\mu(y)\,d\mu(x) \\ &\lesssim \sum_{j=1}^2 |\lambda_j| \sum_{k=3}^\infty \|a_j\|_{L^2(\mu)}[\mu(Q_j)]^{1/2} \frac{\mu(2^{k+1}R)}{[\ell(2^{k-2}R)]^n} \frac{\mu(4R)}{[\ell(2^{k-2}R)]^n} \\ &\lesssim \sum_{j=1}^2 |\lambda_j|.\end{aligned}$$

To estimate I_1, we write

$$\begin{aligned}\mathrm{I}_1 &\le \sum_{j=1}^2 |\lambda_j| \int_{4Q_j} \mathcal{M}_\Phi\left([Ta_j]\chi_{8R}\right)(x)\,d\mu(x) \\ &\quad + \sum_{j=1}^2 |\lambda_j| \int_{4R\setminus 4Q_j} \mathcal{M}_\Phi\left([Ta_j]\chi_{2Q_j}\right)(x)\,d\mu(x) \\ &\quad + \sum_{j=1}^2 |\lambda_j| \int_{4R\setminus 4Q_j} \mathcal{M}_\Phi\left([Ta_j]\chi_{8R\setminus 2Q_j}\right)(x)\,d\mu(x) \\ &=: \mathrm{I}_{1,1} + \mathrm{I}_{1,2} + \mathrm{I}_{1,3}.\end{aligned}$$

By Definition 3.3.1, we see that $\mathcal{M}_\Phi$ is bounded on $L^\infty(\mu)$. This, together with Lemmas 3.3.4 and 5.2.1, implies that $\mathcal{M}_\Phi$ is also bounded on $L^p(\mu)$ for all $p \in$

$(1, \infty)$. From the Hölder inequality and the boundedness of T on $L^2(\mu)$, it follows that

$$\begin{aligned}
\mathrm{I}_{1,1} &\le \sum_{j=1}^{2} |\lambda_j| [\mu(4Q_j)]^{1/2} \left\| \mathcal{M}_\Phi([Ta_j]\chi_{8R}) \right\|_{L^2(\mu)} \\
&\lesssim \sum_{j=1}^{2} |\lambda_j| [\mu(4Q_j)]^{1/2} \|Ta_j\|_{L^2(\mu)} \\
&\lesssim \sum_{j=1}^{2} |\lambda_j| [\mu(4Q_j)]^{1/2} \|a_j\|_{L^2(\mu)} \\
&\lesssim \sum_{j=1}^{2} |\lambda_j|.
\end{aligned}$$

By (ii) of Definition 3.3.1, the Hölder inequality, the boundedness of T on $L^2(\mu)$ and the fact that

$$1 + \delta(Q_j, 4R) \lesssim 1 + \delta(Q_j, R), \tag{5.4.15}$$

we have

$$\begin{aligned}
\mathrm{I}_{1,2} &\le \sum_{j=1}^{2} |\lambda_j| \int_{4R \setminus 4Q_j} \sup_{\varphi \sim x} \left| \int_{2Q_j} Ta_j(y)\varphi(y)\, d\mu(y) \right| d\mu(x) \\
&\lesssim \sum_{j=1}^{2} |\lambda_j| \int_{4R \setminus 4Q_j} \frac{1}{|x - x_j|^n}\, d\mu(x) \int_{2Q_j} |Ta_j(y)|\, d\mu(y) \\
&\lesssim \sum_{j=1}^{2} |\lambda_j| [1 + \delta(Q_j, 4R)] \|Ta_j\|_{L^2(\mu)} [\mu(2Q_j)]^{1/2} \\
&\lesssim \sum_{j=1}^{2} |\lambda_j| [1 + \delta(Q_j, R)] \|a_j\|_{L^2(\mu)} [\mu(2Q_j)]^{1/2} \\
&\lesssim \sum_{j=1}^{2} |\lambda_j|.
\end{aligned}$$

For $j \in \{1, 2\}$, denote $N_{Q_j, 4R}$ simply by N_j. For $\mathrm{I}_{1,3}$, we further write

$$\begin{aligned}
\mathrm{I}_{1,3} &= \sum_{j=1}^{2}|\lambda_j|\sum_{k=2}^{N_j}\int_{2^{k+1}Q_j\setminus 2^kQ_j}\mathcal{M}_\Phi\left([Ta_j]\,\chi_{8R\setminus 2Q_j}\right)(x)\,d\mu(x)\\
&\le \sum_{j=1}^{2}|\lambda_j|\sum_{k=2}^{N_j}\int_{2^{k+1}Q_j\setminus 2^kQ_j}\mathcal{M}_\Phi\left(|Ta_j|\chi_{2^{k+2}Q_j\setminus 2^{k-1}Q_j}\right)(x)\,d\mu(x)\\
&\quad+\sum_{j=1}^{2}|\lambda_j|\sum_{k=2}^{N_j}\int_{2^{k+1}Q_j\setminus 2^kQ_j}\mathcal{M}_\Phi\left(|Ta_j|\chi_{\max\{2^{k+2}Q_j,\,8R\}\setminus 2^{k+2}Q_j}\right)(x)\,d\mu(x)\\
&\quad+\sum_{j=1}^{2}|\lambda_j|\sum_{k=2}^{N_j}\int_{2^{k+1}Q_j\setminus 2^kQ_j}\mathcal{M}_\Phi\left(|Ta_j|\chi_{2^{k-1}Q_j\setminus 2Q_j}\right)(x)\,d\mu(x)\\
&=: \mathrm{E}+\mathrm{F}+\mathrm{G}.
\end{aligned}$$

The boundedness of $\mathcal{M}_\Phi$ on $L^2(\mu)$, (5.3.1) and (3.2.1) tell us that

$$\begin{aligned}
\mathrm{E} &\le \sum_{j=1}^{2}|\lambda_j|\sum_{k=2}^{N_j}[\mu(2^{k+1}Q_j)]^{1/2}\left\|\mathcal{M}_\Phi\left(|Ta_j|\chi_{2^{k+2}Q_j\setminus 2^{k-1}Q_j}\right)\right\|_{L^2(\mu)}\\
&\lesssim \sum_{j=1}^{2}|\lambda_j|\sum_{k=2}^{N_j}[\mu(2^{k+1}Q_j)]^{1/2}\\
&\quad\times\left\{\int_{2^{k+2}Q_j\setminus 2^{k-1}Q_j}\left|\int_{Q_j}K(y,z)a_j(z)\,d\mu(z)\right|^2 d\mu(y)\right\}^{1/2}\\
&\lesssim \sum_{j=1}^{2}|\lambda_j|\sum_{k=2}^{N_j}\frac{\mu(2^{k+2}Q_j)}{[\ell(2^{k-3}Q_j)]^n}\|a_j\|_{L^2(\mu)}[\mu(Q_j)]^{1/2}\\
&\lesssim \sum_{j=1}^{2}|\lambda_j|.
\end{aligned}$$

By Definition 3.3.1, (5.3.1), (5.4.15), (2.1.3) and (3.2.1), we easily see that

$$\begin{aligned}
\mathrm{G} &\le \sum_{j=1}^{2}|\lambda_j|\sum_{k=2}^{N_j}\int_{2^{k+1}Q_j\setminus 2^kQ_j}\sup_{\varphi\sim x}\left[\int_{2^{k-1}Q_j\setminus 2Q_j}|Ta_j(y)|\varphi(y)\,d\mu(y)\right]d\mu(x)\\
&\le \sum_{j=1}^{2}|\lambda_j|\sum_{k=2}^{N_j}\int_{2^{k+1}Q_j\setminus 2^kQ_j}\sum_{l=1}^{k-2}\int_{2^{l+1}Q_j\setminus 2^lQ_j}
\end{aligned}$$

$$\times \left| \int_{Q_j} K(y, z)a_j(z)\, d\mu(z) \right| \frac{1}{|y-x|^n}\, d\mu(y)\, d\mu(x)$$

$$\lesssim \sum_{j=1}^{2} |\lambda_j| \sum_{k=2}^{N_j} \frac{\mu(2^{k+1}Q)}{[\ell(2^{k+1}Q_j)]^n} \sum_{l=1}^{k-2} \frac{\mu(2^{l+1}Q)}{[\ell(2^{l+1}Q_j)]^n} \|a_j\|_{L^2(\mu)} [\mu(Q_j)]^{1/2}$$

$$\lesssim \sum_{j=1}^{2} |\lambda_j| \left[\delta_{Q_j, R}\right]^2 \|a_j\|_{L^2(\mu)} [\mu(Q_j)]^{1/2}$$

$$\lesssim \sum_{j=1}^{2} |\lambda_j|,$$

where, for any cubes $Q \subset R$, $\delta_{Q,R}$ is as in (2.1.2). By an argument similar to that used in the estimate for G, we conclude that

$$\mathrm{F} \lesssim \sum_{j=1}^{2} |\lambda_j|.$$

The estimates for E, F and G imply the desired estimate for $\mathrm{I}_{1,3}$. Combining the estimates for $\mathrm{I}_{1,1}$, $\mathrm{I}_{1,2}$, $\mathrm{I}_{1,3}$ and I_2, we see that

$$\mathrm{I} = \int_{4R} \mathcal{M}_{\Phi}(Tb)(x)\, d\mu(x) \lesssim \sum_{j=1}^{2} |\lambda_j| \sim |b|_{H^{1,2}_{\mathrm{atb},2}(\mu)}. \tag{5.4.16}$$

Now we turn to the estimate for II. Let z_R be the center of the cube R. Invoking that $T^*1 = 0$, we obtain

$$\mathrm{II} = \int_{\mathbb{R}^D \setminus 4R} \sup_{\varphi \sim x} \left| \int_{\mathbb{R}^D} Tb(y)[\varphi(y) - \varphi(z_R)]\, d\mu(y) \right| d\mu(x)$$

$$\leq \int_{\mathbb{R}^D \setminus 4R} \sup_{\varphi \sim x} \left| \int_{2R} Tb(y)[\varphi(y) - \varphi(z_R)]\, d\mu(y) \right| d\mu(x)$$

$$+ \int_{\mathbb{R}^D \setminus 4R} \sup_{\varphi \sim x} \left| \int_{\mathbb{R}^D \setminus 2R} Tb(y)\, [\varphi(y) - \varphi(z_R)]\, d\mu(y) \right| d\mu(x)$$

$$=: \mathrm{II}_1 + \mathrm{II}_2.$$

Notice that, for any $z \in 2R$ and

$$x \in 2^{k+1}R \setminus 2^k R \quad \text{with} \quad k \geq 2,$$

it holds true that $|x-z| \ge \ell(2^{k-2}R)$. This, together with (iii) of Definition 3.3.1 and the mean value theorem, leads to

$$|\varphi(y)-\varphi(z_R)| \lesssim \frac{\ell(R)}{\ell(2^{k-2}R)^{n+1}} \tag{5.4.17}$$

for all $y \in 2R$. By (5.4.17), (5.3.1), the Hölder inequality, the boundedness of T in $L^2(\mu)$ and (3.2.1), we know that

$$\begin{aligned}
\mathrm{II}_1 &\le \sum_{j=1}^{2}|\lambda_j|\sum_{k=2}^{\infty}\int_{2^{k+1}R\setminus 2^kR}\sup_{\varphi\sim x}\left[\int_{2R\setminus 2Q_j}|Ta_j(y)|\,|\varphi(y)-\varphi(z_R)|\,d\mu(y)\right]d\mu(x)\\
&\quad+\sum_{j=1}^{2}|\lambda_j|\sum_{k=2}^{\infty}\int_{2^{k+1}R\setminus 2^kR}\sup_{\varphi\sim x}\left[\int_{2Q_j}|Ta_j(y)||\varphi(y)-\varphi(z_R)|\,d\mu(y)\right]d\mu(x)\\
&\lesssim \sum_{j=1}^{2}|\lambda_j|\sum_{k=2}^{\infty}\int_{2^{k+1}R\setminus 2^kR}\frac{\ell(R)}{\ell(2^{k-2}R)^{n+1}}\\
&\quad\times\sum_{l=1}^{N_j-1}\int_{2^{l+1}Q_j\setminus 2^lQ_j}\int_{Q_j}\frac{|a_j(z)|}{|y-z|^n}\,d\mu(z)\,d\mu(y)\,d\mu(x)\\
&\quad+\sum_{j=1}^{2}|\lambda_j|\sum_{k=2}^{\infty}\int_{2^{k+1}R\setminus 2^kR}\frac{\ell(R)}{\ell(2^{k-2}R)^{n+1}}\left\|(Ta_j)\chi_{2Q_j}\right\|_{L^1(\mu)}d\mu(x)\\
&\lesssim \sum_{j=1}^{2}|\lambda_j|\sum_{k=2}^{\infty}2^{-k}\sum_{l=1}^{N_j-1}\frac{\mu(2^{l+1}Q_j)}{\ell(2^{l+1}Q_j)^n}\|a_j\|_{L^2(\mu)}[\mu(Q_j)]^{1/2}\\
&\quad+\sum_{j=1}^{2}|\lambda_j|\sum_{k=2}^{\infty}2^{-k}\|(Ta_j)\chi_{2Q_j}\|_{L^2(\mu)}[\mu(2Q_j)]^{1/2}\\
&\lesssim \sum_{j=1}^{2}|\lambda_j|\delta_{Q_j,R}\|a_j\|_{L^2(\mu)}[\mu(Q_j)]^{1/2}+\sum_{j=1}^{2}|\lambda_j|\|a_j\|_{L^2(\mu)}[\mu(2Q_j)]^{1/2}\\
&\lesssim \sum_{j=1}^{2}|\lambda_j|.
\end{aligned}$$

We further estimate II_2 by

$$\mathrm{II}_2=\sum_{k=2}^{\infty}\int_{2^{k+1}R\setminus 2^kR}\sup_{\varphi\sim x}\left|\int_{\mathbb{R}^D\setminus 2R}Tb(y)[\varphi(y)-\varphi(z_R)]\,d\mu(y)\right|d\mu(x)$$

$$\begin{aligned}
&\leq \sum_{k=2}^{\infty} \int_{2^{k+1}R\setminus 2^k R} \mathcal{M}_\Phi\left[|Tb|\chi_{2^{k+2}R\setminus 2^{k-1}R}\right](x)\,d\mu(x)\\
&\quad+\sum_{k=2}^{\infty} \int_{2^{k+1}R\setminus 2^k R} \sup_{\varphi\sim x}\left[\int_{2^{k+2}R\setminus 2^{k-1}R} |Tb(y)|\varphi(z_R)\,d\mu(y)\right] d\mu(x)\\
&\quad+\sum_{k=2}^{\infty} \int_{2^{k+1}R\setminus 2^k R} \sup_{\varphi\sim x}\left[\int_{\mathbb{R}^D\setminus 2^{k+2}R} |Tb(y)|\,\{\varphi(y)+\varphi(z_R)\}\,d\mu(y)\right] d\mu(x)\\
&\quad+\sum_{k=2}^{\infty} \int_{2^{k+1}R\setminus 2^k R} \sup_{\varphi\sim x}\left[\int_{2^{k-1}R\setminus 2R} |Tb(y)|\,\{\varphi(y)+\varphi(z_R)\}\,d\mu(y)\right] d\mu(x)\\
&=: \mathrm{II}_{2,1}+\mathrm{II}_{2,2}+\mathrm{II}_{2,3}+\mathrm{II}_{2,4}.
\end{aligned}$$

From the boundedness of $\mathcal{M}_\Phi$ on $L^2(\mu)$, the fact that $\int_{\mathbb{R}^D} b(x)\,d\mu(x)=0$, (5.3.1) and (5.3.8), we deduce that

$$\begin{aligned}
\mathrm{II}_{2,1} &\leq \sum_{k=2}^{\infty} [\mu(2^{k+1}R)]^{1/2}\left\|\mathcal{M}_\Phi\left[|Tb|\chi_{2^{k+2}R\setminus 2^{k-1}R}\right]\right\|_{L^2(\mu)}\\
&\lesssim \sum_{k=2}^{\infty} [\mu(2^{k+1}R)]^{1/2}\left\{\int_{2^{k+2}R\setminus 2^{k-1}R}\right.\\
&\quad\times\left.\left|\int_R [K(y,z)-K(y,z_R)]\,b(z)\,d\mu(z)\right|^2 d\mu(y)\right\}^{1/2}\\
&\lesssim \sum_{k=2}^{\infty} \mu(2^{k+1}R)\frac{[\ell(R)]^\delta}{[\ell(2^kR)]^{n+\delta}}\|b\|_{L^1(\mu)}\\
&\lesssim \sum_{j=1}^{2} |\lambda_j|,
\end{aligned}$$

where we used the fact that

$$\|b\|_{L^1(\mu)} \leq \sum_{j=1}^{2} |\lambda_j|\|a_j\|_{L^1(\mu)} \lesssim \sum_{j=1}^{2} |\lambda_j|.$$

From an argument similar to that used in the estimate for $\mathrm{II}_{2,1}$, we deduce that

$$\mathrm{II}_{2,2} \lesssim \sum_{j=1}^{2} |\lambda_j|.$$

Finally, we estimate $\mathrm{II}_{2,3}$. By the fact that

$$\int_{\mathbb{R}^D} b(x)\,d\mu(x) = 0,$$

(ii) of Definition 3.3.1 and (5.3.8), we find that

$$\begin{aligned}
\mathrm{II}_{2,3} &\le \sum_{k=2}^{\infty} \int_{2^{k+1}R\setminus 2^k R} \sum_{l=k+2}^{\infty} \int_{2^{l+1}R\setminus 2^l R} \int_R |K(y,z) - K(y,z_R)||b(z)|\,d\mu(z) \\
&\quad \times \left[\frac{1}{|y-x|^n} + \frac{1}{|z_R - x|^n}\right] d\mu(y)\,d\mu(x) \\
&\lesssim \sum_{k=2}^{\infty} \sum_{l=k+2}^{\infty} \frac{\mu(2^{k+1}R)}{[\ell(2^{k+1}R)]^n} \frac{\mu(2^{l+1}R)[\ell(R)]^\delta}{[\ell(2^{l+1}R)]^{n+\delta}} \|b\|_{L^1(\mu)} \\
&\lesssim \sum_{j=1}^{2} |\lambda_j|.
\end{aligned}$$

An argument similar to that used in the estimate for $\mathrm{II}_{2,3}$ implies that

$$\mathrm{II}_{2,4} \lesssim \sum_{j=1}^{2} |\lambda_j|.$$

Combining the estimates for $\mathrm{II}_{2,1}$, $\mathrm{II}_{2,2}$, II_{23} and $\mathrm{II}_{2,4}$, we obtain the desired estimate for II_2. By the estimates for II_1 and II_2, we conclude that

$$\mathrm{II} = \int_{\mathbb{R}^D\setminus 4R} \mathcal{M}_\Phi(Tb)(x)\,d\mu(x) \lesssim |b|_{H^{1,2}_{\mathrm{atb},2}(\mu)}. \tag{5.4.18}$$

By the estimates (5.4.16) and (5.4.18), we then obtain (5.4.14).

On the other hand, from an argument similar to that used in the proof of the implication (i) $\Longrightarrow$ (iii) in Theorem 5.4.4, we deduce that

$$\left\|\sum_{i=1}^{\infty} |Tb_i|\right\|_{L^1(\mu)} \le \sum_{i=1}^{\infty} \|Tb_i\|_{L^1(\mu)} \lesssim \sum_{i=1}^{\infty} |b_i|_{H^{1,2}_{\mathrm{atb},2}(\mu)} < \infty. \tag{5.4.19}$$

Observe that, for each $x \in \mathbb{R}^D$ and $\varphi \sim x$, there exists a positive constant M, depending on x, such that, for all $y \in \mathbb{R}^D$,

$$0 \le \varphi(y) \le M.$$

Moreover, by the boundedness of $\mathcal{M}_\Phi$ from $H^1(\mu)$ to $L^1(\mu)$, we conclude that

$$Tf = \sum_{i=1}^{\infty} Tb_i \text{ in } L^1(\mu).$$

These two facts, together with (5.4.19) and the Lebesgue dominated convergence theorem, imply that

$$\begin{aligned}\int_{\mathbb{R}^D} \varphi(y)Tf(y)\,d\mu(y) &= \int_{\mathbb{R}^D} \sum_{i=1}^{\infty} \varphi(y)Tb_i(y)\,d\mu(y)\\ &= \sum_{i=1}^{\infty} \int_{\mathbb{R}^D} \varphi(y)Tb_i(y)\,d\mu(y).\end{aligned}$$

From this, it further follows that, for all $x \in \mathbb{R}^D$,

$$\mathcal{M}_\Phi(Tf)(x) \le \sum_{i=1}^{\infty} \mathcal{M}_\Phi(Tb_i)(x),$$

which, together with the Levi lemma and (5.4.14), implies that

$$\begin{aligned}\|\mathcal{M}_\Phi(Tf)\|_{L^1(\mu)} &\le \left\|\sum_{i=1}^{\infty} \mathcal{M}_\Phi(Tb_i)\right\|_{L^1(\mu)}\\ &\le \sum_{i=1}^{\infty} \|\mathcal{M}_\Phi(Tb_i)\|_{L^1(\mu)}\\ &\lesssim \sum_{i=1}^{\infty} |b_i|_{H^{1,2}_{\mathrm{atb},2}(\mu)}.\end{aligned}$$

This, combined with Definition 3.3.1 and the fact that $H^1(\mu)$ and $H^{1,2}_{\mathrm{atb},2}(\mu)$ coincide with equivalent norms, in turn implies that

$$Tf \in H^1(\mu) \quad \text{and} \quad \|Tf\|_{H^1(\mu)} \lesssim \|f\|_{H^1(\mu)},$$

which completes the proof of Theorem 5.4.6. □

From Theorem 5.4.6, the fact that

$$\mathrm{RBMO}\,(\mu) = \left(H^1(\mu)\right)^*$$

and a standard dual argument, it is easy to deduce the boundedness of the adjoint operator of T in RBMO (μ) as follows.

Corollary 5.4.7. *Let T be the same as in Theorem 5.4.6 and T^* the adjoint operator of T. Then T is bounded on* RBMO (μ).

5.5 Weighted Estimates for Calderón–Zygmund Operators

In the setting of Euclidean spaces, the weighted estimates with $A_p(\mathbb{R}^D)$ weights (the weight function class of Muckenhoupt[3]) for the classical Calderón–Zygmund operators, rely essentially on the following three facts:

(a) the sharp function estimate for Calderón–Zygmund operators;
(b) the reverse Hölder inequality properties of $A_p(\mathbb{R}^D)$ weights;
(c) the prior estimate that, for any bounded function f with compact support, $Tf \in L^p(w)$ whenever $p \in [1, \infty)$ and $w \in A_p(\mathbb{R}^D)$.

To establish the weighted $L^p(u)$ estimates with $u \in A_p^\rho(\mu)$ for singular integral operators, our first motivation is to follow the program which was used in the classical setting. However, for a nonnegative Radon measure μ only satisfying (0.0.1) and a weight $u \in A_p^\rho(\mu)$ with $\rho,\ p \in [1, \infty)$, we do not know whether u enjoys the reverse Hölder inequality properties or not. Thus, we do not expect that T is bounded on $L^p(u)$ for $u \in A_p^\rho(\mu)$. Also, some estimates in establishing the weighted estimates for singular integral operators should be more refined.

The main result in this section is the following weighted weak type estimate for the Calderón–Zygmund operator $T^\sharp$.

Theorem 5.5.1. *Let $\rho \in [1, \infty)$, T be a Calderón–Zygmund operator with kernel K as in the sense of* (5.3.7), *satisfying* (5.3.1) *and* (5.3.8), *and $T^\sharp$ the corresponding maximal operator defined by* (5.3.9). *Then, for any $p \in [1, \infty)$ and $u \in A_p^\rho(\mu)$, both T and $T^\sharp$ are bounded from $L^p(u)$ to $L^{p,\infty}(u)$.*

By Lemma 5.1.7, Theorem 5.5.1 is an easy consequence of the following result.

Theorem 5.5.2. *Under the hypotheses of Theorem 5.5.1, for any $\rho,\ p \in [1, \infty)$ and $u \in A_p^\rho(\mu)$, there exists a positive constant C such that, for any bounded function f with compact support,*

$$\sup_{\lambda \in (0,\infty)} \lambda^p u(\{x \in \mathbb{R}^D : |Tf(x)| > \lambda\})$$
$$\leq C \sup_{\lambda \in (0,\infty)} \lambda^p u(\{x \in \mathbb{R}^D : \mathcal{M}_{(\frac{9}{8}\rho)} f(x) > \lambda\})$$

[3]See [121].

and

$$\begin{aligned}
&\sup_{\lambda\in(0,\infty)} \lambda^p u(\{x \in \mathbb{R}^D : T^\sharp f(x) > \lambda\}) \\
&\quad\le C \sup_{\lambda\in(0,\infty)} \lambda^p u(\{x \in \mathbb{R}^D : \mathcal{M}_{(\frac{9}{8}\rho)} f(x) > \lambda\}).
\end{aligned}$$

To prove Theorem 5.5.2, we need the following sharp function estimate for T and $T^\sharp$.

Lemma 5.5.3. *Let $\rho \in [1,\infty)$ and $r \in (0,1)$. Under the assumptions of Theorem 5.5.1, there exists a positive constant C such that, for all bounded functions f with compact support and μ-almost every $x \in \mathbb{R}^D$,*

$$\mathcal{M}_r^{\rho,\natural}(T^\sharp f)(x) + \mathcal{M}_r^{\rho,\natural}(|Tf|)(x) \le C\mathcal{M}_{(\frac{9}{8}\rho)} f(x). \tag{5.5.1}$$

Proof. We only consider the operator $T^\sharp$, since the estimate for T is similar, the details being omitted. For each cube Q and each bounded function f with compact support, let

$$h_Q := m_Q\left(T^\sharp\left(f\chi_{\mathbb{R}^D\setminus\frac{4}{3}Q}\right)\right).$$

It follows, from Lemma 5.1.1, (5.1.1) and Lemma 5.1.4, that, for any cube Q and $s \in (0, \beta_{2\rho}^{-1}/4)$,

$$\begin{aligned}
&\int_Q |T^\sharp f(y) - m_{T^\sharp f}(\tilde{Q}^\rho)|^r \, d\mu(y) \\
&\le \int_Q |T^\sharp f(y) - h_Q|^r \, d\mu(y) + \left|h_Q - h_{\tilde{Q}^\rho}\right|^r \mu(Q) \\
&\quad + \left|m^\rho_{0,s;\tilde{Q}^\rho}(T^\sharp f) - m_{T^\sharp f}(\tilde{Q}^\rho)\right|^r \mu(Q) \\
&\quad + \left|m^\rho_{0,s;\tilde{Q}^\rho}(T^\sharp f) - h_{\tilde{Q}^\rho}\right|^r \mu(Q) \\
&\le \int_Q |T^\sharp f(y) - h_Q|^r \, d\mu(y) + \left|h_Q - h_{\tilde{Q}^\rho}\right|^r \mu(Q) \\
&\quad + \left[m^\rho_{0,s;\tilde{Q}^\rho}(T^\sharp f - m_{T^\sharp f}(\tilde{Q}^\rho))\right]^r \mu(Q) \\
&\quad + \left[m^\rho_{0,s;\tilde{Q}^\rho}(T^\sharp f - h_{\tilde{Q}^\rho})\right]^r \mu(Q) \\
&\lesssim \int_Q |T^\sharp f(y) - h_Q|^r \, d\mu(y) + \left|h_Q - h_{\tilde{Q}^\rho}\right|^r \mu(Q)
\end{aligned}$$

$$+(3^r s^{-1} + s^{-1})\frac{\mu(Q)}{\mu(\tilde{Q}^\rho)}\int_{\tilde{Q}^\rho}\left|T^\sharp f(y) - h_{\tilde{Q}^\rho}\right|^r d\mu(y).$$

For any two 2ρ-doubling cubes $Q \subset R$, we have

$$\begin{aligned}
&\left|m_{T^\sharp f}(Q) - m_{T^\sharp f}(R)\right| \\
&\le \left|m^\rho_{0,s;Q}(T^\sharp f) - h_Q\right| + \left|h_Q - h_R\right| + \left|m^\rho_{0,s;R}(T^\sharp f) - h_R\right| \\
&\quad + \left|m^\rho_{0,s;Q}(T^\sharp f) - m_{T^\sharp f}(Q)\right| + \left|m^\rho_{0,s;R}(T^\sharp f) - m_{T^\sharp f}(R)\right| \\
&\le m^\rho_{0,s;Q}\left(T^\sharp f - h_Q\right) + |h_Q - h_R| + m^\rho_{0,s;R}\left(T^\sharp f - h_R\right) \\
&\quad + m^\rho_{0,s;Q}\left(T^\sharp f - m_{T^\sharp f}(Q)\right) + m^\rho_{0,s;R}\left(T^\sharp f - m_{T^\sharp f}(R)\right) \\
&\le 4s^{-1/r}\left[\frac{1}{\mu(\frac{3}{2}\rho Q)}\int_Q \left|T^\sharp f(y) - h_Q\right|^r d\mu(y)\right]^{1/r} + |h_Q - h_R| \\
&\quad + 4s^{-1/r}\left[\frac{1}{\mu(\frac{3}{2}\rho R)}\int_R \left|T^\sharp f(y) - h_R\right|^r d\mu(y)\right]^{1/r}.
\end{aligned}$$

Thus, the proof of (5.5.1) is reduced to proving that, for any cube Q,

$$\left[\frac{1}{\mu(\frac{3}{2}\rho Q)}\int_Q \left|T^\sharp f(y) - h_Q\right|^r d\mu(y)\right]^{1/r} \lesssim \inf_{x\in Q}\mathcal{M}_{(\frac{9}{8}\rho)}f(x) \qquad (5.5.2)$$

and, for any two cubes $Q \subset R$ with R a doubling cube,

$$|h_Q - h_R| \lesssim [1 + \delta(Q, R)]\inf_{x\in Q}\mathcal{M}_{(\frac{9}{8}\rho)}f(x). \qquad (5.5.3)$$

We first consider (5.5.2). For any cube Q, write

$$\begin{aligned}
\int_Q \left|T^\sharp f(y) - h_Q\right|^r d\mu(y) &\le \int_Q \left|T^\sharp f(y) - T^\sharp(f\chi_{\mathbb{R}^D\setminus\frac{4}{3}Q})(y)\right|^r d\mu(y) \\
&\quad + \int_Q \left|T^\sharp(f\chi_{\mathbb{R}^D\setminus\frac{4}{3}Q})(y) - h_Q\right|^r d\mu(y) \\
&\le \int_Q \left|T^\sharp(f\chi_{\frac{4}{3}Q})(y)\right|^r d\mu(y) \\
&\quad + \int_Q \left|T^\sharp(f\chi_{\mathbb{R}^D\setminus\frac{4}{3}Q})(y) - h_Q\right|^r d\mu(y).
\end{aligned}$$

Recall that $T^\sharp$ is bounded from $L^1(\mu)$ to $L^{1,\infty}(\mu)$. It follows, from the Kolmogorov inequality, that

$$\left[\frac{1}{\mu(\frac{3}{2}\rho Q)}\int_Q \left|T^\sharp(f\chi_{\frac{4}{3}Q})(y)\right|^r d\mu(y)\right]^{1/r} \lesssim \frac{1}{\mu(\frac{3}{2}\rho Q)}\|f\chi_{\frac{4}{3}Q}\|_{L^1(\mu)}$$
$$\lesssim \inf_{x\in Q}\mathcal{M}_{(\frac{9}{8}\rho)}f(x).$$

On the other hand, by the size condition (5.3.1) and the regularity condition (5.3.8), via a standard computation, we know that, for any $y, z \in Q$,

$$\begin{aligned}
&\left|T^\sharp(f\chi_{\mathbb{R}^D\setminus\frac{4}{3}Q})(y) - T^\sharp(f\chi_{\mathbb{R}^D\setminus\frac{4}{3}Q})(z)\right|\\
&\quad\le \sup_{\epsilon\in(0,\infty)}\left|\int_{|y-w|>\epsilon} K(y,w)f(w)\chi_{\mathbb{R}^D\setminus\frac{4}{3}Q}(w)\,d\mu(w)\right.\\
&\qquad\left.-\int_{|y-w|>\epsilon} K(z,w)f(w)\chi_{\mathbb{R}^D\setminus\frac{4}{3}Q}(w)\,d\mu(w)\right|\\
&\qquad+\sup_{\epsilon\in(0,\infty)}\left|\int_{|y-w|>\epsilon} K(z,w)f(w)\chi_{\mathbb{R}^D\setminus\frac{4}{3}Q}(w)\,d\mu(w)\right.\\
&\qquad\left.-\int_{|z-w|>\epsilon} K(z,w)f(w)\chi_{\mathbb{R}^D\setminus\frac{4}{3}Q}(w)\,d\mu(w)\right|\\
&\quad\lesssim \int_{\mathbb{R}^D\setminus\frac{4}{3}Q}|K(y,w)-K(z,w)||f(w)|\,d\mu(w)\\
&\qquad+\sup_{\epsilon\in(0,\infty)}\epsilon^{-n}\int_{|z-w|<C\epsilon}|f(w)|\chi_{\mathbb{R}^D\setminus\frac{4}{3}Q}(w)\,d\mu(w)\\
&\quad\lesssim \inf_{x\in Q}\mathcal{M}_{(\frac{9}{8}\rho)}f(x).
\end{aligned}\tag{5.5.4}$$

This in turn implies that

$$\left\{\frac{1}{\mu(\frac{3}{2}\rho Q)}\int_Q |T^\sharp(f\chi_{\mathbb{R}^D\setminus\frac{4}{3}Q})(y)-h_Q|^r\,d\mu(y)\right\}^{1/r} \lesssim \inf_{x\in Q}\mathcal{M}_{(\frac{9}{8}\rho)}f(x)$$

and hence the estimate (5.5.2) holds true.

Now we turn our attention to (5.5.3). Denote $N_{Q,R}^\rho + 1$ simply by N. Write

$$\begin{aligned}
|h_Q - h_R| &\le m_Q(T^\sharp(f\chi_{2\rho Q\setminus\frac{4}{3}Q})) + m_Q(T^\sharp(f\chi_{(2\rho)^N Q\setminus 2\rho Q}))\\
&\quad+\left|m_Q(T^\sharp(f\chi_{\mathbb{R}^D\setminus(2\rho)^N Q})) - m_R(T^\sharp(f\chi_{\mathbb{R}^D\setminus(2\rho)^N Q}))\right|
\end{aligned}$$

$$+m_R(T^\sharp(f\chi_{(2\rho)^N Q\setminus\frac{4}{3}R}))$$
$$=: \mathrm{I}_1 + \mathrm{I}_2 + \mathrm{I}_3 + \mathrm{I}_4.$$

The size condition (5.3.1), together with the growth condition (0.0.1), implies that, for any $x,\ y \in Q$,

$$T^\sharp(f\chi_{2\rho Q\setminus\frac{4}{3}Q})(y) \lesssim \frac{1}{[\ell(Q)]^n}\int_{2\rho Q}|f(z)|\,d\mu(z) \lesssim \mathcal{M}_{(\frac{9}{8}\rho)}f(x)$$

and, for any $y \in R$ and $x \in Q$,

$$T^\sharp(f\chi_{(2\rho)^N Q\setminus\frac{4}{3}R})(y) \lesssim \frac{1}{[\ell(R)]^n}\int_{4\rho R}|f(z)|\,d\mu(z) \lesssim \mathcal{M}_{(\frac{9}{8}\rho)}f(x).$$

Therefore,

$$\mathrm{I}_1 + \mathrm{I}_4 \lesssim \inf_{x\in Q}\mathcal{M}_{(\frac{9}{8}\rho)}f(x).$$

For the term I_2, observing that, for any $y \in Q$,

$$\begin{aligned}
T^\sharp(f\chi_{(2\rho)^N Q\setminus 2\rho Q})(y) &\lesssim \sum_{k=1}^{N-1}\int_{(2\rho)^{k+1}Q\setminus(2\rho)^k Q}\frac{|f(z)|}{|y-z|^n}\,d\mu(z)\\
&\lesssim \sum_{k=1}^{N-1}\frac{\mu((2\rho)^{k+2}Q)}{[\ell((2\rho)^k Q)]^n}\inf_{x\in Q}\mathcal{M}_{2\rho}f(x)\\
&\lesssim \delta_{Q,R}^{\rho}\inf_{x\in Q}\mathcal{M}_{(\frac{9}{8}\rho)}f(x),
\end{aligned}$$

we then conclude that

$$\mathrm{I}_2 \lesssim \delta_{Q,R}^{\rho}\inf_{x\in Q}\mathcal{M}_{(\frac{9}{8}\rho)}f(x).$$

Finally, as in the inequality (5.5.4), a familiar argument, invoking the condition (5.3.8), shows that, for any $y \in Q$ and $z \in R$,

$$|T^\sharp(f\chi_{\mathbb{R}^D\setminus(2\rho)^N Q})(y) - T^\sharp(f\chi_{\mathbb{R}^D\setminus(2\rho)^N Q})(z)| \lesssim \inf_{x\in Q}\mathcal{M}_{(\frac{9}{8}\rho)}f(x)$$

and hence

$$\mathrm{I}_3 \lesssim \inf_{x\in Q}\mathcal{M}_{(\frac{9}{8}\rho)}f(x).$$

The inequality (5.5.3) now follows, which completes the proof of Lemma 5.5.3. □

Proof of Theorem 5.5.2. We only consider $T^\sharp$. We first prove that, for any bounded function f with compact support, $\rho,\ p \in [1, \infty)$ and $u \in A_p^\rho(\mu)$,

$$\sup_{\lambda\in(0,R)} \lambda^p u(\{x \in \mathbb{R}^D : |T^\sharp f(x)| > \lambda\}) < \infty. \tag{5.5.5}$$

To see this, let $t \in (2, \infty)$ be large enough such that the support of f is contained in the ball $B(0, t)$. It is obvious that

$$\sup_{\lambda\in(0,R)} \lambda^p u(\{x \in B(0, 2t) : |T^\sharp f(x)| > \lambda\}) \leq R^p u(B(0, 2t)) < \infty.$$

On the other hand, by the size condition (5.3.1), there exists a positive constant $\tilde{C}$ such that, if $x \in \mathbb{R}^D \setminus B(0, 2t)$, then

$$T^\sharp f(x) \leq \int_{\mathbb{R}^D} |K(x, y) f(y)|\, d\mu(y) \leq \frac{\tilde{C}}{|x|^n} \|f\|_{L^1(\mu)}.$$

This, via (i) of Lemma 5.1.8 and (0.0.1), leads to that, if $\lambda \in (0, \tilde{C}\|f\|_{L^1(\mu)}/2]$,

$$\begin{aligned}
&u(\{x \in \mathbb{R}^D \setminus B(0, 2t) : |T^\sharp f(x)| > \lambda\})\\
&\quad\leq u(\{x \in \mathbb{R}^D : |x|^{-n} > \lambda/(\tilde{C}\|f\|_{L^1(\mu)})\})\\
&\quad\leq u\left(B\left(0, \frac{9}{8}\rho(\tilde{C}\|f\|_{L^1(\mu)})^{1/n}\lambda^{-1/n}\right)\right)\\
&\quad\lesssim u(B(0, 1))\left[\frac{\mu(B(0, \frac{9}{8}\rho\tilde{C}\|f\|_{L^1(\mu)}^{1/n}\lambda^{-1/n}))}{\mu(B(0, 1))}\right]^p\\
&\quad\leq C_f \frac{u(B(0, 1))}{[\mu(B(0, 1))]^p}\lambda^{-p},
\end{aligned}$$

where C_f is a positive constant.

Notice that for $\lambda \in (\tilde{C}\|f\|_{L^1(\mu)}/2, \infty)$, there does not exist any point $x \notin B(0, 2t)$ satisfying that $T^\sharp f(x) > \lambda$. Therefore,

$$\begin{aligned}
&\sup_{\lambda\in(0,\infty)} \lambda^p u\left(\{x \in \mathbb{R}^D \setminus B(0, 2t) : \left|T^\sharp f(x)\right| > \lambda\}\right)\\
&\quad= \sup_{\lambda\in(0,\tilde{C}\|f\|_{L^1(\mu)}/2]} \lambda^p u\left(\{x \in \mathbb{R}^D \setminus B(0, 2t) : \left|T^\sharp f(x)\right| > \lambda\}\right)\\
&\quad\leq C_f \frac{u(B(0, 1))}{[\mu(B(0, 1))]^p},
\end{aligned}$$

which shows (5.5.5).

We now conclude the proof of Theorem 5.5.2. If $\mu(\mathbb{R}^D) = \infty$, the desired result follows from (i) of Lemma 5.1.9, Theorem 5.1.12 with $s_1 = \beta_{2\rho}^{-1}/5$ and $p_0 = 1$, (5.1.17) and Lemma 5.5.3.

If $\mu(\mathbb{R}^D) < \infty$, ρ, $p \in [1, \infty)$ and $u \in A_p^\rho(\mu)$, we then know that

$$\begin{aligned} &u\left(\mathbb{R}^D\right)\left[\mu\left(\mathbb{R}^D\right)\right]^{-p} \|T^\sharp f\|_{L^{1,\infty}(\mu)}^p \\ &\lesssim u\left(\mathbb{R}^D\right)\left[\mu\left(\mathbb{R}^D\right)\right]^{-p} \|f\|_{L^1(\mu)}^p \\ &\lesssim u\left(\mathbb{R}^D\right)\left(\inf_{x\in\mathbb{R}^D} \mathcal{M}_{(\frac{9}{8}\rho)} f(x)\right)^p \\ &\lesssim \sup_{\lambda\in(0,\infty)} \left[\lambda^p u\left(\left\{x \in \mathbb{R}^D : \mathcal{M}_{(\frac{9}{8}\rho)} f(x) > \lambda\right\}\right)\right], \end{aligned}$$

where, in the first inequality, we invoked the fact that $T^\sharp$ is bounded from $L^1(\mu)$ to $L^{1,\infty}(\mu)$ and the second inequality follows from the fact that

$$\begin{aligned} \frac{1}{\mu(\mathbb{R}^D)} \int_{\mathbb{R}^D} |f(y)|\, d\mu(y) &= \lim_{\ell(Q)\to\infty} \frac{1}{\mu(\frac{9}{8}\rho Q)} \int_Q |f(y)|\, d\mu(y) \\ &\le \inf_{x\in\mathbb{R}^D} M_{(\frac{9}{8}\rho)} f(x). \end{aligned}$$

The desired results again follows from (i) of Lemma 5.1.9, Theorem 5.1.12 with $s_1 = \beta_{2\rho}^{-1}/5$ and $p_0 = 1$, (5.1.17) and Lemma 5.5.3, which completes the proof of Theorem 5.5.2. □

5.6 Multilinear Commutators of Singular Integrals

In this section, we first consider the boundedness of commutators generated by RBMO (μ) functions and singular integral operators on $H^1(\mu)$. We then prove the boundedness on $L^p(\mu)$, with $p \in (1, \infty)$, and a weak type endpoint estimate for multilinear commutators generated by singular integrals and RBMO(μ) functions or $\mathrm{Osc}_{\exp L^r}(\mu)$ functions, in analogy with the results for multilinear commutators of Calderón–Zygmund operators with μ being a D-dimensional Lebesgue measure.[4]

---

[4] See [108].

5.6.1 Boundedness of Commutators on $H^1(\mu)$

In this subsection, we consider the commutators generated by RBMO (μ) functions and singular integral operators.

Let K be a function on

$$(\mathbb{R}^D \times \mathbb{R}^D) \setminus \{(x, y) : x = y\},$$

which satisfies (5.3.1) and that

$$\sum_{l=1}^{\infty} l \int_{2^l R<|x-y|\leq 2^{l+1} R} [|K(x, y) - K(x, \tilde{y})| \\ +|K(y, x) - K(\tilde{y}, x)|] \, d\mu(x) \lesssim 1 \tag{5.6.1}$$

for any $R \in (0, \infty)$ and $y, \tilde{y} \in \mathbb{R}^D$ such that $|y - \tilde{y}| < R$. For the operator T defined in (5.3.7) and a function $b \in \mathrm{RBMO}(\mu)$, the *commutator* T_b is defined by setting, for all suitable functions f,

$$T_b f := bTf - T(bf). \tag{5.6.2}$$

We establish the weak type $(H^1(\mu), L^1(\mu))$ estimate and the $L^p(\mu)$ estimate, with $p \in (1, \infty)$, for T_b. To this end, we first recall the generalization of the Hölder inequality.[5]

Lemma 5.6.1. *Let $\Phi, \Psi_1, \ldots, \Psi_m$ be Young functions such that, for all $t \in (0, \infty)$,*

$$\Psi_1^{-1}(t) \cdots \Psi_m^{-1}(t) \leq C\Phi^{-1}(t),$$

where C is a positive constant independent of t. Then, there exists a positive constant C such that,

(i) *for all $t_1, \ldots, t_m \in (0, \infty)$,*

$$\Phi(t_1 \cdots t_m) \leq C \sum_{j=1}^{m} \Psi_j(t_j);$$

(ii) *for all suitable functions $f_1, \ldots, f_m$ and cube Q,*

$$\|f_1 \cdots f_m\|_{\Phi, Q, \mu} \leq C \prod_{j=1}^{m} \|f_j\|_{\Psi, Q, \mu}.$$

[5] See [109] for the details.

Theorem 5.6.2. *Let K be a function on*

$$\left(\mathbb{R}^D \times \mathbb{R}^D\right) \setminus \{x = y\}$$

satisfying (5.3.1) and (5.6.1). Suppose that the operator T defined in (5.3.7) is bounded on $L^2(\mu)$. Then, for any $b \in \mathrm{RBMO}(\mu)$, T_b is bounded from $H^1(\mu)$ to $L^{1,\infty}(\mu)$, that is, there exists a positive constant C such that, for all $\lambda \in (0,\infty)$ and functions $f \in H^1(\mu)$,

$$\mu\left(\left\{x \in \mathbb{R}^D : |T_b f(x)| > \lambda\right\}\right) \le C \|b\|_{\mathrm{RBMO}(\mu)} \lambda^{-1} \|f\|_{H^1(\mu)}.$$

Proof. By the homogeneity, we may assume that $\|b\|_{\mathrm{RBMO}(\mu)} = 1$. For each fixed $f \in H^1(\mu)$, by Theorem 3.2.12, we have the decomposition

$$f = \sum_j h_j,$$

where $\{h_j\}_j$ are $(\infty, 2)$-atomic blocks, defined as in Definition 3.2.1, such that

$$\sum_j |h_j|_{H^{1,\infty}_{\mathrm{atb},2}(\mu)} \le 2\|f\|_{H^1(\mu)}.$$

Let R_j be a cube such that $\operatorname{supp} h_j \subset R_j$. Write

$$\begin{aligned} T_b f(x) &= \sum_{j=1}^{\infty} \left[b(x) - m_{\widetilde{R_j}}(b)\right] T h_j(x) + T\left(\sum_{j=1}^{\infty} \left[m_{\widetilde{R_j}}(b) - b\right] h_j\right)(x) \\ &=: T_b^{\mathrm{I}} f(x) + T_b^{\mathrm{II}} f(x). \end{aligned}$$

By Theorem 5.3.4, we see that $T^\sharp$ is bounded from $L^1(\mu)$ to $L^{1,\infty}(\mu)$ and so is T. Therefore,

$$\begin{aligned} &\mu\left(\left\{x \in \mathbb{R}^D : \left|T_b^{\mathrm{II}} f(x)\right| > \lambda\right\}\right) \\ &\quad \lesssim \frac{1}{\lambda} \sum_{j=1}^{\infty} \int_{R_j} \left|b(x) - m_{\widetilde{R_j}}(b)\right| |h_j(x)| \, d\mu(x). \end{aligned} \tag{5.6.3}$$

For each fixed j, decompose h_j as

$$h_j := r_j^1 a_j^1 + r_j^2 a_j^2,$$

where r_j^i, for $i \in \{1, 2\}$, satisfies that

$$|h_j|_{H^{1,\infty}_{\mathrm{atb},2}(\mu)} = |r_j^1| + |r_j^2|,$$

a^i_j, for $i \in \{1, 2\}$, is supported on some cube $Q^i_j \subset R_j$ and satisfies (3.2.1) for $p = \infty$ and $\eta = 4$. Write

$$\begin{aligned}\int_{R_j} \left|b(x) - m_{\widetilde{R_j}}(b)\right| |h_j(x)|\, d\mu(x) &\le |r^1_j| \int_{R_j} \left|b(x) - m_{\widetilde{R_j}}(b)\right| |a^1_j(x)|\, d\mu(x)\\ &\quad + |r^2_j| \int_{R_j} \left|b(x) - m_{\widetilde{R_j}}(b)\right| |a^2_j(x)|\, d\mu(x)\\ &=: \mathrm{E} + \mathrm{F}.\end{aligned}$$

The fact that

$$\left|m_{\widetilde{R_j}}(b) - m_{\widetilde{Q^1_j}}(b)\right| \lesssim [1 + \delta(Q^1_j, R_j)]$$

and a trivial computation lead to that

$$\begin{aligned}\mathrm{E} &\le |r^1_j| \|a^1_j\|_{L^\infty(\mu)} \left[\int_{Q^1_j} \left|b(x) - m_{\widetilde{Q^1_j}}(b)\right| d\mu(x) + \left|m_{\widetilde{R_j}}(b) - m_{\widetilde{Q^1_j}}(b)\right| \mu(Q^1_j)\right]\\ &\lesssim |r^1_j| \left[\mu(4Q^1_j)\right]^{-1} [1 + \delta(Q^1_j, R_j)]^{-2} \left\{\mu(2Q^1_j) + [1 + \delta(Q^1_j, R_j)]\mu(Q^1_j)\right\}\\ &\lesssim |r^1_j|.\end{aligned}$$

Similarly,

$$\mathrm{F} \lesssim |r^2_j|.$$

Combining the estimates for the terms E and F with (5.6.3), we see that

$$\mu\left(\{x \in \mathbb{R}^D : \left|T^{\mathrm{II}}_b f(x)\right| > \lambda\}\right) \lesssim \lambda^{-1} \|f\|_{H^1(\mu)}. \tag{5.6.4}$$

Now we turn our attention to the estimate for $T^{\mathrm{I}}_b f$. Write

$$\begin{aligned}\mu\left(\{x \in \mathbb{R}^D : \left|T^{\mathrm{I}}_b f(x)\right| > \lambda\}\right) &\le \lambda^{-1} \sum_{j=1}^{\infty} \int_{2R_j} \left|b(x) - m_{\widetilde{R_j}}(b)\right| |Th_j(x)|\, d\mu(x)\\ &\quad + \lambda^{-1} \sum_{j=1}^{\infty} \int_{\mathbb{R}^D \setminus 2R_j} \cdots\\ &=: \mathrm{G} + \mathrm{H}.\end{aligned}$$

Denote by x_j the center of R_j. An argument, invoking the fact that

$$\int_{\mathbb{R}^D} h_j(x)\, d\mu(x) = 0,$$

shows that

$$\begin{aligned}\mathrm{H} &\le \lambda^{-1}\sum_{j=1}^{\infty}\int_{\mathbb{R}^D\setminus 2R_j}\left|b(x)-m_{\widetilde{R_j}}(b)\right|\\&\quad\times\left|\int_{R_j}\left[K(x,y)-K(x,x_j)\right]h_j(y)\,d\mu(y)\right|\,d\mu(x)\\&\le \lambda^{-1}\sum_{j=1}^{\infty}\int_{R_j}\int_{\mathbb{R}^D\setminus 2R_j}|h_j(y)|\left|b(x)-m_{\widetilde{R_j}}(b)\right|\\&\quad\times\left|K(x,y)-K(x,x_j)\right|\,d\mu(x)\,d\mu(y).\end{aligned}$$

Write

$$\begin{aligned}&\int_{\mathbb{R}^D\setminus 2R_j}\left|b(x)-m_{\widetilde{R_j}}(b)\right|\left|K(x,y)-K(x,x_j)\right|\,d\mu(x)\\&\quad\le\sum_{k=1}^{\infty}\int_{2^{k+1}R_j\setminus 2^kR_j}\left|b(x)-m_{\widetilde{2^{k+1}R_j}}(b)\right|\left|K(x,y)-K(x,x_j)\right|\,d\mu(x)\\&\quad\quad+\sum_{k=1}^{\infty}\int_{2^{k+1}R_j\setminus 2^kR_j}\left|m_{\widetilde{2^{k+1}R_j}}(b)-m_{\widetilde{R_j}}(b)\right|\left|K(x,y)-K(x,x_j)\right|\,d\mu(x)\\&\quad=:\mathrm{H}_1+\mathrm{H}_2.\end{aligned}$$

Notice that, by Lemma 2.1.3,

$$\delta\left(\widetilde{2^{k+1}R_j},\,\widetilde{R_j}\right)\lesssim 1+\delta(R_j,\,2^{k+1}R_j)\lesssim k.$$

We then see that

$$\mathrm{H}_2\lesssim\sum_{k=1}^{\infty}k\int_{2^{k+1}R_j\setminus 2^kR_j}\left|K(x,y)-K(x,x_j)\right|\,d\mu(x)\lesssim 1.$$

Applying Lemma 5.6.1, we obtain

$$\begin{aligned}\mathrm{H}_1&\lesssim\sum_{k=1}^{\infty}\mu(2^{k+2}R_j)\left\|b-m_{\widetilde{2^{k+1}R_j}}(b)\right\|_{\exp L,\,2^{k+1}R_j,\,\mu}\\&\quad\times\|\{K(\cdot,y)-K(\cdot,x_j)\}\chi_{2^{k+1}R_j\setminus 2^kR_j}(\cdot)\|_{L\log L,\,2^{k+1}R_j,\,\mu}\\&\lesssim\sum_{k=1}^{\infty}\mu(2^{k+2}R_j)\|\{K(\cdot,y)-K(\cdot,x_j)\}\chi_{2^{k+1}R_j\setminus 2^kR_j}(\cdot)\|_{L\log L,\,2^{k+1}R_j,\,\mu},\end{aligned}$$

where the last inequality follows from the inequality

$$\left\| b - m_{\widetilde{2^{k+1}R_j}}(b) \right\|_{\exp L,\, 2^{k+1}R_j,\, \mu} \lesssim 1,$$

which is a simple corollary of the John–Nirenberg inequality. Choose

$$\lambda_k := \left[\mu\left(2^{k+2}R_j\right)\right]^{-1} \left[k \int_{2^{k+1}R_j \setminus 2^k R_j} \left|K(x,\, y) - K(x,\, x_j)\right| d\mu(x) + 2^{-k} \right].$$

We find that, for all $y \in R_j$,

$$\begin{aligned}
&\frac{1}{\mu(2^{k+2}R_j)} \int_{2^{k+1}R_j \setminus 2^k R_j} \frac{\left|K(x,\, y) - K(x,\, x_j)\right|}{\lambda_k} \\
&\quad \times \log\left(2 + \frac{\left|K(x,\, y) - K(x,\, x_j)\right|}{\lambda_k}\right) d\mu(x) \\
&\quad \lesssim \frac{1}{\mu(2^{k+2}R_j)} \int_{2^{k+1}R_j \setminus 2^k R_j} \frac{\left|K(x,\, y) - K(x,\, x_j)\right|}{\lambda_k} \\
&\quad \times \log\left(2 + \frac{1}{\lambda_k |x - y|^n} + \frac{1}{\lambda_k |x - x_j|^n}\right) d\mu(x) \\
&\quad \lesssim \frac{k}{\mu(2^{k+2}R_j)} \int_{2^{k+1}R_j \setminus 2^k R_j} \frac{\left|K(x,\, y) - K(x,\, x_j)\right|}{\lambda_k} d\mu(x) \\
&\quad \lesssim 1,
\end{aligned}$$

which in turn implies that

$$\|\{K(\cdot,\, y) - K(\cdot,\, x_j)\}\chi_{2^{k+1}R_j \setminus 2^k R_j}(\cdot)\|_{L\log L,\, 2^{k+1}R_j,\, \mu} \lesssim \lambda_k.$$

Therefore,

$$\begin{aligned}
\mathrm{H}_1 &\lesssim \sum_{k=1}^{\infty} \mu(2^{k+2}R_j)\lambda_k \\
&\lesssim \sum_{k=1}^{\infty} \left[k \int_{2^{k+1}R_j \setminus 2^k R_j} \left|K(x,\, y) - K(x,\, x_j)\right| d\mu(x) + 2^{-k} \right] \\
&\lesssim 1.
\end{aligned}$$

We then conclude, by combining the estimates for H_1 and H_2, the desired estimate for H that

$$\mathrm{H} \lesssim \lambda^{-1} \sum_{j=1}^{\infty} \|h_j\|_{L^1(\mu)} \lesssim \lambda^{-1} \|f\|_{H^1(\mu)}.$$

It remains to estimate the term G. For each fixed j,

$$\begin{aligned}
&\int_{2R_j} \left|b(x) - m_{\widetilde{R_j}}(b)\right| |Th_j(x)| \, d\mu(x) \\
&\quad \le |r_j^1| \int_{2R_j} \left|b(x) - m_{\widetilde{R_j}}(b)\right| \left|Ta_j^1(x)\right| \, d\mu(x) \\
&\quad\quad + |r_j^2| \int_{2R_j} \left|b(x) - m_{\widetilde{R_j}}(b)\right| \left|Ta_j^2(x)\right| \, d\mu(x) \\
&\quad =: \mathrm{L}_j^1 + \mathrm{L}_j^2.
\end{aligned}$$

We only consider the term L_j^1, the other term L_j^2 can be estimated in a similar way. Write

$$\begin{aligned}
\mathrm{L}_j^1 &\le |r_j^1| \int_{2R_j \setminus 2Q_j^1} \left|b(x) - m_{\widetilde{R_j}}(b)\right| \left|Ta_j^1(x)\right| \, d\mu(x) \\
&\quad + |r_j^1| \int_{2Q_j^1} \left|b(x) - m_{\widetilde{2Q_j^1}}(b)\right| \left|Ta_j^1(x)\right| \, d\mu(x) \\
&\quad + |r_j^1| \left|m_{\widetilde{2Q_j^1}}(b) - m_{\widetilde{R_j}}(b)\right| \int_{2Q_j^1} \left|Ta_j^1(x)\right| \, d\mu(x) \\
&=: \mathrm{U}_j + \mathrm{V}_j + \mathrm{W}_j.
\end{aligned}$$

The Hölder inequality now implies that

$$\begin{aligned}
\mathrm{W}_j &\lesssim |r_j^1| [1 + \delta(Q_j^1, R_j)] \|Ta_j^1\|_{L^2(\mu)} [\mu(2Q_j^1)]^{1/2} \\
&\lesssim |r_j^1| [1 + \delta(Q_j^1, R_j)] \|a_j^1\|_{L^2(\mu)} [\mu(2Q_j^1)]^{1/2} \\
&\lesssim |r_j^1|.
\end{aligned}$$

On the other hand, it follows, from the boundedness of T on $L^2(\mu)$, the Hölder inequality and Corollary 3.1.20, that

$$\mathrm{V}_j \lesssim |r_j^1| \left[\int_{2Q_j^1} \left|b(x) - m_{\widetilde{2Q_j^1}}(b)\right|^2 \, d\mu(x) \right]^{1/2} \|Ta_j^1\|_{L^2(\mu)}$$

$$\lesssim |r_j^1|[\mu(4Q_j^1)]^{1/2}\|a_j^1\|_{L^2(\mu)}$$
$$\lesssim |r_j^1|.$$

Let x_j^1 be the center of Q_j^1. Observe that, for $x \notin 2Q_j^1$,

$$|Ta_j^1(x)| \lesssim \frac{\|a_j^1\|_{L^1(\mu)}}{|x-x_j^1|^n} \lesssim \frac{\|a_j^1\|_{L^\infty(\mu)}\mu(Q_j)}{|x-x_j^1|^n}.$$

A straightforward computation invoking (2.1.3) shows that

$$\begin{aligned}
\mathrm{U}_j &\lesssim |r_j^1|\|a_j^1\|_{L^\infty(\mu)}\mu(Q_j^1)\int_{2R_j\setminus 2Q_j^1}\frac{|b(x)-m_{\widetilde{R_j}}(b)|}{|x-x_j^1|^n}\,d\mu(x)\\
&\lesssim |r_j^1|\|a_j^1\|_{L^\infty(\mu)}\mu(Q_j^1)\\
&\quad\times\left\{\sum_{k=1}^{N_{2Q_j^1,2R_j}+1}\left[\int_{2^{k+1}Q_j^1\setminus 2^kQ_j^1}\frac{|b(x)-m_{\widetilde{2^{k+1}Q_j^1}}(b)|}{|x-x_j^1|^n}\,d\mu(x)\right.\right.\\
&\quad\left.\left.+\left|m_{\widetilde{2^{k+1}Q_j^1}}(b)-m_{\widetilde{R_j}}(b)\right|\int_{2^{k+1}Q_j^1\setminus 2^kQ_j^1}\frac{1}{|x-x_j^1|^n}\,d\mu(x)\right]\right\}\\
&\lesssim |r_j^1|\|a_j^1\|_{L^\infty(\mu)}\mu(Q_j^1)\\
&\quad\times\sum_{k=1}^{N_{2Q_j^1,2R_j}+2}\left\{\frac{\mu(2^{k+1}Q_j^1)}{\ell(2^{k+1}Q_j^1)^n}+[1+\delta(Q_j^1,R_j)]\frac{\mu(2^{k+1}Q_j^1)}{\ell(2^{k+1}Q_j^1)^n}\right\}\\
&\lesssim |r_j^1|\|a_j^1\|_{L^\infty(\mu)}\mu(Q_j^1)[1+\delta(Q_j^1,R_j)]^2\\
&\lesssim |r_j^1|.
\end{aligned}$$

We finally obtain

$$\mu\left(\{x\in\mathbb{R}^D:\ |T_b^I f(x)|>\lambda\}\right)\lesssim\lambda^{-1}\|f\|_{H^1(\mu)},$$

which, together with the estimate (5.6.4), completes the proof of Theorem 5.6.2. □

An interesting application of Theorem 5.6.2 is that it implies the boundedness on $L^p(\mu)$ for commutators of singular integral operators whose kernels satisfy the minimum regularity condition (5.6.1).

Theorem 5.6.3. *Let the kernel K and the operator T be the same as in Theorem 5.6.2. Suppose that, for some fixed p_0 with $p_0 \in (1,\infty)$, the operator T in* (5.3.9) *is bounded on $L^{p_0}(\mu)$. Then, for $b \in \mathrm{RBMO}(\mu)$, the commutator T_b is bounded on $L^p(\mu)$ for any $p \in (1,\infty)$.*

Notice that, by Theorem 5.3.4, we know that T is bounded on $L^{p_0}(\mu)$ for some $p_0 \in (1,\infty)$ implies that T and $T^\sharp$ are bounded on $L^p(\mu)$ for any $p \in (1,\infty)$. Thus, by Theorem 5.2.8 and (5.1.16), the proof of Theorem 5.6.3 is deduced from the following pointwise estimate.

Theorem 5.6.4. *Let the kernel K and the operator T be the same as in Theorem 5.6.2. Suppose that $b \in \mathrm{RBMO}(\mu)$ and T is bounded on $L^2(\mu)$. Then, for any $r \in (1,\infty)$, there exists a positive constant $C_{(r)}$, depending on r, such that, for any bounded function f with compact support,*

$$\mathcal{M}^\sharp(T_b f)(x) \le C_{(r)}\|b\|_{\mathrm{RBMO}(\mu)}\left[\|f\|_{L^\infty(\mu)} + \mathcal{M}_{r,(3/2)}Tf(x) + \mathcal{M}_{r,(9/8)}f(x) + T^\sharp f(x)\right]. \tag{5.6.5}$$

Proof. Without loss of generality, we may assume that $\|b\|_{\mathrm{RBMO}(\mu)} = 1$. We first claim that, for all x and Q with $x \in Q$,

$$\frac{1}{\mu(\frac{3}{2}Q)}\int_Q |T_b f(y) - h_Q|\,d\mu(y) \lesssim \mathcal{M}_{r,(9/8)}f(x) + \mathcal{M}_{r,(3/2)}Tf(x) + \|f\|_{L^\infty(\mu)} \tag{5.6.6}$$

and, for all cubes $Q \subset R$ with $x \in Q$,

$$|h_Q - h_R| \lesssim [1+\delta(Q,R)]^2\left[\mathcal{M}_{r,(9/8)}f(x) + T^\sharp f(x) + \|f\|_{L^\infty(\mu)}\right], \tag{5.6.7}$$

where

$$h_Q := m_Q\left(T\left(\left[b - m_{\tilde{Q}}(b)\right] f\chi_{\mathbb{R}^D\setminus\frac{4}{3}Q}\right)\right)$$

and

$$h_R := m_R\left(T\left(\left[b - m_{\tilde{R}}(b)\right] f\chi_{\mathbb{R}^D\setminus\frac{4}{3}R}\right)\right).$$

Our hypotheses imply h_Q and h_R are both finite.

We first consider (5.6.6). For some fixed cube Q and $x \in Q$, write $T_b f$ in the following way

$$\begin{aligned} T_b f(y) &= [b(y) - m_{\tilde{Q}}(b)]Tf(y) - T\left([b - m_{\tilde{Q}}(b)]f\right)(y) \\ &= [b(y) - m_{\tilde{Q}}(b)]Tf(y) - T\left([b - m_{\tilde{Q}}(b)]f_1\right)(y) \\ &\quad - T\left([b - m_{\tilde{Q}}(b)]f_2\right)(y), \end{aligned}$$

where

$$f_1 := f\chi_{\frac{4}{3}Q} \quad \text{and} \quad f_2 := f - f_1.$$

An applications of the Hölder inequality and the John–Nirenberg inequality implies that, for all $x \in Q$,

$$\begin{aligned}
&\frac{1}{\mu(\frac{3}{2}Q)} \int_Q |b(y) - m_{\tilde{Q}}(b)||Tf(y)|\,d\mu(y) \\
&\leq \left[\frac{1}{\mu(\frac{3}{2}Q)} \int_Q |b(y) - m_{\tilde{Q}}(b)|^{r'}\,d\mu(y)\right]^{1/r'} \left[\frac{1}{\mu(\frac{3}{2}Q)} \int_Q |Tf(y)|^r\,d\mu(y)\right]^{1/r} \\
&\lesssim \mathcal{M}_{r,(3/2)}(Tf)(x).
\end{aligned}$$

On the other hand, if we choose $s \in (1, r)$, then, by the Hölder inequality, Corollary 3.1.20 and the boundedness of T on $L^r(\mu)$, we see that, for all $x \in Q$,

$$\begin{aligned}
&\frac{1}{\mu(\frac{3}{2}Q)} \int_Q |T(b - m_{\tilde{Q}}(b) f_1)(y)|\,d\mu(y) \\
&\leq \left[\frac{1}{\mu(\frac{3}{2}Q)} \int_Q |T(b - m_{\tilde{Q}}(b) f_1)(y)|^s\,d\mu(y)\right]^{1/s} \\
&\lesssim \left[\frac{1}{\mu(\frac{3}{2}Q)} \int_{\frac{4}{3}Q} |b(y) - m_{\tilde{Q}}(b)|^s |f(y)|^s\,d\mu(y)\right]^{1/s} \\
&\lesssim \left[\frac{1}{\mu(\frac{3}{2}Q)} \int_{\frac{4}{3}Q} |f(y)|^r\,d\mu(y)\right]^{1/r} \\
&\lesssim \mathcal{M}_{r,(9/8)} f(x).
\end{aligned}$$

To estimate

$$\left|T\left([b - m_{\tilde{Q}}(b)]f_2\right)(y) - h_Q\right|,$$

we employ the generalization of the Hölder inequality to conclude that, for all $x, y \in Q$,

$$\begin{aligned}
&\left|T\left([b - m_{\tilde{Q}}(b)]f_2\right)(y) - T\left([b - m_{\tilde{Q}}(b)]f_2\right)(x)\right| \\
&\leq \int_{\mathbb{R}^D \setminus \frac{4}{3}Q} |K(y, z) - K(x, z)| \left|b(z) - m_{\tilde{Q}}(b)\right| |f(z)|\,d\mu(z)
\end{aligned}$$

$$
\begin{aligned}
&\leq \|f\|_{L^\infty(\mu)} \sum_{k=1}^{\infty} \int_{2^k \frac{4}{3} Q \setminus 2^{k-1} \frac{4}{3} Q} |K(y,z) - K(x,z)| \left| b(z) - m_{\tilde{Q}}(b) \right| d\mu(z) \\
&\lesssim \|f\|_{L^\infty(\mu)} \sum_{k=1}^{\infty} \left[k \int_{2^k \frac{4}{3} Q \setminus 2^{k-1} \frac{4}{3} Q} |K(y,z) - K(x,z)| \, d\mu(z) + 2^{-k} \right] \\
&\lesssim \|f\|_{L^\infty(\mu)},
\end{aligned}
\tag{5.6.8}
$$

which in turn implies that

$$
\begin{aligned}
&\left|T\left([b - m_{\tilde{Q}}(b)] f_2\right)(y) - h_Q\right| \\
&\quad \leq \left| T\left([b - m_{\tilde{Q}}(b)] f_2\right)(y) - m_Q\left[T\left([b - m_{\tilde{Q}}(b)] f_2\right)\right] \right| \\
&\quad \lesssim \|f\|_{L^\infty(\mu)}.
\end{aligned}
$$

Combining these estimates, we obtain (5.6.6).

We now show (5.6.7). For any cubes $Q \subset R$ with $x \in Q$, we write the difference $|h_Q - h_R|$ in the following way that

$$
\begin{aligned}
&|h_Q - h_R| \\
&\quad = \left| m_Q\left[T\left(\left[b - m_{\tilde{Q}}(b)\right] f \chi_{\mathbb{R}^D \setminus \frac{4}{3} Q}\right)\right] - m_R\left[T\left(\left[b - m_{\tilde{R}}(b)\right] f \chi_{\mathbb{R}^D \setminus \frac{4}{3} R}\right)\right] \right| \\
&\quad \leq \left| m_Q\left[T\left(\left[b - m_{\tilde{Q}}(b)\right] f \chi_{2Q \setminus \frac{4}{3} Q}\right)\right] \right| \\
&\qquad + \left| m_Q\left[T\left(\left[m_{\tilde{Q}}(b) - m_{\tilde{R}}(b)\right] f \chi_{\mathbb{R}^D \setminus 2Q}\right)\right] \right| \\
&\qquad + \left| m_Q\left[T\left(\left[b - m_{\tilde{R}}(b)\right] f \chi_{2^{N_{Q,R}+1} Q \setminus 2Q}\right)\right] \right| \\
&\qquad + \left| m_Q\left[T\left(\left[b - m_{\tilde{R}}(b)\right] f \chi_{\mathbb{R}^D \setminus 2^{N_{Q,R}+1} Q}\right)\right] \right. \\
&\qquad\quad \left. - m_R\left[T\left(\left[b - m_{\tilde{R}}(b)\right] f \chi_{\mathbb{R}^D \setminus 2^{N_{Q,R}+1} Q}\right)\right] \right| \\
&\qquad + \left| m_R\left[T\left(\left[b - m_{\tilde{R}}(b)\right] f \chi_{2^{N_{Q,R}+1} Q \setminus \frac{4}{3} R}\right)\right] \right| \\
&\quad =: \mathrm{M}_1 + \mathrm{M}_2 + \mathrm{M}_3 + \mathrm{M}_4 + \mathrm{M}_5.
\end{aligned}
$$

Using (5.3.1) and the Hölder inequality, we see that, for any $y \in Q$,

$$
\left| T\left(\left[b - m_{\tilde{Q}}(b)\right] f \chi_{2Q \setminus \frac{4}{3} Q}\right)(y) \right| \lesssim \mathcal{M}_{r,(9/8)} f(x)
$$

and hence

$$
\mathrm{M}_1 \lesssim \mathcal{M}_{r,(9/8)} f(x).
$$

Similarly, we have

$$\mathrm{M}_3 \lesssim [1 + \delta(Q, R)]^2 \mathcal{M}_{r,(9/8)} f(x)$$

and

$$\mathrm{M}_5 \lesssim \mathcal{M}_{r,(9/8)} f(x).$$

To estimate M_2, observe that, for $x, y \in Q$,

$$\left|T(f\chi_{\mathbb{R}^D \setminus 2Q})(y)\right| \lesssim \|f\|_{L^\infty(\mu)} + T^\sharp f(x) + \mathcal{M}_{r,(9/8)} f(x).$$

Thus,

$$\begin{aligned}\mathrm{M}_2 &= \left|m_Q\left[T\left(\left[m_{\tilde{Q}}(b) - m_{\tilde{R}}(b)\right] f\chi_{\mathbb{R}^D \setminus 2Q}\right)\right]\right| \\ &\lesssim [1 + \delta(Q, R)]\left[\|f\|_{L^\infty(\mu)} + T^\sharp f(x) + \mathcal{M}_{r,(9/8)} f(x)\right].\end{aligned}$$

Finally, we deal with the term M_4. As in the inequality (5.6.8), we see that, for all $y, z \in R$,

$$\begin{aligned}&\left|T\left(\left[b - m_{\tilde{R}}(b)\right] f\chi_{\mathbb{R}^D \setminus 2^{N_2} Q}\right)(y) - T\left(\left[b - m_{\tilde{R}}(b)\right] f\chi_{\mathbb{R}^D \setminus 2^{N_2} Q}\right)(z)\right| \\ &\quad \lesssim \|f\|_{L^\infty(\mu)}.\end{aligned}$$

Therefore,

$$\mathrm{M}_4 \lesssim \|f\|_{L^\infty(\mu)}.$$

Combining the estimates for M_1, M_2, M_3, M_4 and M_5 implies the inequality (5.6.7).

It remains to prove that our claim implies the conclusion of Theorem 5.6.4. By (5.6.6), we see that, for any doubling cube Q and $x \in Q$,

$$\begin{aligned}|m_Q(T_b f) - h_Q| &\le \frac{1}{\mu(Q)} \int_Q |T_b f(y) - h_Q|\, d\mu(y) \\ &\lesssim \mathcal{M}_{r,(9/8)} f(x) + \mathcal{M}_{r,(3/2)}(Tf)(x) + \|f\|_{L^\infty(\mu)}. \quad (5.6.9)\end{aligned}$$

Also, for any cube Q containing x, by (5.6.6) and (5.6.7), we know that

$$\begin{aligned}&\frac{1}{\mu\left(\frac{3}{2}Q\right)} \int_Q |T_b f(y) - m_{\tilde{Q}}(T_b f)|\, d\mu(y) \\ &\le \frac{1}{\mu\left(\frac{3}{2}Q\right)} \int_Q |T_b f(y) - h_Q|\, d\mu(y) + |h_Q - h_{\tilde{Q}}| + |h_{\tilde{Q}} - m_{\tilde{Q}}(T_b f)| \\ &\lesssim \mathcal{M}_{r,(9/8)} f(x) + \mathcal{M}_{r,(3/2)}(Tf)(x) + T^\sharp f(x) + \|f\|_{L^\infty(\mu)}.\end{aligned}$$

On the other hand, for all doubling cubes Q and R with $Q \subset R$ and $x \in Q$, if $\delta(Q, R) \leq C_4$ with C_4 as in Lemma 3.1.8, it is easy to see that

$$|h_Q - h_R| \lesssim \left[\mathcal{M}_{r,(9/8)}f(x) + T^\sharp f(x) + \|f\|_{L^\infty(\mu)}\right][1 + \delta(Q, R)].$$

Thus, by Lemma 3.1.8, we find that

$$|h_Q - h_R| \lesssim \left[\mathcal{M}_{r,(9/8)}f(x) + T^\sharp f(x) + \|f\|_{L^\infty(\mu)}\right][1 + \delta(Q, R)],$$

provided that $Q \subset R$ are doubling cubes. This, along with (5.6.9), leads to that, for any doubling cubes Q and R with $Q \subset R$ and $x \in Q$,

$$\begin{aligned}
&|m_Q(T_b f) - m_R(T_b f)| \\
&\quad \lesssim \left[\mathcal{M}_{r,(9/8)}f(x) + \mathcal{M}_{r,(3/2)}(Tf)(x) + T^\sharp f(x) + \|f\|_{L^\infty(\mu)}\right][1 + \delta(Q, R)],
\end{aligned}$$

which then completes the proof of Theorem 5.6.4. □

5.6.2 Boundedness of Multilinear Commutators on $L^p(\mu)$

Let T be the Calderón–Zygmund operator defined by (5.3.7) with kernel K satisfying (5.3.1) and (5.3.8), $k \in \mathbb{N}$ and $b_i \in \mathrm{RBMO}(\mu)$, $i \in \{1, \ldots, k\}$, the *multilinear commutator* $T_{\vec{b}}$ is formally defined by setting, for all suitable functions f and $x \in \mathbb{R}^D$,

$$T_{\vec{b}} f(x) := [b_k, [b_{k-1}, \cdots, [b_1, T]\cdots]] f(x), \tag{5.6.10}$$

where $\vec{b} := (b_1, \ldots, b_k)$ and

$$[b_1, T]f(x) := b_1(x)Tf(x) - T(b_1 f)(x). \tag{5.6.11}$$

In what follows, if $k = 1$ and $\vec{b} = b$, we *denote* $T_{\vec{b}} f$ *simply by* $T_b f$. In this section we prove the following conclusion.

Theorem 5.6.5. *Let $k \in \mathbb{N}$ and $b_i \in \mathrm{RBMO}(\mu)$ for $i \in \{1, \ldots, k\}$. Let T and $T_{\vec{b}}$ be as in* (5.3.7) *and* (5.6.10) *with kernel K satisfying* (5.3.1) *and* (5.3.8)*. If T is bounded on $L^2(\mu)$, then the multilinear commutator $T_{\vec{b}}$ in* (5.6.10) *is also bounded on $L^p(\mu)$ with operator norm $C\prod_{i=1}^k \|b_i\|_{\mathrm{RBMO}(\mu)}$, where $p \in (1, \infty)$, namely, there exists a positive constant C such that, for all $f \in L^p(\mu)$,*

$$\|T_{\vec{b}} f\|_{L^p(\mu)} \leq C \prod_{i=1}^k \|b_i\|_{\mathrm{RBMO}(\mu)} \|f\|_{L^p(\mu)}.$$

In order to prove Theorem 5.6.5, we need to recall some maximal operators in Sect. 5.1. Let $\mathcal{M}^\sharp(f)$ and $\mathcal{M}^d(f)$ be as in (5.1.15) and (5.1.20) with $\rho = 1$, respectively. We have the following relation for their $L^p(\mu)$ norms between $\mathcal{M}^d(f)$ and $\mathcal{M}^\sharp(f)$.

Theorem 5.6.6. *Let $f \in L^1_{\mathrm{loc}}(\mu)$ with*

$$\int_{\mathbb{R}^D} f(x)\, d\mu(x) = 0 \quad \text{when} \quad \mu\left(\mathbb{R}^D\right) < \infty.$$

For $p \in (1, \infty)$, if

$$\inf\left\{1,\ \mathcal{M}^d(f)\right\} \in L^p(\mu),$$

then

$$\left\|\mathcal{M}^d(f)\right\|_{L^p(\mu)} \le C \left\|\mathcal{M}^\sharp(f)\right\|_{L^p(\mu)}. \tag{5.6.12}$$

Proof. We may assume that $\mu(\mathbb{R}^D) = \infty$ by similarity. For some fixed $\eta \in (0, 1)$ and all $\epsilon \in (0, \infty)$ we prove that there exists some $\sigma \in (0, \infty)$ such that, for any $\lambda \in (0, \infty)$, the following good-λ inequality holds true:

$$\begin{aligned} &\mu\left(\left\{x \in \mathbb{R}^D : \mathcal{M}^d(f)(x) > (1+\epsilon)\lambda, \mathcal{M}^\sharp(f)(x) \le \sigma\lambda\right\}\right) \\ &\quad \le \eta\mu\left(\left\{x \in \mathbb{R}^D : \mathcal{M}^d(f)(x) > \lambda\right\}\right). \end{aligned} \tag{5.6.13}$$

By this inequality, we obtain

$$\left\|\mathcal{M}^d(f)\right\|_{L^p(\mu)} \lesssim \left\|\mathcal{M}^\sharp(f)\right\|_{L^p(\mu)} \quad \text{if} \quad \inf\left\{1, \mathcal{M}^d(f)\right\} \in L^p(\mu).$$

Let

$$\Omega_\lambda := \left\{x \in \mathbb{R}^D : \mathcal{M}^d(f)(x) > \lambda\right\}$$

and

$$E_\lambda := \left\{x \in \mathbb{R}^D : \mathcal{M}^d(f)(x) > (1+\epsilon)\lambda,\ \mathcal{M}^\sharp(f)(x) \le \sigma\lambda\right\}.$$

For the moment we assume that $f \in L^p(\mu)$. For each $x \in E_\lambda$, among the doubling cubes Q that contain x and such that

$$m_Q(|f|) > (1 + \epsilon/2)\lambda,$$

we consider one cube Q_x which has almost the maximal side length in the sense that, if some doubling cube $\tilde{Q}$ with side length $\ell(\tilde{Q}) \ge 2\ell(Q_x)$, then

$$m_{\tilde{Q}}(|f|) \le (1+\epsilon/2)\lambda.$$

It is easy to show that this maximal cube Q_x exists, because $f \in L^p(\mu)$ and $\mu(\mathbb{R}^D) = \infty$.

Let $S_x := \widetilde{3Q_x}$. Then, assuming σ small enough, we have $m_{S_x}(|f|) > \lambda$ and hence $S_x \subset \Omega_\lambda$. Indeed, by construction, we have $\delta_{Q_x, S_x} \lesssim 1$. Then, since $Q_x \subset S_x$ are doubling cubes containing x, we see that

$$|m_{Q_x}(|f|) - m_{S_x}(|f|)| \le \delta_{Q_x, S_x} \mathcal{M}^\sharp(|f|)(x) \le C_{11} 5\beta_6 \sigma\lambda.$$

Thus, for $\sigma < \epsilon/(C_{11} 10\beta_6)$, it holds true that

$$m_{S_x}(|f|) > (1+\epsilon/2)\lambda - C_{11} 5\beta_6\sigma\lambda.$$

By Theorem 1.1.1, there exist N_D subfamilies $\mathcal{D}_k := \{S_i^k\}_i$, $k \in \{1, \ldots, N_D\}$, of cubes S_x such that they cover E_λ and are centered at points $x_i^k \in E_\lambda$, and each subfamily $\mathcal{D}_k$ is disjoint, where N_D is as in Theorem 1.1.1. Therefore, at least one subfamily $\mathcal{D}_k$ satisfies that

$$\mu\left(\bigcup_i S_i^k\right) \ge \frac{1}{N_D}\mu\left(\bigcup_{i,k} S_i^k\right).$$

Suppose, for example, that it is $\mathcal{D}_1$. We prove that, for each cube S_i^1,

$$\mu\left(S_i^1 \bigcap E_\lambda\right) \le \mu(S_i^1)/(2N_D), \tag{5.6.14}$$

if σ is chosen small enough. By this inequality, we have

$$\mu\left(E_\lambda \bigcap \bigcup_i S_i^1\right) \le \frac{1}{2N_D}\sum_i \mu(S_i^1) \le \frac{1}{2N_D}\mu(\Omega_\lambda).$$

Thus,

$$\begin{aligned}
\mu(E_\lambda) &\le \mu\left(\bigcup_{i,k} S_i^k \setminus \bigcup_i S_i^1\right) + \mu\left(E_\lambda \bigcap \bigcup_i S_i^1\right)\\
&\le \left(1 - \frac{1}{N_D}\right)\mu\left(\bigcup_i S_i^1\right) + \frac{1}{2N_D}\mu(\Omega_\lambda)\\
&\le \left(1 - \frac{1}{2N_D}\right)\mu(\Omega_\lambda).
\end{aligned} \tag{5.6.15}$$

We now prove (5.6.14). Let $y \in S_i^1 \cap E_\lambda$. If $Q \ni y$ is doubling and

$$m_Q(|f|) > (1+\epsilon)\lambda,$$

then $\ell(Q) \le \ell(S_i^1)/8$. Otherwise,

$$\widetilde{30Q} \supset S_i^1 \supset Q_{x_i^1}$$

and, since Q and $\widetilde{30Q}$ are doubling, we have

$$\left| m_Q(|f|) - m_{\widetilde{30Q}}(|f|) \right| \le \delta_{Q,\widetilde{30Q}} \mathcal{M}^\sharp(|f|)(y) \le C_{12}\sigma\lambda \le \frac{\epsilon}{2}\lambda,$$

assuming $C_{12}\sigma < \epsilon/2$, and hence

$$m_{\widetilde{30Q}}(|f|) > (1+\epsilon/2)\lambda,$$

which contradicts the choice of $Q_{x_i^1}$ because

$$\widetilde{30Q} \supset Q_{x_i} \quad \text{and} \quad \ell(\widetilde{30Q}) > 2\ell(Q_{x_i^1}).$$

Thus, $\mathcal{M}^d(f)(y) > (1+\epsilon)\lambda$, which implies that

$$\mathcal{M}^d\left(\chi_{\frac{5}{4}S_i^1} f\right)(y) > (1+\epsilon)\lambda.$$

On the other hand, we also have

$$m_{\widetilde{\frac{5}{4}S_i^1}}(|f|) \le (1+\epsilon/2)\lambda,$$

since $\widetilde{\frac{5}{4}S_i^1}$ is doubling and its side length is larger than $2\ell(Q_{x_i^1})$. Therefore, we see that

$$\mathcal{M}^d\left(\chi_{\frac{5}{4}S_i^1} |f| - m_{\widetilde{\frac{5}{4}S_i^1}}(|f|)\right)(y) > \frac{\epsilon}{2}\lambda$$

and hence, by the weak (1, 1) boundedness of $\mathcal{N}$ and the fact that S_i^1 is doubling, we know that

$$\begin{aligned}\mu\left(S_i^1 \bigcap E_\lambda\right) &\le \mu\left(\left\{y \in \mathbb{R}^D : \mathcal{N}\left(\chi_{\frac{5}{4}S_i^1}\left(|f| - m_{\widetilde{\frac{5}{4}S_i^1}}(|f|)\right)\right)(y) > \frac{\epsilon}{2}\lambda\right\}\right)\\ &\lesssim \frac{1}{\epsilon\lambda}\int_{\frac{5}{4}S_i^1}\left[|f(y)| - m_{\widetilde{\frac{5}{4}S_i^1}}(|f|)\right] d\mu(y)\end{aligned}$$

$$\lesssim \frac{1}{\epsilon\lambda}\mu(2S_i^1)\mathcal{M}^\sharp(|f|)(x_i^1)$$
$$\leq \frac{C_{13}\sigma}{\epsilon}\mu(S_i^1).$$

Thus, (5.6.14) follows from choosing $\sigma < \epsilon/(2N_D C_{13})$, which implies (5.6.14) and, as a consequence, we obtain (5.6.13) and hence (5.6.12) under the assumption $f \in L^p(\mu)$.

Suppose that $f \notin L^p(\mu)$. We consider the function f_q, $q \in \mathbb{N}$, introduced in Lemma 3.1.18. Since, for all functions $g, h \in L^1_{\rm loc}(\mu)$, we have

$$\mathcal{M}^\sharp(g+h) \leq \mathcal{M}^\sharp(g) + \mathcal{M}^\sharp(h) \quad \text{and} \quad \mathcal{M}^\sharp(|g|) \lesssim \mathcal{M}^\sharp(g),$$

arguing as in Lemma 3.1.18, we see that

$$\mathcal{M}^\sharp(f_q) \lesssim \mathcal{M}^\sharp(f).$$

On the other hand, we know that

$$|f_q| \leq q \inf\{1, |f|\} \leq q \inf\{1, \mathcal{M}^d(f)\}$$

and hence $f_q \in L^p(\mu)$. Therefore,

$$\left\|\mathcal{M}^d(f_q)\right\|_{L^p(\mu)} \lesssim \left\|\mathcal{M}^\sharp(f)\right\|_{L^p(\mu)}.$$

Taking the limit as $q \to \infty$, we then obtain (5.6.12). This finishes the proof of Theorem 5.6.6. □

Proof of Theorem 5.6.5. We prove the theorem by induction on k. If $k = 1$, by arguing as in Theorem 5.6.4, we see that, for any $f \in L^p(\mu)$, with $p \in (1,\infty)$, and $x \in \mathbb{R}^D$,

$$\mathcal{M}^\sharp(T_b f)(x) \lesssim \|\vec{b}\|_{\mathrm{RBMO}(\mu)}\left[\mathcal{M}_{r,(3/2)}Tf(x) + \mathcal{M}_{r,(9/8)}f(x) + T^\sharp f(x)\right].$$

This, together with the boundedness of T on $L^p(\mu)$, $\mathcal{M}_{r,(3/2)}$, $\mathcal{M}_{r,(9/8)}$ and $T^\sharp$, implies that $\mathcal{M}^\sharp \circ T_b$ is bounded on $L^p(\mu)$ for $p \in (1,\infty)$, which, combined with Theorem 5.6.6, implies the boundedness of T_b on $L^p(\mu)$ for any $p \in (1,\infty)$.

Now we assume that $k \geq 2$ is an integer and that, for any $i \in \{1,\ldots,k-1\}$ and any subset $\sigma := \{\sigma(1),\ldots,\sigma(i)\}$ of $\{1,\ldots,k\}$, $T_{\vec{b}_\sigma}$ is bounded on $L^p(\mu)$ for any $p \in (1,\infty)$. We first claim that, for any $r \in (1,\infty)$, $T_{\vec{b}}$ satisfies the following sharp function estimate that, for all $x \in \mathbb{R}^D$,

$$\mathcal{M}^\sharp(T_{\vec{b}} f)(x) \lesssim \|\vec{b}\|_{\mathrm{RBMO}(\mu)} \{\mathcal{M}_{r,(3/2)}(Tf)(x) + \mathcal{M}_{r,(9/8)} f(x)\} + \sum_{i=1}^{k-1} \sum_{\sigma \in C_i^k} \|\vec{b}_\sigma\|_{\mathrm{RBMO}(\mu)} \mathcal{M}_{r,(3/2)}(T_{\vec{b}_{\tilde{\sigma}}} f)(x). \tag{5.6.16}$$

Indeed, by homogeneity, we may assume that $\|b_i\|_{\mathrm{RBMO}(\mu)} = 1$ for $i \in \{1, \ldots, k\}$. Then it suffices to show that, for any x and cube Q with $Q \ni x$,

$$\frac{1}{\mu(\frac{3}{2}Q)} \int_Q |T_{\vec{b}} f(y) - h_Q| \, d\mu(y) \lesssim \mathcal{M}_{r,(3/2)}(Tf)(x) + \mathcal{M}_{r,(9/8)} f(x) + \sum_{i=1}^{k-1} \sum_{\sigma \in C_i^k} \mathcal{M}_{r,(3/2)}(T_{\vec{b}_{\tilde{\sigma}}} f)(x) \tag{5.6.17}$$

and, for any cubes $Q \subset R$ with $Q \ni x$,

$$|h_Q - h_R| \lesssim \delta_{Q,R}^{k+1} \{\mathcal{M}_{r,(9/8)} f(x) + \mathcal{M}_{r,(3/2)}(Tf)(x)\} + \delta_{Q,R}^{k+1} \sum_{i=1}^{k-1} \sum_{\sigma \in C_i^k} \mathcal{M}_{r,(3/2)}(T_{\vec{b}_{\tilde{\sigma}}} f)(x), \tag{5.6.18}$$

where Q is an arbitrary cube, R a doubling cube,

$$h_Q := m_Q \left(T \left[\left(m_{\tilde{Q}}(b_1) - b_1 \right) \cdots \left(m_{\tilde{Q}}(b_k) - b_k \right) f \chi_{\mathbb{R}^D \setminus \frac{4}{3} Q} \right] \right)$$

and

$$h_R := m_R \left(T \left[(m_R(b_1) - b_1) \cdots (m_R(b_k) - b_k) f \chi_{\mathbb{R}^D \setminus \frac{4}{3} R} \right] \right).$$

Let us first prove the estimate (5.6.17). With the aid of the formula that, for all $y, z \in \mathbb{R}^D$,

$$\prod_{i=1}^k \left[m_{\tilde{Q}}(b_i) - b_i(z) \right] = \sum_{i=0}^k \sum_{\sigma \in C_i^k} [b(y) - b(z)]_{\tilde{\sigma}} \left[m_{\tilde{Q}}(b) - b(y) \right]_\sigma, \tag{5.6.19}$$

where, if $i = 0$, then $\tilde{\sigma} := \{1, \ldots, k\}$ and $\sigma := \emptyset$, it is easy to see that, for all $y \in \mathbb{R}^D$,

$$T_{\vec{b}} f(y) = T \left(\prod_{i=1}^k \left[m_{\tilde{Q}}(b_i) - b_i \right] f \right)(y) - \sum_{i=1}^k \sum_{\sigma \in C_i^k} \left[m_{\tilde{Q}}(b) - b(y) \right]_\sigma T_{\vec{b}_{\tilde{\sigma}}} f(y),$$

where, if $i = k$, we denote $T_{\vec{b}_{\vec{\sigma}}} f(y)$ by $Tf(y)$. Therefore,

$$\begin{aligned}
&\frac{1}{\mu(\frac{3}{2}Q)} \int_Q \left|T_{\vec{b}} f(y) - h_Q\right| \, d\mu(y) \\
&\leq \frac{1}{\mu(\frac{3}{2}Q)} \int_Q \left| T\left(\prod_{i=1}^{k} \left[m_{\tilde{Q}}(b_i) - b_i\right] f\chi_{\frac{4}{3}Q}\right)(y)\right| \, d\mu(y) \\
&\quad + \sum_{i=1}^{k} \sum_{\sigma \in C_i^k} \frac{1}{\mu(\frac{3}{2}Q)} \int_Q \left|\left[m_{\tilde{Q}}(b) - b(y)\right]_\sigma\right| \left|T_{\vec{b}_{\vec{\sigma}'}} f(y)\right| \, d\mu(y) \\
&\quad + \frac{1}{\mu(\frac{3}{2}Q)} \int_Q \left| T\left(\prod_{i=1}^{k} \left[m_{\tilde{Q}}(b_i) - b_i\right] f\chi_{\mathbb{R}^D \setminus \frac{4}{3}Q}\right)(y) - h_Q\right| \, d\mu(y) \\
&=: \mathrm{I}_1 + \mathrm{I}_2 + \mathrm{I}_3.
\end{aligned}$$

Take $s := \sqrt{r}$ and expand

$$b_i(y) - m_{\tilde{Q}}(b_i) = b_i(y) - m_{\widetilde{\frac{4}{3}Q}}(b_i) + m_{\widetilde{\frac{4}{3}Q}}(b_i) - m_{\tilde{Q}}(b_i)$$

for $i \in \{1, \ldots, k\}$. From Corollary 3.1.20, it is easy to deduce that

$$\int_{\frac{4}{3}Q} \prod_{i=1}^{k} \left|b_i(y) - m_{\tilde{Q}}(b_i)\right|^{ss'} \, d\mu(y) \lesssim \mu\left(\frac{3}{2}Q\right). \tag{5.6.20}$$

By the boundedness of T on $L^s(\mu)$ for $s \in (1, \infty)$, the Hölder inequality and (5.6.20), we conclude that, for all $x \in Q$,

$$\begin{aligned}
\mathrm{I}_1 &\leq \frac{\mu(Q)^{1-1/s}}{\mu(\frac{3}{2}Q)} \left\| T\left(\prod_{i=1}^{k} \left[m_{\tilde{Q}}(b_i) - b_i\right] f\chi_{\frac{4}{3}Q}\right)\right\|_{L^s(\mu)} \\
&\lesssim \frac{\mu(Q)^{1-1/s}}{\mu(\frac{3}{2}Q)} \left\| \prod_{i=1}^{k} \left[m_{\tilde{Q}}(b_i) - b_i\right] f\chi_{\frac{4}{3}Q}\right\|_{L^s(\mu)} \\
&\lesssim \frac{1}{\mu(\frac{3}{2}Q)^{1/s}} \left\{\int_{\frac{4}{3}Q} \prod_{i=1}^{k} \left|b_i(y) - m_{\tilde{Q}}(b_i)\right|^{ss'} \, d\mu(y)\right\}^{1/ss'} \\
&\quad \times \left\{\int_{\frac{4}{3}Q} |f(y)|^r \, d\mu(y)\right\}^{1/r} \\
&\lesssim \mathcal{M}_{r,(9/8)} f(x).
\end{aligned}$$

From the Hölder inequality and Corollary 3.1.20, it follows that, for all $x \in Q$,

$$
\begin{aligned}
\mathrm{I}_2 &\le \sum_{i=1}^{k} \sum_{\sigma \in C_i^k} \left\{ \frac{1}{\mu(\frac{3}{2}Q)} \int_Q \left| \left[b(y) - m_{\tilde{Q}}(b) \right]_\sigma \right|^{r'} d\mu(y) \right\}^{1/r'} \\
&\quad \times \left\{ \frac{1}{\mu(\frac{3}{2}Q)} \int_Q |T_{\vec{b}_{\bar{\sigma}}} f(y)|^r \, d\mu(y) \right\}^{1/r} \\
&\lesssim \sum_{i=1}^{k} \sum_{\sigma \in C_i^k} \mathcal{M}_{r,(3/2)}(T_{\vec{b}_{\bar{\sigma}}} f)(x).
\end{aligned}
$$

To estimate I_3, we need to calculate the difference

$$
\left| T\left(\prod_{i=1}^{k} \left[m_{\tilde{Q}}(b_i) - b_i \right] f \chi_{\mathbb{R}^D \setminus \frac{4}{3}Q} \right) - h_Q \right|.
$$

Take $y,\ y_1 \in Q$. Then

$$
y,\ y_1 \notin \operatorname{supp}\left(f \chi_{\mathbb{R}^D \setminus \frac{4}{3}Q} \right).
$$

Thus, by the condition (5.3.8) and the Hölder inequality, we conclude that, for all $y,\ y_1,\ x \in Q$,

$$
\begin{aligned}
&\left| T\left(\prod_{i=1}^{k} \left[m_{\tilde{Q}}(b_i) - b_i \right] f \chi_{\mathbb{R}^D \setminus \frac{4}{3}Q} \right)(y) - T\left(\prod_{i=1}^{k} \left[m_{\tilde{Q}}(b_i) - b_i \right] f \chi_{\mathbb{R}^D \setminus \frac{4}{3}Q} \right)(y_1) \right| \\
&\lesssim \int_{\mathbb{R}^D \setminus \frac{4}{3}Q} \frac{|y - y_1|^\delta}{|z - y|^{n+\delta}} \prod_{i=1}^{k} \left| b_i(z) - m_{\tilde{Q}}(b_i) \right| |f(z)| \, d\mu(z) \\
&\lesssim \sum_{j=1}^{\infty} \int_{2^j \frac{4}{3}Q \setminus 2^{j-1} \frac{4}{3}Q} \prod_{i=1}^{k} \left[\left| b_i(z) - m_{\widetilde{2^j \frac{4}{3}Q}}(b_i) \right| + \left| m_{\widetilde{2^j \frac{4}{3}Q}}(b_i) - m_{\tilde{Q}}(b_i) \right| \right] \\
&\quad \times \frac{[\ell(Q)]^\delta}{|z - y|^{n+\delta}} |f(z)| \, d\mu(z) \\
&\lesssim \sum_{j=1}^{\infty} \sum_{i=0}^{k} \sum_{\sigma \in C_i^k} 2^{-j\delta} j^{k-i} \frac{1}{[\ell(2^j Q)]^n} \int_{2^j \frac{4}{3}Q} \left| \left[b(z) - m_{\widetilde{2^j \frac{4}{3}Q}}(b) \right]_\sigma \right| |f(z)| \, d\mu(z) \\
&\lesssim \sum_{i=0}^{k} \sum_{\sigma \in C_i^k} \sum_{j=1}^{\infty} 2^{-j\delta} j^{k-i} \mathcal{M}_{r,(9/8)} f(x) \\
&\lesssim \mathcal{M}_{r,(9/8)} f(x),
\end{aligned}
$$

where we used the fact that, for $i \in \{1, \dots, k\}$,

$$\left| m_{\widetilde{2^j \frac{4}{3} Q}}(b_i) - m_{\tilde{Q}}(b_i) \right| \lesssim \delta_{\tilde{Q}, \widetilde{2^j \frac{4}{3} Q}} \lesssim \delta_{Q, 2^j \frac{4}{3} Q} \lesssim j.$$

From the above estimate and the choice of h_Q, we deduce that, for all $x \in Q$,

$$\begin{aligned}
&\left| T\left(\prod_{i=1}^{k}\left[m_{\tilde{Q}}(b_i) - b_i\right] f\chi_{\mathbb{R}^D \setminus \frac{4}{3}Q}\right)(y) - h_Q \right| \\
&= \left| T\left(\prod_{i=1}^{k}\left[m_{\tilde{Q}}(b_i) - b_i\right] f\chi_{\mathbb{R}^D \setminus \frac{4}{3}Q}\right)(y) \right. \\
&\quad \left. - m_Q\left[T\left(\prod_{i=1}^{k}\left[m_{\tilde{Q}}(b_i) - b_i\right] f\chi_{\mathbb{R}^D \setminus \frac{4}{3}Q}\right)\right] \right| \\
&\lesssim \mathcal{M}_{r,(9/8)} f(x)
\end{aligned}$$

and hence

$$\mathrm{I}_3 \lesssim \mathcal{M}_{r,(9/8)} f(x).$$

By the estimates for I_1, I_2 and I_3, we obtain (5.6.17).

Now we turn our attention to the estimate for (5.6.18). For any cubes $Q \subset R$ with $Q \ni x$, where Q is arbitrary cube and R is a doubling cube, we denote $N_{Q,R} + 1$ simply by N. Noticing that R is a doubling cube, we have $R = \tilde{R}$. Write the difference $|h_Q - h_R|$ in the following way that

$$\begin{aligned}
&\left| m_Q\left[T\left(\prod_{i=1}^{k}\left[m_{\tilde{Q}}(b_i) - b_i\right] f\chi_{\mathbb{R}^D \setminus \frac{4}{3}Q}\right)\right] \right. \\
&\quad \left. - m_R\left[T\left(\prod_{i=1}^{k}[m_R(b_i) - b_i]\, f\chi_{\mathbb{R}^D \setminus \frac{4}{3}R}\right)\right] \right| \\
&\le \left| m_R\left[T\left(\prod_{i=1}^{k}\left[m_{\tilde{Q}}(b_i) - b_i\right] f\chi_{\mathbb{R}^D \setminus 2^N Q}\right)\right] \right. \\
&\quad \left. - m_Q\left[T\left(\prod_{i=1}^{k}\left[m_{\tilde{Q}}(b_i) - b_i\right] f\chi_{\mathbb{R}^D \setminus 2^N Q}\right)\right] \right| \\
&\quad + \left| m_R\left[T\left(\prod_{i=1}^{k}[m_R(b_i) - b_i]\, f\chi_{\mathbb{R}^D \setminus 2^N Q}\right)\right] \right.
\end{aligned}$$

$$-m_R\left[T\left(\prod_{i=1}^k\left[m_{\tilde{Q}}(b_i)-b_i\right]f\chi_{\mathbb{R}^D\setminus 2^N Q}\right)\right]\Bigg|$$

$$+\left|m_Q\left[T\left(\prod_{i=1}^k\left[m_{\tilde{Q}}(b_i)-b_i\right]f\chi_{2^N Q\setminus\frac{4}{3}Q}\right)\right]\right|$$

$$+\left|m_R\left[T\left(\prod_{i=1}^k[m_R(b_i)-b_i]f\chi_{2^N Q\setminus\frac{4}{3}R}\right)\right]\right|$$

$$=: \mathrm{L}_1+\mathrm{L}_2+\mathrm{L}_3+\mathrm{L}_4.$$

An estimate similar to that used in the estimate for I_3 tells us that, for all $x\in Q$,

$$\mathrm{L}_1 \lesssim \delta_{Q,R}^k \mathcal{M}_{r,(9/8)} f(x).$$

To estimate L_2, we first have

$$\left|T\left(\prod_{i=1}^k[m_R(b_i)-b_i]f\chi_{\mathbb{R}^D\setminus 2^N Q}\right)(y)-T\left(\prod_{i=1}^k\left[m_{\tilde{Q}}(b_i)-b_i\right]f\chi_{\mathbb{R}^D\setminus 2^N Q}\right)(y)\right|$$

$$=\Bigg|T\left(\prod_{i=1}^k[m_R(b_i)-b_i]f\chi_{\mathbb{R}^D\setminus 2^N Q}\right)(y)$$

$$-\sum_{i=0}^k\sum_{\sigma\in C_i^k}\left[m_{\tilde{Q}}(b)-m_R(b)\right]_{\tilde{\sigma}}T\left([m_R(b)-b]_\sigma f\chi_{\mathbb{R}^D\setminus 2^N Q}\right)(y)\Bigg|$$

$$\lesssim \sum_{i=0}^{k-1}\sum_{\sigma\in C_i^k}\delta_{Q,R}^{k-i}\left|T\left([m_R(b)-b]_\sigma f\chi_{\mathbb{R}^D\setminus 2^N Q}\right)(y)\right|$$

$$\lesssim \sum_{i=0}^{k-1}\sum_{\sigma\in C_i^k}\delta_{Q,R}^{k-i}\left\{\left|T\left([m_R(b)-b]_\sigma f\right)(y)\right|+\left|T\left([m_R(b)-b]_\sigma f\chi_{2^N Q}\right)(y)\right|\right\}$$

$$\lesssim \sum_{i=0}^{k-1}\sum_{\sigma\in C_i^k}\delta_{Q,R}^{k-i}\|\vec{b}_{\tilde{\sigma}}\|_{\mathrm{RBMO}(\mu)}\left\{\sum_{j=0}^{i}\sum_{\eta\in C_j^i}\left|[m_R(b)-b(y)]_{\eta'}\right|\left|T_{\vec{b}_\eta}f(y)\right|\right.$$

$$\left.+\left|T\left([m_R(b)-b]_\sigma f\chi_{2^N Q\setminus\frac{4}{3}R}\right)(y)\right|+\left|T\left([m_R(b)-b]_\sigma f\chi_{\frac{4}{3}R}\right)(y)\right|\right\}.$$

From the Hölder inequality and the fact that R is a doubling cube, it follows that, for all $x\in Q$,

$$\frac{1}{\mu(R)}\int_R |[b(y)-m_R(b)]_{\eta'}||T_{\vec{b}_\eta}f(y)|\,d\mu(y) \lesssim \mathcal{M}_{r,(3/2)}(T_{\vec{b}_\eta}f)(x). \quad (5.6.21)$$

By the Hölder inequality, Corollary 3.1.20 and the condition (0.0.1), it is easy to see that, for all $y \in R$ and $x \in Q$,

$$\begin{aligned}
&\left|T\left([m_R(b)-b]_\sigma f\chi_{2^N Q\setminus\frac{4}{3}R}\right)(y)\right| \\
&\leq \int_{2^N Q\setminus\frac{4}{3}R} |K(y,z)|\,|[m_R(b)-b(z)]_\sigma|\,|f(y)|\,d\mu(y) \\
&\lesssim \frac{1}{[\ell(R)]^n}\int_{2^N Q} |[b(z)-m_R(b)]_\sigma|\,|f(y)|\,d\mu(y) \\
&\lesssim \mathcal{M}_{r,(9/8)}f(x).
\end{aligned}$$

Taking the mean over $y \in R$, we see that, for all $x \in Q$,

$$m_R\left[\left|T\left([m_R(b)-b]_\sigma f\chi_{2^N Q\setminus\frac{4}{3}R}\right)\right|\right] \lesssim \mathcal{M}_{r,(9/8)}f(x). \quad (5.6.22)$$

An argument similar to that used in the estimate for I_1 leads to that, for all $x \in Q$,

$$m_R\left[\left|T\left([m_R(b)-b]_\sigma f\chi_{\frac{4}{3}R}\right)\right|\right] \lesssim \mathcal{M}_{r,(9/8)}f(x). \quad (5.6.23)$$

Combining the estimates (5.6.21), (5.6.22) and (5.6.23), we find that, for all $x \in Q$,

$$\mathrm{L}_2 \lesssim \delta_{Q,R}^k\left\{\sum_{i=1}^{k-1}\sum_{\sigma\in C_i^k}\mathcal{M}_{r,(3/2)}(T_{\vec{b}_\sigma}f)(x)+\mathcal{M}_{r,(3/2)}(Tf)(x)+\mathcal{M}_{r,(9/8)}f(x)\right\}.$$

Now, we deal with L_3. For all $x,\ y \in Q$, we know that

$$\begin{aligned}
&\left|T\left(\prod_{i=1}^k\left[m_{\tilde{Q}}(b_i)-b_i\right]f\chi_{2^N Q\setminus\frac{4}{3}Q}\right)(y)\right| \\
&\lesssim \sum_{j=1}^{N-1}\frac{1}{(\ell(2^j Q))^n}\int_{2^{j+1}Q\setminus 2^j Q}\prod_{i=1}^k\left|b_i(z)-m_{\tilde{Q}}(b_i)\right||f(z)|\,d\mu(z) \\
&\quad+\frac{1}{(\ell(Q))^n}\int_{2Q\setminus\frac{4}{3}Q}\prod_{i=1}^k\left|b_i(z)-m_{\tilde{Q}}(b_i)\right||f(z)|\,d\mu(z) \\
&\lesssim \sum_{j=1}^{N-1}\frac{1}{(\ell(2^j Q))^n}\left\{\int_{2^{j+1}Q}\prod_{i=1}^k\left|b_i(z)-m_{\tilde{Q}}(b_i)\right|^{r'}d\mu(z)\right\}^{1/r'}
\end{aligned}$$

$$\times\left\{\int_{2^{j+1}Q}|f(z)|^r\,d\mu(z)\right\}^{1/r}$$

$$+\frac{1}{(\ell(Q))^n}\left\{\int_{2Q}\prod_{i=1}^k\left|b_i(z)-m_{\tilde{Q}}(b_i)\right|^{r'}\,d\mu(z)\right\}^{1/r'}$$

$$\times\left\{\int_{2Q}|f(z)|^r\,d\mu(z)\right\}^{1/r}$$

$$\lesssim\sum_{j=1}^{N-1}\frac{1}{(\ell(2^jQ))^n}\left\{\int_{2^{j+1}Q}\prod_{i=1}^k\left(\left|b_i(z)-m_{\widetilde{2^{j+1}Q}}(b_i)\right|\right.\right.$$

$$\left.\left.+\left|m_{\widetilde{2^{j+1}Q}}(b_i)-m_{\tilde{Q}}(b_i)\right|\right)^{r'}\,d\mu(z)\right\}^{1/r'}\left\{\int_{2^{j+1}Q}|f(z)|^r\,d\mu(z)\right\}^{1/r}$$

$$+\left\{\frac{1}{(\ell(Q))^n}\int_{2Q}|f(z)|^r\,d\mu(z)\right\}^{1/r}$$

$$\lesssim\delta_{Q,R}^k\sum_{j=1}^{N-1}\frac{\mu(2^{j+2}Q)}{(\ell(2^{j+1}Q))^n}\left\{\frac{1}{\mu(2^{j+2}Q)}\int_{2^{j+1}Q}|f(z)|^r\,d\mu(z)\right\}^{1/r}$$

$$+\mathcal{M}_{r,(9/8)}f(x)$$

$$\lesssim\delta_{Q,R}^{k+1}\mathcal{M}_{r,(9/8)}f(x).$$

Taking the mean over $y\in Q$, we conclude that, for all $x\in Q$,

$$\mathrm{L}_3\lesssim\delta_{Q,R}^{k+1}\mathcal{M}_{r,(9/8)}f(x).$$

Finally, we estimate L_4. It is easy to see that, for all $y\in R$ and $x\in Q$,

$$\left|T\left(\prod_{i=1}^k[m_R(b_i)-b_i]\,f\chi_{2^NQ\setminus\frac{4}{3}R}\right)(y)\right|$$

$$\lesssim\frac{1}{(\ell(R))^n}\int_{2^NQ\setminus\frac{4}{3}R}\prod_{i=1}^k|b_i(y)-m_R(b_i)|\,|f(y)|\,d\mu(y)$$

$$\lesssim\left\{\frac{1}{(\ell(2^NQ))^n}\int_{2^NQ}\prod_{i=1}^k|b_i(y)-m_R(b_i)|^{r'}\,d\mu(y)\right\}^{1/r'}$$

$$\times\left\{\frac{1}{(\ell(2^NQ))^n}\int_{2^NQ}|f(y)|^r\,d\mu(y)\right\}^{1/r}$$

$$\lesssim\mathcal{M}_{r,(9/8)}f(x).$$

Therefore,

$$\mathrm{L}_4 \lesssim \mathcal{M}_{r,(9/8)} f(x).$$

By the estimates for L_1, L_2, L_3 and L_4, we obtain (5.6.18).

Now we have proved that $T_{\vec{b}}$ satisfies (5.6.16). If we choose r such that $1 < r < p < \infty$, by Theorem 5.6.6, the boundedness of $\mathcal{M}_{r,(3/2)}$ on $p \in (r, \infty]$ for $p > r$ and (5.6.16), we obtain

$$\begin{aligned}\|T_{\vec{b}} f\|_{L^p(\mu)} &\lesssim \|\mathcal{M}^\sharp(T_{\vec{b}} f)\|_{L^p(\mu)}\\ &\lesssim \|\mathcal{M}_{r,(3/2)}(Tf)\|_{L^p(\mu)} + \|\mathcal{M}_{r,(9/8)} f\|_{L^p(\mu)}\\ &\quad + \sum_{i=1}^{k-1}\sum_{\sigma\in C_i^k} \left\|\mathcal{M}_{r,(3/2)}(T_{\vec{b}_{\bar{\sigma}}} f)\right\|_{L^p(\mu)}\\ &\lesssim \|f\|_{L^p(\mu)}.\end{aligned}$$

This finishes the proof of Theorem 5.6.5. □

5.6.3 *Endpoint Estimates for Multilinear Commutators*

In this subsection, we consider the endpoint case of Theorem 5.6.10. To this end, we first recall the definition of the following function space of Orlicz type, which is a variant with a non-doubling measure of the space $\mathrm{Osc}_{\exp L^r}(\mathbb{R}^D)$.[6]

Definition 5.6.7. For $r \in [1, \infty)$, a locally integrable function f is said to belong to the *space* $\mathrm{Osc}_{\exp L^r}(\mu)$, if there exists a positive constant $\tilde{C}$ such that,

(i) for any Q,

$$\begin{aligned}&\|f - m_{\tilde{Q}}(f)\|_{\exp L^r,\, Q,\, \mu/\mu(2Q)}\\ &:= \inf\left\{\lambda \in (0, \infty) : \frac{1}{\mu(2Q)} \int_Q \exp\left(\frac{|f - m_{\tilde{Q}}(f)|}{\lambda}\right)^r d\mu \le 2\right\} \le \tilde{C},\end{aligned}$$

(ii) for any doubling cubes $Q_1 \subset Q_2$,

$$|m_{Q_1}(f) - m_{Q_2}(f)| \le \tilde{C}\delta_{Q_1,\,Q_2}.$$

[6] See [108].

The $\mathrm{Osc}_{\exp L^r}(\mu)$ *norm* of f is defined by the minimal constant $\tilde{C}$ satisfying (i) and (ii) and denoted by $\|f\|_{\mathrm{Osc}_{\exp L^r}(\mu)}$.

Obviously, for any $r \in [1, \infty)$,

$$\mathrm{Osc}_{\exp L^r}(\mu) \subset \mathrm{RBMO}(\mu).$$

Moreover, from the John–Nirenberg inequality, it follows that

$$\mathrm{Osc}_{\exp L^1}(\mu) = \mathrm{RBMO}(\mu).$$

If μ is the D-dimensional Lebesgue measure in $\mathbb{R}^D$, the counterpart of the space $\mathrm{Osc}_{\exp L^r}(\mu)$ when $r \in (1, \infty)$ is a proper subspace of the classical space $\mathrm{BMO}(\mathbb{R}^D)$.[7] However, it is still unknown whether the space $\mathrm{Osc}_{\exp L^r}(\mu)$ is a proper subspace of the space $\mathrm{RBMO}(\mu)$ or not, when μ is a non-doubling measure as in (0.0.1).

To state the weak type estimate for the multilinear commutator $T_{\vec{b}}$, we need to introduce the following notation. For $i \in \{1, \ldots, k\}$, we denote by C_i^k the *family of all finite subsets* $\sigma := \{\sigma(1), \ldots, \sigma(i)\}$ *of* $\{1, \ldots, k\}$ *with* i *different elements.* For any $\sigma \in C_i^k$, the *complementary sequence* $\tilde{\sigma}$ is given by $\tilde{\sigma} := \{1, \ldots, k\} \setminus \sigma$. For any

$$\sigma := \{\sigma(1), \ldots, \sigma(i)\} \in C_i^k,$$

we write, for any i–tuple $r := (r_1, \ldots, r_i)$,

$$1/r_\sigma := 1/r_{\sigma(1)} + \cdots + 1/r_{\sigma(i)} \quad \text{and} \quad 1/r_{\tilde{\sigma}} := 1/r - 1/r_\sigma,$$

where

$$1/r := 1/r_1 + \cdots + 1/r_k.$$

Let $\vec{b} := (b_1, \ldots, b_k)$ be a finite family of locally integrable functions. For all $i \in \{1, \ldots, k\}$ and

$$\sigma := \{\sigma(1), \ldots, \sigma(i)\} \in C_i^k,$$

let

$$\vec{b}_\sigma := (b_{\sigma(1)}, \ldots, b_{\sigma(i)})$$

and the product

[7] See [108].

$$b_\sigma := b_{\sigma(1)} \cdots b_{\sigma(i)}.$$

With this notation, we write, for any i-tuple $r := (r_1, \ldots, r_i)$ of positive numbers,

$$\|\vec{b}_\sigma\|_{\mathrm{RBMO}(\mu)} := \|b_{\sigma(1)}\|_{\mathrm{RBMO}(\mu)} \cdots \|b_{\sigma(i)}\|_{\mathrm{RBMO}(\mu)}$$

and

$$\|\vec{b}_\sigma\|_{Osc_{\exp L^{r_\sigma}}(\mu)} := \|b_{\sigma(1)}\|_{Osc_{\exp L^{r_{\sigma(1)}}}(\mu)} \cdots \|b_{\sigma(i)}\|_{Osc_{\exp L^{r_{\sigma(i)}}}(\mu)}.$$

In particular, for $i \in \{1, \ldots, k\}$ and $\sigma := \{\sigma(1), \ldots, \sigma(i)\} \in C_i^k$, we write

$$[b(y) - b(z)]_\sigma := \left[b_{\sigma(1)}(y) - b_{\sigma(1)}(z)\right] \cdots \left[b_{\sigma(i)}(y) - b_{\sigma(i)}(z)\right]$$

and

$$\left[m_{\tilde{Q}}(b) - b(y)\right]_\sigma := \left[m_{\tilde{Q}}(b_{\sigma(1)}) - b_{\sigma(1)}(y)\right] \cdots \left[m_{\tilde{Q}}(b_{\sigma(i)}) - b_{\sigma(i)}(y)\right],$$

where Q is any cube in $\mathbb{R}^D$ and $y, z \in \mathbb{R}^D$. For the product of all the functions, we simply write

$$\|\vec{b}\|_{\mathrm{RBMO}(\mu)} := \|b_1\|_{\mathrm{RBMO}(\mu)} \cdots \|b_k\|_{\mathrm{RBMO}(\mu)}$$

and

$$\|\vec{b}\|_{Osc_{\exp L^r}(\mu)} := \|b_1\|_{Osc_{\exp L^{r_1}}(\mu)} \cdots \|b_m\|_{Osc_{\exp L^{r_k}}(\mu)}.$$

For any $\sigma \in C_i^k$ and suitable function f, let

$$T_{\vec{b}_\sigma} f := \left[b_{\sigma(i)}, \left[b_{\sigma(i-1)}, \cdots, \left[b_{\sigma(1)}, T\right] \cdots\right]\right] f.$$

In particular, when $\sigma := \{1, \ldots, k\}$, we *denote* $T_{\vec{b}_\sigma}$ *simply by* $T_{\vec{b}}$ as in (5.6.10) with (5.6.11).

The main result of this section is as follows.

Theorem 5.6.8. *Let $k \in \mathbb{N}$, $r_i \in [1, \infty)$ and $b_i \in \mathrm{Osc}_{\exp L^{r_i}}(\mu)$ for $i \in \{1, \ldots, k\}$. Let T and $T_{\vec{b}}$ be as in* (5.3.7) *and* (5.6.10) *with kernel K satisfying* (5.3.1) *and* (5.3.8). *If T is bounded on $L^2(\mu)$, then there exists a positive constant C such that, for all $\lambda \in (0, \infty)$ and bounded functions f with compact support,*

$$\begin{aligned}
&\mu\left(\{x \in \mathbb{R}^D : |T_{\vec{b}} f(x)| > \lambda\}\right) \\
&\quad \le C\Phi_{1/r}\left(\|\vec{b}\|_{\mathrm{Osc}_{\exp L^r}(\mu)}\right) \int_{\mathbb{R}^D} \frac{|f(y)|}{\lambda} \log^{1/r}\left(2 + \frac{|f(y)|}{\lambda}\right) d\mu(y),
\end{aligned}$$

where

$$1/r := 1/r_1 + \cdots + 1/r_k$$

and, for $s \in (0, \infty)$ *and all* $t \in (0, \infty)$,

$$\Phi_s(t) := t \log^s(2 + t).$$

Proof. Without loss of generality, we may assume that, for any $i \in \{1, \ldots, k\}$, $\|b_i\|_{\mathrm{Osc}_{\exp L^{r_i}}(\mu)} = 1$. Indeed, let

$$\overline{b}_i := \frac{b_i}{\|b\|_{\mathrm{Osc}_{\exp L^{r_i}}(\mu)}}$$

for $i \in \{1, \ldots, k\}$. The homogeneity tells us that, for any $\lambda \in (0, \infty)$,

$$\begin{aligned}
&\mu\left(\{x \in \mathbb{R}^D : |T_{\vec{b}} f(x)| > \lambda\}\right) \\
&= \mu\left(\left\{x \in \mathbb{R}^D : \left|\left[\overline{b}_k, \left[\overline{b}_{k-1}, \cdots, \left[\overline{b}_1, T\right]\cdots\right]\right] f(x)\right| \right.\right. \\
&\qquad\qquad \left.\left. > \lambda / \|\vec{b}\|_{\mathrm{Osc}_{\exp L^r}(\mu)}\right\}\right). \qquad (5.6.24)
\end{aligned}$$

Noticing that $\|\overline{b}_i\|_{\mathrm{Osc}_{\exp L^{r_i}}(\mu)} = 1$ for $i \in \{1, \ldots, k\}$, if, when $\|b_i\|_{\mathrm{Osc}_{\exp L^{r_i}}(\mu)} = 1$, $i \in \{1, \ldots, k\}$, the theorem is true, by (5.6.24) and the fact that, for any $s \in (0, \infty)$ and $t_1, t_2 \in [0, \infty)$,

$$\Phi_s(t_1 t_2) \lesssim \Phi_s(t_1)\Phi_s(t_2),$$

we easily see that the theorem still holds true for any $b_i \in \mathrm{Osc}_{\exp L^{r_i}}(\mu)$, $i \in \{1, \ldots, k\}$.

We prove the theorem by two steps: $k = 1$ and $k > 1$.

Step I $k = 1$. For each fixed bounded and compact supported function f and each

$$\lambda > 2^{D+1}\|f\|_{L^1(\mu)}/\|\mu\|,$$

applying the Calderón–Zygmund decomposition to f at level λ, we obtain a sequence of cubes, $\{Q_j\}$, with bounded overlaps and functions g and h, where

$$g := f\chi_{\mathbb{R}^D \setminus \bigcup_j Q_j} + \sum_j \phi_j,$$

and

$$h := f - g = \sum_j (w_j f - \phi_j).$$

Notice that $\|g\|_{L^1(\mu)} \lesssim \|f\|_{L^1(\mu)}$. The boundedness of T_b on $L^2(\mu)$ and the fact that $\|g\|_{L^\infty(\mu)} \lesssim \lambda$ show that

$$\mu\left(\{x \in \mathbb{R}^D : |T_b g(x)| > \lambda\}\right) \lesssim \lambda^{-1} \int_{\mathbb{R}^D} |f(y)|\, d\mu(y).$$

By (1.4.1), we conclude that

$$\mu\left(\bigcup_j 2Q_j\right) \lesssim \lambda^{-1} \int_{\mathbb{R}^D} |f(y)|\, d\mu(y).$$

Then the proof of Theorem 5.6.8 is reduced to proving that

$$\begin{aligned}\mu&\left(\left\{x \in \mathbb{R}^D \setminus \bigcup_j 2Q_j : |T_b h(x)| > \lambda\right\}\right)\\ &\lesssim \int_{\mathbb{R}^D} \frac{|f(y)|}{\lambda} \log^{1/r}\left(2 + \frac{|f(y)|}{\lambda}\right) d\mu(y).\end{aligned}$$

For each fixed j, let $b_j := b - m_{\widetilde{Q}_j}(b)$ and $h_j := w_j f - \phi_j$. For all $x \in \mathbb{R}^D$, write

$$\begin{aligned}T_b h(x) &= \sum_j b_j(x) T h_j(x) - \sum_j T(b_j h_j)(x)\\ &=: \mathrm{I}(x) + \mathrm{II}(x).\end{aligned}$$

The boundedness of T from $L^1(\mu)$ to $L^{1,\infty}(\mu)$ implies that

$$\begin{aligned}\mu\left(\{x \in \mathbb{R}^D : |\mathrm{II}(x)| > \lambda\}\right) &\lesssim \lambda^{-1} \sum_j \int_{\mathbb{R}^D} \left|b_j(y) h_j(y)\right| d\mu(y)\\ &\lesssim \lambda^{-1} \sum_j \int_{Q_j} \left|b(y) - m_{\widetilde{Q}_j}(b)\right| |f(y)|\, d\mu(y)\\ &\quad + \lambda^{-1} \sum_j \|\phi_j\|_{L^\infty(\mu)} \int_{R_j} \left|b(y) - m_{\widetilde{Q}_j}(b)\right| d\mu(y)\\ &=: \mathrm{E} + \mathrm{F},\end{aligned}$$

where, for each j, R_j is the smallest $(6, 6^{D+1})$-doubling cube concentric with Q_j. It is obvious that R_j is also $(2, 6^{D+1})$-doubling. Thus, we find that

$$\begin{aligned}
&\int_{R_j} \left|b(y) - m_{\widetilde{Q_j}}(b)\right| d\mu(y)\\
&\quad\le \int_{R_j} \left|b(y) - m_{R_j}(b)\right| d\mu(y)\\
&\qquad + \mu(R_j)\left[\left|m_{\widetilde{6Q_j}}(b) - m_{R_j}(b)\right| + \left|m_{\widetilde{6Q_j}}(b) - m_{\widetilde{Q_j}}(b)\right|\right]\\
&\quad\lesssim \mu(2R_j) + \mu(R_j)\left[\delta_{6Q_j, R_j} + \delta_{Q_j, 6Q_j}\right].
\end{aligned}$$

A trivial computation shows that $\delta_{6Q_j, R_j} \lesssim 1$. This, via the fact

$$\mu(2R_j) \le \mu(6R_j) \le 6^{D+1}\mu(R_j),$$

in turn implies that

$$\mathrm{F} \lesssim \lambda^{-1} \sum_j \|\phi_j\|_{L^\infty(\mu)}\mu(R_j) \lesssim \lambda^{-1} \int_{\mathbb{R}^D} |f(y)|\, d\mu(y).$$

On the other hand, from Lemma 5.6.1, it follows that

$$\begin{aligned}
\mathrm{E} &\lesssim \lambda^{-1} \sum_j \mu(2Q_j) \|f\|_{L(\log L)^{1/r}, Q_j, \mu/\mu(2Q_j)} \|b_j\|_{\exp L^r, Q_j, \mu/\mu(2Q_j)}\\
&\lesssim \lambda^{-1} \sum_j \mu(2Q_j) \|f\|_{L(\log L)^{1/r}, Q_j, \mu/\mu(2Q_j)}\\
&\lesssim \lambda^{-1} \sum_j \mu(2Q_j) \inf\left\{t + \frac{t}{\mu(2Q_j)} \int_{Q_j} \frac{|f(y)|}{t} \log^{1/r}\left(2 + \frac{|f(y)|}{t}\right) d\mu(y)\right\}\\
&\lesssim \int_{\mathbb{R}^D} \frac{|f(y)|}{\lambda} \log^{1/r}\left(2 + \frac{|f(y)|}{\lambda}\right) d\mu(y).
\end{aligned}$$

Now we turn our attention to $\mathrm{I}(x)$. Let x_j be the center of Q_j. Noticing that $\operatorname{supp} h_j \subset R_j$, thus for $x \in \mathbb{R}^D \setminus 2R_j$, using the condition (5.3.8), we write

$$\begin{aligned}
&\int_{\mathbb{R}^D \setminus \bigcup_j 2Q_j} |\mathrm{I}(x)|\, d\mu(x)\\
&\quad\le \sum_j \int_{\mathbb{R}^D \setminus 2R_j} \int_{\mathbb{R}^D} \left|K(x, y) - K(x, x_j)\right| |b_j(x) h_j(y)|\, d\mu(y) d\mu(x)\\
&\qquad + \sum_j \int_{2R_j \setminus 2Q_j} |b_j(x)| |T h_j(x)|\, d\mu(x)
\end{aligned}$$

$$\begin{aligned}
&\lesssim \sum_j [\ell(Q_j)]^\delta \int_{\mathbb{R}^D} |h_j(y)|\, d\mu(y) \int_{\mathbb{R}^D\setminus 2Q_j} \frac{|b_j(x)|}{|x-x_j|^{n+\delta}}\, d\mu(x)\\
&\quad + \sum_j \int_{2R_j\setminus 2Q_j} |b_j(x)||T(w_j f)(x)|\, d\mu(x)\\
&\quad + \sum_j \int_{2R_j} |b_j(x)||T\phi_j(x)|\, d\mu(x)\\
&=: \mathrm{G}+\mathrm{H}+\mathrm{J}.
\end{aligned}$$

Employing the condition (0.0.1), we have

$$\begin{aligned}
&\int_{\mathbb{R}^D\setminus 2Q_j} \frac{|b_j(x)|}{|x-x_j|^{n+\delta}}\, d\mu(x)\\
&\quad \lesssim \sum_{k=1}^{\infty} [2^k \ell(Q_j)]^{-n-\delta} \int_{2^{k+1}Q_j} \left| b(x) - m_{\widetilde{2^{k+1}Q_j}}(b) \right| d\mu(x)\\
&\qquad + \sum_{k=1}^{\infty} [2^k \ell(Q_j)]^{-n-\delta} \mu(2^{k+1}Q_j) \left| m_{\widetilde{Q_j}}(b) - m_{\widetilde{2^{k+1}Q_j}}(b) \right|\\
&\quad \lesssim \sum_{k=1}^{\infty} [2^k \ell(Q_j)]^{-n-\delta} \mu(2^{k+2}Q_j)\\
&\qquad + \sum_{k=1}^{\infty} \delta_{Q_j,2^{k+1}Q_j} [2^k \ell(Q_j)]^{-n-\delta} \mu(2^{k+1}Q_j)\\
&\quad \lesssim [\ell(Q_j)]^{-\delta}.
\end{aligned}$$

Since

$$\|h_j\|_{L^1(\mu)} \lesssim \int_{Q_j} |f(y)|\, d\mu(y),$$

the desired estimate for G follows directly.

On the other hand, the Hölder inequality and the boundedness of T on $L^2(\mu)$ imply that

$$\mathrm{J} \le \sum_j \int_{2R_j} \left| b(x) - m_{\widetilde{2R_j}}(b) \right| |T\phi_j(x)|\, d\mu(x)$$

$$
\begin{aligned}
&+\sum_j \left|m_{\widetilde{Q_j}}(b) - m_{\widetilde{2R_j}}(b)\right| \int_{2R_j} |T\phi_j(x)|\,d\mu(x)\\
&\le \sum_j \left\{\int_{2R_j} \left|b(x) - m_{\widetilde{2R_j}}(b)\right|^2 d\mu(x)\right\}^{1/2} \|T\phi_j(x)\|_{L^2(\mu)}\\
&\quad+\sum_j [\mu(2R_j)]^{1/2}\|T\phi_j(x)\|_{L^2(\mu)} \left|m_{\widetilde{Q_j}}(b) - m_{\widetilde{2R_j}}(b)\right|\\
&\lesssim \sum_j [\mu(4R_j)]^{1/2}\|T\phi_j(x)\|_{L^2(\mu)} \left(1 + \left|m_{\widetilde{Q_j}}(b) - m_{\widetilde{2R_j}}(b)\right|\right)\\
&\lesssim \sum_j [\mu(4R_j)]^{1/2}\|T\phi_j(x)\|_{L^2(\mu)}\\
&\lesssim \int_{\mathbb{R}^D} |f(y)|\,d\mu(y),
\end{aligned}
$$

where the second-to-last inequality follows from the fact that

$$\left|m_{\widetilde{Q_j}}(b) - m_{\widetilde{2R_j}}(b)\right| \le \left|m_{\widetilde{Q_j}}(b) - m_{R_j}(b)\right| + \left|m_{R_j}(b) - m_{\widetilde{2R_j}}(b)\right| \lesssim 1.$$

To estimate H, observe that, for $x \in 2R_j \setminus 2Q_j$,

$$|T(w_j f)(x)| \lesssim \frac{1}{|x - x_j|^n} \int_{Q_j} |f(y)|\,d\mu(y).$$

Therefore, we estimate H by

$$
\begin{aligned}
\mathrm{H} &\lesssim \sum_j \left\{\int_{2R_j\setminus R_j} \frac{|b_j(x)|}{|x - x_j|^n}\,d\mu(x) + \int_{R_j\setminus Q_j} \frac{|b_j(x)|}{|x - x_j|^n}\,d\mu(x)\right\}\\
&\quad\times \int_{Q_j} |f(y)|\,d\mu(y)\\
&\lesssim \sum_j \left\{\frac{\mu(4R_j)}{[\ell(R_j)]^n} + \frac{\mu(2R_j)}{[\ell(R_j)]^n}\left|m_{\widetilde{Q_j}}(b) - m_{\widetilde{2R_j}}(b)\right|\right\} \int_{Q_j} |f(y)|\,d\mu(y)\\
&\quad+\sum_j \sum_{k=0}^{N-1} [\ell(6^k Q_j)]^{-n} \int_{6^{k+1}Q_j\setminus 6^k Q_j} \left|b(x) - m_{\widetilde{6^{k+1}Q_j}}(b)\right| d\mu(x)\\
&\quad\times \int_{Q_j} |f(y)|\,d\mu(y)
\end{aligned}
$$

$$+\sum_j \sum_{k=0}^{N-1} [\ell(6^k Q_j)]^{-n} \mu(6^{k+1} Q_j) \left| m_{\widetilde{Q_j}}(b) - m_{\widetilde{6^{k+1} Q_j}}(b) \right|$$

$$\times \int_{Q_j} |f(y)| \, d\mu(y),$$

where $N \in \mathbb{N}$ is such that $R_j = 6^N Q_j$. Obviously, for each $k \in \{0, \ldots, N-1\}$, $6^{k+1} Q_j \subset R_j$ and hence

$$\left| m_{\widetilde{Q_j}}(b) - m_{\widetilde{6^{k+1} Q_j}}(b) \right| \lesssim \delta_{Q_j, 6^{k+1} Q_j} \lesssim \delta_{Q_j, R_j} \lesssim 1.$$

Consequently,

$$\begin{aligned}
\mathrm{H} &\lesssim \sum_j \int_{Q_j} |f(y)| \, d\mu(y) \\
&\quad + \sum_j \sum_{k=0}^{N-1} [\ell(6^k Q_j)]^{-n} \mu(2 \times 6^{k+1} Q_j) \int_{Q_j} |f(y)| \, d\mu(y) \\
&\quad + \sum_j \sum_{k=0}^{N-1} [\ell(6^k Q_j)]^{-n} \mu(6^{k+1} Q_j) \int_{Q_j} |f(y)| \, d\mu(y) \\
&\lesssim \sum_j \left\{ 1 + \sum_{k=0}^{N-1} [\ell(6^k Q_j)]^{-n} \mu(2 \times 6^{k+1} Q_j) \right\} \int_{Q_j} |f(y)| \, d\mu(y).
\end{aligned}$$

Notice that there does not exist any $(6, 6^{D+1})$-doubling cube between Q_j and R_j. We then see that, for each fixed integer k with $k \in \{0, \ldots, N-1\}$,

$$\mu(6^{k+1} Q_j) \leq \frac{\mu(6^N Q_j)}{6^{(n+1)(N-k-1)}} \lesssim [\ell(6^N Q_j)]^n 6^{-(n+1)(N-k-1)} \sim [\ell(6^k Q_j)]^n 6^{k-N}.$$

We thus conclude that, for each fixed j,

$$\begin{aligned}
&\sum_{k=0}^{N-1} [\ell(6^k Q_j)]^{-n} \mu(6^{k+2} Q_j) \\
&\quad \lesssim \sum_{k=1}^{N-1} [\ell(6^{k-1} Q_j)]^{-n} \mu(6^{k+1} Q_j) + [\ell(6^{N-1} Q_j)]^{-n} \mu(6 R_j) \\
&\quad \lesssim \sum_{l=1}^{\infty} 6^{-l} + 1 \\
&\quad \lesssim 1.
\end{aligned}$$

Combining the estimates for G, H and J above, we see that

$$\int_{\mathbb{R}^D \setminus \bigcup_j 2Q_j} |\mathrm{I}(x)| \, d\mu(x) \lesssim \sum_j \int_{Q_j} |f(y)| \, d\mu(y).$$

This finishes the proof of Theorem 5.6.8 in the case that $k = 1$.

Step II $k \geq 2$. Now we assume that $k \geq 2$ is an integer and that, for any $i \in \{1, \ldots, k-1\}$ and any subset $\sigma := \{\sigma(1), \ldots, \sigma(i)\}$ of $\{1, \ldots, k\}$, Theorem 5.6.8 is true. For each fixed f and

$$\lambda > 2^{D+1} \|f\|_{L^1(\mu)} / \mu\left(\mathbb{R}^D\right),$$

let $Q_j, R_j, \phi_j, w_j, g, h$ and h_j be the same as in Step I. It suffices to show that

$$\mu\left(\left\{x \in \mathbb{R}^D \setminus \bigcup_j 2Q_j : |T_{\vec{b}} h(x)| > \lambda\right\}\right) \lesssim \int_{\mathbb{R}^D} \frac{|f(y)|}{\lambda} \log^{1/r}\left(2 + \frac{|f(y)|}{\lambda}\right) d\mu(y).$$

With the aid of the formula (5.6.19), it is easy to see that

$$\begin{aligned} T_{\vec{b}} h(x) &= \sum_j \prod_{i=1}^k \left[b_i(x) - m_{\widetilde{Q}_j} b_i\right] T h_j(x) - \sum_j T\left(\prod_{i=1}^k \left[b_i - m_{\widetilde{Q}_j}(b_i)\right] h_j\right)(x) \\ &\quad - \sum_j \sum_{i=1}^{k-1} \sum_{\sigma \in C_i^k} T_{\vec{b}_{\bar{\sigma}}}\left(\left[b - m_{\widetilde{Q}_j}(b)\right]_\sigma h_j\right)(x) \\ &=: T_{\vec{b}}^{\mathrm{I}} h(x) - T_{\vec{b}}^{\mathrm{II}} h(x) - \sum_{i=1}^{k-1} \sum_{\sigma \in C_i^k} T_{\vec{b}_{\bar{\sigma}}}^{\mathrm{III}} h(x). \end{aligned}$$

The same argument as that used in Step I shows that

$$\mu\left(\left\{x \in \mathbb{R}^D \setminus \bigcup_j 2Q_j : |T_{\vec{b}}^{\mathrm{I}} h(x)| > \lambda\right\}\right) \lesssim \lambda^{-1} \int_{\mathbb{R}^D} |f(y)| \, d\mu(y).$$

By an argument similar to that used in the estimate for $\mathrm{II}(x)$ in Step I, we know that

$$\mu\left(\left\{x \in \mathbb{R}^D \setminus \bigcup_j 2Q_j : |T_{\vec{b}}^{\mathrm{II}} h(x)| > \lambda\right\}\right) \lesssim \int_{\mathbb{R}^D} \frac{|f(y)|}{\lambda} \log^{1/r}\left(2 + \frac{|f(y)|}{\lambda}\right) d\mu(y).$$

For each fixed i with $i \in \{1, \ldots, k-1\}$, the induction hypothesis now states that

$$\begin{aligned}
&\mu\left(\left\{x \in \mathbb{R}^D : |T^{\mathrm{III}}_{\vec{b}_{\tilde{\sigma}}} h(x)| > \lambda\right\}\right) \\
&\lesssim \int_{\mathbb{R}^D} \Phi_{1/r_{\tilde{\sigma}}}\left(\sum_j \left|\left[b(y) - m_{\widetilde{Q}_j}(b)\right]_\sigma\right| \frac{|h_j(y)|}{\lambda}\right) d\mu(y) \\
&\lesssim \sum_j \int_{\mathbb{R}^D} \Phi_{1/r_{\tilde{\sigma}}}\left(\left|\left[b(y) - m_{\widetilde{Q}_j}(b)\right]_\sigma\right| \frac{|w_j(y) f(y)|}{\lambda}\right) d\mu(y) \\
&\quad + \sum_j \int_{\mathbb{R}^D} \Phi_{1/r_{\tilde{\sigma}}}\left(\left|\left[b(y) - m_{\widetilde{Q}_j}(b)\right]_\sigma\right| \frac{|\phi_j(y)|}{\lambda}\right) d\mu(y) \\
&=: \mathrm{U}_\sigma + \mathrm{V}_\sigma.
\end{aligned}$$

Let

$$\Psi_{\sigma(l)}(t) := \exp t^{r_{\sigma(l)}} - 1 \text{ for } l \in \{1, \ldots, i\} \text{ and } t \in (0, \infty).$$

Notice that

$$\Psi^{-1}_{\sigma(l)}(t) \sim \log^{1/r_{\sigma(l)}}(2+t)$$

and

$$\Phi^{-1}_{1/r_{\tilde{\sigma}}}(t) \sim t \log^{-1/r_{\tilde{\sigma}}}(2+t)$$

for all $t \in (0, \infty)$. Thus, by Lemma 5.6.1, we see that, for any $t_0, t_1, \ldots, t_i \in (0, \infty)$,

$$\Phi_{1/r_{\tilde{\sigma}}}(t_0 t_1 \cdots t_i) \lesssim \Phi_{1/r}(t_0) + \exp t_1^{r_{\sigma(1)}} + \cdots + \exp t_i^{r_{\sigma(i)}}.$$

From this and the assumption that $\|b_i\|_{\mathrm{Osc}_{\exp L^{r_i}}(\mu)} = 1$ for $i \in \{1, \ldots, k\}$, it follows that

$$\begin{aligned}
\mathrm{U}_\sigma &\lesssim \sum_j \int_{\mathbb{R}^D} \Phi_{1/r}\left(\frac{|\chi_{Q_j}(y) f(y)|}{\lambda}\right) d\mu(y) \\
&\quad + \sum_j \sum_{l=1}^{i} \int_{\mathbb{R}^D} \exp\left(\frac{|b_{\sigma(l)}(y) - m_{\widetilde{Q}_j}(b_{\sigma(l)})|}{C} \chi_{Q_j}(y)\right)^{r_{\sigma(l)}} d\mu(y)
\end{aligned}$$

$$\lesssim \int_{\mathbb{R}^D} \frac{|f(y)|}{\lambda} \log^{1/r}\left(2+\frac{|f(y)|}{\lambda}\right) d\mu(y) + \sum_j \mu(2Q_j)$$

$$\lesssim \int_{\mathbb{R}^D} \frac{|f(y)|}{\lambda} \log^{1/r}\left(2+\frac{|f(y)|}{\lambda}\right) d\mu(y).$$

To estimate V_σ, let $r_j := \lambda^{-1}|\phi_j|$. Let $\Lambda \subset \mathbb{N}$ be a *finite index set*. The convexity of $\Phi_{1/r_{\tilde{\sigma}}}$ says that

$$\begin{aligned}
&\Phi_{1/r_{\tilde{\sigma}}}\left(\sum_{j\in\Lambda}\left|\left[b(y)-m_{\widetilde{Q}_j}(b)\right]_\sigma\right|\frac{|\phi_j(y)|}{\lambda}\right)\\
&\le \sum_{j\in\Lambda}\left[\frac{r_j(y)}{\sum_{l\in\Lambda} r_l(y)}\right]\Phi_{1/r_{\tilde{\sigma}}}\left(\left|\left[b(y)-m_{\widetilde{Q}_j}(b)\right]_\sigma\right|\chi_{R_j}(y)\sum_{l\in\Lambda} r_l(y)\right)\\
&\le \left[\frac{1}{\sum_{l\in\Lambda} r_l(y)}\right]\Phi_{1/r_{\tilde{\sigma}}}\left(\sum_{l\in\Lambda} r_l(y)\right)\\
&\quad\times\sum_{j\in\Lambda} r_j(y)\Phi_{1/r_{\tilde{\sigma}}}\left\{\left|\left[b(y)-m_{\widetilde{Q}_j}(b)\right]_\sigma\right|\chi_{R_j}(y)\right\}\\
&\le \log^{1/r_{\tilde{\sigma}}}\left(2+\sum_{l\in\Lambda} r_l(y)\right)\sum_j r_j(y)\Phi_{1/r_{\tilde{\sigma}}}\left\{\left|\left[b(y)-m_{\widetilde{Q}_j}(b)\right]_\sigma\right|\chi_{R_j}(y)\right\}\\
&\le \log^{1/r_{\tilde{\sigma}}}(2+B)\sum_j r_j(y)\Phi_{1/r_{\tilde{\sigma}}}\left\{\left|\left[b(y)-m_{\widetilde{Q}_j}(b)\right]_\sigma\right|\chi_{R_j}(y)\right\},
\end{aligned}$$

where B is as in (1.4.5). This in turn implies that

$$\begin{aligned}
\mathrm{V}_\sigma &\lesssim \lambda^{-1}\sum_j \|\phi_j\|_{L^\infty(\mu)}\int_{R_j}\Phi_{1/r_{\tilde{\sigma}}}\left(\left|\left[b(y)-m_{\widetilde{Q}_j}(b)\right]_\sigma\right|\right)d\mu(y)\\
&\lesssim \lambda^{-1}\sum_j \|\phi_j\|_{L^\infty(\mu)}\mu(R_j)\\
&\lesssim \lambda^{-1}\int_{\mathbb{R}^D}|f(y)|\,d\mu(y).
\end{aligned}$$

Therefore, for $i \in \{1,\ldots,k-1\}$, we have

$$\mu\left(\left\{x\in\mathbb{R}^D : |T^{\mathrm{III}}_{\vec{b}_{\tilde{\sigma}}}h(x)| > \lambda\right\}\right) \lesssim \int_{\mathbb{R}^D}\Phi_{1/r}\left(\frac{|f(y)|}{\lambda}\right)d\mu(y).$$

This gives the desired estimate for

$$\sum_{i=1}^{k-1}\sum_{\sigma\in C_i^k} T^{\mathrm{III}}_{\vec{b}_{\tilde{\sigma}}} h(x)$$

and hence finishes the proof of Theorem 5.6.8. □

5.7 Notes

- The operators $\mathcal{M}_{0,s}^{\rho,d}$ and $\mathcal{M}_{0,s}^{\rho,\natural}$ in the setting of Euclidean spaces were first introduced by John [73] and then rediscovered by Strömberg [123] and Lerner [80,81]. The weight function class $A_p^\rho(\mu)$ was introduced by Komori [76], while for $\sigma\in[0,\infty)$, $A^\rho_{p,(\log L)^\sigma}(\mu)$ was introduced by Hu et al. in [65]. Komori also established the weighted estimates with $A_p^\rho(\mu)$ weights for the operator $\mathcal{M}_{(\eta)}$ with $\eta>\rho$, which is the first result concerning the weighted estimate for the Hardy–Littlewood maximal operator with measure as in (0.0.1). The operators $\mathcal{M}_{0,s}^{\rho,d}$ and $M_{0,s}^{\rho,\natural}$, and the preliminary lemmas in Sect. 5.1 were given by Hu and Yang [64]. Theorem 5.6.6, a first good-λ inequality linking the maximal operator $\mathcal{M}^d$ and the sharp maximal operator $\mathcal{M}^\sharp$, was established by Tolsa in [131]. It seems that Theorem 5.1.11 is more suitable for the study of operators, since the assumption therein is fairly weak.
- The interpolation Theorem 5.2.4 was given by Hu et al. in [50] and Theorem 5.2.5 by Hu et al. in [55]. In [131], Tolsa showed that, for a linear operator T, if T is bounded from $H^1(\mu)$ to $L^1(\mu)$ and from $L^\infty(\mu)$ to RBMO (μ), then T can be extended boundedly to $L^p(\mu)$ for all $p\in(1,\infty)$. See Mateu et al. [94] for another version of the interpolation theorem between $(H^1_{\mathrm{at}}(\mu), L^1(\mu))$ and $(L^\infty(\mu), \mathrm{BMO}(\mu))$, where $H^1_{\mathrm{at}}(\mu)$ and $\mathrm{BMO}(\mu)$ are the atomic Hardy space and BMO-type space introduced in [94].
- When $\mu(\mathbb{R}^D)<\infty$, it is unknown whether Theorems 5.2.4 and 5.2.5 are still true or not.
- In the case that μ is the D-dimensional Lebesgue measure, Theorem 5.3.3 was proved by Grafakos [39]. The proof of Theorem 5.3.3 follows from the ideas of Grafakos and was given by Hu et al. in [57].
- Theorems 5.4.1 and 5.4.3 were given by Da. Yang and Do. Yang in [155].
- If T is a Calderón–Zygmund operator with measure μ as in (0.0.1). Theorem 5.4.4 was given by Tolsa [132]. The proof of Theorem 5.4.4 here follows almost the same line as in [132]. Theorem 5.4.6 is from Chen et al. [11]. The boundedness on $L^p(\mu)$, with $p\in(1,\infty)$, was first considered by Tolsa in [131]. It should be pointed out that the endpoint estimate $(H^1(\mu),\ L^{1,\infty}(\mu))$ for a sublinear operator implies the weak type $L\log L$ estimate for this operator (see Hu and Liang [51]). Theorem 5.6.2 was proved by Hu, Meng and Yang in [55].
- In Chen and Miao [12], a vector-valued commutator theory was established. As an application, they obtained the boundedness of maximal commutators

of Calderón–Zygmund operators with RBMO(μ) functions. See Hu et al. [56, 58, 59] for endpoint estimates for maximal commutators of Calderón–Zygmund operators.

- In [129], Tolsa established a version of the $T(1)$ theorem for the Cauchy transform. In [144], Verdera presented a new proof of the $T(1)$ theorem for the Cauchy integral. Independently, based on random dyadic lattices, Nazarov, Treil and Volberg [102] obtained another $T(1)$ theorem for more general Calderón–Zygmund operators. In [130], Tolsa obtained a version of the $T(1)$ theorem for any general Calderón–Zygmund operator and any measure μ which is non-doubling in general and may contain atoms. In [132], by using the Littlewood–Paley techniques, Tolsa proved a version of the $T(1)$ theorem for Calderón–Zygmund operators on $\mathbb{R}^D$.
- In [22], David established a version of the $T(b)$ theorem for non-doubling measures that solved the Vitushkin conjecture for sets with positive finite one-dimensional Hausdorff measure. In [104], Nazarov, Treil and Volberg established a local $T(b)$ theorem on non-homogeneous spaces. Also, another $T(b)$ theorem for any non-doubling measure on a metric space was established by Nazarov, Treil and Volberg [105], which is closer to the classical one than the ones mentioned above.
- Theorems 5.6.5 and 5.6.8 were established by Hu et al. in [54]. See Meng and Yang [97] for the boundedness of commutators generated by Calderón–Zygmund operators or fractional integrals with Lipschitz functions in the Lebesgue space and the Hardy space.
- In [96], Meng and Yang obtained the boundedness in some Hardy-type spaces of multilinear commutators generated by Calderón–Zygmund operators or fractional integrals with RBMO (μ) functions, where the Hardy-type spaces are some appropriate subspaces, associated to the considered RBMO (μ) functions, of the Hardy space $H^1(\mu)$.
- The $A_p^\rho(\mu)$ weights of Muckenhoupt type in the setting of $\mathbb{R}^D$ with the measure as in (0.0.1) were first introduced by Orobitg and Pérez [106] for $\rho = 1$ and by Komori [76] for $\rho \in [1, \infty)$, and extended to $A^\rho_{p,\,(\log L)^\sigma}(\mu)$ by Hu et al. in [65]. When the measure μ in (0.0.1) satisfies the following additional assumption that

$$\mu(\partial Q) = 0, \tag{5.7.1}$$

namely, the faces (or edges) of any cube have μ measure zero, Orobitg and Pérez [106] proved that, if K satisfies (5.3.1) and (5.3.8) and if T is bounded on $L^2(\mu)$, then T is bounded on $L^p(u)$ for all $p \in (1, \infty)$ and $u \in A_p(\mu)$. It should be pointed out that, for any nonnegative Radon measure μ, there exists an orthonormal system in $\mathbb{R}^D$ such that (5.7.1) holds true (see [94]). However, it is not so clear how the $A_p(\mu)$ weights and singular integrals related to μ as in (0.0.1) depend on different orthonormal systems in $\mathbb{R}^D$. Orobitg and Perez showed that, when μ satisfies (5.7.1), $u \in A_p(\mu)$ implies that $u \in A_{p-\epsilon}(\mu)$ for some $\epsilon \in (0, \infty)$. Notice that, for $\rho \in (1, \infty)$, $A_p^1(\mu) \subset A_p^\rho(\mu)$ and it is still unknown whether a weight $u \in A_p^\rho(\mu)$ enjoys the reverse Hölder inequality or not. Theorem 5.5.1 is from Hu and Yang [64]. For the weighted estimates for commutators of singular integral operators, see [13, 65, 86].

- In [141], Tolsa introduced a maximal operator N, which coincides with the Hardy–Littlewood maximal operator if $\mu(B(x,r)) \sim r^n$ for all $x \in \operatorname{supp} \mu$ and $r \in (0,\infty)$, and showed that, for a fixed $p \in (1,\infty)$, all Calderón–Zygmund operators are bounded on $L^p(w d\mu)$ if and only if N is bounded on $L^p(w d\mu)$. Tolsa also proved that this happens if and only if some conditions of Sawyer type hold true. Tolsa obtained analogous results about the weak (p,p) estimates. Such weights do not satisfy a reverse Hölder inequality, in general, but some kind of self-improving property still holds true. On the other hand, if $f \in \mathrm{RBMO}(\mu)$ and $\epsilon \in (0,\infty)$ is small enough, then $e^{\epsilon f}$ belongs to this class of weights.
- In [53], Hu, Meng and Yang obtained the (L^p, L^q)-boundedness and the weak type endpoint estimate for the multilinear commutators generated by fractional integrals with $\mathrm{RBMO}(\mu)$ functions of Tolsa or with $\mathrm{Osc}_{\exp L^r}(\mu)$ functions for $r \in [1,\infty)$.
- In [60], Hu, Meng and Yang proved that, for a class of linear operators including Riesz potentials on $\mathbb{R}^D$, their boundedness in Lebesgue spaces is equivalent to their boundedness in the Hardy space or its weak type endpoint estimates, respectively. They further obtained several new end estimates for a class of linear operators. Moreover, the result in [60] is new even for the D-dimensional Lebesgue measure.
- Let $\mathcal{X}$ be a set, d a quasidistance on $\mathcal{X}$ and μ a measure satisfying the polynomial growth condition. In [7], Bramanti showed that the Calderón–Zygmund operator T is bounded on $L^2(\mu)$ by establishing the boundedness of T on the Hölder space $C^\alpha(\mathcal{X})$. Then, using the result established by Nazarov et al. in [103], Bramanti [7] further showed that T is bounded on $L^p(\mu)$ for any $p \in (1,\infty)$ and from $L^1(\mu)$ to $L^{1,\infty}(\mu)$.
- Let $(\mathcal{X}, d, \mu)$ be a metric measure space with μ satisfying the polynomial growth condition. In [32], García–Cuerva and Gatto established the boundedness of the fractional integral operator I_α with $\alpha \in (0,n)$ on Lebesgue spaces and Lipschitz spaces, respectively. More general fractional integrals with extra regularity conditions were also considered. When $(\mathcal{X}, d, \mu) := (\mathbb{R}^D, |\cdot|, \mu)$ with μ satisfying (0.0.1), Chen and Lai in [10] obtained the result that the fractional integral I_α of order α is bounded from $H^1(\mu)$ to $L^q(\mu)$ for $1/q = 1 - \alpha/n$.
- For $\alpha \in (0,n)$, the *radial fractional maximal function* M_α is defined by

$$M_\alpha(f)(x) := \sup_{Q \ni x} \frac{1}{[\ell(Q)]^{n-\alpha}} \int_Q |f(y)|\, d\mu(y), \quad \forall x \in \mathbb{R}^D.$$

In [36], García–Cuerva and Martell established a characterization of pairs of weight functions (in terms of Muckenhoupt type conditions) such that M_α is bounded from one weighted L^p space to another weighted L^q space.

- In [52], Hu, Lin and Yang introduced the Marcinkiewicz integral with kernel satisfying the Hörmander-type condition. By assuming that the Marcinkiewicz integral is bounded on $L^2(\mu)$, the authors in [52] then established its boundedness, respectively, from the Lebesgue space $L^1(\mu)$ to the weak Lebesgue space $L^{1,\infty}(\mu)$, from the Hardy space $H^1(\mu)$ to $L^1(\mu)$ and from the Lebesgue space $L^\infty(\mu)$ to the space $\mathrm{RBLO}\,(\mu)$. As a corollary, they obtained the boundedness of the Marcinkiewicz integral in the Lebesgue space $L^p(\mu)$ with $p \in (1,\infty)$. Moreover, they established the boundedness of the commutator generated by the $\mathrm{RBMO}\,(\mu)$ function and the Marcinkiewicz integral with kernel satisfying some slightly stronger Hörmander-type condition, respectively, from $L^p(\mu)$ with $p \in (1,\infty)$ to itself, from the space $L\log L(\mu)$ to $L^{1,\infty}(\mu)$ and from $H^1(\mu)$ to $L^{1,\infty}(\mu)$. Some of the results are also new even for the classical Marcinkiewicz integral.
- In [149], Xu introduced the multilinear singular integral T and obtained its boundedness from $L^{p_1}(\mu) \times L^{p_2}(\mu)$ to $L^p(\mu)$ under the assumption that T is bounded from $L^1(\mu) \times L^1(\mu)$ to $L^{1/2,\infty}(\mu)$, where

$$p_1,\ p_2 \in (1,\infty) \quad \text{and} \quad 1/p := 1/p_1 + /p_2.$$

In [95], Meng obtained the same boundedness of T under the assumption that T is bounded from $H^1(\mu) \times H^1(\mu)$ to $L^{1/2}(\mu)$. See [150] for the boundedness of commutators generated by multilinear singular integrals and $\mathrm{RBMO}\,(\mu)$ functions of Tolsa, and [85] for the boundedness of the maximal multilinear Calderón–Zygmund operators.
- Let $(\mathcal{X}, d, \mu)$ be a metric measure space with μ satisfying the polynomial growth condition. In [34], García–Cuerva and Martell studied weighted inequalities for Calderón–Zygmund operators. Specifically, for $p \in (1,\infty)$, they identified sufficient conditions for the weight on one side, which guarantees the existence of another weight on the other side, so that the weighted L^p inequality holds true. They also dealt with this problem by developing a vector-valued theory for Calderón–Zygmund operators. For the case of the Cauchy integral operator, which is the most important example, they proved that the conditions for the weights are also necessary. Some of these results on weighted inequalities for Calderón–Zygmund operators were extended to the case of the maximal Calderón–Zygmund operators by García–Cuerva and Martell in [35].

Chapter 6
Littlewood–Paley Operators and Maximal Operators Related to Approximations of the Identity

In this chapter, we turn our attention to the boundedness on $L^p(\mu)$, with $p \in (1, \infty)$, and endpoint estimates of operators related to approximations of the identity in Sect. 2.4, including Littlewood–Paley operators and maximal operators.

6.1 Boundedness in Hardy-Type Spaces

For any $k \in \mathbb{Z}$, let S_k be as in Sect. 2.4 and $D_k := S_k - S_{k-1}$ for $k \in \mathbb{Z}$, and we also use D_k to denote the corresponding integral kernel of the operator D_k. The *homogeneous Littlewood–Paley g-function* $\dot{g}(f)$ is then defined by

$$\dot{g}(f)(x) := \left[\sum_{k=-\infty}^{\infty} |D_k f(x)|^2\right]^{1/2}.$$

We first establish the boundedness on $L^p(\mu)$, with $p \in (1, \infty)$, of $\dot{g}(f)$. To this end, for each $k \in \mathbb{Z}$, we let

$$E_k := \sum_{j \in \mathbb{Z}} D_{k+j} D_j \quad \text{and} \quad \Phi_N := \sum_{|k| \le N} E_k$$

for each $N \in \mathbb{N}$. Notice that $D_k 1 = 0$ for all $k \in \mathbb{Z}$ except in the case $k = 1$ with $\mathbb{R}^D$ being an initial cube. In what follows, for any $p \in [1, \infty]$ and operator T, $\|T\|_{p,p}$ stands for the *operator norm in* $L^p(\mu)$.

Lemma 6.1.1. *The following hold true:*

(a) *For all* $j, k \in \mathbb{Z}$ *and some* $\eta \in (0, \infty)$,

$$\|D_j D_k\|_{2,2} \le C 2^{-|j-k|\eta};$$

D. Yang et al., *The Hardy Space H^1 with Non-doubling Measures and Their Applications*, Lecture Notes in Mathematics 2084, DOI 10.1007/978-3-319-00825-7_6,

(b) $\sum_{k\in\mathbb{Z}} D_k = I$ *with strong convergence in* $L^2(\mu)$*;*
(c) *The series* $E_k := \sum_{j\in\mathbb{Z}} D_{k+j} D_j$ *converges strongly in* $L^2(\mu)$ *and*

$$\|E_k\|_{2,2} \le C|k|2^{-|k|\eta}$$

for all $k \in \mathbb{Z}$*;*
(d) $\Phi_N \to I$ *as* $N \to \infty$ *in the operator norm in* $L^2(\mu)$.

Proof. For simplicity we assume that all the cubes $Q_{x,k}$, $x \in \operatorname{supp}\mu$, $k \in \mathbb{Z}$, are transit cubes. Let $p \in [1,\infty]$. To prove (a), it suffices to show that there exist positive constants $C_{(p)}$ and η such that, for all $j,\ k \in \mathbb{Z}$,

$$\left\|D_j D_k\right\|_{p,p} \le C_{(p)} 2^{-|j-k|\eta}. \tag{6.1.1}$$

To this end, assume $j \ge k+2$. The kernel of the operator $D_j D_k$ is given by

$$K_{j,k}(x,y) := \int_{\mathbb{R}^D} D_j(x,z) D_k(z,y)\, d\mu(z),$$

where $x,\ y \in \mathbb{R}^D$. Since $\operatorname{supp}(D_j(x,\cdot)) \subset Q_{x,j-2}$, we have

$$|K_{j,k}(x,y)| \le \int_{Q_{x,j-2}} |D_j(x,z)[D_k(z,y) - D_k(x,y)]|\, d\mu(z).$$

By (b) of Theorem 2.4.4 (taking into account that $Q_{x,j-2} \subset Q_{x,k}$), we see that, for all $x,\ y,\ z \in \mathbb{R}^D$,

$$|D_k(z,y) - D_k(x,y)| \lesssim \frac{\ell(Q_{x,j-2})}{\ell(Q_{x,k})} \frac{1}{[\ell(Q_{x,k}) + \ell(Q_{y,k}) + |x-y|]^n}.$$

By Lemma 2.2.5, we conclude that

$$\ell(Q_{x,j-2}) \lesssim 2^{-\eta|j-k|} \ell(Q_{x,k})$$

for some $\eta \in (0,\infty)$. Therefore, for all $x,\ y \in \mathbb{R}^D$, it holds true that

$$\begin{aligned}
&|K_{j,k}(x,y)| \\
&\quad \lesssim 2^{-\eta|j-k|} \frac{1}{[\ell(Q_{x,k}) + \ell(Q_{y,k}) + |x-y|]^n} \int_{Q_{x,j-2}} |D_j(x,z)|\, d\mu(z) \\
&\quad \lesssim 2^{-\eta|j-k|} \frac{1}{[\ell(Q_{x,k}) + \ell(Q_{y,k}) + |x-y|]^n}.
\end{aligned} \tag{6.1.2}$$

Also, we know that

$$\mathrm{supp}\,(K_{j,k}(x,\cdot)) \subset Q_{x,k-3} \quad \text{and} \quad \mathrm{supp}\,(K_{j,k}(\cdot,y)) \subset Q_{y,k-3}.$$

Indeed, if $K_{j,k}(x,\cdot) \neq 0$ then there exists some z such that $D_j(x,z) \neq 0$ and $D_k(z,y) \neq 0$. Thus,

$$z \in Q_{x,j-2} \subset Q^1_{x,k-2}.$$

Observe that $y \in Q_{z,k-2}$. We see that $z \in Q^1_{y,k-2}$. Therefore, $Q^1_{y,k-2} \cap Q^1_{x,k-2} \neq \emptyset$, which implies that $Q^1_{y,k-2} \subset Q_{x,k-3}$. Thus,

$$\mathrm{supp}\,(K_{j,k}(x,\cdot)) \subset Q_{x,k-3}.$$

Similarly,

$$\mathrm{supp}\,(K_{j,k}(\cdot,y)) \subset Q_{y,k-3}.$$

Thus, we know that, for all $x \in \mathbb{R}^D$,

$$\begin{aligned}\int_{\mathbb{R}^D} |K_{j,k}(x,y)|\,d\mu(y) &\lesssim 2^{-\eta|j-k|} \int_{Q_{x,k-3}} \frac{1}{[\ell(Q_{x,k}) + |x-y|]^n}\,d\mu(y)\\ &\lesssim 2^{-\eta|j-k|}[1 + \delta(Q_{x,k}, Q_{x,k-3})]\\ &\lesssim 2^{-\eta|j-k|}. \end{aligned} \tag{6.1.3}$$

In an analogous way, we conclude that, for all $y \in \mathbb{R}^D$,

$$\int_{\mathbb{R}^D} |K_{j,k}(x,y)|\,d\mu(x) \lesssim 2^{-\eta|j-k|}. \tag{6.1.4}$$

Combining (6.1.3) and (6.1.4) and using the Schur lemma,[1] we obtain (6.1.1) for $j \geq k+2$.

On the other hand, for $k \geq j+2$, operating in a similar way, we also obtain

$$\|D_j D_k\|_{p,p} \lesssim 2^{-|j-k|\eta},$$

and, if $|j-k| \leq 1$, then we have

$$\|D_j D_k\|_{p,p} \leq \|D_j\|_{p,p}\|D_k\|_{p,p} \lesssim 1.$$

Thus, (6.1.1) holds true in any case.

[1] See [40, p.457].

If there exist stopping cubes, then the kernels of the operators S_k satisfy properties which are similar to the ones stated in Theorem 2.4.4 and some estimates as the ones above work. If $\mathbb{R}^D$ is an initial cube, then

$$\int_{\mathbb{R}^D} D_1(x,y)\,d\mu(y) \neq 0$$

in general. However, in the arguments above the equality

$$\int_{\mathbb{R}^D} D_j(x,y)\,d\mu(y) = 0 \tag{6.1.5}$$

is only used to estimate $\|D_j D_k\|_{p,p}$ in the case $j \geq k+2$. When $k \geq 1$ and $j \geq k+2$, (6.1.5) still holds true, and hence (6.1.1) is true in this case. When $k \leq 0$, then $D_k = 0$ and (6.1.1) holds true trivially. Combining these two cases, we then complete the proof of (a).

From (a) and the Cotlar–Knapp–Stein lemma,[2] we deduce that $\sum_{k\in\mathbb{Z}} D_k = I$ with strong convergence in $L^2(\mu)$ and

$$\sum_{k\in\mathbb{Z}} \|D_k f\|_{L^2(\mu)}^2 \lesssim \|f\|_{L^2(\mu)}^2 \tag{6.1.6}$$

for all $f \in L^2(\mu)$, which implies (b).

Moreover, applying the Cotlar–Knapp–Stein lemma and (a) again, we have (c).

Finally, by (c), we see that $\{\Phi_N\}_N$ is convergent in the operator norm in $L^2(\mu)$. On the other hand, from (b), we deduce that

$$I = \sum_{k\in\mathbb{Z}} \sum_{j\in\mathbb{Z}} D_{k+j} D_j$$

with strong convergence in $L^2(\mu)$. By these two facts, we obtain (d), which completes the proof of Lemma 6.1.1. □

As a corollary of Lemma 6.1.1, we have the following conclusion.

Theorem 6.1.2. *There exists a constant $C \in (1,\infty)$ such that, for all $f \in L^2(\mu)$,*

$$C^{-1} \sum_k \|D_k f\|_{L^2(\mu)}^2 \leq \|f\|_{L^2(\mu)}^2 \leq C \sum_k \|D_k f\|_{L^2(\mu)}^2. \tag{6.1.7}$$

Proof. The left inequality of (6.1.7) follows from (6.1.6). To obtain the right inequality in (6.1.7), we operate as follows. By Lemma 6.1.1(c), for N big enough,

$$\|I - \Phi_N\|_{2,2} \leq 1/2$$

[2]See [41, p. 224].

and hence Φ_N is an invertible operator on $L^2(\mu)$. This implies that

$$\|f\|_{L^2(\mu)} \lesssim \|\Phi_N f\|_{L^2(\mu)} \quad \text{for any} \quad f \in L^2(\mu).$$

Therefore, to see that the right inequality of (6.1.7) holds true, we only have to show that

$$\|\Phi_N f\|_{L^2(\mu)}^2 \lesssim \sum_k \|D_k f\|_{L^2(\mu)}^2.$$

This follows from a duality argument. Indeed, if we let

$$D_k^N := \sum_{j:\,|j-k|\le N} D_j$$

then we also have

$$\Phi_N = \sum_{k\in\mathbb{Z}} D_k^N D_k.$$

Given $f \in L^2(\mu)$, we conclude that

$$\begin{aligned}
|\langle \Phi_N f, g\rangle| &= \left|\sum_{k\in\mathbb{Z}} \langle D_k^N D_k f, g\rangle\right| \\
&= \left|\sum_{k\in\mathbb{Z}} \langle D_k f, D_k^{N*} g\rangle\right| \\
&\le \sum_{k\in\mathbb{Z}} \|D_k f\|_{L^2(\mu)} \left\|D_k^{N*} g\right\|_{L^2(\mu)} \\
&\le \left[\sum_k \|D_k f\|_{L^2(\mu)}^2\right]^{1/2} \left[\sum_{k\in\mathbb{Z}} \left\|D_k^{N*} g\right\|_{L^2(\mu)}^2\right]^{1/2}. \qquad (6.1.8)
\end{aligned}$$

From the definition of D_k^N and the left inequality of (6.1.7), we deduce that

$$\sum_{k\in\mathbb{Z}} \left\|D_k^{N*} g\right\|_{L^2(\mu)}^2 \lesssim N^2 \sum_{k\in\mathbb{Z}} \|D_k^* g\|_{L^2(\mu)}^2 \lesssim N^2 \|g\|_{L^2(\mu)}^2. \qquad (6.1.9)$$

Thus, by (6.1.8) and (6.1.9), we obtain the right inequality of (6.1.7), which completes the proof of Theorem 6.1.2. □

Lemma 6.1.3. *The operator Φ_N tends to I in the norm*

$$|||T||| := \|T\|_{2,2} + C$$

as $N \to \infty$, where C is the best constant that appears in (5.3.1) *and* (5.3.2)*. Moreover, Φ_N can be extended boundedly on $L^p(\mu)$, $p \in (1, \infty)$. For N big enough it is invertible in $L^p(\mu)$ (with N depending on p).*

Proof. Observe that $\Phi_N^* = \Phi_N$ for each $N \in \mathbb{N}$. If we can show that $I - \Phi_N$ satisfies the hypotheses of Corollary 5.4.5 such that

$$|||I - \Phi_N||| \to 0 \quad \text{as} \quad N \to \infty,$$

then Lemma 6.1.3 follows from Corollary 5.4.5. We have seen, in Lemma 6.1.1, that $\Phi_N \to I$ as $N \to \infty$ in the operator norm of $L^2(\mu)$. Thus, it only remains to show that $I - \Phi_N$ satisfies Lemma 6.1.1 and the constant C in (5.3.1) and (5.3.2) for $I - \Phi_N$ tends to 0 as $N \to \infty$.

First we deal with the inequality (5.3.1). In (6.1.2), we have shown that, if $k \ge 2$, then the kernel $K_{j+k,j}$ of $D_{j+k}D_j$ satisfies that, for all $x, y \in \operatorname{supp}\mu$,

$$|K_{j+k,j}(x,y)| \lesssim 2^{-\eta k} \frac{1}{[\ell(Q_{x,j}) + \ell(Q_{y,j}) + |x-y|]^n}. \tag{6.1.10}$$

Moreover, just below (6.1.2) we have seen that $K_{j+k,j}(x,y) = 0$ if $y \notin Q_{x,j-3}$ or $x \notin Q_{y,j-3}$. For $x, y \in \operatorname{supp}\mu$, $x \ne y$, let j_0 be the largest integer such that $y \in Q_{x,j_0}$. Since $y \notin Q_{x,j_0+h}$ for $h \in \mathbb{N}$, we obtain $K_{j+k,j}(x,y) = 0$ if $j \ge j_0 + 4$. Observe that, for $j \le j_0$,

$$|x-y| \lesssim \ell(Q_{x,j_0}) \lesssim 2^{-\eta|j-j_0|}\ell(Q_{x,j}).$$

By this and (6.1.10), we have

$$\begin{aligned}
\sum_{j\in\mathbb{Z}} |K_{j+k,j}(x,y)| &\lesssim 2^{-\eta|k|} \sum_{j=-\infty}^{j_0+3} \frac{1}{[\ell(Q_{x,j}) + |x-y|]^n} \\
&\lesssim 2^{-\eta|k|} \left\{ \sum_{j=-\infty}^{j_0} \frac{1}{[\ell(Q_{x,j})]^n} + \frac{1}{|x-y|^n} \right\} \\
&\lesssim 2^{-\eta|k|} \frac{1}{|x-y|^n}.
\end{aligned}$$

An analogous estimate can be obtained for $k \le -2$. Thus, the kernel $K_N(x,y)$ of $I - \Phi_N$ satisfies that, for all $x, y \in \operatorname{supp}\mu$,

$$|K_N(x,y)| \lesssim 2^{-\eta N} \frac{1}{|x-y|^n}. \tag{6.1.11}$$

Now we have to show that the constant $\tilde{C}$ in (5.3.2) corresponding to the kernel of $I - \Phi_N$ tends to 0 as $N \to \infty$. We now deal with the term

$$\mathrm{I}_{j,k} := |K_{j+k,j}(x,y) - K_{j+k,j}(\tilde{x},y)|,$$

assuming $k \ge N \ge 10$. Let h_0 be the largest integer such that $\tilde{x} \in Q_{x,h_0}$. Using (6.1.11), it is easy to show that

$$\sum_{k\ge N}\sum_{j\in\mathbb{Z}}\int_{Q_{x,h_0-10}\setminus B(x,2|x-\tilde{x}|)} \mathrm{I}_{j,k}\,d\mu(y) \lesssim 2^{-\eta N}. \tag{6.1.12}$$

Thus, we only have to estimate the integral

$$\int_{\mathbb{R}^D\setminus Q_{x,h_0-10}} \mathrm{I}_{j,k}\,d\mu(y).$$

Notice that

$$\mathrm{supp}\,(K_{j+k,j}(x,\cdot)) \subset Q_{x,j-3}$$

and

$$\mathrm{supp}\,(K_{j+k,j}(\tilde{x},\cdot)) \subset Q_{\tilde{x},j-3} \subset Q_{x,j-4}\bigcup Q_{x,h_0-10},$$

and hence

$$\mathrm{supp}\,\mathrm{I}_{j,k} \subset Q_{x,j-4}\bigcup Q_{x,h_0-10}.$$

Thus, we may assume $j-4 \le h_0 - 10$. Let us consider the case $j+k > h_0$, that is, $\tilde{x} \notin Q_{x,j+k}$. By (6.1.3), we see that, for N large enough,

$$\begin{aligned}
&\sum_{k\ge N}\ \sum_{\substack{j:\ j-4\le h_0-10\\ j>h_0-k}} \int_{\mathbb{R}^D} \mathrm{I}_{j,k}\,d\mu(y)\\
&\quad\lesssim \sum_{k\ge N}\ \sum_{j=h_0-k+1}^{h_0-6} \int_{\mathbb{R}^D} \left[|K_{j+k,j}(x,y)| + |K_{j+k,j}(\tilde{x},y)|\right] d\mu(y)\\
&\quad\lesssim \sum_{k\ge N}\ \sum_{j=h_0-k+1}^{h_0-6} 2^{-\eta k}\\
&\quad\lesssim 2^{-\eta N/2}.
\end{aligned} \tag{6.1.13}$$

Assume now that $j + k \le h_0$, that is, $\tilde{x} \in Q_{x,j+k}$ (and $k \ge N \ge 10$ too). Observe that

$$\begin{aligned} K_{j+k,j}(x,y) &- K_{j+k,j}(\tilde{x},y) \\ &= \int_{\mathbb{R}^D} [D_{j+k}(x,z) - D_{j+k}(\tilde{x},z)] D_j(z,y)\, d\mu(z) \\ &= \int_{\mathbb{R}^D} [D_{j+k}(x,z) - D_{j+k}(\tilde{x},z)][D_j(z,y) - D_j(x,y)]\, d\mu(z). \end{aligned} \tag{6.1.14}$$

It is easily to show that the integrand above is null unless $z \in Q_{x,j+k-3}$. Since $\tilde{x} \in Q_{x,j+k}$, it follows that

$$|D_{j+k}(x,z) - D_{j+k}(\tilde{x},z)| \lesssim \frac{|x-\tilde{x}|}{\ell(Q_{x,j+k})} \frac{1}{[\ell(Q_{x,j+k}) + |x-z|]^n}.$$

Also, for $z \in Q_{x,j+k-3} \subset Q_{x,j}$, we know that

$$|D_j(z,y) - D_j(x,y)| \lesssim \frac{|x-z|}{\ell(Q_{x,j})} \frac{1}{[\ell(Q_{x,j}) + |x-y|]^n}. \tag{6.1.15}$$

Therefore,

$$\begin{aligned} \mathrm{I}_{j,k} &\lesssim \frac{|x-\tilde{x}|\ell(Q_{x,j+k-3})}{\ell(Q_{x,j})\ell(Q_{x,j+k})[\ell(Q_{x,j}) + |x-y|]^n} \\ &\quad \times \int_{Q_{x,j+k-3}} \frac{1}{[\ell(Q_{x,j+k}) + |x-z|]^n}\, d\mu(z) \\ &\lesssim \frac{|x-\tilde{x}|\ell(Q_{x,j+k-3})}{\ell(Q_{x,j})\ell(Q_{x,j+k})[\ell(Q_{x,j}) + |x-y|]^n}. \end{aligned}$$

Using

$$\ell(Q_{x,j+k-3})/\ell(Q_{x,j}) \lesssim 2^{-\eta k},$$

we obtain

$$\begin{aligned} \int_{\mathbb{R}^D} \mathrm{I}_{j,k}\, d\mu(y) &\lesssim 2^{-\eta k} \frac{|x-\tilde{x}|}{\ell(Q_{x,j+k})} \int_{Q_{x,j-4}} \frac{1}{[\ell(Q_{x,j}) + |x-y|]^n}\, d\mu(y) \\ &\lesssim 2^{-\eta k} \frac{|x-\tilde{x}|}{\ell(Q_{x,j+k})}. \end{aligned}$$

Thus, we see that

$$\begin{aligned}
&\sum_{k\geq N}\sum_{j:\,j+k\leq h_0}\int_{\mathbb{R}^D}|K_{j+k,j}(x,y)-K_{j+k,j}(\tilde{x},y)|\,d\mu(y)\\
&\quad\lesssim\sum_{k\geq N}2^{-\eta k}\sum_{j:\,j+k\leq h_0}\frac{\ell(Q_{x,h_0})}{\ell(Q_{x,j+k})}\\
&\quad\lesssim\sum_{k\geq N}2^{-\eta k}\\
&\quad\lesssim 2^{-\eta N}.
\end{aligned}$$

Let us consider the term

$$\mathrm{J}_{j,k}:=|K_{j+k,j}(y,x)-K_{j+k,j}(y,\tilde{x})|.$$

As (6.1.12), we have

$$\sum_{k\geq N}\sum_{j\in\mathbb{Z}}\int_{Q_{x,h_0-10}\setminus B(x,2|x-\tilde{x}|)}\mathrm{J}_{j,k}\,d\mu(y)\lesssim 2^{-\eta N},$$

and we only have to consider the integral

$$\int_{\mathbb{R}^D\setminus Q_{x,h_0-10}}\mathrm{J}_{j,k}\,d\mu(y).$$

Moreover, it is easily seen that we also have

$$\operatorname{supp}\mathrm{J}_{j,k}\subset\left(Q_{x,j-4}\bigcup Q_{x,h_0-10}\right)$$

in this case.Thus, we may assume $j-4\leq h_0$ again. Operating as (6.1.13), by (6.1.4), we see that

$$\sum_{k\geq N}\sum_{\substack{j:\,j-4\leq h_0-10\\ j>h_0-k}}\int_{\mathbb{R}^D}\mathrm{J}_{j,k}\,d\mu(y)\lesssim 2^{-\eta N/2}.$$

Suppose that $j+k\leq h_0$, that is, $\tilde{x}\in Q_{x,j+k}$. We have

$$\mathrm{J}_{j,k}\leq\int_{\mathbb{R}^D}|D_{j+k}(y,z)||D_j(z,x)-D_j(z,\tilde{x})|\,d\mu(z).$$

Since $\tilde{x} \in Q_{x,h_0} \subset Q_{x,j}$, we conclude that

$$|D_j(z,x) - D_j(z,\tilde{x})| \lesssim \frac{|x-\tilde{x}|}{\ell(Q_{x,j})} \frac{1}{[\ell(Q_{x,j}) + |x-z|]^n}.$$

Thus,

$$\begin{aligned}
\mathrm{J}_{j,k} &\lesssim \frac{|x-\tilde{x}|}{\ell(Q_{x,j})} \int_{Q_{x,j-4}} \frac{|D_{j+k}(y,z)|}{[\ell(Q_{x,j}) + |x-z|]^n} \, d\mu(z) \\
&\lesssim \frac{|x-\tilde{x}|}{\ell(Q_{x,j})} \left[\int_{\substack{z \in Q_{x,j-4} \\ |y-z| \le |x-y|/2}} \frac{|D_{j+k}(y,z)|}{[\ell(Q_{x,j}) + |x-z|]^n} \, d\mu(z) + \int_{\substack{z \in Q_{x,j-4} \\ |y-z| > |x-y|/2}} \cdots \right] \\
&=: \mathrm{J}^1_{j,k} + \mathrm{J}^2_{j,k}.
\end{aligned}$$

Let us estimate $\mathrm{J}^1_{j,k}$ by

$$\begin{aligned}
\mathrm{J}^1_{j,k} &\lesssim \frac{|x-\tilde{x}|}{\ell(Q_{x,j})} \int_{\mathbb{R}^D} \frac{|D_{j+k}(y,z)|}{[\ell(Q_{x,j}) + |x-y|]^n} \, d\mu(z) \\
&\lesssim \frac{|x-\tilde{x}|}{\ell(Q_{x,j})} \frac{1}{[\ell(Q_{x,j}) + |x-y|]^n}.
\end{aligned}$$

We consider $\mathrm{J}^2_{j,k}$ now. On the one hand, we have

$$\begin{aligned}
\mathrm{J}^2_{j,k} &\lesssim \frac{|x-\tilde{x}|}{\ell(Q_{x,j})} \int_{Q_{x,j-4}} \frac{1}{|x-y|^n} \frac{1}{[\ell(Q_{x,j}) + |x-z|]^n} \, d\mu(z) \\
&\lesssim \frac{|x-\tilde{x}|}{\ell(Q_{x,j})|x-y|^n}.
\end{aligned}$$

On the other hand, we see that

$$\mathrm{J}^2_{j,k} \lesssim \frac{|x-\tilde{x}|}{\ell(Q_{x,j})} \int_{\mathbb{R}^D} \frac{|D_{j+k}(y,z)|}{[\ell(Q_{x,j})]^n} \, d\mu(z) \lesssim \frac{|x-\tilde{x}|}{[\ell(Q_{x,j})]^{n+1}}.$$

Thus, we conclude that

$$\mathrm{J}^2_{j,k} \lesssim \frac{|x-\tilde{x}|}{\ell(Q_{x,j})} \frac{1}{[\ell(Q_{x,j}) + |x-y|]^n}$$

in any case. Therefore,

$$\int_{\mathbb{R}^D} \mathrm{J}_{j,k} d\mu(y) \lesssim \frac{|x-\tilde{x}|}{\ell(Q_{x,j})} \int_{Q_{x,j-4}} \frac{1}{[\ell(Q_{x,j}) + |x-y|]^n} \, d\mu(y) \lesssim \frac{\ell(Q_{x,h_0})}{\ell(Q_{x,j})}$$

and hence

$$\begin{aligned}
&\sum_{k\geq N}\sum_{j:\,j+k\leq h_0}\int_{\mathbb{R}^D}|K_{j+k,j}(y,x)-K_{j+k,j}(y,\tilde{x})|\,d\mu(y)\\
&\quad\lesssim\sum_{k\geq N}\sum_{j:\,j+k\leq h_0}\frac{\ell(Q_{x,h_0})}{\ell(Q_{x,j})}\\
&\quad\lesssim\sum_{k\geq N}\sum_{j:\,j+k\leq h_0}\frac{[\ell(Q_{x,j+k})]^{1/2}}{[\ell(Q_{x,j})]^{1/2}}\frac{[\ell(Q_{x,h_0})]^{1/2}}{[\ell(Q_{x,j})]^{1/2}}\\
&\quad\lesssim\sum_{k\geq N}2^{-\eta k/2}\\
&\quad\lesssim 2^{-\eta N/2}.
\end{aligned}$$

When k is negative ($k\leq -N$), we have analogous estimates. Consequently, the kernel of $I-\Phi_N$ satisfies Hörmander's condition (5.3.2) with the positive constant $C\lesssim 2^{-\eta N/2}$, and the proof of Lemma 6.1.3 is completed. □

Corollary 6.1.4. *Let $p\in(1,\infty)$. There exists a constant $\tilde{C}\in(1,\infty)$ such that, if $f\in L^p(\mu)$, then*

$$\tilde{C}^{-1}\,\|\dot{g}(f)\|_{L^p(\mu)}\leq\|f\|_{L^p(\mu)}\leq\tilde{C}\,\|\dot{g}(f)\|_{L^p(\mu)}\,. \tag{6.1.16}$$

Proof. The right inequality follows from the left one (with p' instead of p). Indeed, by an argument similar to that used in the case $p=2$ as in (6.1.8), we find that

$$\|\Phi_N f\|_{L^p(\mu)}\lesssim\|\dot{g}(f)\|_{L^p(\mu)}\,.$$

On the other hand, in Lemma 6.1.3, we see that Φ_N is bounded and invertible in $L^p(\mu)$ and hence

$$\|f\|_{L^p(\mu)}\lesssim\|\Phi_N f\|_{L^p(\mu)}.$$

These two facts imply the right hand side of (6.1.16)

The left inequality in (6.1.16) is proved by using techniques of vector-valued Calderón–Zygmund operators. Let us denote by $L^p(\ell^2,\mu)$ the *Banach space of sequences of functions* $\{g_k\}_{k\in\mathbb{Z}}$, $g_k\in L^1_{\rm loc}(\mu)$, *such that*

$$\left\{\int_{\mathbb{R}^D}\left[\sum_k|g_k(x)|^2\right]^{p/2}d\mu(x)\right\}^{1/p}<\infty.$$

We consider the operator $\tilde{D} : L^p(\mu) \to L^p(\ell^2, \mu)$ given by $\tilde{D} f := \{D_k f\}_{k\in\mathbb{Z}}$. By Theorem 6.1.2, $\tilde{D}$ is bounded from $L^2(\mu)$ into $L^2(\ell^2, \mu)$. If we can prove that the kernel $\tilde{D}(x, y) := \{D_k(x, y)\}_k$ of $\tilde{D}$ satisfies that

(i)

$$\left\|\tilde{D}(x, y)\right\|_{\ell^2} \lesssim 1/|x-y|^n \text{ for all } x \neq y,$$

(ii)

$$\int_{|x-y|\geq 2|x-\tilde{x}|} \left[\left\|\tilde{D}(x, y)-\tilde{D}(\tilde{x}, y)\right\|_{\ell^2} + \left\|\tilde{D}(y, x)-\tilde{D}(y, \tilde{x})\right\|_{\ell^2}\right] d\mu(y) \lesssim 1,$$ [3]

then $\tilde{D}$ is bounded from $L^p(\mu)$ into $L^p(\ell^2, \mu)$, $p \in (1, \infty)$, because $\tilde{D}$ is a vector-valued Calderón–Zygmund operator. Thus, we only have to show that these conditions are satisfied. Let us show that (i) holds true. Given $x, y \in \operatorname{supp} \mu$, $x \neq y$, let $j \in \mathbb{Z}$ be such that $y \in Q_{x,j} \setminus Q_{x,j+1}$. Since $\operatorname{supp}(d_k(x, \cdot)) \subset Q_{x,k-2}$, we have

$$\begin{aligned}
\sum_{k\in\mathbb{Z}} |D_k(x, y)|^2 &\leq \sum_{k\leq j+2} \frac{1}{[\ell(Q_{x,k}) + |x-y|]^{2n}} \\
&\lesssim \frac{1}{|x-y|^{2n}} + \sum_{k\leq j} \frac{1}{[\ell(Q_{x,k})]^{2n}} \\
&\lesssim \frac{1}{|x-y|^{2n}}.
\end{aligned}$$

Thus, (i) holds true.

Now we show that the condition (ii) is also satisfied. Since $\tilde{D}(x, y) = \tilde{D}(y, x)$, we only have to deal with the term $\|\tilde{D}(x, y) - \tilde{D}(\tilde{x}, y)\|_{\ell^2}$. Let $h \in \mathbb{Z}$ be such that $\tilde{x} \in Q_{x,h} \setminus Q_{x,h+1}$ and suppose that $y \in Q_{x,j} \setminus Q_{x,j+1}$ for some $j \leq h - 10$. Notice that, if $k > j + 4$, then

$$D_k(x, y) - D_k(\tilde{x}, y) = 0. \tag{6.1.17}$$

Indeed, we have

$$\operatorname{supp}(D_k(x, \cdot) - D_k(\tilde{x}, \cdot)) \subset Q_{x,k-2} \bigcup Q_{\tilde{x},k-2}.$$

If $k \geq h$, then we see that

$$Q_{x,k-2} \bigcup Q_{\tilde{x},k-2} \subset Q_{x,h-3} \subset Q_{x,j+1}$$

[3] See [34, Theorem 2.7].

and, if $j+4<k<h$, we then have

$$Q_{x,k-2}\bigcup Q_{\tilde{x},k-2}\subset Q_{x,k-3}\subset Q_{x,j+1}.$$

This implies (6.1.17).

Assuming $k\le j+4$, since $\tilde{x}\in Q_{x,h}$ with $h>k$, we find that

$$|D_k(x,y)-D_k(\tilde{x},y)|\lesssim\frac{|x-\tilde{x}|}{\ell(Q_{x,k})}\frac{1}{[\ell(Q_{x,k})+|x-y|]^n}.$$

This, together with (6.1.17), implies that

$$\begin{aligned}\sum_{k\in\mathbb{Z}}|D_k(x,y)-D_k(\tilde{x},y)|^2&\lesssim\sum_{k\le j+4}\left\{\frac{|x-\tilde{x}|}{\ell(Q_{x,k})}\frac{1}{[\ell(Q_{x,k})+|x-y|]^n}\right\}^2\\&\lesssim\left[\frac{|x-\tilde{x}|}{\ell(Q_{x,j+4})|x-y|^n}\right]^2.\end{aligned}$$

From this, the choice of $h\in\mathbb{Z}$ and (0.0.1), we deduce that

$$\begin{aligned}&\int_{|x-y|\ge2|x-\tilde{x}|}\|D(x,y)-D(\tilde{x},y)\|_{\ell^2}\,d\mu(y)\\&\quad\lesssim\int_{Q_{x,h-10}\setminus B(x,2|x-\tilde{x}|)}\frac{1}{|x-y|^n}\,d\mu(y)\\&\qquad+\sum_{i=10}^{\infty}\int_{Q_{x,h-i-1}\setminus Q_{x,h-i}}\frac{|x-\tilde{x}|}{\ell(Q_{x,h-i+4})|x-y|^n}\,d\mu(y)\\&\quad\lesssim1+\sum_{i=10}^{\infty}\frac{|x-\tilde{x}|}{\ell(Q_{x,h-i+4})}\\&\quad\lesssim1,\end{aligned}$$

which completes the proof of Corollary 6.1.4. □

Now we show the boundedness of $\dot{g}(f)$ from $H^1(\mu)$ to $L^1(\mu)$.

Theorem 6.1.5. *There exists a positive constant C such that, for all $f\in H^1(\mu)$,*

$$\|\dot{g}(f)\|_{L^1(\mu)}\le C\|f\|_{H^1(\mu)}.$$

Proof. Let b be any ∞-atomic block as in Definition 3.2.1. To be precise, assume that $b:=\lambda_1a_1+\lambda_2a_2$. By the Fatou lemma, to prove Theorem 6.1.5, it suffices to show that $\dot{g}(b)$ is in $L^1(\mu)$ and

$$\|\dot{g}(b)\|_{L^1(\mu)}\lesssim|\lambda_1|+|\lambda_2|.$$

Assume that $\operatorname{supp}(b) \subset R$ and $\operatorname{supp}(a_j) \subset Q_j$ for $j \in \{1, 2\}$ as in Definition 3.2.1. Since $\dot{g}$ is sublinear, we write

$$\begin{aligned}\int_{\mathbb{R}^D} \dot{g}(b)(x)\,d\mu(x) &= \int_{4R} \dot{g}(b)(x)\,d\mu(x) + \int_{\mathbb{R}^D\setminus 4R} \cdots \\ &\le \sum_{j=1}^{2} |\lambda_j| \int_{2Q_j} \dot{g}(a_j)(x)\,d\mu(x) + \sum_{j=1}^{2} |\lambda_j| \int_{4R\setminus 2Q_j} \cdots \\ &\quad + \int_{\mathbb{R}^D\setminus 4R} \dot{g}(b)(x)\,d\mu(x) \\ &=: \mathrm{I}_1 + \mathrm{I}_2 + \mathrm{I}_3.\end{aligned}$$

Recalling that $\dot{g}$ is bounded on $L^2(\mu)$, by the Hölder inequality and (3.2.1), we then see that

$$\begin{aligned}\mathrm{I}_1 &\le \sum_{j=1}^{2} |\lambda_j| \left\{\int_{2Q_j} [\dot{g}(a_j)(x)]^2\,d\mu(x)\right\}^{\frac{1}{2}} [\mu(2Q_j)]^{\frac{1}{2}} \\ &\lesssim \sum_{j=1}^{2} |\lambda_j| \left\{\int_{Q_j} [a_j(x)]^2\,d\mu(x)\right\}^{\frac{1}{2}} [\mu(2Q_j)]^{\frac{1}{2}} \\ &\lesssim \sum_{j=1}^{2} |\lambda_j| \|a_j\|_{L^\infty(\mu)} \mu(2Q_j) \\ &\le \sum_{j=1}^{2} |\lambda_j|,\end{aligned}$$

which is a desired estimate.

For $j \in \{1, 2\}$, let x_j be the center of Q_j. Notice that, for $x \notin 2Q_j$ and $y \in Q_j$, $|x - y| \sim |x - x_j|$. From this fact, the Hölder inequality, (6.1.17) and (3.2.1), it follows that

$$\begin{aligned}\dot{g}(a_j)(x) &\le \left[\int_{Q_j} \sum_{k=-\infty}^{\infty} |D_k(x, y)|^2 |a_j(y)|^2\,d\mu(y)\right]^{\frac{1}{2}} [\mu(Q_j)]^{\frac{1}{2}} \\ &\lesssim \left[\int_{Q_j} \frac{|a_j(y)|^2}{|x - y|^{2n}}\,d\mu(y)\right]^{\frac{1}{2}} [\mu(Q_j)]^{\frac{1}{2}}. \\ &\lesssim \frac{\|a_j\|_{L^\infty(\mu)}}{|x - x_j|^n} \mu(Q_j) \\ &\lesssim \frac{1}{|x - x_j|^n} \frac{1}{1 + \delta(Q_j, R)}.\end{aligned}$$

By Lemma 2.1.3(d),

$$\delta(2Q_j, 4R) \lesssim 1 + \delta(Q_j, R),$$

which in turn implies that

$$\mathrm{I}_2 \lesssim \sum_{j=1}^{2} \frac{|\lambda_j|}{1 + \delta(Q_j, R)} \delta(2Q_j, 4R) \lesssim \sum_{j=1}^{2} |\lambda_j|.$$

We now estimate I_3. Let $x_0 \in \operatorname{supp} \mu \cap R$. By

$$\int_{\mathbb{R}^D} b(x)\, d\mu(x) = 0,$$

the Minkowski inequality and the Hölder inequality, for $x \notin 4R$, we see that

$$\begin{aligned}
&\left\{\sum_{k=-\infty}^{\infty} |D_k b(x)|^2\right\}^{1/2} \\
&= \left\{\sum_{k=-\infty}^{\infty} \left|\int_R [D_k(x, y) - D_k(x, x_0)] b(y)\, d\mu(y)\right|^2\right\}^{1/2} \\
&\le \left\{\sum_{k=-\infty}^{\infty} \left[\sum_{j=1}^{2} |\lambda_j| \int_{Q_j} |D_k(x, y) - D_k(x, x_0)||a_j(y)|\, d\mu(y)\right]^2\right\}^{1/2} \\
&\lesssim \sum_{j=1}^{2} |\lambda_j| \left[\mu(Q_j)\right]^{1/2} \\
&\quad \times \left\{\int_{Q_j} \left[\sum_{k=-\infty}^{\infty} |D_k(x, y) - D_k(x, x_0)|^2\right] |a_j(y)|^2\, d\mu(y)\right\}^{1/2}.
\end{aligned}$$

Therefore, the proof of Theorem 6.1.5 is reduced to showing that

$$\begin{aligned}
&\int_{\mathbb{R}^D \setminus 4R} \left\{\int_{Q_j} \left[\sum_{k=-\infty}^{\infty} |D_k(x, y) - D_k(x, x_0)|^2\right] |a_j(y)|^2\, d\mu(y)\right\}^{\frac{1}{2}} d\mu(x) \\
&\lesssim [\mu(2Q_j)]^{-1/2}.
\end{aligned}$$

We now claim that, for any $y \in Q_j$, integer $i \ge 3$ and $k \ge H_R^{x_0} - i + 4$, it holds true that

$$\operatorname{supp}\left(D_k(\cdot, y) - D_k(\cdot, x_0)\right) \subset Q_{x_0, H_R^{x_0} - i + 1}. \tag{6.1.18}$$

Indeed, by (c) of Theorem 2.4.4 and the fact that $\{Q_{x,k}\}_k$ is decreasing, we see that

$$\operatorname{supp}(D_k(\cdot, y) - D_k(\cdot, x_0)) \subset Q_{y,k-2} \bigcup Q_{x_0,k-2} \subset Q_{y,\, H_R^{x_0}-i+2} \bigcup Q_{x_0,\, H_R^{x_0}-i+2}.$$

Since $i \geq 3$, it follows that $y \in Q_j$, which, together with the decreasing property of $\{Q_{x_0,k}\}_k$, implies that $y \in Q_{x_0,\, H_R^{x_0}-i+2}$. By this, combined with Lemma 2.3.6, we know that

$$Q_{y,\, H_R^{x_0}-i+2} \subset Q_{x_0,\, H_R^{x_0}-i+1}.$$

Thus, (6.1.18) holds true.

Observe that, for any $y \in Q_j$, we have $y \in Q_{x_0,k}$ for $k \leq H_R^{x_0} - i + 3$. Then the symmetry of S_k and Lemma 2.4.4(e) imply that

$$|D_k(x, y) - D_k(x, x_0)| \lesssim \frac{|x_0 - y|}{\ell(Q_{x_0,k})} \frac{1}{[\ell(Q_{x_0,k}) + |x - x_0|]^n}. \tag{6.1.19}$$

On the other hand, since

$$\ell(Q_{x_0,\, H_R^{x_0}}) \leq \frac{1}{10}\ell(Q_{x_0,\, H_R^{x_0}-1}),$$

we then have $4R \subset Q_{x_0,\, H_R^{x_0}-1}$ and

$$\mathbb{R}^D \setminus (4R) = \left(Q_{x_0,\, H_R^{x_0}-2} \setminus (4R)\right) \bigcup \bigcup_{i=3}^{\infty} \left(Q_{x_0,\, H_R^{x_0}-i} \setminus Q_{x_0,\, H_R^{x_0}-i+1}\right).$$

Suppose that

$$x \in Q_{x_0,\, H_R^{x_0}-i} \setminus Q_{x_0,\, H_R^{x_0}-i+1}$$

for $i \geq 3$. Then (6.1.18) and (6.1.19), along with Lemma 2.2.5, imply that, for any $y \in Q_j$,

$$\begin{aligned}
\sum_{k=-\infty}^{\infty} |D_k(x, x_0) - D_k(x, y)|^2 &= \sum_{k=-\infty}^{H_R^{x_0}-i+3} |D_k(x, x_0) - D_k(x, y)|^2 \\
&\lesssim \sum_{k=-\infty}^{H_R^{x_0}-i+3} \frac{|x_0 - y|^2}{[\ell(Q_{x_0,k})]^2} \frac{1}{[\ell(Q_{x_0,k}) + |x - x_0|]^{2n}}
\end{aligned}$$

$$\lesssim \sum_{k=-\infty}^{H_R^{x_0}-i+3} \frac{[\ell(R)]^2}{[\ell(Q_{x_0,k})]^2} \frac{1}{|x-x_0|^{2n}}$$
$$\lesssim \frac{[\ell(R)]^2}{|x-x_0|^{2n}} \frac{1}{[\ell(Q_{x_0,H_R^{x_0}-i+3})]^2}.$$

Notice that, for any $k \in \mathbb{Z}$ and $x \in \operatorname{supp} \mu$,

$$\delta(Q_{x,k}, Q_{x,k-1}) \lesssim 1. \tag{6.1.20}$$

As a consequence, another application of (3.2.1), together with $R \subset Q_{x_0,H_R^{x_0}}$, shows that

$$\sum_{i=-\infty}^{H_R^{x_0}-3} \int_{Q_{x_0,i} \setminus Q_{x_0,i+1}} \left\{ \int_{Q_j} \left[\sum_{k=-\infty}^{\infty} |D_k(x,y) - D_k(x,x_0)|^2 \right] \right.$$
$$\left. \times |a_j(y)|^2 \, d\mu(y) \right\}^{\frac{1}{2}} d\mu(x)$$
$$\lesssim \sum_{i=-\infty}^{H_R^{x_0}-3} \int_{Q_{x_0,i} \setminus Q_{x_0,i+1}} \left\{ \int_{Q_j} \frac{[\ell(R)]^2}{|x-x_0|^{2n}} \frac{|a_j(y)|^2}{[\ell(Q_{x_0,i+3})]^2} \, d\mu(y) \right\}^{\frac{1}{2}} d\mu(x)$$
$$\lesssim \sum_{i=-\infty}^{H_R^{x_0}-3} \frac{\ell(R)}{\ell(Q_{x_0,i+3})} \|a_j\|_{L^\infty(\mu)} [\mu(Q_j)]^{1/2} \delta(Q_{x_0,i+1}, Q_{x_0,i})$$
$$\lesssim [\mu(2Q_j)]^{-1/2}.$$

On the other hand, since $Q_{x_0,H_R^{x_0}+2} \subset 4R$, it follows, from (3.2.1), (6.1.17) and the fact that for any $x \notin 4R$ and $y \in R$, $|x-x_0| \sim |x-y|$, that

$$\int_{Q_{x_0,H_R^{x_0}-2} \setminus 4R} \left\{ \int_{Q_j} \left[\sum_k |D_k(x,y) - D_k(x,x_0)|^2 \right] |a_j(y)|^2 \, d\mu(y) \right\}^{\frac{1}{2}} d\mu(x)$$
$$\lesssim \int_{Q_{x_0,H_R^{x_0}-2} \setminus 4R} \left[\int_{Q_j} \frac{|a_j(y)|^2}{|x-x_0|^{2n}} \, d\mu(y) \right]^{\frac{1}{2}} d\mu(x)$$
$$\lesssim \sum_{i=H_R^{x_0}-2}^{H_R^{x_0}+1} \|a_j\|_{L^\infty(\mu)} [\mu(Q_j)]^{1/2} \delta(Q_{x_0,i+1}, Q_{x_0,i})$$
$$\lesssim [\mu(2Q_j)]^{-1/2}.$$

Therefore,

$$\mathrm{I}_3 \lesssim \sum_{j=1}^{2} |\lambda_j|,$$

which completes the proof of Theorem 6.1.5. □

Let $g(f)$ be the *inhomogeneous Littlewood–Paley g-function* defined by

$$g(f) := \left[\sum_{k=2}^{\infty} |D_k f(x)|^2\right]^{1/2} + |S_1 f(x)|.$$

We then have the following conclusion.

Theorem 6.1.6. *There exists a positive constant C such that, for all $f \in h^{1,\infty}_{\mathrm{atb}}(\mu)$,*

$$\|g(f)\|_{L^1(\mu)} \leq C\|f\|_{h^{1,\infty}_{\mathrm{atb}}(\mu)}.$$

Proof. By the Fatou lemma, to prove Theorem 6.1.6, it suffices to show that, for any ∞-atomic block or ∞-block $b := \sum_{j=1}^{2} \lambda_j a_j$ as in Definition 3.2.1 or Definition 4.1.3, we have

$$\|g(b)\|_{L^1(\mu)} \lesssim \sum_{j=1}^{2} |\lambda_j|.$$

For any $x \in \operatorname{supp} \mu$,

$$g(b)(x) \leq |S_1(b)(x)| + \left\{\sum_{k=2}^{\infty} |D_k(b)(x)|^2\right\}^{\frac{1}{2}}.$$

By this and (4.3.7), we see that the proof of Theorem 6.1.6 is reduced to showing

$$\int_{\mathbb{R}^D} \left\{\sum_{k=2}^{\infty} |D_k(b)(x)|^2\right\}^{\frac{1}{2}} d\mu(x) \lesssim \sum_{j=1}^{2} |\lambda_j|. \tag{6.1.21}$$

Assume that b is an ∞-atomic block with $\operatorname{supp} b \subset R \notin \mathcal{D}$. By an argument similar to that used in the proof of Theorem 6.1.5, we obtain the estimate (6.1.21).

If b is an ∞-block with $\operatorname{supp} b \subset R \in \mathcal{D}$, we write

$$\begin{aligned}
&\int_{\mathbb{R}^D} \left\{ \sum_{k=2}^{\infty} |D_k(b)(x)|^2 \right\}^{\frac{1}{2}} d\mu(x) \\
&= \int_{4R} \left\{ \sum_{k=2}^{\infty} |D_k(b)(x)|^2 \right\}^{\frac{1}{2}} d\mu(x) + \int_{\mathbb{R}^D \setminus 4R} \cdots \\
&=: \mathrm{I} + \mathrm{II}.
\end{aligned}$$

Using the boundedness of the g-function $g(f)$ in $L^2(\mu)$ and an argument similar to the estimates for I_1 and I_2 in the proof of Theorem 6.1.5 again, we also obtain

$$\mathrm{I} \lesssim \sum_{j=1}^{2} |\lambda_j|.$$

To estimate II, choose any point $x_0 \in R \cap \operatorname{supp} \mu$. From the Hölder inequality, (6.1.17) and (3.2.1), it follows that, for $j \in \{1, 2\}$ and any $x \in Q_{x_0, H_R^{x_0}-2} \setminus (4R)$,

$$\begin{aligned}
\left\{ \sum_{k=2}^{\infty} |D_k(a_j)(x)|^2 \right\}^{\frac{1}{2}} &\le \left[\int_{Q_j} \sum_{k=2}^{\infty} |D_k(x,y)|^2 |a_j(y)|^2 \, d\mu(y) \right]^{\frac{1}{2}} [\mu(Q_j)]^{\frac{1}{2}} \\
&\lesssim \left[\int_{Q_j} \frac{|a_j(y)|^2}{|x-y|^{2n}} \, d\mu(y) \right]^{\frac{1}{2}} [\mu(Q_j)]^{\frac{1}{2}} \\
&\lesssim \frac{1}{|x-x_0|^n},
\end{aligned}$$

where x_j is the center of Q_j and, in the last step, we used the fact that

$$|x - x_j| \sim |x - x_0|$$

for any $x \in Q_{x_0, H_R^{x_0}-2} \setminus (4R)$. On the other hand, since $R \in \mathcal{D}$, by Lemma 4.1.2(d), we have $H_R^{x_0} \le 1$. Thus, for any $k \ge 2$ and $y \in R \cap \operatorname{supp} \mu$, by (c) in Theorem 2.4.4 and Lemma 2.3.6, we have

$$Q_{y,k-2} \subset Q_{y, H_R^{x_0}-1} \subset Q_{x_0, H_R^{x_0}-2}$$

and hence

$$\operatorname{supp}(D_k(b)) \subset Q_{x_0, H_R^{x_0}-2}.$$

Therefore, by Lemma 4.1.2(a), we find that

$$\begin{aligned}
\mathrm{II} &= \int_{Q_{x_0, H_R^{x_0}-2}\setminus 4R} \left\{\sum_{k=2}^{\infty} |D_k(b)(x)|^2\right\}^{\frac{1}{2}} d\mu(x) \\
&\lesssim \sum_{j=1}^{2} |\lambda_j| \int_{Q_{x_0, H_R^{x_0}-2}\setminus Q_{x_0, H_R^{x_0}+1}} \frac{1}{|x-x_0|^n} d\mu(x) \\
&\lesssim \sum_{j=1}^{2} |\lambda_j|,
\end{aligned}$$

which completes the proof of Theorem 6.1.6. □

Let $S := \{S_k\}_{k\in\mathbb{Z}}$ be an approximation of the identity as in Sect. 2.4. We then consider the following maximal operators: for any locally integrable function f and $x \in \mathbb{R}^D$, define

$$\dot{\mathcal{M}}_S(f)(x) := \sup_{k\in\mathbb{Z}} |S_k(f)(x)| \tag{6.1.22}$$

and

$$\mathcal{M}_S(f)(x) := \sup_{k\in\mathbb{N}} |S_k(f)(x)|. \tag{6.1.23}$$

Observe that

$$\mathcal{M}_S(f)(x) \le \dot{\mathcal{M}}_S(f)(x) \lesssim \mathcal{M}_{(2)} f(x)$$

for all $x \in \operatorname{supp} \mu$. Then we have the following lemma.

Lemma 6.1.7. *Let $p \in (1, \infty]$. Then there exists a positive constant $C_{(p)}$ such that, for all $f \in L^p(\mu)$,*

$$\|\mathcal{M}_S(f)\|_{L^p(\mu)} \le \left\|\dot{\mathcal{M}}_S(f)\right\|_{L^p(\mu)} \le C_{(p)} \|f\|_{L^p(\mu)}$$

and there exists a positive constant C such that, for all $f \in L^1(\mu)$ and all $\lambda \in (0,\infty)$,

$$\begin{aligned}
\mu(\{x \in \mathbb{R}^D : \mathcal{M}_S(f)(x) > \lambda\}) &\le \mu\left(\{x \in \mathbb{R}^D : \dot{\mathcal{M}}_S(f)(x) > \lambda\}\right) \\
&\le \frac{C}{\lambda} \|f\|_{L^1(\mu)}.
\end{aligned}$$

The following result further shows that $\dot{\mathcal{M}}_S$ is bounded from $H^1(\mu)$ to $L^1(\mu)$.

Theorem 6.1.8. *There exists a nonnegative constant C such that, for all $f \in H^1(\mu)$,*

$$\|\dot{\mathcal{M}}_S(f)\|_{L^1(\mu)} \le C\|f\|_{H^1(\mu)}.$$

Proof. Let

$$b := \lambda_1 a_1 + \lambda_2 a_2$$

be any ∞-atomic block as in Definition 3.2.1. By the Fatou lemma, to prove Theorem 6.1.8, it suffices to show that

$$\|\dot{\mathcal{M}}_S(b)\|_{L^1(\mu)} \lesssim |\lambda_1| + |\lambda_2|. \tag{6.1.24}$$

Since $\dot{\mathcal{M}}_S$ is sublinear, we write

$$\begin{aligned}
&\int_{\mathbb{R}^D} \dot{\mathcal{M}}_S(b)(x)\,d\mu(x)\\
&= \int_{4R} \dot{\mathcal{M}}_S(b)(x)\,d\mu(x) + \int_{\mathbb{R}^D\setminus 4R} \cdots\\
&\le \sum_{j=1}^2 |\lambda_j| \int_{2Q_j} \dot{\mathcal{M}}_S(a_j)(x)\,d\mu(x) + \sum_{j=1}^2 |\lambda_j| \int_{4R\setminus 2Q_j} \cdots + \sum_{j=1}^2 |\lambda_j| \int_{\mathbb{R}^D\setminus 4R} \cdots\\
&=: \mathrm{I}_1 + \mathrm{I}_2 + \mathrm{I}_3.
\end{aligned}$$

From the Hölder inequality, Lemma 6.1.7, and (3.2.1), it then follows that

$$\begin{aligned}
\mathrm{I}_1 &\le \sum_{j=1}^2 |\lambda_j| \left\{\int_{2Q_j} [\dot{\mathcal{M}}_S(a_j)(x)]^2\,d\mu(x)\right\}^{\frac{1}{2}} [\mu(2Q_j)]^{\frac{1}{2}}\\
&\lesssim \sum_{j=1}^2 |\lambda_j| \left\{\int_{Q_j} [a_j(x)]^2\,d\mu(x)\right\}^{\frac{1}{2}} [\mu(2Q_j)]^{\frac{1}{2}}\\
&\lesssim \sum_{j=1}^2 |\lambda_j| \|a_j\|_{L^\infty(\mu)} \mu(2Q_j)\\
&\lesssim \sum_{j=1}^2 |\lambda_j|,
\end{aligned}$$

which is the desired estimate.

For $j \in \{1, 2\}$, let x_j be the center of Q_j. Notice that, for any $x \notin 2Q_j$ and $y \in Q_j$, it holds true that $|x - y| \sim |x - x_j|$. From this fact, the Hölder inequality, (d) of Theorem 2.4.4 and (3.2.1), it follows that

$$\begin{aligned}\dot{\mathcal{M}}_S(a_j)(x) &\lesssim \int_{Q_j} \frac{|a_j(y)|}{|x - y|^n}\, d\mu(y)\\ &\lesssim \frac{\|a_j\|_{L^\infty(\mu)}\mu(Q_j)}{|x - x_j|^n}\\ &\lesssim \frac{1}{|x - x_j|^n}\frac{1}{1 + \delta(Q_j,\, R)}.\end{aligned}$$

Therefore, by (4.3.1),

$$\mathrm{I}_2 \lesssim \sum_{j=1}^{2} \frac{|\lambda_j|\delta(2Q_j, 4R)}{1 + \delta(Q_j,\, R)} \lesssim \sum_{j=1}^{2} |\lambda_j|.$$

We now estimate I_3. Fix any $x_0 \in R \cap \operatorname{supp} \mu$. It follows, from Lemma 4.1.2(a), that $4R \subset Q_{x_0, H_R^{x_0}-1}$. We then write

$$\begin{aligned}\mathrm{I}_3 &= \int_{\mathbb{R}^D \setminus Q_{x_0, H_R^{x_0}-1}} \dot{\mathcal{M}}_S(b)(x)\, d\mu(x) + \int_{Q_{x_0, H_R^{x_0}-1} \setminus 4R} \cdots\\ &=: \mathrm{F}_1 + \mathrm{F}_2.\end{aligned}$$

By Lemma 4.1.2(a) again, we see that $Q_{x_0, H_R^{x_0}+1} \subset 4R$. From this fact, (d) of Theorem 2.4.4, (3.2.1) and the fact that, for any $x \notin 4R$ and $y \in R$, $|x - x_0| \sim |x - y|$, it follows that

$$\begin{aligned}\mathrm{F}_2 &\lesssim \sum_{j=1}^{2} |\lambda_j| \int_{Q_{x_0, H_R^{x_0}-1} \setminus 4R} \sup_{k \in \mathbb{Z}} \int_{Q_j} \frac{|a_j(y)|}{|x - x_0|^n}\, d\mu(y)\, d\mu(x)\\ &\lesssim \sum_{j=1}^{2} |\lambda_j| \int_{Q_{x_0, H_R^{x_0}-1} \setminus Q_{x_0, H_R^{x_0}+1}} \frac{\|a_j\|_{L^\infty(\mu)}\mu(Q_j)}{|x - x_0|^n}\, d\mu(x)\\ &\lesssim \sum_{j=1}^{2} |\lambda_j| \sum_{i=H_R^{x_0}-1}^{H_R^{x_0}} \delta(Q_{x_0, i+1}, Q_{x_0, i})\\ &\lesssim \sum_{j=1}^{2} |\lambda_j|.\end{aligned}$$

By

$$\int_{\mathbb{R}^D} b(x)\,d\mu(x) = 0,$$

we see that, for any $x \in \mathbb{R}^D \setminus Q_{x_0, H_R^{x_0}-1}$ and $k \in \mathbb{Z}$,

$$\begin{aligned}|S_k(b)(x)| &\le \int_R |S_k(x,y) - S_k(x,x_0)||b(y)|\,d\mu(y)\\ &\le \sum_{j=1}^{2} |\lambda_j| \int_{Q_j} |S_k(x,y) - S_k(x,x_0)||a_j(y)|\,d\mu(y).\end{aligned}$$

We claim that, for any $y \in Q_j$ with $j \in \{1,2\}$, integer $i \ge 2$ and $k \ge H_R^{x_0} - i + 3$, it holds true that

$$\operatorname{supp}(S_k(\cdot,y) - S_k(\cdot,x_0)) \subset Q_{x_0, H_R^{x_0}-i+1}. \tag{6.1.25}$$

Indeed, by (c) of Theorem 2.4.4 and the fact that $\{Q_{x,k}\}_k$ is decreasing in k, we see that

$$\begin{aligned}\operatorname{supp}(S_k(\cdot,y) - S_k(\cdot,x_0)) &\subset \left(Q_{y,k-1} \bigcup Q_{x_0,k-1}\right)\\ &\subset \left(Q_{y, H_R^{x_0}-i+2} \bigcup Q_{x_0, H_R^{x_0}-i+2}\right).\end{aligned}$$

Since $i \ge 2$, it follows that $y \in Q_j$, which, combined with the decreasing property of $\{Q_{x_0,k}\}_k$ in k, implies that $y \in Q_{x_0, H_R^{x_0}-i+2}$. By this fact and Lemma 2.3.3, we conclude that

$$Q_{y, H_R^{x_0}-i+2} \subset Q_{x_0, H_R^{x_0}-i+1}.$$

Thus, the above claim (6.1.25) holds true.

Observe that $Q_j \subset Q_{x_0,k}$ for $k \le H_R^{x_0} - i + 2$, $j \in \{1,2\}$. Then, (a) and (e) of Theorem 2.4.4 imply that, for any $y \in Q_j$,

$$\begin{aligned}|S_k(x,y) - S_k(x,x_0)| &\lesssim \frac{|x_0 - y|}{\ell(Q_{x_0,k})} \frac{1}{[\ell(Q_{x_0,k}) + |x - x_0|]^n}\\ &\lesssim \frac{\ell(R)}{\ell(Q_{x_0, H_R^{x_0}-i+2})} \frac{1}{|x - x_0|^n}.\end{aligned}$$

Therefore, from the fact that

$$\int_{\mathbb{R}^D} b(y)\,d\mu(y) = 0,$$

(6.1.25) and the last inequality above, it follows that

$$\begin{aligned}
\mathrm{F}_1 &= \sum_{i=2}^{\infty} \int_{Q_{x_0,\,H_R^{x_0}-i}\setminus Q_{x_0,\,H_R^{x_0}-i+1}} \sup_{k\in\mathbb{Z}} |S_k(b)(x)|\,d\mu(x) \\
&\lesssim \sum_{j=1}^{2} |\lambda_j| \sum_{i=2}^{\infty} \int_{Q_{x_0,\,H_R^{x_0}-i}\setminus Q_{x_0,\,H_R^{x_0}-i+1}} \sup_{k\le H_R^{x_0}-i+2} \int_{Q_j} |S_k(x,y)-S_k(x,x_0)| \\
&\quad \times |a_j(y)|\,d\mu(y)\,d\mu(x) \\
&\lesssim \sum_{j=1}^{2} |\lambda_j| \sum_{i=2}^{\infty} \int_{Q_{x_0,\,H_R^{x_0}-i}\setminus Q_{x_0,\,H_R^{x_0}-i+1}} \frac{\ell(R)}{\ell(Q_{x_0,\,H_R^{x_0}-i+2})} \frac{1}{|x-x_0|^n}\,d\mu(x) \\
&\lesssim \sum_{j=1}^{2} |\lambda_j| \sum_{i=2}^{\infty} \frac{\ell(R)}{\ell(Q_{x_0,\,H_R^{x_0}-i+2})} \\
&\lesssim \sum_{j=1}^{2} |\lambda_j|.
\end{aligned}$$

Therefore,

$$\mathrm{I}_3 \lesssim \sum_{j=1}^{2} |\lambda_j|,$$

which completes the proof of Theorem 6.1.8. □

We now establish the boundedness of $\mathcal{M}_S$ from $h^1(\mu)$ to $L^1(\mu)$.

Theorem 6.1.9. *There exists a nonnegative constant C such that, for all $f \in h^1(\mu)$,*

$$\|\mathcal{M}_S(f)\|_{L^1(\mu)} \le C\|f\|_{h^1(\mu)}.$$

Proof. By the Fatou lemma, to prove Theorem 6.1.9, it suffices to show that, for any ∞-atomic block or ∞-block $b := \sum_{j=1}^{2} \lambda_j a_j$ as in Definition 4.1.3, we have

$$\|\mathcal{M}_S(b)\|_{L^1(\mu)} \lesssim \sum_{j=1}^{2} |\lambda_j|. \tag{6.1.26}$$

If b is ∞-atomic block as in Definition 4.1.3, then, by the fact that

$$\mathcal{M}_S b(x) \le \dot{\mathcal{M}}_S b(x)$$

for all $x \in \mathbb{R}^D$ and (6.1.24), we see that (6.1.26) holds true. Let b be an ∞-block as in Definition 4.1.3. By Definition 4.1.3, there exists $R \in \mathcal{D}$ such that $\operatorname{supp} b \subset R$. Write

$$\begin{aligned}
&\int_{\mathbb{R}^D} \sup_{k\in\mathbb{N}} |S_k(b)(x)|\, d\mu(x)\\
&\quad\le \sum_{j=1}^{2} |\lambda_j| \int_{2Q_j} \sup_{k\in\mathbb{N}} |S_k(a_j)(x)|\, d\mu(x)\\
&\quad\quad + \sum_{j=1}^{2} |\lambda_j| \int_{4R\setminus 2Q_j} \cdots + \sum_{j=1}^{2} |\lambda_j| \int_{\mathbb{R}^D\setminus 4R} \cdots\\
&\quad =: \mathrm{J}_1 + \mathrm{J}_2 + \mathrm{J}_3.
\end{aligned}$$

Since the argument of estimates for I_1 and I_2 for the proof of Theorem 6.1.8 also works for the present situation, we then find that

$$\mathrm{J}_1 + \mathrm{J}_2 \lesssim \sum_{j=1}^{2} |\lambda_j|.$$

To estimate J_3, fix any $x_0 \in R \cap \operatorname{supp} \mu$. Notice that, for any $x \in \mathbb{R}^D \setminus 4R$ and $y \in Q_j$ with $j \in \{1, 2\}$, it holds true that $|x - y| \sim |x - x_0|$. From this fact, Definition 4.1.3 and (d) of Theorem 2.4.4, it follows that, for $j \in \{1,2\}$ and any $x \in \mathbb{R}^D \setminus 4R$,

$$\sup_{k\in\mathbb{N}} |S_k(a_j)(x)| \lesssim \sup_{k\in\mathbb{N}} \int_{Q_j} \frac{|a_j(y)|}{|x-y|^n}\, d\mu(y) \lesssim \frac{\|a_j\|_{L^\infty(\mu)} \mu(Q_j)}{|x-x_0|^n} \lesssim \frac{1}{|x-x_0|^n}.$$

On the other hand, since $R \in \mathcal{Q}$, by Lemma 4.1.2(d), we conclude that $H_R^{x_0} \le 1$. This observation, together with Lemma 2.3.3, in turn implies that, for any $k \in \mathbb{N}$ and $y \in R \cap \operatorname{supp} \mu$,

$$Q_{y,k-1} \subset Q_{y,\, H_R^{x_0}-1} \subset Q_{x_0,\, H_R^{x_0}-2}.$$

We then conclude that

$$\operatorname{supp}(S_k(b)) \subset Q_{x_0,\, H_R^{x_0}-2}$$

for any $k \in \mathbb{N}$. Moreover, from Lemma 4.1.2(a), we deuce that $Q_{x_0,\, H_R^{x_0}+1} \subset 4R$.

Therefore, we obtain

$$\begin{aligned}
\mathrm{J}_3 &\le \sum_{j=1}^{2} |\lambda_j| \int_{\mathbb{R}^D\setminus 4R} \sup_{k\in\mathbb{N}} |S_k(a_j)(x)|\,d\mu(x)\\
&\lesssim \sum_{j=1}^{2} |\lambda_j| \int_{Q_{x_0,H_R^{x_0}-2}\setminus 4R} \frac{1}{|x-x_0|^n}\,d\mu(x)\\
&\lesssim \sum_{j=1}^{2} |\lambda_j|,
\end{aligned}$$

which completes the proof of Theorem 6.1.9. □

6.2 Boundedness in BMO-Type Spaces

We now consider the uniform boundedness of S_k in RBLO (μ). To this end, we first establish the following lemma, which is a version of Lemma 4.3.5 for RBLO (μ).

Lemma 6.2.1. *There exists a positive constant C such that, for any cubes $Q \subset R$ and $f \in \mathrm{RBLO}\,(\mu)$,*

$$\int_R \frac{|f(y) - \operatorname{ess\,inf}_{y\in\tilde{Q}} f(y)|}{[|y-x_Q| + \ell(Q)]^n}\,d\mu(y) \le C[1+\delta(Q,R)]^2 \|f\|_{\mathrm{RBLO}\,(\mu)}.$$

Proof. Since RBLO $(\mu) \subset$ RBMO (μ), we show Lemma 6.2.1 by Lemma 4.3.5 as below. From Definition 4.4.1, it is easy to see that, for any $f \in \mathrm{RBLO}\,(\mu)$ and cube Q,

$$m_{\tilde{Q}}(f) - \operatorname*{ess\,inf}_{y\in\tilde{Q}} f(y) \le \|f\|_{\mathrm{RBLO}\,(\mu)}.$$

Therefore, an easy computation, involving Lemma 4.3.5 and (0.0.1), implies that

$$\begin{aligned}
&\int_R \frac{|f(y) - \operatorname{ess\,inf}_{y\in\tilde{Q}} f(y)|}{[|y-x_Q| + \ell(Q)]^n}\,d\mu(y)\\
&\quad\le \int_R \frac{|f(y) - m_{\tilde{Q}}(f)|}{[|y-x_Q| + \ell(Q)]^n}\,d\mu(y) + \int_R \frac{m_{\tilde{Q}}(f) - \operatorname{ess\,inf}_{y\in\tilde{Q}} f(y)}{[|y-x_Q| + \ell(Q)]^n}\,d\mu(y)\\
&\quad\lesssim [1+\delta(Q,R)]^2 \|f\|_{\mathrm{RBLO}\,(\mu)},
\end{aligned}$$

which completes the proof of Lemma 6.2.1. □

Theorem 6.2.2. *For any $k \in \mathbb{Z}$, let S_k be as in Sect. 2.4. Then S_k is uniformly bounded on* RBLO (μ)*, namely, there exists a nonnegative constant C, independent of k, such that, for all $f \in$* RBLO (μ)*,*

$$\|S_k(f)\|_{\mathrm{RBLO}\,(\mu)} \le C\|f\|_{\mathrm{RBLO}\,(\mu)}.$$

Proof. Without loss of generality, we may assume that $\|f\|_{\mathrm{RBLO}\,(\mu)} = 1$. We only need to consider the case that $\mathbb{R}^D$ is not an initial cube, since, if $\mathbb{R}^D$ is an initial cube, then, for any $k \in \mathbb{N}$, the argument is similar and, for any $k \le 0$, $S_k = 0$ and hence Theorem 6.2.2 holds true automatically in this case. To this end, it suffices to show that, for any doubling Q,

$$\frac{1}{\mu(Q)}\int_Q \left[S_k(f)(x) - \operatorname*{ess\,inf}_{Q} S_k(f)(y)\right] d\mu(x) \lesssim 1 \tag{6.2.1}$$

and, for any two doubling cubes $Q \subset R$,

$$m_Q(S_k(f)) - m_R(S_k(f)) \lesssim 1 + \delta(Q, R). \tag{6.2.2}$$

To show (6.2.1), let us consider the following two cases:

(i) There exists some $x_0 \in Q \cap \operatorname{supp}\mu$ such that $Q \subset Q_{x_0,k-2}$;
(ii) For any $x \in Q \cap \operatorname{supp}\mu$, $Q \not\subset Q_{x,k-2}$.

In Case (i), for each $x \in Q$, we see that

$$\begin{aligned}
&S_k(f)(x) - \operatorname*{ess\,inf}_{Q} S_k(f)(y)\\
&= \left[S_k(f)(x) - \operatorname*{ess\,inf}_{Q_{x,k}} f(y)\right] + \left[\operatorname*{ess\,inf}_{Q_{x,k}} f(y) - \operatorname*{ess\,inf}_{Q} S_k(f)(y)\right]\\
&=: \mathrm{I}_1 + \mathrm{I}_2.
\end{aligned}$$

It then follows, from (c) and (d) of Theorem 2.4.4 and Lemma 6.2.1, that

$$\mathrm{I}_1 \lesssim \int_{Q_{x,k-1}} \frac{|f(y) - \operatorname{ess\,inf}_{Q_{x,k}} f(y)|}{[|x-y| + \ell(Q_{x,k})]^n}\, d\mu(y) \lesssim 1. \tag{6.2.3}$$

On the other hand, in this case, for any $x,\ y \in Q \cap \operatorname{supp}\mu$, we know that $Q_{x,k}$ and $Q_{y,k}$ are contained in $Q_{x,k-4}$ by Lemma 2.3.3, which, together with (4.4.2) and Lemma 2.1.3, further implies that

$$\begin{aligned}
&\left|\operatorname*{ess\,inf}_{Q_{x,k}} f(z) - \operatorname*{ess\,inf}_{Q_{y,k}} f(z)\right|\\
&\quad\le \left|\operatorname*{ess\,inf}_{Q_{x,k}} f(z) - \operatorname*{ess\,inf}_{Q_{x,k-4}} f(z)\right| + \left|\operatorname*{ess\,inf}_{Q_{x,k-4}} f(z) - \operatorname*{ess\,inf}_{Q_{y,k}} f(z)\right|
\end{aligned}$$

$$
\begin{aligned}
&\lesssim 1+\delta(Q_{x,k},Q_{x,k-4})+\delta(Q_{y,k},Q_{x,k-4})\\
&\lesssim 1+\delta(Q_{y,k},Q_{y,k-3})+\delta(Q_{y,k-3},Q_{x,k-4})\\
&\lesssim 1+\delta(Q_{y,k-3},Q_{y,k-5})\\
&\lesssim 1. \qquad (6.2.4)
\end{aligned}
$$

By this observation, (b), (c) and (d) of Theorem 2.4.4 and Lemma 6.2.1, similar to the proof of (6.2.3), we see that, for any $y \in Q \cap \operatorname{supp}\mu$,

$$
\begin{aligned}
&\left|S_k(f)(y) - \operatorname*{ess\,inf}_{Q_{x,k}} f(z)\right|\\
&\quad\le \int_{Q_{y,k-1}} S_k(y,w)\left|f(w)-\operatorname*{ess\,inf}_{Q_{x,k}} f(z)\right| d\mu(w)\\
&\quad\le \int_{Q_{y,k-1}} S_k(y,w)\left|f(w)-\operatorname*{ess\,inf}_{Q_{y,k}} f(z)\right| d\mu(w) + \left|\operatorname*{ess\,inf}_{Q_{x,k}} f(z) - \operatorname*{ess\,inf}_{Q_{y,k}} f(z)\right|\\
&\quad\lesssim 1.
\end{aligned}
$$

Taking the infimum over all doubling cubes containing y, we have

$$
\mathrm{I}_2 \lesssim 1,
$$

which completes the proof of Case (i).

In Case (ii), it is easy to see that, for any $y \in Q \cap \operatorname{supp}\mu$, $k \ge H_Q^y + 3$. Then, by Lemma 4.1.2(b), we know that, for any $y \in Q \cap \operatorname{supp}\mu$, $Q_{y,k-1} \subset \frac{7}{5}Q$. Therefore, for any $x,\ y \in Q$, it holds true that

$$
\begin{aligned}
S_k(f)(x) - S_k(f)(y) &\le \left[S_k(f)(x) - \operatorname*{ess\,inf}_{\widetilde{\frac{7}{5}Q}} f(z)\right]\\
&\quad + \left[\operatorname*{ess\,inf}_{Q_{y,k}} f(z) - S_k(f)(y)\right]\\
&=: \mathrm{J}_1 + \mathrm{J}_2.
\end{aligned}
$$

From the Tonelli theorem, Theorem 2.4.4, (4.4.1) and the doubling property of Q, it follows that

$$
\frac{1}{\mu(Q)}\int_Q \mathrm{J}_1\,d\mu(x) \le \frac{1}{\mu(Q)}\int_{\frac{7}{5}Q}\left|f(w) - \operatorname*{ess\,inf}_{\widetilde{\frac{7}{5}Q}} f(y)\right| d\mu(w) \lesssim 1.
$$

On the other hand, (6.2.3) implies that

$$
\mathrm{J}_2 \lesssim 1,
$$

which shows (6.2.1).

Now we estimate (6.2.2). As in the proof of (6.2.1), we consider the following three cases:

(i) There exists some $x_0 \in Q \cap \operatorname{supp} \mu$ such that $R \subset Q_{x_0,k-2}$;
(ii) For any $x \in Q \cap \operatorname{supp} \mu$, $Q \not\subset Q_{x,k-2}$;
(iii) For any $x \in Q \cap \operatorname{supp} \mu$, $R \not\subset Q_{x,k-2}$ and there exists some $x_0 \in Q \cap \operatorname{supp} \mu$ such that $Q \subset Q_{x_0,k-2}$.

In Case (i), by (6.2.4) and (6.2.3), we know that

$$
\begin{aligned}
& m_Q(S_k(f)) - m_R(S_k(f)) \\
&= \frac{1}{\mu(Q)} \frac{1}{\mu(R)} \int_Q \int_R [S_k(f)(x) - S_k(f)(y)]\, d\mu(x)\, d\mu(y) \\
&\le \frac{1}{\mu(Q)} \frac{1}{\mu(R)} \int_Q \int_R \left\{ \left| S_k(f)(x) - \operatorname*{ess\,inf}_{z \in Q_{x,k}} f(z) \right| \right. \\
&\quad + \left| \operatorname*{ess\,inf}_{z \in Q_{x,k}} f(z) - \operatorname*{ess\,inf}_{z \in Q_{y,k}} f(z) \right| \\
&\quad \left. + \left| S_k(f)(y) - \operatorname*{ess\,inf}_{z \in Q_{y,k}} f(z) \right| \right\} d\mu(x)\, d\mu(y) \\
&\lesssim 1.
\end{aligned}
$$

In Case (ii), Lemma 4.1.2(b) implies that, for any $x \in Q \cap \operatorname{supp} \mu$, $Q_{x,k-1} \subset \frac{7}{5}Q$. By Lemma 2.1.3, we see that

$$
\begin{aligned}
\left| \operatorname*{ess\,inf}_{z \in \widetilde{\frac{7}{5}Q}} f(z) - \operatorname*{ess\,inf}_{z \in \widetilde{\frac{7}{5}R}} f(z) \right| &\le \left| \operatorname*{ess\,inf}_{z \in \widetilde{\frac{7}{5}Q}} f(z) - \operatorname*{ess\,inf}_{z \in Q} f(z) \right| \\
&\quad + \left| \operatorname*{ess\,inf}_{z \in Q} f(z) - \operatorname*{ess\,inf}_{z \in \widetilde{\frac{7}{5}R}} f(z) \right| \\
&\lesssim 1 + \delta(Q, R).
\end{aligned}
$$

This fact and the Tonelli theorem imply that

$$
\begin{aligned}
& m_Q(S_k(f)) - m_R(S_k(f)) \\
&\le \frac{1}{\mu(Q)} \frac{1}{\mu(R)} \int_Q \int_R |S_k(f)(x) - S_k(f)(y)|\, d\mu(x)\, d\mu(y) \\
&\le \frac{1}{\mu(Q)} \frac{1}{\mu(R)} \int_Q \int_R \left\{ \left| S_k(f)(x) - \operatorname*{ess\,inf}_{z \in \widetilde{\frac{7}{5}Q}} f(z) \right| \right.
\end{aligned}
$$

$$+\left|\widetilde{\operatorname*{ess\,inf}_{z\in\frac{7}{5}Q}} f(z) - \widetilde{\operatorname*{ess\,inf}_{z\in\frac{7}{5}R}} f(z)\right|$$

$$+\left|S_k(f)(y) - \widetilde{\operatorname*{ess\,inf}_{z\in\frac{7}{5}R}} f(z)\right|\Bigg\}\, d\mu(x)\, d\mu(y)$$

$$\lesssim 1+\delta(Q,R).$$

Finally, in Case (iii), by Lemma 2.1.3 and the fact that, for any $x\in Q\cap\operatorname{supp}\mu$,

$$Q_{x,k-1}\subset\frac{7}{5}R \quad\text{and}\quad Q_{x_0,k-2}\subset Q_{x,k-3},$$

we conclude that, for any $x\in Q\cap\operatorname{supp}\mu$,

$$\begin{aligned}
\left|\operatorname*{ess\,inf}_{z\in Q_{x,k}} f(z) - \widetilde{\operatorname*{ess\,inf}_{z\in\frac{7}{5}R}} f(z)\right| &\le 1+\delta\left(Q_{x,k},\widetilde{\frac{7}{5}R}\right)\\
&\lesssim 1+\delta(Q_{x,k},Q_{x_0,k-2})+\delta\left(Q_{x_0,k-2},\widetilde{\frac{7}{5}R}\right)\\
&\lesssim 1+\delta(Q_{x,k},Q_{x,k-3})+\delta\left(Q,\widetilde{\frac{7}{5}R}\right)\\
&\lesssim 1+\delta(Q,R).
\end{aligned}$$

From this, the Tonelli theorem and (6.2.3), we deduce that

$$\begin{aligned}
&m_Q(S_k(f))-m_R(S_k(f))\\
&\quad\le\frac{1}{\mu(Q)}\frac{1}{\mu(R)}\int_Q\int_R\Bigg\{\left|S_k(f)(x)-\operatorname*{ess\,inf}_{z\in Q_{x,k}} f(z)\right|\\
&\qquad+\left|\operatorname*{ess\,inf}_{z\in Q_{x,k}} f(z)-\widetilde{\operatorname*{ess\,inf}_{z\in\frac{7}{5}R}} f(z)\right|\\
&\qquad+\left|\widetilde{\operatorname*{ess\,inf}_{z\in\frac{7}{5}R}} f(z)-S_k(f)(y)\right|\Bigg\}\, d\mu(x)\, d\mu(y)\\
&\quad\lesssim 1+\delta(Q,R),
\end{aligned}$$

which completes the proof of Theorem 6.2.2. □

The following conclusion is a slight variant of Lemma 3.1.8, which can be proved by a slight modification of the proof of Lemma 3.1.8, the details being omitted.

Lemma 6.2.3. *For a positive constant C large enough, the following statement holds true: Let $x \in \mathbb{R}^D$ be a fixed point and $\{f_Q\}_{Q \ni x}$ a collection of numbers. If, for some constant $C_{(x)}$, it holds true that*

$$f_Q - f_R \le C_{(x)}[1 + \delta(Q, R)]$$

for all doubling cubes Q, R with $x \in Q \subset R$ and $\delta(Q, R) \le C$, then

$$f_Q - f_R \le C[1 + \delta(Q, R)]C_{(x)}$$

for all doubling cubes $Q \subset R$ with $x \in Q$.

Theorem 6.2.4. *For any $f \in \mathrm{RBMO}\,(\mu)$, $\dot{g}(f)$ is either infinite everywhere or finite almost everywhere and, in the latter case,*

$$\|[\dot{g}(f)]^2\|_{\mathrm{RBLO}\,(\mu)} \le C\|f\|^2_{\mathrm{RBMO}\,(\mu)}, \tag{6.2.5}$$

where $C \in (0, \infty)$ is independent of f.

Proof. We first claim that, for any $f \in \mathrm{RBMO}\,(\mu)$, if there exists a point $x_0 \in \mathbb{R}^D$ such that $g(f)(x_0) < \infty$, then, for any doubling cube $Q \ni x_0$, it holds true that

$$\frac{1}{\mu(Q)} \int_Q \left\{[\dot{g}(f)(x)]^2 - \operatorname*{ess\,inf}_{y \in Q} [\dot{g}(f)(y)]^2\right\} d\mu(x) \lesssim \|f\|^2_{\mathrm{RBMO}\,(\mu)}. \tag{6.2.6}$$

Without loss of generality, we may assume that $\|f\|_{\mathrm{RBMO}\,(\mu)} = 1$. For any point $x \in \operatorname{supp} \mu \cap Q$, let

$$\left[\dot{g}^{H_Q^x}(f)(x)\right]^2 := \sum_{k=H_Q^x+4}^{\infty} |D_k f(x)|^2$$

and

$$\left[\dot{g}_{H_Q^x}(f)(x)\right]^2 := \sum_{k=-\infty}^{H_Q^x+3} |D_k f(x)|^2.$$

Notice that $Q_{x,j} \subset \frac{4}{3}Q$ when $j \ge H_Q^x + 2$. This fact, together with

$$\operatorname{supp}(D_k(x, \cdot)) \subset Q_{x,k-2}$$

and

$$\int_{\mathbb{R}^D} D_k(x, y)\, d\mu(y) = 0, \tag{6.2.7}$$

implies that, when $k \ge H_Q^x + 4$,

$$D_k f(x) = D_k \left[\left(f - m_{\widetilde{\frac{4}{3}Q}}(f)\right) \chi_{\frac{4}{3}Q}\right](x).$$

It follows, from the doubling property of Q, along with the boundedness of $\dot{g}$ on $L^2(\mu)$ and Corollary 3.1.20, that

$$\begin{aligned}
&\frac{1}{\mu(Q)}\int_Q \left[\dot{g}^{H_Q^x}(f)(x)\right]^2 d\mu(x)\\
&\le \frac{1}{\mu(Q)}\int_Q \left\{\dot{g}\left[\left(f-m_{\widetilde{\frac{4}{3}Q}}(f)\right)\chi_{\frac{4}{3}Q}\right](x)\right\}^2 d\mu(x)\\
&\lesssim \frac{1}{\mu(2Q)}\int_{\frac{4}{3}Q}\left|f(x)-m_{\widetilde{\frac{4}{3}Q}}(f)\right|^2 d\mu(x)\\
&\lesssim 1.
\end{aligned} \tag{6.2.8}$$

Now observe that, for any $x, y \in Q$,

$$\left[\dot{g}_{H_Q^x}(f)(x)\right]^2 - [\dot{g}(y)]^2 \le \left[\dot{g}_{H_Q^x}(f)(x)\right]^2 - \left[\dot{g}_{H_Q^x}(f)(y)\right]^2.$$

Thus, taking (6.2.8) into account, to show (6.2.6), we only need to show that, for μ-almost every $y \in Q$,

$$\left[\dot{g}_{H_Q^x}(f)(x)\right]^2 - \left[\dot{g}_{H_Q^x}(f)(y)\right]^2 \lesssim 1. \tag{6.2.9}$$

We observe that, for each $k \in \mathbb{Z}$ and $z \in \mathbb{R}^D$,

$$|D_k f(z)| \lesssim 1. \tag{6.2.10}$$

Indeed, (d) of Theorem 2.4.4 implies that, for all $z, y \in \operatorname{supp}\mu$,

$$|D_k(z,y)| \lesssim \frac{1}{[\ell(Q_{z,k}) + \ell(Q_{y,k}) + |z-y|]^n}. \tag{6.2.11}$$

Then, since

$$\operatorname{supp}(D_k(z,\cdot)) \subset Q_{z,k-2},$$

by (6.2.7), Lemma 4.3.5 and (6.2.11), we know that, for all $z \in \mathbb{R}^D$,

$$\begin{aligned}
|D_k f(z)| &\le \int_{Q_{z,k-2}} |D_k(z,y)||f(y) - m_{Q_{z,k}}(f)|\, d\mu(y)\\
&\lesssim \int_{Q_{z,k-2}} \frac{|f(y) - m_{Q_{z,k}}(f)|}{[|z-y| + \ell(Q_{z,k})]^n}\, d\mu(z)\\
&\lesssim 1.
\end{aligned}$$

Thus, (6.2.10) holds true. From this observation, we deduce that, for all $x, y \in Q$,

$$\begin{aligned}
&\left[\dot{g}_{H_Q^x}(f)(x)\right]^2 - \left[\dot{g}_{H_Q^x}(f)(y)\right]^2 \\
&\quad \le \sum_{k=-\infty}^{H_Q^x-3} |D_k f(x) - D_k f(y)||D_k f(x) + D_k f(y)| + \sum_{k=H_Q^x-2}^{H_Q^x+3} |D_k f(x)|^2 \\
&\quad \lesssim \sum_{k=-\infty}^{H_Q^x-3} |D_k f(x) - D_k f(y)| + 1.
\end{aligned}$$

By the symmetry of D_k and (6.1.18), we see that, for any fixed integer $i \ge 3$ and $k \ge H_Q^x - i + 4$, and all $z \in Q_{x,H_Q^x-i} \setminus Q_{x,H_Q^x-i+1}$,

$$D_k(x,z) - D_k(y,z) = 0.$$

Therefore, from (6.2.7), we deduce that

$$\begin{aligned}
&\sum_{k=-\infty}^{H_Q^x-3} |D_k f(x) - D_k f(y)| \\
&\quad \le \int_{\mathbb{R}^D} \left[\sum_{k=-\infty}^{H_Q^x-3} |D_k(x,z) - D_k(y,z)|\right] \left|f(z) - m_{Q_{x,H_Q^x}}(f)\right| d\mu(z) \\
&\quad \le \sum_{i=3}^{\infty} \int_{Q_{x,H_Q^x-i} \setminus Q_{x,H_Q^x-i+1}} \left[\sum_{k=-\infty}^{H_Q^x-i+3} |D_k(x,z) - D_k(y,z)|\right] \\
&\qquad \times \left|f(z) - m_{Q_{x,H_Q^x}}(f)\right| d\mu(z) \\
&\qquad + \int_{Q_{x,H_Q^x-2}} \left[\sum_{k=-\infty}^{H_Q^x-3} |D_k(x,z) - D_k(y,z)|\right] \left|f(z) - m_{Q_{x,H_Q^x}}(f)\right| d\mu(z) \\
&\quad =: \mathrm{J}_1 + \mathrm{J}_2.
\end{aligned}$$

Since $x, y \in Q$ implies that $x, y \in Q_{x,k}$ for $k \le H_Q^x$, by Theorem 2.4.4(e) and Lemma 2.2.5, we further conclude that

$$\begin{aligned}
\sum_{k=-\infty}^{H_Q^x-i+3} |D_k(x,z) - D_k(y,z)| &\lesssim \sum_{k=-\infty}^{H_Q^x-i+3} \frac{|x-y|}{\ell(Q_{x,k})[\ell(Q_{x,k}) + |x-z|]^n} \\
&\lesssim \frac{\ell(Q)}{\ell(Q_{x,H_Q^x-i+3})} \frac{1}{|x-z|^n}.
\end{aligned}$$

Moreover, by (4.3.9), we have

$$\begin{aligned}
&\int_{Q_{x,H_Q^x-i}\setminus Q_{x,H_Q^x-i+1}} \frac{|f(z)-m_{Q_{x,H_Q^x-i+1}}(f)|}{|x-z|^n}\,d\mu(z)\\
&\quad\lesssim \left[1+\delta\left(Q_{x,H_Q^x-i+1},\,Q_{x,H_Q^x-i}\right)\right]^2\\
&\quad\lesssim 1.
\end{aligned}$$

Therefore, these facts, together with Definition 3.1.5, (6.1.20) and Lemma 2.2.5, imply that

$$\begin{aligned}
\mathrm{J}_1 &\lesssim \sum_{i=3}^{\infty}\frac{\ell(Q)}{\ell(Q_{x,H_Q^x-i+3})}\int_{Q_{x,H_Q^x-i}\setminus Q_{x,H_Q^x-i+1}} \frac{|f(z)-m_{Q_{x,H_Q^x}}(f)|}{|x-z|^n}\,d\mu(z)\\
&\le \sum_{i=3}^{\infty}\frac{\ell(Q)}{\ell(Q_{x,H_Q^x-i+3})}\left[\int_{Q_{x,H_Q^x-i}\setminus Q_{x,H_Q^x-i+1}} \frac{|f(z)-m_{Q_{x,H_Q^x-i+1}}(f)|}{|x-z|^n}\,d\mu(z)\right.\\
&\quad\left.+\int_{Q_{x,H_Q^x-i}\setminus Q_{x,H_Q^x-i+1}} \frac{|m_{Q_{x,H_Q^x-i+1}}(f)-m_{Q_{x,H_Q^x}}(f)|}{|x-z|^n}\,d\mu(z)\right]\\
&\lesssim \sum_{i=3}^{\infty}\frac{\ell(Q)}{\ell(Q_{x,H_Q^x-i+3})}\left\{1+[1+\delta(Q_{x,H_Q^x},Q_{x,H_Q^x-i+1})]^2\right\}\\
&\lesssim \sum_{i=3}^{\infty}\frac{\ell(Q)}{\ell(Q_{x,H_Q^x-i+3})}(1+i)^2\\
&\lesssim 1.
\end{aligned}$$

Now we turn our attention to J_2. By the estimate (6.2.11), Lemma 2.2.5, (0.0.1), Definition 3.1.5 and (6.1.20), we find that

$$\begin{aligned}
&\int_{Q_{x,H_Q^x-2}} \sum_{k=-\infty}^{H_Q^x-3} |D_k(x,z)|\left|f(z)-m_{Q_{x,H_Q^x}}(f)\right|\,d\mu(z)\\
&\quad\lesssim \int_{Q_{x,H_Q^x-2}} \sum_{k=-\infty}^{H_Q^x-3}\frac{|f(z)-m_{Q_{x,H_Q^x}}(f)|}{[|x-z|+\ell(Q_{x,k})]^n}\,d\mu(z)\\
&\quad\lesssim \int_{Q_{x,H_Q^x-2}} \frac{|f(z)-m_{Q_{x,H_Q^x}}(f)|}{[\ell(Q_{x,H_Q^x-2})]^n}\,d\mu(z)
\end{aligned}$$

$$\lesssim \int_{Q_{x,H_Q^x-2}} \frac{|f(z)-m_{Q_{x,H_Q^x-2}}(f)|}{[\ell(Q_{x,H_Q^x-2})]^n}\,d\mu(z) + \left|m_{Q_{x,H_Q^x-2}}(f) - m_{Q_{x,H_Q^x}}(f)\right|$$
$$\lesssim 1.$$

On the other hand, notice that, by Lemma 2.2.5, for any $z \in Q_{y,H_Q^x-3}$, it holds true that

$$\sum_{k=-\infty}^{H_Q^x-3} \frac{1}{[|y-z|+\ell(Q_{y,k})]^n} \lesssim \frac{1}{[\ell(Q_{y,H_Q^x-3})]^n}.$$

Since $y \in Q \subset Q_{x,H_Q^x}$, it follows that $Q_{x,H_Q^x-2} \subset Q_{y,H_Q^x-3}$ as a result of Lemma 2.3.3. Then, by these observations and (6.2.11), together with Definition 3.1.5, we conclude that

$$\int_{Q_{x,H_Q^x-2}} \sum_{k=-\infty}^{H_Q^x-3} |D_k(y,z)| \left|f(z)-m_{Q_{x,H_Q^x}}(f)\right| d\mu(z)$$
$$\lesssim \int_{Q_{x,H_Q^x-2}} \sum_{k=-\infty}^{H_Q^x-3} \frac{|f(z)-m_{Q_{x,H_Q^x}}(f)|}{[|y-z|+\ell(Q_{y,k})]^n}\,d\mu(z)$$
$$\lesssim \int_{Q_{x,H_Q^x-2}} \frac{|f(z)-m_{Q_{x,H_Q^x}}(f)|}{[\ell(Q_{y,H_Q^x-3})]^n}\,d\mu(z)$$
$$\lesssim \int_{Q_{x,H_Q^x-2}} \frac{|f(z)-m_{Q_{x,H_Q^x}}(f)|}{[\ell(Q_{x,H_Q^x-2})]^n}\,d\mu(z)$$
$$\lesssim 1.$$

Combining these estimates above, we find that

$$\mathrm{J}_2 \le \int_{Q_{x,H_Q^x-2}} \left\{\sum_{k=-\infty}^{H_Q^x-3} [|D_k(x,z)|+|D_k(y,z)|]\right\} \left|f(z)-m_{Q_{x,H_Q^x}}(f)\right| d\mu(z) \lesssim 1.$$

Thus, (6.2.9) holds true.

To finish the proof of Theorem 6.2.4, by Lemma 6.2.3, it suffices to show that, for any doubling cubes $Q \subset R$,

$$\left[\dot{g}(f)^2\right]_Q - \left[\dot{g}(f)^2\right]_R \lesssim [1+\delta(Q,R)]^4. \tag{6.2.12}$$

For any $x \in \operatorname{supp}\mu \cap Q$, we first consider the case that $H_Q^x \geq H_R^x + 10$ by writing

$$\begin{aligned}
&[\dot{g}(f)^2]_Q - [\dot{g}(f)^2]_R \\
&\leq \frac{1}{\mu(Q)} \int_Q \left[\dot{g}^{H_Q^x} f(x)\right]^2 d\mu(x) + \frac{1}{\mu(Q)} \int_Q \sum_{k=H_R^x+4}^{H_Q^x+3} |D_k f(x)|^2 \, d\mu(x) \\
&\quad + \frac{1}{\mu(Q)} \frac{1}{\mu(R)} \int_Q \int_R \left([\dot{g}_{H_R^x} f(x)]^2 - [\dot{g}_{H_R^x} f(y)]^2 \right) d\mu(y)\, d\mu(x).
\end{aligned}$$

By (6.2.9) with Q replaced by R, we see that

$$\frac{1}{\mu(Q)} \frac{1}{\mu(R)} \int_Q \int_R \left([\dot{g}_{H_R^x} f(x)]^2 - [\dot{g}_{H_R^x} f(y)]^2 \right) d\mu(y)\, d\mu(x) \lesssim 1.$$

Therefore by (6.2.8), the estimate (6.2.12) is reduced to proving that

$$\frac{1}{\mu(Q)} \int_Q \sum_{k=H_R^x+4}^{H_Q^x-1} |D_k f(x)|^2 \, d\mu(x) \lesssim [1 + \delta(Q, R)]^4. \tag{6.2.13}$$

By splitting

$$Q_{x,k-2} = (Q_{x,k-2} \setminus Q_{x,k-1}) \bigcup (Q_{x,k-1} \setminus Q_{x,k}) \bigcup Q_{x,k},$$

it follows, from the fact that

$$\operatorname{supp}(D_k(x, \cdot)) \subset Q_{x,k-2},$$

(6.2.7) and (6.2.11), that

$$\begin{aligned}
\sum_{k=H_R^x+4}^{H_Q^x-1} |D_k f(x)| &\leq \sum_{k=H_R^x+4}^{H_Q^x-1} \int_{Q_{x,k-2}} \frac{|f(z) - m_{Q_{x,H_Q^x-1}}(f)|}{[|x-z| + \ell(Q_{x,k})]^n} \, d\mu(z) \\
&\leq 2 \int_{Q_{x,H_R^x+2} \setminus Q_{x,H_Q^x-1}} \frac{|f(z) - m_{Q_{x,H_Q^x-1}}(f)|}{|x-z|^n} \, d\mu(z) \\
&\quad + \sum_{k=H_R^x+4}^{H_Q^x-1} \int_{Q_{x,k}} \frac{|f(z) - m_{Q_{x,H_Q^x-1}}(f)|}{[|x-z| + \ell(Q_{x,k})]^n} \, d\mu(z) \\
&=: \mathrm{L}_1 + \mathrm{L}_2.
\end{aligned}$$

By

$$Q \subset Q_{x, H_Q^x - 1}, \ Q_{x, H_R^x + 2} \subset 2R$$

and Lemma 2.1.3, we see that

$$\delta\left(Q_{x, H_Q^x - 1}, Q_{x, H_R^x + 2}\right) \lesssim 1 + \delta(Q, R). \tag{6.2.14}$$

Thus, by (6.2.14) and (4.3.9), we have

$$\mathrm{L}_1 \lesssim \left[1 + \delta\left(Q_{x, H_Q^x - 1}, Q_{x, H_R^x + 2}\right)\right]^2 \lesssim [1 + \delta(Q, R)]^2.$$

To estimate L_2, by Lemma 2.2.5, we first know that, for any integer

$$k \in \left[H_R^x + 4, H_Q^x - 1\right],$$

there exists a unique integer

$$j_k \in \left[0, N_{Q_{x, H_Q^x - 1}, Q_{x, H_R^x + 4}}\right]$$

such that

$$2^{j_k} Q_{x, H_Q^x - 1} \subset Q_{x,k} \subset 2^{j_k + 1} Q_{x, H_Q^x - 1}$$

and, for different k, j_k is different. It then follows, from Definition 3.1.5, the decreasing property of $Q_{x,k}$, (6.2.14) and (2.1.3), that

$$\begin{aligned}
\mathrm{L}_2 &\le \sum_{k=H_R^x+4}^{H_Q^x - 1} \int_{Q_{x,k}} \frac{|f(z) - m_{Q_{x,k}}(f)|}{[\ell(Q_{x,k})]^n} \, d\mu(z) \\
&\quad + \sum_{k=H_R^x+4}^{H_Q^x - 1} \frac{\mu(Q_{x,k})}{[\ell(Q_{x,k})]^n} \left| m_{Q_{x,k}}(f) - m_{Q_{x, H_Q^x - 1}}(f) \right| \\
&\lesssim \sum_{k=H_R^x+4}^{H_Q^x - 1} \frac{\mu(Q_{x,k})}{[\ell(Q_{x,k})]^n} + \sum_{k=H_R^x+4}^{H_Q^x - 1} \frac{\mu(Q_{x,k})}{[\ell(Q_{x,k})]^n} [1 + \delta(Q, R)] \\
&\lesssim \sum_{k=H_R^x+4}^{H_Q^x - 1} \frac{\mu(2^{j_k+1} Q_{x, H_Q^x - 1})}{[\ell(2^{j_k} Q_{x, H_Q^x - 1})]^n} [1 + \delta(Q, R)]
\end{aligned}$$

$$\lesssim \delta_{Q,2R}\,[1+\delta\,(Q,R)]$$
$$\lesssim [1+\delta\,(Q,R)]^2\,.$$

Consequently, (6.2.13) follows from combining the estimates for L_1 and L_2.

If $H_R^x \leq H_Q^x \leq H_R^x + 9$, then, by the estimates (6.2.8) through (6.2.10), we also see that (6.2.12) holds true, which completes the proof of Theorem 6.2.4. □

From Theorem 6.2.4, we can easily deduce the following result.

Corollary 6.2.5. *For any $f \in \mathrm{RBMO}\,(\mu)$, $\dot{g}(f)$ is either infinite everywhere or finite almost everywhere and, in the latter case,*

$$\|\dot{g}(f)\|_{\mathrm{RBLO}\,(\mu)} \leq C\|f\|_{\mathrm{RBMO}\,(\mu)},$$

where $C \in (0,\infty)$ is independent of f.

Proof. First, with the aid of (6.2.6) and the inequality that, for any $a, b \in \mathbb{Z}_+$,

$$a - b \leq \left|a^2 - b^2\right|^{1/2}, \tag{6.2.15}$$

it is easy to see that, if

$$\operatorname*{ess\,inf}_{y\in Q} \dot{g}(f)(y) < \infty,$$

then

$$\frac{1}{\mu(Q)}\int_Q \left[\dot{g}(f)(x) - \operatorname*{ess\,inf}_{y\in Q} \dot{g}(f)(y)\right] d\mu(x) \lesssim \|f\|_{\mathrm{RBMO}\,(\mu)}.$$

Moreover, in the argument of (6.2.12), we see that, for any doubling cubes $Q \subset R$, $x \in Q$ and $y \in R$,

$$\left|\left[\dot{g}(f)(x)\right]^2 - \left[\dot{g}(f)(y)\right]^2\right| \leq \left[\dot{g}^{H_Q^x} f(x)\right]^2 + \left[\dot{g}^{H_R^x} f(y)\right]^2 + \sum_{k=H_R^x+4}^{H_Q^x+3} |D_k f(x)|^2$$
$$+ \left|\left[\dot{g}_{H_R^x} f(x)\right]^2 - \left[\dot{g}_{H_R^x} f(y)\right]^2\right|.$$

From this fact, together with (6.2.8) through (6.2.10), (6.2.13) and (6.2.15), we deduce that, for any doubling cubes $Q \subset R$,

$$m_Q[\dot{g}(f)] - m_R[\dot{g}(f)] \lesssim [1+\delta(Q,R)]^2\|f\|_{\mathrm{RBMO}\,(\mu)}.$$

An application of Lemma 6.2.3 leads to the conclusion of Corollary 6.2.5, which completes the proof of Corollary 6.2.5. □

The following conclusion is a local variant of Lemma 3.1.8, which is an immediate corollary of Lemma 6.2.3.

Lemma 6.2.6. *There exists some constant P_0 (big enough) depending on C_0 and n such that, if $x \in \mathbb{R}^D$ is some fixed point and $\{f_Q\}_{Q \ni x}$ a collection of numbers such that*

$$f_Q - f_R \leq [1 + \delta(Q, R)]C_{(x)}$$

for all doubling cubes $Q \subset R$ with $x \in Q$ and $Q \notin \mathcal{D}$ such that

$$1 + \delta(Q, R) \leq P_0,$$

and $|f_R| \leq C_{(x)}$ for all doubling cubes $R \in \mathcal{D}$ with $R \ni x$, then

$$f_Q - f_R \leq C[1 + \delta(Q, R)]C_{(x)}$$

for all doubling cubes $Q \subset R$ with $x \in Q$ and $Q \notin \mathcal{D}$, where C depends on C_0, n and P_0.

Analogous to Theorem 6.2.4 for the boundedness of the homogeneous Littlewood–Paley g-function $\dot{g}(f)$ in RBMO (μ), we have Theorem 6.2.7 below for the boundedness of the inhomogeneous Littlewood–Paley g-function $g(f)$ in rbmo (μ). However, unlike Theorem 6.2.4, if $f \in \mathrm{rbmo}\,(\mu)$, then $g(f)$ is finite almost everywhere.

Theorem 6.2.7. *There exists a positive constant C such that, for all $f \in \mathrm{rbmo}\,(\mu)$,*

$$\|[g(f)]^2\|_{\mathrm{rblo}\,(\mu)} \leq C\|f\|^2_{\mathrm{rbmo}\,(\mu)}.$$

Proof. By the homogeneity of $g(f)$, we may assume that $\|f\|_{\mathrm{rbmo}\,(\mu)} = 1$. We first consider the case that $\mathbb{R}^D$ is an initial cube. In this case, to show Theorem 6.2.7, we first prove that, for any cube $Q \in \mathcal{D}$,

$$\frac{1}{\mu(2Q)} \int_Q [g(f)(x)]^2\, d\mu(x) \lesssim 1. \tag{6.2.16}$$

For any $x \in Q \cap \operatorname{supp} \mu$, we write

$$[g(f)(x)]^2 = |S_1(f)(x)|^2 + \sum_{k=2}^{H_Q^x+3} |D_k(f)(x)|^2 + \left[g^{H_Q^x}(f)(x)\right]^2,$$

where

$$\left[g^{H_Q^x}(f)(x)\right]^2 := \sum_{k=H_Q^x+4}^{\infty} |D_k(f)(x)|^2.$$

By the fact that $Q_{x,k} \subset \frac{7}{5}Q$ for $k \geq H_Q^x + 2$, the boundedness of g on $L^2(\mu)$ and Corollary 4.2.9, we see that

$$\begin{aligned}\frac{1}{\mu(2Q)}\int_Q \left[g^{H_Q^x}(f)(x)\right]^2 d\mu(x) &\leq \frac{1}{\mu(2Q)}\int_Q \left[g\left(f\chi_{\frac{7}{5}Q}\right)(x)\right]^2 d\mu(x)\\ &\lesssim \frac{1}{\mu(2Q)}\int_{\frac{7}{5}Q} |f(x)|^2\, d\mu(x)\\ &\lesssim 1. \qquad (6.2.17)\end{aligned}$$

Moreover, from the fact that $\mathrm{rbmo}\,(\mu) \subset \mathrm{RBMO}\,(\mu)$, it follows that, for any $f \in \mathrm{rbmo}\,(\mu)$, $k \geq 2$ and $z \in \operatorname{supp}\mu$,

$$|D_k(f)(z)| \lesssim 1. \qquad (6.2.18)$$

Thus, (6.2.18) holds true, which, together with the fact that $0 \leq H_Q^x \leq 1$, implies that

$$\sum_{k=2}^{H_Q^x+3} |D_k(f)(x)|^2 \lesssim 1. \qquad (6.2.19)$$

The estimates (6.2.17) and (6.2.19), together with (4.3.10), imply that the estimate (6.2.16) holds true.

From (6.2.16), it follows that, for any doubling cube Q,

$$\operatorname*{ess\,inf}_Q [g(f)]^2 \lesssim 1.$$

Therefore, to complete the proof of Theorem 6.2.7, by Proposition 4.4.15, (6.2.16) and Lemma 6.2.6, we still need to show that, for any doubling cube $Q \notin \mathcal{D}$ and $y \in Q$,

$$\frac{1}{\mu(Q)}\int_Q \{[g(f)(x)]^2 - [g(f)(y)]^2\}\, d\mu(x) \lesssim 1 \qquad (6.2.20)$$

and, for any doubling cubes $Q \subset R$ with $Q \notin \mathcal{D}$,

$$m_Q\left[g(f)^2\right] - m_R\left[g(f)^2\right] \lesssim [1 + \delta(Q, R)]^4. \qquad (6.2.21)$$

We first establish (6.2.20). For any doubling cube $Q \notin \mathcal{D}$ and $x \in Q \cap \operatorname{supp}\mu$, if $0 \leq H_Q^x \leq 5$, we use the following trivial estimate that

$$\begin{aligned}
&[g(f)(x)]^2 - [g(f)(y)]^2 \\
&\quad \le [g(f)(x)]^2 \\
&\quad = |S_1(f)(x)|^2 + \sum_{k=2}^{H_Q^x+3} |D_k(f)(x)|^2 + \left[g^{H_Q^x}(f)(x)\right]^2.
\end{aligned} \tag{6.2.22}$$

If $\frac{7}{5}Q \in \mathcal{D}$, then (6.2.20) is deduced from (6.2.22), (6.2.17), (6.2.18) and (4.3.10) directly. If $\frac{7}{5}Q \notin \mathcal{D}$, notice that

$$\int_{\mathbb{R}^D} D_k(x,y)\,d\mu(y) = 0 \quad \text{when} \quad 0 \le H_Q^x \le 5 \quad \text{and} \quad k \ge H_Q^x + 4.$$

Moreover, by Lemma 4.1.2(b), we have

$$\operatorname{supp}(D_k(x,\cdot)) \subset Q_{x,k-2} \subset \frac{7}{5}Q.$$

These facts imply that, for all $x \in \mathbb{R}^D$,

$$D_k f(x) = D_k\left[\left(f - m_{\widetilde{\frac{7}{5}Q}}(f)\right)\chi_{\frac{7}{5}Q}\right](x).$$

On the other hand, since $g^{H_Q^x}(f) \le g(f)$, by the doubling property of Q, together with the boundedness of g on $L^2(\mu)$ and Corollary 4.2.9, we find that

$$\begin{aligned}
&\frac{1}{\mu(Q)}\int_Q \left[g^{H_Q^x}(f)(x)\right]^2 d\mu(x) \\
&\quad \le \frac{1}{\mu(Q)}\int_Q \left[g\left(\left[f - m_{\widetilde{\frac{7}{5}Q}}(f)\right]\chi_{\frac{7}{5}Q}\right)(x)\right]^2 d\mu(x) \\
&\quad \lesssim \frac{1}{\mu(2Q)}\int_{\frac{7}{5}Q} \left|f(x) - m_{\widetilde{\frac{7}{5}Q}}(f)\right|^2 d\mu(x) \\
&\quad \lesssim 1.
\end{aligned} \tag{6.2.23}$$

This, together with (6.2.22), (4.3.10) and (6.2.18), implies (6.2.20).

Now suppose that $H_Q^x \ge 6$. We then have

$$\begin{aligned}
[g(f)(x)]^2 - [g(f)(y)]^2 &\le |S_1(f)(x)|^2 + \sum_{k=2}^{H_Q^x-3} \left[|D_k(f)(x)|^2 - |D_k(f)(y)|^2\right] \\
&\quad + \sum_{k=H_Q^x-2}^{H_Q^x+3} |D_k(f)(x)|^2 + \left[g^{H_Q^x}(f)(x)\right]^2.
\end{aligned}$$

Using (6.2.23), (6.2.18) and (4.3.10) again, we see that the estimate (6.2.20) is reduced to showing that, for any $x, y \in Q \cap \operatorname{supp} \mu$,

$$\sum_{k=2}^{H_Q^x - 3} \left[|D_k(f)(x)|^2 - |D_k(f)(y)|^2 \right] \lesssim 1. \tag{6.2.24}$$

An application of (6.2.18) implies that

$$\sum_{k=2}^{H_Q^x - 3} \left[|D_k(f)(x)|^2 - |D_k(f)(y)|^2 \right] \lesssim \sum_{k=2}^{H_Q^x - 3} |D_k(f)(x) - D_k(f)(y)| .$$

For $y \in Q \cap \operatorname{supp} \mu$ and $2 \le k \le H_Q^x - 3$, we know that $Q \subset Q_{x,k}$, which, together with Lemma 2.3.3, implies that $Q_{y,k-2} \subset Q_{x,k-3}$. By this fact, together with (6.2.7), (e) of Theorem 2.4.4 and Lemma 4.3.5, we further see that

$$\begin{aligned}
&|D_k(f)(x) - D_k(f)(y)| \\
&\quad = \left| \int_{Q_{x,k-3}} [D_k(x,z) - D_k(y,z)] \left[f(z) - m_{Q_{x,k}}(f) \right] d\mu(z) \right| \\
&\quad \lesssim \int_{Q_{x,k-3}} \frac{|x-y||f(z) - m_{Q_{x,k}}(f)|}{\ell(Q_{x,k})[|x-z| + \ell(Q_{x,k})]^n} \, d\mu(z) \\
&\quad \lesssim \frac{|x-y|}{\ell(Q_{x,k})}.
\end{aligned}$$

Therefore from the fact that $|x-y| \lesssim \ell(Q)$ and Lemma 2.2.5, it follows that

$$\sum_{k=2}^{H_Q^x - 3} \left[|D_k(f)(x)|^2 - |D_k(f)(y)|^2 \right] \lesssim \sum_{k=2}^{H_Q^x - 3} \frac{|x-y|}{\ell(Q_{x,k})} \lesssim \frac{\ell(Q)}{\ell(Q_{x,H_Q^x - 3})} \lesssim 1,$$

since $Q \subset Q_{x,H_Q^x - 3}$.

We now prove (6.2.21). For any doubling cubes $Q \subset R$ with $Q \notin \mathcal{D}$, $x \in Q \cap \operatorname{supp} \mu$ and $y \in R \cap \operatorname{supp} \mu$, we first consider the case that $H_Q^x \ge H_R^x + 10$. In this case, if $H_R^x \ge 6$, we write

$$\begin{aligned}
[g(f)(x)]^2 - [g(f)(y)]^2 &\le |S_1(f)(x)|^2 + \sum_{k=2}^{H_R^x + 3} \left[|D_k(f)(x)|^2 - |D_k(f)(y)|^2 \right] \\
&\quad + \sum_{k=H_R^x + 4}^{H_Q^x + 3} |D_k(f)(x)|^2 + \left[g^{H_Q^x}(f)(x) \right]^2 .
\end{aligned}$$

Observe that, for $k \geq H_R^x + 4$ and $x \in \operatorname{supp} \mu$,

$$\operatorname{supp}(D_k(x, \cdot)) \subset Q_{x,k-2}$$

and hence, by Lemma 4.1.2(e), it holds true that $Q_{x,k-2} \notin \mathcal{D}$. Therefore, using (6.2.18) and repeating the argument of (6.2.12), we obtain

$$\sum_{k=H_R^x+4}^{H_Q^x+3} |D_k(f)(x)| \lesssim 1 + \sum_{k=H_R^x+4}^{H_Q^x-1} |D_k(f)(x)| \lesssim [1 + \delta(Q, R)]^2. \tag{6.2.25}$$

Consequently, (6.2.21) follows from (6.2.23), (4.3.10), (6.2.24) and (6.2.25).

If $0 \leq H_R^x \leq 5$, by writing

$$\begin{aligned}
&[g(f)(x)]^2 - [g(f)(y)]^2 \\
&\leq |S_1(f)(x)|^2 + \sum_{k=2}^{H_R^x+3} |D_k(f)(x)|^2 + \sum_{k=H_R^x+4}^{H_Q^x+3} |D_k(f)(x)|^2 + \left[g^{H_Q^x}(f)(x)\right]^2,
\end{aligned}$$

we see that (6.2.21) follows from (6.2.23), (6.2.18), (4.3.10) and (6.2.25).

Similarly, if $H_R^x \leq H_Q^x \leq H_R^x + 9$, (6.2.21) follows from (6.2.23), (6.2.18), (4.3.10) and (6.2.24), which completes the proof of the case that $\mathbb{R}^D$ is an initial cube.

If $\mathbb{R}^D$ is not an initial cube, we can also show that (6.2.16), (6.2.20) and (6.2.21) hold true, the details being omitted. This finishes the proof of Theorem 6.2.7. □

From an argument similar to that used in the proof of Corollary 6.2.5, we deduce the following conclusion, the details being omitted again.

Corollary 6.2.8. *There exists a positive constant C such that, for all $f \in \mathrm{rbmo}(\mu)$,*

$$\|g(f)\|_{\mathrm{rblo}(\mu)} \leq C\|f\|_{\mathrm{rbmo}(\mu)}.$$

Now we consider the boundedness of the maximal operators $\dot{\mathcal{M}}_S$ and $\mathcal{M}_S$ in (6.1.22) and (6.1.23) from RBMO (μ) to RBLO (μ).

Theorem 6.2.9. *If $\mathbb{R}^D$ is not an initial cube, then, for any $f \in \mathrm{RBMO}(\mu)$, $\dot{\mathcal{M}}_S(f)$ is either infinite everywhere or finite almost everywhere and, in the latter case, there exists a positive constant C, independent of f, such that*

$$\|\dot{\mathcal{M}}_S(f)\|_{\mathrm{RBLO}(\mu)} \leq C\|f\|_{\mathrm{RBMO}(\mu)}.$$

If $\mathbb{R}^D$ is an initial cube, the same conclusions as above are true if $\dot{\mathcal{M}}_S$ is replaced by $\mathcal{M}_S$.

Proof. By homogeneity, we may assume that

$$\|f\|_{\mathrm{RBMO}\,(\mu)} = 1.$$

Moreover, when $\mathbb{R}^D$ is an initial cube, by the convention, we have $S_k = 0$ when $k \le 0$. Thus, using this convention, we can also write $\mathcal{M}_S$ into $\dot{\mathcal{M}}_S$.

We first claim that, if there exists a point $x_0 \in \mathbb{R}^D$ such that $\dot{\mathcal{M}}_S(f)(x_0) < \infty$, then, for any doubling cube $Q \ni x_0$,

$$\frac{1}{\mu(Q)}\int_Q \left[\dot{\mathcal{M}}_S(f)(x) - \inf_Q \dot{\mathcal{M}}_S(f)\right] d\mu(x) \lesssim 1. \tag{6.2.26}$$

Indeed, for any cube Q and $f \in L^1_{\mathrm{loc}}(\mu)$, define

$$\dot{\mathcal{M}}_{S,Q,1}(f)(x) := \sup_{k \ge H_Q^x + 1} |S_k(f)(x)|, \tag{6.2.27}$$

$$\dot{\mathcal{M}}_{S,Q,2}(f)(x) := \sup_{k \le H_Q^x} |S_k(f)(x)|, \tag{6.2.28}$$

$$Q_1 := \{x \in Q : \dot{\mathcal{M}}_{S,Q,1}(f)(x) \ge \dot{\mathcal{M}}_{S,Q,2}(f)(x)\} \quad \text{and} \quad Q_2 := Q \setminus Q_1.$$

We then have

$$\dot{\mathcal{M}}_S(f) = \max\left\{\dot{\mathcal{M}}_{S,Q,1}(f),\ \dot{\mathcal{M}}_{S,Q,2}(f)\right\}.$$

Write

$$f_1 := [f - m_Q(f)]\chi_{\frac{4}{3}Q} \quad \text{and} \quad f_2 := [f - m_Q(f)]\chi_{\mathbb{R}^D \setminus \frac{4}{3}Q}.$$

Since $\dot{\mathcal{M}}_{S,Q,1}$ is sublinear, we see that

$$\begin{aligned}
&\frac{1}{\mu(Q)}\int_Q \left[\dot{\mathcal{M}}_S(f)(x) - \inf_Q \dot{\mathcal{M}}_S(f)\right] d\mu(x)\\
&\le \frac{1}{\mu(Q)}\int_{Q_1} \left[\dot{\mathcal{M}}_{S,Q,1}(f_1)(x) + \dot{\mathcal{M}}_{S,Q,1}(f_2)(x)\right] d\mu(x)\\
&\quad + \left[|m_Q(f)| - \inf_Q \dot{\mathcal{M}}_S(f)\right]\\
&\quad + \frac{1}{\mu(Q)}\int_{Q_2} \left[\dot{\mathcal{M}}_{S,Q,2}(f)(x) - \inf_Q \dot{\mathcal{M}}_S(f)\right] d\mu(x)\\
&=: \mathrm{E}_1 + \mathrm{E}_2 + \mathrm{E}_3.
\end{aligned}$$

By Lemma 6.1.7, $\dot{\mathcal{M}}_S$ is bounded on $L^2(\mu)$. From this fact, the Hölder inequality, the doubling property of Q, Lemma 2.1.3, Proposition 3.1.4 and Corollary 3.1.20, it follows that

$$\begin{aligned}
&\frac{1}{\mu(Q)}\int_{Q_1}\dot{\mathcal{M}}_{S,Q,1}(f_1)(x)\,d\mu(x)\\
&\lesssim\left\{\frac{1}{\mu(Q)}\int_{\frac{4}{3}Q}\left|f(x)-m_Q(f)\right|^2\,d\mu(x)\right\}^{1/2}\\
&\lesssim\left\{\frac{1}{\mu(Q)}\int_{\frac{4}{3}Q}\left|f(x)-m_{\widetilde{\frac{4}{3}Q}}(f)\right|^2\,d\mu(x)\right\}^{1/2}+\left|m_Q(f)-m_{\widetilde{\frac{4}{3}Q}}(f)\right|\\
&\lesssim 1.
\end{aligned}$$

By this inequality, the estimate for E_1 is reduced to showing that

$$\frac{1}{\mu(Q)}\int_{Q_1}\dot{\mathcal{M}}_{S,Q,1}(f_2)(x)\,d\mu(x)\lesssim 1. \tag{6.2.29}$$

By applying Lemma 4.3.5 and (6.1.20), we see that

$$\int_{Q_{x,k-1}}\frac{|f(z)-m_{Q_{x,k}}(f)|}{[|z-x|+\ell(Q_{x,k})]^n}\,d\mu(z)\lesssim[1+\delta(Q_{x,k},Q_{x,k-1})]^2\lesssim 1. \tag{6.2.30}$$

Moreover, if $k\geq H_Q^x+4$, then $Q_{x,k-1}\subset\frac{4}{3}Q$, which is deduced from applying Lemma 4.1.2(b), together with the fact that

$$\ell(Q_{x,k-1})\leq\frac{1}{10}\ell(Q_{x,k-2})\quad\text{for any}\quad x\in\operatorname{supp}\mu\quad\text{and}\quad k\in\mathbb{Z}.$$

Then, by this fact, (b) through (d) of Theorem 2.4.4, (6.2.30), Proposition 3.1.4, Lemma 2.1.3(d) and Lemma 4.1.2(e), we conclude that, for any $x\in Q_1$,

$$\begin{aligned}
\dot{\mathcal{M}}_{S,Q,1}(f_2)(x)&=\sup_{H_Q^x+1\leq k\leq H_Q^x+3}|S_k(f_2)(x)|\\
&\leq\sup_{H_Q^x+1\leq k\leq H_Q^x+3}\left[\left|S_k\left((f-m_{Q_{x,k}}(f))\chi_{\mathbb{R}^D\setminus\frac{4}{3}Q}\right)(x)\right|\right.\\
&\qquad\left.+\left|m_{Q_{x,k}}(f)-m_{Q_{x,H_Q^x+1}}(f)\right|+\left|m_{Q_{x,H_Q^x+1}}(f)-m_Q(f)\right|\right]\\
&\lesssim 1.
\end{aligned}$$

This implies (6.2.29).

Now we estimate E_2. From (b) through (d) of Theorem 2.4.4 and (6.2.30), it follows that, for any $k \in \mathbb{Z}$ and $x \in \operatorname{supp} \mu$,

$$|S_k(f)(x) - m_{Q_{x,k}}(f)| \lesssim 1. \tag{6.2.31}$$

Then, applying this, together with Lemma 4.1.2(e) and Proposition 3.1.4, we see that, for any $y \in Q$,

$$\begin{aligned}|m_Q(f)| - \dot{\mathcal{M}}_S(f)(y) &\le \left|m_Q(f) - S_{H_Q^y+1}(f)(y)\right| \\ &\le \left|m_Q(f) - m_{Q_{y,H_Q^y+1}}(f)\right| \\ &\quad + \left|m_{Q_{y,H_Q^y+1}}(f) - S_{H_Q^y+1}(f)(y)\right| \\ &\lesssim 1.\end{aligned}$$

Thus, we have

$$E_2 \lesssim 1.$$

On the other hand, for any $x,\ y \in Q$ and $k \le H_Q^x$, Lemma 2.3.3 implies that

$$Q_{y,k} \subset Q_{x,k-1} \subset Q_{y,k-2}.$$

Then, by (d) and (e) of Lemma 2.1.3, we see that $\delta(Q_{y,k}, Q_{x,k}) \lesssim 1$. Therefore, it follows, from Proposition 3.1.4 and (6.2.31), that

$$\begin{aligned}&|S_k f(x)| - \dot{\mathcal{M}}_S(f)(y) \\ &\quad \le |S_k f(x) - S_k f(y)| \\ &\quad \le |S_k f(x) - m_{Q_{x,k}}(f)| + |m_{Q_{x,k}}(f) - m_{Q_{y,k}}(f)| + |m_{Q_{y,k}}(f) - S_k f(y)| \\ &\quad \lesssim 1,\end{aligned}$$

which implies that

$$E_3 \lesssim 1.$$

Combining estimates for E_1 through E_3, we obtain (6.2.26).

By (6.2.26), if there exists a point $x_0 \in \mathbb{R}^D$ such that $\dot{\mathcal{M}}_S(f)(x_0) < \infty$, then $\dot{\mathcal{M}}_S(f)(x)$ is finite almost everywhere and, for any doubling cube Q,

$$\frac{1}{\mu(Q)} \int_Q \left[\dot{\mathcal{M}}_S(f)(x) - \operatorname*{ess\,inf}_Q \dot{\mathcal{M}}_S(f)\right] d\mu(x) \lesssim 1.$$

To complete the proof of Theorem 6.2.9, it suffices to show that, for any doubling cube $Q \subset R$,

$$m_Q[\dot{\mathcal{M}}_S(f)] - m_R[\dot{\mathcal{M}}_S(f)] \lesssim 1 + \delta(Q, R).$$

Let $\dot{\mathcal{M}}_{S,R,1}(f)$ and $\dot{\mathcal{M}}_{S,R,2}(f)$ be as in (6.2.27) and (6.2.28) with Q replaced by R,

$$Q_1 := \{x \in Q : \dot{\mathcal{M}}_{S,R,1}(f)(x) \ge \dot{\mathcal{M}}_{S,R,2}(f)(x)\}$$

and $Q_2 := Q \setminus Q_1$. Split

$$\begin{aligned} f &= [f - m_R(f)]\chi_{\frac{4}{3}Q} + [f - m_R(f)]\chi_{\mathbb{R}^D \setminus \frac{4}{3}Q} + m_R(f) \\ &=: f_1 + f_2 + m_R(f). \end{aligned}$$

From the fact that $\dot{\mathcal{M}}_{S,R,1}$ is sublinear, it follows that

$$\begin{aligned} & m_Q[\dot{\mathcal{M}}_S(f)] - m_R[\dot{\mathcal{M}}_S(f)] \\ & \le \frac{1}{\mu(Q)} \int_{Q_1} \{\dot{\mathcal{M}}_{S,R,1}(f_1)(x) + \dot{\mathcal{M}}_{S,R,1}(f_2)(x)\}\, d\mu(x) \\ & \quad + \left[|m_R(f)| - m_R[\dot{\mathcal{M}}_S(f)]\right] \\ & \quad + \frac{1}{\mu(Q)} \int_{Q_2} \{\dot{\mathcal{M}}_{S,R,2}(f)(x) - m_R[\dot{\mathcal{M}}_S(f)]\}\, d\mu(x) \\ & =: \mathrm{F}_1 + \mathrm{F}_2 + \mathrm{F}_3. \end{aligned}$$

By the boundedness of $\dot{\mathcal{M}}_S$ in $L^2(\mu)$, the Hölder inequality, the doubling property of Q, Lemma 2.1.3, Proposition 3.1.4 and Corollary 3.1.20, we conclude that

$$\begin{aligned} & \frac{1}{\mu(Q)} \int_{Q_1} \dot{\mathcal{M}}_{S,R,1}(f_1)(x)\, d\mu(x) \\ & \lesssim \left\{ \frac{1}{\mu(Q)} \int_{\frac{4}{3}Q} \left| f(x) - m_{\widetilde{\frac{4}{3}Q}}(f) \right|^2 d\mu(x) \right\}^{1/2} \\ & \quad + \left| m_{\widetilde{\frac{4}{3}Q}}(f) - m_Q(f) \right| + |m_Q(f) - m_R(f)| \\ & \lesssim 1 + \delta(Q, R). \end{aligned} \tag{6.2.32}$$

From the fact that

$$Q_{x,k-1} \subset \frac{4}{3}Q \quad \text{for} \quad k \ge H_Q^x + 4,$$

(b) through (d) of Theorem 2.4.4, (6.2.30), Lemma 2.1.3, Proposition 3.1.4 and (a) and (e) of Lemma 4.1.2, we deduce that, for any $x \in Q$,

$$\begin{aligned}\mathcal{M}_{S,R,1}(f_2)(x) &\leq \sup_{H_R^x+1\leq k\leq H_Q^x+3} \left\{S_k\left[\left|f - m_{Q_{x,k}}(f)\right|\right](x)\right.\\&\quad\left.+\left|m_{Q_{x,k}}(f) - m_{\widetilde{3R}}(f)\right| + \left|m_{\widetilde{3R}}(f) - m_R(f)\right|\right\}\\&\lesssim 1 + \delta(Q_{x,H_Q^x+3}, \widetilde{3R})\\&\lesssim 1 + \delta(Q, R).\end{aligned}$$

By this and (6.2.32), we find that

$$\mathrm{F}_1 \lesssim 1 + \delta(Q, R).$$

On the other hand, Lemma 4.1.2(e), Proposition 3.1.4 and (6.2.31) imply that, for any $y \in R$,

$$\begin{aligned}|m_R(f)| - \dot{\mathcal{M}}_S(f)(y) &\leq \left|m_R(f) - S_{H_R^y+1}(f)(y)\right|\\&\leq \left|m_R(f) - m_{Q_{y,H_R^y+1}}(f)\right|\\&\quad+\left|m_{Q_{y,H_R^y+1}}(f) - S_{H_R^y+1}(f)(y)\right|\\&\lesssim 1.\end{aligned}$$

Taking average over $y \in R$, we then see that

$$\mathrm{F}_2 \lesssim 1.$$

Observe that, for any $x, y \in R$ and $k \leq H_R^x$, it holds true that $\delta(Q_{y,k}, Q_{x,k}) \lesssim 1$. Then it follows, from Proposition 3.1.4 and (6.2.31), that

$$\begin{aligned}&|S_k f(x)| - \dot{\mathcal{M}}_S(f)(y)\\&\quad\leq |S_k f(x) - m_{Q_{x,k}}(f)| + |m_{Q_{x,k}}(f) - m_{Q_{y,k}}(f)| + |m_{Q_{y,k}}(f) - S_k f(y)|\\&\quad\lesssim 1,\end{aligned}$$

which implies that

$$\mathrm{F}_3 \lesssim 1$$

and hence completes the proof of Theorem 6.2.9. □

The following theorem is a local version of Theorem 6.2.9.

Theorem 6.2.10. *There exists a positive constant C such that, for all $f \in$ rbmo (μ),*

$$\|\mathcal{M}_S(f)\|_{\mathrm{rblo}\,(\mu)} \le C\|f\|_{\mathrm{rbmo}\,(\mu)}.$$

Proof. By homogeneity, we may assume that $\|f\|_{\mathrm{rbmo}\,(\mu)} = 1$. We first consider the case that $\mathbb{R}^D$ is an initial cube. In this case, we claim that, for any cube $Q \in \mathcal{D}$,

$$\frac{1}{\mu(2Q)} \int_Q \mathcal{M}_S(f)(x)\,d\mu(x) \lesssim 1. \tag{6.2.33}$$

By Lemma 6.1.7, we know that $\mathcal{M}_S$ is bounded on $L^2(\mu)$. From this fact, together with the Hölder inequality and Corollary 4.2.9, we deduce that

$$\frac{1}{\mu(2Q)} \int_Q \mathcal{M}_S\left[f\chi_{\frac{4}{3}Q}\right](x)\,d\mu(x) \lesssim \left\{\frac{1}{\mu(2Q)} \int_{\frac{4}{3}Q} |f(x)|^2\,d\mu(x)\right\}^{1/2} \lesssim 1.$$

On the other hand, by Lemma 4.1.2(c), $0 \le H_Q^x \le 1$ for any $x \in Q$, which in turn implies that $k \ge H_Q^x + 4$ for $k \ge 5$. Then we have $Q_{x,k-1} \subset \frac{4}{3}Q$. Moreover, Lemma 4.3.5 implies that (6.2.30) holds true for any $f \in$ rbmo (μ). From these facts, together with (b) of Theorem 2.4.4, it follows that, for any $x \in Q$,

$$\begin{aligned}
&\mathcal{M}_S\left[f\chi_{\mathbb{R}^D\setminus\frac{4}{3}Q}\right](x)\\
&\quad= \sup_{1\le k\le 4}\left|\int_{Q_{x,k-1}} S_k(x,y)f(y)\chi_{\mathbb{R}^D\setminus\frac{4}{3}Q}(y)\,d\mu(y)\right|\\
&\quad\le \sup_{1\le k\le 4}\left\{\int_{Q_{x,k-1}} \frac{|f(y)-m_{Q_{x,k}}(f)|}{[|x-y|+\ell(Q_{x,k})]^n}\,d\mu(y) + \left|m_{Q_{x,k}}(f)\right|\right\}\\
&\quad\lesssim 1,
\end{aligned}$$

where in the last inequality,by Definition 4.2.1, $\left|m_{Q_{x,k}}(f)\right| \le 1$ if $k = 1$; and

$$\left|m_{Q_{x,k}}(f)\right| \le \left|m_{Q_{x,k}}(f) - m_{Q_{x,1}}(f)\right| + \left|m_{Q_{x,1}}(f)\right| \lesssim 1$$

if $2 \le k \le 4$. Therefore (6.2.33) follows and $\mathcal{M}_S(f)$ is finite almost everywhere.

We now prove that, for any doubling cube $Q \notin \mathcal{D}$,

$$\frac{1}{\mu(Q)} \int_Q \left[\mathcal{M}_S(f)(x) - \operatorname*{ess\,inf}_Q \mathcal{M}_S(f)\right] d\mu(x) \lesssim 1. \tag{6.2.34}$$

To this end, let

$$\mathcal{M}_{S,Q,1}(f)(x) := \sup_{k \geq H_Q^x + 1} |S_k(f)(x)|,$$

$$\mathcal{M}_{S,Q,2}(f)(x) := \sup_{1 \leq k \leq H_Q^x} |S_k(f)(x)|$$

(If $H_Q^x = 0$, then $\mathcal{M}_{S,Q,2}(f)$ disappears),

$$Q_1 := \{x \in Q : \mathcal{M}_{S,Q,1}(f)(x) \geq \mathcal{M}_{S,Q,2}(f)(x)\}$$

and $Q_2 := Q \setminus Q_1$. Moreover, write

$$\begin{aligned} f &= [f - m_Q(f)]\chi_{\frac{4}{3}Q} + [f - m_Q(f)]\chi_{\mathbb{R}^D \setminus \frac{4}{3}Q} + m_Q(f) \\ &=: f_1 + f_2 + m_Q(f). \end{aligned}$$

Then we have

$$\begin{aligned} &\frac{1}{\mu(Q)} \int_Q \left[\mathcal{M}_S(f)(x) - \operatorname*{ess\,inf}_Q \mathcal{M}_S(f)\right] d\mu(x) \\ &\quad \leq \frac{1}{\mu(Q)} \int_{Q_1} [\mathcal{M}_{S,Q,1}(f_1)(x) + \mathcal{M}_{S,Q,1}(f_2)(x)]\, d\mu(x) \\ &\qquad + \left[|m_Q(f)| - \operatorname*{ess\,inf}_Q \mathcal{M}_S(f)\right] \\ &\qquad + \frac{1}{\mu(Q)} \int_{Q_2} \left[\mathcal{M}_{S,Q,2}(f)(x) - \operatorname*{ess\,inf}_Q \mathcal{M}_S(f)\right] d\mu(x) \\ &\quad =: \mathrm{G}_1 + \mathrm{G}_2 + \mathrm{G}_3. \end{aligned}$$

We claim that

$$\frac{1}{\mu(Q)} \int_{Q_1} \mathcal{M}_{S,Q,1}(f_1)(x)\, d\mu(x) \lesssim 1. \tag{6.2.35}$$

Indeed, it is obvious that Proposition 3.1.4 holds true for any $f \in \mathrm{rbmo}\,(\mu)$ with $\|f\|_{\mathrm{rbmo}\,(\mu)} = 1$. Then, by the Hölder inequality, the boundedness of $\mathcal{M}_S$ in $L^2(\mu)$, the doubling property of Q, Corollary 4.2.9 and Lemma 2.1.3, we know that, if $\frac{4}{3}Q \in \mathcal{D}$, then

$$\begin{aligned} &\frac{1}{\mu(Q)} \int_{Q_1} \mathcal{M}_{S,Q,1}(f_1)(x)\, d\mu(x) \\ &\lesssim \left\{\frac{1}{\mu(Q)} \int_{\frac{4}{3}Q} |f(x) - m_Q(f)|^2\, d\mu(x)\right\}^{1/2} \end{aligned}$$

$$\lesssim \left\{\frac{1}{\mu(Q)}\int_{\frac{4}{3}Q}|f(x)|^2\,d\mu(x)\right\}^{1/2}+\left|m_Q(f)-m_{\widetilde{\frac{4}{3}Q}}(f)\right|+\left|m_{\widetilde{\frac{4}{3}Q}}(f)\right|$$
$$\lesssim 1$$

and, if $\frac{4}{3}Q\notin\mathcal{D}$, then

$$\frac{1}{\mu(Q)}\int_{Q_1}\mathcal{M}_{S,Q,1}(f_1)(x)\,d\mu(x)$$
$$\leq\left\{\frac{1}{\mu(Q)}\int_{\frac{4}{3}Q}\left|f(x)-m_{\widetilde{\frac{4}{3}Q}}(f)\right|^2\,d\mu(x)\right\}^{1/2}+\left|m_{\widetilde{\frac{4}{3}Q}}(f)-m_Q(f)\right|$$
$$\lesssim 1.$$

Therefore, (6.2.35) follows.

From (b) through (d) of Theorem 2.4.4, Proposition 3.1.4, (6.2.30), Lemma 4.1.2 (e) and the fact that $Q_{x,k-1}\subset\frac{4}{3}Q$ for $k\geq H_Q^x+4$, it follows that, for any $x\in Q_1$,

$$\begin{aligned}\mathcal{M}_{S,Q,1}(f_2)(x)\leq&\sup_{H_Q^x+1\leq k\leq H_Q^x+3}\Big\{S_k\left[\left|f-m_{Q_{x,k}}(f)\right|\right](x)\\&+\left|m_{Q_{x,k}}(f)-m_{Q_{x,H_Q^x+1}}(f)\right|\\&+\left|m_{Q_{x,H_Q^x+1}}(f)-m_Q(f)\right|\Big\}\\\lesssim&\,1.\end{aligned}$$

This and (6.2.35) lead to

$$\mathrm{G}_1\lesssim 1.$$

Observe that Lemma 4.3.5 implies that (6.2.31) also holds true for any $f\in\mathrm{rbmo}(\mu)$. Similar to the estimate for E_2 in the proof of Theorem 6.2.9, by Lemma 4.1.2(e) and (6.2.31), we conclude that, for any $y\in Q$,

$$|m_Q(f)|-\mathcal{M}_S(f)(y)\leq\left|m_Q(f)-S_{H_Q^y+1}(f)(y)\right|\lesssim 1,$$

which implies that

$$\mathrm{G}_2\lesssim 1.$$

On the other hand, it follows, from Proposition 3.1.4 and (6.2.31), that, for any $x,\,y\in Q$ and $1\leq k\leq H_Q^x$, it holds true that

$$\begin{aligned}
&|S_k f(x)| - \mathcal{M}_S(f)(y) \\
&\quad \le |S_k f(x) - m_{Q_{x,k}}(f)| + |m_{Q_{x,k}}(f) - m_{Q_{y,k}}(f)| \\
&\qquad + |m_{Q_{y,k}}(f) - S_k f(y)| \\
&\quad \lesssim 1.
\end{aligned}$$

Thus, we have

$$\mathrm{G}_3 \lesssim 1.$$

Combining the estimates for G_1 through G_3, we obtain (6.2.34).

Notice that, by (6.2.33), we see that, for any cube $Q \in \mathcal{D}$,

$$\operatorname*{ess\,inf}_{\tilde{Q}} \mathcal{M}_S(f) \lesssim \frac{1}{\mu(2\tilde{Q})} \int_{\tilde{Q}} \mathcal{M}_S(f)(x)\, d\mu(x) \lesssim 1.$$

Then, by (6.2.33) and (6.2.34), to complete the proof of Theorem 6.2.10 in the case that $\mathbb{R}^D$ is an initial cube, it suffices to prove that, for any doubling cubes $Q \subset R$ with $Q \notin \mathcal{D}$,

$$m_Q(\mathcal{M}(f)) - m_R(\mathcal{M}(f)) \lesssim 1 + \delta(Q, R).$$

Let

$$\mathcal{M}_{S,R,1}(f)(x) := \sup_{k \ge H_R^x + 1} |S_k(f)(x)|,$$

$$\mathcal{M}_{S,R,2}(f)(x) := \sup_{1 \le k \le H_R^x} |S_k(f)(x)|$$

(If $H_R^x = 0$, then $\mathcal{M}_{S,R,2}(f)$ disappears),

$$Q_1 := \{x \in Q : \mathcal{M}_{S,R,1}(f)(x) \ge \mathcal{M}_{S,R,2}(f)(x)\}$$

and $Q_2 := Q \setminus Q_1$. Since $\mathcal{M}_S$ is sublinear, it follows that

$$\begin{aligned}
&m_Q(\mathcal{M}_S(f)) - m_R(\mathcal{M}_S(f)) \\
&\quad \le \frac{1}{\mu(Q)} \int_{Q_1} \{\mathcal{M}_{S,R,1}(f_1)(x) + \mathcal{M}_{S,R,1}(f_2)(x)\}\, d\mu(x) \\
&\qquad + [|m_R(f)| - m_R(\mathcal{M}_S(f))] \\
&\qquad + \frac{1}{\mu(Q)} \int_{Q_2} \{\mathcal{M}_{S,R,2}(f)(x) - m_R(\mathcal{M}_S(f))\}\, d\mu(x) \\
&\quad =: \mathrm{H}_1 + \mathrm{H}_2 + \mathrm{H}_3,
\end{aligned}$$

where

$$f_1 := [f - m_R(f)]\chi_{\frac{4}{3}Q} \quad \text{and} \quad f_2 := [f - m_R(f)]\chi_{\mathbb{R}^D \setminus \frac{4}{3}Q}.$$

By the boundedness of $\mathcal{M}_S$ in $L^2(\mu)$, the Hölder inequality, the doubling property of Q, Corollary 4.2.9, Proposition 3.1.4 and Lemma 2.1.3, we see that, if $\frac{4}{3}Q \in \mathcal{D}$, then

$$\begin{aligned}
&\frac{1}{\mu(Q)}\int_{Q_1} \mathcal{M}_{S,Q,1}(f_1)(x)\,d\mu(x)\\
&\lesssim \left\{\frac{1}{\mu(Q)}\int_{\frac{4}{3}Q} |f(x)-m_R(f)|^2\,d\mu(x)\right\}^{1/2}\\
&\lesssim \left\{\frac{1}{\mu(Q)}\int_{\frac{4}{3}Q} |f(x)|^2\,d\mu(x)\right\}^{1/2} + |m_R(f)-m_Q(f)|\\
&\quad + \left|m_Q(f) - m_{\widetilde{\frac{4}{3}Q}}(f)\right| + \left|m_{\widetilde{\frac{4}{3}Q}}(f)\right|\\
&\lesssim 1 + \delta(Q,R)
\end{aligned}$$

and, if $\frac{4}{3}Q \notin \mathcal{D}$, then

$$\begin{aligned}
\frac{1}{\mu(Q)}\int_{Q_1} \mathcal{M}_{S,Q,1}(f_1)(x)\,d\mu(x) &\lesssim \left\{\frac{1}{\mu(Q)}\int_{\frac{4}{3}Q} \left|f(x) - m_{\widetilde{\frac{4}{3}Q}}(f)\right|^2 d\mu(x)\right\}^{1/2}\\
&\quad + \left|m_{\widetilde{\frac{4}{3}Q}}(f) - m_Q(f)\right| + |m_Q(f) - m_R(f)|\\
&\lesssim 1 + \delta(Q,R).
\end{aligned}$$

On the other hand, from the fact that $Q_{x,k-1} \subset \frac{4}{3}Q$ for $k \geq H_Q^x + 4$, (b), (c) and (d) of Theorem 2.4.4, (6.2.30), Lemmas 2.1.3 and 4.1.2, we deduce that, for any $x \in Q$,

$$\begin{aligned}
\mathcal{M}_{S,R,1}(f_2)(x) &= \sup_{H_R^x+1\leq k\leq H_Q^x+3} |S_k(f_2)(x)|\\
&\leq \sup_{H_R^x+1\leq k\leq H_Q^x+3} \int_{Q_{x,k-1}} S_k(x,z)|f(z) - m_R(f)|\,d\mu(z)\\
&\leq \sup_{H_R^x+1\leq k\leq H_Q^x+3} \left[\int_{Q_{x,k-1}} S_k(x,z)|f(z) - m_{Q_{x,k}}(f)|\,d\mu(z)\right.\\
&\quad \left. + \left|m_{Q_{x,k}}(f) - m_{\widetilde{3R}}(f)\right| + \left|m_{\widetilde{3R}}(f) - m_R(f)\right|\right]
\end{aligned}$$

$$\lesssim 1 + \delta(Q_{x, H_Q^x+3}, \widetilde{3R})$$
$$\lesssim 1 + \delta(Q, R).$$

Then we have

$$\frac{1}{\mu(Q)} \int_{Q_1} \mathcal{M}_{S, Q, 1}(f_2)(x)\, d\mu(x) \lesssim 1 + \delta(Q, R),$$

which, together with the estimate for $\mathcal{M}_{S, Q, 1}(f_1)(x)$, implies that

$$\mathrm{H}_1 \lesssim 1 + \delta(Q, R).$$

By Lemma 4.1.2(e), Proposition 3.1.4 and (6.2.31), we conclude that, for any $y \in R$,

$$\begin{aligned} |m_R(f)| - \mathcal{M}_S(f)(y) &\le \left| m_R(f) - S_{H_R^y+1}(f)(y) \right| \\ &\le \left| m_R(f) - m_{Q_{y, H_R^y+1}}(f) \right| \\ &\quad + \left| m_{Q_{y, H_R^y+1}}(f) - S_{H_R^y+1}(f)(y) \right| \\ &\lesssim 1. \end{aligned}$$

This further implies that

$$\mathrm{H}_2 \lesssim 1.$$

Moreover, for any $x, y \in R$ and $1 \le k \le H_R^x$, from the fact that $\delta(Q_{y,k}, Q_{x,k}) \lesssim 1$, Proposition 3.1.4 and (6.2.31), it follows that

$$\begin{aligned} &|S_k f(x)| - \mathcal{M}_S(f)(y) \\ &\quad \le |S_k f(x) - m_{Q_{x,k}}(f)| + |m_{Q_{x,k}}(f) - m_{Q_{y,k}}(f)| \\ &\qquad + |m_{Q_{y,k}}(f) - S_k f(y)| \\ &\quad \lesssim 1. \end{aligned}$$

Therefore, the proof of Theorem 6.2.10 in the case that $\mathbb{R}^D$ is an initial cube is completed.

When $\mathbb{R}^D$ is not an initial cube, the proof is similar, the details being omitted, which completes the proof of Theorem 6.2.10. □

6.3 Boundedness in Morrey-Type Spaces

In this section, we consider the boundedness of the Littlewood–Paley g-function in the Morrey-type spaces.[4]

Definition 6.3.1. Let $\eta,\ \rho \in (1,\infty)$, $p \in [1,\infty)$, $\beta_\rho := \rho^{D+1}$ and $\alpha \in (-\infty, 0]$. A function $f \in L^p_{\mathrm{loc}}(\mu)$ is said to belong to the *Morrey-type space* $\mathcal{C}^{\alpha,\,p}(\mu)$ if there exists some constant $\tilde{C} \in [0,\infty)$ such that, for any cube Q,

$$\left\{\frac{1}{\mu(\eta Q)}\int_Q \left|f(x) - m_{\tilde{Q}^\rho}(f)\right|^p d\mu(x)\right\}^{\frac{1}{p}} \le \tilde{C}[\mu(\eta Q)]^{\frac{\alpha}{n}} \tag{6.3.1}$$

and, for any two (ρ,β_ρ)-doubling cubes $Q \subset R$,

$$\left|m_Q(f) - m_R(f)\right| \le \tilde{C}[1+\delta(Q,\,R)][\mu(Q)]^{\frac{\alpha}{n}}. \tag{6.3.2}$$

The minimal constant $\tilde{C}$ as above is defined to be the *norm* of f in the space $\mathcal{C}^{\alpha,\,p}(\mu)$ and denoted by $\|f\|_{\mathcal{C}^{\alpha,\,p}(\mu)}$.

When $\alpha = 0$, the space $\mathcal{C}^{\alpha,\,p}(\mu)$ is just the space $\mathrm{RBMO}(\mu)$. Moreover, the space $\mathcal{C}^{\alpha,\,p}(\mu)$ has the following property.

Lemma 6.3.2. *Let $\alpha \in (-\infty,0]$, $\eta \in (1,\infty)$, $p \in [1,\infty)$ and $\rho \in [\eta,\infty)$. There exist positive constants C and $\tilde{C}$ such that,*

(i) *for any (ρ,ρ^{D+1})-doubling cubes Q and R,*

$$\left|m_Q(f) - m_R(f)\right| \le C\left(\max\left\{[\mu(Q)]^{\frac{\alpha}{n}},\ [\mu(R)]^{\frac{\alpha}{n}}\right\}\right) \times[1+\delta(Q,R)]\|f\|_{\mathcal{C}^{\alpha,\,p}(\mu)};$$

(ii) *for any cubes $Q \subset R$,*

$$\left|m_{\tilde{Q}^\rho}(f) - m_{\tilde{R}^\rho}(f)\right| \le \tilde{C}[\mu(\eta Q)]^{\frac{\alpha}{n}}[1+\delta(Q,R)]\|f\|_{\mathcal{C}^{\alpha,\,p}(\mu)}.$$

Proof. Since (ii) follows from (i), the fact that

$$\mu(\eta Q) \lesssim \min\left\{\mu(\tilde{Q}^\rho),\ \mu(\tilde{R}^\rho)\right\},$$

(b) and (e) of Lemma 2.1.3, we only need to prove (i). Without loss of generality, we may assume that $\ell(R_Q) \ge \ell(Q_R)$. Then $Q_R \subset 2R_Q$. Let $R_0 := \widetilde{2R_Q}^\rho$. From Lemma 2.1.3, we deduce that

$$\delta(R,R_0) = \delta(R,R_Q) + \delta(R_Q,R_0) \lesssim 1+\delta(Q,R).$$

[4] See also [117].

Thus, by (6.3.2), we find that

$$|m_R(f) - m_{R_0}(f)| \lesssim [\mu(R)]^{\frac{\alpha}{n}}[1 + \delta(Q, R)]\|f\|_{\mathcal{C}^{\alpha,p}(\mu)}. \tag{6.3.3}$$

On the other hand, we also have

$$\begin{aligned}\delta(Q, R_0) &\lesssim 1 + \delta(Q, 3R_Q) + \delta(3R_Q, R_0)\\ &\lesssim 1 + \delta(Q, Q_R) + \delta(Q_R, 3R_Q)\\ &\lesssim 1 + \delta(Q, R) + \delta(R, 3R_Q)\\ &\lesssim 1 + \delta(Q, R).\end{aligned}$$

By (6.3.2), we see that

$$|m_Q(f) - m_{R_0}(f)| \lesssim [\mu(Q)]^{\frac{\alpha}{n}}[1 + \delta(Q, R)]\|f\|_{\mathcal{C}^{\alpha,p}(\mu)}. \tag{6.3.4}$$

Therefore, (i) follows from (6.3.3) and (6.3.4), which completes the proof of Lemma 6.3.2. □

Proposition 6.3.3. *Let $\alpha \in (-\infty, 0]$, $\eta \in (1,\infty)$, $p \in [1,\infty)$ and $\rho \in [\eta,\infty)$.*

(i) *The space $\mathcal{C}^{\alpha,p}(\mu)$ is independent of the choice of η;*
(ii) *The space $\mathcal{C}^{\alpha,p}(\mu)$ is independent of the choice of ρ;*
(iii) *For any $f \in \mathcal{C}^{\alpha,p}(\mu)$, $|f| \in \mathcal{C}^{\alpha,p}(\mu)$ and there exists a positive constant C, independent of f, such that*

$$\||f|\|_{\mathcal{C}^{\alpha,p}(\mu)} \le C\|f\|_{\mathcal{C}^{\alpha,p}(\mu)};$$

(iv) *If $f, g \in \mathcal{C}^{\alpha,p}(\mu)$ are real-valued, then $\min\{f, g\}, \max\{f, g\} \in \mathcal{C}^{\alpha,p}(\mu)$ and there exists a positive constant $\tilde{C}$, independent of f and g, such that*

$$\|\min\{f, g\}\|_{\mathcal{C}^{\alpha,p}(\mu)},\ \|\max\{f, g\}\|_{\mathcal{C}^{\alpha,p}(\mu)} \le \tilde{C}\left[\|f\|_{\mathcal{C}^{\alpha,p}(\mu)} + \|g\|_{\mathcal{C}^{\alpha,p}(\mu)}\right].$$

Proof. Because (iii) is easy to show and (iv) follows from (iii) directly, we only need to prove (i) and (ii).

We first consider (i). To this end, assume that

$$\frac{\alpha}{n} + \frac{1}{p} > 0 \quad \text{and} \quad \eta_1 < \eta_2 \le \rho.$$

Then the inclusion $\mathcal{C}^{\alpha,p}_{\eta_1,\rho}(\mu) \subset \mathcal{C}^{\alpha,p}_{\eta_2,\rho}(\mu)$ is obvious. Conversely, let $f \in \mathcal{C}^{\alpha,p}_{\eta_2,\rho}(\mu)$ and Q be a cube. We have to estimate

$$[\mu(\eta_1 Q)]^{-\frac{\alpha}{n}-\frac{1}{p}}\left\{\int_Q \left|f(x) - m_{\tilde{Q}^\rho}(f)\right|^p d\mu(x)\right\}^{\frac{1}{p}}.$$

Observe that there exist some cubes $Q_1, \ldots, Q_N$ with the same sidelength such that

$$Q \subset \bigcup_{i=1}^{N} Q_i, \ \eta_2 Q_i \subset \eta_1 Q \quad \text{for all} \quad i \in \{1, \ldots, N\},$$

where N depends on D, η_1 and η_2. Arguing as in (3.1.12) and applying Lemma 6.3.2, we see that, for all $i \in \{1, \ldots, N\}$,

$$\left| m_{\tilde{Q}_i^\rho}(f) - m_{\tilde{Q}^\rho}(f) \right| \lesssim \|f\|_{\mathcal{C}^{\alpha,p}_{\eta_2,\rho}(\mu)} [\mu(\eta_2 Q_i)]^{\frac{\alpha}{n}}.$$

Using this fact, we easily obtain

$$\begin{aligned}
&[\mu(\eta_1 Q)]^{-\frac{\alpha}{n}-\frac{1}{p}} \left\{ \int_Q \left| f(x) - m_{\tilde{Q}^\rho}(f) \right|^p d\mu(x) \right\}^{\frac{1}{p}} \\
&\quad \leq \sum_{i=1}^{N} [\mu(\eta_2 Q_i)]^{-\frac{\alpha}{n}-\frac{1}{p}} \left\{ \left[\int_{Q_i} \left| f(x) - m_{\tilde{Q}_i^\rho}(f) \right|^p d\mu(x) \right]^{\frac{1}{p}} \right. \\
&\quad\quad \left. + [\mu(Q_i)]^{\frac{1}{p}} \left| m_{\tilde{Q}_i^\rho}(f) - m_{\tilde{Q}^\rho}(f) \right| \right\} \\
&\quad \lesssim \|f\|_{\mathcal{C}^{\alpha,p}_{\eta_2,\rho}(\mu)}.
\end{aligned}$$

Assume that $\frac{\alpha}{n} + \frac{1}{p} \leq 0$ and let $\eta_1 < \eta_2 \leq \rho$. It is trivial to see that, for any $f \in \mathcal{C}^{\alpha,p}_{\eta_2,\rho}(\mu)$, then

$$f \in \mathcal{C}^{\alpha,p}_{\eta_1,\rho}(\mu) \quad \text{and} \quad \|f\|_{\mathcal{C}^{\alpha,p}_{\eta_1,\rho}(\mu)} \leq \|f\|_{\mathcal{C}^{\alpha,p}_{\eta_2,\rho}(\mu)}.$$

Conversely, let $f \in \mathcal{C}^{\alpha,p}_{\eta_1,\rho}(\mu)$. For any cube Q, let $Q_0 := \frac{\eta_2}{\eta_1} Q$. By the facts that

$$\delta(\tilde{Q}^\rho, \widetilde{Q_0}^\rho) \lesssim 1 + \delta(Q, Q_0) \lesssim 1$$

and

$$\mu(\eta_2 Q) \lesssim \min\left\{ \mu\left(\tilde{Q}^\rho\right), \mu\left(\widetilde{Q_0}^\rho\right) \right\},$$

and Lemma 6.3.2(a), we have

$$\begin{aligned}
&\left\{ \int_Q \left| f(x) - m_{\tilde{Q}^\rho}(f) \right|^p d\mu(x) \right\}^{\frac{1}{p}} \\
&\quad \leq \left\{ \int_{Q_0} \left| f(x) - m_{\widetilde{Q_0}^\rho}(f) \right|^p d\mu(x) \right\}^{\frac{1}{p}} + [\mu(Q_0)]^{\frac{1}{p}} \left| m_{\widetilde{Q_0}^\rho}(f) - m_{\tilde{Q}^\rho}(f) \right|
\end{aligned}$$

$$\lesssim [\mu(\eta_1 Q_0)]^{\frac{\alpha}{n}+\frac{1}{p}} \|f\|_{\mathcal{C}^{\alpha,p}_{\eta_1,\rho}(\mu)} + [\mu(\eta_2 Q)]^{\frac{\alpha}{n}+\frac{1}{p}} \left[1+\delta\left(\tilde{Q}^{\rho}, \widetilde{Q_0}^{\rho}\right)\right] \|f\|_{\mathcal{C}^{\alpha,p}_{\eta_1,\rho}(\mu)}$$

$$\lesssim [\mu(\eta_2 Q)]^{\frac{\alpha}{n}+\frac{1}{p}} \|f\|_{\mathcal{C}^{\alpha,p}_{\eta_1,\rho}(\mu)},$$

which implies (i).

To show (ii), assume that $\eta \in (1,\infty)$ is fixed, $\rho_1, \rho_2 \in [\eta,\infty)$ and $f \in \mathcal{C}^{\alpha,p}_{\eta,\rho_1}(\mu)$. From (6.3.1), (6.3.2), Lemma 6.3.2(b), the Hölder inequality and the fact that ρ_1, $\rho_2 \geq \eta$, it follows that

$$\begin{aligned}
&\left\{\frac{1}{\mu(\eta Q)}\int_Q \left|f(x) - m_{\tilde{Q}^{\rho_2}}(f)\right|^p d\mu(x)\right\}^{\frac{1}{p}}\\
&\quad\leq \left\{\frac{1}{\mu(\eta Q)}\int_Q \left|f(x) - m_{\tilde{Q}^{\rho_1}}(f)\right|^p d\mu(x)\right\}^{\frac{1}{p}} + \left|m_{\tilde{Q}^{\rho_1}}(f) - m_{\tilde{Q}^{\rho_2}}(f)\right|\\
&\quad\leq \|f\|_{\mathcal{C}^{\alpha,p}_{\eta,\rho_1}(\mu)}[\mu(\eta Q)]^{\frac{\alpha}{n}} + \left|m_{\tilde{Q}^{\rho_1}}(f) - m_{\widetilde{\tilde{Q}^{\rho_2}}^{\rho_1}}(f)\right|\\
&\qquad + \left|m_{\widetilde{\tilde{Q}^{\rho_2}}^{\rho_1}}(f) - m_{\tilde{Q}^{\rho_2}}(f)\right|\\
&\quad\lesssim \|f\|_{\mathcal{C}^{\alpha,p}_{\eta,\rho_1}(\mu)}[\mu(\eta Q)]^{\frac{\alpha}{n}} + [1+\delta(Q,\tilde{Q}^{\rho_2})]\|f\|_{\mathcal{C}^{\alpha,p}_{\eta,\rho_1}(\mu)}[\mu(\eta Q)]^{\frac{\alpha}{n}}\\
&\qquad + \frac{1}{\mu(\tilde{Q}^{\rho_2})}\int_{\tilde{Q}^{\rho_2}} \left|f(x) - m_{\widetilde{\tilde{Q}^{\rho_2}}^{\rho_1}}(f)\right| d\mu(x)\\
&\quad\lesssim \|f\|_{\mathcal{C}^{\alpha,p}_{\eta,\rho_1}(\mu)}[\mu(\eta Q)]^{\frac{\alpha}{n}}.
\end{aligned} \tag{6.3.5}$$

On the other hand, for any (ρ_2, ρ_2^{D+1})-doubling cubes $Q \subset R$, Lemma 6.3.2(ii), together with Lemma 2.1.3, implies that

$$\begin{aligned}
|m_Q(f) - m_R(f)| &\leq \left|m_Q(f) - m_{\tilde{Q}^{\rho_1}}(f)\right| + \left|m_{\tilde{Q}^{\rho_1}}(f) - m_{\tilde{R}^{\rho_1}}(f)\right|\\
&\quad + \left|m_{\tilde{R}^{\rho_1}}(f) - m_R(f)\right|\\
&\lesssim [1+\delta(Q,R)]\|f\|_{\mathcal{C}^{\alpha,p}_{\eta,\rho_1}(\mu)}[\mu(Q)]^{\frac{\alpha}{n}}.
\end{aligned} \tag{6.3.6}$$

Combining (6.3.5) and (6.3.6), we see that $f \in \mathcal{C}^{\alpha,p}_{\eta,\rho_2}(\mu)$ and

$$\|f\|_{\mathcal{C}^{\alpha,p}_{\eta,\rho_2}(\mu)} \lesssim \|f\|_{\mathcal{C}^{\alpha,p}_{\eta,\rho_1}(\mu)}.$$

Since the converse holds true by symmetry, we complete the proof of (ii) and hence Proposition 6.3.3. □

Remark 6.3.4. As a result of (i) and (ii) of Proposition 6.3.3, unless explicitly pointed out, in what follows, when we mention $\mathcal{C}^{\alpha,\,p}(\mu)$, we *always take* $\rho = \eta = 2$ *in Definition 6.3.1.*

Definition 6.3.5. Let $\eta,\,\rho \in (1,\infty)$, $p \in [1,\infty)$, $\beta_\rho := \rho^{D+1}$ and $\alpha \in (-\infty, 0]$. A function $f \in L^p_{\rm loc}(\mu)$ is said to belong to the *space* $\mathcal{C}^{\alpha,\,p}_*(\mu)$ if there exists some constant $\tilde{C} \in [0,\infty)$ such that, for any cube Q,

$$\left\{\frac{1}{\mu(\eta Q)}\int_Q \left[f(x) - \operatorname*{ess\,inf}_{\tilde{Q}^\rho} f\right]^p d\mu(x)\right\}^{\frac{1}{p}} \le \tilde{C}[\mu(\eta Q)]^{\frac{\alpha}{n}}$$

and, for any two (ρ, ρ^{D+1})-doubling cubes $Q \subset R$,

$$m_Q(f) - m_R(f) \le \tilde{C}[1 + \delta(Q, R)][\mu(Q)]^{\frac{\alpha}{n}}.$$

The minimal constant $\tilde{C}$ as above is defined to be the *norm* of f in the space $\mathcal{C}^{\alpha,\,p}_*(\mu)$ and denoted by $\|f\|_{\mathcal{C}^{\alpha,\,p}_*(\mu)}$.

If $p = 1$ and $\alpha = 0$, then $\mathcal{C}^{\alpha,\,p}_*(\mu)$ is RBLO(μ). It is obvious that

$$\mathcal{C}^{\alpha,\,p}_*(\mu) \subset \mathcal{C}^{\alpha,\,p}(\mu).$$

Moreover, the space $\mathcal{C}^{\alpha,\,p}_*(\mu)$ enjoys the following properties that are easy to prove, the details being omitted.

Proposition 6.3.6. *Let* $\alpha \in (-\infty, 0]$, $\eta \in (1,\infty)$, $p \in [1,\infty)$ *and* $\rho \in [\eta,\infty)$.

(i) *The space* $\mathcal{C}^{\alpha,\,p}_*(\mu)$ *is independent of the choice of* η.
(ii) *The definition of* $\mathcal{C}^{\alpha,\,p}_*(\mu)$ *is independent of the choice of* ρ.

Remark 6.3.7. As a result of (i) and (ii) of Proposition 6.3.6, unless explicitly pointed out, in what follows, when we mention $\mathcal{C}^{\alpha,\,p}_*(\mu)$, we *always take* $\rho = \eta = 2$ *in Definition 6.3.5.*

We now establish the boundedness from $\mathcal{C}^{\alpha,\,p}(\mu)$ to $\mathcal{C}^{\alpha,\,p}_*(\mu)$ of the homogeneous Littlewood–Paley g-function $\dot{g}(f)$. To this end, we need the following estimate.

Lemma 6.3.8. *There exists a positive constant* C *such that, for any cubes* $Q \subset R$ *and* $f \in \mathcal{C}^{\alpha,\,p}(\mu)$,

$$\int_R \frac{|f(y) - m_{\tilde{Q}}(f)|}{[|y - z_Q| + \ell(Q)]^n}\,d\mu(y) \le C[1 + \delta(Q, R)]^2\|f\|_{\mathcal{C}^{\alpha,\,p}(\mu)}[\mu(2Q)]^{\frac{\alpha}{n}}.$$

Proof. From the Hölder inequality, (0.0.1) and Definition 6.3.1, it follows that

$$\int_Q \frac{|f(y)-m_{\tilde{Q}}(f)|}{[|y-z_Q|+\ell(Q)]^n}\,d\mu(y) \le \frac{1}{[\ell(Q)]^n}\int_Q \left|f(y)-m_{\tilde{Q}}(f)\right|\,d\mu(y)$$
$$\lesssim [\mu(2Q)]^{\frac{\alpha}{n}}.$$

Therefore, to show Lemma 6.3.8, it suffices to show that

$$\int_{R\setminus Q} \frac{|f(y)-m_{\tilde{Q}}(f)|}{|y-z_Q|^n}\,d\mu(y) \lesssim [1+\delta(Q,R)]^2[\mu(2Q)]^{\frac{\alpha}{n}}. \tag{6.3.7}$$

By (0.0.1), the Hölder inequality, Lemmas 6.3.2 and 2.1.3, and Definition 6.3.1, we conclude that

$$\begin{aligned}
&\int_{R\setminus Q} \frac{|f(y)-m_{\tilde{Q}}(f)|}{|y-z_Q|^n}\,d\mu(y)\\
&\quad\lesssim \sum_{k=0}^{N_{Q,R}} \frac{1}{[\ell(2^{k+1}Q)]^n}\int_{2^{k+1}Q\setminus 2^kQ} \left|f(y)-m_{\tilde{Q}}(f)\right|\,d\mu(y)\\
&\quad\lesssim \sum_{k=0}^{N_{Q,R}} \frac{1}{[\ell(2^{k+1}Q)]^n}\int_{2^{k+1}Q\setminus 2^kQ} \left|f(y)-m_{\widetilde{2^{k+1}Q}}(f)\right|\,d\mu(y)\\
&\qquad+\sum_{k=0}^{N_{Q,R}} \frac{\mu\left(2^{k+1}Q\right)}{[\ell(2^{k+1}Q)]^n}\left|m_{\widetilde{2^{k+1}Q}}(f)-m_{\tilde{Q}}(f)\right|\\
&\quad\lesssim \sum_{k=1}^{N_{Q,R}+1} \frac{\mu(2^kQ)}{[\ell\left(2^kQ\right)]^n}[\mu(2^kQ)]^{\frac{\alpha}{n}}\\
&\qquad+\sum_{k=1}^{N_{Q,R}+1} \frac{\mu(2^kQ)}{[\ell\left(2^kQ\right)]^n}\left[1+\delta\left(Q,2^kQ\right)\right][\mu(2Q)]^{\frac{\alpha}{n}}\\
&\quad\lesssim \{\delta_{Q,R}+\delta_{Q,R}[1+\delta(Q,\,R)]\}[\mu(2Q)]^{\frac{\alpha}{n}}\\
&\quad\lesssim [1+\delta(Q,\,R)]^2[\mu(2Q)]^{\frac{\alpha}{n}},
\end{aligned}$$

which completes the proof of Lemma 6.3.8. □

Theorem 6.3.9. *Let $\alpha\in(-\infty,0]$ and $p\in(1,\infty)$. If $\mathbb{R}^D$ is not an initial cube, then, for any $f\in \mathcal{C}^{\alpha,\,p}(\mu)$, $\dot{g}(f)$ is either infinite everywhere or finite almost everywhere and, in the latter case, there exists a positive constant C, independent of f, such that*

$$\|\dot{g}(f)\|_{\mathcal{C}_*^{\alpha,\,p}(\mu)} \le C\|f\|_{\mathcal{C}^{\alpha,\,p}(\mu)}.$$

Proof. By the homogeneity, we may assume $\|f\|_{\mathcal{C}^{\alpha,p}(\mu)} = 1$. We first prove that, if there exists $x_0 \in \mathbb{R}^D$ satisfying $\dot{g}(f)(x_0) < \infty$, then, for any cube $Q \ni x_0$,

$$\left\{\frac{1}{\mu(2Q)}\int_Q \left|\dot{g}(f)(x) - \inf_{\tilde{Q}} \dot{g}(f)\right|^p d\mu(x)\right\}^{\frac{1}{p}} \lesssim [\mu(2Q)]^{\frac{\alpha}{n}}. \tag{6.3.8}$$

Write

$$\left[\dot{g}^{H_{\tilde{Q}}^x}(f)(x)\right]^2 := \sum_{k=H_{\tilde{Q}}^x+4}^{\infty} |D_k(f)(x)|^2$$

and

$$\left[\dot{g}_{H_{\tilde{Q}}^x}(f)(x)\right]^2 := \sum_{k=-\infty}^{H_{\tilde{Q}}^x-4} |D_k(f)(x)|^2.$$

Then we have

$$\begin{aligned}
&\left\{\frac{1}{\mu(2Q)}\int_Q \left|\dot{g}(f)(x) - \inf_{\tilde{Q}} \dot{g}(f)\right|^p d\mu(x)\right\}^{\frac{1}{p}} \\
&\le \left\{\frac{1}{\mu(2Q)}\int_Q \left[\dot{g}^{H_{\tilde{Q}}^x}(f)(x)\right]^p d\mu(x)\right\}^{\frac{1}{p}} \\
&\quad + \left\{\frac{1}{\mu(2Q)}\int_Q \left[\sum_{k=H_{\tilde{Q}}^x-3}^{H_{\tilde{Q}}^x+3} |D_k f(x)|^2\right]^{\frac{p}{2}} d\mu(x)\right\}^{\frac{1}{p}} \\
&\quad + \left\{\frac{1}{\mu(2Q)}\int_Q \left\{\dot{g}_{H_{\tilde{Q}}^x}(f)(x) - \inf_{\tilde{Q}} \dot{g}_{H_{\tilde{Q}}^x}(f)\right\}^p d\mu(x)\right\}^{\frac{1}{p}} \\
&=: \mathrm{F}_1 + \mathrm{F}_2 + \mathrm{F}_3.
\end{aligned}$$

We first estimate F_1. By (6.2.7) and Lemma 4.1.2(a), we know that, for $k \ge H_{\tilde{Q}}^x + 4$ and $x \in \mathbb{R}^D$,

$$D_k(f)(x) = D_k([f - m_{\widetilde{\frac{7}{5}Q}}(f)]\chi_{\frac{7}{5}Q})(x).$$

Then, from the boundedness of $\dot{g}(f)$ in $L^p(\mu)$ for $p \in (1,\infty)$ and Proposition 6.3.3(i), it follows that

$$\begin{aligned}
\mathrm{F}_1 &\le \left[\frac{1}{\mu(2Q)}\int_Q \left\{\dot{g}\left(\left[f - m_{\widetilde{\frac{7}{5}Q}}(f)\right]\chi_{\frac{7}{5}Q}\right)(x)\right\}^p d\mu(x)\right]^{\frac{1}{p}}\\
&\lesssim \left\{\frac{1}{\mu(2Q)}\int_{\frac{7}{5}Q}\left|f(x) - m_{\widetilde{\frac{7}{5}Q}}(f)\right|^p d\mu(x)\right\}^{\frac{1}{p}}\\
&\lesssim [\mu(2Q)]^{\frac{\alpha}{n}}.
\end{aligned}$$

Similarly, applying the $L^p(\mu)$ boundedness of $\dot{g}(f)$, we have

$$\begin{aligned}
&\left[\frac{1}{\mu(2Q)}\int_Q\left\{\sum_{k=H_Q^x-3}^{H_Q^x+3}\left|D_k\left\{\left(f - m_{\widetilde{\frac{7}{5}Q}}(f)\right)\chi_{\frac{7}{5}Q}\right\}(x)\right|^2\right\}^{p/2} d\mu(x)\right]^{\frac{1}{p}}\\
&\quad\lesssim \left\{\frac{1}{\mu(2Q)}\int_{\frac{7}{5}Q}\left|f(x) - m_{\widetilde{\frac{7}{5}Q}}(f)\right|^p d\mu(x)\right\}^{\frac{1}{p}}\\
&\quad\lesssim [\mu(2Q)]^{\frac{\alpha}{n}}.
\end{aligned}$$

Therefore, by (6.2.7), the estimate of F_2 is reduced to showing that

$$\begin{aligned}
&\left[\frac{1}{\mu(2Q)}\int_Q\left\{\sum_{k=H_Q^x-3}^{H_Q^x+3}\left|D_k\left[\left(f - m_{\widetilde{\frac{7}{5}Q}}(f)\right)\chi_{\mathbb{R}^D\setminus\frac{7}{5}Q}\right](x)\right|^2\right\}^{p/2} d\mu(x)\right]^{\frac{1}{p}}\\
&\quad\lesssim [\mu(2Q)]^{\frac{\alpha}{n}}.
\end{aligned}$$

To this end, we only need to show that, for any cubes $Q \subset R$ and $x \in Q$,

$$\sum_{k=H_R^x-3}^{H_Q^x+3}\left|D_k\left[\left(f - m_{\widetilde{\frac{7}{5}Q}}(f)\right)\chi_{\mathbb{R}^D\setminus\frac{7}{5}Q}\right](x)\right| \lesssim [1+\delta(Q,R)]^2[\mu(2Q)]^{\frac{\alpha}{n}}. \tag{6.3.9}$$

Indeed, for each $x \in Q$ and $k \in [H_R^x - 3, H_Q^x + 3]$, write

$$Q_{x,k-2} = (Q_{x,k-2}\setminus Q_{x,k-1})\bigcup(Q_{x,k-1}\setminus Q_{x,k})\bigcup Q_{x,k}.$$

It follows, from (c) and (d) of Theorem 2.4.4, that

$$\begin{aligned}
&\sum_{k=H_R^x-3}^{H_Q^x+3} \left| D_k \left[\left(f - m_{\widetilde{\frac{7}{5}Q}}(f) \right) \chi_{\mathbb{R}^D \setminus \frac{7}{5}Q} \right](x) \right| \\
&\lesssim \sum_{k=H_R^x-3}^{H_Q^x+3} \int_{Q_{x,k-2}} \frac{|f(z) - m_{\widetilde{\frac{7}{5}Q}}(f)|}{[|x-z| + \ell(Q_{x,k})]^n} \chi_{\mathbb{R}^D \setminus \frac{7}{5}Q}(z)\, d\mu(z) \\
&\lesssim \int_{Q_{x,H_R^x-5} \setminus \frac{7}{5}Q} \frac{|f(z) - m_{\widetilde{\frac{7}{5}Q}}(f)|}{|x-z|^n}\, d\mu(z) \\
&\quad + \sum_{k=H_R^x-3}^{H_Q^x+3} \int_{Q_{x,k} \setminus \frac{7}{5}Q} \frac{|f(z) - m_{\widetilde{\frac{7}{5}Q}}(f)|}{[|x-z| + \ell(Q_{x,k})]^n}\, d\mu(z) \\
&=: \mathrm{J}_1 + \mathrm{J}_2.
\end{aligned}$$

Notice that

$$\frac{7}{5}Q \subset Q_{x,H_R^x-5} \quad \text{and} \quad |z_Q - z| \lesssim |x-z| \quad \text{for any} \quad z \notin \frac{7}{5}Q,$$

where z_Q is the center of Q. By (6.3.7), Lemmas 4.1.2 and 2.1.3, and the fact that $\alpha \le 0$, we have

$$\begin{aligned}
\mathrm{J}_1 &\lesssim \left[1 + \delta\left(\frac{7}{5}Q, Q_{x,H_R^x-5}\right)\right]^2 \left[\mu\left(\frac{14}{5}Q\right)\right]^{\frac{\alpha}{n}} \\
&\lesssim [1 + \delta(Q,R)]^2 [\mu(2Q)]^{\frac{\alpha}{n}}.
\end{aligned}$$

Observe that, for each $x \in Q$, $\{Q_{x,k}\}_{k\in\mathbb{Z}}$ is decreasing in k and $Q_{x,H_Q^x+3} \subset \frac{7}{5}Q$. Let

$$k_0 \in \left[H_R^x - 3, H_Q^x + 3\right]$$

be the largest integer such that $Q_{x,k_0} \not\subset \frac{7}{5}Q$. Then, for each $k \in \left[H_R^x - 3, k_0\right]$, it holds true that $2Q \subset 30Q_{x,k}$. Let

$$N_0 := N_{Q_{x,k_0}, Q_{x,H_R^x-3}}.$$

By Lemma 2.2.5, we see that, for any integer $k \in \left[H_R^x - 3, k_0\right]$, there exists a unique integer $j_k \in [0, N_0]$ such that

$$2^{j_k} Q_{x,k_0} \subset Q_{x,k} \subset 2^{j_k+1} Q_{x,k_0}$$

and, for different k, j_k is different. On the other hand, by (e) and (d) of Lemma 2.1.3 and Lemma 4.1.2(e), for any $x \in Q$ and $k \in [H_R^x - 3, H_Q^x + 3]$, we see that

$$\delta(Q, Q_{x,k}) + \delta(Q_{x,k}, R) \lesssim 1 + \delta(Q, R). \tag{6.3.10}$$

It then follows, from Definition 6.3.1, the doubling property of $Q_{x,k}$, Lemmas 6.3.2 and 2.1.3, and (6.3.10), that

$$\begin{aligned}
\mathrm{J}_2 &\leq \sum_{k=H_R^x-3}^{k_0} \int_{15Q_{x,k}} \frac{|f(z) - m_{\widetilde{15Q_{x,k}}}(f)|}{[\ell(Q_{x,k})]^n} \, d\mu(z) \\
&\quad + \sum_{k=H_R^x-3}^{k_0} \frac{\mu(15Q_{x,k})}{[\ell(Q_{x,k})]^n} \left| m_{\widetilde{15Q_{x,k}}}(f) - m_{\widetilde{\frac{7}{5}Q}}(f) \right| \\
&\lesssim \sum_{k=H_R^x-3}^{k_0} \frac{\mu(30Q_{x,k})}{[\ell(Q_{x,k})]^n} [\mu(30Q_{x,k})]^{\frac{\alpha}{n}} \\
&\quad + \sum_{k=H_R^x-3}^{k_0} \frac{\mu(30Q_{x,k})}{[\ell(Q_{x,k})]^n} \left[1 + \delta\left(\frac{7}{5}Q, 15Q_{x,k}\right) \right] [\mu(2Q)]^{\frac{\alpha}{n}} \\
&\lesssim \sum_{j_k=0}^{N_0} \frac{\mu\left(2^{j_k+1} 30 Q_{x,k_0}\right)}{\left[\ell\left(2^{j_k} 30 Q_{x,k_0}\right)\right]^n} \left[1 + \delta\left(Q, R\right)\right] [\mu(2Q)]^{\frac{\alpha}{n}} \\
&\lesssim \left[1 + \delta\left(Q, R\right)\right]^2 [\mu(2Q)]^{\frac{\alpha}{n}}.
\end{aligned}$$

Consequently, (6.3.9) follows from the combination of estimates for J_1 and J_2.

Now we estimate F_3. For any $x \in Q$ and $k \leq H_{\tilde{Q}}^x$, it holds true that $Q \subset Q_{x,k}$. By (c) and (d) of Theorem 2.4.4, Lemma 6.3.8 and the doubling property of $Q_{x,k}$, we know that

$$\begin{aligned}
|D_k(f)(x)| &\lesssim \int_{Q_{x,k-2}} \frac{|f(z) - m_{Q_{x,k}}(f)|}{[\ell(Q_{x,k}) + |x - z|]^n} \, d\mu(z) \\
&\lesssim [1 + \delta(Q_{x,k}, Q_{x,k-2})]^2 [\mu(2Q_{x,k})]^{\frac{\alpha}{n}} \\
&\lesssim [\mu(2Q)]^{\frac{\alpha}{n}}.
\end{aligned} \tag{6.3.11}$$

From this, we deduce that, for any $x \in Q$ and $y \in \tilde{Q}$,

$$\dot{g}_{H_Q^x}(f)(x) - \dot{g}_{H_Q^x}(f)(y)$$
$$\leq \left\{ \sum_{k=-\infty}^{H_{\tilde{Q}}^x - 4} |D_k(f)(x) - D_k(f)(y)||D_k(f)(x) + D_k(f)(y)| \right\}^{\frac{1}{2}}$$
$$\lesssim [\mu(2Q)]^{\frac{\alpha}{2n}} \left\{ \sum_{k=-\infty}^{H_{\tilde{Q}}^x - 4} |D_k(f)(x) - D_k(f)(y)| \right\}^{\frac{1}{2}}.$$

We now claim that, for any $y \in \tilde{Q}$, $i \geq 4$ and $k \geq H_{\tilde{Q}}^x - i + 5$,

$$\operatorname{supp}(D_k(\cdot, y) - D_k(\cdot, x)) \subset Q_{x, H_{\tilde{Q}}^x - i + 1}.$$

This can be seen by the fact that

$$\begin{aligned}\operatorname{supp}(D_k(\cdot, y) - D_k(\cdot, x)) &\subset \left(Q_{y,k-3} \bigcup Q_{x,k-3}\right) \\ &\subset Q_{x,k-4} \\ &\subset Q_{x, H_{\tilde{Q}}^x - i + 1}.\end{aligned}$$

From this fact, we deduce that

$$\sum_{k=-\infty}^{H_{\tilde{Q}}^x - 4} |D_k(f)(x) - D_k(f)(y)|$$
$$\leq \sum_{i=4}^{\infty} \int_{Q_{x, H_{\tilde{Q}}^x - i} \setminus Q_{x, H_{\tilde{Q}}^x - i + 1}} \sum_{k=-\infty}^{H_{\tilde{Q}}^x - i + 4} |D_k(x,z) - D_k(y,z)|$$
$$\times \left| f(z) - m_{Q_{x, H_{\tilde{Q}}^x}}(f) \right| d\mu(z)$$
$$+ \int_{Q_{x, H_{\tilde{Q}}^x - 3}} \sum_{k=-\infty}^{H_{\tilde{Q}}^x - 4} |D_k(x,z) - D_k(y,z)| \left| f(z) - m_{Q_{x, H_{\tilde{Q}}^x}}(f) \right| d\mu(z)$$
$$=: \mathrm{G}_1 + \mathrm{G}_2.$$

Let us estimate G_1. Observe that, for each $i \geq 4$ and $k \leq H^x_{\tilde{Q}} - i + 4$, it holds true that $x, y \in Q_{x,k}$. Then, by (e) of Theorem 2.4.4 and the fact that

$$\ell(Q_{x,k}) \leq \frac{1}{10}\ell(Q_{x,k-1}) \quad \text{for any} \quad x \in \operatorname{supp}\mu \quad \text{and} \quad k \in \mathbb{Z},$$

we conclude that

$$\begin{aligned}\sum_{k=-\infty}^{H^x_{\tilde{Q}}-i+4} |D_k(x,z) - D_k(y,z)| &\lesssim \sum_{k=-\infty}^{H^x_{\tilde{Q}}-i+4} \frac{|x-y|}{\ell(Q_{x,k})[\ell(Q_{x,k}) + |x-z|]^n} \\ &\lesssim \frac{\ell(\tilde{Q})}{\ell(Q_{H^x_{\tilde{Q}}-i+4})|x-z|^n}.\end{aligned}$$

This, together with (6.3.7), implies that

$$\begin{aligned}\mathrm{G}_1 &\lesssim \sum_{i=4}^{\infty} \frac{\ell(\tilde{Q})}{\ell(Q_{x,H^x_{\tilde{Q}}-i+4})} \int_{Q_{x,H^x_{\tilde{Q}}-i} \setminus Q_{x,H^x_{\tilde{Q}}-i+1}} \frac{|f(z) - m_{Q_{x,H^x_{\tilde{Q}}}}(f)|}{|x-z|^n}\,d\mu(z) \\ &\lesssim \sum_{i=4}^{\infty} \frac{\ell(\tilde{Q})}{\ell(Q_{x,H^x_{\tilde{Q}}-i+4})} \int_{Q_{x,H^x_{\tilde{Q}}-i} \setminus Q_{x,H^x_{\tilde{Q}}-i+1}} \frac{|f(z) - m_{Q_{x,H^x_{\tilde{Q}}-i+1}}(f)|}{|x-z|^n}\,d\mu(z) \\ &\quad + \sum_{i=4}^{\infty} \frac{\ell(\tilde{Q})}{\ell(Q_{x,H^x_{\tilde{Q}}-i+4})} \left| m_{Q_{x,H^x_{\tilde{Q}}-i+1}}(f) - m_{Q_{x,H^x_{\tilde{Q}}}}(f) \right| \\ &\quad \times \left[1 + \delta\left(Q_{x,H^x_{\tilde{Q}}-i+1}, Q_{x,H^x_{\tilde{Q}}-i}\right)\right] \\ &\lesssim \sum_{i=4}^{\infty} \frac{\ell(\tilde{Q})}{\ell(Q_{x,H^x_{\tilde{Q}}-i+4})} \left[\mu(Q_{x,H^x_{\tilde{Q}}-i+1})\right]^{\frac{\alpha}{n}} \left[1 + \delta\left(Q_{x,H^x_{\tilde{Q}}-i+1}, Q_{x,H^x_{\tilde{Q}}-i}\right)\right]^2 \\ &\quad + \sum_{i=4}^{\infty} \frac{\ell(\tilde{Q})}{\ell(Q_{x,H^x_{\tilde{Q}}-i+4})} i \left[\mu(Q_{x,H^x_{\tilde{Q}}})\right]^{\frac{\alpha}{n}} \\ &\lesssim [\mu(Q)]^{\frac{\alpha}{n}}.\end{aligned}$$

To estimate G_2, by (d) of Theorem 2.4.4, Lemma 2.1.3, the facts that $\tilde{Q} \subset Q_{x,H^x_{\tilde{Q}}-3}$ for any $x \in Q$ and that

$$\ell(Q_{x,k}) \leq \frac{1}{10}\ell(Q_{x,k-1})$$

for any $k \in \mathbb{Z}$, we first see that

$$\int_{Q_{x,H_{\tilde{Q}}^x-3}} \sum_{k=-\infty}^{H_{\tilde{Q}}^x-4} |D_k(x,z)| \left| f(z) - m_{Q_{x,H_{\tilde{Q}}^x}}(f) \right| d\mu(z)$$

$$\lesssim \int_{Q,H_{\tilde{Q}}^x-3} \sum_{k=-\infty}^{H_{\tilde{Q}}^x-4} \frac{|f(z) - m_{Q_{x,H_{\tilde{Q}}^x}}(f)|}{[|x-z| + \ell(Q_{x,k})]^n} d\mu(z)$$

$$\lesssim \int_{Q_{x,H_{\tilde{Q}}^x-3}} \frac{|f(z) - m_{Q_{x,H_{\tilde{Q}}^x}}(f)|}{[\ell(Q_{x,H_{\tilde{Q}}^x-3})]^n} d\mu(z)$$

$$\lesssim \int_{Q_{x,H_{\tilde{Q}}^x-3}} \frac{|f(z) - m_{Q_{x,H_{\tilde{Q}}^x-3}}(f)|}{[\ell(Q_{x,H_{\tilde{Q}}^x-3})]^n} d\mu(z) + \left| m_{Q_{x,H_{\tilde{Q}}^x-3}}(f) - m_{Q_{x,H_{\tilde{Q}}^x}}(f) \right|$$

$$\lesssim [\mu(Q)]^{\frac{\alpha}{n}}.$$

On the other hand, for any $z \in Q_{y,H_{\tilde{Q}}^x-4}$, we know that

$$\sum_{k=-\infty}^{H_{\tilde{Q}}^x-4} \frac{1}{[|y-z| + \ell(Q_{y,k})]^n} \lesssim \frac{1}{[\ell(Q_{x,H_{\tilde{Q}}^x-3})]^n}.$$

This, together with Theorem 2.4.4(d), implies that

$$\int_{Q_{x,H_{\tilde{Q}}^x-3}} \sum_{k=-\infty}^{H_{\tilde{Q}}^x-4} |D_k(y,z)| \left| f(z) - m_{Q_{x,H_{\tilde{Q}}^x}}(f) \right| d\mu(z)$$

$$\lesssim \int_{Q_{x,H_{\tilde{Q}}^x-3}} \sum_{k=-\infty}^{H_{\tilde{Q}}^x-4} \frac{|f(z) - m_{Q_{x,H_{\tilde{Q}}^x}}(f)|}{[|y-z| + \ell(Q_{y,k})]^n} d\mu(z)$$

$$\lesssim \int_{Q_{x,H_{\tilde{Q}}^x-3}} \frac{|f(z) - m_{Q_{x,H_{\tilde{Q}}^x-3}}(f)|}{[\ell(Q_{x,H_{\tilde{Q}}^x-3})]^n} d\mu(z)$$

$$\lesssim [\mu(Q)]^{\frac{\alpha}{n}}.$$

Therefore, we see that

$$\mathrm{G}_2 \lesssim \int_{Q_{x,H_{\tilde{Q}}^x-3}} \left\{ \sum_{k=-\infty}^{H_{\tilde{Q}}^x-4} [|D_k(x,z)| + |D_k(y,z)|] \left| f(z) - m_{Q_{x,H_{\tilde{Q}}^x}}(f) \right| \right\} d\mu(z)$$

$$\lesssim [\mu(Q)]^{\frac{\alpha}{n}},$$

which, together with the estimate for G_1, implies that

$$\mathrm{F}_3 \lesssim [\mu(Q)]^{\frac{\alpha}{n}}.$$

Combining estimates for F_1 through F_3, we obtain (6.3.8).

By Lemma 6.2.3, to complete the proof of Theorem 6.3.9, it suffices to prove that, for any doubling cubes $Q \subset R$,

$$m_Q(\dot{g}(f)) - m_R(\dot{g}(f)) \lesssim [1 + \delta(Q,R)]^4 [\mu(Q)]^{\frac{\alpha}{n}}. \tag{6.3.12}$$

For any $x \in Q$, assume that $H_{\tilde{Q}}^x \geq H_{\tilde{R}}^x + 10$. We then write

$$\begin{aligned}
&m_Q(\dot{g}(f)) - m_R(\dot{g}(f)) \\
&\leq \frac{1}{\mu(Q)} \int_Q \dot{g}^{H_{\tilde{Q}}^x}(f)(x)\, d\mu(x) + \frac{1}{\mu(Q)} \int_Q \left\{ \sum_{k=H_{\tilde{R}}^x-3}^{H_{\tilde{Q}}^x+3} |D_k f(x)|^2 \right\}^{\frac{1}{2}} d\mu(x) \\
&\quad + \frac{1}{\mu(Q)} \frac{1}{\mu(R)} \int_Q \int_R \{[\dot{g}_{H_{\tilde{R}}^x}(f)(x)] - [\dot{g}_{H_{\tilde{R}}^x}(f)(y)]\}\, d\mu(y)\, d\mu(x) \\
&=: \mathrm{H}_1 + \mathrm{H}_2 + \mathrm{H}_3.
\end{aligned}$$

Similar to the estimate for F_1, we have

$$\mathrm{H}_1 \lesssim [\mu(Q)]^{\frac{\alpha}{n}}.$$

Moreover, arguing as in the estimate of F_3, we conclude that

$$\mathrm{H}_3 \lesssim [\mu(R)]^{\frac{\alpha}{n}} \lesssim [\mu(Q)]^{\frac{\alpha}{n}}.$$

Therefore, the proof of (6.3.12) is reduced to proving that

$$\mathrm{H}_2 \lesssim [1 + \delta(Q,R)]^2 [\mu(Q)]^{\frac{\alpha}{n}}.$$

By the Hölder inequality and the $L^p(\mu)$ boundedness of $\dot{g}(f)$, we see that

$$\frac{1}{\mu(Q)}\int_Q \left\{\sum_{k=H_R^x-3}^{H_Q^x+3} \left|D_k\left[\left(f - m_{\widetilde{\frac{7}{5}Q}}(f)\right)\chi_{\frac{7}{5}Q}\right](x)\right|^2\right\}^{\frac{1}{2}} d\mu(x)$$

$$\leq \left\{\frac{1}{\mu(Q)}\int_Q \left[\sum_{k=H_R^x-3}^{H_Q^x+3} \left|D_k\left[\left(f - m_{\widetilde{\frac{7}{5}Q}}(f)\right)\chi_{\frac{7}{5}Q}\right](x)\right|^2\right]^{\frac{p}{2}} d\mu(x)\right\}^{\frac{1}{p}}$$

$$\lesssim [\mu(Q)]^{\frac{\alpha}{n}}.$$

By this and (6.3.9), we find that

$$\mathrm{H}_2 \lesssim [\mu(Q)]^{\frac{\alpha}{n}}.$$

Consequently, (6.3.12) follows from the combination of estimates for H_1 through H_3.

If $H_R^x \leq H_Q^x \leq H_R^x + 9$, by an analogous argument, we also see that (6.3.12) holds true, which completes the proof of Theorem 6.3.9. □

Remark 6.3.10. There is not any result for the boundedness of the Littlewood–Paley g-function $g(f)$ on $\mathcal{C}^{\alpha,p}(\mu)$. It is reasonable to conjecture that $g(f)$ is bounded from a local version of $\mathcal{C}^{\alpha,p}(\mu)$ to a local version of $\mathcal{C}_*^{\alpha,p}(\mu)$.

Now we consider the boundedness of the maximal operator $\dot{\mathcal{M}}_S(f)$ in (6.1.22) from $\mathcal{C}^{\alpha,p}(\mu)$ to $\mathcal{C}_*^{\alpha,p}(\mu)$. Recall that $\dot{\mathcal{M}}_S(f)$ is bounded on $L^p(\mu)$ for all $p \in (1,\infty)$.

Theorem 6.3.11. *Let $\alpha \in (-\infty, 0]$ and $p \in (1,\infty)$. If $\mathbb{R}^D$ is not an initial cube, for any $f \in \mathcal{C}^{\alpha,p}(\mu)$, $\dot{\mathcal{M}}_S(f)$ is either infinite everywhere or finite almost everywhere and, in the latter case, there exists a positive constant C, independent of f, such that*

$$\left\|\dot{\mathcal{M}}_S(f)\right\|_{\mathcal{C}_*^{\alpha,p}(\mu)} \leq C\|f\|_{\mathcal{C}^{\alpha,p}(\mu)}.$$

Proof. By the homogeneity, we may assume $\|f\|_{\mathcal{C}^{\alpha,p}(\mu)} = 1$. We first show that, if there exists $x_0 \in \mathbb{R}^D$ such that $\dot{\mathcal{M}}_S(f)(x_0) < \infty$, then, for any cube $Q \ni x_0$,

$$\left\{\frac{1}{\mu(2Q)}\int_Q \left|\dot{\mathcal{M}}_S(f)(x) - \inf_{\tilde{Q}} \dot{\mathcal{M}}_S(f)\right|^p d\mu(x)\right\}^{\frac{1}{p}} \lesssim [\mu(2Q)]^{\frac{\alpha}{n}}. \qquad (6.3.13)$$

To this end, write

$$f = \left[f - m_{\widetilde{\frac{7}{5}Q}}(f)\right]\chi_{\frac{7}{5}Q} + \left[f - m_{\widetilde{\frac{7}{5}Q}}(f)\right]\chi_{\mathbb{R}^D\setminus\frac{7}{5}Q} + m_{\widetilde{\frac{7}{5}Q}}(f)$$
$$=: f_1 + f_2 + m_{\widetilde{\frac{7}{5}Q}}(f).$$

For any $x \in Q$, let

$$\dot{\mathcal{M}}_{S,Q,1}(f)(x) := \sup_{k \geq H_{\tilde{Q}}^x} |S_k(f)(x)|,$$

$$\dot{\mathcal{M}}_{S,Q,2}(f)(x) := \sup_{k \leq H_{\tilde{Q}}^x - 1} |S_k(f)(x)|,$$

$$Q_1 := \{x \in Q : \dot{\mathcal{M}}_{S,Q,1}(f)(x) \geq \dot{\mathcal{M}}_{S,Q,2}(f)(x)\} \quad \text{and} \quad Q_2 := Q \setminus Q_1.$$

Then, for any $x \in Q$,

$$\dot{\mathcal{M}}_S(f)(x) = \max\left\{\dot{\mathcal{M}}_{S,Q,1}(f)(x),\ \dot{\mathcal{M}}_{S,Q,2}(f)(x)\right\}.$$

Moreover, by the sublinearity of $\dot{\mathcal{M}}_S$, we know that

$$\left\{\frac{1}{\mu(2Q)}\int_Q \left|\dot{\mathcal{M}}_S(f)(x) - \inf_{\tilde{Q}} \dot{\mathcal{M}}_S(f)\right|^p d\mu(x)\right\}^{\frac{1}{p}}$$
$$\leq \left\{\frac{1}{\mu(2Q)}\int_{Q_1} \left|\dot{\mathcal{M}}_S(f)(x) - \inf_{\tilde{Q}} \left|S_{H_{\tilde{Q}}^x - 1}(f)\right|\right|^p d\mu(x)\right\}^{\frac{1}{p}}$$
$$+ \left\{\frac{1}{\mu(2Q)}\int_{Q_2} \left|\dot{\mathcal{M}}_S(f)(x) - \inf_{\tilde{Q}} \dot{\mathcal{M}}_{S,Q,2}(f)\right|^p d\mu(x)\right\}^{\frac{1}{p}}$$
$$\leq \left\{\frac{1}{\mu(2Q)}\int_{Q_1} \left[\dot{\mathcal{M}}_S(f_1)(x) + \dot{\mathcal{M}}_S(f_2)(x)\right.\right.$$
$$\left.\left. + \left|\left|m_{\widetilde{\frac{7}{5}Q}}(f)\right| - \inf_{\tilde{Q}} \left|S_{H_{\tilde{Q}}^x - 1}(f)\right|\right|\right]^p d\mu(x)\right\}^{\frac{1}{p}}$$
$$+ \left\{\frac{1}{\mu(2Q)}\int_{Q_2} \left|\dot{\mathcal{M}}_S(f)(x) - \inf_{\tilde{Q}} \dot{\mathcal{M}}_{S,Q,2}(f)\right|^p d\mu(x)\right\}^{\frac{1}{p}}$$
$$=: \mathrm{L}_1 + \mathrm{L}_2.$$

By the fact that $\dot{\mathcal{M}}_S$ is bounded on $L^p(\mu)$ for all $p \in (1,\infty)$ and Proposition 6.3.3(i), we find that

$$\left\{\frac{1}{\mu(2Q)}\int_{Q_1}\left|\dot{\mathcal{M}}_S(f_1)(x)\right|^p\,d\mu(x)\right\}^{\frac{1}{p}} \lesssim [\mu(2Q)]^{\frac{\alpha}{n}}. \tag{6.3.14}$$

We now show that, for any $x \in Q_1$,

$$\dot{\mathcal{M}}_S(f_2)(x) \lesssim [\mu(2Q)]^{\frac{\alpha}{n}}. \tag{6.3.15}$$

Recall that, for $x \in Q_1$,

$$\dot{\mathcal{M}}(f)(x) = \dot{\mathcal{M}}_{S,Q,1}(f)(x).$$

If $k \geq H_Q^x + 3$, then, by applying Lemma 4.1.2(b), we know that $Q_{x,k-1} \subset \frac{7}{5}Q$. This fact, together with (c) of Theorem 2.4.4, implies that, for any $x \in Q_1$,

$$\dot{\mathcal{M}}_{S,Q,1}(f_2)(x) = \sup_{H_{\tilde{Q}}^x \leq k \leq H_Q^x+2} |S_k(f_2)(x)|.$$

For each $k \in [H_{\tilde{Q}}^x, H_Q^x + 2]$, by Lemmas 4.1.2 and 2.1.3, we find that

$$\delta\left(\frac{7}{5}Q, Q_{x,k}\right) \lesssim 1.$$

If $Q_{x,k-1} \not\subset \frac{7}{5}Q$, then $\frac{7}{5}Q \subset 15Q_{x,k-1}$. Since, for any $z \in (\mathbb{R}^D \setminus \frac{7}{5}Q)$ and $x \in Q$, $|z_Q - z| \lesssim |x - z|$, from (6.3.7), Lemmas 6.3.2(ii) and 2.1.3, and (0.0.1), it follows that

$$\begin{aligned}
|S_k(f_2)(x)| &\lesssim \int_{Q_{x,k-1}\setminus\frac{7}{5}Q} \frac{|f(z) - m_{\widetilde{\frac{7}{5}Q}}(f)|}{[|z-x| + \ell(Q_{x,k})]^n}\,d\mu(z)\\
&\lesssim \int_{15Q_{x,k-1}\setminus\frac{7}{5}Q} \frac{|f(z) - m_{\tilde{Q}}(f)|}{|z - z_Q|^n}\,d\mu(z)\\
&\quad + \left|m_{\tilde{Q}}(f) - m_{\widetilde{\frac{7}{5}Q}}(f)\right| \int_{15Q_{x,k-1}\setminus\frac{7}{5}Q} \frac{1}{|z-z_Q|^n}\,d\mu(z)\\
&\lesssim \left[1 + \delta\left(\frac{7}{5}Q, 15Q_{x,k-1}\right) + \delta(Q, \frac{7}{5}Q)\right]^2 [\mu(2Q)]^{\frac{\alpha}{n}}\\
&\lesssim [\mu(2Q)]^{\frac{\alpha}{n}}.
\end{aligned} \tag{6.3.16}$$

If $Q_{x,k-1} \subset \frac{7}{5}Q$, then $|S_k(f_2)(x)| = 0$ and hence (6.3.16) also holds true. This implies (6.3.15).

Notice that, by Lemma 4.1.2(c), for any $x \in Q$, $y \in \tilde{Q}$, it holds true that

$$|H_{\tilde{Q}}^x - H_{\tilde{Q}}^y| \le 1.$$

Then we have

$$Q \subset Q_{y,H_{\tilde{Q}}^x - 1} \quad \text{and} \quad \delta\left(Q, Q_{y,H_{\tilde{Q}}^x - 1}\right) \lesssim 1.$$

Moreover, both $2\widetilde{\frac{7}{5}Q}$ and $2Q_{y,H_{\tilde{Q}}^x - 1}$ contain $2Q$. On the other hand, by Lemma 6.3.8, and (b), (c) and (d) of Theorem 2.4.4, we see that, for any $k \in \mathbb{Z}$ and $x \in \operatorname{supp}\mu$,

$$\begin{aligned} |S_k(f)(x) - m_{Q_{x,k}}(f)| &\lesssim \int_{Q_{x,k-1}} \frac{|f(y) - m_{Q_{x,k}}(f)|}{[|x-y| + \ell(Q_{x,k})]^n}\, d\mu(y) \\ &\lesssim [\mu(Q_{x,k})]^{\frac{\alpha}{n}}. \end{aligned} \tag{6.3.17}$$

From these facts and Lemma 6.3.2(i), we deduce that, for any $y \in \tilde{Q}$,

$$\begin{aligned} &\left| m_{\widetilde{\frac{7}{5}Q}}(f) - S_{H_{\tilde{Q}}^x - 1}(f)(y) \right| \\ &\le \left| m_{\widetilde{\frac{7}{5}Q}}(f) - m_{Q_{y,H_{\tilde{Q}}^x - 1}}(f) \right| + \left| m_{Q_{y,H_{\tilde{Q}}^x - 1}}(f) - S_{H_{\tilde{Q}}^x - 1}(f)(y) \right| \\ &\lesssim \left[1 + \delta\left(\widetilde{\frac{7}{5}Q}, Q_{y,H_{\tilde{Q}}^x - 1} \right) \right] [\mu(2Q)]^{\frac{\alpha}{n}} \\ &\quad + \int_{Q_{y,H_{\tilde{Q}}^x - 2}} \frac{|f(z) - m_{Q_{y,H_{\tilde{Q}}^x - 1}}(f)|}{[\ell(Q_{y,H_{\tilde{Q}}^x - 1}) + |y-z|]^n}\, d\mu(z) \\ &\lesssim [\mu(2Q)]^{\frac{\alpha}{n}}. \end{aligned}$$

Combining this estimate with (6.3.14) and (6.3.15), we find that

$$\mathrm{L}_1 \lesssim [\mu(2Q)]^{\frac{\alpha}{n}}.$$

To estimate L_2, we first claim that, for any cube Q, $x, y \in \operatorname{supp}\mu \cap Q$ and $k \le H_Q^x$, it holds true that

$$\delta(Q_{x,k}, Q_{y,k}) \lesssim 1. \tag{6.3.18}$$

Indeed, by the facts that $y \in Q$ and $k \le H_Q^x$, we see that $y \in Q_{x,k}$. By Lemma 2.3.3, we further know that

$$Q_{x,k} \subset Q_{y,k-1} \subset Q_{x,k-2}.$$

Then, (e) and (d) of Lemma 2.1.3 imply that

$$\begin{aligned}\delta(Q_{x,k}, Q_{y,k}) &\lesssim \delta(Q_{x,k}, Q_{y,k-1}) + \delta(Q_{y,k}, Q_{y,k-1}) + 1\\ &\lesssim \delta(Q_{x,k}, Q_{x,k-2}) + 1\\ &\lesssim 1,\end{aligned}$$

which implies (6.3.18).

Observe that, by Lemma 4.1.2(c), we conclude that, for any $x \in Q$, $y \in \tilde{Q}$ and $k \le H_{\tilde{Q}}^{x} - 1$,

$$\tilde{Q} \subset Q_{y,\, H_{\tilde{Q}}^{y}} \subset Q_{y,\, H_{\tilde{Q}}^{x}-1} \subset Q_{y,k}.$$

This, together with the fact that $\tilde{Q} \subset Q_{x,k}$, Lemma 6.3.2(i) and (6.3.18), implies that

$$|m_{Q_{x,k}}(f) - m_{Q_{y,k}}(f)| \lesssim [\mu(\tilde{Q})]^{\frac{\alpha}{n}},$$

from which and (6.3.17) it follows that, for any $x \in Q$ and $y \in \tilde{Q}$,

$$\begin{aligned}&\sup_{k\le H_{\tilde{Q}}^{x}-1} |S_k(f)(x)| - \sup_{k\le H_{\tilde{Q}}^{x}-1} |S_k(f)(y)|\\ &\quad\le \sup_{k\le H_{\tilde{Q}}^{x}-1} |S_k(f)(x) - S_k(f)(y)|\\ &\quad\le \sup_{k\le H_{\tilde{Q}}^{x}-1} \Big[|S_k(f)(x) - m_{Q_{x,k}}(f)| + |m_{Q_{x,k}}(f) - m_{Q_{y,k}}(f)|\\ &\qquad + |m_{Q_{y,k}}(f) - S_k(f)(y)|\Big]\\ &\quad\lesssim [\mu(2Q)]^{\frac{\alpha}{n}}.\end{aligned}$$

This leads to

$$\mathrm{L}_2 \lesssim [\mu(2Q)]^{\frac{\alpha}{n}}.$$

Combining estimates for L_1 and L_2, we obtain (6.3.13).

By (6.3.13) and Lemma 6.2.3, to finish the proof of Theorem 6.3.11, it suffices to show that, for any doubling cubes $Q \subset R$,

$$m_Q(\dot{\mathcal{M}}_S(f)) - m_R(\dot{\mathcal{M}}_S(f)) \lesssim [1 + \delta(Q, R)]^2 [\mu(Q)]^{\frac{\alpha}{n}}. \tag{6.3.19}$$

For any $x \in Q$, let

$$\dot{\mathcal{M}}_{S,R,1}(f)(x) := \sup_{k \geq H_R^x} |S_k(f)(x)|,$$

$$\dot{\mathcal{M}}_{S,R,2}(f)(x) := \sup_{k \leq H_R^x - 1} |S_k(f)(x)|,$$

$$Q_1 := \{x \in Q : \dot{\mathcal{M}}_{S,R,1}(f)(x) \geq \dot{\mathcal{M}}_{S,R,2}(f)(x)\} \quad \text{and} \quad Q_2 := Q \setminus Q_1.$$

Split

$$\begin{aligned} f &= [f - m_{\widetilde{\frac{7}{5}R}}(f)]\chi_{\frac{7}{5}Q} + [f - m_{\widetilde{\frac{7}{5}R}}(f)]\chi_{\mathbb{R}^D \setminus \frac{7}{5}Q} + m_{\widetilde{\frac{7}{5}R}}(f) \\ &=: f_1 + f_2 + m_{\widetilde{\frac{7}{5}R}}(f). \end{aligned}$$

From the fact that $\dot{\mathcal{M}}_{S,R,1}$ is sublinear, it follows that

$$\begin{aligned} &m_Q\left(\dot{\mathcal{M}}_S(f)\right) - m_R\left(\dot{\mathcal{M}}_S(f)\right) \\ &\leq \frac{1}{\mu(Q)} \int_{Q_1} \{\dot{\mathcal{M}}_{S,R,1}(f_1)(x) \\ &\quad + \dot{\mathcal{M}}_{S,R,1}(f_2)(x) + \left[\left|m_{\widetilde{\frac{7}{5}R}}(f)\right| - m_R\left(\dot{\mathcal{M}}_S(f)\right)\right]\} \, d\mu(x) \\ &\quad + \frac{1}{\mu(Q)} \int_{Q_2} \{\dot{\mathcal{M}}_{S,R,2}(f)(x) - m_R\left(\dot{\mathcal{M}}_S(f)\right)\} \, d\mu(x) \\ &=: \mathrm{F}_1 + \mathrm{F}_2. \end{aligned}$$

By the Hölder inequality, the boundedness of $\dot{\mathcal{M}}_S$ in $L^p(\mu)$, Lemmas 6.3.2(ii) and 2.1.3, and the doubling property of Q, we see that

$$\begin{aligned} &\frac{1}{\mu(Q)} \int_{Q_1} \dot{\mathcal{M}}_{S,R,1}(f_1)(x) \, d\mu(x) \\ &\lesssim \left\{\frac{1}{\mu(Q)} \int_{\frac{7}{5}Q} \left|f(x) - m_{\widetilde{\frac{7}{5}Q}}(f)\right|^p d\mu(x)\right\}^{1/p} + \left|m_{\widetilde{\frac{7}{5}Q}}(f) - m_{\widetilde{\frac{7}{5}R}}(f)\right| \\ &\lesssim [1 + \delta(Q, R)][\mu(Q)]^{\frac{\alpha}{n}}. \end{aligned} \tag{6.3.20}$$

By (c) of Theorem 2.4.4 and the fact that $Q_{x,k-1} \subset \frac{7}{5}Q$ for $k \geq H_Q^x + 3$ and any $x \in Q \cap \operatorname{supp} \mu$, to estimate $\mathcal{M}_{S,R,1}(f_2)(x)$, we only need to consider the case when $k \in [H_R^x, H_Q^x + 2]$. If $Q_{x,k-1} \subset \frac{7}{5}Q$, then by another application of Theorem 2.4.4(c), we know that $S_k(f_2)(x) = 0$. If $Q_{x,k-1} \not\subset \frac{7}{5}Q$, then $\frac{7}{5}Q \subset 15Q_{x,k-1}$. Observe that, for any $y \in (\mathbb{R}^D \setminus \frac{7}{5}Q)$ and $x \in Q$, it holds true that $|z_Q - y| \lesssim |x - y|$. From (b), (c) and (d) of Theorem 2.4.4, (6.3.7), (6.3.10), Lemmas 6.3.2(ii) and 2.1.3, we deduce that, for each $k \in [H_R^x, H_Q^x + 2]$ and $x \in Q$,

$$\begin{aligned}|S_k(f_2)(x)| &\le S_k\left[\left|f - m_{\widetilde{\frac{7}{5}Q}}(f)\right| \chi_{\mathbb{R}^D\setminus\frac{7}{5}Q}\right](x) + \left|m_{\widetilde{\frac{7}{5}Q}}(f) - m_{\widetilde{\frac{7}{5}R}}(f)\right| \\ &\lesssim \int_{15Q_{x,k-1}\setminus\frac{7}{5}Q} \frac{|f(y) - m_{\widetilde{\frac{7}{5}Q}}(f)|}{|z_Q - y|^n}\, d\mu(y) + [1 + \delta(Q, R)][\mu(Q)]^{\frac{\alpha}{n}} \\ &\lesssim \left[1 + \delta\left(\frac{7}{5}Q, 15Q_{x,k-1}\right) + \delta(Q, R)\right]^2 [\mu(Q)]^{\frac{\alpha}{n}} \\ &\lesssim [1 + \delta(Q, R)]^2[\mu(Q)]^{\frac{\alpha}{n}}.\end{aligned}$$

On the other hand, Lemmas 6.3.2 and 4.1.2(e), and (6.3.17) imply that, for any $y \in R$,

$$\begin{aligned}\left|m_{\widetilde{\frac{7}{5}R}}(f)\right| - \dot{\mathcal{M}}_S(f)(y) &\le \left|m_{\widetilde{\frac{7}{5}R}}(f) - S_{H_R^y}(f)(y)\right| \\ &\le \left|m_{\widetilde{\frac{7}{5}R}}(f) - m_{Q_{y,H_R^y}}(f)\right| \\ &\quad + \left|m_{Q_{y,H_R^y}}(f) - S_{H_R^y}(f)(y)\right| \\ &\lesssim [\mu(R)]^{\frac{\alpha}{n}}.\end{aligned}$$

Taking average over $y \in R$, we see that

$$\left|m_{\widetilde{\frac{7}{5}R}}(f)\right| - m_R\left(\dot{\mathcal{M}}_S(f)\right) \lesssim [\mu(R)]^{\frac{\alpha}{n}}.$$

We then conclude that

$$\mathrm{F}_1 \lesssim [1 + \delta(Q, R)]^2[\mu(Q)]^{\frac{\alpha}{n}}.$$

Observe that, for any $x,\ y \in R$ and $k \le H_R^x - 1$, it holds true that

$$\delta(Q_{y,k}, Q_{x,k}) \lesssim 1 \quad \text{and} \quad Q \subset R \subset Q_{y,k}.$$

By this, together with Lemma 6.3.2 and (6.3.17), we see that

$$\begin{aligned}&|S_k(f)(x)| - \dot{\mathcal{M}}_S(f)(y) \\ &\quad \le |S_k(f)(x) - m_{Q_{x,k}}(f)| + |m_{Q_{x,k}}(f) - m_{Q_{y,k}}(f)| \\ &\quad\quad + |m_{Q_{y,k}}(f) - S_k(f)(y)| \\ &\quad \lesssim [\mu(R)]^{\frac{\alpha}{n}},\end{aligned}$$

which implies that

$$\mathrm{F}_3 \lesssim [\mu(Q)]^{\frac{\alpha}{n}}$$

and hence completes the proof of Theorem 6.3.11. □

Remark 6.3.12. There is not any result for the boundedness of the maximal function $\mathcal{M}_S$ on $\mathcal{C}^{\alpha,p}(\mu)$. It is reasonable to conjecture that $\mathcal{M}_S$ is bounded from a local version of $\mathcal{C}^{\alpha,p}(\mu)$ to a local version of $\mathcal{C}_*^{\alpha,p}(\mu)$.

Finally, we consider the boundedness of approximations of the identity on $\mathcal{C}^{\alpha,p}(\mu)$.

Theorem 6.3.13. *Let $p \in (1,\infty)$ and $\alpha \in (-\infty,0]$. For any $k \in \mathbb{Z}$, let S_k be as in Sect. 2.4. Then $\{S_k\}_k$ is uniformly bounded on $\mathcal{C}^{\alpha,p}(\mu)$, namely, there exists a nonnegative constant C, independent of k, such that, for all $f \in \mathcal{C}^{\alpha,p}(\mu)$,*

$$\|S_k(f)\|_{\mathcal{C}^{\alpha,p}(\mu)} \leq C\|f\|_{\mathcal{C}^{\alpha,p}(\mu)}.$$

Proof. Without loss of generality, we may assume that $\|f\|_{\mathcal{C}^{\alpha,p}(\mu)} = 1$. We only need to consider the case that $\mathbb{R}^D$ is not an initial cube. Indeed, if $\mathbb{R}^D$ is an initial cube, the argument when $k \in \mathbb{N}$ is similar to that used in the case that $\mathbb{R}^D$ is not an initial cube and, since $S_k = 0$ when $k \leq 0$, the conclusion of Theorem 6.3.13 holds true automatically in this case.

To prove the conclusion of Theorem 6.3.13 in the case that $\mathbb{R}^D$ is not an initial cube, by Lemma 3.1.8, it suffices to show that, for any cube Q,

$$\left\{\frac{1}{\mu(2Q)}\int_Q \left[S_k(f)(x) - m_{\tilde{Q}}(S_k(f))\right]^p \, d\mu(x)\right\}^{\frac{1}{p}} \lesssim [\mu(Q)]^{\frac{\alpha}{n}} \tag{6.3.21}$$

and, for any two doubling cubes $Q \subset R$,

$$\left|m_Q(S_k(f)) - m_R(S_k(f))\right| \lesssim [1+\delta(Q,R)]^2[\mu(Q)]^{\frac{\alpha}{n}}. \tag{6.3.22}$$

To show (6.3.21), let us consider the following three cases:

(i) there exists some point $x_0 \in Q \cap \operatorname{supp}\mu$ such that $k \geq H_Q^{x_0} + 4$;
(ii) there exists some point $x_0 \in Q \cap \operatorname{supp}\mu$ such that $k \leq H_{\tilde{Q}}^{x_0} - 2$;
(iii) for any point $x \in Q$,

$$k \in [H_{\tilde{Q}}^x - 1, H_Q^x + 3].$$

In Case (i), by (c) and (a) of Lemma 4.1.2, we see that, for any $x \in Q \cap \operatorname{supp}\mu$,

$$k \geq H_Q^x + 3 \quad \text{and} \quad Q_{x,k-1} \subset \frac{7}{5}Q.$$

Recall that $\{S_k\}_k$ is uniformly bounded on $L^p(\mu)$ for $p \in (1, \infty)$. From (b) and (c) of Theorem 2.4.4, the Hölder inequality and Proposition 6.3.3(i), it follows that

$$\begin{aligned}
&\left\{\frac{1}{\mu(2Q)} \int_Q \left[S_k(f)(x) - m_{\tilde{Q}}(S_k(f))\right]^p d\mu(x)\right\}^{\frac{1}{p}} \\
&\leq \left\{\frac{1}{\mu(2Q)} \int_Q \left|S_k\left(\left[f - m_{\widetilde{\frac{7}{5}Q}}(f)\right]\right)(x)\right|^p d\mu(x)\right\}^{\frac{1}{p}} \\
&\quad + \left|m_{\widetilde{\frac{7}{5}Q}}(f) - m_{\tilde{Q}}(S_k(f))\right| \\
&\lesssim \left\{\frac{1}{\mu(2Q)} \int_Q \left|S_k\left(\left[f - m_{\widetilde{\frac{7}{5}Q}}(f)\right]\chi_{\frac{7}{5}Q}\right)\right|^p d\mu(x)\right\}^{\frac{1}{p}} \\
&\quad + \left\{\frac{1}{\mu(2\tilde{Q})} \int_{\tilde{Q}} \left|S_k\left(\left[f - m_{\widetilde{\frac{7}{5}Q}}(f)\right]\chi_{\frac{7}{5}Q}\right)\right|^p d\mu(x)\right\}^{\frac{1}{p}} \\
&\lesssim \left\{\frac{1}{\mu(2Q)} \int_{\frac{7}{5}Q} \left|f(x) - m_{\widetilde{\frac{7}{5}Q}}(f)\right|^p d\mu(x)\right\}^{\frac{1}{p}} \\
&\lesssim [\mu(2Q)]^{\frac{\alpha}{n}}.
\end{aligned}$$

In Case (ii), by Lemma 4.1.2(a), we see that, for any $x,\ y \in Q \cap \operatorname{supp} \mu$,

$$k \leq H_{\tilde{Q}}^x - 1, \quad 2Q \subset Q_{x,k} \quad \text{and} \quad \tilde{Q} \subset Q_{y,k} \bigcap Q_{x,k}.$$

Thus, by Theorem 2.4.4(b), together with (6.3.17), Lemma 6.3.2 and (6.3.18), we know that

$$\begin{aligned}
&\left|S_k(f)(x) - m_{\tilde{Q}}(S_k(f))\right| \\
&\leq \left|S_k\left(\left[f - m_{Q_{x,k}}(f)\right]\right)(x)\right| + \left|m_{\tilde{Q}}\left(S_k(f) - m_{Q_{x,k}}(f)\right)\right| \\
&\lesssim [\mu(Q_{x,k})]^{\frac{\alpha}{n}} \\
&\quad + \frac{1}{\mu(\tilde{Q})} \int_{\tilde{Q}} \left[\left|S_k(f)(y) - m_{Q_{y,k}}(f)\right| + \left|m_{Q_{y,k}}(f) - m_{Q_{x,k}}(f)\right|\right] d\mu(y) \\
&\lesssim [\mu(Q_{x,k})]^{\frac{\alpha}{n}} + \frac{1}{\mu(\tilde{Q})} \int_{\tilde{Q}} \left\{[\mu(Q_{y,k})]^{\frac{\alpha}{n}} + [\mu(\tilde{Q})]^{\frac{\alpha}{n}}\right\} d\mu(y) \\
&\lesssim [\mu(2Q)]^{\frac{\alpha}{n}},
\end{aligned}$$

which implies (6.3.21).

In Case (iii), Lemmas 2.1.3(e) and 4.1.2(e) lead to that, for any $x \in Q$,

$$\delta(Q, Q_{x,k}) \lesssim 1 + \delta(Q, Q_{x,H_Q^x}) + \delta(Q_{x,H_Q^x}, Q_{x,k}) \lesssim 1. \tag{6.3.23}$$

On the other hand, by writing

$$\begin{aligned} \left|S_k(f)(x) - m_{\tilde{Q}}(S_k(f))\right| &\le \left|S_k\left(\left[f - m_{\widetilde{\frac{7}{5}Q}}(f)\right]\right)(x)\right| \\ &\quad + \left|m_{\tilde{Q}}\left(S_k\left[f - m_{\widetilde{\frac{7}{5}Q}}(f)\right]\right)\right|, \end{aligned}$$

we have

$$\begin{aligned} &\left\{\frac{1}{\mu(2Q)}\int_Q \left[S_k(f)(x) - m_{\tilde{Q}}(S_k(f))\right]^p d\mu(x)\right\}^{\frac{1}{p}} \\ &\lesssim \left\{\frac{1}{\mu(2Q)}\int_Q \left|S_k\left(\left[f - m_{\widetilde{\frac{7}{5}Q}}(f)\right]\chi_{\frac{7}{5}Q}\right)(x)\right|^p d\mu(x)\right\}^{\frac{1}{p}} \\ &\quad + \left\{\frac{1}{\mu(2Q)}\int_Q \left|S_k\left(\left[f - m_{\widetilde{\frac{7}{5}Q}}(f)\right]\chi_{\mathbb{R}^D\setminus\frac{7}{5}Q}\right)(x)\right|^p d\mu(x)\right\}^{\frac{1}{p}} \\ &\quad + \left\{\frac{1}{\mu(2\tilde{Q})}\int_{\tilde{Q}} \left|S_k\left(\left[f - m_{\widetilde{\frac{7}{5}Q}}(f)\right]\chi_{\frac{7}{5}\tilde{Q}}\right)(x)\right|^p d\mu(x)\right\}^{\frac{1}{p}} \\ &\quad + \left\{\frac{1}{\mu(2\tilde{Q})}\int_{\tilde{Q}} \left|S_k\left(\left[f - m_{\widetilde{\frac{7}{5}Q}}(f)\right]\chi_{\mathbb{R}^D\setminus\frac{7}{5}\tilde{Q}}\right)(x)\right|^p d\mu(x)\right\}^{\frac{1}{p}} \\ &=: \mathrm{H}_1 + \mathrm{H}_2 + \mathrm{H}_3 + \mathrm{H}_4. \end{aligned}$$

The uniform boundedness of $\{S_k\}_k$ in $L^p(\mu)$, together with Proposition 6.3.3, implies that

$$\mathrm{H}_1 \lesssim [\mu(2Q)]^{\frac{\alpha}{n}}.$$

Similarly, by the uniform $L^p(\mu)$ boundedness of $\{S_k\}_k$, together with Proposition 6.3.3, Lemmas 6.3.2(b) and 2.1.3, and the fact that

$$\mu(2Q) \lesssim \min\left\{\mu\left(\widetilde{\frac{7}{5}Q}\right), \mu\left(\widetilde{\frac{7}{5}\tilde{Q}}\right)\right\},$$

we conclude that

$$\begin{aligned}
\mathrm{H}_3 &\lesssim \left\{\frac{1}{\mu(2\tilde{Q})}\int_{\frac{7}{5}\tilde{Q}}\left|f(x)-m_{\widetilde{\frac{7}{5}\tilde{Q}}}(f)\right|^p\,d\mu(x)\right\}^{\frac{1}{p}}+\left|m_{\widetilde{\frac{7}{5}\tilde{Q}}}(f)-m_{\widetilde{\frac{7}{5}Q}}(f)\right| \\
&\lesssim [\mu(2\tilde{Q})]^{\frac{\alpha}{n}}+[\mu(2Q)]^{\frac{\alpha}{n}}\left[1+\delta\left(\frac{7}{5}Q,\frac{7}{5}\tilde{Q}\right)\right] \\
&\lesssim [\mu(2Q)]^{\frac{\alpha}{n}}.
\end{aligned}$$

For any $x \in Q \cap \operatorname{supp}\mu$, if $Q_{x,k-1} \subset \frac{7}{5}Q$, then H_2 is 0 by Theorem 2.4.4(c). Assume that $Q_{x,k-1} \not\subset \frac{7}{5}Q$. Applying (c) and (d) of Theorem 2.4.4, (6.3.7) and (6.3.23), we see that, for any $x \in Q \cap \operatorname{supp}\mu$,

$$\begin{aligned}
\left|S_k\left(\left[f-m_{\widetilde{\frac{7}{5}Q}}(f)\right]\chi_{\mathbb{R}^D\setminus\frac{7}{5}Q}\right)(x)\right| &\lesssim \int_{15Q_{x,k-1}\setminus\frac{7}{5}Q}\frac{|f(y)-m_{\widetilde{\frac{7}{5}Q}}(f)|}{|x_Q-y|^n}\,d\mu(y) \\
&\lesssim \left[1+\delta\left(\frac{7}{5}Q,15Q_{x,k-1}\right)\right]^2[\mu(2Q)]^{\frac{\alpha}{n}} \\
&\lesssim [\mu(2Q)]^{\frac{\alpha}{n}}.
\end{aligned}$$

This implies that

$$\mathrm{H}_2 \lesssim [\mu(2Q)]^{\frac{\alpha}{n}}.$$

Analogously, to estimate H_4, we only need to consider the case that $x \in \tilde{Q}$ and $Q_{x,k-1} \not\subset \frac{7}{5}\tilde{Q}$. By the facts that, for any $x \in \tilde{Q}$ and $y \in \mathbb{R}^D \setminus \frac{7}{5}\tilde{Q}$,

$$\frac{7}{5}\tilde{Q} \subset 15Q_{x,k-1} \quad \text{and} \quad |x_Q-y| \lesssim |x-y|,$$

and that

$$|m_{\widetilde{\frac{7}{5}\tilde{Q}}}(f)-m_{\widetilde{\frac{7}{5}Q}}(f)| \lesssim [\mu(2Q)]^{\frac{\alpha}{n}},$$

(c) and (d) of Theorem 2.4.4, (6.3.7) and (6.3.23), we see that, for any $x \in \tilde{Q} \cap \operatorname{supp}\mu$,

$$\begin{aligned}
&\left|S_k\left(\left[f-m_{\widetilde{\frac{7}{5}Q}}(f)\right]\chi_{\mathbb{R}^D\setminus\frac{7}{5}\tilde{Q}}\right)(x)\right| \\
&\quad\lesssim \int_{15Q_{x,k-1}\setminus\frac{7}{5}\tilde{Q}}\frac{|f(y)-m_{\widetilde{\frac{7}{5}\tilde{Q}}}(f)|}{|z_Q-y|^n}\,d\mu(y) \\
&\quad\quad+\int_{15Q_{x,k-1}\setminus\frac{7}{5}\tilde{Q}}\frac{|m_{\widetilde{\frac{7}{5}\tilde{Q}}}(f)-m_{\widetilde{\frac{7}{5}Q}}(f)|}{|z_Q-y|^n}\,d\mu(y)
\end{aligned}$$

$$\lesssim \left[1+\delta\left(\frac{7}{5}\tilde{Q}, 15Q_{x,k-1}\right)\right]^2 [\mu(2Q)]^{\frac{\alpha}{n}}$$
$$\lesssim [\mu(2Q)]^{\frac{\alpha}{n}},$$

which implies that

$$\mathrm{H}_4 \lesssim [\mu(2Q)]^{\frac{\alpha}{n}}.$$

This, combining the estimates for H_1 through H_3, finishes the proof of (6.3.21).

Let us estimate (6.3.22). As in the proof of (6.3.21), we consider the following three cases:

(i) there exists some point $x_0 \in Q \cap \operatorname{supp}\mu$ such that $k \geq H_Q^{x_0} + 4$;
(ii) there exists some point $x_0 \in Q \cap \operatorname{supp}\mu$ such that $k \leq H_R^{x_0} - 1$;
(iii) for any point $x \in Q$, $k \in [H_R^x, H_Q^x + 3]$.

In Case (i), for any $x \in Q \cap \operatorname{supp}\mu$,

$$k \geq H_Q^x + 3 \quad \text{and} \quad Q_{x,k-1} \subset \frac{7}{5}Q \subset \frac{7}{5}R.$$

From Theorem 2.4.4(b), the Hölder inequality, Lemmas 6.3.2 and 2.1.3, the boundedness of S_k in $L^p(\mu)$, Proposition 6.3.3 and the doubling property of Q and R, it follows that

$$\begin{aligned}
&\left|m_Q(S_k(f)) - m_R(S_k(f))\right| \\
&\quad \leq \left|m_Q(S_k(f)) - m_{\widetilde{\frac{7}{5}Q}}(f)\right| + \left|m_{\widetilde{\frac{7}{5}Q}}(f) - m_{\widetilde{\frac{7}{5}R}}(f)\right| \\
&\qquad + \left|m_{\widetilde{\frac{7}{5}R}}(f) - m_R(S_k(f))\right| \\
&\quad \lesssim \left\{\frac{1}{\mu(Q)}\int_Q \left|S_k\left(\left[f - m_{\widetilde{\frac{7}{5}Q}}(f)\right]\chi_{\frac{7}{5}Q}\right)\right|^p d\mu(x)\right\}^{\frac{1}{p}} \\
&\qquad + [1+\delta(Q,R)][\mu(Q)]^{\frac{\alpha}{n}} \\
&\qquad + \left\{\frac{1}{\mu(R)}\int_R \left|S_k\left(\left[f - m_{\widetilde{\frac{7}{5}R}}(f)\right]\chi_{\frac{7}{5}R}\right)\right|^p d\mu(x)\right\}^{\frac{1}{p}} \\
&\quad \lesssim [1+\delta(Q,R)][\mu(Q)]^{\frac{\alpha}{n}}.
\end{aligned}$$

In Case (ii), for any $x \in Q \cap \operatorname{supp}\mu$, then

$$k \leq H_R^x \quad \text{and} \quad Q \subset R \subset Q_{x,k}.$$

By these facts, Lemma 6.3.2, (6.3.17) and (6.3.18), we know that

$$\begin{aligned}
&\left|m_Q(S_k(f)) - m_R(S_k(f))\right| \\
&\quad \le \frac{1}{\mu(R)}\frac{1}{\mu(Q)}\int_R\int_Q \left[\left|S_k(f)(x) - m_{Q_{x,k}}(f)\right| + \left|m_{Q_{x,k}}(f) - m_{Q_{y,k}}(f)\right|\right. \\
&\qquad \left. + \left|m_{Q_{y,k}}(f) - S_k(f)(y)\right|\right] d\mu(x)\, d\mu(y) \\
&\quad \lesssim \frac{1}{\mu(R)}\frac{1}{\mu(Q)}\int_R\int_Q \left\{[\mu(Q_{x,k})]^{\frac{\alpha}{n}} + [\mu(R)]^{\frac{\alpha}{n}} + [\mu(Q_{y,k})]^{\frac{\alpha}{n}}\right\} d\mu(x)\, d\mu(y) \\
&\quad \lesssim [\mu(Q)]^{\frac{\alpha}{n}}.
\end{aligned}$$

In Case (iii), by the Hölder inequality, the boundedness of S_k in $L^p(\mu)$ for $p \in (1,\infty)$, (c) and (d) of Theorem 2.4.4, (6.3.7), (6.3.10), Lemmas 6.3.2 and 2.1.3, we see that

$$\begin{aligned}
&\left|m_Q(S_k(f)) - m_R(S_k(f))\right| \\
&\quad \le \left|m_Q(S_k(f)) - m_{\widetilde{\frac{7}{5}Q}}(f)\right| + \left|m_{\widetilde{\frac{7}{5}Q}}(f) - m_{\widetilde{\frac{7}{5}R}}(f)\right| + \left|m_{\widetilde{\frac{7}{5}R}}(f) - m_R(S_k(f))\right| \\
&\quad \lesssim \left|m_Q\left(S_k\left(\left[f - m_{\widetilde{\frac{7}{5}Q}}(f)\right]\chi_{\frac{7}{5}Q}\right)\right)\right| + \left|m_Q\left(S_k\left(\left[f - m_{\widetilde{\frac{7}{5}Q}}(f)\right]\chi_{\mathbb{R}^D\setminus\frac{7}{5}Q}\right)\right)\right| \\
&\qquad + [1+\delta(Q,R)][\mu(Q)]^{\frac{\alpha}{n}} + \left|m_R\left(S_k\left(\left[f - m_{\widetilde{\frac{7}{5}R}}(f)\right]\chi_{\frac{7}{5}R}\right)\right)\right| \\
&\qquad + \left|m_R\left(S_k\left(\left[f - m_{\widetilde{\frac{7}{5}R}}(f)\right]\chi_{\mathbb{R}^D\setminus\frac{7}{5}R}\right)\right)\right| \\
&\quad \lesssim [1+\delta(Q,R)]^2[\mu(Q)]^{\frac{\alpha}{n}}.
\end{aligned}$$

Therefore, (6.3.22) holds true, which completes the proof of Theorem 6.3.13. □

6.4 Notes

- Theorem 6.1.2 was proved by Tolsa in [132].
- Theorem 6.1.5 was proved by Da. Yang and Do. Yang in [153].
- Theorem 6.1.6 was proved by Hu et al. in [66]. If we define the *inhomogeneous Littlewood–Paley g-function* $\tilde{g}(f)$ as follows,

$$\tilde{g}(f)(x) := \left[|S_0(f)(x)|^2 + \sum_{k=1}^{\infty} |D_k(f)(x)|^2\right]^{1/2}, \quad \forall x \in \mathbb{R}^D,$$

then Theorems 6.1.6 and 6.2.7, and Corollary 6.2.8 are still true. Notice that, when $\mathbb{R}^D$ is an initial cube, then $S_0 := 0$ and $\tilde{g}(f)$ degenerates into $g(f)$.

- Theorems 6.1.8 and 6.1.9 were proved by Da. Yang and Do. Yang in [152].
- Theorem 6.2.2 was proved in [152].
- Theorem 6.2.4 was proved in [153].
- Theorem 6.2.7 was proved in [66].
- Theorem 6.2.9 was proved by Da. Yang and Do. Yang in [154] (see also Bennett et al. [3]).
- Theorem 6.2.10 was proved in [154].
- It is interesting whether $H^1(\mu)$ can be characterized by the homogeneous Littlewood–Paley function $\dot{g}(f)$ or not. To be precise, let

$$H^1_{\dot{g}}(\mu) := \left\{ f \in L^1(\mu) : \int_{\mathbb{R}^D} f(x)\,d\mu(x) = 0,\ \dot{g}(f) \in L^1(\mu) \right\}$$

endowed with the *norm*

$$\|f\|_{H^1_{\dot{g}}(\mu)} := \|f\|_{L^1(\mu)} + \|\dot{g}(f)\|_{L^1(\mu)}.$$

It is *unclear* whether the spaces $H^1(\mu)$ and $H^1_{\dot{g}}(\mu)$ coincide with equivalent norms or not.
- It is interesting whether $H^1(\mu)$ can be characterized by the homogeneous maximal function $\dot{\mathcal{M}}_S(f)$ or not. To be precise, let

$$H^1_{\dot{\mathcal{M}}_S}(\mu) := \left\{ f \in L^1(\mu) : \int_{\mathbb{R}^D} f(x)\,d\mu(x) = 0,\ \dot{\mathcal{M}}_S(f) \in L^1(\mu) \right\}$$

endowed with the *norm*

$$\|f\|_{H^1_{\dot{\mathcal{M}}_S}(\mu)} := \|f\|_{L^1(\mu)} + \left\|\dot{\mathcal{M}}_S(f)\right\|_{L^1(\mu)}.$$

It is *unclear* whether the spaces $H^1(\mu)$ and $H^1_{\dot{\mathcal{M}}_S}(\mu)$ coincide with equivalent norms or not.
- It is interesting whether $h^1(\mu)$ can be characterized by the inhomogeneous Littlewood–Paley function $g(f)$ or not. To be precise, let

$$h^1_g(\mu) := \{ f \in L^1(\mu) :\ g(f) \in L^1(\mu) \}$$

endowed with the *norm*

$$\|f\|_{h^1_g(\mu)} := \|f\|_{L^1(\mu)} + \|g(f)\|_{L^1(\mu)}.$$

It is *unclear* whether the spaces $h^1(\mu)$ and $h^1_g(\mu)$ coincide with equivalent norms or not.

- It is interesting whether $h^1(\mu)$ can be characterized by the inhomogeneous maximal function $\mathcal{M}_S(f)$ or not. To be precise, let

$$h^1_{\mathcal{M}_S}(\mu) := \{f \in L^1(\mu) : \mathcal{M}_S(f) \in L^1(\mu)\}$$

endowed with the *norm*

$$\|f\|_{h^1_{\mathcal{M}_S}(\mu)} := \|f\|_{L^1(\mu)} + \|\mathcal{M}_S(f)\|_{L^1(\mu)}.$$

It is *unclear* whether the spaces $h^1(\mu)$ and $h^1_{\mathcal{M}_S}(\mu)$ coincide with equivalent norms or not.
- In [84], Lin and Meng proved that, for suitable $\rho \in (0,\infty)$ and $\lambda \in (1,\infty)$, the parameterized Littlewood–Paley function $g_\lambda^{*,\rho}(f)$ is bounded on $L^p(\mu)$ for $p \in [2,\infty)$ under the assumption that the kernel of $g_\lambda^{*,\rho}(f)$ satisfies some Hörmander-type condition, and bounded from $L^1(\mu)$ into weak $L^1(\mu)$ under the assumption that the kernel satisfies some slightly stronger Hörmander-type condition. As a corollary, $g_\lambda^{*,\rho}(f)$ with the kernel satisfying the above stronger Hörmander-type condition is bounded on $L^p(\mu)$ for $p \in (1,2)$. Moreover, Lin and Meng [84] proved that, for suitable ρ and λ, $g_\lambda^{*,\rho}(f)$ is bounded from $L^\infty(\mu)$ into RBLO (μ) if the kernel satisfies the Hörmander-type condition, and from the Hardy space $H^1(\mu)$ into $L^1(\mu)$ if the kernel satisfies the above stronger Hörmander-type condition. The corresponding properties for the parameterized area integral are also presented in [84].
- For the space $H^1(\mu)$, it is still *unknown* whether there exists an area integral $S(f)$ or $g_\lambda^*(f)$ which characterizes $H^1(\mu)$ or not.
- In [48], Han and Yang established a theory of Triebel-Lizorkin spaces $\dot{F}^s_{pq}(\mu)$ for $p \in (1,\infty)$, $q \in [1,\infty]$ and $s \in (-\theta,\theta)$, where $\theta \in (0,\infty)$ depends on μ, C_0, n and D as in (0.0.1). Moreover, in [48], the method, without using the vector-valued maximal function inequality of Fefferman and Stein, is new even for the classical case. As applications, the lifting properties of these spaces by using the Riesz potential operators and the dual spaces were given.
- In [25], Deng, Han and Yang established a theory of Besov spaces $\dot{B}^s_{pq}(\mu)$ for $p,\ q \in [1,\infty]$ and $s \in (-\theta,\theta)$, where $\theta \in (0,\infty)$ depends on μ, C_0, n and D as in (0.0.1); See also Deng and Han [24].
- To the best of our knowledge, there is not any result for the boundedness of the Littlewood–Paley g-function $g(f)$ on $\mathcal{C}^{\alpha,\,p}(\mu)$. It is reasonable to guess that $g(f)$ is bounded from a local version of $\mathcal{C}^{\alpha,\,p}(\mu)$ to a local version of $\mathcal{C}^{\alpha,\,p}_*(\mu)$.
- To the best of our knowledge, there is not any result for the boundedness of the maximal function $\mathcal{M}_S$ on $\mathcal{C}^{\alpha,\,p}(\mu)$. It is reasonable to guess that $g(f)$ is bounded from a local version of $\mathcal{C}^{\alpha,\,p}(\mu)$ to a local version of $\mathcal{C}^{\alpha,\,p}_*(\mu)$.
- Let $\rho \in (1,\infty)$ and $1 \le q \le p < \infty$. The Morrey space $M^p_q(\rho,\mu)$ is defined by

$$M^p_q(\rho,\mu) := \{f \in L^q_{\mathrm{loc}}(\mu) : \|f\|_{M^p_q(\rho,\mu)} < \infty\},$$

where

$$\|f\|_{M_q^p(\rho,\mu)} := \sup_Q [\mu(\rho Q)]^{1/p-1/q} \left[\int_Q |f(x)|^q \, d\mu(x)\right]^{1/q}.$$

In [116], Sawano and Takana introduced $M_q^p(\rho,\mu)$ and showed that the space $M_q^p(\rho,\mu)$ is independent of the choice of $\rho \in (1,\infty)$. Then, the authors in [116] considered several classical operators, e. g. the maximal operator or the fractional integral operator, and investigated the boundedness of these operators in the Morrey spaces. Sawano and Takana in [117] further investigated the connection between RBMO (μ) and the Morrey spaces, and in [120] investigated the predual of the Morrey spaces.

- In [119], Sawano and Takana established a sharp maximal inequality for Morrey spaces and obtained the boundedness of commutators generated by singular integral or fractional integrals with RBMO (μ) functions in Morrey spaces. In [113], Sawano extended these results to a vector-valued setting.
- In [115], Sawano and Shirai considered the multi-commutators on the Morrey spaces generated by functions in RBMO (μ) and singular integral operators or by functions in RBMO (μ) and fractional integral operators, and they showed that the multi-commutators are compact if one of the RBMO (μ) functions can be approximated by compactly supported smooth functions.
- In [114], Sawano further studied the generalized Morrey space which consists of all functions $f \in L^p_{\rm loc}(\mu)$ such that

$$\sup_Q \left[\frac{1}{\varphi(\mu(kQ))}\int_Q |f(y)|^p \, d\mu(y)\right]^{1/p} < \infty,$$

where $\varphi : (0,\infty) \to (0,\infty)$ is increasing and the supremum is taken over all cubes Q with $\mu(Q) > 0$.
- In [46], Gunawan, Sawano and Sihwaningrum established the boundedness of the fractional integral operator and its generalized version on a version of generalized Morrey spaces on $\mathbb{R}^D$ with the measure as in (0.0.1).

Part II
Non-homogeneous Spaces $(\mathcal{X}, \nu)$

The classical theory of Hardy spaces and singular integrals has been well developed into a large branch of analysis on spaces of homogeneous type in the sense of Coifman and Weiss [18, 19]. Recall that a (quasi-)metric space $(\mathcal{X}, d)$ equipped with a nonnegative measure μ is called a *space of homogeneous type* if $(\mathcal{X}, d, \mu)$ satisfies the *measure doubling condition*: there exists a positive constant $C_{(\mu)}$ such that, for any ball $B(x,r) := \{y \in \mathcal{X} : d(x,y) < r\}$ with $x \in \mathcal{X}$ and $r \in (0, \infty)$,

$$\mu(B(x,2r)) \leq C_{(\mu)}\mu(B(x,r)).$$ [1]

Typical examples of spaces of homogeneous type include Euclidean spaces, Euclidean spaces with weighted measures satisfying the doubling property, Heisenberg groups, connected and simply connected nilpotent Lie groups and the boundary of an unbounded model polynomial domain in $\mathbb{C}^N$ or, more generally, Carnot–Carathéodory spaces with doubling measures. Since the 1970s, there have been a lot of fruitful results on the theory of Hardy spaces and singular integral operators on spaces of homogeneous type. It is now well known that the space of homogeneous type is a natural setting for the theory of function spaces and singular integrals; see, for example, [1, 18, 19, 24, 47, 99–101].

On the other hand, substantial progress in the study of the theory on function spaces and singular integrals with non-doubling measures disproved the long held belief of the decades of the 1970s and the 1980s that the doubling property of the measures is indispensable in the theory of harmonic analysis.

However, as pointed out by Hytönen in [68], the measures satisfying (0.0.1) do not include the doubling measures as special cases. In [68], Hytönen introduced a new class of metric measure spaces satisfying the so-called geometrically doubling and the upper doubling conditions. This new class of metric measure spaces, which are called *non-homogeneous spaces*, includes both the spaces of homogeneous type and metric spaces with polynomial growth measures as special

---

[1] We restrict ourselves to a metric space throughout this book.

cases. Recently, many classical results have been proved still valid if the underlying spaces are replaced by the non-homogeneous spaces (see, for example, [9, 30, 66, 68–70, 89]). It is now also known that the theory of the singular integral operators on non-homogeneous spaces arises naturally in the study of complex and harmonic analysis questions in several complex variables (see [70, 147]). The purpose of this part is to introduce the theory of the Hardy space H^1 and singular integrals in non-homogeneous spaces.

This part consists of two chapters, namely, Chaps. 7 and 8. In Chap. 7, we introduce the non-homogeneous space $(\mathcal{X}, d, \nu)$ and present some basic properties. Based on these properties, we further introduce the atomic Hardy space $H^1(\mathcal{X}, \nu)$ and its dual space, the BMO-type space $\mathrm{RBMO}\,(\mathcal{X}, \nu)$ in this setting, establishing the John–Nirenberg inequality for $\mathrm{RBMO}\,(\mathcal{X}, \nu)$ and some equivalent characterizations of $\mathrm{RBMO}\,(\mathcal{X}, \nu)$ and $H^1(\mathcal{X}, \nu)$, respectively. As applications of Chap. 7, in Chap. 8, we discuss the boundedness of Calderón–Zygmund operators over non-homogeneous spaces $(\mathcal{X}, \nu)$. By establishing the Calderón–Zygmund decomposition, we first show that the Calderón–Zygmund operator T is bounded from $H^1(\mathcal{X}, \nu)$ to $L^1(\mathcal{X}, \nu)$. We then establish the molecular characterization for $H^1(\mathcal{X}, \nu)$ and its variant, $\tilde{H}^1(\mathcal{X}, \nu)$, which is a subspace of $H^1(\mathcal{X}, \nu)$, and obtain the boundedness of T on $\tilde{H}^1(\mathcal{X}, \nu)$. We also prove that the boundedness of T on $L^p(\mathcal{X}, \nu)$, with $p \in (1, \infty)$, is equivalent to its various estimates, and establish some weighted estimates involving the John–Strömberg maximal operators and the John–Strömberg sharp maximal operators, and some weighted norm inequalities for the multilinear Calderón–Zygmund operators. In addition, the boundedness of multilinear commutators of Calderón–Zygmund operators on Orlicz spaces is also presented.

We now make some necessary conventions on notation. As in Part I, throughout this part, we use C, $\tilde{C}$, c and $\tilde{c}$ to denote *positive constants* which are independent of the main parameters, but may change their values at different occurrences. *Constants with subscripts*, such as C_1 and c_1, retain their values at different occurrences throughout this part. Furthermore, $C_{(\rho, \gamma, \ldots)}$ stands for a *positive constant* depending on the parameter ρ, γ, ... Also, the *symbol* $Y \lesssim Z$ means that $Y \le CZ$ for some positive constant C, and $Y \sim Z$ means that $Y \lesssim Z \lesssim Y$.

In this part, unless explicitly pointed out, a *ball* means an open set

$$B := B(x_B, r_B) := \{y \in \mathcal{X} : \ d(x_B, y) < r_B\}$$

with $x_B \in \mathcal{X}$ and $r_B \in (0, \infty)$. For any ball $B := B(x_B, r_B)$ and $\varrho \in (0, \infty)$,

$$\varrho B := B(x_B, \varrho r_B).$$

Finally, in this part, we assume that ν is a nonnegative Borel measure on $\mathcal{X}$ and let $\|\nu\| := \nu(\mathcal{X})$. For any set $E \subset \mathcal{X}$, we denote by χ_E the *characteristic function of* E. Moreover, let

$$\mathrm{diam}\,(\mathcal{X}) := \sup\{d(x, y) : \ x, \ y \in \mathcal{X}\}.$$

For any $f \in L^1_{\mathrm{loc}}(\mathcal{X}, \nu)$ and ball B, $m_B(f)$ denotes the *mean* of f over B, that is,

$$m_B(f) := \frac{1}{\mu(B)} \int_B f(x)\, d\mu(x).$$

For any $p \in [1, \infty]$, in this part, $L^p_b(\mathcal{X}, \nu)$ stands for the *space of functions in* $L^p(\mathcal{X}, \nu)$ *with bounded support* and $L^p_{b,0}(\mathcal{X}, \nu)$ the *space of functions in* $L^p_b(\mathcal{X}, \nu)$ *having integral 0*. We also use $C_b(\mathcal{X})$ to denote the *space of all continuous functions with bounded support*.

Chapter 7
The Hardy Space $H^1(\mathcal{X}, \nu)$ and Its Dual Space RBMO$(\mathcal{X}, \nu)$

In this chapter, we introduce and study a class of metric measure spaces $(\mathcal{X}, d, \nu)$, which include both Euclidean spaces with nonnegative Radon measures satisfying the polynomial growth condition and spaces of homogeneous type as special cases. We also introduce the BMO-type space RBMO $(\mathcal{X}, \nu)$ and the atomic Hardy space $H^1(\mathcal{X}, \nu)$ in this setting, establish the John–Nirenberg inequality for RBMO $(\mathcal{X}, \nu)$ and some equivalent characterizations of RBMO $(\mathcal{X}, \nu)$ and $H^1(\mathcal{X}, \nu)$, respectively, and show that the dual space of $H^1(\mathcal{X}, \nu)$ is RBMO $(\mathcal{X}, \nu)$.

7.1 Upper Doubling Metric Measure Spaces and Geometrically Doubling Spaces

In this section, we introduce the class of metric measure spaces $(\mathcal{X}, d, \nu)$ and study some basic geometric properties. We begin with the notion of upper doubling spaces.

Definition 7.1.1. A metric measure space $(\mathcal{X}, d, \nu)$ is said to be *upper doubling* if ν is a Borel measure on $\mathcal{X}$ and there exists a dominating function

$$\lambda : \ \mathcal{X} \times (0, \infty) \to (0, \infty)$$

and a positive constant $C_{(\lambda)}$ such that, for each $x \in \mathcal{X}$, $r \to \lambda(x, r)$ is non-decreasing and, for all $x \in \mathcal{X}$ and $r \in (0, \infty)$,

$$\nu(B(x, r)) \leq \lambda(x, r) \leq C_{(\lambda)} \lambda(x, r/2). \tag{7.1.1}$$

We write

$$\varsigma := \log_2 C_{(\lambda)},$$

which can be thought of as a dimension of the measure.

D. Yang et al., *The Hardy Space H^1 with Non-doubling Measures and Their Applications*, Lecture Notes in Mathematics 2084, DOI 10.1007/978-3-319-00825-7_7,

Example 7.1.2. A metric space $(\mathcal{X}, d)$ equipped with a non-negative Borel measure ν is called a *space of homogeneous type* if $(\mathcal{X}, d, \nu)$ satisfies the following *doubling condition* that there exists a positive constant $C_{(\nu)}$ such that, for any ball $B \subset \mathcal{X}$,

$$0 < \nu(2B) \le C_{(\nu)}\nu(B). \tag{7.1.2}$$

A space of homogeneous type is a special case of upper doubling spaces, where we take the dominating function $\lambda(x, r) := \nu(B(x, r))$.

Example 7.1.3. Let μ be a non-negative Radon measure on $\mathbb{R}^D$ which only satisfies the polynomial growth condition as in (0.0.1). By taking $\lambda(x, r) := Cr^n$, we see that $(\mathbb{R}^D, |\cdot|, \mu)$ is also an upper doubling measure space.

Example 7.1.4. Let $\mathcal{X}$ be the *unit ball* $\mathbb{B}_{2D}$ of $\mathbb{C}^D$ and ν the measure in $\mathbb{B}_{2D}$ for which the analytic Besov–Sobolev space $B_2^\sigma(\mathbb{B}_{2D})$ continuously embeds into $L^2(\mathcal{X}, \nu)$. The measure ν satisfies the upper power bound $\nu(B(x, r)) \le r^m$ with $m \in (0, 2D]$, except possibly when $B(x, r) \subset H$, where H is a fixed open set. However, in the exceptional case there it holds true that

$$r \le \delta(x) := d(x, \mathbb{C}^D \setminus H)$$

and hence

$$\nu(B(x, r)) \le \lim_{\epsilon \to 0} \nu(B(x, \delta(x) + \epsilon)) \le \lim_{\epsilon \to 0}[\delta(x) + \epsilon]^m = [\delta(x)]^m.$$

Thus, ν is actually upper doubling with

$$\nu(B(x, r)) \le \max\{[\delta(x)]^m, r^m\} =: \lambda(x, r).$$

It is not difficult to show that $\lambda(\cdot, \cdot)$ satisfies the conditions of Definition 7.1.1.

Let $(\mathcal{X}, d, \nu)$ be an upper doubling space and λ a dominating function on $\mathcal{X} \times (0, \infty)$ as in Definition 7.1.1. The function λ does not need to satisfy the *additional property* that there exists a positive constant C such that, for all $x, y \in \mathcal{X}$ with $d(x, y) \le r$,

$$\lambda(x, r) \le C\lambda(y, r). \tag{7.1.3}$$

However, by the proposition below, we see that there always exists another dominating function related to λ satisfying (7.1.3).

Proposition 7.1.5. *Let $(\mathcal{X}, d, \nu)$ be an upper doubling space and λ a dominating function with the positive constant $C_{(\lambda)}$ as in Definition 7.1.1. Then there exists another dominating function $\tilde{\lambda}$ with a positive constant $C_{(\tilde{\lambda})} \le C_{(\lambda)}$ such that $\tilde{\lambda} \le \lambda$ and $\tilde{\lambda}$ satisfies* (7.1.3).

Proof. Define $\tilde{\lambda}(x, r)$ by setting, for all $x \in \mathcal{X}$ and $r \in (0, \infty)$,

$$\tilde{\lambda}(x, r) := \inf_{z \in \mathcal{X}} \lambda(z, r + d(x, z)).$$

Clearly, $\tilde{\lambda}$ is non-decreasing in r and $\tilde{\lambda} \leq \lambda$. By (7.1.1), we see that, if $x, y, z \in \mathcal{X}$ with $d(x, y) \leq r$, then

$$\nu(B(x, r)) \leq \nu(B(z, r + d(x, z))) \leq \lambda(z, r + d(x, z)),$$

$$\lambda(z, r + d(x, z)) \leq \lambda\left(z, 2\left[\frac{1}{2}r + d(x, z)\right]\right) \leq C_{(\lambda)}\lambda\left(z, \frac{1}{2}r + d(x, z)\right)$$

and

$$\begin{aligned} \lambda(z, r + d(x, z)) &\leq \lambda(z, r + d(x, y) + d(y, z)) \\ &\leq \lambda(z, 2[r + d(y, z)]) \\ &\leq C_{(\lambda)}\lambda(z, r + d(y, z)). \end{aligned}$$

Taking the infimum over $z \in \mathcal{X}$ of these inequalities, we conclude, respectively, that

$$\nu(B(x, r)) \leq \tilde{\lambda}(x, r),$$

$$\tilde{\lambda}(x, r) \leq C_{(\lambda)}\tilde{\lambda}(x, r/2) \quad \text{and} \quad \tilde{\lambda}(x, r) \leq C_{(\lambda)}\tilde{\lambda}(y, r),$$

which complete the proof of Proposition 7.1.5. □

Remark 7.1.6. By Proposition 7.1.5, in what follows, we *always assume that the dominating function* λ *satisfies* (7.1.3).

Recall that the doubling condition (7.1.2) is not assumed in our context. As a substitute of (7.1.2) in the present investigation, the following geometrically doubling condition is also well known in analysis on metric spaces.

Definition 7.1.7. A metric space $(\mathcal{X}, d)$ is said to be *geometrically doubling* if there exists some $N_0 \in \mathbb{N}$ such that, for any ball $B(x, r) \subset \mathcal{X}$, there exists a finite ball covering $\{B(x_i, r/2)\}_i$ of $B(x, r)$ such that the cardinality of this covering is at most N_0.

Proposition 7.1.8. *Let* $(\mathcal{X}, d)$ *be a metric space. The following statements are mutually equivalent:*

(i) $(\mathcal{X}, d)$ *is geometrically doubling;*
(ii) *For any* $\epsilon \in (0, 1)$ *and any ball* $B(x, r) \subset \mathcal{X}$, *there exists a finite ball covering* $\{B(x_i, \epsilon r)\}_i$ *of* $B(x, r)$ *such that the cardinality of this covering is at most* $N_0\epsilon^{-n_0}$, *here and in what follows,* N_0 *is as in Definition 7.1.7 and*

$$n_0 := \log_2 N_0;$$

(iii) *For every* $\epsilon \in (0, 1)$, *any ball* $B(x, r) \subset \mathcal{X}$ *contains at most* $N_0\epsilon^{-n_0}$ *centers* $\{x_i\}_i$ *of disjoint balls with radius* ϵr;

(iv) *There exists* $M \in \mathbb{N}$ *such that any ball* $B(x, r) \subset \mathcal{X}$ *contains at most* M *centers* $\{x_i\}_i$ *of disjoint balls* $\{B(x_i, r/4)\}_{i=1}^M$.

Proof. (i) $\Longrightarrow$ (ii). Let $\epsilon \in [2^{-k}, 2^{1-k})$ with $k \in \mathbb{Z}_+$. By iterating (i), we see that $B(x, r)$ is covered by at most N_0^k balls

$$B(x_i, 2^{-k}r) \subset B(x_i, \epsilon r),$$

here

$$N_0^k = 2^{k \log_2 N_0} \le (2\epsilon^{-1})^{\log_2 N_0} = N_0\epsilon^{-\log_2 N_0}.$$

(ii) $\Longrightarrow$ (iii). Suppose that $\{y_j\}_{j\in J} \subset B(x, r)$ are centers of disjoint balls

$$\{B(y_j, \epsilon r)\}_{j\in J},$$

and choose a covering of $B(x, r)$ consisting of balls $\{B(x_i, \epsilon r)\}_{i\in I}$, where $|I| \le N_0\epsilon^{-n_0}$. Then every y_j belongs to some $B(x_i, \epsilon r)$, and no two $y_j \ne y_k$ can belong to the same $B(x_i, \epsilon r)$, for otherwise

$$x_i \in B(y_j, \epsilon r) \cap B(y_k, \epsilon r) = \emptyset.$$

Thus,

$$|J| \le |I| \le N_0\epsilon^{-n_0}.$$

(iii) $\Longrightarrow$ (iv) is obvious.

(iv) $\Longrightarrow$ (i). Keep selecting disjoint balls $B(y_j, r/4)$ with $y_j \in B(x, r)$ as long as it is possible; the process terminates after at most M steps by assumption. Then every $y \in B(x, r)$ belongs to some $B(y_j, r/2)$, for otherwise the ball $B(y, r/4)$ could still have been chosen, which completes the proof. □

It is well known that spaces of homogeneous type are geometrically doubling spaces.[1] Conversely, if $(\mathcal{X}, d)$ is a complete geometrically doubling metric space, then there exists a Borel measure ν on $\mathcal{X}$ such that $(\mathcal{X}, d, \nu)$ is a space of homogeneous type.[2] However, the point of view taken in the investigation is that the measure is given by a particular problem, and not something that one is free to choose or construct. Thus, even if there exist some doubling measures on the metric space of interest, one might still work with a non-doubling one.

---

[1] See [18, p. 67].

[2] See [92] and [148].

Lemma 7.1.9. *In a geometrically doubling metric space, any disjoint collection of balls is at most countable.*

Proof. Let $x_0 \in \mathcal{X}$ be a fixed reference point. By Proposition 7.1.8, any ball $B(x_0, k)$ contains at most finitely many centers of disjoint balls of radius bigger than a given j^{-1}, where $j, k \in \mathbb{Z}_+$. Since every ball has its center in some $B(x_0, k)$ and radius bigger than some j^{-1}, we then conclude the desired conclusion and hence complete the proof of Lemma 7.1.9. □

Using Lemma 7.1.9, we next prove that $\mathcal{X}$ is separable as follows.

Proposition 7.1.10. *Let $(\mathcal{X}, d)$ be a metric space which satisfies the geometrically doubling condition as in Definition 7.1.7. Then $\mathcal{X}$ is separable.*

Proof. It suffices to prove that there exists a countable subset which is dense in $\mathcal{X}$. Let $\varepsilon \in (0, \infty)$. A collection of points, $\{x_\beta\}_\beta \subset \mathcal{X}$, is said to be *$\varepsilon$-separated*, if, for any two points $x_\alpha, x_\beta \in \{x_\gamma\}_\gamma$ with $\alpha \neq \beta$, it holds that

$$d(x_\alpha, x_\beta) \geq \varepsilon.$$

Fix a small parameter $\delta \in (0, 1)$. By the geometrically doubling condition and the Zorn lemma, we see that there exists a collection of maximal 1-separated points in $\mathcal{X}$, denoted by $\mathcal{X}^0 := \{x_\alpha^0\}_\alpha$. Inductively, for $k \in \mathbb{N}$, there exist a maximal δ^k-separated set in $\mathcal{X}$, denoted by

$$\mathcal{X}^k := \{x_\alpha^k\}_\alpha \supset \mathcal{X}^{k-1},$$

and a maximal δ^{-k}-separated set in $\mathcal{X}^{-(k-1)}$, denoted by

$$\mathcal{X}^{-k} := \{x_\alpha^{-k}\}_\alpha \subset \mathcal{X}^{-(k-1)}.$$ [3]

We claim that, for all $k \in \mathbb{Z}$ and $x \in \mathcal{X}$, it holds true that, for any $\alpha \neq \beta$,

$$d(x_\alpha^k, x_\beta^k) \geq \delta^k \tag{7.1.4}$$

and

$$d(x, \mathcal{X}^k) := \min_\alpha \{d(x, x_\alpha^k)\} < 2\delta^k. \tag{7.1.5}$$

Indeed, the separation property (7.1.4) is part of the construction. We now prove (7.1.5). From the maximality of $\{x_\alpha^k\}_\alpha$, it follows that, for all $x \in \mathcal{X}$ and $k \in \mathbb{Z}_+$,

$$d(x, \mathcal{X}^k) := \min_\alpha \{d(x, x_\alpha^k)\} < \delta^k,$$

[3] See, for example, [1, 14].

namely, (7.1.5) holds true when $k \in \mathbb{Z}_+$. Moreover, for $k \in \mathbb{N}$ and given $x \in \mathcal{X}$, we can recursively find points, $x^0_{\alpha_0}$, $x^{-1}_{\alpha_1}, \ldots, x^{-k}_{\alpha_k}$, such that $x^{-i}_{\alpha_i} \in \mathcal{X}^{-i}$ for $i \in \{0, \ldots, k\}$ and

$$d(x, x^0_{\alpha_0}) < 1, \; d(x^0_{\alpha_0}, x^{-1}_{\alpha_1}) < \delta^{-1}, \; \ldots, \; d(x^{-k+1}_{\alpha_{k-1}}, x^{-k}_{\alpha_k}) < \delta^{-k}.$$

By this fact, we conclude that

$$d(x, x^{-k}_{\alpha_k}) \le d(x, x^0_{\alpha_0}) + \sum_{j=0}^{k-1} d(x^{-j}_{\alpha_j}, x^{-j-1}_{\alpha_{j+1}}) < 2\delta^{-k},$$

which implies that (7.1.5) also holds true for $k \in \mathbb{Z} \setminus \mathbb{Z}_+$. Thus, the above claim holds true.

From (7.1.5), we deduce that $\mathcal{Y} := \cup_{k\in\mathbb{Z}} \mathcal{X}^k$ is dense in $\mathcal{X}$. Moreover, for any $k \in \mathbb{Z}$, by the separability of $\mathcal{X}^k$ (7.1.4), the geometrically doubling condition and Lemma 7.1.9, we conclude that $\mathcal{X}^k$ is at most countable for each $k \in \mathbb{Z}$. Thus, $\mathcal{Y}$ is countable, which completes the proof. □

As in Part I, although the doubling measure condition (7.1.2) is not assumed uniformly for all balls in the space $(\mathcal{X}, d, \nu)$, it makes sense to ask whether such an inequality is true for a given particular ball or not.

Definition 7.1.11. Let $\alpha, \beta \in (1, \infty)$. A ball $B \subset \mathcal{X}$ is said to be *(α, β)-doubling* if $\nu(\alpha B) \le \beta \nu(B)$.

The following two lemmas tell us that when $(\mathcal{X}, d, \nu)$ satisfies the upper doubling condition and the geometrically doubling condition, there exist still many small and large balls that have the following (α, β)-doubling property.

Lemma 7.1.12. *Let the metric measure space $(\mathcal{X}, d, \nu)$ be upper doubling and*

$$\beta > C_{(\lambda)}^{\log_2 \alpha} = \alpha^{\varsigma}.$$

Then, for every ball $B \subset \mathcal{X}$ there exists $j \in \mathbb{N}$ such that $\alpha^j B$ is (α, β)-doubling.

Proof. Assume contrary to the claim that none of the balls $\alpha^j B$, $j \in \mathbb{N}$, is (α, β)-doubling, namely, $\nu(\alpha^{j+1} B) > \beta \nu(\alpha^j B)$ for all $j \in \mathbb{N}$. From this, it then follows that

$$\begin{aligned}
\nu(B) &\le \beta^{-1} \nu(\alpha B) \\
&\le \cdots \\
&\le \beta^{-j} \nu(\alpha^{-j} B) \\
&\le \beta^{-j} \lambda(x_B, \alpha^j r_B) \\
&\le \beta^{-j} C_{(\lambda)}^{(j \log_2 \alpha)+1} \lambda(x_B, r_B) \\
&= C_{(\lambda)} \left(\frac{\alpha^{\varsigma}}{\beta} \right)^j \lambda(x_B, r_B) \to 0,
\end{aligned}$$

as $j \to \infty$. Hence $\nu(B) = 0$. But the same argument also holds true with αB in place of B, leading to $\nu(\alpha B) = 0$. Then B is (α, β)-doubling after all, which is a contradiction and completes the proof of Lemma 7.1.12. □

Lemma 7.1.13. *Let $(\mathcal{X}, d)$ be geometrically doubling and $\beta > \alpha^{n_0}$, where n_0 is as in Proposition 7.1.8. If ν is a Borel measure on $\mathcal{X}$ which is finite on bounded sets, then, for ν-almost every $x \in \mathcal{X}$, there exist arbitrarily small (α, β)-doubling balls centered at x. Indeed, their radius may be chosen to be of the form $\alpha^{-j} r$, $j \in \mathbb{N}$, for any preassigned number $r \in (0, \infty)$.*

Proof. Consider a fixed ball $B := B(x_0, r)$. It suffices to prove Lemma 7.1.13 for ν-almost all $x \in B$.

For $x \in B$ and $k \in \mathbb{N}$, let

$$B_x^k := B(x, \alpha^{-k} r).$$

The point x is said to be k-*bad* if none of the balls $\alpha^j B_x^k$, $j \in \{0, \ldots, k\}$, is (α, β)-doubling. Notice that

$$\alpha^k B_x^k = B(x, r) \subset 3B$$

and hence, for every k-bad point x, it holds true that

$$\nu(B_x^k) \le \beta^{-k} \nu(\alpha^k B_x^k) \le \beta^{-k} \nu(3B).$$

Among the k-bad points, choose a maximal $\alpha^{-k} r$-separated family Y. Hence the balls $\{B_y^k\}_{y \in Y}$ cover all the bad points. On the other hand, the balls

$$\{2^{-1} B_y^k\}_{y \in Y} = \{B(y, 2^{-1} \alpha^{-k} r)\}_{y \in Y}$$

are disjoint with their centers contained in $B := B(x_0, r)$, and hence there exist at most $N_0 (2^{-1} \alpha^{-k})^{-n_0} = N_0 2^{n_0} \alpha^{k n_0}$ of them. Thus,

$$\begin{aligned}
\nu(\{x \in B : x \text{ is } k - \text{bad}\}) &\le \nu\left(\bigcup_{y \in Y} B_y^k\right) \\
&\le \sum_{y \in Y} \nu(B_y^k) \\
&\le \sum_{y \in Y} \beta^{-k} \nu(3B) \\
&\le N_0 2^{n_0} \nu(3B) \left(\frac{\alpha^{n_0}}{\beta}\right)^k \to 0,
\end{aligned}$$

as $k \to \infty$. Hence only a zero-set of points can be k-bad for all $k \in \mathbb{N}$ and hence this finishes the proof of Lemma 7.1.13. □

Lemma 7.1.14. *For $\alpha \in (1, \infty)$, $\beta \in (\alpha^{3\varsigma}, \infty)$ and any ball B, there exists $j \in \mathbb{N}$ such that $\alpha^j B$, $\alpha^{j+1} B$ and $\alpha^{j+2} B$ are three consecutive (α, β)-doubling balls.*

Proof. If $\beta > \alpha^{3\varsigma}$, then, by Proposition 7.1.12, for any ball $B \subset \mathcal{X}$, there exists $j \in \mathbb{N}$ such that $\alpha^j B$ is (α^3, β)-doubling. That is,

$$\nu(\alpha^3 \alpha^j B) \le \beta \nu(\alpha^j B).$$

This implies that

$$\nu(\alpha\alpha^j B) \le \beta\nu(\alpha^j B), \quad \nu(\alpha\alpha^{j+1} B) \le \beta\nu(\alpha^{j+1} B) \quad \text{and}$$
$$\nu(\alpha\alpha^{j+2} B) \le \beta\nu(\alpha^{j+2} B).$$

Therefore, the three balls $\alpha^j B$, $\alpha^{j+1} B$ and $\alpha^{j+2} B$ are three (α, β)-doubling balls, which completes the proof of Lemma 7.1.14. □

In what follows, for any $\alpha \in (1, \infty)$ and ball B, $\tilde{B}^\alpha$ denotes the *smallest* (α, β_α)-*doubling* ball of the form $\alpha^j B$ with $j \in \mathbb{Z}_+$, where

$$\beta_\alpha := \max\left\{\alpha^{3n_0}, \alpha^{3\varsigma}\right\} + 30^{n_0} + 30^{\varsigma} = \alpha^{\max\{3n_0, 3\varsigma\}} + 30^{n_0} + 30^{\varsigma}. \qquad (7.1.6)$$

The following coefficient $\delta(B, S)$ for any balls B and S is an analogue of $\delta(Q, R)$ in Part I.

Definition 7.1.15. For all balls $B \subset S$, let

$$\delta(B, S) := \int_{(2S)\setminus B} \frac{1}{\lambda(x_B, d(x, x_B))}\, d\nu(x).$$

For the coefficient $\delta(B, S)$, we also have the following useful properties similar to Lemma 2.1.3.

Lemma 7.1.16. (i) *For all balls $B \subset R \subset S$,*

$$\delta(B, R) \le \delta(B, S).$$

(ii) *For any balls $B \subset S$,*

$$\delta(B, S) \le C_{(\lambda)} \log_2 \frac{6r_S}{r_B}.$$

Moreover, for any $\rho \in [1, \infty)$, there exists a positive constant C, depending on ρ, such that,

$$\text{for all balls} \quad B \subset S \quad \text{with} \quad r_S \le \rho r_B, \quad \delta(B, S) \le C.$$

(iii) *For any* $\alpha \in (1, \infty)$, *there exists a positive constant* $\tilde{C}$, *depending on* α, *such that,*

$$\text{for all balls} \quad B, \quad \delta(B, \tilde{B}^{\alpha}) \le \tilde{C}.$$

Moreover, for any two concentric balls $B \subset S$ *such that there does not exist any* (α, β)*-doubling ball in the form of* $\alpha^k B$, *with* $k \in \mathbb{N}$, *such that*

$$B \subset \alpha^k B \subset S, \quad \delta(B, S) \le \tilde{C}.$$

(iv) *There exists a positive constant* c *such that, for all balls* $B \subset R \subset S$,

$$\delta(B, S) \le \delta(B, R) + c\delta(R, S).$$

In particular, if B *and* R *are concentric, then* $c = 1$.

(v) *There exists a positive constant* $\tilde{c}$ *such that, for all balls* $B \subset R \subset S$,

$$\delta(R, S) \le \tilde{c}[1 + \delta(B, S)].$$

Moreover, if B *and* R *are concentric, then*

$$\delta(R, S) \le \delta(B, S).$$

Proof. (i) is obvious.

To show (ii), let N be the smallest integer such that $2S \subset 2^N B$. Then we see that $N \le \log_2 \frac{6r_S}{r_B}$. From (7.1.1), it follows that

$$\begin{aligned}\delta(B, S) &\le \sum_{k=1}^{N} \int_{2^k B \setminus 2^{k-1} B} \frac{1}{\lambda(x_B, d(x, x_B))} d\nu(x) \\ &\le C_{(\lambda)} N \\ &\le C_{(\lambda)} \log_2 \frac{6r_S}{r_B}.\end{aligned}$$

Now we prove (iii). Let $j \in \mathbb{N}$ be such that $\tilde{B}^{\alpha} = \alpha^j B$. Using (7.1.1) again, we have

$$\begin{aligned}\delta\left(B, \tilde{B}^{\alpha}\right) &\le \int_{2\tilde{B}^{\alpha} \setminus \tilde{B}^{\alpha}} \frac{1}{\lambda(x_B, d(x, x_B))} d\nu(x) + \sum_{i=1}^{j} \int_{\alpha^i B \setminus \alpha^{i-1} B} \cdots \\ &\le \frac{\nu(2\tilde{B}^{\alpha})}{\lambda(x_B, \alpha^j r_B)} + \sum_{i=1}^{j} \frac{\nu(\alpha^i B)}{\lambda(x_B, \alpha^{i-1} r_B)}\end{aligned}$$

$$\begin{aligned}
&\le C_{(\lambda)} + \sum_{i=1}^{j} \frac{\beta^{i-j}\nu(\alpha^j B)}{\alpha^{\varsigma(i-j-1)}\lambda(x_B, \alpha^j r_B)} \\
&\lesssim 1 + \sum_{i=1}^{j} \left(\frac{\alpha^\varsigma}{\beta}\right)^{j-i} \\
&\lesssim 1.
\end{aligned}$$

This implies (iii).

From (7.1.3) and (7.1.1), we deduce that, for any ball B, $x \notin 2B$ and $y \in B$,

$$\lambda(x_B, d(x, x_B)) \sim \lambda(x, d(x, x_B)) \sim \lambda(x, d(x, y)) \sim \lambda(y, d(x, y)). \quad (7.1.7)$$

By this, there exists a positive constant c such that

$$\begin{aligned}
\delta(B, S) &= \int_{(2S)\setminus(2R)} \frac{1}{\lambda(x_B, d(x, x_B))}\, d\nu(x) + \int_{(2R)\setminus B} \cdots \\
&\le c \int_{(2S)\setminus(2R)} \frac{1}{\lambda(x_R, d(x, x_R))}\, d\nu(x) + \delta(B, R) \\
&\le c\delta(R, S) + \delta(B, R).
\end{aligned}$$

Moreover, if B and R are concentric, then we take $c = 1$. This finishes the proof of (iv).

Finally, by (7.1.7) and (ii) of this lemma, we see that, for all balls $B \subset R \subset S$,

$$\begin{aligned}
\delta(R, S) &= \int_{(2S)\setminus(2R)} \frac{1}{\lambda(x_R, d(x, x_R))}\, d\nu(x) + \delta(R, R) \\
&\lesssim 1 + \int_{(2S)\setminus(2R)} \frac{1}{\lambda(x_B, d(x, x_B))}\, d\nu(x) \\
&\lesssim 1 + \delta(B, S).
\end{aligned}$$

On the other hand, it is easy to see that, if B and R are concentric and $B \subset R \subset S$, then $\delta(R, S) \le \delta(B, S)$. This shows (v), and hence finishes the proof of Lemma 7.1.16. □

We now recall the basic cover lemma in metric measure spaces, which is of importance for applications.[4]

Lemma 7.1.17. *Every family $\mathcal{F}$ of balls of uniformly bounded diameter in a metric space $\mathcal{X}$ contains a disjointed subfamily $\mathcal{G}$ such that*

---

[4]See [49, Theorem 1.2].

$$\bigcup_{B\in\mathcal{F}} B \subset \bigcup_{B\in\mathcal{G}} 5B.$$

Indeed, every ball B from $\mathcal{F}$ meets a ball from $\mathcal{G}$ with radius at least half that of B.

From Lemma 7.1.17, we immediately deduce the following cover lemma.

Lemma 7.1.18. *Every family of balls $\{B_i\}_{i\in F}$ of uniformly bounded diameter in a metric space $\mathcal{X}$ contains a disjoint subfamily $\{B_i\}_{i\in E}$ with $E \subset F$ such that*

(i) $\bigcup_{i\in F} B_i \subset \bigcup_{i\in E} 6B_i$;
(ii) *For each $x \in \mathcal{X}$,*

$$\sum_{i\in E} \chi_{6B_i}(x) < \infty.$$

We remark that in (ii) of Lemma 7.1.18, the sum $\sum_{i\in E} \chi_{6B_i} < \infty$ at each x but these sums are not necessarily uniformly bounded on $\mathcal{X}$.

Proof of Lemma 7.1.18. By Lemma 7.1.17, we pick a disjoint subfamily

$$\{B_i : \ B_i := B(x_{B_i}, r_{B_i})\}_{i\in E}$$

with $E \subset F$ satisfying (i). If, for $i,\ j \in E$, $6B_i \subset 6B_j$, we then remove i from E. Thus, we may further assume that for $i,\ j \in E$, neither $6B_i \subset 6B_j$ nor $6B_j \subset 6B_i$.

To prove (ii), we assume in contradiction that there exists some $x \in \mathcal{X}$ such that there exist infinite number of balls $\{B_i : i \in I_x \subset E\}$ such that $x \in 6B_i$ for all $i \in I_x$. We show that

$$\liminf_{i\in I_x} r_{B_i} > 0.$$

Otherwise, for any $\epsilon \in (0,\infty)$ there exists $i_\epsilon \in I_x$ such that $r_{B_{i_\epsilon}} < \epsilon$. Therefore, if B_0 is any ball in the family $\{B_i : \ i \in I_x\}$, there exists $r \in (0,\infty)$ such that $B(x,r) \subset 6B_0$. For $\epsilon = r/30$, we have $x \in 6B_{i_\epsilon}$ and $r_{6B_{i_\epsilon}} < r/4$. This implies that

$$6B_{i_\epsilon} \subset B(x,r) \subset 6B_0,$$

which is a contradiction.

Thus $\liminf_{i\in I_x} r_{B_i} > 0$. This, together with the uniform boundedness of diameter of the family of balls, implies that there exist $m,\ M \in (0,\infty)$ such that $m < r_{B_i} < M$ for all $i \in I_x$. Obviously,

$$\bigcup_{i\in I_x} B_i \subset B(x, 2M).$$

By Definition 7.1.7, there exists a finite family $\{B(z_j, m/30)\}_{j=1}^K$ of balls such that

$$B(x, 2M) \subset \bigcup_{j=1}^{K} B(z_j, m/30).$$

On the other hand, because the number of the balls $\{B_i\}_{i\in I_x}$ is infinite, so is that of $\{\frac{1}{6}B_i\}_{i\in I_x}$. Therefore, from this and the fact that

$$\bigcup_{i\in I_x} \frac{1}{6}B_i \subset \bigcup_{j=1}^{K} B(z_j, m/30),$$

we see that there exists a ball

$$B_k \in \left\{B\left(z_j, \frac{m}{30}\right) : j \in \{1, \ldots, K\}\right\},$$

and at least two balls B_1 and B_2 in $\{B_i : i \in I_x\}$ such that $B_k \cap \frac{1}{6}B_1 \neq \emptyset$ and $B_k \cap \frac{1}{6}B_2 \neq \emptyset$. Since

$$\min\left\{r_{\frac{1}{6}B_1}, r_{\frac{1}{6}B_2}\right\} > \frac{1}{6}m = 5r_{B_k},$$

we have $B_k \subset B_1 \cap B_2$. This is a contradiction, because the family of balls $\{B_i : i \in I_x\}$ is pairwise disjoint. This finishes the proof of Lemma 7.1.18. □

If let $C_b(\mathcal{X})$ be the *space of all continuous functions with bounded support*, then we have the following conclusion.

Proposition 7.1.19. *Let $(\mathcal{X}, d)$ be a geometrically doubling metric space and ν a Borel measure on $\mathcal{X}$ which is finite on bounded sets. Then continuous, boundedly supported functions are dense in $L^p(\mathcal{X}, \nu)$ for $p \in [1, \infty)$.*

Proof. It suffices to approximate the characterization function of a Borel set E of finite measure in the $L^p(\mathcal{X}, \nu)$ norm by a continuous, boundedly supported function. Since, for any given $x_0 \in \mathcal{X}$,

$$\nu(E) = \lim_{r\to\infty} \nu(E \cap B(x_0, r)),$$

without loss of generality, we assume E is bounded. By a general result concerning Borel measures on metric spaces, there exists a closed set $F \subset E$ and an open set $\Omega \supset E$ such that $\nu(\Omega \setminus F) < \epsilon$. Since Ω may be replaced by $\Omega \cap B$, where B is any ball containing E, we take Ω to be bounded.

Let $\beta > 6^n$, as required in Lemma 7.1.13. For each $x \in F$, choose a $(6, \beta)$-doubling ball B_x of radius $r_x \le 1$ centered at x with $6B_x \subset \Omega$. By Lemma 7.1.17, extract a disjoint subcollection

$$\{B^i\}_{i=1}^{\infty} := \{B_{x_i}\}_{i=1}^{\infty}$$

such that $F \subset \cup_{i=1}^{\infty} 5B^i$. Since

$$\sum_{i=1}^{\infty} \nu(\overline{5B^i}) \le \sum_{i=1}^{\infty} \nu(6B^i) \le \beta \sum_{i=1}^{\infty} \nu(B^i) = \beta \nu\left(\bigcup_{i=1}^{\infty} B^i\right) \le \beta \nu(\Omega) < \infty,$$

it follows that

$$\lim_{j\to\infty} \nu\left(\bigcup_{i>j} \overline{5B^i}\right) = 0.$$

Thus,

$$\nu(F) = \lim_{j\to\infty} \nu\left(F \bigcap \left(\bigcup_{i=1}^{j} \overline{5B^i}\right)\right),$$

and hence F can be replaced by the closed set $F \cap (\cup_{i=1}^{j} \overline{5B^i})$ for some large $j \in \mathbb{N}$. Since $6B^i \subset \Omega$, it follows that

$$d(\overline{5B^i}, \mathcal{X} \setminus \Omega) \ge r_{B^i}$$

and hence the new set F satisfies $d(F, \mathcal{X} \setminus \Omega) > 0$. Thus, the function

$$\varphi(x) := \frac{d(x, \mathcal{X} \setminus \Omega)}{d(x, \mathcal{X} \setminus \Omega) + d(x, F)}, \quad \forall x \in \mathcal{X}$$

is continuous as the quotient of continuous functions, with denominator bounded away from zero, and satisfies that $\chi_F \le \varphi \le \chi_\Omega$, where Ω is a bounded set. Thus,

$$|\chi_E - \varphi| \le \chi_{\Omega\setminus F}$$

and hence

$$\|\chi_E - \varphi\|_{L^p(\mathcal{X},\nu)}^p \le \nu(\Omega \setminus F) < \epsilon,$$

which completes the proof of Proposition 7.1.19. □

Let $\eta \in [5, \infty)$. In what follows, let $L^1_{\mathrm{loc}}(\mathcal{X}, \nu)$ denote the *set of all* ν*-locally integrable functions on* $\mathcal{X}$. For any $f \in L^1_{\mathrm{loc}}(\mathcal{X}, \nu)$, consider the following *Hardy–Littlewood maximal function* $\mathcal{M}_\eta$, defined by setting

$$\mathcal{M}_\eta(f)(x) := \sup_{B\ni x} \frac{1}{\nu(\eta B)} \int_B |f(y)|\, d\nu(y), \quad \forall x \in \mathcal{X}, \tag{7.1.8}$$

where the supremum is taken over all balls B containing x. For any ν-measurable function f, the maximal function $\mathcal{M}_\eta(f)$ is lower semi-continuous and hence Borel measurable.

Theorem 7.1.20. *Let $\eta \in [5, \infty)$. If $(\mathcal{X}, d)$ is geometrically doubling and ν a Borel measure on $\mathcal{X}$ which is finite on bounded sets, then, for all $f \in L^1(\mathcal{X}, \nu)$ and $t \in (0, \infty)$, it holds true that*

$$\nu\left(\{x \in \mathcal{X} : \mathcal{M}_\eta(f)(x) > t\}\right) \le \frac{1}{t}\|f\|_{L^1(\mathcal{X},\nu)}.$$

Proof. For all $R \in (0, \infty)$, define the *operator* $\mathcal{M}_\eta^R$ by setting, for all $x \in \mathcal{X}$,

$$\mathcal{M}_\eta^R(f)(x) := \sup_{B \ni x,\ r_B < R} \frac{1}{\nu(\eta B)} \int_B |f(y)|\, d\nu(y). \tag{7.1.9}$$

Then $\mathcal{M}_\eta^R(f)$ is lower semi-continuous. Let

$$\Omega_R := \left\{y \in \mathcal{X} : \mathcal{M}_\eta^R(f)(y) > t\right\}.$$

Then, for every $x \in \Omega_R$, there exists a ball B_x of radius at most R such that

$$\nu(\eta B_x)^{-1} \int_{B_x} |f|\, d\nu > t.$$

In particular, the balls B_x of uniformly bounded radius cover the set Ω_R. By Lemma 7.1.17, among these balls we pick a disjoint subcollection $\{B_i\}_{i\in I}$ such that the balls $\{5B_i\}_{i\in I}$ still cover Ω_R. Thus,

$$\nu(\Omega_R) \le \sum_{i\in I} \nu(5B^i) \le \frac{1}{t}\sum_{i\in I} \int_{B_i} |f(y)|\, d\nu(y) \le \frac{1}{t}\|f\|_{L^1(\mathcal{X},\nu)}.$$

Since $\mathcal{M}_\eta^R(f) \uparrow \mathcal{M}_\eta(f)$ as $R \to \infty$, the desired result follows from the dominated convergence theorem, which completes the proof of Theorem 7.1.20. □

Corollary 7.1.21. *Let $(\mathcal{X}, d)$ be a geometrically doubling metric space and ν a Borel measure on $\mathcal{X}$ which is finite on bounded sets. Let $\eta \in [5, \infty)$ and $\beta \in (\eta^{n_0}, \infty)$. Then, for all $f \in L^1_{\rm loc}(\mathcal{X}, \nu)$ and ν-almost every $x \in \mathcal{X}$,*

$$f(x) = \lim_{\substack{B \downarrow x \\ (\eta,\beta)-\text{doubling}}} \frac{1}{\nu(B)} \int_B f(y)\, d\nu(y).$$

Proof. By Lemma 7.1.13, there exist arbitrarily small (η, β)-doubling balls containing x such that the limit makes sense for ν-almost all $x \in \mathcal{X}$. By a standard localization, it suffices to consider $f \in L^1(\mathcal{X}, \nu)$. The assertion is furthermore clear for continuous boundedly supported functions, which are dense in $L^1(\mathcal{X}, \nu)$ by Proposition 7.1.19. For $f \in L^1(\mathcal{X}, \nu)$ and a continuous boundedly supported function g, we see that

$$\begin{aligned}
&\limsup_{\substack{B\downarrow x\\ (\eta,\beta)-\text{doubling}}} \frac{1}{\nu(B)}\int_B |f(y)-f(x)|\,d\nu(y)\\
&\quad\le \sup_{B\ni x}\frac{\beta}{\nu(\eta B)}\int_B |f(y)-g(y)|\,d\nu(y)+|g(x)-f(x)|\\
&\quad=\beta\mathcal{M}_\eta(f-g)(x)+|g(x)-f(x)|.
\end{aligned}$$

By Theorem 7.1.20, we conclude that, for a given $\epsilon\in(0,\infty)$,

$$\nu(\{x\in\mathcal{X}:\ \mathcal{M}_\eta(f-g)(x)>\epsilon\})\lesssim \epsilon^{-1}\|g-f\|_{L^1(\mathcal{X},\nu)}.$$

If we choose g such that

$$\|g-f\|_{L^1(\mathcal{X},\nu)}<\epsilon^2,$$

then these two inequalities imply the desired conclusion, which completes the proof of Corollary 7.1.21. □

7.2 The BMO Space RBMO($\mathcal{X}$, ν)

In this section, we introduce the space RBMO($\mathcal{X}$, ν) in this setting and investigate its basic properties.

Definition 7.2.1. Let $\rho\in(1,\infty)$. A function $f\in L^1_{\rm loc}(\mathcal{X},\nu)$ is said to be in the *space* RBMO($\mathcal{X}$, ν) if there exists a positive constant C and, for all balls B, a complex number f_B such that

$$\frac{1}{\nu(\rho B)}\int_B |f(y)-f_B|\,d\nu(y)\le C \tag{7.2.1}$$

and, for all balls $B\subset S$,

$$|f_B-f_S|\le C[1+\delta(B,S)]. \tag{7.2.2}$$

Moreover, the RBMO($\mathcal{X}$, ν) *norm* of f is defined to be the minimal constant C as above and denoted by $\|f\|_{{\rm RBMO}(\mathcal{X},\nu)}$.

Proposition 7.2.2. *The following properties hold true:*

(i) RBMO($\mathcal{X}$, ν) *is a Banach space;*
(ii) $L^\infty(\mathcal{X},\nu)\subset$ RBMO ($\mathcal{X}$, ν). *Moreover, for all* $f\in L^\infty(\mathcal{X},\nu)$,

$$\|f\|_{{\rm RBMO}(\mathcal{X},\nu)}\le 2\|f\|_{L^\infty(\mathcal{X},\nu)};$$

(iii) *The* RBMO $(\mathcal{X}, \nu)$ *space is independent of the choice of the parameter* $\rho \in (1, \infty)$*;*

(iv) *If* $f \in$ RBMO$(\mathcal{X}, \nu)$*, then* $|f| \in$ RBMO$(\mathcal{X}, \nu)$ *and there exists a positive constant* C *such that, for all* $f \in$ RBMO$(\mathcal{X}, \nu)$*,*

$$\||f|\|_{\mathrm{RBMO}(\mathcal{X},\nu)} \leq C\|f\|_{\mathrm{RBMO}(\mathcal{X},\nu)};$$

(v) *If real-valued functions* $f, g \in$ RBMO$(\mathcal{X}, \nu)$*, then*

$$\min\{f, g\}, \max\{f, g\} \in \mathrm{RBMO}(\mathcal{X}, \nu)$$

and there exists a positive constant C *such that, for all* $f, g \in$ RBMO $(\mathcal{X}, \nu)$*,*

$$\|\min\{f, g\}\|_{\mathrm{RBMO}(\mathcal{X},\nu)} \leq C[\|f\|_{\mathrm{RBMO}(\mathcal{X},\nu)} + \|g\|_{\mathrm{RBMO}(\mathcal{X},\nu)}]$$

and

$$\|\max\{f, g\}\|_{\mathrm{RBMO}(\mathcal{X},\nu)} \leq C[\|f\|_{\mathrm{RBMO}(\mathcal{X},\nu)} + \|g\|_{\mathrm{RBMO}(\mathcal{X},\nu)}].$$

Proof. To prove (i), it is routinely to show that RBMO $(\mathcal{X}, \nu)$ is a linear space and $\|\cdot\|_{\mathrm{RBMO}(\mathcal{X},\nu)}$ a norm when any two functions, whose difference is ν-almost everywhere equal to a constant, are identified. To prove completeness, first fix a reference ball B_0 and replace each function $f^k \in$ RBMO $(\mathcal{X}, \nu)$, where

$$\sum_{k=1}^{\infty} \|f^k\|_{\mathrm{RBMO}(\mathcal{X},\nu)} < \infty, \tag{7.2.3}$$

by the function $f^k - f^k_{B_0}$ from the same equivalence class. Also replace the constant f^k_B by $f^k_B - f^k_{B_0}$. Keep denoting this new function still by f^k such that now $f^k_{B_0} = 0$. From (7.2.2), it follows that, for every ball B,

$$|f^k_B| \leq C_{(B)}\|f^k\|_{\mathrm{RBMO}(\mathcal{X},\nu)},$$

where $C_{(B)}$ is a positive constant depending only on B. Thus, by (7.2.3), the series $\sum_{k=1}^{\infty} f^k_B$ converges to a number f_B for each ball B. Using these numbers in the definition of the RBMO $(\mathcal{X}, \nu)$ space, it is easy to show that $\sum_{k=1}^{\infty} f^k$ converges ν-almost every $x \in \mathcal{X}$ and in the norm of RBMO $(\mathcal{X}, \nu)$ to a function f and

$$\|f\|_{\mathrm{RBMO}(\mathcal{X},\nu)} \leq \sum_{k=1}^{\infty} \|f^k\|_{\mathrm{RBMO}(\mathcal{X},\nu)},$$

which completes the proof of (i).

Observe that (ii) is obvious, (iv) follows from (iii) and (v) from (iv), to finish the proof of Proposition 7.2.2, then it suffices to prove (iii). To this end, denote RBMO $(\mathcal{X}, \nu)$ with parameter ρ temporarily by $\mathrm{RBMO}_\rho(\mathcal{X}, \nu)$, and let $\rho > \sigma > 1$. It is obvious that

$$\mathrm{RBMO}_\sigma(\mathcal{X}, \nu) \subset \mathrm{RBMO}_\rho(\mathcal{X}, \nu),$$

where the inclusion map has norm at most 1, hence only the converse direction requires proof.

Let $\delta := (\sigma - 1)/\rho$ and consider a fixed ball B_0. Then there exist balls

$$\{B_i := B(x_i, \delta r) : \ x_i \in B_0, \ i \in I\},$$

which cover B_0, where $|I| \le N_0 \delta^{-n_0}$. Moreover,

$$\rho B_i := B(x_i, \delta\rho r) \subset B(x_0, \sigma r) = \sigma B_0,$$

since $r + \delta\rho r = \sigma r$. From this, it then follows that

$$\begin{aligned}|f_{B_i} - f_{B_0}| &\le |f_{B_i} - f_{\sigma B_0}| + |f_{\sigma B_0} - f_{B_0}| \\ &\lesssim \|f\|_{\mathrm{RBMO}_\rho(\mathcal{X},\nu)} \left[1 + \log_2 \frac{6\sigma r_{B_0}}{r_{B_i}} + \log_2 6\sigma\right] \\ &\lesssim \|f\|_{\mathrm{RBMO}_\rho(\mathcal{X},\nu)}.\end{aligned}$$

Thus,

$$\begin{aligned}\int_{B_0} |f(x) - f_{B_0}|\, d\nu(x) &\le \sum_{i\in I} \int_{B_i} |f(x) - f_{B_0}|\, d\nu(x) \\ &\le \sum_{i\in I} \left\{\int_{B_i} |f(x) - f_{B_i}|\, d\nu(x) + \nu(B_i)|f_{B_i} - f_{B_0}|\right\} \\ &\lesssim \sum_{i\in I} \|f\|_{\mathrm{RBMO}_\rho(\mathcal{X},\nu)} \nu(\rho B_i) \\ &\lesssim \|f\|_{\mathrm{RBMO}_\rho(\mathcal{X},\nu)} \nu(\sigma B_0).\end{aligned}$$

Hence,

$$\|f\|_{\mathrm{RBMO}_\sigma(\mathcal{X},\nu)} \lesssim \|f\|_{\mathrm{RBMO}_\rho(\mathcal{X},\nu)},$$

and the same numbers f_B work in the definition of both spaces. This finishes the proof of (iii) and hence Proposition 7.2.2. □

When $(\mathcal{X}, d, \nu)$ is a space of homogeneous type as in Example 7.1.2 and

$$\lambda(x, r) := \nu(B(x, r))$$

for all $x \in \mathcal{X}$ and $r \in (0, \infty)$. A function $f \in L^1_{\rm loc}(\mathcal{X}, \nu)$ is said to belong to the *space* BMO $(\mathcal{X}, \nu)$ if there exists a positive constant C such that

$$\sup_B \frac{1}{\mu(B)} \int_B |f(x) - m_B(f)| \, d\mu(x) \le C,$$

where the supremum is taken over all balls in $\mathcal{X}$. Moreover, the BMO$(\mathcal{X}, \nu)$ *norm* of f is defined to be the minimal constant C as above and denoted by $\|f\|_{{\rm BMO}(\mathcal{X}, \nu)}$.[5] Then, when $\nu(\mathcal{X}) = \infty$, we have the following conclusion that RBMO $(\mathcal{X}, \nu)$ and BMO $(\mathcal{X}, \nu)$ coincide with equivalent norms. This is false when $\nu(\mathcal{X}) < \infty$, namely, for a general doubling measure ν with $\nu(\mathcal{X}) < \infty$, it may happen that

$$\text{RBMO}\,(\mathcal{X}, \nu) \subsetneqq \text{BMO}\,(\mathcal{X}, \nu)$$

(see Example 3.1.14).

Proposition 7.2.3. *If ν is a doubling measure, $\nu(\mathcal{X}) = \infty$ and, for all $x \in \mathcal{X}$ and $r \in (0, \infty)$,*

$$\lambda(x, r) := \nu(B(x, r)),$$

then the spaces RBMO $(\mathcal{X}, \nu)$ *and* BMO $(\mathcal{X}, \nu)$ *coincide with equivalent norms.*

Proof. If ν is doubling, then (7.2.1) is equivalent to the usual BMO conditions and, if this condition holds true for some f_B, it also holds true with $f_B := m_B(f)$, here and in what follows, $m_B(f)$ denotes the *mean* of f over B, namely,

$$m_B(f) := \frac{1}{\nu(B)} \int_B f(y) \, d\nu(y).$$

Hence it remains to investigate the other condition (7.2.2) in this case. We claim that

$$\begin{aligned} &|m_B(f) - m_{B_1}(f)| \\ &\quad \lesssim \|f\|_{\text{BMO}\,(\mathcal{X}, \nu)} \left[1 + \log_2 \frac{\nu(B_1)}{\nu(B)}\right] \\ &\quad \lesssim \|f\|_{\text{BMO}\,(\mathcal{X}, \nu)} \left\{1 + \int_{(2B_1 \setminus B)} \frac{1}{\lambda(B(x_B, d(x, x_B)))} \, d\nu(x)\right\}, \quad (7.2.4) \end{aligned}$$

which proves the assertion.

For $B \subset B_1$, define inductively $B^0 := B$ and B^i to be the smallest $2^k B^{i-1}$, $k \in \mathbb{N}$, such that

---

[5] See [19].

$$\nu(2^k B^{i-1}) > 2\nu(B^{i-1}) \tag{7.2.5}$$

(the assumption that $\nu(\mathcal{X}) = \infty$ guarantees the existence of B^i); hence the doubling condition of ν implies that

$$\nu(B^i) = \nu(2^k B^{i-1}) \le C_{(\nu)}\nu(2^{k-1} B^{i-1}) \le 2C_{(\nu)}\nu(B^{i-1}). \tag{7.2.6}$$

Let i_0 be the first index such that $B^{i_0} \not\subset 2B_1$. Then $r_{B^{i_0}} > r_{B_1}$, hence

$$B_1 \subset 2B^{i_0} \subset B^{i_0+1}$$

and, therefore,

$$\begin{aligned}\nu(B_1) &\le \nu(B^{i_0+1})\\ &\le 2C_{(\nu)}\nu(B^{i_0})\\ &\le 4C_{(\nu)}^2\nu(B^{i_0-1})\\ &\le 4C_{(\nu)}^2\nu(2B_1)\\ &\le 4C_{(\nu)}^3\nu(B_1).\end{aligned}$$

On the other hand, (7.2.5) and (7.2.6) imply that

$$2^{i_0}\nu(B) < \nu(B^{i_0}) \le (2C_{(\nu)})^{i_0}\nu(B).$$

Combining these two chains of inequalities, we see that

$$2^{i_0-1}C_{(\nu)}^{-2} \le \frac{\nu(B_1)}{\nu(B)} \le (2C_{(\nu)})^{i_0+1}. \tag{7.2.7}$$

Thus, it holds true that

$$\begin{aligned}|m_B(f) - m_{B_1}(f)| &\le \sum_{i=1}^{i_0+1} |m_{B^i}(f) - m_{B^{i-1}}(f)| + |m_{B^{i_0+1}}(f) - m_{B_1}(f)|\\ &\le \sum_{i=1}^{i_0+1} \frac{1}{\nu(B^{i-1})}\int_{B^{i-1}} |f(y) - m_{B^i}(f)|\,d\nu(y)\\ &\quad + \frac{1}{\nu(B_1)}\int_{B_1} |f(y) - m_{B^{i_0+1}}(f)|\,d\nu(y)\\ &\le \sum_{i=1}^{i_0+1} \frac{\nu(B^i)}{\nu(B^{i-1})}\frac{1}{\nu(B^i)}\int_{B^i} |f(y) - m_{B^i}(f)|\,d\nu(y)\end{aligned}$$

$$+\frac{\nu(B^{i_0+1})}{\nu(B_1)} \frac{1}{\nu(B^{i_0+1})} \int_{B^{i_0+1}} |f(y) - m_{B^{i_0+1}}(f)| \, d\nu(y)$$
$$\lesssim \sum_{i=1}^{i_0+1} \|f\|_{\mathrm{RBMO}\,(\mathcal{X},\,\nu)} + \|f\|_{\mathrm{RBMO}\,(\mathcal{X},\,\nu)}$$
$$\lesssim (1 + i_0)\|f\|_{\mathrm{RBMO}\,(\mathcal{X},\,\nu)}$$
$$\lesssim \left[1 + \log_2 \frac{\nu(B_1)}{\nu(B)}\right] \|f\|_{\mathrm{RBMO}\,(\mathcal{X},\,\nu)}.$$

On the other hand, (7.2.7), together with (7.1.1), implies that the quantity on the right of (7.2.2) is minorized by

$$\int_{2B_1 \setminus B} \frac{d\nu(x)}{\lambda(x_B, d(x, x_B))} \geq \sum_{i=1}^{i_0-1} \int_{B^i \setminus B^{i-1}} \frac{d\nu(x)}{\lambda(x_B, d(x, x_B))}$$
$$\gtrsim \sum_{i=1}^{i_0-1} \frac{\nu(B^i \setminus B^{i-1})}{\nu(B^i)}$$
$$\gtrsim i_0 - 1$$
$$\gtrsim \log_2 \frac{\nu(B_1)}{\nu(B)}.$$

This finishes the proof of (7.2.4) and hence Proposition 7.2.3. □

Analogous to RBMO$(\mathcal{X}, \nu)$, in the case $(\mathcal{X}, d, \nu) := (\mathbb{R}^D, |\cdot|, \mu)$ with μ satisfying the polynomial growth condition (0.0.1), the space RBMO$(\mathcal{X}, \nu)$ has the following generalized form.

Definition 7.2.4. Let $\rho \in (1, \infty)$ and $\gamma \in [1, \infty)$. A function $f \in L^1_{\mathrm{loc}}(\mathcal{X}, \nu)$ is said to be in the *space* RBMO$_\gamma(\mathcal{X}, \nu)$ if there exists a positive constant C and, for all balls B, a complex number f_B such that (7.2.1) holds true and that, for all balls $B \subset S$,

$$|f_B - f_S| \leq C[1 + \delta(B, S)]^\gamma. \tag{7.2.8}$$

Moreover, the RBMO$_\gamma(\mathcal{X}, \nu)$ *norm* of f is defined to be the minimal constant C as above and denoted by $\|f\|_{\mathrm{RBMO}_\gamma(\mathcal{X},\nu)}$.

If $\gamma = 1$, the space RBMO$_1(\mathcal{X}, \nu)$ is just RBMO$(\mathcal{X}, \nu)$. From the following proposition, we see that the space RBMO$_\gamma(\mathcal{X}, \nu)$ is independent of $\gamma \in [1, \infty)$.

Proposition 7.2.5. *Let* $\gamma \in (1, \infty)$. *Then* RBMO$_\gamma(\mathcal{X}, \nu)$ *and* RBMO$(\mathcal{X}, \nu)$ *coincide with equivalent norms.*

The proof of Proposition 7.2.5 depends on the following lemmas, which are analogues of Lemmas 3.1.7 and 3.1.8 whose proofs are similar and hence omitted.

Lemma 7.2.6. *Whenever $B_1 \subset B_2 \subset \cdots \subset B_m$ are concentric balls with*

$$\delta(B_i, B_{i+1}) \geq 3C_{(\lambda)}$$

for all $i \in \{1, \ldots, m-1\}$, it holds true that

$$\sum_{i=1}^{m-1}[1 + \delta(B_i, B_{i+1})] \leq 3\delta(B_1, B_m).$$

Lemma 7.2.7. *For a large positive constant D_1, the following statement holds true: let $x \in \mathcal{X}$ be a fixed point and $\{f_B\}_{B \ni x}$ some collection of numbers. If, for some constant $C_{(x)}$, it holds true that*

$$|f_B - f_S| \leq C_{(x)}[1 + \delta(B, S)]$$

for all balls B, S with $x \in B \subset S$ and $\delta(B, S) \leq D_1$, then there exists a positive constant C, independent of x, such that

$$|f_B - f_S| \leq CC_{(x)}[1 + \delta(B, S)]$$

for all balls B, S with $x \in B \subset S$.

Proof of Proposition 7.2.5. Obviously,

$$\mathrm{RBMO}\,(\mathcal{X},\ \nu) \subset \mathrm{RBMO}_\gamma\,(\mathcal{X},\ \nu).$$

To see the converse, assume that $f \in \mathrm{RBMO}_\gamma\,(\mathcal{X},\ \nu)$. Then f satisfies (7.2.1) with C replaced by $\|f\|_{\mathrm{RBMO}_\gamma\,(\mathcal{X},\nu)}$. To show that f satisfies (7.2.2), let $x \in \mathcal{X}$ and $B \subset S$ with $x \in B$ such that $\delta(B, S) \leq D_1$, where D_1 is as in the statement of Lemma 7.2.7. Then $f \in \mathrm{RBMO}_\gamma\,(\mathcal{X},\ \nu)$ implies that

$$\begin{aligned}|f_B - f_S| &\leq [1 + \delta(B, S)]^\gamma \|f\|_{\mathrm{RBMO}_\gamma\,(\mathcal{X},\nu)}\\ &\leq [1 + \delta(B, S)](1 + D_1)^{\gamma-1}\|f\|_{\mathrm{RBMO}_\gamma\,(\mathcal{X},\nu)},\end{aligned}$$

which is as the assumption of Lemma 7.2.7 with $C_{(x)} := (1 + D_1)^{\gamma-1}$. By Lemma 7.2.7, we see that, for all balls $B \subset S$ with $x \in B$,

$$|f_B - f_S| \lesssim [1 + \delta(B, S)](1 + D_1)^{\gamma-1}\|f\|_{\mathrm{RBMO}_\gamma\,(\mathcal{X},\nu)}.$$

This, together with (7.2.1), implies that $f \in \mathrm{RBMO}\,(\mathcal{X},\ \nu)$ and

$$\|f\|_{\mathrm{RBMO}\,(\mathcal{X},\nu)} \lesssim \|f\|_{\mathrm{RBMO}_\gamma\,(\mathcal{X},\nu)},$$

and hence finishes the proof of Proposition 7.2.5. □

Remark 7.2.8. By Proposition 7.2.5, unless explicitly pointed out, in what follows, we always assume that the *constant γ in Definition 7.2.4 is equal to 1.*

We now establish a characterization of $\mathrm{RBMO}_\gamma(\mathcal{X}, \nu)$ in terms of the average of functions on $(6, \beta_6)$-doubling balls, where β_6 is as in (7.1.6) with $\alpha := 6$.

Proposition 7.2.9. *Let $\rho \in (1, \infty)$, $\gamma \in [1, \infty)$ and $f \in L^1_{\mathrm{loc}}(\mathcal{X}, \nu)$. The following statements are equivalent:*

(a) $f \in \mathrm{RBMO}_\gamma(\mathcal{X}, \nu)$;
(b) *There exists $\tilde{C} \in (0, \infty)$ such that, for any $(6, \beta_6)$-doubling ball B,*

$$\int_B |f(x) - m_B(f)|\, d\nu(x) \le \tilde{C}\nu(B) \tag{7.2.9}$$

and, for all $(6, \beta_6)$-doubling balls $B \subset S$,

$$|m_B(f) - m_S(f)| \le \tilde{C}[1 + \delta(B, S)]^\gamma. \tag{7.2.10}$$

Moreover, let $\|f\|_\circ$ be the minimal constant $\tilde{C}$ in (b). Then there exists a constant $C \in [1, \infty)$ such that

$$\|f\|_\circ / C \le \|f\|_{\mathrm{RBMO}_\gamma(\mathcal{X}, \nu)} \le C\|f\|_\circ.$$

Proof. By Proposition 7.2.5 and Proposition 7.2.2(iii), we take $\rho := 6/5$ in Definition 7.2.1.

We first show that (a) implies (b). To this end, assume that $f \in \mathrm{RBMO}_\gamma(\mathcal{X}, \nu)$. If B is $(6, \beta_6)$-doubling, then we have

$$\begin{aligned} |f_B - m_B(f)| &= \left| \frac{1}{\nu(B)} \int_B [f(x) - f_B]\, d\nu(x) \right| \\ &\le \|f\|_{\mathrm{RBMO}_\gamma(\mathcal{X}, \nu)} \frac{\nu(\frac{6}{5}B)}{\nu(B)} \\ &\lesssim \|f\|_{\mathrm{RBMO}_\gamma(\mathcal{X}, \nu)}, \end{aligned} \tag{7.2.11}$$

which, together with (7.2.1) and B is $(6, \beta_6)$-doubling, implies that

$$\begin{aligned} &\frac{1}{\nu(B)} \int_B |f(x) - m_B(f)|\, d\nu(x) \\ &\le \frac{1}{\nu(B)} \int_B |f(x) - f_B|\, d\nu(x) + |f_B - m_B(f)| \\ &\lesssim \|f\|_{\mathrm{RBMO}_\gamma(\mathcal{X}, \nu)}. \end{aligned}$$

Moreover, by (7.2.2) and (7.2.11), we see that, if $B \subset S$ are both $(6, \beta_6)$-doubling balls, then

$$|m_B(f) - m_S(f)| \le |m_B(f) - f_B| + |f_B - f_S| + |f_S - m_S(f)|$$
$$\lesssim [1 + \delta(B, S)]^\gamma \|f\|_{\mathrm{RBMO}_\gamma(\mathcal{X},\nu)}.$$

Thus, (b) holds true.

We now show that (b) implies (a). Let f satisfy (7.2.9) and (7.2.10). We show that

$$f \in \mathrm{RBMO}_\gamma(\mathcal{X}, \nu) \text{ with } f_B := m_{\tilde{B}^6}(f) \text{ for each ball } B.$$

Let B be any ball which is not $(6, \beta_6)$-doubling. For ν-almost every $x \in B$, let B_x be the biggest $(30, \beta_6)$-doubling ball with center x and radius $30^{-k} r_B$ for some $k \in \mathbb{N}$. Recall that such ball exists by Lemma 7.1.13, since $\beta_6 > 30^{\max\{n_0, \varsigma\}}$. We claim that

$$B_x \subset (6/5)B \subset \tilde{B}^6$$

and

$$\left|m_{B_x}(f) - m_{\tilde{B}^6}(f)\right| \lesssim \tilde{C}. \tag{7.2.12}$$

Obviously, since B is not $(6, \beta_6)$-doubling, then $\tilde{B}^6$ has the radius at least $6r_B$. From this, it follows that

$$B_x \subset (6/5)B \subset \tilde{B}^6.$$

Let A_x be the smallest $(30, \beta_6)$-doubling ball of the form $30^k B_x$, $k \in \mathbb{N}$, which exists by Lemma 7.1.12. Then, by the choices of A_x and B_x, $r_{A_x} \ge r_B$. To show (7.2.12), we consider the following two cases.

Case i) $r_{\tilde{B}^6} \le r_{A_x}$. In this case, $\tilde{B}^6 \subset 2A_x$. From (ii), (iii), (iv) and (v) of Lemma 7.1.16, we deduce that

$$\delta\left(\tilde{B}^6, \widetilde{2A_x}^6\right) \lesssim 1 + \delta\left(B_x, \widetilde{2A_x}^6\right) \lesssim 1.$$

This, combined with (7.2.10) and the fact that B_x is also $(6, \beta_6)$-doubling, implies that

$$\left|m_{B_x}(f) - m_{\tilde{B}^6}(f)\right| \le \left|m_{B_x}(f) - m_{\widetilde{2A_x}^6}(f)\right| + \left|m_{\widetilde{2A_x}^6}(f) - m_{\tilde{B}^6}(f)\right|$$
$$\lesssim \tilde{C}\left[1 + \delta\left(B_x, \widetilde{2A_x}^6\right) + \delta\left(\tilde{B}^6, \widetilde{2A_x}^6\right)\right]^\gamma$$
$$\lesssim \tilde{C}.$$

Case ii) $r_{\tilde{B}^6} > r_{A_x}$. In this case, $B \subset 2A_x \subset 3\tilde{B}^6$. From this, together with (7.2.10), the fact that B_x is also $(6, \beta_6)$-doubling and (ii) through (v) of Lemma 7.1.16, we deduce that

$$\begin{aligned}
\left|m_{B_x}(f) - m_{\tilde{B}^6}(f)\right| &\le \left|m_{B_x}(f) - m_{\widetilde{3\tilde{B}^6}^6}(f)\right| + \left|m_{\widetilde{3\tilde{B}^6}^6}(f) - m_{\tilde{B}^6}(f)\right| \\
&\lesssim \tilde{C}\left[1 + \delta\left(B_x, \widetilde{3\tilde{B}^6}^6\right) + \delta\left(\tilde{B}^6, \widetilde{3\tilde{B}^6}^6\right)\right]^\gamma \\
&\lesssim \tilde{C}\left[1 + \delta\left(B_x, 2A_x\right) + \delta\left(2A_x, \widetilde{3\tilde{B}^6}^6\right)\right]^\gamma \\
&\lesssim \tilde{C}\left[1 + \delta\left(B, \widetilde{3\tilde{B}^6}^6\right)\right]^\gamma \\
&\lesssim \tilde{C}.
\end{aligned}$$

Thus, (7.2.12) holds true. That is, the claim is true.

Now, by Lemma 7.1.18, there exists a countable disjoint subfamily $\{B_i\}_i$ of $\{B_x\}_x$ such that, for every $x \in B$, $x \in \cup_i 5B_i$. Moreover, since for any i, $5B_i$ is $(6, \beta_6)$-doubling, by (7.2.9), (7.2.12) and Lemma 7.1.16(ii), we have

$$\begin{aligned}
&\int_B \left|f(x) - m_{\tilde{B}^6}(f)\right| d\nu(x) \\
&\quad \le \sum_i \int_{5B_i} \left|f(x) - m_{\tilde{B}^6}(f)\right| d\nu(x) \\
&\quad \lesssim \sum_i \int_{5B_i} |f(x) - m_{5B_i}(f)|\, d\nu(x) \\
&\qquad + \sum_i |m_{5B_i}(f) - m_{B_i}(f)|\, \nu(5B_i) \\
&\qquad + \sum_i \left|m_{B_i}(f) - m_{\tilde{B}^6}(f)\right| \nu(5B_i) \\
&\quad \lesssim \tilde{C} \sum_i \nu(5B_i) \\
&\quad \lesssim \tilde{C} \sum_i \nu(B_i) \\
&\quad \lesssim \tilde{C}\nu\left(\frac{6}{5}B\right).
\end{aligned} \tag{7.2.13}$$

By an argument similar to that used in the proof of (7.2.12), we conclude that, for any balls $B \subset S$,

$$\left|m_{\tilde{B}^6}(f) - m_{\tilde{S}^6}(f)\right| \lesssim \tilde{C}[1 + \delta(B, S)]^\gamma.$$

This, together with (7.2.13), implies that $f \in \mathrm{RBMO}_\gamma(\mathcal{X}, \nu)$ and

$$\|f\|_{\mathrm{RBMO}_\gamma(\mathcal{X},\nu)} \lesssim \tilde{C},$$

which completes the proof of Proposition 7.2.9. □

We now establish another characterization of $\mathrm{RBMO}_\gamma(\mathcal{X}, \nu)$ which is useful in applications. To be precise, let $\gamma \in [1, \infty)$ and $f \in L^1_{\mathrm{loc}}(\mathcal{X}, \nu)$. The *median value* of f on any ball B, denoted by $\alpha_f(B)$, is defined as follows. If f is real-valued, then, for any ball B, let $\alpha_B(f)$ be some *real number* such that $\inf_{\alpha \in \mathbb{R}} m_B(|f - \alpha|)$ is attained. It is known that $\alpha_B(f)$ satisfies that

$$\nu(\{x \in B : f(x) > \alpha_B(f)\}) \le \nu(B)/2 \tag{7.2.14}$$

and

$$\nu(\{x \in B : f(x) < \alpha_B(f)\}) \le \nu(B)/2. \tag{7.2.15}$$

For all balls B with $\nu(B) = 0$, let $\alpha_B(f) := 0$. If f is complex-valued, we take

$$\alpha_B(f) := [\alpha_B(\Re f)] + i[\alpha_B(\Im f)], \tag{7.2.16}$$

where $i^2 = -1$, and, for any complex number z, $\Re z$ and $\Im z$ denote the *real part* and the *imaginary part* of z, respectively. Furthermore, for any $f \in L^1_{\mathrm{loc}}(\mathcal{X}, \nu)$, we denote by $\|f\|_*$ the *minimal nonnegative constant* $\tilde{C}$ such that, for any ball B,

$$\frac{1}{\nu(3B)} \int_B \left|f(x) - \alpha_{\tilde{B}^6}(f)\right| d\nu(x) \le \tilde{C} \tag{7.2.17}$$

and, for all $(6, \beta_6)$-doubling balls $B \subset S$,

$$|\alpha_B(f) - \alpha_S(f)| \le \tilde{C}[1 + \delta(B, S)]^\gamma. \tag{7.2.18}$$

Then we have the following conclusion.

Proposition 7.2.10. *$\|\cdot\|_*$ is an equivalent norm with $\|\cdot\|_{\mathrm{RBMO}_\gamma(\mathcal{X},\nu)}$.*

Proof. By Proposition 7.2.2, we may take $\rho := 3$ in Definition 7.2.1. Observe that, for any complex-valued function $f \in \mathrm{RBMO}_\gamma(\mathcal{X}, \nu)$, both $\Re f$ and $\Im f$ are in $\mathrm{RBMO}_\gamma(\mathcal{X}, \nu)$. By this fact, without loss of generality, we may assume that f is real-valued, since if f is complex-valued, we may consider its real part and imaginary part, respectively.

Assume that f satisfies (7.2.17) and (7.2.18). We need to show that

$$f \in \mathrm{RBMO}_\gamma(\mathcal{X}, \nu) \quad \text{and} \quad \|f\|_{\mathrm{RBMO}_\gamma(\mathcal{X},\nu)} \lesssim \|f\|_*.$$

Indeed, for any ball $B \subset \mathcal{X}$, let $f_B := \alpha_{\widetilde{B}^6}(f)$. Obviously, (7.2.1) holds true with C replaced by $\|f\|_*$. Moreover, by arguing as the estimate of (7.2.12), we see that, for all balls $B \subset S$,

$$\left|\alpha_{\widetilde{B}^6}(f) - \alpha_{\widetilde{S}^6}(f)\right| \lesssim [1 + \delta(B, S)]^\gamma \|f\|_*.$$

Thus,

$$f \in \mathrm{RBMO}_\gamma(\mathcal{X}, \nu) \quad \text{and} \quad \|f\|_{\mathrm{RBMO}_\gamma(\mathcal{X},\nu)} \lesssim \|f\|_*.$$

Now, for any $f \in \mathrm{RBMO}_\gamma(\mathcal{X}, \nu)$, we prove that f satisfies (7.2.17) and (7.2.18) and that

$$\|f\|_* \lesssim \|f\|_{\mathrm{RBMO}_\gamma(\mathcal{X},\nu)}.$$

To this end, assume that B is $(6, \beta_6)$-doubling. By the definition of $\alpha_B(f)$, we have

$$\begin{aligned}
|\alpha_B(f) - f_B| &\le \frac{1}{\nu(B)} \int_B [|f(x) - f_B| + |f(x) - \alpha_B(f)|]\, d\nu(x) \\
&\le \frac{2}{\nu(B)} \int_B |f(x) - f_B|\, d\nu(x) \\
&\lesssim \|f\|_{\mathrm{RBMO}_\gamma(\mathcal{X},\nu)}.
\end{aligned} \tag{7.2.19}$$

Then, from this, together with (7.2.2) and Lemma 7.1.16(ii), it follows that, for any ball B,

$$|f_B - \alpha_{\widetilde{B}^6}(f)| \le \left|f_B - f_{\widetilde{B}^6}\right| + \left|f_{\widetilde{B}^6} - \alpha_{\widetilde{B}^6}(f)\right| \lesssim \|f\|_{\mathrm{RBMO}_\gamma(\mathcal{X},\nu)}.$$

By this and (7.2.1), we have

$$\begin{aligned}
&\frac{1}{\nu(3B)} \int_B \left|f(x) - \alpha_{\widetilde{B}^6}(f)\right| d\nu(x) \\
&\quad\le \frac{1}{\nu(3B)} \int_B |f(x) - f_B|\, d\nu(x) + \left|f_B - \alpha_{\widetilde{B}^6}(f)\right| \\
&\quad\lesssim \|f\|_{\mathrm{RBMO}_\gamma(\mathcal{X},\nu)}.
\end{aligned} \tag{7.2.20}$$

Moreover, from (7.2.2) and (7.2.19), we deduce that, for all $(6, \beta_6)$-doubling balls $B \subset S$,

$$\begin{aligned}
|\alpha_B(f) - \alpha_S(f)| &\le |\alpha_B(f) - f_B| + |f_B - f_S| + |f_S - \alpha_S(f)| \\
&\lesssim [1 + \delta(B, S)]^\gamma \|f\|_{\mathrm{RBMO}_\gamma(\mathcal{X},\nu)}.
\end{aligned}$$

This, combined with (7.2.20), implies that

$$\|f\|_* \lesssim \|f\|_{\mathrm{RBMO}_\gamma(\mathcal{X},\nu)}$$

and hence finishes the proof of Proposition 7.2.10. □

The space RBMO ($\mathcal{X}$, ν) also enjoys the following John–Nirenberg inequality.

Theorem 7.2.11. *For every $\rho \in (1, \infty)$, there exists a positive constant c such that, for every $f \in \mathrm{RBMO}(\mathcal{X}, \nu)$ and every ball $B_0 := B(x_0, r)$,*

$$\nu(\{x \in B_0 : |f(x) - f_{B_0}| > t\}) \le 2\nu(\rho B_0)e^{-ct/\|f\|_{\mathrm{RBMO}(\mathcal{X},\nu)}}.$$

Proof. Let $\alpha := 5\rho$ and β be large enough as required in Lemmas 7.1.12 and 7.1.13, and L a large positive constant to be chosen. For every $x \in B_0$, let $\tilde{B}_x^\alpha$ be the maximal (α, β)-doubling ball of the form $\tilde{B}_x^\alpha := B(x, \alpha^{-i} r)$, $i \in \mathbb{N}$, such that

$$\tilde{B}_x^\alpha \subset \sqrt{\rho} B_0$$

and $|f_{\tilde{B}_x^\alpha} - f_{B_0}| > L$, if any exists. Notice that, for ν-almost every $x \in B_0$, if $|f(x) - f_{B_0}| > 2L$, then there exist arbitrarily small doubling balls $B := B(x, \alpha^{-i} r)$ such that $|m_B f - f_{B_0}| > 2L$. Thus, from this and the fact that

$$|m_B f - f_B| \le \beta\|f\|_{\mathrm{RBMO}(\mathcal{X},\nu)} \text{ for every } (\alpha, \beta) - \text{doubling ball } B, \tag{7.2.21}$$

it follows that

$$|f_B - f_{B_0}| > 2L - |m_B f - f_B| > L,$$

provided that $L > \beta\|f\|_{\mathrm{RBMO}(\mathcal{X},\nu)}$. Therefore, for all $x \in B_0$ with $|f(x) - f_{B_0}| > 2L$, a ball $\tilde{B}_x^\alpha$ is found. Observe that

$$\begin{aligned}
\frac{1}{\nu(\tilde{B}_x^\alpha)} \int_{\tilde{B}_x^\alpha} |f(y) - f_{B_0}|\, d\nu(y) &\ge \left|m_{\tilde{B}_x^\alpha} f - f_{B_0}\right| \\
&> L - \left|m_{\tilde{B}_x^\alpha} f - f_{\tilde{B}_x^\alpha}\right| \\
&\ge L - \beta\|f\|_{\mathrm{RBMO}(\mathcal{X},\nu)} \\
&> L/2
\end{aligned} \tag{7.2.22}$$

by (7.2.21), provided that $L > 2\beta\|f\|_{\mathrm{RBMO}(\mathcal{X},\nu)}$.

From the maximality of $\tilde{B}_x^\alpha$, it follows that $\widetilde{\tilde{B}_x^\alpha}^\alpha := (\alpha \tilde{B}_x^\alpha)^\alpha$ satisfies that

$$\widetilde{\tilde{B}_x^\alpha}^\alpha \not\subset \sqrt{\rho} B_0 \quad \text{or} \quad |f_{\widetilde{\tilde{B}_x^\alpha}^\alpha} - f_{B_0}| \le L.$$

In the first case, let $\alpha^i \tilde{B}_x^\alpha$, $i \in \mathbb{Z}_+$, be the smallest expansion of $\tilde{B}_x^\alpha$, with $\alpha^i \tilde{B}_x^\alpha \not\subset \sqrt{\rho}B_0$, such that

$$r_{\alpha^i \tilde{B}_x^\alpha} \sim r_{B_0} \quad \text{and} \quad \widetilde{\tilde{B}_x^\alpha}^\alpha = \widetilde{(\alpha^i \tilde{B}_x^\alpha)}^\alpha.$$

Hence by Definition 7.2.1 and Lemma 7.1.16, we see that there exists a positive constant c_1 such that

$$\left| f_{\widetilde{\tilde{B}_x^\alpha}^\alpha} - f_{B_0} \right| \le \left| f_{\widetilde{\tilde{B}_x^\alpha}^\alpha} - f_{\alpha^i \tilde{B}_x^\alpha} \right| + \left| f_{\alpha^i \tilde{B}_x^\alpha} - f_{B_0} \right| \le c_1 \|f\|_{\mathrm{RBMO}(\mathcal{X},\nu)}.$$

But this means that actually $|f_{\widetilde{\tilde{B}_x^\alpha}^\alpha} - f_{B_0}| \le L$ in any case, provided that

$$L \ge c_1 \|f\|_{\mathrm{RBMO}(\mathcal{X},\nu)}.$$

Hence since

$$\left| f_{\tilde{B}_x^\alpha} - f_{\widetilde{\tilde{B}_x^\alpha}^\alpha} \right| \le c_2 \|f\|_{\mathrm{RBMO}(\mathcal{X},\nu)}$$

for some positive constant c_2, if $L \ge 2c_2 \|f\|_{\mathrm{RBMO}(\mathcal{X},\nu)}$, then it follows that

$$L < \left| f_{\tilde{B}_x^\alpha} - f_{B_0} \right| \le \left| f_{\tilde{B}_x^\alpha} - f_{\widetilde{\tilde{B}_x^\alpha}^\alpha} \right| + \left| f_{\widetilde{\tilde{B}_x^\alpha}^\alpha} - f_{B_0} \right| \le c_2 \|f\|_{\mathrm{RBMO}(\mathcal{X},\nu)} + L \le 3L/2.$$

By Lemma 7.1.17, among the balls $\tilde{B}_x^\alpha$, one now choose disjoint balls B_i, $i \in I$, such that the expanded balls $5B_i$ cover all the original $\tilde{B}_x^\alpha$. Let c_3 be a positive constant such that

$$|f_{B_i} - f_{5B_i}| \le c_3 \|f\|_{\mathrm{RBMO}(\mathcal{X},\nu)}.$$

If $x \in 5B_i$ and $|f(x) - f_{B_0}| > nL$, then

$$\begin{aligned} |f(x) - f_{5B_i}| &\ge |f(x) - f_{B_0}| - |f_{B_0} - f_{B_i}| - |f_{B_i} - f_{5B_i}| \\ &> nL - 3L/2 - c_3 \|f\|_{\mathrm{RBMO}(\mathcal{X},\nu)} \\ &\ge (n-2)L, \end{aligned}$$

provided $L \ge 2c_3 \|f\|_{\mathrm{RBMO}(\mathcal{X},\nu)}$. For $n \ge 2$, it thus follows that

$$\begin{aligned} \{x \in B_0 : |f(x) - f_{B_0}| > nL\} &\subset \bigcup_{\substack{x \in B_0 \\ |f(x) - f_{B_0}| > nL}} \{y \in \tilde{B}_x^\alpha : |f(y) - f_{B_0}| > nL\} \\ &\subset \bigcup_{i \in I} \{y \in 5B_i : |f(y) - f_{5B_i}| > (n-2)L\}. \end{aligned}$$

Using (7.2.22) and the fact that the balls $\{B_i\}_{i\in I} := \{\tilde{B}^{\alpha}_{x_i}\}_{i\in I}$ are (α, β)-doubling, disjoint, and contained in $\sqrt{\rho}B_0$, we then conclude that

$$\begin{aligned}
\sum_{i\in I} \nu(\rho 5B_i) &= \sum_{i\in I} \nu(\alpha B_i) \\
&\le \beta \sum_{i\in I} \nu(B_i) \\
&\le \frac{2\beta}{L} \sum_{i\in I} \int_{B_i} |f(y) - f_{B_0}|\, d\nu(y) \\
&\le \frac{2\beta}{L} \int_{\sqrt{\rho}B_0} |f(y) - f_{B_0}|\, d\nu(y) \\
&\le \frac{2\beta}{L} \left[\int_{\sqrt{\rho}B_0} |f(y) - f_{\sqrt{\rho}B_0}|\, d\nu(y) + \nu(\sqrt{\rho}B_0)|f_{\sqrt{\rho}B_0} - f_{B_0}| \right] \\
&\le \frac{c_4}{L} \nu(\rho B_0) \|f\|_{\mathrm{RBMO}(\mathcal{X}, \nu)} \\
&\le \frac{1}{2} \nu(\rho B_0),
\end{aligned}$$

provided $L \ge 2c_4 \|f\|_{\mathrm{RBMO}(\mathcal{X}, \nu)}$ for some positive constant c_4.

Write $B^i := 5B_i$, the above conclusions are summarized as

$$\{x \in B_0 : |f(x) - f_{B_0}| > nL\} \subset \bigcup_{i\in I} \{x \in B^i : |f(x) - f_{B^i}| > (n-2)L\}$$

and

$$\sum_{i\in I} \nu(\rho B^i) \le \frac{1}{2} \nu(\rho B_0).$$

This contains the essence of the matter, for now we iterate, with the balls B^i in place of B_0, to the result that

$$\begin{aligned}
&\{x \in B_0 : |f(x) - f_{B_0}| > 2nL\} \\
&\quad \subset \bigcup_{i_1} \{x \in B^{i_1} : |f(x) - f_{B^{i_1}}| > 2(n-1)L\} \\
&\quad \subset \bigcup_{i_1, i_2} \{x \in B^{i_1, i_2} : |f(x) - f_{B^{i_1, i_2}}| > 2(n-2)L\} \\
&\quad \subset \cdots \\
&\quad \subset \bigcup_{i_1, \ldots, i_n} \{x \in B^{i_1, \ldots, i_n} : |f(x) - f_{B^{i_1, \ldots, i_n}}| > 0\},
\end{aligned}$$

and hence

$$\begin{aligned}\nu(\{x \in B_0 : |f(x) - f_{B_0}| > 2nL\}) &\le \sum_{i_1,\dots,i_n} \nu(B^{i_1,\dots,i_n}) \\ &\le \sum_{i_1,\dots,i_{n-1}} \sum_{i_n} \nu(\rho B^{i_1,\dots,i_{n-1},i_n}) \\ &\le \sum_{i_1,\dots,i_{n-1}} \frac{1}{2}\nu(\rho B^{i_1,\dots,i_{n-1}}) \\ &\le \cdots \\ &\le \frac{1}{2^n}\nu(\rho B_0).\end{aligned}$$

Take

$$\tilde{C} := \max\{2\beta, c_1, 2c_2, 2c_3, 2c_4\}$$

and $L \ge \tilde{C}\|f\|_{\mathrm{RBMO}(\mathcal{X},\nu)}$, and choose $n \in \mathbb{N}$ such that

$$2nL \le t < 2(n+1)L.$$

Then

$$\begin{aligned}\nu(\{x \in B_0 : |f(x) - f_{B_0}| > t\}) &\le \nu(\{x \in B_0 : |f(x) - f_{B_0}| > 2nL\}) \\ &\le 2^{-n}\nu(\rho B_0) \\ &\le 2^{-(2L)^{-1}t+1}\nu(\rho B_0) \\ &= 2\exp(-ct/\|f\|_{\mathrm{RBMO}(\mathcal{X},\nu)})\nu(\rho B_0),\end{aligned}$$

which completes the proof of Theorem 7.2.11. □

From Theorem 7.2.11 and Proposition 7.2.9, we immediately deduce the following conclusion.

Corollary 7.2.12. (i) *For every $\rho \in (1,\infty)$ and $p \in [1,\infty)$, there exists a positive constant C such that, for every $f \in$ RBMO$(\mathcal{X}, \nu)$ and every ball B,*

$$\left[\frac{1}{\nu(\rho B)}\int_B |f(x) - f_B|^p \, d\nu(x)\right]^{1/p} \le C\|f\|_{\mathrm{RBMO}(\mathcal{X},\nu)}. \tag{7.2.23}$$

Moreover, the infimum of all positive constants C satisfying (7.2.23) *and* (7.2.10) *is an equivalent* RBMO$(\mathcal{X}, \nu)$ *norm of f.*

(ii) *The conclusion of (i) still holds true if f_B is replaced by $m_{\tilde{B}^6} f$ for any ball B in* (7.2.10) *and* (7.2.23).

7.3 An Equivalent Characterization of RBMO $(\mathcal{X}, \nu)$ Via the Local Sharp Maximal Operator

This section is devoted to an equivalent characterization of RBMO $(\mathcal{X}, \nu)$. To this end, we introduce the notion of the John–Strömberg sharp maximal function. Let $s \in (0, 1)$ and $\varrho \in (1, \infty)$. For any fixed ball B and ν-measurable function f, define $m^{\varrho}_{0,s;B}(f)$ by setting

$$m^{\varrho}_{0,s;B}(f) := \inf\{t \in (0, \infty) : \nu(\{y \in B : |f(y)| > t\}) < s\nu(\varrho B)\}$$

when $\nu(B) > 0$, and $m^{\varrho}_{0,s;B}(f) := 0$ when $\nu(B) = 0$. For any ν-measurable function f, the *John–Strömberg sharp maximal function* $M^{\varrho,\sharp}_{0,s}(f)$ is defined by setting, for all $x \in \mathcal{X}$,

$$M^{\varrho,\sharp}_{0,s}(f)(x) := \sup_{B \ni x} m^{\varrho}_{0,s;B}\left(f - \alpha_{\widetilde{B}^{6\varrho^2}}(f)\right) + \sup_{\substack{x \in B \subset S \\ B,\, S\ (6\varrho^2,\, \beta_{6\varrho^2})\text{-doubling}}} \frac{|\alpha_B(f) - \alpha_S(f)|}{1 + \delta(B, S)},$$

where $\alpha_B(f)$ and $\alpha_S(f)$ are the median values of f, respectively, on the balls B and S as in (7.2.16).

Using $M^{\varrho,\sharp}_{0,s}$, we introduce the space $\mathrm{RBMO}_{0,s}(\mathcal{X}, \nu)$ as follows.

Definition 7.3.1. Let $s \in (0, 1)$ and $\varrho \in (1, \infty)$. A ν-measurable function f is said to belong to the *space* $\mathrm{RBMO}_{0,s}(\mathcal{X}, \nu)$ if $M^{\varrho,\sharp}_{0,s}(f) \in L^{\infty}(\mathcal{X}, \nu)$. Moreover, $\|M^{\varrho,\sharp}_{0,s}(f)\|_{L^{\infty}(\mathcal{X},\nu)}$ is defined to be the $\mathrm{RBMO}_{0,s}(\mathcal{X}, \nu)$ *norm* of f and denoted by $\|f\|_{\mathrm{RBMO}_{0,s}(\mathcal{X},\nu)}$.

The main result of this section is as follows.

Theorem 7.3.2. *Let $\varrho \in (1, \infty)$ and $s \in (0, \beta^{-1}_{6\varrho^2}/4)$. Then* RBMO $(\mathcal{X}, \nu)$ *and* $\mathrm{RBMO}_{0,s}(\mathcal{X}, \nu)$ *coincide with equivalent norms.*

The remainder of this section is devoted to the proof of Theorem 7.3.2. To this end, we first establish the corresponding John–Nirenberg inequality for the space $\mathrm{RBMO}_{0,s}(\mathcal{X}, \nu)$ with $\varrho \in (1, \infty)$ and $s \in (0, \beta^{-2}_{6\varrho^2}/4)$, which is a variant of Theorem 7.2.11. It plays an important role in the proof of Theorem 7.3.2.

Proposition 7.3.3. *For any $\varrho \in (1, \infty)$ and $s \in (0, \beta^{-2}_{6\varrho^2}/4)$, there exist two positive constant $C_{(\varrho)}$ and $c_{(\varrho)}$ such that, for all $f \in \mathrm{RBMO}_{0,s}(\mathcal{X}, \nu)$, balls $B_0 \subset \mathcal{X}$ and $t \in (0, \infty)$,*

$$\nu\left(\left\{x \in B_0 : \left|f(x) - \alpha_{\widetilde{B_0}^{6\varrho^2}}(f)\right| > t\right\}\right) \le C_{(\varrho)} e^{-\frac{c_{(\varrho)} t}{\|f\|_{\mathrm{RBMO}_{0,s}(\mathcal{X},\nu)}}} \nu\left(\varrho^2 B_0\right).$$

To prove Proposition 7.3.3, we need some technical lemmas. Let $\varrho \in (1, \infty)$. The *doubling maximal operator* $\mathcal{N}$ and the *doubling local maximal operator* $M^d_{0,s}$ are, respectively, defined by setting, for all $f \in L^1_{\mathrm{loc}}(\mathcal{X}, \nu)$ and $x \in \mathcal{X}$,

$$\mathcal{N}(f)(x) := \sup_{\substack{B \ni x \\ B\ (6\varrho^2,\ \beta_{6\varrho^2})-\text{doubling}}} \frac{1}{\nu(B)} \int_B |f(y)|\, d\nu(y)$$

and, for all ν-measurable functions f and $x \in \mathcal{X}$,

$$M^d_{0,s}(f)(x) := \sup_{\substack{B \ni x \\ B\ (6\varrho^2,\ \beta_{6\varrho^2})-\text{doubling}}} m^{\varrho}_{0,s;\,B}(f).$$

Lemma 7.3.4. *Let $\varrho \in (1, \infty)$ and $s \in (0, \beta^{-1}_{6\varrho^2})$. If f is a ν-measurable function, then, for all $t \in (0, \infty)$,*

$$\nu(\{x \in \mathcal{X} : |f(x)| > t\}) \le \nu\left(\left\{x \in \mathcal{X} : M^d_{0,s}(f)(x) \ge t\right\}\right). \tag{7.3.1}$$

Proof. It is easy to see that, for all $t \in (0, \infty)$,

$$\{x \in \mathcal{X} : |f(x)| > t\} = \left\{x \in \mathcal{X} : \chi_{\{y \in \mathcal{X}:\ |f(y)|>t\}}(x) = 1\right\},$$

which, combined with Corollary 7.1.21 and the fact $s \in (0, \beta^{-1}_{6\varrho^2})$, implies that, for ν-almost every $x \in \mathcal{X}$ satisfying $|f(x)| > t$,

$$\mathcal{N}\left(\chi_{\{y \in \mathcal{X}:\ |f(y)|>t\}}\right)(x) \ge \chi_{\{y \in \mathcal{X}:\ |f(y)|>t\}}(x) = 1 > s\beta_{6\varrho^2}.$$

This means that

$$\{x \in \mathcal{X} : |f(x)| > t\} \subset \left\{x \in \mathcal{X} : \mathcal{N}\left(\chi_{\{y \in \mathcal{X}:\ |f(y)|>t\}}\right)(x) > s\beta_{6\varrho^2}\right\} \bigcup \Theta,$$

where $\nu(\Theta) = 0$. By Corollary 7.1.21 again, we see that, for any $x \in \mathcal{X}$ satisfying

$$\mathcal{N}(\chi_{\{y \in \mathcal{X}:\ |f(y)|>t\}})(x) > s\beta_{6\varrho^2},$$

there exists a $(6\varrho^2, \beta_{6\varrho^2})$-doubling ball B containing x such that

$$\frac{1}{\nu(B)} \int_B \chi_{\{y \in \mathcal{X}:\ |f(y)|>t\}}(y)\, d\nu(y) > s\beta_{6\varrho^2}.$$

This means that

$$\nu(\{y \in B : |f(y)| > t\}) > s\beta_{6\varrho^2}\nu(B).$$

Notice that

$$\nu(\varrho B) \le \nu(6\varrho^2 B) \le \beta_{6\varrho^2}\nu(B).$$

Hence,

$$\nu(\{y \in B : |f(y)| > t\}) > s\nu(2\varrho B).$$

On the other hand, by the definition of $m^{\varrho}_{0,s;B}(f)$, we easily conclude that, for any $r \in (m^{\varrho}_{0,s;B}(f), \infty)$,

$$\nu(\{y \in B : |f(y)| > r\}) < s\nu(2\varrho B).$$

Therefore, $m^{\varrho}_{0,s;B}(f) \geq t$ and hence $M^{d}_{0,s}(f)(x) \geq t$, which implies that

$$\{x \in \mathcal{X} : |f(x)| > t\} \subset \left\{x \in \mathcal{X} : M^{d}_{0,s}(f)(x) \geq t\right\} \bigcup \Theta.$$

The desired conclusion (7.3.1) then follows directly, which completes the proof of Lemma 7.3.4. □

Lemma 7.3.5. *Let $\varrho \in (1, \infty)$, $s \in (0, \beta^{-1}_{6\varrho^2}/2]$ and B be a $(6\varrho^2, \beta_{6\varrho^2})$-doubling ball. Then, for any ν-measurable real-valued function f, it holds true that*

$$|\alpha_B(f)| \leq m^{\varrho}_{0,s;B}(f).$$

The proof of Lemma 7.3.5 is similar to that of Lemma 5.1.2. We omit the details here. Moreover, by Lemma 7.3.5, we easily conclude that, for all ν-measurable complex-valued functions f and all $(6\varrho^2, \beta_{6\varrho^2})$-doubling balls $B \subset \mathcal{X}$,

$$\begin{aligned} |\alpha_B(f)| &\leq |\alpha_B(\Re f)| + |\alpha_B(\Im f)| \\ &\leq m^{\varrho}_{0,s;B}(\Re f) + m^{\varrho}_{0,s;B}(\Im f) \\ &\leq 2m^{\varrho}_{0,s;B}(f), \end{aligned} \tag{7.3.2}$$

which is used in Sect. 8.4.

Lemma 7.3.6. *Let $\rho \in (1, \infty)$ and $\beta_{6\rho^2} \in (C_{(\lambda)}^{\log_2 6\rho^2}, \infty)$. If $f \in \mathrm{RBMO}\,(\mathcal{X}, \nu)$, then, for all balls B,*

$$\frac{1}{\nu(\rho B)} \int_B \left|f(y) - m_{\widetilde{B}^{6\rho^2}}(f)\right| d\nu(y) \lesssim \|f\|_{\mathrm{RBMO}\,(\mathcal{X},\nu)}$$

and that, for all $(6\rho^2, \beta_{6\rho^2})$-doubling balls $B \subset S$,

$$|m_B(f) - m_S(f)| \lesssim [1 + \delta(B, S)]\|f\|_{\mathrm{RBMO}\,(\mathcal{X},\nu)}.$$

Lemma 7.3.6 is a variant of Proposition 7.2.9. The proof is still valid. We omit the details here.

Proof of Proposition 7.3.3. It suffices to prove that there exist two positive constant $\tilde{C}_\varrho$ and $\tilde{c}_\varrho$ such that, for any real-valued function $f \in \mathrm{RBMO}_{0,s}(\mathcal{X}, \nu)$, ball $B_0 \subset \mathcal{X}$ and $t \in (0, \infty)$,

$$\nu\left(\left\{x \in B_0 : \left|f(x) - \alpha_{\tilde{B}_0^{6\varrho^2}}(f)\right| > t\right\}\right) \le \tilde{C}_\varrho e^{-\frac{\tilde{c}_\varrho t}{\|f\|_{\mathrm{RBMO}_{0,s}(\mathcal{X},\nu)}}} \nu\left(\varrho^2 B_0\right). \quad (7.3.3)$$

Indeed, for any complex-valued function f, we write

$$f := f_1 + i f_2,$$

where f_1 and f_2 are, respectively, the *real* and the *imaginary parts* of f. Notice that $\|f_1\|_{\mathrm{RBMO}_{0,s}(\mathcal{X},\nu)}$ and $\|f_2\|_{\mathrm{RBMO}_{0,s}(\mathcal{X},\nu)}$ are both not greater than $\|f\|_{\mathrm{RBMO}_{0,s}(\mathcal{X},\nu)}$. Therefore, if the inequality (7.3.3) holds true for the real-valued functions f_1 and f_2, then

$$\begin{aligned}
&\nu\left(\left\{x \in B_0 : \left|f(x) - \alpha_{\tilde{B}_0^{6\varrho^2}}(f)\right| > t\right\}\right) \\
&\quad\le \nu\left(\left\{x \in B_0 : \left|f_1(x) - \alpha_{\tilde{B}_0^{6\varrho^2}}(f_1)\right| > t/2\right\}\right) \\
&\qquad + \nu\left(\left\{x \in B_0 : \left|f_2(x) - \alpha_{\tilde{B}_0^{6\varrho^2}}(f_2)\right| > t/2\right\}\right) \\
&\quad\le \tilde{C}_\varrho \left[e^{-\frac{\tilde{c}_\varrho t}{2\|f_1\|_{\mathrm{RBMO}_{0,s}(\mathcal{X},\nu)}}} + e^{-\frac{\tilde{c}_\varrho t}{2\|f_2\|_{\mathrm{RBMO}_{0,s}(\mathcal{X},\nu)}}}\right] \nu\left(\varrho^2 B_0\right) \\
&\quad\le \tilde{C}_\varrho \left[e^{-\frac{\tilde{c}_\varrho t}{2\|f\|_{\mathrm{RBMO}_{0,s}(\mathcal{X},\nu)}}} + e^{-\frac{\tilde{c}_\varrho t}{2\|f\|_{\mathrm{RBMO}_{0,s}(\mathcal{X},\nu)}}}\right] \nu\left(\varrho^2 B_0\right) \\
&\quad\le 2\tilde{C}_\varrho e^{-\frac{\tilde{c}_\varrho t}{2\|f\|_{\mathrm{RBMO}_{0,s}(\mathcal{X},\nu)}}} \nu\left(\varrho^2 B_0\right).
\end{aligned}$$

Therefore, Proposition 7.3.3 holds true for all complex-valued functions with

$$C_{(\varrho)} := 2\tilde{C}_{(\varrho)} \quad \text{and} \quad c_{(\varrho)} := \tilde{c}_{(\varrho)}/2.$$

To show (7.3.3), without loss of generality, we may assume that

$$\|f\|_{\mathrm{RBMO}_{0,s}(\mathcal{X},\nu)} > 0.$$

Otherwise, by the definition of $\|f\|_{\mathrm{RBMO}_{0,s}(\mathcal{X},\nu)}$, we easily conclude that, for all $(6\varrho^2, \beta_{6\varrho^2})$-doubling balls $B \subset S$, $\alpha_B(f) = \alpha_S(f)$ and

$$\sup_{B \subset \mathcal{X}} m^{\varrho}_{0,s;B}\left(f - \alpha_{\tilde{B}^{6\varrho^2}}(f)\right) = 0.$$

Thus, there exists a constant M such that, for any $(6\varrho^2, \beta_{6\varrho^2})$-doubling ball B, $\alpha_B(f) = M$ and hence $m_{0,s;B}(f - M) = 0$. This further implies that, for all $x \in \mathcal{X}$,

$$M^d_{0,s}(f - M)(x) = 0.$$

From this and Lemma 7.3.4, it follows that $f(x) = M$ for ν-almost every $x \in \mathcal{X}$, which implies that, for any ball $B_0 \subset \mathcal{X}$ and $t \in (0, \infty)$,

$$\nu\left(\left\{x \in B_0 : \left|f(x) - \alpha_{\tilde{B}_0^{6\varrho^2}}(f)\right| > t\right\}\right) = 0.$$

Therefore, the inequality (7.3.3) holds true in this case.

Denote by L a large positive constant which is determined later. Choose $\gamma \in (2\beta_{6\varrho^2}, \infty)$ such that $\gamma s < \beta_{6\varrho^2}^{-1}/2$. It is easy to see that, for all $x \in B_0$ satisfying $|f(x) - \alpha_{\tilde{B}_0^{6\varrho^2}}(f)| > 2L$,

$$\chi_{\{y\in B_0:\ |f(y)-\alpha_{\tilde{B}_0^{6\varrho^2}}(f)|>2L\}}(x) = 1 > \gamma s.$$

On the other hand, from Corollary 7.1.21, it follows that, for ν-almost every $x \in B_0$ satisfying $|f(x) - \alpha_{\tilde{B}_0^{6\varrho^2}}(f)| > 2L$,

$$\begin{aligned}&\chi_{\{y\in B_0:\ |f(y)-\alpha_{\tilde{B}_0^{6\varrho^2}}(f)|>2L\}}(x)\\ &\quad= \lim_{\substack{B\downarrow x\\ (6\varrho^2,\beta_{6\varrho^2})-\text{doubling}}} \frac{1}{\nu(B)}\int_B \chi_{\{y\in B_0:\ |f(y)-\alpha_{\tilde{B}_0^{6\varrho^2}}(f)|>2L\}}(y)\,d\nu(y).\end{aligned}$$

Therefore, for ν-almost every $x \in B_0$ satisfying $|f(x) - \alpha_{\tilde{B}_0^{6\varrho^2}}(f)| > 2L$, there exists an arbitrarily small $(6\varrho^2, \beta_{6\varrho^2})$-doubling ball $B := B(x, (6\varrho^2)^{-k}r)$ such that

$$\frac{1}{\nu(B)}\int_B \chi_{\{y\in\mathcal{X}:\ |f(y)-\alpha_{\tilde{B}_0^{6\varrho^2}}(f)|>2L\}}(y)\,d\nu(y) > \gamma s.$$

This means that

$$\nu\left(\left\{y \in B : \left|f(y) - \alpha_{\tilde{B}_0^{6\varrho^2}}(f)\right| > 2L\right\}\right) > \gamma s \nu(B). \tag{7.3.4}$$

Let B_x^* be the *maximal* $(6\varrho^2, \beta_{6\varrho^2})$*-doubling ball* of the form $B(x, (6\varrho^2)^{-k}r)$ with $k \in \mathbb{N}$ satisfying $B_x^* \subset \varrho B_0$ and (7.3.4).

Denote by B_x^{**} the *smallest* $(6\varrho^2, \beta_{6\varrho^2})$*-doubling ball of the form* $(6\varrho^2)^k B_x^*$ *with* $k \in \mathbb{N}$. We claim that

$$\left|\alpha_{B_x^{**}}(f) - \alpha_{\tilde{B}_0^{6\varrho^2}}(f)\right| \le 2L. \tag{7.3.5}$$

To show (7.3.5), we consider the following three cases.

Case A). $B_x^{**} \not\subset \varrho B_0$ and $r(B_x^{**}) \le r(\tilde{B}_0^{6\varrho^2})$. In this case, $B_x^{**} \subset 6\varrho^2 \tilde{B}_0^{6\varrho^2}$. By Definition 7.3.1, we conclude that

$$\left|\alpha_{\tilde{B}_0^{6\varrho^2}}(f) - \alpha_{\widetilde{6\varrho^2 \tilde{B}_0^{6\varrho^2}}^{6\varrho^2}}(f)\right| \le \left[1 + \delta\left(\tilde{B}_0^{6\varrho^2}, \widetilde{6\varrho^2 \tilde{B}_0^{6\varrho^2}}^{6\varrho^2}\right)\right] \|f\|_{\mathrm{RBMO}_{0,s}(\mathcal{X},\nu)}. \tag{7.3.6}$$

On the other hand, it holds true that

$$\left|\alpha_{B_x^{**}}(f) - \alpha_{\widetilde{6\varrho^2 \tilde{B}_0^{6\varrho^2}}^{6\varrho^2}}(f)\right| \le \left[1 + \delta\left(B_x^{**}, \widetilde{6\varrho^2 \tilde{B}_0^{6\varrho^2}}^{6\varrho^2}\right)\right] \|f\|_{\mathrm{RBMO}_{0,s}(\mathcal{X},\nu)}. \tag{7.3.7}$$

Let $(6\varrho^2)^{k_0} B_x^*$ be the *smallest expansion of* B_x^* *such that* $(6\varrho^2)^k B_x^* \not\subset \varrho B_0$ *with* $k \in \mathbb{N}$. Then,

$$r((6\varrho^2)^{k_0} B_x^*) \sim r(B_0), \quad (6\varrho^2)^{k_0-1} B_x^* \subset \varrho B_0 \quad \text{and} \quad (6\varrho^2)^{k_0-1} B_x^* \subset B_x^{**}.$$

Thus, by (iv), (v), (ii) and (iii) of Lemma 7.1.16, we see that there exists a positive constant $C_{1,1}$, depending on ϱ and ν, such that

$$\begin{aligned}
&2 + \delta\left(\tilde{B}_0^{6\varrho^2}, \widetilde{6\varrho^2 \tilde{B}_0^{6\varrho^2}}^{6\varrho^2}\right) + \delta\left(B_x^{**}, \widetilde{6\varrho^2 \tilde{B}_0^{6\varrho^2}}^{6\varrho^2}\right)\\
&\quad\le 2 + \delta\left(\tilde{B}_0^{6\varrho^2}, 6\varrho^2 \tilde{B}_0^{6\varrho^2}\right) + \delta\left(6\varrho^2 \tilde{B}_0^{6\varrho^2}, \widetilde{6\varrho^2 \tilde{B}_0^{6\varrho^2}}^{6\varrho^2}\right)\\
&\qquad + \delta\left(\left(6\varrho^2\right)^{k_0-1} B_x^*, 6\varrho^2 B_0\right)\\
&\qquad + c\left[\delta\left(6\varrho^2 B_0, 6\varrho^2 \tilde{B}_0^{6\varrho^2}\right) + \delta\left(6\varrho^2 \tilde{B}_0^{6\varrho^2}, \widetilde{6\varrho^2 \tilde{B}_0^{6\varrho^2}}^{6\varrho^2}\right)\right]\\
&\quad\le C_{1,1}.
\end{aligned}$$

From this, the estimates (7.3.6) and (7.3.7), it follows that

$$\left|\alpha_{B_x^{**}}(f) - \alpha_{\tilde{B}_0^{6\varrho^2}}(f)\right| \le \left|\alpha_{B_x^{**}}(f) - \alpha_{\widetilde{6\varrho^2 \tilde{B}_0^{6\varrho^2}}^{6\varrho^2}}(f)\right|$$

$$+\left|\alpha_{\widetilde{6\varrho^2 \tilde{B}_0^{6\varrho^2}}^{6\varrho^2}}(f) - \alpha_{\tilde{B}_0^{6\varrho^2}}(f)\right|$$

$$\le C_{1,1}\|f\|_{\mathrm{RBMO}_{0,s}(\mathcal{X},\nu)}.$$

Case B). $B_x^{**} \not\subset \varrho B_0$ and $r(B_x^{**}) > r(\tilde{B}_0^{6\varrho^2})$. In this case, $\tilde{B}_0^{6\varrho^2} \subset 6\varrho^2 B_x^{**}$. It follows, from Definition 7.3.1, that

$$\left|\alpha_{B_x^{**}}(f) - \alpha_{\widetilde{(6\varrho^2)^2 B_x^{**}}^{6\varrho^2}}(f)\right|$$
$$\le \left[1 + \delta\left(B_x^{**}, \widetilde{(6\varrho^2)^2 B_x^{**}}^{6\varrho^2}\right)\right]\|f\|_{\mathrm{RBMO}_{0,s}(\mathcal{X},\nu)} \tag{7.3.8}$$

and

$$\left|\alpha_{\tilde{B}_0^{6\varrho^2}}(f) - \alpha_{\widetilde{(6\varrho^2)^2 B_x^{**}}^{6\varrho^2}}(f)\right|$$
$$\le \left[1 + \delta\left(\tilde{B}_0^{6\varrho^2}, \widetilde{(6\varrho^2)^2 B_x^{**}}^{6\varrho^2}\right)\right]\|f\|_{\mathrm{RBMO}_{0,s}(\mathcal{X},\nu)}. \tag{7.3.9}$$

Since $\tilde{B}_0^{6\varrho^2} \subset 6\varrho^2 B_x^{**}$, it follows that there exists a positive constant $m \in \mathbb{N}$ such that

$$r(\tilde{B}_0^{6\varrho^2}) \ge r((6\varrho^2)^m B_x^*)/(6\varrho^2)^2$$

and

$$\tilde{B}_0^{6\varrho^2} \subset (6\varrho^2)^m B_x^* \subset \widetilde{(6\varrho^2)^2 B_x^{**}}^{6\varrho^2}.$$

Thus,

$$r(\tilde{B}_0^{6\varrho^2}) \sim r((6\varrho^2)^m B_x^*)$$

and hence, by (iv), (v), (ii) and (iii) of Lemma 7.1.16, we see that there exists a positive constant $C_{1,2}$, depending on ϱ and ν, such that

$$
\begin{aligned}
&2+\delta\left(B_x^{**}, \widetilde{(6\varrho^2)^2 B_x^{**}}^{6\varrho^2}\right)+\delta\left(\tilde{B}_0^{6\varrho^2}, \widetilde{(6\varrho^2)^2 B_x^{**}}^{6\varrho^2}\right)\\
&\quad\le 3+\delta\left(B_x^{**}, (6\varrho^2)^2 B_x^{**}\right)+\delta\left((6\varrho^2)^2 B_x^{**}, \widetilde{(6\varrho^2)^2 B_x^{**}}^{6\varrho^2}\right)\\
&\quad\quad+c\left[\delta\left(\tilde{B}_0^{6\varrho^2}, (6\varrho^2)^m B_x^{*}\right)+\delta\left(B_x^{*}, (6\varrho^2)^2 B_x^{**}\right)\right.\\
&\quad\quad\left.+\delta\left((6\varrho^2)^2 B_x^{**}, \widetilde{(6\varrho^2)^2 B_x^{**}}^{6\varrho^2}\right)\right]\\
&\quad\le C_{1,2}.
\end{aligned}
$$

This, combined with (7.3.8) and (7.3.9), implies that

$$
\begin{aligned}
&\left|\alpha_{B_x^{**}}(f)-\alpha_{\tilde{B}_0^{6\varrho^2}}(f)\right|\\
&\quad\le\left|\alpha_{B_x^{**}}(f)-\alpha_{\widetilde{(6\varrho^2)^2 B_x^{**}}^{6\varrho^2}}(f)\right|+\left|\alpha_{\widetilde{(6\varrho^2)^2 B_x^{**}}^{6\varrho^2}}(f)-\alpha_{\tilde{B}_0^{6\varrho^2}}(f)\right|\\
&\quad\le C_{1,2}\|f\|_{\mathrm{RBMO}_{0,s}(\mathcal{X},\nu)}.
\end{aligned}
$$

Case C). $B_x^{**}\subseteq\varrho B_0$. Recall the fact $\gamma s<\beta_{6\varrho^2}^{-1}/2$. Then we choose $\eta\in(0,\infty)$ such that $\gamma s+\eta<\beta_{6\varrho^2}^{-1}/2$. From the choice of B_x^{*}, it follows that (7.3.4) does not hold true for the ball B_x^{**}, that is,

$$
\nu\left(\left\{y\in B_x^{**}:\left|f(y)-\alpha_{\tilde{B}_0^{6\varrho^2}}(f)\right|>2L\right\}\right)\le\gamma s\nu\left(B_x^{**}\right)<(\gamma s+\eta)\nu\left(B_x^{**}\right).
$$

This means that

$$
m^{\varrho}_{0,\gamma s+\eta;\,B_x^{**}}\left(f-\alpha_{\tilde{B}_0^{6\varrho^2}}(f)\right)\le 2L,
$$

which, along with Lemma 7.3.5, implies that

$$
\begin{aligned}
\left|\alpha_{B_x^{**}}(f)-\alpha_{\tilde{B}_0^{6\varrho^2}}(f)\right|&=\left|\alpha_{B^{**}}\left(f-\alpha_{\tilde{B}_0^{6\varrho^2}}(f)\right)\right|\\
&\le m^{\varrho}_{0,\gamma s+\eta;\,B_x^{**}}\left(f-\alpha_{\tilde{B}_0^{6\varrho^2}}(f)\right)\\
&\le 2L,
\end{aligned}
$$

where we used the fact that, for any ball B, $c\in\mathbb{C}$ and ν-measurable function h,

$$
\alpha_B(h)-c=\alpha_B(h-c). \tag{7.3.10}
$$

Let $C_1 := \max\{C_{1,1}, C_{1,2}\}$. Choose

$$L \geq \frac{C_1}{2} \|f\|_{\mathrm{RBMO}_{0,s}(\mathcal{X},\nu)}.$$

Then (7.3.5) is true.

Let $C_2 \in (1, \infty)$ be a constant, depending on ϱ and ν, such that

$$1 + \delta\left(B_x^*, B_x^{**}\right) \leq C_2.$$

Then, if $L \geq 2C_2\|f\|_{\mathrm{RBMO}_{0,s}(\mathcal{X},\nu)}$, by Definition 7.3.1, (7.3.5) and Lemma 7.1.16 (iii), we conclude that

$$\begin{aligned}
&\left|\alpha_{B_x^*}(f) - \alpha_{\tilde{B}_0^{6\varrho^2}}(f)\right| \\
&\quad \leq \left|\alpha_{B_x^*}(f) - \alpha_{B_x^{**}}(f)\right| + \left|\alpha_{B_x^{**}}(f) - \alpha_{\tilde{B}_0^{6\varrho^2}}(f)\right| \\
&\quad \leq \left[1 + \delta\left(B_x^*, B_x^{**}\right)\right] \|f\|_{\mathrm{RBMO}_{0,s}(\mathcal{X},\nu)} + 2L \\
&\quad \leq C_2\|f\|_{\mathrm{RBMO}_{0,s}(\mathcal{X},\nu)} + 2L \\
&\quad \leq \frac{5}{2}L. \qquad (7.3.11)
\end{aligned}$$

By Lemma 7.1.17, we choose disjoint balls $\{B_{x_i}^*\}_i$ among the balls $\{B_x^*\}_{x\in B_0}$ so that the expanded balls $\{5B_{x_i}^*\}_i$ cover all the original B_x^*. It follows, from Definition 7.3.1, (iv), (ii) and (iii) of Lemma 7.1.16, that there exists a constant $C_3 \in (1, \infty)$, depending on ϱ and ν, such that

$$\begin{aligned}
&\left|\alpha_{B_{x_i}^*}(f) - \alpha_{\widetilde{5B_{x_i}^*}^{6\varrho^2}}(f)\right| \\
&\quad \leq \left[1 + \delta\left(B_{x_i}^*, \widetilde{5B_{x_i}^*}^{6\varrho^2}\right)\right] \|f\|_{\mathrm{RBMO}_{0,s}(\mathcal{X},\nu)} \\
&\quad \leq \left[1 + \delta\left(B_{x_i}^*, 5B_{x_i}^*\right) + \delta\left(5B_{x_i}^*, \widetilde{5B_{x_i}^*}^{6\varrho^2}\right)\right] \|f\|_{\mathrm{RBMO}_{0,s}(\mathcal{X},\nu)} \\
&\quad \leq C_3\|f\|_{\mathrm{RBMO}_{0,s}(\mathcal{X},\nu)},
\end{aligned}$$

which, together with (7.3.11), implies that, if $x \in 5B_{x_i}^*$ satisfying that

$$\left|f(x) - \alpha_{\tilde{B}_0^{6\varrho^2}}(f)\right| > kL$$

with $k \geq 3$ and $L \geq 2C_3\|f\|_{\mathrm{RBMO}_{0,s}(\mathcal{X},\nu)}$, then

$$
\begin{aligned}
&\left|f(x) - \alpha_{\widetilde{5B_{x_i}^*}^{6\varrho^2}}(f)\right| \\
&\geq \left|f(x) - \alpha_{\tilde{B}_0^{6\varrho^2}}(f)\right| - \left|\alpha_{\tilde{B}_0^{6\varrho^2}}(f) - \alpha_{B_{x_i}^*}(f)\right| - \left|\alpha_{B_{x_i}^*}(f) - \alpha_{\widetilde{5B_{x_i}^*}^{6\varrho^2}}(f)\right| \\
&> kL - \frac{5}{2}L - C_3\|f\|_{\mathrm{RBMO}_{0,s}(\mathcal{X},\nu)} \\
&\geq (k-3)L.
\end{aligned}
$$

Therefore,

$$
\begin{aligned}
&\left\{x \in B_0 : \left|f(x) - \alpha_{\tilde{B}_0^{6\varrho^2}}(f)\right| > kL\right\} \\
&\quad \subset \bigcup_{\{x \in B_0:\, |f(x) - \alpha_{\tilde{B}_0^{6\varrho^2}}(f)| > kL\}} \left\{y \in B_x^* : \left|f(y) - \alpha_{\tilde{B}_0^{6\varrho^2}}(f)\right| > kL\right\} \\
&\quad \subset \bigcup_i \left\{y \in 5B_{x_i}^* : \left|f(y) - \alpha_{\widetilde{5B_{x_i}^*}^{6\varrho^2}}(f)\right| > (k-3)L\right\}.
\end{aligned}
$$

Using Definition 7.3.1, and (iv), (ii) and (iii) of Lemma 7.1.16, we see that there exists a constant $C_4 \in (1, \infty)$, depending on ϱ and ν, such that

$$
\begin{aligned}
&\left|\alpha_{\widetilde{\varrho B_0}^{6\varrho^2}}(f) - \alpha_{\tilde{B}_0^{6\varrho^2}}(f)\right| \\
&\quad \leq \left|\alpha_{\widetilde{\varrho B_0}^{6\varrho^2}}(f) - \alpha_{B_0}(f)\right| + \left|\alpha_{B_0}(f) - \alpha_{\tilde{B}_0^{6\varrho^2}}(f)\right| \\
&\quad \leq \left[2 + \delta\left(B_0, \widetilde{\varrho B_0}^{6\varrho^2}\right) + \delta\left(B_0, \tilde{B}_0^{6\varrho^2}\right)\right] \|f\|_{\mathrm{RBMO}_{0,s}(\mathcal{X},\nu)} \\
&\quad \leq C_4 \|f\|_{\mathrm{RBMO}_{0,s}(\mathcal{X},\nu)}.
\end{aligned} \tag{7.3.12}
$$

Take L such that $L \geq C_4\|f\|_{\mathrm{RBMO}_{0,s}(\mathcal{X},\nu)}$. Then, by the facts that $\{B_{x_i}^*\}_i$ are $(6\varrho^2, \beta_{6\varrho^2})$-doubling and disjoint, which are contained in ϱB_0, (7.3.12) and (7.3.4), we conclude that

$$
\begin{aligned}
\sum_i \nu\left(5\varrho^2 B_{x_i}^*\right) &\leq \sum_i \nu\left(6\varrho^2 B_{x_i}^*\right) \\
&\leq \beta_{6\varrho^2} \sum_i \nu\left(B_{x_i}^*\right)
\end{aligned}
$$

$$
\begin{aligned}
&\le \frac{\beta_{6\varrho^2}}{\gamma s} \sum_i \nu\left(\left\{y \in B_{x_i}^* : \left|f(y) - \alpha_{\tilde{B}_0^{6\varrho^2}}(f)\right| > 2L\right\}\right) \\
&\le \frac{\beta_{6\varrho^2}}{\gamma s} \nu\left(\left\{y \in \varrho B_0 : \left|f(y) - \alpha_{\tilde{B}_0^{6\varrho^2}}(f)\right| > 2L\right\}\right) \\
&\le \frac{\beta_{6\varrho^2}}{\gamma s} \nu\left(\left\{y \in \varrho B_0 : \left|f(y) - \alpha_{\widetilde{\varrho B_0}^{6\varrho^2}}(f)\right|\right.\right. \\
&\quad \left.\left. + \left|\alpha_{\widetilde{\varrho B_0}^{6\varrho^2}}(f) - \alpha_{\tilde{B}_0^{6\varrho^2}}(f)\right| > 2L\right\}\right) \\
&\le \frac{\beta_{6\varrho^2}}{\gamma s} \nu\left(\left\{y \in \varrho B_0 : \left|f(y) - \alpha_{\widetilde{\varrho B_0}^{6\varrho^2}}(f)\right| > L\right\}\right) \\
&< \frac{\beta_{6\varrho^2}}{\gamma} \nu\left(\varrho^2 B_0\right).
\end{aligned}
$$

Therefore,

$$
\begin{aligned}
&\left\{x \in B_0 : \left|f(x) - \alpha_{\tilde{B}_0^{6\varrho^2}}(f)\right| > kL\right\} \\
&\quad \subset \bigcup_i \left\{y \in 5B_{x_i}^* : \left|f(y) - \alpha_{\widetilde{5B_{x_i}^*}^{6\varrho^2}}(f)\right| > (k-3)L\right\}
\end{aligned}
$$

and

$$
\sum_i \nu\left(5\varrho^2 B_{x_i}^*\right) \le \frac{\beta_{6\varrho^2}}{\gamma} \nu\left(\varrho^2 B_0\right).
$$

Denote $5B_{x_i}^*$ simply by B^i. Let $n \in \mathbb{N}$. Iterating n times with the balls B^i in place of B_0, we see that

$$
\begin{aligned}
&\left\{x \in B_0 : \left|f(x) - \alpha_{\tilde{B}_0^{6\varrho^2}}(f)\right| > 3nL\right\} \\
&\quad \subset \bigcup_{i_1} \left\{y \in B^{i_1} : \left|f(y) - \alpha_{\widetilde{B^{i_1}}^{6\varrho^2}}(f)\right| > 3(n-1)L\right\} \\
&\quad \subset \bigcup_{i_1, i_2} \left\{y \in B^{i_1, i_2} : \left|f(y) - \alpha_{\widetilde{B^{i_1, i_2}}^{6\varrho^2}}(f)\right| > 3(n-2)L\right\} \\
&\quad \subset \cdots \\
&\quad \subset \bigcup_{i_1, \ldots, i_n} \left\{y \in B^{i_1, \ldots, i_n} : \left|f(y) - \alpha_{\widetilde{B^{i_1, \ldots, i_n}}^{6\varrho^2}}(f)\right| > 0\right\}
\end{aligned}
$$

and hence

$$\begin{aligned}
&\nu\left(\left\{x \in B_0 : \left|f(x) - \alpha_{\widetilde{B}_0^{6\varrho^2}}(f)\right| > 3nL\right\}\right) \\
&\leq \sum_{i_1, \ldots, i_n} \nu\left(B^{i_1, \ldots, i_n}\right) \\
&\leq \sum_{i_1, \ldots, i_{n-1}} \sum_{i_n} \nu\left(B^{i_1, \ldots, i_{n-1}, i_n}\right) \\
&\leq \frac{\beta_{6\varrho^2}}{\gamma} \sum_{i_1, \ldots, i_{n-1}} \nu\left(\varrho^2 B^{i_1, \ldots, i_{n-1}}\right) \\
&\leq \cdots \\
&\leq \left(\frac{\beta_{6\varrho^2}}{\gamma}\right)^n \nu\left(\varrho^2 B_0\right).
\end{aligned}$$

Take

$$L := C_5 \|f\|_{\mathrm{RBMO}_{0,s}(\mathcal{X},\nu)} \quad \text{with} \quad C_5 := \max\{C_1/2, 2C_2, 2C_3, C_4\}$$

and choose $n \in \mathbb{N}$ such that $t \in [3nL, 3(n+1)L)$. We then know that

$$\begin{aligned}
&\nu\left(\left\{x \in B_0 : \left|f(x) - \alpha_{\widetilde{B}_0^{6\varrho^2}}(f)\right| > t\right\}\right) \\
&\leq \nu\left(\left\{x \in B_0 : \left|f(x) - \alpha_{\widetilde{B}_0^{6\varrho^2}}(f)\right| > 3nL\right\}\right) \\
&\leq \left(\frac{\beta_{6\varrho^2}}{\gamma}\right)^n \nu\left(\varrho^2 B_0\right) \\
&\leq \left(\frac{\beta_{6\varrho^2}}{\gamma}\right)^{\frac{t}{3L}-1} \nu\left(\varrho^2 B_0\right) \\
&\leq 2e^{-\frac{(\ln 2)t}{3C_5\|f\|_{\mathrm{RBMO}_{0,s}(\mathcal{X},\nu)}}} \nu\left(\varrho^2 B_0\right).
\end{aligned}$$

This means that (7.3.3) is true for any $t \in [3L, \infty)$.

On the other hand, it is easy to show that, for any $t \in (0, 3L)$,

$$\begin{aligned}
&\nu\left(\left\{x \in B_0 : \left|f(x) - \alpha_{\widetilde{B}_0^{6\varrho^2}}(f)\right| > t\right\}\right) \\
&\leq \nu(\varrho^2 B_0) \\
&= e^{\frac{(\ln 2)t}{3C_5\|f\|_{\mathrm{RBMO}_{0,s}(\mathcal{X},\nu)}}} e^{-\frac{(\ln 2)t}{3C_5\|f\|_{\mathrm{RBMO}_{0,s}(\mathcal{X},\nu)}}} \nu\left(\varrho^2 B_0\right) \\
&\leq 2e^{-\frac{(\ln 2)t}{3C_5\|f\|_{\mathrm{RBMO}_{0,s}(\mathcal{X},\nu)}}} \nu\left(\varrho^2 B_0\right).
\end{aligned}$$

Thus, (7.3.3) still holds true for any $t \in (0, 3L)$, which completes the proof of Proposition 7.3.3. □

Based on Proposition 7.3.3, we now prove Theorem 7.3.2.

Proof of Theorem 7.3.2. We first show that, if $f \in \mathrm{RBMO}(\mathcal{X}, \nu)$, then

$$f \in \mathrm{RBMO}_{0,s}(\mathcal{X}, \nu).$$

Let $\varrho \in (1, \infty)$. For any ball $B \subset \mathcal{X}$, from the definition of $m_{0,s;B}^{\varrho}(f - \alpha_{\widetilde{B}^{6\varrho^2}}(f))$, we deduce that, for any $t \in (0, m_{0,s;B}^{\varrho}(f - \alpha_{\widetilde{B}^{6\varrho^2}}(f)))$,

$$\nu\left(\left\{y \in B : \left|f(y) - \alpha_{\widetilde{B}^{6\varrho^2}}(f)\right| > t\right\}\right) \geq s\nu(\varrho B),$$

which implies that

$$t \leq \frac{1}{s\nu(\varrho B)} \int_B \left|f(x) - \alpha_{\widetilde{B}^{6\varrho^2}}(f)\right| d\nu(x).$$

Letting $t \to m_{0,s;B}^{\varrho}(f - \alpha_{\widetilde{B}^{6\varrho^2}}(f))$, we then conclude that

$$m_{0,s;B}^{\varrho}\left(f - \alpha_{\widetilde{B}^{6\varrho^2}}(f)\right) \leq \frac{1}{s\nu(\varrho B)} \int_B \left|f(x) - \alpha_{\widetilde{B}^{6\varrho^2}}(f)\right| d\nu(x).$$

Choose $\rho = \varrho$ in Lemma 7.3.6. Then the above estimate, combined with (7.3.10) and Lemmas 7.3.5 and 7.3.6, implies that

$$\begin{aligned}
&m_{0,s;B}^{\varrho}\left(f - \alpha_{\widetilde{B}^{6\varrho^2}}(f)\right)\\
&\quad\leq \frac{1}{s\nu(\varrho B)} \int_B \left|f(x) - m_{\widetilde{B}^{6\varrho^2}}(f)\right| d\nu(x) + \frac{\nu(B)}{s\nu(\varrho B)} \left|m_{\widetilde{B}^{6\varrho^2}}(f) - \alpha_{\widetilde{B}^{6\varrho^2}}(f)\right|\\
&\quad\leq \frac{1}{s\nu(\varrho B)} \int_B \left|f(x) - m_{\widetilde{B}^{6\varrho^2}}(f)\right| d\nu(x) + \left|\alpha_{\widetilde{B}^{6\varrho^2}}\left(f - m_{\widetilde{B}^{6\varrho^2}}(f)\right)\right|\\
&\quad\leq \frac{1}{s\nu(\varrho B)} \int_B \left|f(x) - m_{\widetilde{B}^{6\varrho^2}}(f)\right| d\nu(x) + m_{0,s;\widetilde{B}^{6\varrho^2}}^{\varrho}\left(f - m_{\widetilde{B}^{6\varrho^2}}(f)\right)\\
&\quad\leq \frac{1}{s\nu(\varrho B)} \int_B \left|f(x) - m_{\widetilde{B}^{6\varrho^2}}(f)\right| d\nu(x)\\
&\qquad + \frac{1}{s\nu(\varrho \widetilde{B}^{6\varrho^2})} \int_{\widetilde{B}^{6\varrho^2}} \left|f(x) - m_{\widetilde{B}^{6\varrho^2}}(f)\right| d\nu(x)\\
&\quad\leq s^{-1}\|f\|_{\mathrm{RBMO}(\mathcal{X},\nu)}.
\end{aligned}$$

On the other hand, similarly, we conclude that, for all $(6\varrho^2, \beta_{6\varrho^2})$-doubling balls $B \subset S \subset \mathcal{X}$,

$$\begin{aligned}
&|\alpha_B(f) - \alpha_S(f)|\\
&\quad\leq |\alpha_B(f) - m_B(f)| + |m_B(f) - m_S(f)| + |m_S(f) - \alpha_S(f)|
\end{aligned}$$

$$\begin{aligned}
&\le |\alpha_B(f - m_B(f))| + |m_B(f) - m_S(f)| + |\alpha_S(f - m_S(f))| \\
&\le m^{\varrho}_{0,s;B}(f - m_B(f)) + [1 + \delta(B, S)]\, \|f\|_{\mathrm{RBMO}(\mathcal{X},\nu)} \\
&\quad + m^{\varrho}_{0,s;S}(f - m_S(f)) \\
&\le \frac{1}{s\nu(\varrho B)} \int_B |f(x) - m_B(f)|\, d\nu(x) + [1 + \delta(B, S)]\, \|f\|_{\mathrm{RBMO}(\mathcal{X},\nu)} \\
&\quad + \frac{1}{s\nu(\varrho S)} \int_S |f(x) - m_S(f)|\, d\nu(x) \\
&\le (2s^{-1} + 1)\|f\|_{\mathrm{RBMO}(\mathcal{X},\nu)}.
\end{aligned}$$

Therefore, for any $f \in \mathrm{RBMO}(\mathcal{X}, \nu)$, we see that $f \in \mathrm{RBMO}_{0,s}(\mathcal{X}, \nu)$ and

$$\|f\|_{\mathrm{RBMO}_{0,s}(\mathcal{X},\nu)} \lesssim \|f\|_{\mathrm{RBMO}(\mathcal{X},\nu)}.$$

Now we prove that, if $f \in \mathrm{RBMO}_{0,s}(\mathcal{X}, \nu)$, then $f \in \mathrm{RBMO}(\mathcal{X}, \nu)$ and

$$\|f\|_{\mathrm{RBMO}(\mathcal{X},\nu)} \lesssim \|f\|_{\mathrm{RBMO}_{0,s}(\mathcal{X},\nu)}. \tag{7.3.13}$$

To prove (7.3.13), we consider the following two cases.

Case I. $\|f\|_{\mathrm{RBMO}_{0,s}(\mathcal{X},\nu)} = 0$. Just as in the proof of Proposition 7.3.3, we know that there exists a constant M such that $f(x) = M$ for ν-almost every $x \in \mathcal{X}$, which implies that $\|f\|_{\mathrm{RBMO}(\mathcal{X},\nu)} = 0$.

Case II. $\|f\|_{\mathrm{RBMO}_{0,s}(\mathcal{X},\nu)} > 0$.

We now show (7.3.13). Indeed, by Definition 7.3.1 and Lemma 7.3.6, to prove (7.3.13), it suffices to show that

$$\sup_{B \ni x} \frac{1}{\nu(\varrho^2 B)} \int_B \left|f(x) - m_{\widetilde{B}^{6\varrho^2}}(f)\right| d\nu(x) \lesssim \|f\|_{\mathrm{RBMO}_{0,s}(\mathcal{X},\nu)} \tag{7.3.14}$$

and that

$$\sup_{\substack{x \in B \subset S \\ B,S\ (6\varrho^2, \beta_{6\varrho^2})-\text{doubling}}} |m_B(f) - m_S(f)| \lesssim [1 + \delta(B, S)]\|f\|_{\mathrm{RBMO}_{0,s}(\mathcal{X},\nu)}. \tag{7.3.15}$$

With the aid of Proposition 7.3.3, we easily see that, for all balls $B \subset \mathcal{X}$,

$$\begin{aligned}
&\frac{1}{\nu(\varrho^2 B)} \int_B \left|f(x) - m_{\widetilde{B}^{6\varrho^2}}(f)\right| d\nu(x) \\
&\quad \le \frac{1}{\nu(\varrho^2 B)} \int_B \left|f(x) - \alpha_{\widetilde{B}^{6\varrho^2}}(f)\right| d\nu(x) + \left|\alpha_{\widetilde{B}^{6\varrho^2}}(f) - m_{\widetilde{B}^{6\varrho^2}}(f)\right|
\end{aligned}$$

$$\begin{aligned}
&\le \frac{1}{\nu(\varrho^2 B)} \int_B \left|f(x) - \alpha_{\tilde{B}^{6\varrho^2}}(f)\right| d\nu(x) \\
&\quad + \frac{1}{\nu(\tilde{B}^{6\varrho^2})} \int_{\tilde{B}^{6\varrho^2}} \left|f(x) - \alpha_{\tilde{B}^{6\varrho^2}}(f)\right| d\nu(x) \\
&= \frac{1}{\nu(\varrho^2 B)} \int_0^\infty \nu\left(\left\{x \in B : \left|f(x) - \alpha_{\tilde{B}^{6\varrho^2}}(f)\right| > t\right\}\right) dt \\
&\quad + \frac{1}{\nu(\tilde{B}^{6\varrho^2})} \int_0^\infty \nu\left(\left\{x \in \tilde{B}^{6\varrho^2} : \left|f(x) - \alpha_{\tilde{B}^{6\varrho^2}}(f)\right| > t\right\}\right) dt \\
&\lesssim \frac{1}{\nu(\varrho^2 B)} \int_0^\infty \exp\left(-\frac{c_\varrho t}{\|f\|_{\mathrm{RBMO}_{0,s}(\mathcal{X},\nu)}}\right) \nu\left(\varrho^2 B\right) dt \\
&\quad + \frac{1}{\nu(\tilde{B}^{6\varrho^2})} \int_0^\infty \exp\left(-\frac{c_\varrho t}{\|f\|_{\mathrm{RBMO}_{0,s}(\mathcal{X},\nu)}}\right) \nu\left(\tilde{B}^{6\varrho^2}\right) dt \\
&\lesssim \|f\|_{\mathrm{RBMO}_{0,s}(\mathcal{X},\nu)},
\end{aligned}$$

which implies (7.3.14).

On the other hand, applying Proposition 7.3.3 again, we conclude that, for all $(6\varrho^2, \beta_{6\varrho^2})$-doubling balls $B \subset S \subset \mathcal{X}$,

$$\begin{aligned}
&|m_B(f) - m_S(f)| \\
&\quad \le |m_B(f) - \alpha_B(f)| + |\alpha_B(f) - \alpha_S(f)| + |\alpha_S(f) - m_S(f)| \\
&\quad \le \frac{1}{\nu(B)} \int_B |f(x) - \alpha_B(f)|\, d\nu(x) + [1 + \delta(B, S)]\|f\|_{\mathrm{RBMO}_{0,s}(\mathcal{X},\nu)} \\
&\qquad + \frac{1}{\nu(S)} \int_S |f(x) - \alpha_S(f)|\, d\nu(x) \\
&\quad = \frac{1}{\nu(B)} \int_0^\infty \nu(\{x \in B : |f(x) - \alpha_B(f)| > t\})\, dt \\
&\qquad + [1 + \delta(B, S)]\|f\|_{\mathrm{RBMO}_{0,s}(\mathcal{X},\nu)} \\
&\qquad + \frac{1}{\nu(S)} \int_0^\infty \nu(\{x \in S : |f(x) - \alpha_S(f)| > t\})\, dt \\
&\quad \lesssim \frac{1}{\nu(B)} \int_0^\infty \exp\left(-\frac{c_\varrho t}{\|f\|_{\mathrm{RBMO}_{0,s}(\mathcal{X},\nu)}}\right) \nu\left(\varrho^2 B\right) dt \\
&\qquad + [1 + \delta(B, S)]\|f\|_{\mathrm{RBMO}_{0,s}(\mathcal{X},\nu)} \\
&\qquad + \frac{1}{\nu(S)} \int_0^\infty \exp\left(-\frac{c_\varrho t}{\|f\|_{\mathrm{RBMO}_{0,s}(\mathcal{X},\nu)}}\right) \nu(\varrho^2 S)\, dt \\
&\quad \lesssim [1 + \delta(B, S)]\|f\|_{\mathrm{RBMO}_{0,s}(\mathcal{X},\nu)},
\end{aligned}$$

which implies (7.3.15). This finishes the proof of Theorem 7.3.2. □

Let φ be a *strictly increasing and nonnegative continuous function* on $[0, \infty)$ such that

$$\lim_{t\to\infty} \varphi(t) = \infty.$$

Denote by φ^{-1} the *inverse function* of φ. Notice that, for all balls $B \subset \mathcal{X}$,

$$m^{\varrho}_{0,s;B}\left(f - \alpha_{\tilde{B}^{6\varrho^2}}(f)\right) \le \varphi^{-1}\left(\frac{1}{s\nu(2\varrho B)}\int_B \varphi\left(\left|f(x) - \alpha_{\tilde{B}^{6\varrho^2}}(f)\right|\right) d\nu(x)\right).$$

Then, by Theorem 7.3.2, we obtain the following conclusion.

Corollary 7.3.7. *Let $\varrho \in (1, \infty)$ and φ be as above. If $f \in L^1_{\rm loc}(\mathcal{X}, \nu)$ and there exists a positive constant C such that, for all balls $B \subset \mathcal{X}$,*

$$\frac{1}{\nu(2\varrho B)}\int_B \varphi\left(\left|f(x) - \alpha_{\tilde{B}^{6\varrho^2}}(f)\right|\right) d\nu(x) \le C$$

and that, for all $(6\varrho^2, \beta_{6\varrho^2})$-doubling balls $B \subset S$,

$$|\alpha_B(f) - \alpha_S(f)| \le C[1 + \delta(B, S)],$$

then $f \in$ RBMO$(\mathcal{X}, \nu)$.

A typical example of φ satisfying Corollary 7.3.7 is

$$\varphi(r) := r^p$$

for all $r \in [0, \infty)$ with $p \in (0, \infty)$. We remark that, if $p \in [1, \infty)$, the conclusion that a ν-locally integrable function satisfying the hypothesis of Corollary 7.3.7 belongs to RBMO$(\mathcal{X}, \nu)$ can be deduced from the Hölder inequality. However, if $p \in (0, 1)$, this conclusion cannot be deduced from the John–Nirenberg inequality anymore. Other typical examples of φ satisfying the hypothesis of Corollary 7.3.7 are

$$\varphi(r) := \underbrace{\log(\cdots\log}_{k}(e^k + r)\cdots)$$

with $k \in \mathbb{N}$.

7.4 The Atomic Hardy Space $H^1(\mathcal{X}, \nu)$

In this section, we study the atomic Hardy space $H^1(\mathcal{X}, \nu)$. We begin with the notion of $H^{1,q}_{{\rm atb},\gamma}(\mathcal{X}, \nu)$.

Definition 7.4.1. Let $\rho \in (1,\infty)$, $\gamma \in [1,\infty)$ and $q \in (1,\infty]$. A function $b \in L^1_{\rm loc}(\mathcal{X}, \nu)$ is called a *$(q, \gamma)_\lambda$-atomic block* if

(i) there exists some ball B such that $\operatorname{supp}(b) \subset B$,
(ii)

$$\int_{\mathcal{X}} b(x)\, d\nu(x) = 0,$$

(iii) for any $j \in \{1,2\}$, there exist a function a_j supported on a ball $B_j \subset B$ and a complex number λ_j such that

$$b = \lambda_1 a_1 + \lambda_2 a_2 \tag{7.4.1}$$

and

$$\|a_j\|_{L^q(\mathcal{X},\nu)} \le [\nu(\rho B_j)]^{1/q-1}[1 + \delta(B_j, B)]^{-\gamma}. \tag{7.4.2}$$

Moreover, let

$$|b|_{H^{1,q}_{\rm atb,\gamma}(\mathcal{X},\nu)} := |\lambda_1| + |\lambda_2|. \tag{7.4.3}$$

A function $f \in L^1(\mathcal{X}, \nu)$ is said to belong to the *space* $H^{1,q}_{\rm atb,\gamma}(\mathcal{X},\nu)$ if there exist $(q,\gamma)_\lambda$-atomic blocks $\{b_i\}_{i\in\mathbb{N}}$ such that

$$f = \sum_{i=1}^{\infty} b_i$$

and

$$\sum_{i=1}^{\infty} |b_i|_{H^{1,q}_{\rm atb,\gamma}(\mathcal{X},\nu)} < \infty.$$

The $H^{1,q}_{\rm atb,\gamma}(\mathcal{X}, \nu)$ *norm* of f is defined by

$$\|f\|_{H^{1,q}_{\rm atb,\gamma}(\mathcal{X},\nu)} := \inf\left\{\sum_{i=1}^{\infty} |b_i|_{H^{1,q}_{\rm atb,\gamma}(\mathcal{X},\nu)}\right\},$$

where the infimum is taken over all the possible decompositions of f as above.

Notice that the coefficients $\delta(B_j, B)$ for $j \in \{1,2\}$ in Definition 7.4.1 depend on the choice of the dominating function λ; that is why λ is included in the name "$(q,\gamma)_\lambda$-atomic block". When $\gamma = 1$, we *write* $H^{1,q}_{\rm atb,\gamma}(\mathcal{X}, \nu)$ *simply as* $H^{1,q}_{\rm atb}(\mathcal{X}, \nu)$. For the spaces $H^{1,q}_{\rm atb,\gamma}(\mathcal{X}, \nu)$, we have the following conclusions.

Proposition 7.4.2. (i) *Let* $q \in (1, \infty]$, $\rho \in (1, \infty)$ *and* $\gamma \in [1, \infty)$. *The space* $H^{1,q}_{\mathrm{atb},\gamma}(\mathcal{X}, \nu)$ *is a Banach space;*

(ii) *For each fixed* $q \in (1, \infty]$ *and* $\gamma \in [1, \infty)$, *the space* $H^{1,q}_{\mathrm{atb},\gamma}(\mathcal{X}, \nu)$ *is independent of the choice of* $\rho \in (1, \infty)$;

(iii) *For all* $q \in (1, \infty]$, $\rho \in (1, \infty)$ *and* $\gamma \in [1, \infty)$,

$$H^{1,\infty}_{\mathrm{atb},\gamma}(\mathcal{X}, \nu) \subset H^{1,q}_{\mathrm{atb},\gamma}(\mathcal{X}, \nu) \subset L^1(\mathcal{X}, \nu).$$

Moreover, $H^{1,\infty}_{\mathrm{atb},\gamma}(\mathcal{X}, \nu)$ *is dense in* $H^{1,q}_{\mathrm{atb},\gamma}(\mathcal{X}, \nu)$.

Proof. The proof for (i) is standard, the details being omitted.

To prove (ii), assume that $\rho_1 > \rho_2 > 1$. For $i \in \{1, 2\}$, write the Hardy spaces corresponding to ρ_i as $H^{1,q}_{\mathrm{atb},\gamma,\rho_i}(\mathcal{X}, \nu)$ for the moment. Clearly,

$$H^{1,q}_{\mathrm{atb},\gamma,\rho_1}(\mathcal{X}, \nu) \subset H^{1,q}_{\mathrm{atb},\gamma,\rho_2}(\mathcal{X}, \nu).$$

Conversely, let

$$b := \sum_{j=1}^{2} \lambda_j a_j \in H^{1,q}_{\mathrm{atb},\gamma,\rho_2}(\mathcal{X}, \nu)$$

be a $(q, \gamma)_\lambda$-atomic block, where for any $j \in \{1, 2\}$, $\operatorname{supp} a_j \subset B_j \subset B$ for some balls B_j and B as in Definition 7.4.1. By (iv) and (ii) of Lemma 7.1.16, without loss of generality, we may further assume that B is (ρ, β_ρ)-doubling with $\rho \geq \rho_1$ and

$$\beta_\rho := \rho^{\max\{3n_0, 3\varsigma\}} + 30^{n_0} + 30^{\varsigma}.$$

Then, for each j, we know that

$$\|a_j\|_{L^q(\mathcal{X},\nu)} \leq [\nu(\rho_2 B_j)]^{1/q-1}[1 + \delta(B_j, B)]^{-\gamma}. \tag{7.4.4}$$

From Proposition 7.1.8, it follows that there exists a sequence $\{B_{j,k}\}_{k=1}^N$ of balls such that

$$B_j \subset \bigcup_{k=1}^{N} B_{j,k} := \bigcup_{k=1}^{N} B\left(x_{B_{j,k}}, \frac{\rho_2 - 1}{10(\rho_1 + 1)} r_{B_j}\right)$$

and $x_{B_{j,k}} \in B_j$ for all $k \in \{1, \ldots, N\}$. Observe that $\rho_1 B_{j,k} \subset \rho_2 B_j$. For any $k \in \{1, \ldots, N\}$, define

$$a_{j,k} := a_j \frac{\chi_{B_{j,k}}}{\sum_{k=1}^{N} \chi_{B_{j,k}}}$$

and $\lambda_{j,k} := \lambda_j$. Then we have

$$b = \sum_{j=1}^{2} \lambda_j a_j = \sum_{j=1}^{2} \sum_{k=1}^{N} \lambda_{j,k} a_{j,k}.$$

Moreover, by (7.4.4), together with (i), (ii) and (iv) of Lemma 7.1.16, we see that

$$\begin{aligned}\|a_{j,k}\|_{L^q(\mathcal{X},\nu)} &\leq \left[\nu(\rho_2 B_j)\right]^{1/q-1} [1+\delta(B_j, B)]^{-\gamma} \\ &\leq \left[\nu(\rho_1 B_{j,k})\right]^{1/q-1} [1+\delta(B_j, B)]^{-\gamma} \\ &\lesssim \left[\nu(\rho_1 B_{j,k})\right]^{1/q-1} [1+\delta(B_{j,k}, \rho B)]^{-\gamma}. \end{aligned} \tag{7.4.5}$$

Let

$$C_{j,k} := \lambda_{j,k}(a_{j,k} + \gamma_{j,k}\chi_B),$$

where

$$\gamma_{j,k} := -\frac{1}{\nu(B)} \int_{\mathcal{X}} a_{j,k}(x)\, d\nu(x).$$

We claim that $C_{j,k}$ is a $(q, \gamma)_\lambda$-atomic block. Indeed,

$$\operatorname{supp} C_{j,k} \subset \rho B \quad \text{and} \quad \int_{\mathcal{X}} C_{j,k}(x)\, d\nu(x) = 0.$$

Moreover, since B is (ρ, β_ρ)-doubling and $B_{j,k} \subset \rho B$, by the Hölder inequality and (7.4.5), we conclude that

$$\begin{aligned}\left\|\gamma_{j,k}\chi_B\right\|_{L^q(\mathcal{X},\nu)} &\lesssim [\nu(B)]^{1/q-1}[\nu(B_{j,k})]^{1-1/q}[\nu(\rho_1 B_{j,k})]^{1/q-1} \\ &\quad \times [1+\delta(B_{j,k}, \rho B)]^{-\gamma} \\ &\lesssim [\nu(\rho_1 B)]^{1/q-1}.\end{aligned}$$

This, together with (7.4.5), implies that

$$|C_{j,k}|_{H^{1,q}_{\mathrm{atb},\gamma,\rho_1}(\mathcal{X},\nu)} \lesssim |\lambda_{j,k}|.$$

Thus, the claim is true.

By the claim, we see that

$$b = \sum_{j=1}^{2} \sum_{k=1}^{N} C_{j,k} \in H^{1,q}_{\mathrm{atb},\gamma,\rho_1}(\mathcal{X}, \nu)$$

and

$$\|b\|_{H^{1,q}_{\mathrm{atb},\gamma,\rho_1}(\mathcal{X},\nu)} \lesssim \sum_{j=1}^{2} |\lambda_j|.$$

Thus, we know that

$$H^{1,q}_{\mathrm{atb},\gamma,\rho_2}(\mathcal{X}, \nu) \subset H^{1,q}_{\mathrm{atb},\gamma,\rho_1}(\mathcal{X}, \nu),$$

which shows (ii).

To prove (iii), let $q \in (1, \infty]$, $\rho \in (1, \infty)$ and $\gamma \in [1, \infty)$. The inclusions

$$H^{1,\infty}_{\mathrm{atb},\gamma}(\mathcal{X}, \nu) \subset H^{1,q}_{\mathrm{atb},\gamma}(\mathcal{X}, \nu) \subset L^1(\mathcal{X}, \nu)$$

are obvious. Thus, to show (iii), it suffices to prove that $H^{1,\infty}_{\mathrm{atb},\gamma}(\mathcal{X}, \nu)$ is dense in $H^{1,q}_{\mathrm{atb},\gamma}(\mathcal{X}, \nu)$. By Definition 7.4.1, for any $f \in H^{1,q}_{\mathrm{atb},\gamma}(\mathcal{X}, \nu)$ and $\epsilon \in (0, \infty)$, there exist $N \in \mathbb{N}$ and $g := \sum_{j=1}^{N} b_j$ such that

$$\|f - g\|_{H^{1,q}_{\mathrm{atb},\gamma}(\mathcal{X},\nu)} < \epsilon,$$

where, for any $j \in \{1, \ldots, N\}$, b_j is a $(q, \gamma)_\lambda$-atomic block supported in some ball B_j as in Definition 7.4.1. By Proposition 7.1.19, there exists $h_j \in L^\infty(\mathcal{X}, \nu)$ such that $\operatorname{supp} h_j \subset B_j$ and

$$\|h_j - b_j\|_{L^q(\mathcal{X},\nu)} < \frac{\epsilon}{2^{N+1}[\lambda(x_{B_j}, \rho r_{B_j})]^{1-1/q}}. \tag{7.4.6}$$

Let

$$D_j := \int_{\mathcal{X}} h_j(x)\, d\nu(x).$$

Then, by

$$\int_{\mathcal{X}} b_j(x)\, d\nu(x) = 0,$$

the Hölder inequality and (7.4.6), we see that

$$\begin{aligned}
|D_j| &\le \int_{\mathcal{X}} \left|h_j(x) - b_j(x)\right| d\nu(x) \\
&\le [\nu(B_j)]^{1-1/q} \left\|h_j - b_j\right\|_{L^q(\mathcal{X},\nu)} \\
&\le \frac{\epsilon[\nu(B_j)]^{1-1/q}}{2^{N+1}[\lambda(x_{B_j}, \rho r_{B_j})]^{1-1/q}}.
\end{aligned}$$

For any $j \in \{1, \ldots, N\}$, let

$$\tilde{h}_j := [\nu(B_j)]^{-1} \chi_{B_j} D_j \quad \text{and} \quad \tilde{b}_j := h_j - \tilde{h}_j.$$

Then $\tilde{b}_j \in H^{1,\infty}_{\mathrm{atb},\gamma}(\mathcal{X}, \nu)$ with $\mathrm{supp}\,(\tilde{b}_j) \subset 2B_j$. Moreover, we claim that

$$\|\tilde{b}_j - b_j\|_{H^{1,q}_{\mathrm{atb},\gamma}(\mathcal{X},\nu)} \lesssim \epsilon/2^N.$$

Indeed, from the definition of D_j, the Hölder inequality, (7.4.6), (7.1.1) and Lemma 7.1.16(ii), it follows that

$$\left\|\tilde{h}_j\right\|_{L^q(\mathcal{X},\nu)} \le \frac{\epsilon}{2^{N+1}[\lambda(x_{B_j}, 2\rho r_{B_j})]^{1-1/q}}$$

and

$$\begin{aligned}\left\|\tilde{b}_j - b_j\right\|_{L^q(\mathcal{X},\nu)} &\le \|h_j - b_j\|_{L^q(\mathcal{X},\nu)} + \left\|\tilde{h}_j\right\|_{L^q(\mathcal{X},\nu)}\\ &< \frac{\epsilon}{2^N[\lambda(x_{B_j}, \rho r_{B_j})]^{1-1/q}}\\ &\lesssim \frac{\epsilon}{2^N[\nu(2\rho B_j)]^{1-1/q}[1+\delta(2B_j, 2B_j)]^\gamma}.\end{aligned}$$

This, via Definition 7.4.1, implies that

$$\|\tilde{b}_j - b_j\|_{H^{1,q}_{\mathrm{atb},\gamma}(\mathcal{X},\nu)} \lesssim \epsilon/2^N.$$

Let

$$\tilde{g} := \sum_{j=1}^N \tilde{b}_j.$$

Then $\tilde{g} \in H^{1,\infty}_{\mathrm{atb},\gamma}(\mathcal{X}, \nu)$ and

$$\|f - \tilde{g}\|_{H^{1,q}_{\mathrm{atb},\gamma}(\mathcal{X},\nu)} \le \|f - g\|_{H^{1,q}_{\mathrm{atb},\gamma}(\mathcal{X},\nu)} + \|g - \tilde{g}\|_{H^{1,q}_{\mathrm{atb},\gamma}(\mathcal{X},\nu)} \lesssim \epsilon.$$

This finishes the proof of Proposition 7.4.2(iii). □

Remark 7.4.3. (i) By Proposition 7.4.4(ii), unless explicitly pointed out, we *always assume that $\rho = 2$ in Definition 7.4.1.*

(ii) Let $q \in (1, \infty]$. Instead of Definition 7.4.1, we can also define the atomic block b in the following way: b satisfies (i) and (ii) of Definition 7.4.1 and

$$b = \sum_{j=1}^{\infty} \lambda_j a_j, \tag{7.4.7}$$

where $\{a_j\}_{j=1}^{\infty}$ and $\{\lambda_j\}_{j=1}^{\infty}$ satisfy (7.4.2) and

$$|b|_{\tilde{H}^{1,q}_{\mathrm{atb},\gamma}(\mathcal{X},\nu)} := \sum_{j=1}^{\infty} |\lambda_j| < \infty, \tag{7.4.8}$$

respectively. Correspondingly, we obtain an *atomic Hardy space*, temporarily denoted by $\tilde{H}^{1,q}_{\mathrm{atb},\gamma}(\mathcal{X}, \nu)$. Arguing as in the proof of Proposition 3.2.3, we see that $\tilde{H}^{1,q}_{\mathrm{atb},\gamma}(\mathcal{X}, \nu)$ and $H^{1,q}_{\mathrm{atb},\gamma}(\mathcal{X}, \nu)$ coincide with equivalent norms.

(iii) Let $(\mathcal{X}, d, \nu)$ be the space of homogeneous type as in Example 7.1.2,

$$\lambda(x, r) := \nu(B(x, r)) \text{ for all } x \in \mathcal{X} \text{ and } r \in (0, \infty),$$

and $q \in (1, \infty]$. A function a is called a *q-atom* associated with a ball B if

(i) $\operatorname{supp} a \subset B$;
(ii) $\|a\|_{L^q(\mathcal{X},\nu)} \le [\mu(B)]^{1/q-1}$;
(iii) $\int_{\mathcal{X}} a(x)\, d\nu(x) = 0$.

A function $f \in L^1(\mathcal{X}, \nu)$ is said to belong to the *space* $H^{1,q}(\mathcal{X}, \nu)$ if there exist q-atoms $\{a_i\}_{i\in\mathbb{N}}$ such that

$$f = \sum_{i=1}^{\infty} \lambda_i a_i \quad \text{and} \quad \sum_{i=1}^{\infty} |\lambda_i| < \infty.$$

The $H^{1,q}(\mathcal{X}, \nu)$ *norm* of f is defined by

$$\|f\|_{H^{1,q}(\mathcal{X},\nu)} := \inf\left\{\sum_{i=1}^{\infty} |\lambda_i|\right\},$$

where the infimum is taken over all the possible decompositions of f as above.[6]

To show that, for any $q \in (1, \infty)$ and $\gamma \in [1, \infty)$,

$$\left(H^{1,q}_{\mathrm{atb},\gamma}(\mathcal{X}, \nu)\right)^* = \mathrm{RBMO}_\gamma(\mathcal{X}, \nu),$$

we first have the following conclusion.

[6] See [19].

Lemma 7.4.4. *Let $\gamma \in [1, \infty)$. Then*

$$\mathrm{RBMO}_\gamma\,(\mathcal{X}, \nu) \subset \left(H^{1,\infty}_{\mathrm{atb},\gamma}(\mathcal{X}, \nu)\right)^*.$$

That is, for any fixed $g \in \mathrm{RBMO}_\gamma\,(\mathcal{X}, \nu)$, the linear functional,

$$\mathcal{L}_g(f) := \int_{\mathcal{X}} f(x)g(x)\,d\nu(x)$$

defined over bounded functions f with bounded support, can be extended to a continuous linear functional $\mathcal{L}_g$ over $H^{1,\infty}_{\mathrm{atb},\gamma}(\mathcal{X}, \nu)$ and

$$\|\mathcal{L}_g\|_{(H^{1,\infty}_{\mathrm{atb},\gamma}(\mathcal{X},\nu))^*} \sim \|g\|_{\mathrm{RBMO}_\gamma\,(\mathcal{X},\nu)}$$

with the implicit equivalent positive constants independent of g.

Proof. Without loss of generality, we may assume that g is real-valued.

We first prove that

$$\|\mathcal{L}_g\|_{(H^{1,\infty}_{\mathrm{atb},\gamma}(\mathcal{X},\nu))^*} \lesssim \|g\|_{\mathrm{RBMO}_\gamma\,(\mathcal{X},\nu)}. \tag{7.4.9}$$

To this end, since $H^{1,\infty}_{\mathrm{atb},\gamma}(\mathcal{X}, \nu)$ and $\mathrm{RBMO}_\gamma\,(\mathcal{X}, \nu)$ are both independent of the choice of the parameter ρ, without loss of generality, we may assume that $\rho = 3$ in Definitions 7.2.4 and 7.4.1 in this case. Obviously, to show (7.4.9), it suffices to show that, if b is an $(\infty, \gamma)_\lambda$-atomic block and $g \in \mathrm{RBMO}_\gamma\,(\mathcal{X}, \nu)$, then

$$\left|\int_{\mathcal{X}} b(x)g(x)\,d\nu(x)\right| \lesssim |b|_{H^{1,\infty}_{\mathrm{atb},\gamma}(\mathcal{X},\nu)}\|g\|_{\mathrm{RBMO}_\gamma\,(\mathcal{X},\nu)}. \tag{7.4.10}$$

Assume that $b := \sum_{j=1}^2 \lambda_j a_j$, where, for any $j \in \{1, 2\}$, $\operatorname{supp} a_j \subset B_j \subset B$ for some balls B_j and B as in Definition 7.4.1. Then, from the fact that $\int_{\mathcal{X}} b(x)\,d\nu(x) = 0$, it follows that

$$\begin{aligned}
\mathrm{H} := \left|\int_{\mathcal{X}} b(x)g(x)\,d\nu(x)\right| &= \left|\int_B b(x)[g(x) - g_B]\,d\nu(x)\right| \\
&\le \sum_{j=1}^2 |\lambda_j|\|a_j\|_{L^\infty(\mathcal{X},\nu)} \int_{B_j} |g(x) - g_B|\,d\nu(x).
\end{aligned}$$

Since $f \in \mathrm{RBMO}_\gamma\,(\mathcal{X}, \nu)$, we see that, for each j,

$$\begin{aligned}\int_{B_j} |g(x) - g_B|\, d\nu(x) &\le \int_{B_j} |g(x) - g_{B_j}|\, d\nu(x) + |g_B - g_{B_j}|\, \nu(B_j) \\ &\le \|g\|_{\mathrm{RBMO}_\gamma(\mathcal{X},\nu)} \left\{\nu(3B_j) + [1 + \delta(B_j, B)]^\gamma \nu(B_j)\right\} \\ &\lesssim \|g\|_{\mathrm{RBMO}_\gamma(\mathcal{X},\nu)} [1 + \delta(B_j, B)]^\gamma \nu(3B_j),\end{aligned}$$

which implies that

$$\mathrm{H} \lesssim \sum_{j=1}^{2} |\lambda_j| \|g\|_{\mathrm{RBMO}_\gamma(\mathcal{X},\nu)} \sim |b|_{H^{1,\infty}_{\mathrm{atb},\gamma}(\mathcal{X},\nu)} \|g\|_{\mathrm{RBMO}_\gamma(\mathcal{X},\nu)}.$$

Thus, (7.4.10) holds true.

To prove the converse, by Proposition 7.2.10, it suffices to show that there exists $f \in H^{1,\infty}_{\mathrm{atb},\gamma}(\mathcal{X}, \nu)$ such that

$$|\mathcal{L}_g(f)| \gtrsim \|g\|_* \|f\|_{H^{1,\infty}_{\mathrm{atb},\gamma}(\mathcal{X},\nu)}, \tag{7.4.11}$$

where $\|g\|_*$ is as in Sect. 7.2. Moreover, since $H^{1,\infty}_{\mathrm{atb},\gamma}(\mathcal{X}, \nu)$ is independent of the choice of ρ, without loss of generality, we may assume that $\rho = 3/2$ in Definition 7.4.1 in this case.

Let $\epsilon \in (0, 1/8)$. We show (7.4.11) by considering the following two cases.

Case (i) Assume that there exists some $(6, \beta_6)$-doubling ball B such that

$$\int_B |g(x) - \alpha_B(g)|\, d\nu(x) \ge \epsilon \|g\|_* \nu(B). \tag{7.4.12}$$

In this case, we take f such that $\int_{\mathcal{X}} f(x)\, d\nu(x) = 0$. Indeed, we let $f(x) = 0$ for all $x \notin B$, $f(x) = 1$ for all $x \in B \cap \{x \in \mathcal{X} :\ g(x) > \alpha_B(g)\}$, $f(x) = -1$ for all

$$x \in B \bigcap \{x \in \mathcal{X} :\ g(x) < \alpha_B(g)\},$$

and $f(x) = \alpha$ for all $x \in B \cap \{x \in \mathcal{X} :\ g(x) = \alpha_B(g)\}$, where $\alpha = 1$ or -1. By (7.2.14) and (7.2.15), we see such an f exists. From the definition of f and (7.4.12), it follows that

$$\begin{aligned}\left|\int_{\mathcal{X}} g(x) f(x)\, d\nu(x)\right| &= \left|\int_{\mathcal{X}} [g(x) - \alpha_B(g)] f(x)\, d\nu(x)\right| \\ &= \int_B |g(x) - \alpha_B(g)|\, d\nu(x) \\ &\ge \epsilon \|g\|_* \nu(B).\end{aligned}$$

Since f is an $(\infty, \gamma)_\lambda$-atomic block and B is $(6, \beta_6)$-doubling, from Lemma 7.1.16 (ii), we deduce that

$$\|f\|_{H^{1,\infty}_{\mathrm{atb},\gamma}(\mathcal{X},\nu)} \le |f|_{H^{1,\infty}_{\mathrm{atb},\gamma}(\mathcal{X},\nu)} \lesssim \nu(3B)[1+\delta(2B,2B)]^\gamma \lesssim \nu(B).$$

Thus, we have

$$|\mathcal{L}_g(f)| = \left|\int_{\mathcal{X}} f(x)g(x)\,d\nu(x)\right| \gtrsim \epsilon\|g\|_*\|f\|_{H^{1,\infty}_{\mathrm{atb},\gamma}(\mathcal{X},\nu)}.$$

Namely, (7.4.11) holds true in this case.

Case (ii) Assume that, for any $(6, \beta_6)$-doubling ball B, (7.4.12) does not hold true. In this case, we further consider two subcases.

Subcase (a) Assume that, for any two $(6, \beta_6)$-doubling balls $B \subset S$,

$$|\alpha_B(g) - \alpha_S(g)| \le \frac{1}{2}[1+\delta(B,S)]^\gamma \|g\|_*.$$

In this case, by the definition of $\|g\|_*$, there exists a ball B such that

$$\int_B \left|g(x) - \alpha_{\tilde{B}^6}(g)\right| d\nu(x) > \frac{1}{2}\|g\|_*\nu(3B). \tag{7.4.13}$$

Let $f := a_1 + a_2$, where

$$a_1 := \chi_{B\cap\{x\in\mathcal{X}:\ g(x)>\alpha_{\tilde{B}^6}(g)\}} - \chi_{B\cap\{x\in\mathcal{X}:\ g(x)\le\alpha_{\tilde{B}^6}(g)\}}$$

and $a_2 := C_{\tilde{B}^6}\chi_{\tilde{B}^6}$ for some constant $C_{\tilde{B}^6}$ such that

$$\int_{\mathcal{X}} f(x)\,d\nu(x) = 0.$$

We point out that (7.4.13) implies that $\nu(\tilde{B}^6) \ne 0$, which further guarantees the existence of f. Then we have

$$\|a_2\|_{L^\infty(\mathcal{X},\nu)}\nu(\tilde{B}^6) = \left|\int_{\mathcal{X}} a_2(x)\,d\nu(x)\right| = \left|\int_{\mathcal{X}} a_1(x)\,d\nu(x)\right| \le \nu(B). \tag{7.4.14}$$

Observe that $\operatorname{supp} a_1 \subset 2B$ and $\operatorname{supp} a_2 \subset 2\tilde{B}^6$. Since $\tilde{B}^6$ is $(6, \beta_6)$-doubling, from (7.4.14), together with (i) through (iv) of Lemma 7.1.16, it follows that

$$\begin{aligned}\|f\|_{H^{1,\infty}_{\mathrm{atb},\gamma}(\mathcal{X},\nu)} &\le \|a_1\|_{L^\infty(\mathcal{X},\nu)}\nu(3B)\left[1+\delta\left(2B,2\tilde{B}^6\right)\right]^\gamma \\ &\quad + \|a_2\|_{L^\infty(\mathcal{X},\nu)}\nu(3\tilde{B}^6)\left[1+\delta\left(2\tilde{B}^6,2\tilde{B}^6\right)\right]^\gamma \\ &\lesssim \nu(3B). \end{aligned} \tag{7.4.15}$$

Now we conclude that

$$\begin{aligned}\mathcal{L}_g(f) &= \int_{\mathcal{X}} g(x)f(x)\,d\nu(x)\\ &= \int_{\tilde{B}^6} [g - \alpha_{\tilde{B}^6}(g)]\, f(x)\,d\nu(x)\\ &= \int_{\tilde{B}^6} [g - \alpha_{\tilde{B}^6}(g)]\, a_1(x)\,d\nu(x)\\ &\quad + \int_{\tilde{B}^6} [g - \alpha_{\tilde{B}^6}(g)]\, a_2(x)\,d\nu(x). \qquad (7.4.16)\end{aligned}$$

From the definition of a_1 and (7.4.13), we deduce that

$$\begin{aligned}\left|\int_{\tilde{B}^6} [g - \alpha_{\tilde{B}^6}(g)]\, a_1(x)\,d\nu(x)\right| &= \int_B \left|g(x) - \alpha_{\tilde{B}^6}(g)\right| d\nu(x)\\ &> \frac{1}{2}\|g\|_* \nu(3B). \qquad (7.4.17)\end{aligned}$$

On the other hand, since (7.4.12) does not hold true for all $(6, \beta_6)$-doubling balls, by (7.4.14), we see that

$$\begin{aligned}\left|\int_{\tilde{B}^6} [g(x) - \alpha_{\tilde{B}^6}(g)]\, a_2(x)\,d\nu(x)\right| &\le \frac{\nu(B)}{\nu(\tilde{B}^6)} \int_{\tilde{B}^6} \left|g(x) - \alpha_{\tilde{B}^6}(g)\right| d\nu(x)\\ &< \epsilon\|g\|_* \nu(B).\end{aligned}$$

By using this, (7.4.16), (7.4.17), (7.4.15) and $\epsilon \in (0, 1/8)$, we further have

$$\left|\mathcal{L}_g(f)\right| \gtrsim \|g\|_* \nu(3B) \gtrsim \|g\|_* \|f\|_{H^{1,\infty}_{\mathrm{atb},\gamma}(\mathcal{X},\nu)}.$$

Subcase (b) Assume that there exist $(6, \beta_6)$-doubling balls $B \subset S$ such that

$$|\alpha_B(g) - \alpha_S(g)| > \frac{1}{2}[1 + \delta(B, S)]^\gamma \|g\|_*. \qquad (7.4.18)$$

Let B, S be such balls. Since in Case (ii), (7.4.12) does not hold true for any $(6, \beta_6)$-doubling ball, we conclude that $\nu(S) \ge \nu(B) > 0$. Let

$$f := \frac{1}{\nu(S)}\chi_S - \frac{1}{\nu(B)}\chi_B.$$

Then f is an $(\infty, \gamma)_\lambda$-atomic block and

$$\|f\|_{H^{1,\infty}_{\mathrm{atb},\gamma}(\mathcal{X},\nu)} \lesssim [1 + \delta(B, S)]^\gamma.$$

Moreover,

$$\begin{aligned}
\mathcal{L}_g(f) &= \int_{\mathcal{X}} [g(x) - \alpha_S(g)] f\, d\nu(x) \\
&= \frac{1}{\nu(S)} \int_S [g(x) - \alpha_S(g)]\, d\nu(x) \\
&\quad - \frac{1}{\nu(B)} \int_B [g(x) - \alpha_S(g)]\, d\nu(x) \\
&= \frac{1}{\nu(S)} \int_S [g(x) - \alpha_S(g)]\, d\nu(x) \\
&\quad - \frac{1}{\nu(B)} \int_B [g(x) - \alpha_B(g)]\, d\nu(x) \\
&\quad + [\alpha_B(g) - \alpha_S(g)]. \qquad (7.4.19)
\end{aligned}$$

By $\epsilon \in (0, 1/8)$, (7.4.18), (7.4.19) and the fact that (7.4.12) does not hold true for any $(6, \beta_6)$-doubling ball, we have

$$|\mathcal{L}_g(f)| \geq \frac{1}{4}[1 + \delta(B, S)]^{\gamma} \|g\|_* \gtrsim \|f\|_{H^{1,\infty}_{\mathrm{atb},\gamma}(\mathcal{X},\nu)} \|g\|_*.$$

This finishes the proof of (7.4.10) and hence Lemma 7.4.4. □

We now show that

$$\left(H^{1,q}_{\mathrm{atb},\gamma}(\mathcal{X}, \nu)\right)^* = \mathrm{RBMO}_{\gamma}(\mathcal{X}, \nu)$$

for any $q \in (1, \infty)$ and $\gamma \in [1, \infty)$. To this end, let $q \in (1, \infty]$. For each ball B, we denote by $L^q(B, \nu)$ the *subspace of* $L^q(\mathcal{X}, \nu)$ *in which functions are supported in* B and

$$L^q_0(B, \nu) := L^q_{b,0}(\mathcal{X}, \nu) \bigcap L^q(B, \nu).$$

Obviously,

$$\bigcup_B L^q_0(B, \nu) = L^q_{b,0}(\mathcal{X}, \nu) \quad \text{and} \quad \bigcup_B L^q(B, \nu) = L^q_b(\mathcal{X}, \nu).$$

Proposition 7.4.5. *Let* $q \in (1, \infty)$ *and* $\gamma \in [1, \infty)$. *Then*

$$\left(H^{1,q}_{\mathrm{atb},\gamma}(\mathcal{X}, \nu)\right)^* = \mathrm{RBMO}_{\gamma}(\mathcal{X}, \nu).$$

We first formulate an intermediate result needed in the proof as a separate lemma, since it is used again in the proof of Theorem 8.1.5 in Sect. 8.1.

Lemma 7.4.6. *Let $q \in (1, \infty)$ and $K \in (0, \infty)$. If $g \in L^{q'}_{\rm loc}(\mathcal{X}, \nu)$ satisfies the estimate that, for all $(q, \gamma)_\lambda$-atomic blocks f,*

$$\left|\int_{\mathcal{X}} g(x) f(x) d\nu(x)\right| \le K |f|_{H^{1,q}_{\rm atb,\gamma}(\mathcal{X},\nu)},$$

then $g \in {\rm RBMO}_\gamma(\mathcal{X}, \nu)$ and there exists a positive constant C such that

$$\|g\|_{{\rm RBMO}_\gamma(\mathcal{X},\nu)} \le CK.$$

Proof. By Proposition 7.4.2(ii), without loss of generality, we may assume that $\rho = 3$ in Definition 7.4.1. The assumption implies in particular that

$$\left|\int_{\mathcal{X}} g(x) f(x) d\nu\right| \lesssim K\nu(3B)^{1/q'} \|f\|_{L^q(\mathcal{X},\nu)}$$

for all $f \in L^q_0(B, \nu)$. Hence for all $f \in L^q(B, \nu)$,

$$\begin{aligned}\left|\int_{\mathcal{X}} [g(x) - m_B(g)] f(x) d\nu(x)\right| &= \left|\int_{\mathcal{X}} g(x)[f(x) - m_B(f)] d\nu(x)\right| \\ &\lesssim K\nu(3B)^{1/q'} \|\chi_B[f - m_B(f)]\|_{L^q(\mathcal{X},\nu)} \\ &\lesssim K\nu(3B)^{1/q'} \|f\|_{L^q(\mathcal{X},\nu)},\end{aligned}$$

which implies that

$$\left[\int_B |g(x) - m_B(g)|^{q'} d\nu(x)\right]^{1/q'} \lesssim K\nu(3B)^{1/q'};$$

in particular,

$$\sup_{(6,\beta_6)-\text{doubling ball } B} \left[\frac{1}{\nu(B)} \int_B |g(x) - m_B(g)|^{q'}\, d\nu(x)\right]^{1/q'} \lesssim K. \tag{7.4.20}$$

By Proposition 7.2.9, to show that $\|g\|_{{\rm RBMO}_\gamma(\mathcal{X},\nu)} \lesssim K$, it remains to show that, for all $(6, \beta_6)$-doubling balls $B \subset S$,

$$|m_B(g) - m_S(g)| \lesssim [1 + \delta(B, S)]^\gamma K. \tag{7.4.21}$$

To prove (7.4.21), we only need to consider the case that

$$\nu(\{x \in B : g(x) \ne m_S(g)\}) > 0;$$

for otherwise we have $m_B(g) = m_S(g)$ and (7.4.21) automatically holds true. Let

$$b := a_1 + a_2,$$

where

$$a_1 := \frac{|g - m_S(g)|^{q'}}{[g - m_S(g)]} \chi_{\{x \in B:\ g \neq m_S(g)\}},$$

$a_2 := C_S \chi_S$ and C_S is a constant such that

$$\int_{\mathcal{X}} b(x)\, d\nu(x) = 0.$$

Then, by (i), (ii) and (iv) of Lemma 7.1.16, we see that b is a $(q, \gamma)_\lambda$-atomic block and

$$\begin{aligned} |b|_{H^{1,q}_{\mathrm{atb},\gamma}(\mathcal{X},\nu)} &\leq \|a_1\|_{L^q(\mathcal{X},\nu)} [\nu(6B)]^{1/q'} [1 + \delta(2B, 2S)]^\gamma + |C_S| \nu(6S) \\ &\lesssim \left[\int_B |g(x) - m_S(g)|^{q'}\, d\nu(x) \right]^{1/q} [\nu(B)]^{1/q'} [1 + \delta(B, S)]^\gamma. \end{aligned}$$

By this, (7.4.20) for S, and the assumption of the lemma, we obtain

$$\begin{aligned} &\int_B |g(x) - m_S(g)|^{q'}\, d\nu(x) \\ &\quad = \int_{\mathcal{X}} [g(x) - m_S(g)] a_1(x)\, d\nu(x) \\ &\quad \leq \left| \int_{\mathcal{X}} g(x) b(x)\, d\nu(x) \right| + |C_S| \int_S |g(x) - m_S(g)|\, d\nu(x) \\ &\quad \lesssim K \left[|b|_{H^{1,q}_{\mathrm{atb},\gamma}(\mathcal{X},\nu)} + |C_S| \nu(S) \right] \\ &\quad \lesssim K \left[\int_B |g(x) - m_S(g)|^{q'}\, d\nu(x) \right]^{1/q} [\nu(B)]^{1/q'} [1 + \delta(B, S)]^\gamma, \end{aligned}$$

which implies that

$$\left[\frac{1}{\nu(B)} \int_B |g(x) - m_S(g)|^{q'}\, d\nu(x) \right]^{1/q'} \lesssim K [1 + \delta(B, S)]^\gamma.$$

From this, the Hölder inequality and (7.4.20) for B, it then follows that

$$\begin{aligned}|m_B(g) - m_S(g)| &\le \frac{1}{\nu(B)} \int_B |g(x) - m_B(g)|\, d\nu(x) \\ &\quad + \frac{1}{\nu(B)} \int_B |g(x) - m_S(g)|\, d\nu(x) \\ &\lesssim K[1 + \delta(B, S)]^\gamma.\end{aligned}$$

Thus, (7.4.21) holds true, which, combined with (7.4.20), completes the proof of Lemma 7.4.6. □

Proof of Proposition 7.4.5. To prove that

$$\mathrm{RBMO}_\gamma\,(\mathcal{X}, \nu) \subset \left(H^{1,q}_{\mathrm{atb},\gamma}(\mathcal{X}, \nu)\right)^*,$$

we recall, from Corollary 7.2.12, that the space

$$\mathrm{RBMO}_\gamma\,(\mathcal{X}, \nu) = \mathrm{RBMO}\,(\mathcal{X}, \nu)$$

satisfies the John–Nirenberg inequality, namely,

$$\left[\frac{1}{\nu(\rho B)} \int_B |f(x) - f_B|^{q'}\, d\nu(x)\right]^{1/q'} \lesssim \|f\|_{\mathrm{RBMO}\,(\mathcal{X},\nu)} \sim \|f\|_{\mathrm{RBMO}_\gamma\,(\mathcal{X},\nu)}.$$

Thus,

$$\mathrm{RBMO}_\gamma\,(\mathcal{X}, \nu) \subset L^{q'}_{\mathrm{loc}}(\mathcal{X}, \nu)$$

and hence the action of the functional $\mathcal{L}_g$, defined for $g \in \mathrm{RBMO}_\gamma\,(\mathcal{X}, \nu)$ as in Lemma 7.4.4, makes sense on $(q, \gamma)_\lambda$-atomic blocks. After these observations, the proof of

$$\mathrm{RBMO}_\gamma\,(\mathcal{X}, \nu) \subset \left(H^{1,q}_{\mathrm{atb},\gamma}(\mathcal{X}, \nu)\right)^*$$

is a slight modification of the proof of Lemma 7.4.4. We omit the details.

To show

$$\left(H^{1,q}_{\mathrm{atb},\gamma}(\mathcal{X}, \nu)\right)^* \subset \mathrm{RBMO}_\gamma\,(\mathcal{X}, \nu),$$

we assume that $\mathcal{L} \in (H^{1,q}_{\mathrm{atb},\gamma}(\mathcal{X}, \nu))^*$. By Proposition 7.2.2 and Proposition 7.4.2(ii), we take $\rho = 3$ in Definitions 7.2.4 and 7.4.1. If $f \in L^q_0(B, \nu)$, then $f \in H^{1,q}_{\mathrm{atb},\gamma}(\mathcal{X}, \nu)$ and

$$\|f\|_{H^{1,q}_{\mathrm{atb},\gamma}(\mathcal{X},\nu)} \le [\nu(3B)]^{1/q'}[1+\delta(B,B)]^{\gamma}\|f\|_{L^q(\mathcal{X},\nu)}, \tag{7.4.22}$$

where $q' := q/(q-1)$. Consider the restriction of $\mathcal{L}$ on $L^q_0(B,\nu)$. For all $f \in L^q_0(B,\nu)$, by (7.4.22) and Lemma 7.1.16(ii), we have

$$\begin{aligned} |\mathcal{L}(f)| &\le \|\mathcal{L}\|_{(H^{1,q}_{\mathrm{atb},\gamma}(\mathcal{X},\nu))^*}|f|_{H^{1,q}_{\mathrm{atb},\gamma}(\mathcal{X},\nu)} \\ &\lesssim \|\mathcal{L}\|_{(H^{1,q}_{\mathrm{atb},\gamma}(\mathcal{X},\nu))^*}[\nu(3B)]^{1/q'}\|f\|_{L^q(\mathcal{X},\nu)}. \end{aligned} \tag{7.4.23}$$

Therefore, $\mathcal{L}$ defines a linear functional on $L^q_0(B,\nu)$. By the Riesz representation theorem, together with (7.4.23), there exist

$$[g] \in (L^q_0(B,\nu))^* = L^{q'}(B,\nu)/\mathbb{C}$$

and $g \in [g]$ such that, for all $f \in L^q_0(B,\nu)$,

$$\mathcal{L}(f) = \int_B f(x)g(x)\,d\nu(x). \tag{7.4.24}$$

Here $[g]$ represents the *equivalence class* of $L^{q'}(B,\nu)/\mathbb{C}$ determined by g.

Now let B_0 be a fixed ball and $B_j := 2^j B_0$ for all $j \in \mathbb{N}$. By (7.4.24), we know that, for all $f \in L^q_0(B_0,\nu)$ and $j \in \mathbb{N}$,

$$\int_{B_0} f(x)g_0(x)\,d\nu(x) = \int_{B_0} f(x)g_j(x)\,d\nu(x),$$

where g_j is the function in (7.4.24) associated with B_j for any $j \in \mathbb{Z}_+$. This implies that, for ν-almost every $x \in B_0$, $g_j - g_0 = C_j$ for some constant C_j. Moreover, we conclude that, for all $j, l \in \mathbb{Z}_+$ with $j \le l$ and ν-almost every $x \in B_j$,

$$g_j(x) - C_j = g_l(x) - C_l.$$

Define $g := g_j - C_j$ on B_j for any $j \in \mathbb{Z}_+$. Then g is well defined, and it holds true that

$$\mathcal{L}(f) = \int_{\mathcal{X}} g(x)f(x)\,d\nu(x)$$

and

$$\left|\int_{\mathcal{X}} g(x)f(x)\,d\nu(x)\right| \le \|\mathcal{L}\|_{(H^{1,q}_{\mathrm{atb},\gamma}(\mathcal{X},\nu))^*}|f|_{H^{1,q}_{\mathrm{atb},\gamma}(\mathcal{X},\nu)}$$

for all $f \in L^q_{b,0}(\mathcal{X}, \nu)$. Thus, the functional $\mathcal{L}$ is represented by g, and Lemma 7.4.6 shows that

$$\|g\|_{\mathrm{RBMO}_\gamma(\mathcal{X},\nu)} \lesssim \|\mathcal{L}\|_{(H^{1,q}_{\mathrm{atb},\gamma}(\mathcal{X},\nu))^*},$$

which completes the proof of Proposition 7.4.5. □

Lemma 7.4.7. *Let $(\mathcal{X}, \|\cdot\|_{\mathcal{X}})$ and $(\mathcal{Y}, \|\cdot\|_{\mathcal{Y}})$ be Banach spaces. Assume that $\mathcal{X}$ is a subspace of $\mathcal{Y}$ and the spaces $\mathcal{X}^*$ and $\mathcal{Y}^*$ coincide with equivalent norms. Then the spaces $\mathcal{X}$ and $\mathcal{Y}$ coincide with equivalent norms.*

Proof. Since $\mathcal{X} \subset \mathcal{Y}$, we then define $i : \mathcal{X} \to \mathcal{Y}$ by setting, for all $x \in \mathcal{X}$, $i(x) := x$. By the assumption that the spaces $\mathcal{X}^*$ and $\mathcal{Y}^*$ coincide with equivalent norms, we conclude that,[7] for all $x \in \mathcal{X}$,

$$\begin{aligned}\|x\|_{\mathcal{X}} &= \sup\{|f(x)| : f \in \mathcal{X}^*, \|f\|_{\mathcal{X}^*} \le 1\}\\ &\sim \sup\{|f(x)| : f \in \mathcal{Y}^*, \|f\|_{\mathcal{Y}^*} \le 1\}\\ &\sim \|x\|_{\mathcal{Y}},\end{aligned}$$

which implies that i is continuous and injective. Define the map $i^* : \mathcal{Y}^* \to \mathcal{X}^*$ by setting, for all $x \in \mathcal{X}$ and $f \in \mathcal{Y}^*$,

$$\langle i^*(f), x\rangle := \langle f, i(x)\rangle,$$

which is actually $i^*(f) := f|_{\mathcal{X}}$ for all $f \in \mathcal{Y}^*$; moreover, for all $f \in \mathcal{Y}^*$,

$$\|f\|_{\mathcal{Y}^*} \sim \|f\|_{\mathcal{X}^*} = \|f|_{\mathcal{X}}\|_{\mathcal{X}^*} = \|i^*(f)\|_{\mathcal{X}^*}.$$

This fact implies that i is surjective.[8] Since i is continuous, injective and surjective, it follows, from the open mapping theorem,[9] that i^{-1} is also continuous. This implies that the spaces $\mathcal{X}$ and $\mathcal{Y}$ coincide with equivalent norms, which completes the proof of Lemma 7.4.7. □

Theorem 7.4.8. (i) *Let $\gamma \in [1, \infty)$ and $q \in (1, \infty)$. Then the spaces $H^{1,q}_{\mathrm{atb},\gamma}(\mathcal{X}, \nu)$ and $H^{1,\infty}_{\mathrm{atb},\gamma}(\mathcal{X}, \nu)$ coincide with equivalent norms and*

$$\left(H^{1,\infty}_{\mathrm{atb},\gamma}(\mathcal{X}, \nu)\right)^* = \mathrm{RBMO}_\gamma(\mathcal{X}, \nu).$$

---

[7] See [110, Theorem 4.3].

[8] See [110, Theorem 4.13].

[9] See [110, Corollary 2.12 (b)].

(ii) *Let* $\gamma \in [1,\infty)$ *with* $\gamma \in (1,\infty)$ *and* $q \in (1,\infty]$. *Then the spaces* $H^{1,q}_{\mathrm{atb},\gamma}(\mathcal{X}, \nu)$ *and* $H^{1,q}_{\mathrm{atb}}(\mathcal{X}, \nu)$ *coincide with equivalent norms.*

Proof. If (i) holds true, then, by Proposition 7.2.5 and Lemma 7.4.7, we have (ii). Thus, to prove Theorem 7.4.8, it suffices to show (i). By Lemma 7.4.6, we conclude that, for all $q \in (1,\infty)$ and $\gamma \in [1,\infty)$,

$$\left(H^{1,q}_{\mathrm{atb},\gamma}(\mathcal{X}, \nu)\right)^* = \mathrm{RBMO}_\gamma\,(\mathcal{X}, \nu).$$

On the other hand, from Proposition 7.4.2(iii), it follows that, if $f \in (H^{1,q}_{\mathrm{atb},\gamma}(\mathcal{X}, \nu))^*$, then $f \in (H^{1,\infty}_{\mathrm{atb},\gamma}(\mathcal{X}, \nu))^*$. Consider the maps

$$i:\ H^{1,\infty}_{\mathrm{atb},\gamma}(\mathcal{X}, \nu) \to H^{1,q}_{\mathrm{atb},\gamma}(\mathcal{X}, \nu)$$

and

$$i^*:\ \mathrm{RBMO}_\gamma(\mathcal{X}, \nu) = \left(H^{1,q}_{\mathrm{atb},\gamma}(\mathcal{X}, \nu)\right)^* \to \left(H^{1,\infty}_{\mathrm{atb},\gamma}(\mathcal{X}, \nu)\right)^*.$$

Notice that the map i is an inclusion and i^* the canonical injection from the space $\mathrm{RBMO}_\gamma\,(\mathcal{X}, \nu)$ to $(H^{1,\infty}_{\mathrm{atb},\gamma}(\mathcal{X}, \nu))^*$ (with $g \equiv L_g$ for $g \in \mathrm{RBMO}_\gamma\,(\mathcal{X}, \nu)$). By Lemma 7.4.4, we know that $i^*(\mathrm{RBMO}_\gamma\,(\mathcal{X}, \nu))$ is closed in $(H^{1,\infty}_{\mathrm{atb},\gamma}(\mathcal{X}, \nu))^*$. As an application of the Banach closed range theorem, we conclude that $H^{1,\infty}_{\mathrm{atb},\gamma}(\mathcal{X}, \nu)$ is closed in $H^{1,q}_{\mathrm{atb},\gamma}(\mathcal{X}, \nu)$, which, together with Proposition 7.4.2(iii), implies that

$$H^{1,\infty}_{\mathrm{atb},\gamma}(\mathcal{X}, \nu) = H^{1,q}_{\mathrm{atb},\gamma}(\mathcal{X}, \nu)$$

as a set. Thus, i maps $H^{1,\infty}_{\mathrm{atb},\gamma}(\mathcal{X}, \nu)$ onto $H^{1,q}_{\mathrm{atb},\gamma}(\mathcal{X}, \nu)$. Since both $H^{1,\infty}_{\mathrm{atb},\gamma}(\mathcal{X}, \nu)$ and $H^{1,q}_{\mathrm{atb},\gamma}(\mathcal{X}, \nu)$ are Banach spaces, by the corollary of the open mapping theorem, we further see that

$$H^{1,\infty}_{\mathrm{atb},\gamma}(\mathcal{X}, \nu) = H^{1,q}_{\mathrm{atb},\gamma}(\mathcal{X}, \nu)$$

with an equivalent norm. This finishes the proof of (i) and hence Theorem 7.4.8. □

Remark 7.4.9. (i) If ν satisfies (7.1.2) and

$$\lambda(x,r) := \nu(B(x,r))$$

for all $x \in \mathcal{X}$ and $r \in (0,\infty)$, by Theorem 7.4.8, Proposition 7.2.3 and the dual property between $H^1(\mathcal{X}, \nu)$ and $\mathrm{BMO}\,(\mathcal{X}, \nu)$ on spaces of homogeneous type,[10]

[10]See [19, p. 593, Theorem B].

we know that, for all $q \in (1, \infty]$,

$$\left(H_{\mathrm{atb}}^{1,q}(\mathcal{X}, \nu)\right)^* = \mathrm{RBMO}\,(\mathcal{X}, \nu) = \mathrm{BMO}\,(\mathcal{X}, \nu) = \left(H^{1,q}(\mathcal{X}, \nu)\right)^*,$$

which, together with Lemma 7.4.7, further implies that, for all $q \in (1, \infty]$,

$$H_{\mathrm{atb}}^{1,q}(\mathcal{X}, \nu) = H^{1,q}(\mathcal{X}, \nu),$$

where $H^{1,q}(\mathcal{X}, \nu)$ is the classical Hardy space introduced by Coifman and Weiss.[11]

(ii) By Theorem 7.4.8 and Proposition 7.4.2(ii), we *denote the atomic Hardy space* $H_{\mathrm{atb},\,\gamma}^{1,q}(\mathcal{X}, \nu)$ *simply by* $H^1(\mathcal{X}, \nu)$ and *always assume that* $\rho = 2 = q$ *and* $\gamma = 1$ *in Definition 7.4.1.*

7.5 Notes

- The class of metric measure spaces satisfying the upper doubling condition and the geometrically doubling condition was introduced by Hytönen in [68].
- Examples 7.1.2 and 7.1.3 were given by Hytönen in [68] and Example 7.1.4 was given by Hytönen and Martikainen in [70].
- Theorem 7.1.20 was originally proved by Hytönen in [68].
- The space RBMO $(\mathcal{X}, \nu)$ was introduced by Hytönen when $\gamma = 1$ in [68] and by Hytönen et al. in [71] when $\gamma \in [1, \infty)$.
- Theorem 7.2.11 was established in [68].
- Theorem 7.3.2 was proved by Hu et al. in [61]. If $(\mathcal{X}, d, \nu) := (\mathbb{R}^D, |\cdot|, dx)$, Strömberg [123] proved that

  $$\mathrm{BMO}(\mathbb{R}^D) = \mathrm{BMO}_{0,\,s}(\mathbb{R}^D)$$

  if and only if $s \in (0, 1/2]$. Moreover, if ν is an absolutely continuous measure on $\mathbb{R}^D$, namely, there exists a weight w such that $d\nu = w dx$, Lerner [79] established the John–Strömberg characterization of BMO(w). Furthermore, if $(\mathcal{X}, d, \nu) := (\mathbb{R}^D, |\cdot|, \mu)$ with μ satisfying (0.0.1), it was proved by Hu et al. in [67] that

  $$\mathrm{RBMO}\,(\mu) = \mathrm{RBMO}_{0,\,s}(\mu)$$

  for $s \in (0, \beta_D^{-2}/2)$ with $\beta_D \in (2^D, \infty)$.
- The space $H_{\mathrm{atb},\,\gamma}^{1,q}(\mathcal{X}, \nu)$ in Definition 7.4.1 was introduced in [71]. The space $\tilde{H}_{\mathrm{atb},\,\gamma}^{1,q}(\mathcal{X}, \nu)$ with $\gamma = 1$ in (7.4.8) was introduced by Bui and Duong in [9].

[11] See [19].

- Theorem 7.4.8 was established in [71]. Also, Bui and Duong [9] showed that, for all $q \in (1, \infty)$,

$$\tilde{H}^{1,q}_{\mathrm{atb},\,1}(\mathcal{X}, \nu) = \tilde{H}^{1,\infty}_{\mathrm{atb},\,1}(\mathcal{X}, \nu)$$

and

$$\left(\tilde{H}^{1,q}_{\mathrm{atb},\,1}(\mathcal{X}, \nu)\right)^* = \left(\tilde{H}^{1,\infty}_{\mathrm{atb},\,1}(\mathcal{X}, \nu)\right)^* = \mathrm{RBMO}\,(\mathcal{X},\, \nu).$$

- An interesting *problem* is whether the atomic Hardy space $H^1(\mathcal{X}, \nu)$ has a characterization in terms of the maximal function similar to Theorem 3.3.3 or not.

Chapter 8
Boundedness of Operators over $(\mathcal{X}, \nu)$

In this chapter, we consider the boundedness of Calderón–Zygmund operators over non-homogeneous spaces $(\mathcal{X}, \nu)$. We first show that the Calderón–Zygmund operator T is bounded from $H^1(\mathcal{X}, \nu)$ to $L^1(\mathcal{X}, \nu)$. We then establish the molecular characterization of a version of the atomic Hardy space $\tilde{H}^{1,p}_{\mathrm{atb}}(\mathcal{X}, \nu)$, which is a subspace of $H^{1,p}_{\mathrm{atb}}(\mathcal{X}, \nu)$, and obtain the boundedness of T on $\tilde{H}^{1,p}_{\mathrm{atb}}(\mathcal{X}, \nu)$. We also prove that the boundedness of T on $L^p(\mathcal{X}, \nu)$ with $p \in (1, \infty)$ is equivalent to its various estimates and establish some weighted estimates involving the John–Strömberg maximal operators and the John–Strömberg sharp maximal operators, and some weighted norm inequalities for the multilinear Calderón–Zygmund operators. In addition, the boundedness of multilinear commutators of Calderón–Zygmund operators on Orlicz spaces is also presented.

8.1 Behaviors of Operators on $H^1(\mathcal{X}, \nu)$

This section has twofold purposes. The first is to study the interpolation property of $H^1(\mathcal{X}, \nu)$ for sublinear operators, and the second is to establish the boundedness of Calderón–Zygmund operators on $H^1(\mathcal{X}, \nu)$. To start with, we first establish the following Calderón–Zygmund decomposition.

Theorem 8.1.1. *Let $p \in [1, \infty)$, $f \in L^p(\mathcal{X}, \nu)$ and $t \in (0, \infty)$ (if $\nu(\mathcal{X}) < \infty$, then*

$$t > \zeta \|f\|_{L^p(\mathcal{X},\nu)}/[\nu(\mathcal{X})]^{1/p},$$

where ζ is any fixed positive constant satisfying that $\zeta > \max\{6^{3\varsigma}, 6^{3n_0}\}$). Then

D. Yang et al., *The Hardy Space H^1 with Non-doubling Measures and Their Applications*, Lecture Notes in Mathematics 2084, DOI 10.1007/978-3-319-00825-7_8,

(a) *there exists an almost disjoint family* $\{6B_i\}_i$ *of balls such that* $\{B_i\}_i$ *is pairwise disjoint,*

$$\frac{1}{\nu(6^2 B_i)} \int_{B_i} |f(x)|^p \, d\nu(x) > \frac{t^p}{\zeta} \text{ for all } i, \tag{8.1.1}$$

$$\frac{1}{\nu(6^2 \eta B_i)} \int_{\eta B_i} |f(x)|^p \, d\nu(x) \le \frac{t^p}{\zeta} \text{ for all } i \text{ and all } \eta \in (2, \infty), \tag{8.1.2}$$

and

$$|f(x)| \le t \text{ for } \nu\text{-almost every } x \in \mathcal{X} \setminus \left(\bigcup_i 6B_i\right); \tag{8.1.3}$$

(b) *for each* i, *let* R_i *be a* $(3 \times 6^2, 6^{3\varsigma})$*-doubling ball concentric with* B_i *satisfying that* $r_{R_i} > 6^2 r_{B_i}$, *and*

$$\omega_i := \chi_{6B_i} / \left(\sum_k \chi_{6B_k}\right).$$

Then there exists a family $\{\varphi_i\}_i$ *of functions such that, for each* i, $\operatorname{supp} \varphi_i \subset R_i$, φ_i *has a constant sign on* R_i,

$$\int_{\mathcal{X}} \varphi_i(x) \, d\nu(x) = \int_{6B_i} f(x)\omega_i(x) \, d\nu(x) \tag{8.1.4}$$

and

$$\sum_i |\varphi_i(x)| \le \gamma t \text{ for } \nu\text{-almost every } x \in \mathcal{X}, \tag{8.1.5}$$

where γ *is some positive constant depending only on* $(\mathcal{X}, \nu)$, *and there exists a positive constant* C, *independent of* f, t *and* i, *such that, when* $p = 1$, *it holds true that*

$$\|\varphi_i\|_{L^\infty(\mathcal{X}, \nu)} \nu(R_i) \le C \int_{\mathcal{X}} |f(x)\omega_i(x)| \, d\nu(x) \tag{8.1.6}$$

and, when $p \in (1, \infty)$, *it holds true that*

$$\left\{\int_{R_i} |\varphi_i(x)|^p \, d\nu(x)\right\}^{1/p} [\nu(R_i)]^{1/p'}$$
$$\le \frac{C}{t^{p-1}} \int_{\mathcal{X}} |f(x)\omega_i(x)|^p \, d\nu(x); \tag{8.1.7}$$

(c) *for* $p \in (1, \infty)$, *if, for any* i, *choosing* R_i *in (b) to be the smallest* $(3 \times 6^2, 6^{3\varsigma})$-*doubling ball of* $\{(3 \times 6^2)^k B_i\}_{k \in \mathbb{N}}$, *then*

$$h := \sum_i (f\omega_i - \varphi_i) \in H_{\rm atb}^{1,p}(\mathcal{X}, \nu)$$

and there exists a positive constant C, *independent of* f *and* t, *such that*

$$\|h\|_{H_{\rm atb}^{1,p}(\mathcal{X},\nu)} \le \frac{C}{t^{p-1}} \|f\|_{L^p(\mathcal{X},\nu)}^p. \tag{8.1.8}$$

Proof. For the sake of simplicity, we only give the proof in the case $p = 1$ for (a) and (b). We deduce the conclusion, when $p \in (1, \infty)$, from letting $g := |f|^p \in L^1(\mathcal{X}, \nu)$ and, with a simple modification, we obtain (8.1.7) instead of (8.1.6).

(a) Let

$$E := \{x \in \mathcal{X} : |f(x)| > t\}.$$

For ν-almost every $x \in E$, there exists some ball B_x such that

$$\frac{1}{\nu(6^2 B_x)} \int_{B_x} |f(y)|\, d\nu(y) > \frac{t}{\zeta} \tag{8.1.9}$$

and, if $\tilde{B}_x$ is centered at x with $r_{\tilde{B}_x} > 2r_{B_x}$, then

$$\frac{1}{\nu(6^2 \tilde{B}_x)} \int_{\tilde{B}_x} |f(y)|\, d\nu(y) \le \frac{t}{\zeta}.$$

Now we apply Lemma 7.1.18 to obtain a family of balls, $\{B_i\}_i \subset \{B_x\}_x$, such that $\sum_i \chi_{6B_i}(x) < \infty$ for all $x \in \mathcal{X}$ and (8.1.1), (8.1.2) and (8.1.3) are satisfied.

(b) Assume that the family of balls, $\{B_i\}_i$, is finite. Without loss of generality, we may suppose that $r_{R_i} \le r_{R_{i+1}}$. The function φ_i is constructed of the form $\varphi_i := \alpha_i \chi_{A_i}$ with $A_i \subset R_i$.

First, let $A_1 := R_1$ and $\varphi_1 := \alpha_1 \chi_{A_1}$ such that

$$\int \varphi_1(x)\, d\nu(x) = \int_{6B_1} f(x)\omega_1(x)\, d\nu(x).$$

Assume that $\{\varphi_1, \ldots, \varphi_{k-1}\}$ have been constructed satisfying (8.1.4) and

$$\sum_{i=1}^{k-1} \varphi_i \le \vartheta t, \tag{8.1.10}$$

where ϑ is some positive constant which is fixed later. There are two cases.

Case (i) There exists some $i \in \{1, \dots, k-1\}$ such that $R_i \cap R_k \neq \emptyset$. Let $\{R_{s_1}, \dots, R_{s_m}\}$ be the subfamily of $\{R_1, \dots, R_{k-1}\}$ such that $R_{s_j} \cap R_k \neq \emptyset$. Since $r_{R_{s_j}} \leq r_{R_k}$, it follows that $R_{s_j} \subset 3R_k$. By using the $(3 \times 6^2, 6^{3\varsigma})$-doubling property of R_k and (8.1.2), we conclude that

$$\begin{aligned}\sum_{j=1}^{m} \int_{\mathcal{X}} |\varphi_{s_j}(x)| \, d\nu(x) &\leq \sum_{j=1}^{m} \int_{\mathcal{X}} |f(x)\omega_{s_j}(x)| \, d\nu(x) \\ &\lesssim \int_{3R_k} |f(x)| \, d\nu(x) \\ &\lesssim t\nu(R_k).\end{aligned}$$

From this, it follows that there exists a positive constant c_0 such that

$$\nu\left(\left\{x \in \mathcal{X} : \sum_{j=1}^{m} |\varphi_{s_j}(x)| > 2c_0 t\right\}\right) \leq \frac{\nu(R_k)}{2}.$$

Let

$$A_k := R_k \bigcap \left\{x \in \mathcal{X} : \sum_{j} |\varphi_{s_j}(x)| \leq 2c_0 t\right\}.$$

Then

$$\nu(A_k) \geq \frac{\nu(R_k)}{2}.$$

The constant α_k is chosen such that

$$\int_{\mathcal{X}} \varphi_k(x) \, d\nu(x) = \int_{B_k} f(x)\omega_k(x) \, d\nu(x),$$

where $\varphi_k := \alpha_k \chi_{A_k}$. Then we see that

$$|\alpha_k| \lesssim \frac{1}{\nu(A_k)} \int_{\mathcal{X}} |f(x)|\omega_k(x) \, d\nu(x) \lesssim \frac{1}{\nu(R_k)} \int_{\frac{1}{6}R_k} |f(x)| \, d\nu(x) \leq \tilde{C}t.$$

If we choose $\vartheta := 2c_0 + \tilde{C}$, then (8.1.5) holds true.

Case (ii) $R_i \cap R_k = \emptyset$ for all $i \in \{1, \dots, k-1\}$. Let $A_k := R_k$ and $\varphi_k := \alpha_k \chi_{R_k}$ such that

$$\int_{\mathcal{X}} \varphi_k(x) \, d\nu(x) = \int_{B_k} f(x)\omega_k(x) \, d\nu(x).$$

Using an argument similar to that used in Case (i), and applying (8.1.10), we also obtain (8.1.5).

By the construction of the function φ_i, it is easy to see that $\nu(R_i) \le 2\nu(A_i)$. Hence,

$$\|\varphi_i\|_{L^\infty(\mathcal{X},\nu)}\nu(R_i) \lesssim |\alpha_i|\nu(A_i) \lesssim \int_{\mathcal{X}} |f(x)\omega_i(x)|\, d\nu(x).$$

When the collection of the balls, $\{B_i\}$, is not finite, we can argue as in Theorem 1.4.1. The proofs of (a) and (b) are completed.

(c) Since R_i is the smallest $(3 \times 6^2, 6^{3\varsigma})$-doubling ball of the form $3 \times 6^2 B_i$, $i \in \mathbb{N}$, we then know that $\delta(B_i, R_i) \lesssim 1$. For each i, considering the atomic block $h_i := f\omega_i - \varphi_i$ supported in R_i, by (8.1.1) and (8.1.7), we have

$$|h_i|_{H^{1,p}_{\mathrm{atb},\gamma}(\mathcal{X},\nu)} \lesssim \frac{1}{t^{p-1}} \int_{\mathcal{X}} |f(x)\omega_i(x)|^p\, d\nu(x),$$

which implies that

$$\|h\|_{H^{1,p}_{\mathrm{atb},\gamma}(\mathcal{X},\nu)} \lesssim \frac{1}{t^{p-1}} \int_{\mathcal{X}} \sum_i |f(x)\omega_i(x)|^p\, d\nu(x) \sim \frac{1}{t^{p-1}} \int_{\mathcal{X}} |f(x)|^p\, d\nu(x).$$

This finishes the proof of Theorem 8.1.1. □

Applying Theorem 8.1.1, we obtain the following interpolation theorem.

Theorem 8.1.2. *Let $p_0 \in (1,\infty)$, $p \in (1, p_0)$ and T be a sublinear operator. Assume that T is bounded from $L^{p_0}(\mathcal{X}, \nu)$ to $L^{p_0,\infty}(\mathcal{X}, \nu)$ and from $H^1(\mathcal{X}, \nu)$ to $L^{1,\infty}(\mathcal{X}, \nu)$. Then there exists a positive constant $C_{(p)}$, depending on p, such that, for all $f \in L^p(\mathcal{X}, \nu)$,*

$$\|Tf\|_{L^p(\mathcal{X},\nu)} \le C_{(p)}\|f\|_{L^p(\mathcal{X},\nu)}.$$

Proof. By the Marcinkiewicz interpolation theorem, we only need to prove that, for all $p \in (1, p_0)$, $f \in L^p(\mathcal{X}, \nu)$ and $t > \zeta\|f\|_{L^p(\mathcal{X},\nu)}/[\nu(\mathcal{X})]^{1/p}$,

$$\nu(\{x \in \mathcal{X} : |Tf(x)| > t\}) \lesssim \frac{\|f\|^p_{L^p(\mathcal{X},\nu)}}{t^p}. \tag{8.1.11}$$

To show (8.1.11), for any given $f \in L^p(\mathcal{X}, \nu)$ and

$$t > \zeta\|f\|_{L^p(\mathcal{X},\nu)}/[\nu(\mathcal{X})]^{1/p},$$

by applying Theorem 8.1.1, we conclude that, with the notation same as in Theorem 8.1.1, it holds true that $f = g + h$, where

$$g := f\chi_{\mathcal{X}\setminus\cup_j(6B_j)} + \sum_j \varphi_j$$

and

$$h := \sum_j (\omega_j f - \varphi_j).$$

By Theorem 8.1.1(c), we see that

$$h \in H^1(\mathcal{X}, \nu) \quad \text{and} \quad \|h\|_{H^1(\mathcal{X},\nu)} \lesssim t^{1-p}\|f\|_{L^p(\mathcal{X},\nu)}^p.$$

Moreover, from (a) and (b) of Theorem 8.1.1, together with the Hölder inequality, we deduce that

$$\left\|f\chi_{\mathcal{X}\setminus\cup_j(6B_j)}\right\|_{L^{p_0}(\mathcal{X},\nu)} \le t^{1-\frac{p}{p_0}}\|f\|_{L^p(\mathcal{X},\nu)}^{\frac{p}{p_0}}$$

and

$$\begin{aligned}\left\|\sum_j \varphi_j\right\|_{L^{p_0}(\mathcal{X},\nu)} &\lesssim \left[t^{p_0-1}\sum_j \int_{\mathcal{X}} |\varphi_j(x)|\,d\nu(x)\right]^{\frac{1}{p_0}} \\ &\lesssim \left[t^{p_0-p}\|f\|_{L^p(\mathcal{X},\nu)}^p\right]^{\frac{1}{p_0}} \\ &\sim t^{1-\frac{p}{p_0}}\|f\|_{L^p(\mathcal{X},\nu)}^{\frac{p}{p_0}}.\end{aligned}$$

These facts imply that $g \in L^{p_0}(\mathcal{X}, \nu)$ and

$$\|g\|_{L^{p_0}(\mathcal{X},\nu)} \lesssim t^{1-\frac{p}{p_0}}\|f\|_{L^p(\mathcal{X},\nu)}^{\frac{p}{p_0}}.$$

By the estimates of g and h, together with the boundedness of T from $L^{p_0}(\mathcal{X}, \nu)$ to $L^{p_0,\infty}(\mathcal{X}, \nu)$ and from $H^1(\mathcal{X}, \nu)$ to $L^{1,\infty}(\mathcal{X}, \nu)$, we further see that, for any $t \in (0, \infty)$

$$\begin{aligned}&\nu(\{x \in \mathcal{X} : |T(f)(x)| > t\}) \\ &\quad\le \nu(\{x \in \mathcal{X} : |T(g)(x)| > t/2\}) + \nu(\{x \in \mathcal{X} : |T(h)(x)| > t/2\}) \\ &\quad\lesssim t^{-p_0}\|g\|_{L^{p_0}(\mathcal{X},\nu)}^{p_0} + t^{-1}\|h\|_{H^1(\mathcal{X},\nu)} \\ &\quad\lesssim t^{-p}\|f\|_{L^p(\mathcal{X},\nu)}^p,\end{aligned}$$

which implies (8.1.11) and hence completes the proof of Theorem 8.1.2. □

To state our second interpolation theorem, we first consider the *doubling maximal function*, $\tilde{N} f(x)$, and the *sharp maximal function*, $\mathcal{M}^\sharp f(x)$, which are respectively defined by setting, for $f \in L^1_{\rm loc}(\mathcal{X}, \nu)$ and $x \in \mathcal{X}$,

$$\tilde{N}(f)(x) := \sup_{\substack{B \ni x \\ B\,(6,\beta_6)-\text{doubling}}} \frac{1}{\nu(B)} \int_B |f(y)|\,d\nu(y) \tag{8.1.12}$$

and

$$\begin{aligned} \mathcal{M}^\sharp(f)(x) := \sup_{B \ni x} \frac{1}{\nu(5B)} \int_B |f(y) - m_{\tilde{B}^6}(f)|\,d\nu(y) \\ + \sup_{\substack{x \in B \subset S \\ B,\,S\,(6,\beta_6)-\text{doubling}}} \frac{|m_B(f) - m_S(f)|}{1 + \delta(B, S)}. \end{aligned} \tag{8.1.13}$$

Observe that

$$\mathcal{M}^\sharp(f) \lesssim \mathcal{M}_\eta(f) \quad \text{and} \quad \tilde{N}(f) \lesssim \mathcal{M}_\eta(f),$$

where $\mathcal{M}_\eta(f)$ is as in (7.1.8) with $\eta \in [5, \infty)$. Thus, both $\mathcal{M}^\sharp$ and $\tilde{N}$ are bounded on $L^p(\mathcal{X}, \nu)$ for $p \in (1, \infty)$ and from $L^1(\mathcal{X}, \nu)$ to $L^{1,\infty}(\mathcal{X}, \nu)$.

Lemma 8.1.3. *For any $f \in L^1_{\rm loc}(\mathcal{X}, \nu)$, with $\int_{\mathcal{X}} f(x)\,d\nu(x) = 0$ if $\nu(\mathcal{X}) < \infty$, if $\inf\{1, \tilde{N}(f)\} \in L^{p_0}(\mathcal{X}, \nu)$ for some $p_0 \in (1, \infty)$, then, for any $p \in [p_0, \infty)$, there exists a positive constant $C_{(p)}$, depending on p but independent of f, such that*

$$\left\|\tilde{N}(f)\right\|_{L^p(\mathcal{X},\nu)} \le C_{(p)} \left\|\mathcal{M}^\sharp(f)\right\|_{L^p(\mathcal{X},\nu)}. \tag{8.1.14}$$

Proof. Since the proof for $\nu(\mathcal{X}) < \infty$ is similar, we only prove Lemma 8.1.3 in the case that $\nu(\mathcal{X}) = \infty$. Furthermore, we first prove the following good-λ inequality, namely, for some $\eta \in (0, 1)$ and for all $\epsilon \in (0, \infty)$, there exists some $\vartheta \in (0, \infty)$, independent of ϵ, such that, for all $t \in (0, \infty)$,

$$\begin{aligned} &\nu\left(\left\{x \in \mathcal{X} : \tilde{N}(f)(x) > (1+\epsilon)t,\ \mathcal{M}^\sharp(f)(x) \le \vartheta t\right\}\right) \\ &\quad \le \eta\nu\left(\left\{x \in \mathcal{X} : \tilde{N}(f)(x) > t\right\}\right). \end{aligned} \tag{8.1.15}$$

Indeed, for $f \in L^p(\mathcal{X}, \nu)$, setting

$$E_t := \left\{x \in \mathcal{X} : \tilde{N}(f)(x) > (1+\epsilon)t,\ \mathcal{M}^\sharp(f)(x) \le \vartheta t\right\}$$

and

$$\Omega_t := \left\{x \in \mathcal{X} : \tilde{N}(f)(x) > t\right\}.$$

For each $x \in E_t$, we can choose a $(6, \beta_6)$-doubling ball B_x containing x which satisfies that $m_{B_x}(|f|) > (1+\epsilon)t$ and, if B is any $(6, \beta_6)$-doubling ball containing x with $r_B > 2r_{B_x}$, then

$$m_B(|f|) \le (1+\epsilon/2)t.$$

Such ball B_x exists due to $f \in L^p(\mathcal{X}, \nu)$.

Let S_x be the smallest $(6^3, \beta_6)$-doubling ball of the form $6^{3j} B_x$, $j \in \mathbb{N}$. Then, by Lemmas 7.1.14 and 7.1.16, we see that S_x, $6S_x$ and $6^2 S_x$ are $(6, \beta_6)$-doubling balls and $\delta(B_x, 6S_x) \lesssim 1$, from which we deduce that

$$|m_{B_x}(|f|) - m_{6S_x}(|f|)| \le [1 + \delta(B_x, 6S_x)]\mathcal{M}^\sharp(|f|)(x) \lesssim \mathcal{M}^\sharp(f)(x) \lesssim \vartheta t.$$

This implies that, for sufficiently small ϑ, it holds that $m_{6S_x}(|f|) > t$ and hence $6S_x \subset \Omega_t$.

Without loss of generality, we may assume that $\sup_{x\in E_t} r_{S_x} < \infty$. Now by Lemma 7.1.17, we pick a disjoint collection $\{S_{x_i}\}_{i\in I}$ with $x_i \in E_t$ and $E_t \subset \cup_{i\in I} 6S_{x_i}$. We claim that, for all $i \in I$,

$$\nu\left(6S_{x_i} \bigcap E_t\right) \le \eta\nu(6S_{x_i})/\beta_6. \tag{8.1.16}$$

If (8.1.16) holds true, by the doubling property and the disjoint property of $\{S_{x_i}\}_i$, we see that

$$\nu(E_t) \le \sum_{i\in I} \nu\left(6S_{x_i} \bigcap E_t\right) \le \eta \sum_{i\in I} \nu(6S_{x_i})/\beta_6 \le \eta \sum_{i\in I} \nu(S_{x_i}) \le \eta\nu(\Omega_t),$$

and hence (8.1.15) holds true.

Now we prove (8.1.16). Let $W_{x_i} := 6S_{x_i}$ and $y \in W_{x_i} \cap E_t$. For any $(6, \beta_6)$-doubling ball $B \ni y$ satisfying $m_B(|f|) > (1+\epsilon)t$, we know that $r_B \le r_{W_{x_i}}/8$. Otherwise, if $r_B > r_{W_{x_i}}/8$, then we have

$$B_{x_i} \subset W_{x_i} \subset \widetilde{16B}^6$$

and, for sufficiently small ϑ,

$$\left|m_B(|f|) - m_{\widetilde{16B}^6}(|f|)\right| \le \left[1 + \delta\left(B, \widetilde{16B}^6\right)\right]\mathcal{M}^\sharp(|f|)(y) \lesssim \vartheta t \le \epsilon t/2,$$

which implies that $m_{\widetilde{16B}^6}(|f|) > (1+\epsilon/2)t$. This contradicts to the choice of B_{x_i}, from which we deduce that $r_B \le r_{W_{x_i}}/8$ and $B \subset \frac{5}{4}W_{x_i}$. By this fact, together with $m_B(|f|) > (1+\epsilon)t$, we conclude that

$$\tilde{N}(f\chi_{\frac{5}{4}W_{x_i}})(y) > (1+\epsilon)t$$

and

$$m_{\widetilde{\frac{5}{4}W_{x_i}}^6}(|f|) \leq (1 + \epsilon/2)t.$$

Therefore,

$$\tilde{N}\left(\chi_{\frac{5}{4}W_{x_i}}\left(|f| - m_{\widetilde{\frac{5}{4}W_{x_i}}^6}(|f|)\right)\right)(y) > \frac{\epsilon}{2}t.$$

Observe that $\tilde{N}$ is bounded from $L^1(\mathcal{X}, \nu)$ to $L^{1,\infty}(\mathcal{X}, \nu)$ due to Theorem 7.1.20. By these facts, we see that

$$\begin{aligned}
\nu\left(W_{x_i} \bigcap E_t\right) &\leq \nu\left(\left\{x \in \mathcal{X} : \tilde{N}\left(\chi_{\frac{5}{4}W_{x_i}}\left(|f| - m_{\widetilde{\frac{5}{4}W_{x_i}}^6}(|f|)\right)\right)(x) > \frac{\epsilon}{2}t\right\}\right) \\
&\lesssim \frac{1}{\epsilon t}\int_{\frac{5}{4}W_{x_i}}\left||f(x)| - m_{\widetilde{\frac{5}{4}W_{x_i}}^6}(|f|)\right| d\nu(x) \\
&\lesssim \frac{1}{\epsilon t}\nu\left(\frac{15}{2}W_{x_i}\right)\mathcal{M}^\sharp(|f|)(x_i) \\
&\leq \frac{\tilde{C}\vartheta}{\epsilon}\nu(6S_{x_i}),
\end{aligned}$$

where $\tilde{C}$ is a positive constant depending only on $(\mathcal{X}, \nu)$. Thus, (8.1.16) holds true provided $\vartheta < \eta\epsilon/(\beta_6\tilde{C})$.

Now suppose that $f \notin L^p(\mathcal{X}, \nu)$ for $p \in [p_0, \infty)$, but $\inf\{1, \tilde{N}(f)\} \in L^{p_0}(\mathcal{X}, \nu)$. For each $k \in \mathbb{N}$, let

$$f_k(x) := \begin{cases} f(x), & |f(x)| \leq k; \\ \dfrac{kf(x)}{|f(x)|}, & |f(x)| > k. \end{cases}$$

Then, by Corollary 7.1.21, we see that

$$|f_k| \leq k \inf\{1, |f|\} \leq k \inf\{1, \tilde{N}(f)\}$$

ν-almost everywhere on $\mathcal{X}$. Thus, from the assumption $\inf\{1, \tilde{N}(f)\} \in L^{p_0}(\mathcal{X}, \nu)$, it follows that $f_k \in L^{p_0}(\mathcal{X}, \nu)$ for all $k \in \mathbb{N}$. On the other hand, for each $k \in \mathbb{N}$, $f_k \in L^\infty(\mathcal{X}, \nu)$, and hence $f_k \in L^p(\mathcal{X}, \nu)$ for all $p \in [p_0, \infty)$. Therefore, using these facts, (8.1.14) with f replaced by f_k, and $\mathcal{M}^\sharp(f_k) \lesssim \mathcal{M}^\sharp(f)$, we conclude that, for all $p \in [p_0, \infty)$,

$$\left\|\tilde{N}(f_k)\right\|_{L^p(\mathcal{X},\nu)} \lesssim \left\|\mathcal{M}^\sharp(f_k)\right\|_{L^p(\mathcal{X},\nu)} \lesssim \left\|\mathcal{M}^\sharp(f)\right\|_{L^p(\mathcal{X},\nu)},$$

which, together with the Fatou lemma, completes the proof of Lemma 8.1.3. □

Theorem 8.1.4. *Let T be a linear operator which is bounded from $H^1(\mathcal{X}, \nu)$ into $L^1(\mathcal{X}, \nu)$ and from $L^\infty(\mathcal{X}, \nu)$ into* RBMO$(\mathcal{X}, \nu)$*. Then T can be extended to a bounded linear operator on $L^p(\mathcal{X}, \nu)$ for all $p \in (1, \infty)$.*

Proof. For simplicity we may assume that $\nu(\mathcal{X}) = \infty$. Let $f \in L^\infty_{b,0}(\mathcal{X}, \nu)$. The set of all such functions is dense in $L^p(\mathcal{X}, \nu)$ for all $p \in (1, \infty)$. For such functions f, we see that $f \in H^1(\mathcal{X}, \nu)$ and $Tf \in L^1(\mathcal{X}, \nu)$, then

$$\inf\{1, \tilde{N}(Tf)\} \in L^p(\mathcal{X}, \nu)$$

for any $p \in (1, \infty)$. Therefore, by Lemma 8.1.3 and Corollary 7.1.21, it suffices to show that, for all $p \in (1, \infty)$,

$$\left\|\mathcal{M}^\sharp(Tf)\right\|_{L^p(\mathcal{X},\nu)} \lesssim \|f\|_{L^p(\mathcal{X},\nu)}. \tag{8.1.17}$$

As in Theorem 8.1.1, for such a function f and $t \in (0, \infty)$, we decompose f as $f = g + h$ such that g and h satisfy that $\|g\|_{L^\infty(\mathcal{X},\nu)} \lesssim t$ and

$$\|h\|_{H^{1,p}_{\rm atb}(\mathcal{X},\nu)} \lesssim \frac{1}{t^{p-1}}\|f\|^p_{L^p(\mathcal{X},\nu)}.$$

Since T is bounded from $L^\infty(\mathcal{X}, \nu)$ to RBMO$(\mathcal{X}, \nu)$, it follows that

$$\left\|\mathcal{M}^\sharp(Tg)\right\|_{L^\infty(\mathcal{X},\nu)} \le Ct$$

for some positive constant C. Therefore,

$$\begin{aligned}&\nu\left(\left\{x \in \mathcal{X} : \mathcal{M}^\sharp(Tf)(x) > (C+1)t\right\}\right)\\ &\quad\le \nu\left(\left\{x \in \mathcal{X} : \mathcal{M}^\sharp(Th)(x) > t\right\}\right).\end{aligned} \tag{8.1.18}$$

On the other hand, from the weak type (1, 1) of $\mathcal{M}^\sharp$, we deduce that

$$\nu(\{x \in \mathcal{X} : \mathcal{M}^\sharp(Th)(x) > t\}) \lesssim \frac{\|Th\|_{L^1(\mathcal{X},\nu)}}{t}. \tag{8.1.19}$$

By the boundedness of T from $H^1(\mathcal{X}, \nu)$ to $L^1(\mathcal{X}, \nu)$, we see that

$$\|Th\|_{L^1(\mathcal{X},\nu)} \lesssim \|h\|_{H^{1,p}_{\rm atb}(\mathcal{X},\nu)} \lesssim \frac{1}{t^{p-1}}\|f\|^p_{L^p(\mathcal{X},\nu)}.$$

This, together with (8.1.19) and (8.1.18), implies that

$$\nu\left(\left\{x \in \mathcal{X} : \mathcal{M}^\sharp(Tf)(x) > (C+1)t\right\}\right) \lesssim \frac{1}{t^p}\|f\|^p_{L^p(\mathcal{X},\nu)}.$$

Thus, the sublinear operator $\mathcal{M}^\sharp \circ T$ is of weak type (p, p) for all $p \in (1, \infty)$. By the Marcinkiewicz interpolation theorem, we conclude that $\mathcal{M}^\sharp \circ T$ is bounded for all $p \in (1, \infty)$, which implies (8.1.17) and hence completes the proof of Theorem 8.1.4. □

We now turn to the boundedness of Calderón–Zygmund operators on $H^1(\mathcal{X}, \nu)$, we first establish a boundedness criterion for linear operators from $H^1(\mathcal{X}, \nu)$ to a Banach space, which is of independent interest.

Theorem 8.1.5. *Let $\rho \in (1, \infty)$, $\gamma \in [1, \infty)$, $q \in (1, \infty)$, T be a linear operator defined on $L^q_{b,0}(\mathcal{X}, \nu)$, and $\mathcal{Y}$ a Banach space. Then T can be extended to a bounded linear operator from $H^1(\mathcal{X}, \nu)$ to $\mathcal{Y}$ if and only if there exists a nonnegative constant C such that, for all $(q, \gamma)_\lambda$-atomic blocks b,*

$$\|Tb\|_{\mathcal{Y}} \le C|b|_{H^{1,q}_{\mathrm{atb},\gamma}(\mathcal{X},\nu)}. \tag{8.1.20}$$

Proof. Obviously, if T can be extended to a bounded linear operator from $H^1(\mathcal{X}, \nu)$ to $\mathcal{Y}$, then (8.1.20) holds true.

To show the sufficiency of Theorem 8.1.5, by similarity, we only prove Theorem 8.1.5 for the case $q = 2$. Moreover, via Theorem 7.4.8 and Proposition 7.4.2(ii), without loss of generality, we may assume that $\rho = 2$ and $\gamma = 1$ in Definition 7.4.1. Let B be a fixed ball. If $f \in L^2_0(B, \nu)$, then f is a $(2, 1)_\lambda$-atomic block and

$$|f|_{H^{1,2}_{\mathrm{atb}}(\mathcal{X},\nu)} \le \|f\|_{L^2(\mathcal{X},\nu)}[\nu(2B)]^{1/2}[1 + \delta(B, B)]. \tag{8.1.21}$$

Moreover, from this and (8.1.20), it follows that, for any sequence $\{B_j\}_j$ of increasing concentric balls with $\mathcal{X} = \cup_j B_j$, T is bounded from $L^2_0(B_j, \nu)$ to $\mathcal{Y}$ for all j, which implies that the *adjoint operator* T^* of T is bounded from the dual space $\mathcal{Y}^*$ of $\mathcal{Y}$ to $(L^2_0(B_j, \nu))^*$ for all j. Moreover, this, together with (8.1.20), implies that, for all $f \in \mathcal{Y}^*$ and $(2, 1)_\lambda$-atomic blocks b,

$$\left|\int_{\mathcal{X}} b(x) T^* f(x)\, d\nu(x)\right| = |\langle Tb, f\rangle| \lesssim \|f\|_{\mathcal{Y}^*} |b|_{H^{1,2}_{\mathrm{atb}}(\mathcal{X},\nu)}. \tag{8.1.22}$$

By Lemma 7.4.6, this implies that $T^* f \in \mathrm{RBMO}(\mathcal{X}, \nu)$ and

$$\|T^* f\|_{\mathrm{RBMO}(\mathcal{X},\nu)} \lesssim \|f\|_{\mathcal{Y}^*}.$$

Let $H^{1,2}_{\mathrm{fin}}(\mathcal{X}, \nu)$ be the subspace of all finite linear combinations of $(2, 1)_\lambda$-atomic blocks. Then $H^{1,2}_{\mathrm{fin}}(\mathcal{X}, \nu)$ is dense in $H^{1,2}_{\mathrm{atb}}(\mathcal{X}, \nu)$. On the other hand, $H^{1,2}_{\mathrm{fin}}(\mathcal{X}, \nu)$ coincides with $L^2_{b,0}(\mathcal{X}, \nu)$ as vector spaces. Then, by Theorem 7.4.8(i) and the claim above, we know that, for all $g \in H^{1,2}_{\mathrm{fin}}(\mathcal{X}, \nu)$ and $f \in \mathcal{Y}^*$ with $\|f\|_{\mathcal{Y}^*} = 1$,

$$|\langle Tg, f\rangle| = |\langle g, T^* f\rangle| \lesssim \|g\|_{H^{1,2}_{\rm atb}(\mathcal{X},\nu)}.$$

From this and (8.1.20), it follows that

$$Tg \in \mathcal{Y} \quad \text{and} \quad \|Tg\|_{\mathcal{Y}} \lesssim \|g\|_{H^{1,2}_{\rm atb}(\mathcal{X},\nu)},$$

which, together with the density of $H^{1,2}_{\rm fin}(\mathcal{X},\nu)$ in $H^{1,2}_{\rm atb}(\mathcal{X}, \nu)$, then completes the proof of Theorem 8.1.5. □

As an application of Theorem 8.1.5, we consider the boundedness of Calderón–Zygmund operators from $H^1(\mathcal{X}, \nu)$ to $L^1(\mathcal{X}, \nu)$. Define

$$\Delta := \{(x, x) : \ x \in \mathcal{X}\}.$$

A *standard kernel* is a mapping $K : \ (\mathcal{X} \times \mathcal{X}) \setminus \Delta \to \mathbb{C}$ satisfying that there exist some positive constants τ and C such that, for all $x, y \in \mathcal{X}$ with $x \neq y$,

$$|K(x, y)| \le C \frac{1}{\lambda(x, d(x, y))} \tag{8.1.23}$$

and, for all $x, \tilde{x}, y \in \mathcal{X}$ with $d(x, y) \ge 2d(x, \tilde{x})$,

$$\begin{aligned} &|K(x, y) - K(\tilde{x}, y)| + |K(y, x) - K(y, \tilde{x})| \\ &\quad \le C \frac{[d(x, \tilde{x})]^\tau}{[d(x, y)]^\tau \lambda(x, d(x, y))}. \end{aligned} \tag{8.1.24}$$

A linear operator T is called a *Calderón–Zygmund operator* with kernel K satisfying (8.1.23) and (8.1.24), if, for all $f \in L^\infty_b(\mathcal{X}, \nu)$ and $x \notin \operatorname{supp} f$,

$$Tf(x) := \int_{\mathcal{X}} K(x, y) f(y)\, d\nu(y). \tag{8.1.25}$$

Theorem 8.1.6. *Let T be a Calderón–Zygmund operator as in* (8.1.25) *with kernel K satisfying* (8.1.23) *and* (8.1.24). *If T is bounded on $L^2(\mathcal{X}, \nu)$, then there exists a positive constant C such that, for all $f \in H^1(\mathcal{X}, \nu)$,*

$$\|Tf\|_{L^1(\mathcal{X},\nu)} \le C \|f\|_{H^1(\mathcal{X},\nu)}.$$

Proof. By Theorem 7.4.8, we take $q = 2$ and $\gamma = 1$ in Definition 7.4.1. Moreover, by Theorem 8.1.5, it suffices to show that, for all $(2, 1)_\lambda$-atomic blocks b,

$$\|Tb\|_{L^1(\mathcal{X},\nu)} \lesssim |b|_{H^{1,2}_{\rm atb}(\mathcal{X},\nu)}. \tag{8.1.26}$$

Let $b := \sum_{j=1}^2 \lambda_j a_j$ be a $(2, 1)_\lambda$-atomic block, where, for any $j \in \{1, 2\}$,

$$\operatorname{supp} a_j \subset B_j \subset B$$

for some B_j and B as in Definition 7.4.1. We write

$$\begin{aligned}
\|Tb\|_{L^1(\mathcal{X},\nu)} &= \int_{2B} |Tb(x)|\,d\nu(x) + \int_{\mathcal{X}\setminus(2B)} \cdots \\
&\le \sum_{j=1}^{2} |\lambda_j| \int_{2B_j} |Ta_j(x)|\,d\nu(x) + \sum_{j=1}^{2} |\lambda_j| \int_{(2B)\setminus(2B_j)} \cdots \\
&\quad + \int_{\mathcal{X}\setminus(2B)} |Tb(x)|\,d\nu(x) \\
&=: \sum_{i=1}^{3} \mathrm{I}_i.
\end{aligned}$$

The Hölder inequality and the boundedness of T on $L^2(\mathcal{X}, \nu)$ imply that

$$\mathrm{I}_1 \lesssim \sum_{j=1}^{2} |\lambda_j| [\nu(2B_j)]^{1/2} \|a_j\|_{L^2(\mathcal{X},\nu)} \lesssim \sum_{j=1}^{2} |\lambda_j|.$$

On the other hand, from (7.1.7), (8.1.23) and the Hölder inequality, it follows that

$$\begin{aligned}
\mathrm{I}_2 &\lesssim \sum_{j=1}^{2} |\lambda_j| \int_{(2B)\setminus(2B_j)} \int_{B_j} \frac{|a_j(y)|}{\lambda(x, d(x,y))}\,d\nu(y)\,d\nu(x) \\
&\lesssim \sum_{j=1}^{2} |\lambda_j| \int_{(2B)\setminus(2B_j)} \int_{B_j} \frac{|a_j(y)|}{\lambda(x_{B_j}, d(x, x_{B_j}))}\,d\nu(y)\,d\nu(x) \\
&\lesssim \sum_{j=1}^{2} |\lambda_j| \delta(B_j, B) \|a_j\|_{L^1(\mathcal{X},\nu)} \\
&\lesssim \sum_{j=1}^{2} |\lambda_j|.
\end{aligned}$$

Moreover, by $\int_{\mathcal{X}} b(x)\,d\nu(x) = 0$, (8.1.24), (7.1.1) and (7.1.7), we see that

$$\begin{aligned}
\mathrm{I}_3 &\le \int_{\mathcal{X}\setminus(2B)} \int_{\mathcal{X}} |K(x,y) - K(x, x_B)||b(y)|\,d\nu(y)\,d\nu(x) \\
&\lesssim \int_{\mathcal{X}\setminus(2B)} \int_{\mathcal{X}} \left[\frac{d(y, x_B)}{d(x, x_B)}\right]^{\tau} \frac{|b(y)|}{\lambda(x, d(x,y))}\,d\nu(y)\,d\nu(x)
\end{aligned}$$

$$\lesssim \sum_{j=1}^{2} |\lambda_j| \int_{\mathcal{X}\setminus(2B)} \int_{B_j} \left[\frac{r_B}{d(x, x_B)}\right]^{\tau} \frac{|a_j(y)|}{\lambda(x_B, d(x, x_B))} \, d\nu(y) \, d\nu(x)$$

$$\lesssim \sum_{j=1}^{2} |\lambda_j| \sum_{k=1}^{\infty} \int_{(2^{k+1}B)\setminus(2^k B)} \left(\frac{r_B}{2^{k+1} r_B}\right)^{\tau} \frac{d\nu(x)}{\lambda(x_B, d(x, x_B))}$$

$$\lesssim \sum_{j=1}^{2} |\lambda_j|.$$

This, combined with the estimates for I_1 and I_2, implies (8.1.26), which completes the proof of Theorem 8.1.6. □

Example 8.1.7. Let ν be the measure in $\mathbb{B}_{2D}$ as in Example 7.1.4. Consider the kernel

$$K(x, y) := (1 - \bar{x} \cdot y)^{-m} \tag{8.1.27}$$

for $x,\ y \in \overline{\mathbb{B}}_{2D} \subset \mathbb{C}^D$. Here $\bar{x}$ stands for the *componentwise complex conjugation*, and $\bar{x} \cdot y$ stands for the *usual dot product of D-vector $\bar{x}$ and y*. Moreover, $\overline{\mathbb{B}}_{2D} \setminus \frac{1}{2}\mathbb{B}_{2D}$ is equipped with the *regular quasi-distance*:

$$d(x, y) := ||x| - |y|| + \left|1 - \frac{\bar{x} \cdot y}{|x||y|}\right|.$$

Finally, the set H related to the exceptional balls is the open unit ball $\mathbb{B}_{2D}$. Then $\delta(x) = 1 - |x|$. Observe that,[1] for any $x,\ y \in \mathbb{B}_{2D}$,

$$|1 - \bar{x} \cdot y| \geq 3^{-1} d(x, y) \quad \text{and} \quad |1 - \bar{x} \cdot y| \geq \delta(x).$$

Thus, the kernel $K(x, y)$ defined by (8.1.27) satisfies (8.1.23) and (8.1.24).

8.2 Boundedness of Calderón–Zygmund Operators: Equivalent Characterizations I

In this section, besides the upper doubling condition and the geometrically doubling condition, we also assume that $(\mathcal{X}, d, \mu)$ satisfies the *non-atomic condition* that $\mu(\{x\}) = 0$ for all $x \in \mathcal{X}$. The goal of this section is to show that the boundedness of a Calderón–Zygmund operator T on $L^2(\mathcal{X}, \nu)$ is equivalent to that of T on $L^p(\mathcal{X}, \nu)$

[1] See [127] or [70].

for some $p \in (1,\infty)$, or that of T from $L^1(\mathcal{X},\nu)$ to $L^{1,\infty}(\mathcal{X},\nu)$. We also prove that, if T is a Calderón–Zygmund operator bounded on $L^2(\mathcal{X},\nu)$, then its maximal operator is bounded on $L^p(\mathcal{X},\nu)$ for all $p \in (1,\infty)$ and from $\mathcal{M}(\mathcal{X})$, the *space of all complex-valued Borel measures on* $\mathcal{X}$, to $L^{1,\infty}(\mathcal{X},\nu)$.

For a measure $\omega \in \mathcal{M}(\mathcal{X})$, we denote by

$$\|\omega\| := \int_{\mathcal{X}} |d\omega(x)|$$

the *total variation of* ω and $\operatorname{supp}\omega$ the *smallest closed set* $F \subset \mathcal{X}$ *for which* ω *vanishes on* $\mathcal{X}\setminus F$ (such a smallest closed set always exists, since, by Proposition 7.1.10, $\mathcal{X}$ is separable). Assume that T is a Calderón–Zygmund operator as in (8.1.25) with K satisfying (8.1.23) and (8.1.24). For any $\omega \in \mathcal{M}(\mathcal{X})$ with bounded support and $x \in \mathcal{X} \setminus \operatorname{supp}\omega$, define

$$T\omega(x) := \int_{\mathcal{X}} K(x,y)\,d\omega(y).$$

Moreover, the *maximal operator* $T^\sharp$ associated with T is defined as follows. For every $f \in L_b^\infty(\mathcal{X},\nu)$ and $\omega \in \mathcal{M}(\mathcal{X})$, we let, for all $x \in \mathcal{X}$,

$$T^\sharp f(x) := \sup_{\epsilon\in(0,\infty)} |T_\epsilon f(x)| \tag{8.2.1}$$

and

$$T^\sharp \omega(x) := \sup_{\epsilon\in(0,\infty)} |T_\epsilon \omega(x)|,$$

where, for every $\epsilon \in (0,\infty)$,

$$T_\epsilon f(x) := \int_{d(x,y)>\epsilon} K(x,y)f(y)\,d\nu(y) \tag{8.2.2}$$

and

$$T_\epsilon \omega(x) := \int_{d(x,y)>\epsilon} K(x,y)\,d\omega(y).$$

The main result of this section reads as follows.

Theorem 8.2.1. *Let T be a Calderón–Zygmund operator with kernel K satisfying the conditions* (8.1.23) *and* (8.1.24). *Then the following statements are equivalent:*

(i) *T is bounded on $L^2(\mathcal{X},\nu)$; namely, there exists a positive constant C such that, for all $f \in L^2(\mathcal{X},\nu)$,*

$$\|Tf\|_{L^2(\mathcal{X},\nu)} \le C\|f\|_{L^2(\mathcal{X},\nu)}.$$

(ii) *T is bounded on $L^p(\mathcal{X}, \nu)$ for some $p \in (1,\infty)$; namely, there exists a positive constant $C_{(p)}$, depending on p, such that, for all $f \in L^p(\mathcal{X}, \nu)$,*

$$\|Tf\|_{L^p(\mathcal{X},\nu)} \le C_{(p)}\|f\|_{L^p(\mathcal{X},\nu)}.$$

(iii) *T is bounded from $L^1(\mathcal{X}, \nu)$ to $L^{1,\infty}(\mathcal{X}, \nu)$; namely, there exists a positive constant $\tilde{C}$ such that, for all $f \in L^1(\mathcal{X}, \nu)$,*

$$\|Tf\|_{L^{1,\infty}(\mathcal{X},\nu)} \le \tilde{C}\|f\|_{L^1(\mathcal{X},\nu)}. \tag{8.2.3}$$

To prove Theorem 8.2.1, we make some preliminaries in Sect. 8.2.1, including a Whitney-type covering lemma and a Hörmander-type inequality. In Sect. 8.2.2, we first establish a Cotlar type inequality and an endpoint estimate for T in terms of the so-called *elementary measures*. As an application of these estimates and the non-atomic assumption, we further obtain

$$\text{(i)} \Longrightarrow \text{(iii)}, \quad \text{(iii)} \Longrightarrow \text{(ii)} \quad \text{and} \quad \text{(ii)} \Longrightarrow \text{(iii)}$$

of Theorem 8.2.1. Section 8.2.3 is devoted to the proof of (iii) $\Longrightarrow$ (i) of Theorem 8.2.1, while the proof of Theorem 8.2.17 is presented in Sect. 8.2.4.

8.2.1 Preliminaries

We start with the following Whitney type covering lemma.

Lemma 8.2.2. *Let $\Omega \subsetneqq \mathcal{X}$ be a bounded open set. Then there exists a sequence $\{B_i\}_i$ of balls such that*

(w)$_{\rm i}$ *$\Omega = \cup_i B_i$ and $2B_i \subset \Omega$ for all i;*
(w)$_{\rm ii}$ *there exists a positive constant C such that, for all $x \in \mathcal{X}$,*

$$\sum_i \chi_{B_i}(x) \le C;$$

(w)$_{\rm iii}$ *for all i,*

$$(3B_i)\bigcap(\mathcal{X} \setminus \Omega) \ne \emptyset.$$

Proof. For any $x \in \Omega$, let

$$\hat{r}(x) := \frac{1}{10}\,\text{dist}\,(x, \mathcal{X} \setminus \Omega),$$

where above and in what follows, for any y and set E,

$$\operatorname{dist}(y, E) := \inf_{z \in E} d(y, z).$$

The function $\hat{r}(x)$ is strictly positive because Ω is open and the balls centered at x form a basis of neighborhood of x. Then, by Lemma 7.1.17, there exists a sequence

$$\left\{\hat{B}_i\right\}_i := \{B(x_i, \hat{r}(x_i))\}_i$$

of balls with $\{x_i\}_i \subset \Omega$ satisfying that $\{\hat{B}_i\}_i$ are pairwise disjoint and

$$\{B_i\}_i := \left\{5\hat{B}_i\right\}_i$$

forms a covering of Ω. Moreover, for each i, let $r_i := 5\hat{r}(x_i)$. Then, for any i and $y \in 2B_i$, since $\mathcal{X} \setminus \Omega$ is closed, it follows that

$$\operatorname{dist}(y, \mathcal{X} \setminus \Omega) \geq \operatorname{dist}(x_i, \mathcal{X} \setminus \Omega) - d(y, x_i) > \operatorname{dist}(x_i, \mathcal{X} \setminus \Omega) - 2r_i = 0.$$

This implies that $y \in \Omega$ and hence $2B_i \subset \Omega$, which shows $(\mathrm{w})_i$.

On the other hand, by the definition of r_i, we see that

$$3r_i = \frac{3}{2} \operatorname{dist}(x_i, \mathcal{X} \setminus \Omega)$$

and hence

$$(3B_i) \bigcap (\mathcal{X} \setminus \Omega) \neq \emptyset,$$

which shows $(\mathrm{w})_{\mathrm{iii}}$.

It remains to show $(\mathrm{w})_{\mathrm{ii}}$. To this end, we claim that, for any i and $x \in B_i \cap \Omega$,

$$\frac{1}{3} \operatorname{dist}(x, \mathcal{X} \setminus \Omega) < r_i < \operatorname{dist}(x, \mathcal{X} \setminus \Omega). \tag{8.2.4}$$

Indeed, by the fact that $\mathcal{X} \setminus \Omega$ is closed, we have

$$\operatorname{dist}(x_i, \mathcal{X} \setminus \Omega) \leq \operatorname{dist}(x, \mathcal{X} \setminus \Omega) + d(x, x_i),$$

which further implies that

$$\operatorname{dist}(x_i, \mathcal{X} \setminus \Omega) - r_i < \operatorname{dist}(x, \mathcal{X} \setminus \Omega). \tag{8.2.5}$$

Observe that, by the definition of r_i, it holds true that

$$\mathrm{dist}\,(x_i, \mathcal{X} \setminus \Omega) = 2r_i.$$

From this, together with (8.2.5), we deduce that

$$r_i < \mathrm{dist}\,(x, \mathcal{X} \setminus \Omega). \tag{8.2.6}$$

On the other hand, by this, we also have

$$\mathrm{dist}\,(x, \mathcal{X} \setminus \Omega) \le d(x, x_i) + \mathrm{dist}\,(x_i, \mathcal{X} \setminus \Omega) < 3r_i,$$

which, combined with (8.2.6), implies (8.2.4), and hence the claim holds true.

Now let $x \in \Omega$ and B_i contain x. Then, by (8.2.4), we see that

$$B_i \subset B(x, 2\,\mathrm{dist}\,(x, \mathcal{X} \setminus \Omega)).$$

Observe that $\{\frac{1}{5}B_i\}_i = \{\hat{B}_i\}_i$ are mutually disjoint. This, together with another application of (8.2.4), implies that

$$\left\{B\left(x_i, \frac{1}{15}\,\mathrm{dist}\,(x, \mathcal{X} \setminus \Omega)\right)\right\}_i$$

are also pairwise disjoint. From this and Proposition 7.1.8, we deduce that the cardinality of

$$\left\{B\left(x_i, \frac{1}{15}\,\mathrm{dist}\,(x, \mathcal{X} \setminus \Omega)\right)\right\}_i$$

contained in $B(x, 2\,\mathrm{dist}\,(x, \mathcal{X} \setminus \Omega))$ is at most $N_0 30^n$, and so is the cardinality of $\{B_i\}_i$ containing x. Thus, $(\mathrm{w})_{\mathrm{ii}}$ holds true, which completes the proof of Lemma 8.2.2. □

Let $p \in (0, \infty)$, $f \in L^p_{\mathrm{loc}}(\mathcal{X}, \nu)$ and $\omega \in \mathscr{M}(\mathcal{X})$. The *centered maximal functions* $\mathcal{M}^c_p f$ and $\mathcal{M}^c \omega$ are defined by setting, for all $x \in \mathcal{X}$,

$$\mathcal{M}^c_p f(x) := \sup_{r \in (0,\infty)} \left[\frac{1}{\nu(\overline{B}(x, 5r))} \int_{B(x,r)} |f(y)|^p \, d\nu(y)\right]^{\frac{1}{p}}$$

and

$$\mathcal{M}^c \omega(x) := \sup_{r \in (0,\infty)} \frac{\omega(\overline{B}(x, r))}{\omega(\overline{B}(x, 5r))}.$$

If $p = 1$, we *denote* $\mathcal{M}^c_1$ *simply by* $\mathcal{M}^c$, which is called the *centered Hardy–Littlewood maximal operator*.

Lemma 8.2.3. *The following statements hold true:*

(i) *Let $p \in [1, \infty)$. Then $\mathcal{M}_p^c$ is bounded on $L^q(\mathcal{X}, \nu)$ for all $q \in (p, \infty]$ and from $L^p(\mathcal{X}, \nu)$ to $L^{p,\infty}(\mathcal{X}, \nu)$;*
(ii) *Let $p \in (0, 1)$. Then $\mathcal{M}_p^c$ is bounded on $L^{1,\infty}(\mathcal{X}, \nu)$;*
(iii) *There exists a positive constant C such that, for all $\omega \in \mathcal{M}(\mathcal{X})$,*

$$\mathcal{M}^c\omega \in L^{1,\infty}(\mathcal{X}, \nu) \quad \text{and} \quad \|\mathcal{M}^c\omega\|_{L^{1,\infty}(\mathcal{X},\nu)} \leq C\|\omega\|.$$

Proof. We first show (i). Let $p \in [1, \infty)$. By Lemma 7.1.9, any disjoint collection of open balls is at most countable, so is any disjoint collection of closed balls. Moreover, by an argument similar to that used in the proof of Theorem 7.1.20, we see that $\mathcal{M}_p^c$ is bounded on $L^q(\mathcal{X}, \nu)$ for all $q \in (p, \infty]$ and bounded from $L^p(\mathcal{X}, \nu)$ to $L^{p,\infty}(\mathcal{X}, \nu)$, which implies (i).

Because the proof of (iii) is similar to that of (i), it suffices to prove (ii). To this end, fix $f \in L^{1,\infty}(\mathcal{X}, \nu)$ and split $f := f_t + f^t$, where

$$f_t := f\chi_{\{y\in\mathcal{X}:\, |f(y)|\leq t\}} \quad \text{and} \quad f^t := f\chi_{\{y\in\mathcal{X}:\, |f(y)|>t\}}.$$

Clearly

$$\left\|\mathcal{M}_p^c(f_t)\right\|_{L^\infty(\mathcal{X},\nu)} \leq \|f_t\|_{L^\infty(\mathcal{X},\nu)} \leq t$$

and

$$[\mathcal{M}_p^c(f)]^p \leq \left[\mathcal{M}_p^c(f_t)\right]^p + \left[\mathcal{M}_p^c(f^t)\right]^p.$$

From these facts and the weak type (1,1) of $\mathcal{M}^c$, we deduce that

$$\begin{aligned}
&\nu\left(\left\{x \in \mathcal{X} : |\mathcal{M}_p^c(f)(x)| > 2^{\frac{1}{p}}t\right\}\right)\\
&\quad\leq \nu\left(\left\{x \in \mathcal{X} : |\mathcal{M}_p^c(f^t)(x)|^p > t^p\right\}\right)\\
&\qquad+\nu\left(\left\{x \in \mathcal{X} : |\mathcal{M}_p^c(f_t)(x)|^p > t^p\right\}\right)\\
&\quad\leq \nu\left(\left\{x \in \mathcal{X} : |\mathcal{M}_p^c(f^t)(x)|^p > t^p\right\}\right)\\
&\quad\leq \nu\left(\{x \in \mathcal{X} : |\mathcal{M}^c(|f^t|^p)(x)| > t^p\}\right)\\
&\quad\lesssim \frac{1}{t^p}\int_{\mathcal{X}} \left|f^t(x)\right|^p\, d\nu(x).
\end{aligned}$$

On the other hand, we have

$$\begin{aligned}\int_{\mathcal{X}} |f^t(x)|^p \, d\nu(x) &= p \int_0^t s^{p-1} \nu\left(\left\{x \in \mathcal{X} : \left|f^t(x)\right| > s\right\}\right) ds \\ &\quad + p \int_t^\infty s^{p-1} \nu\left(\left\{x \in \mathcal{X} : \left|f^t(x)\right| > s\right\}\right) ds \\ &= t^p \nu\left(\{x \in \mathcal{X} : \ |f(x)| > t\}\right) \\ &\quad + \int_t^\infty p s^{p-1} \nu\left(\{x \in \mathcal{X} : |f(x)| > s\}\right) ds \\ &\le t^{p-1} \|f\|_{L^{1,\infty}(\mathcal{X},\nu)} + \|f\|_{L^{1,\infty}(\mathcal{X},\nu)} \int_t^\infty p s^{p-2} \, ds \\ &\lesssim t^{p-1} \|f\|_{L^{1,\infty}(\mathcal{X},\nu)}.\end{aligned}$$

By these facts, we conclude that

$$t\nu\left(\left\{x \in \mathcal{X} : \left|\mathcal{M}_p^c(f)(x)\right| > 2^{\frac{1}{p}} t\right\}\right) \lesssim \|f\|_{L^{1,\infty}(\mathcal{X},\nu)}.$$

Taking supremum over $t \in (0,\infty)$ on both sides, we know that

$$\left\|\mathcal{M}_p^c(f)\right\|_{L^{1,\infty}(\mathcal{X},\nu)} \lesssim \|f\|_{L^{1,\infty}(\mathcal{X},\nu)},$$

which implies (ii), and hence completes the proof of Lemma 8.2.3. □

Lemma 8.2.4. *Let $\eta \in \mathscr{M}(\mathcal{X})$ satisfy that $\eta(\mathcal{X}) = 0$ and $\operatorname{supp} \eta \subset \overline{B}(x,\rho)$ for some $\rho \in (0,\infty)$ and $x \in \mathcal{X}$, and T be a Calderón–Zygmund operator with kernel K satisfying* (8.1.23) *and* (8.1.24)*. Then there exists a positive constant C, independent of η, x and ρ, such that, for all nonnegative Borel measures ν on $\mathcal{X}$,*

$$\int_{\mathcal{X}\setminus B(x,2\rho)} |T\eta(y)| \, d\nu(y) \le C \|\eta\| \mathcal{M}^c \nu(x). \tag{8.2.7}$$

Moreover, for any $p \in [1,\infty)$ and $f \in L^p_{\mathrm{loc}}(\mathcal{X},\nu)$, it holds true that

$$\int_{\mathcal{X}\setminus B(x,2\rho)} |T\eta(y)||f(y)| \, d\nu(y) \le C \|\eta\| \mathcal{M}_p^c f(x) \tag{8.2.8}$$

and

$$\int_{\mathcal{X}\setminus B(x,2\rho)} |T\eta(y)| \, d\nu(y) \le C \|\eta\|, \tag{8.2.9}$$

where C is a positive constant, independent of η, x, ρ and f.

Proof. By similarity, we only prove (8.2.7). By $\eta(\mathcal{X}) = 0$, $\operatorname{supp} \eta \subset \overline{B}(x, \rho)$ and (8.1.24), we conclude that, for any $y \in \mathcal{X} \setminus B(x, 2\rho)$,

$$\begin{aligned} |T\eta(y)| &= \left| \int_{\overline{B}(x,\rho)} K(y, \tilde{x})\, d\eta(\tilde{x}) \right| \\ &= \left| \int_{\overline{B}(x,\rho)} [K(y, \tilde{x}) - K(y, x)]\, d\eta(\tilde{x}) \right| \\ &\le \|\eta\| \sup_{\tilde{x} \in \overline{B}(x,\rho)} |K(y, \tilde{x}) - K(y, x)| \\ &\lesssim \|\eta\| \left[\frac{\rho}{d(x, y)} \right]^\tau \frac{1}{\lambda(x, d(x, y))}. \end{aligned}$$

Therefore, by (7.1.1), we know that

$$\begin{aligned} \int_{\mathcal{X} \setminus B(x, 2\rho)} |T\eta(y)|\, d\nu(y) &\lesssim \|\eta\| \int_{\mathcal{X} \setminus B(x, 2\rho)} \left[\frac{\rho}{d(x, y)} \right]^\tau \frac{1}{\lambda(x, d(x, y))}\, d\nu(y) \\ &\lesssim \|\eta\| \sum_{k=1}^\infty \int_{B(x, 2^{k+1}\rho) \setminus B(x, 2^k \rho)} \frac{1}{2^{k\tau}} \frac{1}{\lambda(x, 2^k \rho)}\, d\nu(y) \\ &\lesssim \|\eta\| \sum_{k=1}^\infty \frac{1}{2^{k\tau}} \frac{\nu(B(x, 2^{k+1}\rho))}{\nu(B(x, 5 \cdot 2^{k+1}\rho))} \\ &\lesssim \|\eta\| \sum_{k=1}^\infty \frac{1}{2^{k\tau}} \mathcal{M}^c \nu(x) \\ &\lesssim \|\eta\| \mathcal{M}^c \nu(x), \end{aligned}$$

which completes the proof of Lemma 8.2.4. □

Arguing as in the proof of Proposition 5.3.1, we have the following conclusion, the details being omitted.

Proposition 8.2.5. *Let $p \in (1, \infty)$. If $\{T_\epsilon\}_{\epsilon \in (0,\infty)}$ as in (8.2.2) is bounded on $L^p(\mathcal{X}, \nu)$ uniformly on $\epsilon \in (0, \infty)$, then there exists an operator $\tilde{T}$ which is the weak limit as $\epsilon \to 0$ of some subsequence of the uniformly bounded operators $\{T_\epsilon\}_{\epsilon \in (0,\infty)}$. Moreover, the operator $\tilde{T}$ is also bounded on $L^p(\mathcal{X}, \nu)$ and satisfies that, for $f \in L^p(\mathcal{X}, \nu)$ with compact support and ν-almost every $x \notin \operatorname{supp} f$,*

$$\tilde{T} f(x) = \int_{\mathcal{X}} K(x, y) f(y)\, d\nu(y).$$

8.2.2 *Proof of Theorem 8.2.1, Part I*

This section is devoted to the proof of the implicity

$$(\mathrm{i}) \Longrightarrow (\mathrm{iii}), \quad (\mathrm{iii}) \Longrightarrow (\mathrm{ii}) \quad \text{and} \quad (\mathrm{ii}) \Longrightarrow (\mathrm{iii})$$

of Theorem 8.2.1. To this end, we first establish an endpoint estimate for T via the so-called elementary measures which are finite linear combinations of unit point masses with positive coefficients.

Lemma 8.2.6. *Let T be a Calderón–Zygmund operator with kernel K satisfying* (8.1.23) *and* (8.1.24)*, which is bounded on $L^2(\mathcal{X}, \nu)$. Then there exist positive constants C and c such that, for any $f \in L_b^\infty(\mathcal{X}, \nu)$ and $x \in \operatorname{supp}\nu$,*

$$T^\sharp(f)(x) \le C\mathcal{M}^c(Tf)(x) + c\mathcal{M}_2^c(f)(x). \tag{8.2.10}$$

Proof. Let $x \in \operatorname{supp}\nu$, $r \in (0, \infty)$, and

$$r_j := 5^j r \quad \text{and} \quad \nu_j := \nu(\overline{B}(x, r_j))$$

for all $j \in \mathbb{Z}_+$. We claim that there exists some $j \in \mathbb{N}$ such that $\nu_{j+1} \le 4C_{(\lambda)}^6 \nu_{j-1}$. For otherwise, by (7.1.1), we would conclude that, for every $j \in \mathbb{N}$,

$$\begin{aligned}
\nu_0 &< \left(4C_{(\lambda)}^6\right)^{-j} \nu_{2j} \\
&= \left(4C_{(\lambda)}^6\right)^{-j} \nu\left(\overline{B}\left(x, r_{2j}\right)\right) \\
&\lesssim \left(4C_{(\lambda)}^6\right)^{-j} \lambda\left(x, 5^{2j} r\right) \\
&\lesssim 5^{-j} \lambda(x, r).
\end{aligned}$$

Letting $j \to 0$, we know that $\nu(\overline{B}(x, r)) = 0$, which contradicts to the fact that $\nu(\overline{B}(x, r)) > 0$ for each $r \in (0, \infty)$ and each $x \in \operatorname{supp}\nu$. Thus, the claim holds true.

Let $k \in \mathbb{N}$ be the *smallest integer* such that

$$\nu_{k+1} \le 4C_{(\lambda)}^6 \nu_{k-1} \quad \text{and} \quad R := r_{k-1} := 5^{k-1} r.$$

Then we find that

$$\nu\left(\overline{B}\left(x, 25R\right)\right) \lesssim \nu\left(\overline{B}\left(x, R\right)\right). \tag{8.2.11}$$

Observe that, for all $j \in \{1, \ldots, k\}$, we know that

$$\nu_{j+1} \le (2C_{(\lambda)}^3)^{j+2-k} \nu_k \quad \text{and} \quad \lambda(x, r_k) \le [C_{(\lambda)}]^{\max\{0, k-j-1\}} \lambda(x, r_{j+1}).$$

Let $f \in L_b^\infty(\mathcal{X}, \nu)$. From this, (8.1.23), (7.1.1) and the Hölder inequality, we then deduce that

$$\begin{aligned}
|T_r f(x) - T_{5R} f(x)| &\le \int_{\overline{B}(x,5R)\setminus\overline{B}(x,r)} |K(x,y)||f(y)|\,d\nu(y)\\
&= \sum_{j=1}^{k} \int_{\overline{B}(x,r_j)\setminus\overline{B}(x,r_{j-1})} |K(x,y)||f(y)|\,d\nu(y)\\
&\lesssim \sum_{j=1}^{k} \frac{\nu(\overline{B}(x,r_{j+1}))}{\lambda(x,r_{j+1})} \mathcal{M}^c(f)(x)\\
&\lesssim \sum_{j=1}^{k} 2^{j-k} \mathcal{M}^c(f)(x)\\
&\lesssim \mathcal{M}^c(f)(x).
\end{aligned} \tag{8.2.12}$$

Let

$$V_R(x) := \frac{1}{\nu(\overline{B}(x,R))} \int_{\overline{B}(x,R)} Tf(y)\,d\nu(y).$$

Then we have

$$|V_R(x)| \lesssim \mathcal{M}^c(Tf)(x). \tag{8.2.13}$$

On the other hand, observe that

$$\begin{aligned}
T_{5R} f(x) &= \int_{\mathcal{X}\setminus\overline{B}(x,5R)} K(x,y) f(y)\,d\nu(y)\\
&= \int_{\mathcal{X}} K(x,y)\chi_{\mathcal{X}\setminus\overline{B}(x,5R)}(y) f(y)\,d\nu(y)\\
&= T\left(f\chi_{\mathcal{X}\setminus\overline{B}(x,5R)}\right)(x)\\
&= \left\langle \delta_x, T\left(f\chi_{\mathcal{X}\setminus\overline{B}(x,5R)}\right)\right\rangle\\
&= \left\langle T^*\delta_x, f\chi_{\mathcal{X}\setminus\overline{B}(x,5R)}\right\rangle\\
&= \int_{\mathcal{X}\setminus\overline{B}(x,5R)} T^*\delta_x(y) f(y)\,d\nu(y),
\end{aligned}$$

where above and in what follows, δ_x denotes the *Dirac measure at* x. By writing

$$\begin{aligned}V_R(x) &= \frac{1}{\nu(\overline{B}(x,R))}\int_{\mathcal{X}} \chi_{\overline{B}(x,R)}(y)T(f)(y)\,d\nu(y)\\ &= \frac{1}{\nu(\overline{B}(x,R))}\int_{\mathcal{X}} \chi_{\overline{B}(x,R)}(y)T\left(f\chi_{\overline{B}(x,5R)}\right)(y)\,d\nu(y)\\ &\quad+\int_{\mathcal{X}} T^*\left(\frac{\chi_{\overline{B}(x,R)}}{\nu(\overline{B}(x,R))}\right)(y)f(y)\chi_{\mathcal{X}\setminus\overline{B}(x,5R)}(y)\,d\nu(y),\end{aligned}$$

we then see that

$$\begin{aligned}&|T_{5R}f(x) - V_R(x)|\\ &\quad\le \left|\int_{\mathcal{X}\setminus\overline{B}(x,5R)} T^*\left(\delta_x - \frac{\chi_{\overline{B}(x,R)}}{\nu(\overline{B}(x,R))}\,d\nu\right)(y)f(y)\,d\nu(y)\right|\\ &\qquad+\left|\frac{1}{\nu(\overline{B}(x,R))}\int_{\mathcal{X}}\left[Tf\chi_{\overline{B}(x,5R)}(y)\right]\chi_{\overline{B}(x,R)}(y)\,d\nu(y)\right|\\ &\quad=: \mathrm{L}_1 + \mathrm{L}_2.\end{aligned} \tag{8.2.14}$$

By (8.2.8), we have

$$\mathrm{L}_1 \lesssim \mathcal{M}^c(f)(x).$$

From the Hölder inequality, the boundedness of T on $L^2(\mathcal{X}, \nu)$ and (8.2.11), we further deduce that

$$\begin{aligned}\mathrm{L}_2 &\le \left[\nu\left(\overline{B}(x,R)\right)\right]^{-\frac12}\left[\int_{\mathcal{X}}\left|T\left(f\chi_{\overline{B}(x,5R)}\right)(y)\right|^2 d\nu(y)\right]^{\frac12}\\ &\lesssim \left[\nu\left(\overline{B}(x,R)\right)\right]^{-\frac12}\left[\int_{\overline{B}(x,5R)}|f(y)|^2\,d\nu(y)\right]^{\frac12}\\ &\lesssim \mathcal{M}_2^c(f)(x).\end{aligned}$$

Then, combining the estimates for L_1 and L_2, and using (8.2.14), (8.2.13) and (8.2.12), we conclude that, for any $r \in (0, \infty)$,

$$\begin{aligned}|T_r f(x)| &\le |T_r f(x) - T_{5R}f(x)| + |T_{5R}f(x) - V_R(x)| + |V_R(x)|\\ &\lesssim \mathcal{M}_2^c(f)(x) + \mathcal{M}^c(Tf)(x).\end{aligned}$$

Taking the supremum over $r \in (0, \infty)$, we obtain (8.2.10), and hence complete the proof of Lemma 8.2.6. □

Remark 8.2.7. We point out that, if we replace the boundedness of T on $L^2(\mathcal{X}, \nu)$ in Lemma 8.2.6 by the boundedness of T on $L^q(\mathcal{X}, \nu)$ for some $q \in (1, \infty)$, then (8.2.10) still holds true with $\mathcal{M}_2^c$ replaced by $\mathcal{M}_q^c$.

Definition 8.2.8. A subset A of a measure space $(\mathcal{X}, \nu)$ is called an *atom* if $\nu(A) > 0$ and each $B \subset A$ has measure either equal to zero or equal to $\nu(A)$. A measure space $(\mathcal{X}, \nu)$ is said to be *non-atomic* if it does not contain any atom.

We know, from Definition 8.2.8, that $\mathcal{X}$ is non-atomic if and only if, for any $A \subset \mathcal{X}$ with $\nu(A) > 0$, there exists a proper subset $B \subsetneqq A$ with $\nu(B) > 0$ and $\nu(A \setminus B) > 0$. By this, it is straightforward that if $\nu(\{x\}) = 0$ for any $x \in \mathcal{X}$, then $(\mathcal{X}, \nu)$ is a non-atomic space. Moreover, it is known that, if $(\mathcal{X}, \nu)$ is a non-atomic measure space, then, for any sets $A_0 \subset A_1 \subset \mathcal{X}$ such that $0 < \nu(A_1) < \infty$ and $\nu(A_0) \le t \le \nu(A_1)$ for some $t \in (0, \infty)$, there exists a set E such that $A_0 \subset E \subset A_1$ and $\nu(E) = t$.[2]

We say that ω is an *elementary measure* if it is of the form

$$\omega := \sum_{i=1}^{N} \alpha_i \delta_{x_i},$$

where $N \in \mathbb{N}$, δ_{x_i} is the Dirac measure at some $x_i \in \mathcal{X}$ and $\alpha_i \in (0, \infty)$ for $i \in \{1, \ldots, N\}$. To prove Theorem 8.2.1, we first establish an endpoint estimate for T on these elementary measures.

Theorem 8.2.9. *Let T be a Calderón–Zygmund operator with kernel K satisfying* (8.1.23) *and* (8.1.24)*, which is bounded on $L^2(\mathcal{X}, \nu)$. Then there exist positive constants $\tilde{c}_1$ and $\tilde{c}_2$ such that, for all elementary measures ω,*

$$\|T\omega\|_{L^{1,\infty}(\mathcal{X},\nu)} \le \left[\tilde{c}_1 + \tilde{c}_2 \|T\|_{L^2(\mathcal{X},\nu)\to L^2(\mathcal{X},\nu)}\right] \|\omega\|. \tag{8.2.15}$$

Proof. Without loss of generality, we may normalize ω such that

$$\|\omega\| = \sum_{i=1}^{N} \alpha_i = 1,$$

and hence we only need to prove

$$\|T\omega\|_{L^{1,\infty}(\mathcal{X},\nu)} \le \tilde{c}_1 + \tilde{c}_2 \|T\|_{L^2(\mathcal{X},\nu)\to L^2(\mathcal{X},\nu)}. \tag{8.2.16}$$

Since, for $t \in (0, 1/\nu(\mathcal{X})]$, we have

$$t\nu(\{x \in \mathcal{X} : |T\omega(x)| > t\}) \le t\nu(\mathcal{X}) \le 1,$$

---

[2]See [40, p. 65].

it remains to consider the case $t \in (1/\nu(\mathcal{X}), \infty)$. Let $\overline{B}(x_1, \rho_1)$ be the *smallest closed ball* such that

$$\nu(\overline{B}(x_1, \rho_1)) \geq \alpha_1/t.$$

Indeed, since the function $\rho \to \nu(\overline{B}(x, \rho))$ is increasing and continuous from the right, and

$$\nu\left(\overline{B}(x, \rho)\right) \geq 1/t \geq \alpha_1/t \text{ for sufficiently large } \rho \in (0, \infty),$$

it follows that such ρ_1 exists and is strictly positive. Then

$$\nu(B(x_1, \rho_1)) = \lim_{\rho \to \rho_1 - 0} \nu(\overline{B}(x_1, \rho)) \leq \frac{\alpha_1}{t}.$$

Since $(\mathcal{X}, \nu)$ is non-atomic, we find a Borel set E_1 such that

$$B(x_1, \rho_1) \subset E_1 \subset \overline{B}(x_1, \rho_1)$$

and $\nu(E_1) = \frac{\alpha_1}{t}$.

Let $\overline{B}(x_2, \rho_2)$ be the *smallest closed ball* such that

$$\nu(\overline{B}(x_2, \rho_2) \setminus E_1) \geq \alpha_2/t.$$

Similarly, for the corresponding open ball $B(x_2, \rho_2)$, we have

$$\nu(B(x_2, \rho_2) \setminus E_1) \leq \alpha_2/t$$

and henceforth find a Borel set E_2 with the property:

$$(B(x_2, \rho_2) \setminus E_1) \subset E_2 \subset \left(\overline{B}(x_2, \rho_2) \setminus E_1\right)$$

and $\nu(E_2) = \frac{\alpha_2}{t}$.

Repeating the process, for $i \in \{3, \ldots, N\}$, we have $\overline{B}(x_i, \rho_i)$ and E_i such that $\overline{B}(x_i, \rho_i)$ is the *smallest closed ball* satisfying that

$$\nu\left(\overline{B}(x_i, \rho_i) \setminus \bigcup_{l=1}^{i-1} E_l\right) \geq \alpha_i/t,$$

$$\left(B(x_i, \rho_i) \setminus \bigcup_{l=1}^{i-1} E_l\right) \subset E_i \subset \left(\overline{B}(x_i, \rho_i) \setminus \bigcup_{l=1}^{i-1} E_l\right)$$

and $\nu(E_i) = \frac{\alpha_i}{t}$. Let

$$E = \bigcup_{i=1}^{N} E_i.$$

Then, by the fact that $\sum_{i=1}^{N} \alpha_i = 1$, together with the choices of $\{B(x_i, \rho_i)\}_{i=1}^{N}$ and $\{E_i\}_{i=1}^{N}$, we see that

$$\bigcup_{i=1}^{N} B(x_i, \rho_i) \subset E \subset \bigcup_{i=1}^{N} \overline{B}(x_i, \rho_i)$$

and $\nu(E) = \frac{1}{t}$.

Outside E, let us compare $T\omega$ to $t\sigma$, where

$$\sigma := \sum_{i=1}^{N} \chi_{\mathcal{X}\setminus\overline{B}(x_i, 2\rho_i)} T(\chi_{E_i}\, d\nu).$$

We have

$$\begin{aligned} T\omega - t\sigma &= T\left(\sum_{i=1}^{N} \alpha_i \delta_{x_i}\right) - t \sum_{i=1}^{N} \chi_{\mathcal{X}\setminus\overline{B}(x_i, 2\rho_i)} T(\chi_{E_i}\, d\nu) \\ &= \sum_{i=1}^{N} \left[\alpha_i\, T\delta_{x_i} - t\chi_{\mathcal{X}\setminus\overline{B}(x_i, 2\rho_i)} T(\chi_{E_i}\, d\nu)\right] \\ &=: \sum_{i=1}^{N} \varphi_i. \end{aligned} \tag{8.2.17}$$

Notice that, for any i,

$$\begin{aligned} &\int_{\mathcal{X}\setminus E} |\varphi_i(x)|\, d\nu(x) \\ &= \int_{\mathcal{X}\setminus \bigcup_{i=1}^{N} E_i} \left|\alpha_i\, T\delta_{x_i}(x) - t\chi_{\mathcal{X}\setminus\overline{B}(x_i, 2\rho_i)}(x) T(\chi_{E_i}\, d\nu)(x)\right| d\nu(x) \\ &\le \int_{\mathcal{X}\setminus\overline{B}(x_i, 2\rho_i)} \left|\alpha_i\, T\delta_{x_i}(x) - t\chi_{\mathcal{X}\setminus\overline{B}(x_i, 2\rho_i)}(x) T(\chi_{E_i}\, d\nu)(x)\right| d\nu(x) \\ &\quad + \int_{\overline{B}(x_i, 2\rho_i)\setminus B(x_i, \rho_i)} \cdots \end{aligned}$$

$$
\begin{aligned}
&= \int_{\mathcal{X}\setminus\overline{B}(x_i,2\rho_i)} |T(\alpha_i\delta_{x_i} - t\chi_{E_i}\,d\nu)(x)|\,d\nu(x) \\
&\quad + \int_{\overline{B}(x_i,2\rho_i)\setminus B(x_i,\rho_i)} \alpha_i|T\delta_{x_i}(x)|\,d\nu(x) \\
&=: \mathrm{J}_1 + \mathrm{J}_2.
\end{aligned} \tag{8.2.18}
$$

For each i, using (8.2.9) and $\nu(E_i) = \frac{\alpha_i}{t}$, we find that

$$
\mathrm{J}_1 \lesssim \|\alpha_i\delta_{x_i} - t\chi_{E_i}\,d\nu\| \lesssim \alpha_i.
$$

Moreover, from (8.1.23), (7.1.3) and (7.1.1), we deduce that

$$
\begin{aligned}
\mathrm{J}_2 &\lesssim \int_{\overline{B}(x_i,2\rho_i)\setminus B(x_i,\rho_i)} \frac{\alpha_i}{\lambda(x, d(x,x_i))}\,d\nu(x) \\
&\lesssim \int_{\overline{B}(x_i,2\rho_i)\setminus B(x_i,\rho_i)} \frac{\alpha_i}{\lambda(x_i, d(x,x_i))}\,d\nu(x) \\
&\lesssim \alpha_i \frac{\nu(\overline{B}(x_i, 2\rho_i))}{\lambda(x_i,\rho_i)} \\
&\lesssim \alpha_i.
\end{aligned}
$$

By the estimates for J_1 and J_2, and (8.2.18), we conclude that

$$
\int_{\mathcal{X}\setminus E} |\varphi_i(x)|\,d\nu(x) \lesssim \alpha_i,
$$

which, together with (8.2.17) and the fact that $\sum_{i=1}^N \alpha_i = 1$, further implies that there exists a positive constant $\tilde{c}_3$ such that

$$
\int_{\mathcal{X}\setminus E} |T\omega(x) - t\sigma(x)|\,d\nu(x) \le \sum_{i=1}^N \int_{\mathcal{X}\setminus E} |\varphi_i(x)|\,d\nu(x) \le \tilde{c}_3. \tag{8.2.19}
$$

Via (8.2.19), to accomplish the proof of Theorem 8.2.9, it suffices to show that there exist positive constants C_4 and C_5 such that

$$
C_6 := C_4 + C_5\|T\|_{L^2(\mathcal{X},\nu)\to L^2(\mathcal{X},\nu)}
$$

satisfying that

$$
\nu(\{x \in \mathcal{X} : |\sigma(x)| > C_6\}) \le \frac{2}{t}. \tag{8.2.20}
$$

Indeed, assume that (8.2.20) holds true for the moment. Then, from $\nu(E) = \frac{1}{t}$, (8.2.19) and (8.2.20), we deduce that

$$\begin{aligned}
&\nu(\{x \in \mathcal{X} : |T\omega(x)| > (\tilde{c}_3 + C_6)t\}) \\
&\quad \le \nu(\{x \in \mathcal{X} \setminus E : |T\omega(x)| > (\tilde{c}_3 + C_6)t\}) + \nu(E) \\
&\quad \le \nu(\{x \in \mathcal{X} \setminus E : |T\omega(x) - t\sigma(x)| > \tilde{c}_3 t\}) \\
&\qquad + \nu(\{x \in \mathcal{X} : |\sigma(x)| > C_6\}) + \nu(E) \\
&\quad \le \frac{4}{t}.
\end{aligned}$$

This implies (8.2.16), and hence finishes the proof of Theorem 8.2.9, up to the verification of (8.2.20), which we do in the following lemma. □

Lemma 8.2.10. *The estimate* (8.2.20) *holds true.*

Proof. We first claim that there exist C_4 and C_5 such that, for any set F with $\nu(F) = \frac{1}{t}$,

$$\left| \int_{\mathcal{X}} \sigma(x)\chi_F(x)\,d\nu(x) \right| \le \frac{C_6}{t}. \tag{8.2.21}$$

Indeed, let F be such a set. Then the definition of σ implies that

$$\begin{aligned}
&\int_{\mathcal{X}} \sigma(x)\chi_F(x)\,d\nu(x) \\
&\quad = \sum_{i=1}^N \int_{\mathcal{X}} T\chi_{E_i}(x)\chi_{F\setminus\overline{B}(x_i,2\rho_i)}(x)\,d\nu(x) \\
&\quad = \sum_{i=1}^N \int_{\mathcal{X}} \chi_{E_i}(x)T^*\chi_{F\setminus\overline{B}(x_i,2\rho_i)}(x)\,d\nu(x).
\end{aligned} \tag{8.2.22}$$

By (7.1.3) and (7.1.1), it is easy to see that, for all $x \in E_i \subset \overline{B}(x_i, \rho_i)$ and $y \in \overline{B}(x_i, 2\rho_i) \setminus \overline{B}(x, \rho_i)$,

$$\lambda(x_i, \rho_i) \lesssim \lambda(y, d(x, y)),$$

which, together with (8.1.23) and (7.1.3), further implies that, for all $x \in E_i \subset \overline{B}(x_i, \rho_i)$,

$$\begin{aligned}
\left| T^*\chi_{F\setminus\overline{B}(x_i,2\rho_i)}(x) - T^*\chi_{F\setminus\overline{B}(x,\rho_i)}(x) \right| &\le \int_{\overline{B}(x_i,2\rho_i)\setminus\overline{B}(x,\rho_i)} |K(y,x)|\,d\nu(y) \\
&\lesssim \int_{\overline{B}(x_i,2\rho_i)\setminus\overline{B}(x,\rho_i)} \frac{1}{\lambda(y, d(x,y))}\,d\nu(y) \\
&\lesssim \frac{\nu(\overline{B}(x_i, 2\rho_i))}{\lambda(x_i, \rho_i)} \\
&\lesssim 1.
\end{aligned}$$

By this, combined with the fact that

$$|T^*\chi_{F\setminus\overline{B}(x,\rho_i)}(x)| \le (T^*)^\sharp\chi_F(x)$$

and Lemma 8.2.6, we conclude that, for all $x \in E_i \subset \overline{B}(x_i, \rho_i)$,

$$\begin{aligned}
&\left|T^*\chi_{F\setminus\overline{B}(x_i,2\rho_i)}(x)\right| \\
&\quad\le \left|T^*\chi_{F\setminus\overline{B}(x_i,2\rho_i)}(x) - T^*\chi_{F\setminus\overline{B}(x,\rho_i)}(x)\right| + \left|T^*\chi_{F\setminus\overline{B}(x,\rho_i)}(x)\right| \\
&\quad\lesssim 1 + (T^*)^\sharp\chi_F(x) \\
&\quad\lesssim 1 + \mathcal{M}^c(T^*\chi_F)(x).
\end{aligned}$$

Furthermore, by this, (8.2.22), $E = \cup_{i=1}^N E_i$ (disjoint union) and $\nu(E) = \frac{1}{t}$, we know that

$$\begin{aligned}
\left|\int_{\mathcal{X}} \sigma(x)\chi_F(x)\,d\nu(x)\right| &\le \sum_{i=1}^N \left|\int_{\mathcal{X}} \chi_{E_i}(x)\left[T^*\chi_{F\setminus\overline{B}(x_i,2\rho_i)}\right](x)\,d\nu(x)\right| \\
&\lesssim \sum_{i=1}^N \int_{\mathcal{X}} \chi_{E_i}(x)[1 + \mathcal{M}^c(T^*\chi_F)(x)]\,d\nu(x) \\
&\sim \frac{1}{t} + \int_{\mathcal{X}} \chi_E(x)\mathcal{M}^c(T^*\chi_F)(x)\,d\nu(x). \qquad (8.2.23)
\end{aligned}$$

Since T is bounded on $L^2(\mathcal{X}, \nu)$, by duality, we see that T^* is also bounded on $L^2(\mathcal{X}, \nu)$ and

$$\|T^*\|_{L^2(\mathcal{X},\nu)\to L^2(\mathcal{X},\nu)} = \|T\|_{L^2(\mathcal{X},\nu)\to L^2(\mathcal{X},\nu)}.$$

From this fact, Lemma 8.2.3(i), $\nu(F) = \frac{1}{t} = \nu(E)$ and the Hölder inequality, we further deduce that

$$\begin{aligned}
&\int_{\mathcal{X}} \chi_E(x)\mathcal{M}^c(T^*\chi_F)(x)\,d\nu(x) \\
&\quad\le \|\chi_E\|_{L^2(\mathcal{X},\nu)}\|\mathcal{M}^c(T^*\chi_F)\|_{L^2(\mathcal{X},\nu)} \\
&\quad\le \|\chi_E\|_{L^2(\mathcal{X},\nu)}\|\mathcal{M}^c\|_{L^2(\mathcal{X},\nu)\to L^2(\mathcal{X},\nu)}\|T^*\|_{L^2(\mathcal{X},\nu)\to L^2(\mathcal{X},\nu)}\|\chi_F\|_{L^2(\mathcal{X},\nu)} \\
&\quad= \frac{1}{t}\|\mathcal{M}^c\|_{L^2(\mathcal{X},\nu)\to L^2(\mathcal{X},\nu)}\|T\|_{L^2(\mathcal{X},\nu)\to L^2(\mathcal{X},\nu)},
\end{aligned}$$

which, together with (8.2.23), implies that there exist C_4 and C_5 satisfying (8.2.21). Therefore the claim (8.2.21) holds true.

Suppose that

$$\nu(\{x \in \mathcal{X} : |\sigma(x)| > C_6\}) > 2/t.$$

Then either

$$\nu(\{x \in \mathcal{X} : \sigma(x) > C_6\}) > \frac{1}{t} \tag{8.2.24}$$

or

$$\nu(\{x \in \mathcal{X} : \sigma(x) < -C_6\}) > \frac{1}{t}.$$

Without loss of generality, we may only consider (8.2.24) by similarity. Pick some set $F \subset \mathcal{X}$ with $\nu(F) = 1/t$ such that $\sigma(x) > C_6$ everywhere on F (such F exists because of the statement below Definition 8.2.8). Then, apparently,

$$\int_{\mathcal{X}} \sigma(x)\chi_F(x)\,d\nu(x) > \frac{C_6}{t}. \tag{8.2.25}$$

Thus, we obtain a contradiction by combining (8.2.21) with (8.2.25), which implies (8.2.20), and hence completes the proof of Lemma 8.2.10. □

Remark 8.2.11. (i) Theorem 8.2.9 also holds true with finite linear combinations of Dirac measures with arbitrary real coefficients. Indeed, every such measure ω can be represented as $\omega = \omega_+ - \omega_-$, where ω_+ and ω_- are finite linear combinations of Dirac measures with positive coefficients and

$$\|\omega\| = \|\omega_+\| + \|\omega_-\|.$$

Therefore,

$$\|T\omega\|_{L^{1,\infty}(\mathcal{X},\nu)} \le 2(\tilde{c}_1 + \tilde{c}_2\|T\|_{L^2(\mathcal{X},\nu)\to L^2(\mathcal{X},\nu)})\|\omega\|.$$

(ii) If we replace the assumption of Theorem 8.2.9 that T is bounded on $L^2(\mathcal{X}, \nu)$ by that T is bounded on $L^q(\mathcal{X}, \nu)$ for some $q \in (1, \infty)$, then via a slight modification of the proof Theorem 8.2.9, we have (8.2.15) with $\|T\|_{L^2(\mathcal{X},\nu)\to L^2(\mathcal{X},\nu)}$ replaced by $\|T\|_{L^q(\mathcal{X},\nu)\to L^q(\mathcal{X},\nu)}$.

Proof of Theorem 8.2.1, Part I. In this part, we show that (i) of Theorem 8.2.1 implies (ii) and (iii) of Theorem 8.2.1 and that (ii) of Theorem 8.2.1 implies (iii) of Theorem 8.2.1.

We first assume that (i) holds true and show that (ii) and (iii) hold true. Then, by the Marcinkiewicz interpolation theorem and a duality argument, we obtain (ii) via (iii). Therefore, we only need to prove (iii). To this end, observe that, for any $f \in L^1(\mathcal{X}, \nu)$, it holds true that $f = f^+ - f^-$, where

$$f^+ := \max\{f, 0\} \ge 0 \quad \text{and} \quad f^- := \max\{-f, 0\} \ge 0.$$

Moreover, by Proposition 7.1.19, we see that, for any $f \in L^1(\mathcal{X}, \nu)$ and $f \geq 0$, there exist $\{f_j\}_{j\in\mathbb{N}} \subset C_b(\mathcal{X})$ and $f_j \geq 0$ for all $j \in \mathbb{N}$ such that

$$\|f_j - f\|_{L^1(\mathcal{X},\nu)} \to 0$$

as $j \to \infty$. By these observations, combined with the linear property of T, we see that, to show (iii), it suffices to prove that (8.2.3) holds true for all $f \in C_b(\mathcal{X})$ and $f \geq 0$.

Let $t \in (0, \infty)$,

$$G := \{x \in \mathcal{X} : f(x) > t\}, \quad f^t := f\chi_G \quad \text{and} \quad f_t := f\chi_{\mathcal{X}\setminus G}.$$

Then $Tf = Tf^t + Tf_t$. Notice that

$$\int_{\mathcal{X}} [f_t(x)]^2 \, d\nu(x) \leq t \int_{\mathcal{X}} f_t(x) \, d\nu(x) \leq t\|f\|_{L^1(\mathcal{X},\nu)}.$$

By this and the boundedness of T on $L^2(\mathcal{X}, \nu)$, we find that

$$\int_{\mathcal{X}} |Tf_t(x)|^2 \, d\nu(x) \leq \|T\|^2_{L^2(\mathcal{X},\nu)\to L^2(\mathcal{X},\nu)} t\|f\|_{L^1(\mathcal{X},\nu)},$$

which implies that

$$\nu\left(\{x \in \mathcal{X} : |Tf_t(x)| > t\|T\|_{L^2(\mathcal{X},\nu)\to L^2(\mathcal{X},\nu)}\}\right) \leq \frac{\|f\|_{L^1(\mathcal{X},\nu)}}{t}. \tag{8.2.26}$$

We now estimate Tf^t. Since, by $f \in C_b(\mathcal{X})$, G is a bounded open set, we know that, by Lemma 8.2.2, there exists a sequence $\{B_i\}_i$ of balls with finite overlap such that $G = \cup_i B_i$ and $2B_i \subset G$ for all i. Without loss of generality, we may assume the cardinality of $\{B_i\}_i$ is just $\mathbb{N}$. Then the fact that $\{B_i\}_{i\in\mathbb{N}}$ has the finite overlap implies that

$$f^t = \sum_{i\in\mathbb{N}} f \frac{\chi_{B_i}}{\sum_{j\in\mathbb{N}} \chi_{B_j}} =: \sum_{i\in\mathbb{N}} f_i.$$

It is easy to see that $f_i \geq 0$ for all $i \in \mathbb{N}$. For any $N \in \mathbb{N}$ and $i \in \{1, \ldots, N\}$, define

$$f^{(N)} := \sum_{i=1}^{N} f_i \quad \text{and} \quad \alpha_i := \int_{\mathcal{X}} f_i(y) \, d\nu(y).$$

Then $\alpha_i \geq 0$ for all $i \in \mathbb{N}$. By $G = \cup_{i\in\mathbb{N}} B_i$ and the finite overlap property of $\{B_i\}_{i\in\mathbb{N}}$, we have

$$\sum_{i=1}^{\infty} \alpha_i \leq \sum_{i=1}^{\infty} \int_{B_i} f(y) \, d\nu(y) \lesssim \int_G f(y) \, d\nu(y) \lesssim \|f\|_{L^1(\mathcal{X},\nu)}. \tag{8.2.27}$$

Pick $x_i \in B_i$ and define

$$\nu^{(N)} := \sum_{i=1}^{N} \alpha_i \delta_{x_i}.$$

Then we see that

$$\|\nu^{(N)}\| = \sum_{i=1}^{N} \alpha_i.$$

By (8.2.27), the fact that $2B_i \subset G$ for all $i \in \mathbb{N}$ and (8.2.9), there exists a positive constant C_7 such that

$$\begin{aligned}
&\int_{\mathcal{X}\setminus G} \left|Tf^{(N)}(x) - T\nu^{(N)}(x)\right| d\nu(x) \\
&= \int_{\mathcal{X}\setminus G} \left| T\left(\sum_{i=1}^{N} [f_i\, d\nu - \alpha_i \delta_{x_i}]\right)(x)\right| d\nu(x) \\
&\le \sum_{i=1}^{N} \int_{\mathcal{X}\setminus 2B_i} |T(f_i\, d\nu - \alpha_i \delta_{x_i})(x)|\, d\nu(x) \\
&\lesssim \sum_{i=1}^{N} \alpha_i \\
&\le C_7 \|f\|_{L^1(\mathcal{X},\nu)}.
\end{aligned} \tag{8.2.28}$$

On the other hand, by Theorem 8.2.9, we find that

$$\begin{aligned}
&\nu\left(\left\{x \in \mathcal{X} : \left|T\nu^{(N)}(x)\right| > (\tilde{c}_1 + \tilde{c}_2 \|T\|_{L^2(\mathcal{X},\nu)\to L^2(\mathcal{X},\nu)})t\right\}\right) \\
&\le \frac{1}{t} \|\nu^{(N)}\| \\
&\le \frac{1}{t} \|f\|_{L^1(\mathcal{X},\nu)},
\end{aligned}$$

from which, together with (8.2.28), we deduce that

$$\begin{aligned}
&\nu\left(\left\{x \in \mathcal{X}\setminus G : \left|Tf^{(N)}(x)\right| > \left(C_7 + \tilde{c}_1 + \tilde{c}_2 \|T\|_{L^2(\mathcal{X},\nu)\to L^2(\mathcal{X},\nu)}\right) t\right\}\right) \\
&\le \nu\left(\left\{x \in \mathcal{X}\setminus G : \left|Tf^{(N)}(x) - T\nu^{(N)}(x)\right| > C_7 t\right\}\right) \\
&\quad + \nu\left(\left\{x \in \mathcal{X}\setminus G : \left|T\nu^{(N)}(x)\right| > (\tilde{c}_1 + \tilde{c}_2 \|T\|_{L^2(\mathcal{X},\nu)\to L^2(\mathcal{X},\nu)})t\right\}\right) \\
&\le \frac{2}{t} \|f\|_{L^1(\mathcal{X},\nu)}.
\end{aligned}$$

This, combined with the fact that $\nu(G) \le \|f\|_{L^1(\mathcal{X},\nu)}/t$, implies that

$$\begin{aligned}&\nu\left(\left\{x \in \mathcal{X} : \left|Tf^{(N)}(x)\right| > \left(C_7 + \tilde{c}_1 + \tilde{c}_2\|T\|_{L^2(\mathcal{X},\nu)\to L^2(\mathcal{X},\nu)}\right) t\right\}\right)\\ &\quad\le \frac{3}{t}\|f\|_{L^1(\mathcal{X},\nu)}.\end{aligned} \tag{8.2.29}$$

Observe that $f^{(N)} \to f^t$ in $L^2(\mathcal{X}, \nu)$ as $N \to \infty$. From the boundedness of T on $L^2(\mathcal{X}, \nu)$, we then deduce that $Tf^{(N)} \to Tf^t$ also in $L^2(\mathcal{X}, \nu)$ as $N \to \infty$. By this fact and (8.2.29), we have

$$\nu\left(\left\{x \in \mathcal{X} : \left|Tf^t(x)\right| > \left(C_7 + \tilde{c}_1 + \tilde{c}_2\|T\|_{L^2(\mathcal{X},\nu)\to L^2(\mathcal{X},\nu)}\right) t\right\}\right) \le \frac{3}{t}\|f\|_{L^1(\mathcal{X},\nu)},$$

from which together with (8.2.26), it follows that there exist positive constants C_8 and C_9 such that

$$\sup_{t\in(0,\infty)} t\,\nu(\{x \in \mathcal{X} : |Tf(x)| > t\}) \le \left(C_8 + C_9\|T\|_{L^2(\mathcal{X},\nu)\to L^2(\mathcal{X},\nu)}\right)\|f\|_{L^1(\mathcal{X},\nu)}.$$

This implies (8.2.3), and hence finishes the proof of the implicity (i) $\Longrightarrow$ (iii).

Now assume that (ii) holds true. Then, by Remark 8.2.11(ii) and a similar proof of (i) $\Longrightarrow$ (iii), we see that (iii) holds true. We omit the details, which completes the proof of Part I of Theorem 8.2.1. □

8.2.3 *Proof of Theorem 8.2.1, Part II*

This section is devoted to the proof of (iii) $\Longrightarrow$ (i) of Theorem 8.2.1. To do so, we first establish the boundedness of $T^\sharp$ from $L^1(\mathcal{X}, \nu)$ to $L^{1,\infty}(\mathcal{X}, \nu)$, which implies that $\{T_r\}_{r\in(0,\infty)}$ is uniformly bounded from $L^1(\mathcal{X}, \nu)$ to $L^{1,\infty}(\mathcal{X}, \nu)$. By restricting ν to ν_M, where ν_M is the *restriction of* ν *to a given ball* $\overline{B}(x_0, M)$ *for some* $x_0 \in \mathcal{X}$ *and* $M \in (0,\infty)$, we prove that, for any $r \in (0,\infty)$ and $p \in (1,\infty)$, T_r is bounded on $L^p(\mathcal{X}, \nu_M)$. Then, using a smooth truncation argument, we further show that $\{T_r\}_{r\in(0,\infty)}$ is uniformly bounded from $L^2(\mathcal{X}, \nu)$ to $L^2(\mathcal{X}, \nu_M)$ with the constant independent of M. By letting $M \to \infty$, $\{T_r\}_{r\in(0,\infty)}$ is uniformly bounded on $L^2(\mathcal{X}, \nu)$.

Theorem 8.2.12. *Let T be a Calderón–Zygmund operator with kernel K satisfying* (8.1.23) *and* (8.1.24)*, which is bounded from $L^1(\mathcal{X}, \nu)$ to $L^{1,\infty}(\mathcal{X}, \nu)$. Then there exists a positive constant C such that, for any $f \in L^1(\mathcal{X}, \nu)$,*

$$\left\|T^\sharp f\right\|_{L^{1,\infty}(\mathcal{X},\nu)} \le C\|f\|_{L^1(\mathcal{X},\nu)}.$$

Proof. Let $p \in (0,1)$. By (i) and (ii) of Lemma 8.2.3 and the boundedness of T from $L^1(\mathcal{X}, \nu)$ to $L^{1,\infty}(\mathcal{X}, \nu)$, to show Theorem 8.2.12, we only need to prove that, for any $f \in L_b^\infty(\mathcal{X}, \nu)$ and $x \in \mathcal{X}$,

$$[T^\sharp f(x)]^p \lesssim [\mathcal{M}_p^c Tf(x)]^p + [\mathcal{M}^c f(x)]^p.$$

Moreover, it suffices to prove that, for any $r \in (0, \infty)$, $f \in L_b^\infty(\mathcal{X}, \nu)$ and $x \in \mathcal{X}$,

$$|T_r f(x)|^p \lesssim \left[\mathcal{M}_p^c Tf(x)\right]^p + [\mathcal{M}^c f(x)]^p. \tag{8.2.30}$$

To this end, for any $j \in \mathbb{N}$, let

$$r_j := 5^j r \quad \text{and} \quad \nu_j := \nu(\overline{B}(x, r_j))$$

be as in the proof of Lemma 8.2.6. Again let k be the *smallest positive integer* such that

$$\nu_{k+1} \leq 4[C_{(\lambda)}]^6 \nu_{k-1} \quad \text{and} \quad R := r_{k-1} = 5^{k-1} r.$$

Then, similar to the proof of (8.2.12), we conclude that

$$|T_r f(x) - T_{5R} f(x)| \lesssim \mathcal{M}^c f(x). \tag{8.2.31}$$

Let

$$f_1 := f \chi_{\overline{B}(x,5R)} \quad \text{and} \quad f_2 := f - f_1.$$

For any $u \in \overline{B}(x, R)$, if K is the kernel associated with T, then, by (8.1.24) and (7.1.1), we see that

$$\begin{aligned}|Tf_2(x) - Tf_2(u)| &\leq \int_{d(x,y)>5R} |K(x, y) - K(u, y)|\, |f(y)|\, d\nu(y)\\ &\lesssim \sum_{k=1}^{\infty} \left[\frac{d(x,u)}{5^k R}\right]^\tau \int_{\overline{B}(x,5^{k+1}R)} \frac{|f(y)|}{\lambda(x, 5^k R)}\, d\nu(y)\\ &\lesssim \mathcal{M}^c f(x).\end{aligned}$$

This, combined with (8.2.31) and the fact that

$$Tf_2(x) = \int_{\mathcal{X}} K(x, y) f_2(y)\, d\nu(y) = T_{5R} f(x),$$

implies that

$$\begin{aligned}|T_r f(x)| &\leq |T_r f(x) - T_{5R} f(x)| + |T_{5R} f(x) - Tf_2(u)| + |Tf_2(u)|\\ &\lesssim \mathcal{M}^c f(x) + |Tf(u)| + |Tf_1(u)|,\end{aligned}$$

from which and $p \in (0, 1)$, it further follows that, for all $u \in \overline{B}(x, R)$,

$$|T_r f(x)|^p \lesssim [\mathcal{M}^c f(x)]^p + |Tf(u)|^p + |Tf_1(u)|^p. \tag{8.2.32}$$

Since T is bounded from $L^1(\mathcal{X}, \nu)$ to $L^{1,\infty}(\mathcal{X}, \nu)$, by the Kolmogorov inequality, we conclude that

$$\begin{aligned} &\frac{1}{\nu(\overline{B}(x, R))} \int_{\overline{B}(x, R)} |Tf_1(u)|^p \, d\nu(u) \\ &\quad \lesssim \frac{1}{[\nu(\overline{B}(x, R))]^p} \left[\int_{\overline{B}(x, R)} |f_1(u)| \, d\nu(u)\right]^p. \end{aligned} \tag{8.2.33}$$

Taking the average on the variable u over $\overline{B}(x, R)$ on both sides of (8.2.32), and using (8.2.33), the Hölder inequality and (8.2.11), we see that

$$\begin{aligned} &|T_r f(x)|^p \\ &\quad \lesssim [\mathcal{M}^c f(x)]^p + \left[\mathcal{M}_p^c(Tf)(x)\right]^p + \frac{1}{\nu(\overline{B}(x, R))} \int_{\overline{B}(x, R)} |Tf_1(u)|^p \, d\nu(u) \\ &\quad \lesssim [\mathcal{M}^c f(x)]^p + \left[\mathcal{M}_p^c(Tf)(x)\right]^p + \frac{1}{[\nu(\overline{B}(x, 25R))]^p} \left[\int_{\overline{B}(x, 5R)} |f(u)| \, d\nu(u)\right]^p \\ &\quad \lesssim [\mathcal{M}^c f(x)]^p + \left[\mathcal{M}_p^c(Tf)(x)\right]^p, \end{aligned}$$

which implies (8.2.30), and hence completes the proof Theorem 8.2.12. □

Let $x_0 \in \mathcal{X}$ and $M \in (0, \infty)$. We now obtain the uniform boundedness of the truncated operators $\{T_r\}_{r \in (0,\infty)}$ on $L^p(\mathcal{X}, \nu_M)$ for all $p \in (1, \infty)$. Notice that the set $\mathcal{X} \setminus \overline{B}(x_0, M)$ has ν_M-measure zero by definition, and hence we may agree that any $f \in L^p(\mathcal{X}, \nu_M)$ satisfies $f|_{\mathcal{X} \setminus \overline{B}(x_0, M)} = 0$. With this agreement, observe that

$$T_r f(x) = \int_{d(x, y) > r} K(x, y) f(y) \, d\nu(y) = \int_{d(x, y) > r} K(x, y) f(y) \, d\nu_M(y)$$

for $f \in L^p(\mathcal{X}, \nu_M)$, so we may also replace ν by ν_M in the formula of $T_r f$ when considering functions $f \in L^p(\mathcal{X}, \nu_M)$. Finally, observing that ν_M also satisfies the upper doubling condition, with the same dominating function λ, we then see that all results shown for ν apply equally well to ν_M, with constants uniform with respect to M.

Lemma 8.2.13. *Let $p \in (1, \infty)$ and $r \in (0, \infty)$. Let $M \in (0, \infty)$ and ν_M be as above. Then there exists a positive constant $\tilde{C}$, depending on M and r, such that, for all $f \in L^p(\mathcal{X}, \nu_M)$,*

$$\|T_r f\|_{L^p(\mathcal{X}, \nu_M)} \le \tilde{C} \|f\|_{L^p(\mathcal{X}, \nu_M)}.$$

Proof. We first claim that there exists a positive constant C such that, for all $x \in \overline{B}(x_0, M)$,

$$|T_r f(x)| \leq C[\lambda(x,r)]^{-1/p} \|f\|_{L^p(\mathcal{X},\nu_M)}. \tag{8.2.34}$$

To this end, let $B_0 := B(x,r)$. Then (8.1.23), together with the Hölder inequality, implies that

$$|T_r f(x)| \lesssim \left[\int_{\mathcal{X}\setminus B_0} \frac{d\nu(y)}{[\lambda(x,d(x,y))]^{p'}}\right]^{\frac{1}{p'}} \|f\|_{L^p(\mathcal{X},\nu_M)}. \tag{8.2.35}$$

To prove the claim, we inductively construct an auxiliary sequence $\{r_0, r_1, \ldots\}$ of radii such that $r_0 = r$ and r_{i+1} is the smallest $2^k r_i$ with $k \in \mathbb{N}$ satisfying

$$\lambda(x, 2^k r_i) > 2\lambda(x, r_i), \tag{8.2.36}$$

whenever such a k exists. We consider the following two cases.

Case (i) For each $i \in \mathbb{Z}_+$, there exists $k \in \mathbb{N}$ such that (8.2.36) holds true. In this case, r_{i+1} is the smallest $2^k r_i$ satisfying (8.2.36) and

$$\{B_i\}_{i\in\mathbb{N}} := \{B(x,r_i)\}_{i\in\mathbb{N}}.$$

Now by the choice of r_i, (7.1.1) and the fact that $2^i\lambda(x,r) \leq \lambda(x,r_i)$ for all $i \in \mathbb{Z}_+$, we know that

$$\begin{aligned}\int_{\mathcal{X}\setminus B_0} \frac{1}{[\lambda(x,d(x,y))]^{p'}}\, d\nu(y) &\lesssim \sum_{i=0}^{\infty} \frac{\nu(B_{i+1})}{[\lambda(x,r_{i+1})]^{p'}}\\ &\lesssim \sum_{i=0}^{\infty} \frac{1}{[\lambda(x,r_{i+1})]^{p'-1}}\\ &\lesssim \sum_{i=0}^{\infty} \frac{1}{[2^i\lambda(x,r)]^{p'-1}}\\ &\sim \frac{1}{[\lambda(x,r)]^{p'-1}}\end{aligned} \tag{8.2.37}$$

and hence

$$\left[\int_{\mathcal{X}\setminus B_0} \frac{1}{[\lambda(x,d(x,y))]^{p'}}\, d\nu(y)\right]^{\frac{1}{p'}} \lesssim [\lambda(x,r)]^{-\frac{1}{p}},$$

which, combined with (8.2.35), implies (8.2.34) and the claim holds true in this case.

Case (ii) For some $i_0 \in \mathbb{Z}_+$, (8.2.36) holds true for all $i < i_0$ but does not hold true for i_0. In this case, if $i_0 \in \mathbb{N}$, we let $\{B_i\}_{i=1}^{i_0}$ be as in Case (i), $r_{i_0+1} := \infty$ and $B_{i_0+1} := \mathcal{X}$; otherwise, if $i_0 = 0$, we then let $r_1 := \infty$ and $B_1 := \mathcal{X}$. Then we see that $\lambda(x, 2^k r_{i_0}) \leq 2\lambda(x, r_{i_0})$ for all $k \in \mathbb{N}$ and

$$\nu(\mathcal{X}) := \lim_{t\to\infty} \nu(B(x,t)) \leq \lim_{t\to\infty} \lambda(x,t) =: \lambda(x,\infty) \leq 2\lambda(x, r_{i_0}),$$

which, together with (7.1.1) and the fact that $2^i\lambda(x,r) \leq \lambda(x,r_i)$ for all $i \leq i_0$, implies (8.2.37) in this case, and the claim holds true.

If $x \in \operatorname{supp} \nu_M = \overline{B}(x_0, M)$, then

$$\operatorname{supp} \nu_M \subset B(x, 3M).$$

By this and the definition of $\operatorname{supp} \nu_M$, we conclude that

$$\nu_M(\mathcal{X}) = \nu_M(B(x,3M)) \leq \lambda(x,3M) \leq [C_{(\lambda)}]^{1+\log_2(3M/r)}\lambda(x,r)$$

and hence

$$\frac{1}{\lambda(x,r)} \leq \frac{[C_{(\lambda)}]^{3+\log_2(M/r)}}{\nu_M(\mathcal{X})}.$$

By this fact, we see that

$$\int_{\mathcal{X}} \frac{1}{\lambda(x,r)}\, d\nu_M(x) \leq \frac{[C_{(\lambda)}]^{3+\log_2(M/r)}}{\nu_M(\mathcal{X})} \int_{\mathcal{X}} d\nu_M(x) \leq [C_{(\lambda)}]^{3+\log_2(M/r)}.$$

From this and (8.2.34), it follows that

$$\begin{aligned}\|T_r f\|_{L^p(\mathcal{X},\nu_M)} &\lesssim \|f\|_{L^p(\mathcal{X},\nu_M)} \left[\int_{\mathcal{X}} \frac{1}{\lambda(x,r)}\, d\nu_M(x)\right]^{\frac{1}{p}} \\ &\lesssim \|f\|_{L^p(\mathcal{X},\nu_M)} \left[C_{(\lambda)}\right]^{\frac{3+\log_2(M/r)}{p}} \\ &= \tilde{C}_{(M,r)}\|f\|_{L^p(\mathcal{X},\nu_M)}.\end{aligned}$$

This finishes the proof of Lemma 8.2.13. □

We need the following result which shows that two bounded Calderón–Zygmund operators having the same kernel can at most differ by a multiplication operator.

Proposition 8.2.14. *Let T and $\tilde{T}$ be Calderón–Zygmund operators which have the same kernel satisfying* (8.1.23) *and* (8.1.24) *and are both bounded from $L^p(\mathcal{X}, \nu)$ to $L^{p,\infty}(\mathcal{X}, \nu)$ for some $p \in [1,\infty)$. Then there exists $b \in L^\infty(\mathcal{X}, \nu)$ such that, for all $f \in L^p(\mathcal{X}, \nu)$,*

$$Tf - \tilde{T}f = bf \quad \text{and} \quad \|b\|_{L^\infty(\mathcal{X},\nu)} \le \|T - \tilde{T}\|_{L^p(\mathcal{X},\nu)\to L^{p,\infty}(\mathcal{X},\nu)}.$$

The proof relies on the following lemma.

Lemma 8.2.15. *For a suitable $\delta \in (0,1)$, there exists a sequence of countable Borel partitions, $\{Q_\alpha^k\}_{\alpha\in\mathscr{A}_k}$, $k \in \mathbb{Z}$, of $\mathcal{X}$ with the following properties:*

(i) *For some $x_\alpha^k \in \mathcal{X}$ and constants $0 < c_1 < c_2 < \infty$,*

$$B(x_\alpha^k, c_1\delta^k) \subset Q_\alpha^k \subset B(x_\alpha^k, c_2\delta^k);$$

(ii) *$\{Q_\alpha^{k+1}\}_{\alpha\in\mathscr{A}_{k+1}}$ is a refinement of $\{Q_\alpha^k\}_{\alpha\in\mathscr{A}_k}$.*

Moreover, it may be arranged so that

$$\nu\left(\bigcup_{k\in\mathbb{Z},\,\alpha\in\mathscr{A}_k} \partial Q_\alpha^k\right) = 0, \tag{8.2.38}$$

where, for a set Q,

$$\partial Q := \{x \in \mathcal{X} : d(x,Q) = d(x, \mathcal{X}\setminus Q) = 0\}$$

is the boundary.

Proof. Let $\{Q_\alpha^k\}_{k\in\mathbb{Z},\,\alpha\in\mathscr{A}_k}$ be the *random dyadic cubes* constructed in [70], so indeed $Q_\alpha^k := Q_\alpha^k(\zeta)$, where ζ is a point of an underlying *probability space* Ω. We use $\mathbb{P}$ to denote a *probability measure* on Ω (as constructed in [70]) such that $\mathbb{P}(A)$ is the *probability of the event* $A \subset \Omega$. By the construction given in [70], these sets automatically satisfy the other claims for all $\zeta \in \Omega$, and it remains to show that we can choose $\zeta \in \Omega$ so as to also satisfy (8.2.38).

The "side-length" of Q_α^k is defined by $\ell(Q_\alpha^k) := \delta^k$, where $\delta \in (0,1)$ is a fixed parameter entering the construction. For $\varepsilon \in (0,\infty)$, let

$$\delta_\varepsilon Q := \{x : d(x,Q) \le \varepsilon\ell(Q)\} \bigcap \{x : d(x,\mathcal{X}\setminus Q) \le \varepsilon\ell(Q)\}.$$

It was shown[3] that there exists an $\eta \in (0,\infty)$ such that, for any fixed $x \in \mathcal{X}$ and $k \in \mathbb{Z}$,

$$\mathbb{P}\left(x \in \bigcup_{\alpha\in\mathscr{A}_k} \delta_\varepsilon Q_\alpha^k\right) \lesssim \varepsilon^\eta.$$

[3] See [70, Lemma 10.1].

In particular, by taking the limit as $\varepsilon \to 0$, we see that

$$\mathbb{P}\left(x \in \bigcup_{\alpha \in \mathscr{A}_k} \partial Q_\alpha^k\right) = 0.$$

Then it is possible to sum the zero probabilities over $k \in \mathbb{Z}$ to deduce

$$\mathbb{P}\left(x \in \bigcup_{k \in \mathbb{Z}, \alpha \in \mathscr{A}_k} \partial Q_\alpha^k\right) = 0.$$

Now we compute (the integration variable of the $d\mathbb{P}$-integrals is $\zeta \in \Omega$, the random variable implicit in the random dyadic cubes $Q_\alpha^k := Q_\alpha^k(\zeta)$):

$$\begin{aligned}
\int_\Omega \nu\left(\bigcup_{k \in \mathbb{Z}, \alpha \in \mathscr{A}_k} \partial Q_\alpha^k\right) d\mathbb{P} &= \int_\Omega \int_{\mathcal{X}} \chi_{\bigcup_{k \in \mathbb{Z}, \alpha \in \mathscr{A}_k} \partial Q_\alpha^k}(x)\, d\nu(x)\, d\mathbb{P} \\
&= \int_{\mathcal{X}} \int_\Omega \chi_{\bigcup_{k \in \mathbb{Z}, \alpha \in \mathscr{A}_k} \partial Q_\alpha^k}(x)\, d\mathbb{P}\, d\nu(x) \\
&= \int_{\mathcal{X}} \mathbb{P}\left(x \in \bigcup_{k \in \mathbb{Z}, \alpha \in \mathscr{A}_k} \partial Q_\alpha^k\right) d\nu(x) \\
&= 0.
\end{aligned}$$

Thus, the integral of $\nu(\bigcup_{k \in \mathbb{Z}, \alpha \in \mathscr{A}_k} \partial Q_\alpha^k(\zeta)) \geq 0$ is zero. This means that

$$\nu\left(\bigcup_{k \in \mathbb{Z}, \alpha \in \mathscr{A}_k} \partial Q_\alpha^k(\zeta)\right) = 0$$

for $\mathbb{P}$-almost every $\zeta \in \Omega$. Now we just fix one such ζ and, for this choice, the boundaries of the corresponding dyadic cubes $Q_\alpha^k := Q_\alpha^k(\zeta)$ have ν-measure zero. This implies (8.2.38) and hence finishes the proof of Lemma 8.2.15. □

Proof of Proposition 8.2.14. Let $S := T - \tilde{T}$. Then S is bounded from $L^p(\mathcal{X}, \nu)$ to $L^{p,\infty}(\mathcal{X}, \nu)$ for some $p \in [1, \infty)$ as in the proposition and it has kernel 0. We prove that, for all $M \in \mathbb{N}$, $f \in L^p(\mathcal{X}, \nu)$ with

$$\operatorname{supp} f \subset B_M := \overline{B}(x_0, M),$$

and ν-almost every $x \in \mathcal{X}$, it holds true that

$$Sf(x) = f(x) S(\chi_{B_M})(x) =: f(x) b_M(x) \tag{8.2.39}$$

and

$$\|b_M\|_{L^\infty(\mathcal{X},\nu_M)} \le \|S\|_{L^p(\mathcal{X},\nu)\to L^{p,\infty}(\mathcal{X},\nu)}, \tag{8.2.40}$$

where $\nu_M := \nu|_{B_M}$.

Suppose for the moment that (8.2.39) and (8.2.40) are already proved. If $M < \tilde{M}$, then, for all $f \in L^p(\mathcal{X}, \nu)$ with $\operatorname{supp} f \subset B_M \subset B_{\tilde{M}}$, we have

$$fb_M = Sf = fb_{\tilde{M}}$$

almost everywhere on B_M. Since this is true for all such f, we have $b_{\tilde{M}} = b_M$ on B_M, and hence we can unambiguously define $b(x)$ for all $x \in \mathcal{X}$ by letting $b(x) := b_M(x)$ for $x \in B_M$. The uniform bound (8.2.40) implies that

$$\|b\|_{L^\infty(\mathcal{X},\nu)} \le \|S\|_{L^p(\mathcal{X},\nu)\to L^{p,\infty}(\mathcal{X},\nu)},$$

and we have $Sf = bf$ for all $f \in L^p(\mathcal{X}, \nu)$ with bounded support. Finally, by density this holds true for all $f \in L^p(\mathcal{X}, \nu)$. Thus, proving (8.2.39) and (8.2.40) proves the proposition, and we turn to this task.

Now we prove (8.2.39). Let us consider functions of the form

$$\sum_\alpha x_\alpha^k \chi_{Q_\alpha^k \cap B_M}, \tag{8.2.41}$$

where $\{Q_\alpha^k\}_{\alpha,k}$ are the dyadic cubes with zero-measure boundaries, as provided by Lemma 8.2.15. Since $(\mathcal{X}, d)$ is geometrically doubling and B_M is bounded, we see that only finitely many Q_α^k intersect B_M, and hence the sum in (8.2.41) may taken to be finite.

We claim that, for ν-almost every $x \in \mathcal{X}$,

$$S\left(\chi_{Q_\alpha^k \cap B_M}\right)(x) = \chi_{Q_\alpha^k \cap B_M}(x) \cdot S\left(\chi_{B_M}\right)(x). \tag{8.2.42}$$

Indeed, observe first that, for ν-almost every $x \in \mathcal{X}$,

$$S\left(\chi_{B_M}\right)(x) = S\left(\sum_\beta \chi_{Q_\beta^k \cap B_M}\right)(x) = \sum_\beta S\left(\chi_{Q_\beta^k \cap B_M}\right)(x). \tag{8.2.43}$$

On the other hand, the assumption that S has kernel 0 means that, for any $f \in L_b^\infty(\mathcal{X}, \nu)$ and ν-almost every $x \notin \operatorname{supp} f$,

$$Sf(x) = \int_{\mathcal{X}} 0 f(y)\, d\nu(y) = 0.$$

This implies that

$$\begin{aligned}\operatorname{supp}\left(S\left(\chi_{Q_\beta^k\cap B_M}\right)\right) &\subset \operatorname{supp}\chi_{Q_\beta^k\cap B_M}\\ &= \overline{Q_\beta^k\bigcap B_M}\\ &\subset \overline{Q_\beta^k}\bigcup \overline{B}_M\\ &= \left(Q_\beta^k\bigcap \overline{B}_M\right)\bigcup\left(\partial Q_\beta^k\bigcap \overline{B}_M\right).\end{aligned}$$

Recall that Q_α^k and Q_β^k are disjoint if $\alpha\neq\beta$, which, together with (8.2.38), implies that almost every $x\in Q_\alpha^k\cap B_M$ is outside $\operatorname{supp}(S(\chi_{Q_\beta^k\cap B_M}))$. Hence

$$S\left(\chi_{Q_\beta^k\cap B_M}\right)(x)=0$$

for ν-almost every $x\in Q_\alpha^k\cap B_M$ and thus, for ν-almost every $x\in\mathcal{X}$,

$$\begin{aligned}\chi_{Q_\alpha^k\cap B_M}(x)S\left(\chi_{Q_\beta^k\cap B_M}\right)(x) &= \delta_{\alpha\beta}\chi_{Q_\alpha^k\cap B_M}(x)S\left(\chi_{Q_\alpha^k\cap B_M}\right)(x)\\ &= \delta_{\alpha\beta}S\left(\chi_{Q_\alpha^k\cap B_M}\right)(x),\end{aligned}$$

where $\delta_{\alpha\beta}:=1$ if $\alpha=\beta$ and $\delta_{\alpha\beta}:=0$ otherwise, and the last equality follows from the fact that $\chi_{Q_\alpha^k\cap B_M}(x)=1$ for ν-almost every $x\in\operatorname{supp}(S(\chi_{Q_\alpha^k\cap B_M}))$. Multiplying (8.2.43) by $\chi_{Q_\alpha^k\cap B_M}$, we then see that

$$\begin{aligned}\chi_{Q_\alpha^k\cap B_M}(x)S\left(\chi_{B_M}\right)(x) &= \sum_\beta \chi_{Q_\alpha^k\cap B_M}(x)S\left(\chi_{Q_\beta^k\cap B_M}\right)(x)\\ &= S\left(\chi_{Q_\alpha^k\cap B_M}\right)(x),\end{aligned}$$

which is precisely (8.2.42).

Now it is easy to complete the proof of (8.2.39). For any f of the form (8.2.41), it follows, from (8.2.42), that

$$Sf=\sum_\alpha x_\alpha^k S\left(\chi_{Q_\alpha^k\cap B_M}\right)=\sum_\alpha x_\alpha^k\chi_{Q_\alpha^k\cap B_M}S\left(\chi_{B_M}\right)=fS\left(\chi_{B_M}\right). \quad (8.2.44)$$

Recall that the martingale convergence implies that, for any $f\in L^p(\mathcal{X},\nu)$,

$$\mathbb{E}_k f:=\sum_\alpha m_{Q_\alpha^k}(f)\chi_{Q_\alpha^k}\to f$$

for ν-almost every $x\in\mathcal{X}$ and in $L^p(\mathcal{X},\nu)$ as $k\to\infty$. If $f\in L^p(\mathcal{X},\nu)$ is general, apply (8.2.44) to $\chi_{B_M}\mathbb{E}_k f$. Then, as $k\to\infty$, we have

$$\chi_{B_M}\mathbb{E}_k f \to \chi_{B_M} f$$

in $L^p(\mathcal{X}, \nu)$, hence

$$S(\chi_{B_M}\mathbb{E}_k f) \to S(\chi_{B_M} f)$$

in $L^{p,\infty}(\mathcal{X}, \nu)$, and thus almost everywhere for a subsequence. Also, by (8.2.44), we see that

$$S\left(\chi_{B_M}\mathbb{E}_k f\right) = \chi_{B_M}\left(\mathbb{E}_k f\right) S\left(\chi_{B_M}\right) \to \chi_{B_M} f S\left(\chi_{B_M}\right)$$

for ν-almost every $x \in \mathcal{X}$. As a result, for all $f \in L^p(\mathcal{X}, \nu)$,

$$S(\chi_{B_M} f) = (\chi_{B_M} f)\, S\left(\chi_{B_M}\right) =: \chi_{B_M} f b_M,$$

where

$$b_M := S\left(\chi_{B_M}\right) \in L^{p,\infty}(\mathcal{X}, \nu)$$

since $\chi_{B_M} \in L^p(\mathcal{X}, \nu)$. Thus, (8.2.39) holds true for all $f \in L^p(\mathcal{X}, \nu)$ with $\operatorname{supp} f \subset B_M$.

It remains to prove (8.2.40). Let $t \in (0, \infty)$,

$$f := \chi_{\{|b_M|>t\}\cap B_M} \quad \text{and} \quad B := \|S\|_{L^p(\mathcal{X},\nu)\to L^{p,\infty}(\mathcal{X},\nu)}.$$

Then

$$\|f\|_{L^p(\mathcal{X},\nu)} = \left[\nu\left(\{x \in \mathcal{X} : |b_M(x)| > t\}\bigcap B_M\right)\right]^{1/p}.$$

By this, (8.2.39) and the boundedness of S from $L^p(\mathcal{X}, \nu)$ to $L^{p,\infty}(\mathcal{X}, \nu)$, we see that

$$\begin{aligned}
&t\left[\nu\left(\{x \in \mathcal{X} : |b_M(x)| > t\}\bigcap B_M\right)\right]^{1/p}\\
&\quad = t\left[\nu(\{x \in \mathcal{X} : |b_M(x) f(x)| > t\})\right]^{1/p}\\
&\quad = t\left[\nu(\{x \in \mathcal{X} : |Sf(x)| > t\})\right]^{1/p}\\
&\quad \le \|Sf\|_{L^{p,\infty}(\mathcal{X},\nu)}\\
&\quad \le B\|f\|_{L^p(\mathcal{X},\nu)}\\
&\quad = B\left[\nu\left(\{x \in \mathcal{X} : |b_M(x)| > t\}\bigcap B_M\right)\right]^{1/p}.
\end{aligned}$$

This means that either

$$\nu\left(\{x \in \mathcal{X} : |b_M(x)| > t\}\bigcap B_M\right) = 0$$

or $t \leq B$, which is the same as

$$\|b_M\|_{L^\infty(\mathcal{X},\nu_M)} \leq B.$$

This implies (8.2.40), and hence finishes the proof of Proposition 8.2.14. □

From Proposition 8.2.14, we easily deduce the following consequence.

Lemma 8.2.16. *Let T and $\tilde{T}$ be Calderón–Zygmund operators, which are bounded from $L^1(\mathcal{X}, \nu)$ to $L^{1,\infty}(\mathcal{X}, \nu)$ and have the same kernel satisfying* (8.1.23) *and* (8.1.24). *If $\tilde{T}$ is bounded on $L^2(\mathcal{X}, \nu)$, then T is also bounded on $L^2(\mathcal{X}, \nu)$.*

Proof. By Proposition 8.2.14, we have $Tf = \tilde{T}f + bf$, where $b \in L^\infty(\mathcal{X}, \nu)$. Hence

$$\begin{aligned}\|Tf\|_{L^2(\mathcal{X},\nu)} &\leq \left\|\tilde{T}f\right\|_{L^2(\mathcal{X},\nu)} + \|bf\|_{L^2(\mathcal{X},\nu)}\\ &\leq \left[\left\|\tilde{T}\right\|_{L^2(\mathcal{X},\nu)\to L^2(\mathcal{X},\nu)} + \|b\|_{L^\infty(\mathcal{X},\nu)}\right]\|f\|_{L^2(\mathcal{X},\nu)},\end{aligned}$$

which completes the proof of Lemma 8.2.16. □

Proof of Theorem 8.2.1, Part II. In this part, we show that (iii) of Theorem 8.2.1 implies its (i). Let $\nu_M := \nu|_{\overline{B}(x_0, M)}$ be as before. The assumption clearly implies that T is bounded from $L^1(\mathcal{X}, \nu_M)$ to $L^{1,\infty}(\mathcal{X}, \nu_M)$ with a norm bound independent of M. We then prove that T is bounded on $L^2(\mathcal{X}, \nu_M)$, still with a bound independent of M. By the density of boundedly supported $L^2_{\rm loc}(\mathcal{X}, \nu)$-functions in $L^2(\mathcal{X}, \nu)$ and the monotone convergence, this suffices to conclude the proof of (iii) $\Longrightarrow$ (i) of Theorem 8.2.1. Thus, from now on, we work with the measure ν_M. Recall that ν_M satisfies, uniformly in M, the same assumptions as ν such that everything shown for ν above equally well applies to ν_M.

By Theorem 8.2.12, we see that $T^\sharp$ is bounded from $L^1(\mathcal{X}, \nu_M)$ to $L^{1,\infty}(\mathcal{X}, \nu_M)$, which implies that the operators $\{T_r\}_{r\in(0,\infty)}$ is uniformly bounded from $L^1(\mathcal{X}, \nu_M)$ to $L^{1,\infty}(\mathcal{X}, \nu_M)$, and the bound (denoted by N_1) depends only on the norm of T as the operator from $L^1(\mathcal{X}, \nu)$ to $L^{1,\infty}(\mathcal{X}, \nu)$.

Let $p \in (1,\infty)$. It follows, from Lemma 8.2.13, that, for any $r \in (0,\infty)$, T_r is bounded on $L^p(\mathcal{X}, \nu_M)$ with $p \in (1,\infty)$, but with the norm a priori depending on M and r. We claim, however, that $\{T_r\}_{r\in(0,\infty)}$ is uniformly bounded on $L^2(\mathcal{X}, \nu_M)$. That is, if we denote the corresponding norm by $N_p(r, M)$, then we know that there exists a positive constant C depending on N_1, but not on r or M, such that

$$N_2(r, M) \leq C. \tag{8.2.45}$$

To this end, we define, for any $r \in (0,\infty)$ and $x \in \mathcal{X}$, that

$$T_r^\psi f(x) := \int_{\mathcal{X}} K(x,y)\psi\left(\frac{d(x,y)}{r}\right) f(y)\,d\nu(y),$$

where ψ is a smooth function on $(0, \infty)$ such that $\operatorname{supp}\psi \subset [1/2, \infty)$, $\psi(t) \in [0, 1]$ for all $t \in (0, \infty)$, and $\psi(t) := 1$ when $t \in [1, \infty)$, and K is the kernel of T. It follows, from the definition of T_r^ψ, (8.1.23) and (7.1.1), that, for any $x \in \mathcal{X}$,

$$\begin{aligned}\left|T_r f(x) - T_r^\psi f(x)\right| &\le \int_{\overline{B}(x,r)\setminus B(x,r/2)} |K(x,y)||f(y)|\,d\nu(y)\\ &\lesssim \int_{\overline{B}(x,r)} \frac{|f(y)|}{\lambda(x,r/2)}\,d\nu(y)\\ &\lesssim \mathcal{M}^c f(x).\end{aligned}$$

This fact, together with Lemma 8.2.3(i), implies that the boundedness of T_r on $L^p(\mathcal{X}, \nu_M)$ for $p \in (1,\infty)$ or from $L^1(\mathcal{X}, \nu_M)$ to $L^{1,\infty}(\mathcal{X}, \nu_M)$ is equivalent to that of T_r^ψ. Moreover, if the sequence $\{T_r\}_{r\in(0,\infty)}$ is uniformly bounded on $L^p(\mathcal{X}, \nu_M)$ or from $L^1(\mathcal{X}, \nu_M)$ to $L^{1,\infty}(\mathcal{X}, \nu_M)$, then so is $\{T_r^\psi\}_{r\in(0,\infty)}$; and vice verse.

Now we denote by $\tilde{N}_p(r, M)$ the norm of T_r^ψ on $L^p(\mathcal{X}, \nu_M)$, with $p \in (1,\infty)$, and by $\tilde{N}_1$ the (finite) supremum, over r and M, of the norms of T_r^ψ from $L^1(\mathcal{X}, \nu_M)$ to $L^{1,\infty}(\mathcal{X}, \nu_M)$. Then, to show (8.2.45), we only need to prove that

$$\tilde{N}_2(r, M) \le \tilde{C} \tag{8.2.46}$$

for some positive constant $\tilde{C}$ independent of r and M.

We now prove (8.2.46). Observe that, for each r, T_r^ψ is bounded on $L^2(\mathcal{X}, \nu_M)$ and from $L^1(\mathcal{X}, \nu_M)$ to $L^{1,\infty}(\mathcal{X}, \nu_M)$. Then, from the Marcinkiewicz interpolation theorem, we deduce that T_r^ψ is bounded on $L^{\frac{4}{3}}(\mathcal{X}, \nu_M)$ and

$$\tilde{N}_{\frac{4}{3}}(r, M) \lesssim \tilde{N}_1^{\frac{1}{2}}\left[\tilde{N}_2(r, M)\right]^{\frac{1}{2}}.$$

By duality, the right hand side implies also the bound for the norm of $(T_r^\psi)^*$ on $L^4(\mathcal{X}, \nu_M)$. Observe that, for all $x \in \mathcal{X}$,

$$(T_r^\psi)^*(g)(x) = \int_{\mathcal{X}} \overline{K(y,x)\psi\left(\frac{d(x,y)}{r}\right)}g(y)\,d\nu_M(y).$$

Then $(T_r^\psi)^*$ is also a Calderón–Zygmund operator. Thus, $(T_r^\psi)^*$ is bounded from $L^1(\mathcal{X}, \nu_M)$ to $L^{1,\infty}(\mathcal{X}, \nu_M)$ and the norm is bounded by $c\tilde{N}_1^{\frac{1}{2}}[\tilde{N}_2(r, M)]^{\frac{1}{2}} + \tilde{c}$ for some positive constants c and $\tilde{c}$. Another application of the Marcinkiewicz interpolation theorem implies that the norm of $(T_r^\psi)^*$ on $L^{\frac{4}{3}}(\mathcal{X}, \nu_M)$ is also bounded by $c\tilde{N}_1^{\frac{1}{2}}[\tilde{N}_2(r, M)]^{\frac{1}{2}} + \tilde{c}$. By duality, we further see that

$$\tilde{N}_4(r, M) \le c\tilde{N}_1^{\frac{1}{2}}[\tilde{N}_2(r, M)]^{\frac{1}{2}} + \tilde{c}.$$

Using interpolation again, we conclude that

$$\tilde{N}_2(r, M) \le c\tilde{N}_1^{\frac{1}{2}}[\tilde{N}_2(r, M)]^{\frac{1}{2}} + \tilde{c},$$

from which (8.2.46) follows. Thus, (8.2.45) holds true and the claim is true.

By (8.2.45), we see that $\{T_r\}_{r\in(0,\infty)}$ is uniformly bounded on $L^2(\mathcal{X}, \nu_M)$, with bounds also uniform in M. By letting $M \to \infty$, we see that $\{T_r\}_{r\in(0,\infty)}$ is uniformly bounded on $L^2(\mathcal{X}, \nu)$. Then there exists a weak limit $\tilde{T}$ bounded on $L^2(\mathcal{X}, \nu)$ and some sequence $r_i \to 0$ as $i \to \infty$. That is, for all $f \in L^2(\mathcal{X}, \nu)$ and $g \in L^2(\mathcal{X}, \nu)$,

$$\langle g, \tilde{T} f\rangle = \lim_{r_i\to 0}\langle g, T_{r_i} f\rangle.$$

By a standard argument, it is easy to show that $\tilde{T}$ is a Calderón–Zygmund operator with the same kernel K as T. It follows, from (i) $\Longrightarrow$ (iii) of Theorem 8.2.1 for the operator $\tilde{T}$, that $\tilde{T}$ is also bounded from $L^1(\mathcal{X}, \nu)$ to $L^{1,\infty}(\mathcal{X}, \nu)$. Applying Lemma 8.2.16, we know that T is also bounded on $L^2(\mathcal{X}, \nu)$. This finishes the proof of (iii) $\Longrightarrow$ (i) of Theorem 8.2.1 and hence the proof of Theorem 8.2.1. □

8.2.4 Boundedness of Maximal Calderón–Zygmund Operators

As an application of Theorem 8.2.1, we obtain the following boundedness of the maximal operators associated with the Calderón–Zygmund operators.

Theorem 8.2.17. *Let T be a Calderón–Zygmund operator with kernel K satisfying* (8.1.23) *and* (8.1.24)*, which is bounded on $L^2(\mathcal{X}, \nu)$, and $T^\sharp$ the maximal operator associated with T. Then the following statements hold true:*

(i) *Let $p \in (1, \infty)$. Then there exists a positive constant c such that, for all $f \in L^p(\mathcal{X}, \nu)$,*

$$\|T^\sharp f\|_{L^p(\mathcal{X},\nu)} \le c\|f\|_{L^p(\mathcal{X},\nu)};$$

(ii) *There exists a positive constant $\tilde{c}$ such that, for all $\omega \in \mathcal{M}(\mathcal{X})$,*

$$\|T^\sharp \omega\|_{L^{1,\infty}(\mathcal{X},\nu)} \le \tilde{c}\|\omega\|. \tag{8.2.47}$$

Moreover, for all $f \in L^1(\mathcal{X}, \nu)$,

$$\|T^\sharp f\|_{L^{1,\infty}(\mathcal{X},\nu)} \le \tilde{c}\|f\|_{L^1(\mathcal{X},\nu)}. \tag{8.2.48}$$

We begin the proof of Theorem 8.2.17 with the following inequality for $T^\sharp$ on the elementary measures.

Lemma 8.2.18. *Let* $p \in (0,1)$ *and* T *be a Calderón–Zygmund operator with kernel* K *satisfying* (8.1.23) *and* (8.1.24), *which is bounded on* $L^2(\mathcal{X}, \nu)$. *Then there exist positive constants* C *and* $C_{(p)}$ *such that, for all elementary measures* $\omega = \sum_i \alpha_i \delta_{x_i}$ *and* $x \in \operatorname{supp} \nu$,

$$\left[T^{\sharp}\omega(x)\right]^p \le C\left[\mathcal{M}_p^c T\omega(x)\right]^p + C_{(p)}[\mathcal{M}^c\omega(x)]^p. \tag{8.2.49}$$

Proof. As in Lemma 8.2.6, let $r \in (0,\infty)$, $r_j := 5^{j}r$,

$$\nu_j := \nu(\overline{B}(x, r_j)) \quad \text{for} \quad j \in \mathbb{Z}_+,$$

k be the *smallest positive integer* such that

$$\nu_{k+1} \le 4[C_{(\lambda)}]^6 \nu_{k-1} \quad \text{and} \quad R := r_{k-1} = 5^{k-1}r.$$

Similar to the proof of (8.2.12), we have

$$|T_r\omega(x) - T_{5R}\omega(x)| \lesssim \mathcal{M}^c\omega(x). \tag{8.2.50}$$

Now decompose the measure ω as $\omega = \omega_1 + \omega_2$, where

$$\omega_1 := \sum_{i:\ x_i \in \overline{B}(x,5R)} \alpha_i \delta_{x_i}$$

and

$$\omega_2 := \sum_{i:\ x_i \notin \overline{B}(x,5R)} \alpha_i \delta_{x_i}.$$

Applying (8.2.7) to T^*, we see that, for any $\tilde{x} \in \overline{B}(x, R)$,

$$\begin{aligned}
|T_{5R}\omega(x) - T\omega_2(\tilde{x})| &= \left|\int_{\mathcal{X}} K(x,y)\chi_{\mathcal{X}\setminus\overline{B}(x,5R)}(y)\,d\omega(y) - T\omega_2(\tilde{x})\right| \\
&= \left|\int_{\mathcal{X}} K(x,y)\,d\omega_2(y) - T\omega_2(\tilde{x})\right| \\
&= |T\omega_2(x) - T\omega_2(\tilde{x})| \\
&= |\langle\delta_x, T\omega_2\rangle - \langle\delta_{\tilde{x}}, T\omega_2\rangle| \\
&\le \int_{\mathcal{X}} |T^*(\delta_x - \delta_{\tilde{x}})(y)|\,d\omega_2(y) \\
&\le \int_{\mathcal{X}\setminus\overline{B}(x,5R)} |T^*(\delta_x - \delta_{\tilde{x}})(y)|\,d\omega(y) \\
&\lesssim \mathcal{M}^c\omega(x).
\end{aligned}$$

This implies that

$$\begin{aligned} \mathrm{H}_1 &:= \frac{1}{\nu(\overline{B}(x,R))} \int_{\overline{B}(x,R)} |T_{5R}\omega(x) - T\omega_2(\tilde{x})|^p \, d\omega(\tilde{x}) \\ &\lesssim [\mathcal{M}^c\omega(x)]^p. \end{aligned} \tag{8.2.51}$$

On the other hand, write

$$\begin{aligned} \mathrm{H}_2 &:= \frac{1}{\nu(\overline{B}(x,R))} \int_{\overline{B}(x,R)} |T\omega_2(\tilde{x}) - T\omega(\tilde{x})|^p \, d\nu(\tilde{x}) \\ &= \frac{1}{\nu(\overline{B}(x,R))} \int_{\overline{B}(x,R)} |T\omega_1(\tilde{x})|^p \, d\nu(\tilde{x}) \\ &= \frac{1}{\nu(\overline{B}(x,R))} \int_0^\infty ps^{p-1} \nu\left(\{\tilde{x} \in \overline{B}(x,R) : |T\omega_1(\tilde{x})| > s\}\right) ds. \end{aligned}$$

Since T is bounded on $L^2(\mathcal{X}, \nu)$, by Theorem 8.2.9, we know that, for every $s \in (0, \infty)$,

$$\nu\left(\{\tilde{x} \in \overline{B}(x,R) : |T\omega_1(\tilde{x})| > s\}\right) \lesssim \min\left\{\nu\left(\overline{B}(x,R)\right), \frac{\|\omega_1\|}{s}\right\}. \tag{8.2.52}$$

Observe that

$$\|\omega_1\| = \omega(\overline{B}(x, 5R)).$$

By this, together with (8.2.52), the definition of $\mathcal{M}^c\omega$ and (8.2.11), we see that

$$\begin{aligned} \nu\left(\{\tilde{x} \in \overline{B}(x,R) : |T\omega_1(\tilde{x})| > s\}\right) &\lesssim \nu\left(\overline{B}(x,R)\right) \min\left\{1, \frac{1}{s} \frac{\omega(\overline{B}(x,5R))}{\nu(\overline{B}(x,R))}\right\} \\ &\lesssim \nu\left(\overline{B}(x,R)\right) \min\left\{1, \frac{1}{s}\mathcal{M}^c\omega(x)\right\}, \end{aligned}$$

which further implies that

$$\begin{aligned} \mathrm{H}_2 &\lesssim \int_0^\infty ps^{p-1} \min\left\{1, \frac{1}{s}\mathcal{M}^c\omega(x)\right\} ds \\ &\sim \int_0^{\mathcal{M}^c\omega(x)} ps^{p-1} \, ds + \int_{\mathcal{M}^c\omega(x)}^\infty ps^{p-2}\mathcal{M}^c\omega(x) \, ds \\ &\lesssim [\mathcal{M}^c\omega(x)]^p. \end{aligned}$$

From this, combined with (8.2.51), we deduce that

$$\frac{1}{\nu(\overline{B}(x,R))}\int_{\overline{B}(x,R)} |T_{5R}\omega(x) - T\omega(\tilde{x})|^p \, d\nu(\tilde{x}) \lesssim \mathrm{H}_1 + \mathrm{H}_2 \lesssim [\mathcal{M}^c\omega(x)]^p.$$

Using this and (8.2.50), we see that

$$\begin{aligned}
|T_r\omega(x)|^p &= \frac{1}{\nu(\overline{B}(x, R))}\int_{\overline{B}(x, R)} |T_r\omega(x)|^p \, d\nu(\tilde{x}) \\
&\le \frac{1}{\nu(\overline{B}(x, R))}\int_{\overline{B}(x, R)} [|T_r\omega(x) - T_{5R}\omega(x)|^p \\
&\quad +|T_{5R}\omega(x) - T\omega(\tilde{x})|^p + |T\omega(\tilde{x})|^p] \, d\nu(\tilde{x}) \\
&\lesssim [\mathcal{M}^c\omega(x)]^p + \frac{1}{\nu(\overline{B}(x, R))}\int_{\overline{B}(x, R)} |T\omega(\tilde{x})|^p \, d\nu(\tilde{x}) \\
&\lesssim [\mathcal{M}^c\omega(x)]^p + \left[\mathcal{M}^c_p T\omega(x)\right]^p.
\end{aligned}$$

Taking the supremum over $r \in (0, \infty)$, we find that (8.2.49) holds true, which completes the proof of Lemma 8.2.18. □

As a result of Lemma 8.2.18, by Theorem 8.2.9 and (i) and (ii) of Lemma 8.2.3, we immediately obtain the following conclusion.

Proposition 8.2.19. *Let T be a Calderón–Zygmund operator with kernel K satisfying* (8.1.23) *and* (8.1.24)*, which is bounded on $L^2(\mathcal{X}, \nu)$. Then there exists a positive constant C such that, for all elementary measures $\omega \in \mathcal{M}(\mathcal{X})$,*

$$\left\|T^\sharp\omega\right\|_{L^{1,\infty}(\mathcal{X},\nu)} \le C\,\|\omega\|.$$

Proof of Theorem 8.2.17. By Theorem 8.2.1, Remark 8.2.7, Lemma 8.2.3(i) and a density argument, we have (i).

To prove (ii), it suffices to prove (8.2.47), since for any $f \in L^1(\mathcal{X}, \nu)$, if we define

$$d\omega := f d\nu,$$

then we see that $\omega \in \mathcal{M}(\mathcal{X})$ and (8.2.48) follows from (8.2.47). Moreover, recall that, for any complex measure $\omega \in \mathcal{M}(\mathcal{X})$, $|\omega|(\mathcal{X}) < \infty$.[4] Then, by considering the Jordan decompositions of real and imaginary parts of ω, we only need to prove (8.2.47) for any finite nonnegative measure.

To this end, assume that ω is a finite nonnegative measure and fix $t \in (0, \infty)$. We show that

[4] See [111, Theorem 6.4].

$$\nu\left(\left\{x \in \mathcal{X} : \left|T^{\sharp}\omega(x)\right| > t\right\}\right) \lesssim \frac{\|\omega\|}{t}.$$

Let $R \in (0, \infty)$ and consider the truncated maximal operator

$$T_R^{\sharp}\omega := \sup_{r>R} |T_r\omega|.$$

Since $T_R^{\sharp}\omega(x)$ increases to $T^{\sharp}\omega(x)$ pointwise on $\mathcal{X}$ as $R \to 0$, it suffices to show that there exists a positive constant C such that, for every $R \in (0, \infty)$,

$$\nu\left(\left\{x \in \mathcal{X} : \left|T_R^{\sharp}\omega(x)\right| > t\right\}\right) \leq \frac{C\,\|\omega\|}{t}. \tag{8.2.53}$$

In what follows, we use $\mathbb{P}$ to denote a *probability measure* on a *probability space* Ω, $\mathbb{P}(A)$ the *probability of the event* $A \subset \Omega$, $\mathbb{E}(\xi)$ the *mathematical expectation of a random variable* $\xi \in L^1(\mathbb{P})$ and

$$\mathbb{V}(\xi) := \mathbb{E}[(\xi - \mathbb{E}\xi)^2] = \mathbb{E}\xi^2 - (\mathbb{E}\xi)^2$$

the *variance of* $\xi \in L^2(\mathbb{P})$.

For each $N \in \mathbb{N}$, consider the *random elementary measure*

$$\omega_N := \frac{\|\omega\|}{N} \sum_{i=1}^{N} \delta_{x_i},$$

where the random points $\{x_i\}_{i=1}^N \subset \mathcal{X}$ are independent and satisfy that

$$\mathbb{P}(\{x_i \in E\}) = \omega(E)/\|\omega\|$$

for every Borel set $E \subset \mathcal{X}$. This immediately implies that

$$\mathbb{E}f(x_i) = \frac{1}{\|\omega\|} \int_{\mathcal{X}} f(z)\,d\omega(z)$$

for $f = \chi_E$ by definition, for simple functions f by linearity, and finally for all $f \in L^1(\mathcal{X}, \nu)$ by approximation. From this, we deduce that, for every $x \in \mathcal{X}$ and $r > R$,

$$\mathbb{E}[(T_r\delta_{x_i})(x)] = \frac{1}{\|\omega\|} T_r\omega(x). \tag{8.2.54}$$

Indeed,

$$\|\omega\|\mathbb{E}[(T_r\delta_{x_i})(x)] = \int_{\mathcal{X}} (T_r\delta_z)(x)\,d\omega(z)$$

$$= \int_{\mathcal{X}} \int_{d(y,z)>r} K(x,y)d\delta_z(y)d\omega(z)$$

$$= \int_{\mathcal{X}} \chi_{d(x,z)>r} K(x,z)d\omega(z) = T_r\omega(x).$$

Thus, (8.2.54) holds true.

Fix some $x_0 \in \mathcal{X}$ and $M \in (R, \infty)$. On the other hand, from (7.1.3) and (7.1.1), we deduce that, for any $x \in \overline{B}(x_0, M)$,

$$\lambda(x_0, M) \lesssim \lambda(x, M) \lesssim [C_{(\lambda)}]^{1+\log_2(M/R)}\lambda(x, R).$$

By this, the fact that $r > R$, (8.2.54) and (8.1.23), we conclude that, for any point $x \in \overline{B}(x_0, M)$,

$$\begin{aligned} \mathbb{V}[T_r\delta_{x_i}(x)] &\leq \mathbb{E}\left[|T_r\delta_{x_i}(x)|^2\right] \\ &= \int_{\Omega} \left[\int_{\mathcal{X}} K(x,y)\, d\delta_{x_i}(y)\right]^2 d\mathbb{P} \\ &= \int_{\Omega} [K(x,x_i)]^2 \chi_{\mathcal{X}\setminus\overline{B}(x,r)}(x_i)\, d\mathbb{P} \\ &\lesssim \frac{1}{[\lambda(x,r)]^2} \\ &\lesssim \frac{[C_{(\lambda)}]^{2[1+\log_2(M/R)]}}{[\lambda(x_0,M)]^2}. \end{aligned} \tag{8.2.55}$$

Moreover, by (8.2.54), we see that

$$\mathbb{E}[(T_r\omega_N)(x)] = \sum_{i=1}^{N} \frac{\|\omega\|}{N} \mathbb{E}[(T_r\delta_{x_i})(x)] = T_r\omega(x). \tag{8.2.56}$$

This, together with the Cauchy inequality and (8.2.55), implies that there exists a positive constant c, independent of x_0, M, r, R and N, such that

$$\begin{aligned} \mathbb{V}[T_r\omega_N(x)] &= \frac{\|\omega\|^2}{N^2}\mathbb{V}\left[\sum_{i=1}^{N} T_r\delta_{x_i}(x)\right] \\ &\leq \frac{\|\omega\|^2}{N}\sum_{i=1}^{N} \mathbb{V}\,[T_r\delta_{x_i}(x)] \\ &\leq c\frac{\|\omega\|^2}{N}\frac{[C_{(\lambda)}]^{2[1+\log_2(M/R)]}}{[\lambda(x_0,M)]^2}. \end{aligned}$$

Fix a number $\gamma \in (0, \infty)$ small enough. From the fact above, the Chebyshev inequality and (8.2.56), we deduce that, for every point $x \in \overline{B}(x_0, M)$ satisfying that $|T_r\omega(x)| > t$, it holds true that

$$\begin{aligned}
\mathbb{P}(\{|T_r\omega_N(x)| \le (1-\gamma)t\}) \\
&\le \mathbb{P}(\{|T_r\omega_N(x) - T_r\omega(x)| > \gamma t\}) \\
&\le \frac{\mathbb{V}(T_r\omega_N)(x)}{\gamma^2 t^2} \\
&\le c\frac{1}{\gamma^2 t^2}\frac{\|\omega\|^2}{N}\frac{[C_{(\lambda)}]^{2[1+\log_2(M/R)]}}{[\lambda(x_0, M)]^2} \\
&\le \gamma,
\end{aligned}$$

provided that

$$N \ge c\frac{\|\omega\|^2}{\gamma^3 t^2}\frac{[C_{(\lambda)}]^{2[1+\log_2(M/R)]}}{[\lambda(x_0, M)]^2}.$$

Since $r > R$ is arbitrary, we deduce that, for each $x \in \mathcal{X}$ satisfying $T_R^\sharp\omega(x) > t$,

$$\mathbb{P}\left(\left\{T_R^\sharp\omega_N(x) \le (1-\gamma)t\right\}\right) \le \gamma.$$

Let E be any given Borel set with $\nu(E) < \infty$ such that $T_R^\sharp\omega(x) > t$ for every $x \in E$. Then

$$\begin{aligned}
&\mathbb{E}\left(\nu\left(\left\{x \in E : T_R^\sharp\omega_N(x) \le (1-\gamma)t\right\}\right)\right) \\
&\quad = \int_E \mathbb{P}\left(\left\{T_R^\sharp\omega_N(x) \le (1-\gamma)t\right\}\right) d\nu(x) \\
&\quad \le \gamma\nu(E).
\end{aligned}$$

Thus there exists at least one choice of points $\{x_i\}_{i=1}^N$ such that

$$\nu(\{x \in E : T_R^\sharp\omega_N(x) \le (1-\gamma)t\}) \le \gamma\nu(E)$$

and therefore

$$\nu(\{x \in E : T_R^\sharp\omega_N(x) > (1-\gamma)t\}) \ge (1-\gamma)\nu(E).$$

From this, together with Proposition 8.2.19, it follows that

$$\nu(E) \le \frac{1}{1-\gamma}\nu\left(\left\{x \in E : T_R^\sharp\omega_N(x) > (1-\gamma)t\right\}\right)$$

$$\le \frac{1}{(1-\gamma)^2 t} \left\| T_R^\sharp \omega_N \right\|_{L^{1,\infty}(\mathcal{X},\nu)}$$
$$\lesssim \frac{1}{(1-\gamma)^2 t} \|\omega_N\|$$
$$\lesssim \frac{1}{(1-\gamma)^2 t} \|\omega\|.$$

Since $\gamma \in (0,\infty)$ is arbitrary, we see that $\nu(E) \lesssim \frac{\|\omega\|}{t}$. As E is an arbitrary subset of finite measure of the set of the points $x \in \mathcal{X}$ for which $T_R^\sharp \omega(x) > t$, we obtain (8.2.53), which completes the proof of Theorem 8.2.17. □

Remark 8.2.20. If we replace the assumption of Theorem 8.2.17 that T is bounded on $L^2(\mathcal{X},\nu)$ by that T is bounded on $L^q(\mathcal{X},\nu)$ for some $q \in (1,\infty)$, then Theorem 8.2.17 still holds true.

8.3 Boundedness of Calderón–Zygmund Operators: Equivalent Characterizations II

In this section, we prove that the boundedness of a Calderón–Zygmund operator on $L^2(\mathcal{X},\nu)$ is equivalent to either its boundedness from the atomic Hardy space $H^1(\mathcal{X},\nu)$ to $L^{1,\infty}(\mathcal{X},\nu)$ or from $H^1(\mathcal{X},\nu)$ to $L^1(\mathcal{X},\nu)$.

Let T be a linear operator as in (8.1.25) with kernel K satisfying (8.1.23) and the *Hörmander condition* that there exists a positive constant C such that, for all $x \ne \tilde{x}$,

$$\int_{d(x,y)\ge 2d(x,\tilde{x})} [|K(x,y) - K(\tilde{x},y)| + |K(y,x) - K(y,\tilde{x})|]\, d\mu(y) \le C. \tag{8.3.1}$$

This integral in (8.1.25) may not be convergent even for nice functions. For this reason, we consider the truncated operator T_ϵ in (8.2.2) for any $\epsilon \in (0,\infty)$.

We say that T is *bounded* on $L^p(\mathcal{X},\nu)$ for $p \in (1,\infty)$ if $\{T_\epsilon\}_{\epsilon\in(0,\infty)}$ is bounded on $L^p(\mathcal{X},\nu)$ uniformly in $\epsilon \in (0,\infty)$, and T is *bounded* from a Banach space $\mathcal{Y}$ to $L^{p,\infty}(\mathcal{X},\nu)$ for $p \in [1,\infty)$ if $\{T_\epsilon\}_{\epsilon\in(0,\infty)}$ is bounded from $\mathcal{Y}$ to $L^{p,\infty}(\mathcal{X},\nu)$ uniformly in $\epsilon \in (0,\infty)$.

Lemma 8.3.1. *Let $\epsilon \in (0,\infty)$, $r \in (0,1)$, $\rho \in (1,\infty)$, T and T_ϵ be respectively as in* (8.1.25) *and* (8.2.2) *with kernel K satisfying* (8.1.23) *and* (8.3.1). *If T is bounded from $H^1(\mathcal{X},\nu)$ to $L^{1,\infty}(\mathcal{X},\nu)$, then there exists a positive constant C, depending on r, such that, for all $\epsilon \in (0,\infty)$, $\rho \in (1,\infty)$, balls B and functions $a \in L^\infty(\mathcal{X},\nu)$ supported on B,*

$$\frac{1}{\nu(\rho B)} \int_B |T_\epsilon(a)(x)|^r\, d\nu(x) \le C \|a\|^r_{L^\infty(\mathcal{X},\nu)}.$$

Proof. By similarity, without loss of generality, we may assume $\rho = 2$. For any given ball $B := B(x_B, r_B)$,we consider the following two cases on r_B.

Case (i) $r_B \leq \operatorname{diam}(\operatorname{supp}\nu)/40$. We first claim that there exists a $j_0 \in \mathbb{N}$ such that

$$\nu(6^{j_0} B \setminus 2B) > 0. \tag{8.3.2}$$

Indeed, if for all $j \in \mathbb{N}$, $\nu(6^j B \setminus 2B) = 0$, then we see that $\nu(\mathcal{X} \setminus 2B) = 0$, which implies that $\operatorname{supp}\nu \subset \overline{2B}$, the *closure* of $2B$. This contradicts to that $r_B \leq \operatorname{diam}(\operatorname{supp}\nu)/40$ and thus the claim holds true.

Now assume that S is the smallest ball of the form $6^j B$ such that (8.3.2) holds true. We then have that $\nu(6^{-1}S \setminus 2B) = 0$ and $\nu(S \setminus 2B) > 0$. Thus,

$$\nu\left(S \setminus \left(6^{-1}S \bigcup 2B\right)\right) > 0.$$

By this and Lemma 7.1.13, we choose $x_0 \in S \setminus (6^{-1}S \cup 2B)$ such that the ball centered at x_0 with the radius $6^{-k} r_S$ for some $k \geq 2$ is $(6, \beta_6)$-doubling. Let B_0 be the biggest ball of this form. Then we see that $B_0 \subset 2S$ and $d(B_0, B) \gtrsim r_B$. We now claim that

$$\delta(B, 2S) \lesssim 1. \tag{8.3.3}$$

Indeed, if $S = 6B$, then, by Lemma 7.1.16(ii), we have (8.3.3). If $S \supset 6^2 B$, then $\frac{1}{12}S \supset 3B$. Notice that, in this case, $\nu(6^{-1}S \setminus 2B) = 0$ implies that $\delta(2B, \frac{1}{12}S) = 0$. Thus, by this, together with (ii) and (iv) of Lemma 7.1.16, we further have

$$\begin{aligned}
\delta(B, 2S) &\leq \delta(B, 2B) + \delta\left(2B, \frac{1}{12}S\right) + \delta\left(\frac{1}{12}S, 2S\right) \\
&= \delta(B, 2B) + \delta\left(\frac{1}{12}S, 2S\right) \\
&\lesssim 1.
\end{aligned}$$

Thus, (8.3.3) also holds true in this case. This shows the claim.

Moreover, by the definition of B_0, we see that

$$r_{6(\widetilde{6B_0}^6)} \geq r_S \quad \text{and} \quad 2S \subset 24(\widetilde{6B_0}^6).$$

Therefore, by (i) through (iv) of Lemma 7.1.16, we conclude that

$$\begin{aligned}
\delta(B_0, 2S) &\leq \delta(B_0, 24(\widetilde{6B_0}^6)) \\
&\leq \delta(B_0, \widetilde{6B_0}^6) + \delta(\widetilde{6B_0}, 24(\widetilde{6B_0}^6)) \\
&\lesssim 1.
\end{aligned} \tag{8.3.4}$$

For any $a \in L^\infty(\mathcal{X}, \nu)$ supported on B, we define an atomic block b, supported on $2S$, by $b := a + c_{B_0}\chi_{B_0}$, where c_{B_0} is a constant such that $\int_{\mathcal{X}} b(x)\,d\nu(x) = 0$. Clearly, we know that

$$\|b\|_{H^{1,\infty}_{\rm atb}(\mu)} \le [1+\delta(B,2S)]\|a\|_{L^\infty(\mathcal{X},\nu)}\nu(2B) + [1+\delta(B_0,2S)]|c_{B_0}|\nu(2B_0).$$

From the choice of c_{B_0}, the doubling property of B_0 and the assumption of a, we deduce that

$$c_{B_0} = -\int_{\mathcal{X}} a(x)\,d\nu(x)/\nu(B_0)$$

and

$$|c_{B_0}|\nu(2B_0) \lesssim \|a\|_{L^1(\mathcal{X},\nu)} \lesssim \|a\|_{L^\infty(\mathcal{X},\nu)}\nu(2B). \tag{8.3.5}$$

This, together with (8.3.3) and (8.3.4), implies that

$$\|b\|_{H^{1,\infty}_{\rm atb}(\mu)} \lesssim \|a\|_{L^\infty(\mathcal{X},\nu)}\nu(2B). \tag{8.3.6}$$

By (8.1.23), we conclude that, for any $x \in B$,

$$\begin{aligned}
|T_\epsilon(c_{B_0}\chi_{B_0})(x)| &= \left|\int_{d(x,y)>\epsilon} K(x,y)c_{B_0}\chi_{B_0}(y)\,d\nu(y)\right| \\
&\le |c_{B_0}|\int_{B_0} |K(x,y)|\,d\nu(y) \\
&\lesssim |c_{B_0}|\int_{B_0} \frac{1}{\lambda(x,d(x,y))}\,d\nu(y).
\end{aligned}$$

Notice that, for any $x \in B$ and $y \in B_0$, $d(x,y) \gtrsim r_B$. Then, by this, together with (7.1.1), (7.1.3) and (8.3.5), we have

$$\begin{aligned}
|T_\epsilon(c_{B_0}\chi_{B_0})(x)| &\lesssim |c_{B_0}|\frac{\nu(B_0)}{\lambda(x,r_B)} \\
&\lesssim \frac{\|a\|_{L^\infty(\mathcal{X},\nu)}\nu(2B)}{\lambda(x_B,r_B)} \\
&\lesssim \|a\|_{L^\infty(\mathcal{X},\nu)}.
\end{aligned} \tag{8.3.7}$$

On the other hand, by the boundedness from $H^1(\mathcal{X}, \nu)$ to $L^{1,\infty}(\mathcal{X}, \nu)$ of T, for all $f \in H^1(\mathcal{X}, \nu)$ and balls B, we know that

$$\begin{aligned}
&\int_B |T_\epsilon f(x)|^r \, d\nu(x)\\
&= r\int_0^\infty \nu(\{x \in B : |T_\epsilon f(x)| > \lambda\})\, \lambda^{r-1}\, d\lambda\\
&= r\int_0^{\frac{\|f\|_{H^1(\mathcal{X},\nu)}}{\nu(B)}} \nu(\{x \in B : |T_\epsilon f(x)| > \lambda\})\, \lambda^{r-1}\, d\lambda + r\int_{\frac{\|f\|_{H^1(\mathcal{X},\nu)}}{\nu(B)}}^\infty \cdots\\
&\lesssim \nu(B)\int_0^{\frac{\|f\|_{H^1(\mathcal{X},\nu)}}{\nu(B)}} \lambda^{r-1}\, d\lambda + \int_{\frac{\|f\|_{H^1(\mathcal{X},\nu)}}{\nu(B)}}^\infty \|f\|_{H^1(\mathcal{X},\nu)}\lambda^{r-2}\, d\lambda\\
&\lesssim \frac{\|f\|^r_{H^1(\mathcal{X},\nu)}}{[\nu(B)]^{r-1}}.
\end{aligned}$$

Thus, this estimate, together with (8.3.6) and (8.3.7), further implies that

$$\begin{aligned}
&\int_B |T_\epsilon(a)(x)|^r \, d\nu(x)\\
&\quad\le \int_B |T_\epsilon(b)(x)|^r \, d\nu(x) + \int_B |T_\epsilon(c_{B_0}\chi_{B_0})(x)|^r \, d\nu(x)\\
&\quad\lesssim \frac{\|b\|^r_{H^1(\mathcal{X},\nu)}}{[\nu(B)]^{r-1}} + \nu(B)\|a\|^r_{L^\infty(\mathcal{X},\nu)}\\
&\quad\lesssim \nu(2B)\|a\|^r_{L^\infty(\mathcal{X},\nu)}.
\end{aligned} \tag{8.3.8}$$

Case (ii) $r_B >$ diam (supp $\nu)/40$. In this case, without loss of generality, we may assume that $r_B \le 8\,\mathrm{diam}\,(\mathrm{supp}\,\nu)$. Then, by Proposition 7.1.8, we know that $B \cap \mathrm{supp}\,\nu$ is covered by finite number balls $\{B_j\}_{j=1}^N$ with radius $r_B/800$, where $N \in \mathbb{N}$. For $j \in \{1, \ldots, N\}$ and a as in Lemma 8.3.1, we define

$$a_j := \frac{\chi_{B_j}}{\sum_{k=1}^N \chi_{B_k}} a.$$

From Case (i), we deduce that (8.3.8) is true if we replace B by $2B_j$ which contains the support of a_j. This, together with (8.1.23) and (8.3.8), implies that

$$\begin{aligned}
\int_B |T_\epsilon(a)(x)|^r \, d\nu(x) &\le \sum_{j=1}^N \int_{2B_j} |T_\epsilon(a_j)(x)|^r \, d\nu(x)\\
&\lesssim \sum_{j=1}^N \nu(4B_j)\|a_j\|^r_{L^\infty(\mathcal{X},\nu)}\\
&\lesssim \|a\|^r_{L^\infty(\mathcal{X},\nu)}\nu(2B),
\end{aligned}$$

which, combined with Case (i), completes the proof of Lemma 8.3.1. □

For each $r \in (0, \infty)$, $\rho \in [5, \infty)$, any $f \in L^r_{\rm loc}(\mathcal{X}, \nu)$ and $x \in \mathcal{X}$, define

$$\mathcal{M}_{r,\rho}(f)(x) := \sup_{B \ni x} \left\{ \frac{1}{\nu(\rho B)} \int_B |f(y)|^r \, d\nu(y) \right\}^{\frac{1}{r}}, \tag{8.3.9}$$

and

$$\mathcal{M}^\sharp_r(f)(x) := \{\mathcal{M}^\sharp(|f|^r)(x)\}^{\frac{1}{r}},$$

where $\mathcal{M}^\sharp$ is as in (8.1.13). As in (7.1.8), when $r = 1$, we *write* $\mathcal{M}_{r,\rho}(f)$ *simply by* $\mathcal{M}_\rho(f)$; also, when $r = 1$ and $\rho = 5$, we *write* $\mathcal{M}_{r,\rho}(f)$ *simply by* $\mathcal{M}(f)$.

Applying Lemma 8.3.1 and Theorem 8.1.1, we then have the following uniform boundedness of $\mathcal{M}^\sharp_r \circ T_\epsilon$ on $L^p(\mathcal{X}, \nu)$, with $p \in (1, \infty]$, for all $\epsilon \in (0, \infty)$.

Lemma 8.3.2. *Let $\epsilon \in (0, \infty)$, T and T_ϵ be as in Lemma 8.3.1. Then the following two statements hold true:*

(i) *Let $r \in (0, 1]$. Then there exists a positive constant $C_{(r)}$, depending on r, such that, for all $\epsilon \in (0, \infty)$ and $f \in L^\infty_b(\mathcal{X}, \nu)$,*

$$\left\|\mathcal{M}^\sharp_r(T_\epsilon f)\right\|_{L^\infty(\mathcal{X},\nu)} \le C_{(r)} \|f\|_{L^\infty(\mathcal{X},\nu)}. \tag{8.3.10}$$

(ii) *Let $r \in (0, 1)$ and $p \in (1, \infty)$. Then there exists a positive constant $C_{(p,r)}$, depending on p and r, such that, for all $\epsilon \in (0, \infty)$, $f \in L^\infty_b(\mathcal{X}, \nu)$ and $t \in (0, \infty)$,*

$$\nu\left(\{x \in \mathcal{X} : \mathcal{M}^\sharp_r(T_\epsilon f)(x) > t\}\right) \le \frac{C_{(p,r)}}{t^p} \|f\|^p_{L^p(\mathcal{X},\nu)}.$$

Proof. Let $\tilde r \in (0, 1)$. Applying Lemma 8.3.1, we conclude that, for any $\epsilon \in (0, \infty)$ and ball B,

$$\int_B \left|T_\epsilon\left(f\chi_{\mathcal{X}\setminus\frac{4}{3}B}\right)(x)\right|^{\tilde r} d\nu(x) < \infty.$$

Thus, for all balls B and ν-almost every $x \in \mathcal{X}$,

$$\left|T_\epsilon\left(f\chi_{\mathcal{X}\setminus\frac{4}{3}B}\right)(x)\right| < \infty.$$

To show (i), it suffices to prove that, for all balls $B \subset S$,

$$\begin{aligned} \mathrm{H}_1 &:= \frac{1}{\nu(5B)} \int_B \left||T_\epsilon f(x)|^r - m_B\left(\left|T_\epsilon\left(f\chi_{\mathcal{X}\setminus\frac{4}{3}B}\right)\right|^r\right)\right| d\nu(x) \\ &\lesssim \|f\|^r_{L^\infty(\mathcal{X},\nu)} \end{aligned} \tag{8.3.11}$$

and

$$\begin{aligned}\mathrm{H}_2 :&= \left|m_B\left(\left|T_\epsilon\left(f\chi_{\mathcal{X}\setminus\frac{4}{3}B}\right)\right|^r\right) - m_S\left(\left|T_\epsilon\left(f\chi_{\mathcal{X}\setminus\frac{4}{3}S}\right)\right|^r\right)\right| \\ &\lesssim [1+\delta(B,S)]\|f\|^r_{L^\infty(\mathcal{X},\nu)}. \end{aligned} \tag{8.3.12}$$

We first prove (8.3.12). Let N be the smallest integer such that $S \subset 2^{N-2}B$. We then claim that

$$r_S \geq 2^{N-4}r_B \quad \text{and} \quad 2^{N-4}B \subset 2S. \tag{8.3.13}$$

Indeed, let

$$B := B(x_B, r_B) \quad \text{and} \quad S := B(x_S, r_S).$$

If $r_S < 2^{N-4}r_B$, then, for any $z \in S$, by $x_B \in S$ and $B \subset S$, we have

$$d(z, x_B) \leq d(z, x_S) + d(x_S, x_B) < 2r_S < 2^{N-3}r_B.$$

Thus, $S \subset 2^{N-3}B$, which contradicts to the choice of N. Therefore, $r_S \geq 2^{N-4}r_B$ which, combined with $B \subset S$, further implies that $2^{N-4}B \subset 2S$. Thus, (8.3.13) holds true.

Obviously, we have

$$\begin{aligned}\mathrm{H}_2 \leq\ & \left|m_B\left(\left|T_\epsilon\left(f\chi_{\mathcal{X}\setminus\frac{4}{3}B}\right)\right|^r\right) - m_B\left(\left|T_\epsilon\left(f\chi_{\mathcal{X}\setminus 2^N B}\right)\right|^r\right)\right| \\ &+ \left|m_B\left(\left|T_\epsilon\left(f\chi_{\mathcal{X}\setminus 2^N B}\right)\right|^r\right) - m_S\left(\left|T_\epsilon\left(f\chi_{\mathcal{X}\setminus 2^N B}\right)\right|^r\right)\right| \\ &+ \left|m_S\left(\left|T_\epsilon\left(f\chi_{\mathcal{X}\setminus\frac{4}{3}S}\right)\right|^r\right) - m_S\left(\left|T_\epsilon\left(f\chi_{\mathcal{X}\setminus 2^N B}\right)\right|^r\right)\right| \\ \leq\ & m_B\left(\left|T_\epsilon\left(f\chi_{2^N B\setminus\frac{4}{3}B}\right)\right|^r\right) \\ &+ \left|m_B\left(\left|T_\epsilon\left(f\chi_{\mathcal{X}\setminus 2^N B}\right)\right|^r\right) - m_S\left(\left|T_\epsilon\left(f\chi_{\mathcal{X}\setminus 2^N B}\right)\right|^r\right)\right| \\ &+ m_S\left(\left|T_\epsilon\left(f\chi_{2^N B\setminus\frac{4}{3}S}\right)\right|^r\right) \\ =:\ & \mathrm{I}_1 + \mathrm{I}_2 + \mathrm{I}_3.\end{aligned}$$

To estimate I_3, by (7.1.3), we see that, for any $y \in S$ and $z \in (2^N B) \setminus \frac{4}{3}S$,

$$\lambda(y, d(y,z)) \sim \lambda(z, d(y,z)) \sim \lambda(z, d(x_S,z)) \sim \lambda(x_S, d(x_S,z)) \gtrsim \lambda(x_S, r_S),$$

which, via (8.1.23) and (8.3.13), implies that, for all $y \in S$,

$$\begin{aligned}\left|T_\epsilon\left(f\chi_{2^N B\setminus\frac{4}{3}S}\right)(y)\right| &= \left|\int_{d(y,z)>\epsilon} K(y,z)f(z)\chi_{2^N B\setminus\frac{4}{3}S}(z)\,d\nu(z)\right| \\ &\lesssim \int_{d(y,z)\geq\frac{1}{3}r_S} \frac{|f(z)|\chi_{2^N B\setminus\frac{4}{3}S}(z)}{\lambda(y,d(y,z))}\,d\nu(z) \\ &\lesssim \|f\|_{L^\infty(\mathcal{X},\nu)}\frac{\nu(2^N B)}{\lambda(x_S,r_S)} \\ &\lesssim \|f\|_{L^\infty(\mathcal{X},\nu)}.\end{aligned}$$

Thus,

$$\mathrm{I}_3 \lesssim \|f\|^r_{L^\infty(\mathcal{X},\nu)}.$$

A trivial computation, involving (8.1.23) and (8.3.1), shows that, for any $x \in B$ and $y \in S$, we have

$$\begin{aligned}&\left||T_\epsilon(f\chi_{\mathcal{X}\setminus 2^N B})(x)|^r - |T_\epsilon(f\chi_{\mathcal{X}\setminus 2^N B})(y)|^r\right| \\ &\quad\leq \left|T_\epsilon(f\chi_{\mathcal{X}\setminus 2^N B})(x) - T_\epsilon(f\chi_{\mathcal{X}\setminus 2^N B})(y)\right|^r \\ &\quad\leq \left[\|f\|_{L^\infty(\mathcal{X},\nu)}\int_{\mathcal{X}\setminus 2^N B}|K(x,z)-K(y,z)|\,d\nu(z)\right]^r \\ &\quad\lesssim \|f\|^r_{L^\infty(\mathcal{X},\nu)}.\end{aligned}$$

This, via the fact that

$$\mathrm{I}_2 \leq \frac{1}{\nu(B)\nu(S)}\int_B\int_S \left||T_\epsilon(f\chi_{\mathcal{X}\setminus 2^N B})(x)|^r - |T_\epsilon(f\chi_{\mathcal{X}\setminus 2^N B})(y)|^r\right|\,d\nu(x)\,d\nu(y),$$

implies the desired estimate that

$$\mathrm{I}_2 \lesssim \|f\|^r_{L^\infty(\mathcal{X},\nu)}.$$

On the other hand, from (8.3.13), (8.1.23), and (i) and (ii) of Lemma 7.1.16, we deduce that, for all $y \in B$,

$$\begin{aligned}&\left|T_\epsilon(f\chi_{2^N B\setminus\frac{4}{3}B})(y)\right| \\ &\quad\lesssim \|f\|_{L^\infty(\mathcal{X},\nu)} \\ &\qquad\times\left[\int_{2^N B\setminus 2^{N-4}B}\frac{1}{\lambda(y,d(y,z))}\,d\nu(z) + \int_{2^{N-4}B\setminus 2B}\cdots + \int_{2B\setminus\frac{4}{3}B}\cdots\right] \\ &\quad\lesssim \|f\|_{L^\infty(\mathcal{X},\nu)}[1+\delta(B,S)],\end{aligned}$$

and hence

$$\mathrm{I}_1 \lesssim \|f\|_{L^\infty(\mathcal{X},\nu)}^r [1 + \delta(B, S)]^r \lesssim \|f\|_{L^\infty(\mathcal{X},\nu)}^r [1 + \delta(B, S)].$$

Combining the estimates for I_1 through I_3, we obtain (8.3.12).

To prove (8.3.11), for a fixed ball B and any $f \in L_b^\infty(\mathcal{X}, \nu)$, decompose f as

$$f = f\chi_{\frac{4}{3}B} + f\chi_{\mathcal{X}\setminus\frac{4}{3}B} =: f_1 + f_2$$

and let

$$h_B := m_B(|T_\epsilon f_2|^r).$$

Then, for any $x, y \in B$, an easy computation, involving (8.1.23) and (8.3.1), shows that

$$\begin{aligned} ||T_\epsilon f_2(x)|^r - |T_\epsilon f_2(y)|^r| &\le |T_\epsilon f_2(x) - T_\epsilon f_2(y)|^r \\ &\le \left[\int_{\mathcal{X}\setminus\frac{4}{3}B} |K(x,z) - K(y,z)|\, d\nu(z)\right]^r \|f\|_{L^\infty(\mathcal{X},\nu)}^r \\ &\lesssim \|f\|_{L^\infty(\mathcal{X},\nu)}^r. \end{aligned}$$

Thus, we see that, for any $x \in B$,

$$\begin{aligned} ||T_\epsilon f_2(x)|^r - h_B| &= \frac{1}{\nu(B)} \left| \int_B (|T_\epsilon f_2(y)|^r - |T_\epsilon f_2(x)|^r)\, d\nu(y) \right| \\ &\lesssim \|f\|_{L^\infty(\mathcal{X},\nu)}^r, \end{aligned}$$

which implies that

$$\frac{1}{\nu(5B)} \int_B ||T_\epsilon f_2(x)|^r - h_B|\, d\nu(x) \lesssim \|f\|_{L^\infty(\mathcal{X},\nu)}^r. \tag{8.3.14}$$

By this fact and Lemma 8.3.1, we conclude that

$$\begin{aligned} &\frac{1}{\nu(5B)} \int_B ||T_\epsilon f(x)|^r - h_B|\, d\nu(x) \\ &\lesssim \frac{1}{\nu(5B)} \int_B ||T_\epsilon f(x)|^r - |T_\epsilon f_2(x)|^r|\, d\nu(x) \\ &\quad + \frac{1}{\nu(5B)} \int_B ||T_\epsilon f_2(x)|^r - h_B|\, d\nu(x) \end{aligned}$$

$$\lesssim \frac{1}{\nu(5B)} \int_B |T_\epsilon f_1(x)|^r \, d\nu(x) + \|f\|^r_{L^\infty(\mathcal{X},\nu)}$$
$$\lesssim \|f\|^r_{L^\infty(\mathcal{X},\nu)}.$$

Then, from (8.3.14), we deduce that

$$\begin{aligned}
\mathrm{H}_1 &\leq \frac{1}{\nu(5B)} \int_B \left||T_\epsilon f(x)|^r - h_B\right| d\nu(x) + \frac{\nu(B)}{\nu(5B)} |h_B - m_B(|T_\epsilon f_2|^r)| \\
&\lesssim \|f\|^r_{L^\infty(\mathcal{X},\nu)} + \frac{1}{\nu(5B)} \int_B \left||T_\epsilon f_2(x)|^r - h_B\right| d\nu(x) \\
&\lesssim \|f\|^r_{L^\infty(\mathcal{X},\nu)}.
\end{aligned}$$

This, together with (8.3.11), completes the proof of (i).

To prove (ii), for any fixed $t \in (0,\infty)$ and $f \in L^\infty_b(\mathcal{X},\nu)$, applying Theorem 8.1.1, we conclude that, with the notation same as in the proof of Theorem 8.1.1, $f = g + h$. Moreover, by Theorem 8.1.1, we see that

$$\|g\|_{L^\infty(\mathcal{X},\nu)} \lesssim t, \; h \in H^1(\mathcal{X},\nu) \quad \text{and} \quad \|h\|_{H^1(\mathcal{X},\nu)} \lesssim t^{1-p} \|f\|^p_{L^p(\mathcal{X},\nu)}.$$

The estimate (8.3.10) implies that there exists a positive constant C, independent of ϵ and g, such that

$$\left\|\mathcal{M}^\sharp_r(T_\epsilon g)\right\|_{L^\infty(\mathcal{X},\nu)} \lesssim \|g\|_{L^\infty(\mathcal{X},\nu)} \leq Ct. \tag{8.3.15}$$

By the definitions of $\mathcal{M}_{r,5}$ and $\mathcal{M}^\sharp_r$, we see that there exists a positive constant c, independent of ϵ, f, g and h, such that

$$\mathcal{M}^\sharp_r(T_\epsilon(f)) \leq c\left[\mathcal{M}^\sharp_r(T_\epsilon g) + \mathcal{M}_{r,5}(T_\epsilon h)\right]. \tag{8.3.16}$$

From (8.3.16) and (8.3.15), we deduce that

$$\begin{aligned}
&\nu\left(\{x \in \mathcal{X} : \mathcal{M}^\sharp_r(T_\epsilon f)(x) > c(C+1)t\}\right) \\
&\quad \leq \nu\left(\{x \in \mathcal{X} : \mathcal{M}_{r,5}(T_\epsilon h)(x) > t\}\right).
\end{aligned} \tag{8.3.17}$$

Obviously, we have

$$\begin{aligned}
\{\mathcal{M}_{r,5}(T_\epsilon h)\}^r &\leq \left\{\mathcal{M}_{r,5}\left([T_\epsilon h]\chi_{\{x\in\mathcal{X}:\, |T_\epsilon h(x)| \leq t/2^{\frac{1}{r}}\}}\right)\right\}^r \\
&\quad + \left\{\mathcal{M}_{r,5}\left([T_\epsilon h]\chi_{\{x\in\mathcal{X}:\, |T_\epsilon h(x)| > t/2^{\frac{1}{r}}\}}\right)\right\}^r
\end{aligned}$$

and

$$\begin{aligned}&\left\|\mathcal{M}_{r,5}\left([T_\epsilon h]\chi_{\{x\in\mathcal{X}:\ |T_\epsilon h(x)|\le t/2^{\frac{1}{r}}\}}\right)\right\|_{L^\infty(\mathcal{X},\nu)}\\&\quad\le\left\|[T_\epsilon h]\,\chi_{\{x\in\mathcal{X}:\ |T_\epsilon h(x)|\le t/2^{\frac{1}{r}}\}}\right\|_{L^\infty(\mathcal{X},\nu)}\\&\quad\le\frac{t}{2^{\frac{1}{r}}}.\end{aligned}$$

From these two estimates, together with the boundedness of $\mathcal{M}$ from $L^1(\mathcal{X}, \nu)$ to $L^{1,\infty}(\mathcal{X}, \nu)$ and the boundedness of T_ϵ from $H^1(\mathcal{X}, \nu)$ to $L^{1,\infty}(\mathcal{X}, \nu)$, it follows that

$$\begin{aligned}&\nu\left(\{x\in\mathcal{X}:\ \mathcal{M}_{r,5}(T_\epsilon h)(x)>t\}\right)\\&\quad\le\nu\left(\left\{x\in\mathcal{X}:\ \mathcal{M}_5\left(|T_\epsilon h|^r\chi_{\{x\in\mathcal{X}:\ |(T_\epsilon h)(x)|>t/2^{\frac{1}{r}}\}}\right)>\frac{t^r}{2}\right\}\right)\\&\quad\lesssim t^{-r}\int_{\mathcal{X}}\left|(T_\epsilon h)(x)\chi_{\{x\in\mathcal{X}:\ |T_\epsilon h(x)|>t/2^{\frac{1}{r}}\}}(x)\right|^r d\nu(x)\\&\quad\lesssim t^{-r}\nu(\{x\in\mathcal{X}:\ |T_\epsilon h(x)|>t/2^{\frac{1}{r}}\})\int_0^{t/2^{\frac{1}{r}}}s^{r-1}\,ds\\&\qquad+t^{-r}\int_{t/2^{\frac{1}{r}}}^\infty s^{r-1}\nu(\{x\in\mathcal{X}:\ |T_\epsilon h(x)|>s\})\,ds\\&\quad\lesssim\nu(\{x\in\mathcal{X}:|T_\epsilon h(x)|>t/2^{\frac{1}{r}}\})+\frac{1}{t}\sup_{s\ge t/2^{\frac{1}{r}}}s\nu(\{x\in\mathcal{X}:\ |T_\epsilon h(x)|>s\})\\&\quad\lesssim\frac{\|h\|_{H^1(\mathcal{X},\nu)}}{t}\\&\quad\lesssim t^{-p}\|f\|_{L^p(\mathcal{X},\nu)}^p.\end{aligned}$$

This, together with (8.3.17), implies (ii) and hence finishes the proof of Lemma 8.3.2. □

The following lemma is a weak version of Lemma 8.1.3, whose proof is similar and hence omitted.

Lemma 8.3.3. *For any $f\in L^1_{\mathrm{loc}}(\mathcal{X}, \nu)$, with $\int_{\mathcal{X}} f(x)\,d\nu(x) = 0$ when $\nu(\mathcal{X}) < \infty$, if $\inf\{1, \tilde{N}(f)\}\in L^{p_0}(\mathcal{X}, \nu)$ for some $p_0\in(1,\infty)$, then, for any $p\in[p_0,\infty)$, there exists a positive constant $C_{(p)}$, depending on p but independent of f, such that*

$$\begin{aligned}&\sup_{t\in(0,\infty)}t^p\nu\left(\{x\in\mathcal{X}:\ \tilde{N}(f)(x)>t\}\right)\\&\qquad\le C\sup_{t\in(0,\infty)}t^p\nu\left(\{x\in\mathcal{X}:\ \mathcal{M}^\sharp(f)(x)>t\}\right).\end{aligned}$$

The main result of this section is as follows.

Theorem 8.3.4. *Let T be a Calderón–Zygmund operator as in* (8.1.25) *with kernel K satisfying* (8.1.23) *and* (8.3.1). *If $\nu(\mathcal{X}) = \infty$, then the boundedness of T on $L^2(\mathcal{X}, \nu)$ is equivalent to either of the following two statements:*

(i) *T is bounded from $H^1(\mathcal{X}, \nu)$ to $L^1(\mathcal{X}, \nu)$;*
(ii) *T is bounded from $H^1(\mathcal{X}, \nu)$ to $L^{1,\infty}(\mathcal{X}, \nu)$.*

Proof. By an argument similar to that of Theorem 8.1.6, we see that, if T is bounded on $L^2(\mathcal{X}, \nu)$, then T is also bounded from $H^1(\mathcal{X}, \nu)$ to $L^1(\mathcal{X}, \nu)$, and hence T is bounded from $H^1(\mathcal{X}, \nu)$ to $L^{1,\infty}(\mathcal{X}, \nu)$. Thus, to prove Theorem 8.3.4, we only need to prove that, if T is bounded from $H^1(\mathcal{X}, \nu)$ to $L^{1,\infty}(\mathcal{X}, \nu)$, then T is also bounded on $L^2(\mathcal{X}, \nu)$.

Assume now that T is bounded from $H^1(\mathcal{X}, \nu)$ into $L^{1,\infty}(\mathcal{X}, \nu)$. Choose $r \in (0, 1)$. For each $f \in L_b^\infty(\mathcal{X}, \nu)$ satisfying that

$$\int_{\mathcal{X}} f(x)\, d\nu(x) = 0,$$

it is easy to show that $f \in H^1(\mathcal{X}, \nu)$. By the boundedness of T from $H^1(\mathcal{X}, \nu)$ to $L^{1,\infty}(\mathcal{X}, \nu)$, we see that $T_\epsilon f \in L^{1,\infty}(\mathcal{X}, \nu)$, which implies that

$$\inf\left\{1, \tilde{N}(|T_\epsilon f|^r)\right\} \in L^{\frac{2}{r}}(\mathcal{X}, \nu).$$

Indeed, if, for all $r \in (0, 1)$ and $f \in L^1_{\rm loc}(\mathcal{X}, \nu)$, let

$$\tilde{N}_r(f) := \left[\tilde{N}(|f|^r)\right]^{\frac{1}{r}},$$

then

$$\begin{aligned}
&\int_{\mathcal{X}} \left[\inf\left\{1, \tilde{N}\left(|T_\epsilon f(x)|^r\right)\right\}\right]^{\frac{2}{r}} d\nu(x) \\
&\quad = \int_{\mathcal{X}} \left[\inf\{1, \tilde{N}_r(|T_\epsilon f(x)|)\}\right]^2 d\nu(x) \\
&\quad = 2\int_0^2 t\nu(\{x \in \mathcal{X} : \inf\{1, \tilde{N}_r(T_\epsilon f)(x)\} > t\})\, dt \\
&\qquad +2\int_2^\infty t\nu(\{x \in \mathcal{X} : \inf\{1, \tilde{N}_r(T_\epsilon f)(x)\} > t\})\, dt \\
&\quad \lesssim \int_0^2 t\nu\left(\{x \in \mathcal{X} : \inf\{1, \tilde{N}_r(T_\epsilon f)(x)\} > t\}\right) dt \\
&\quad \lesssim \left\|\tilde{N}_r(T_\epsilon f)\right\|_{L^{1,\infty}(\mathcal{X},\nu)}
\end{aligned}$$

$$\lesssim \|T_\epsilon f\|_{L^{1,\infty}(\mathcal{X},\nu)}$$
$$< \infty,$$

where in the last inequality, we used the boundedness of $\tilde{N}_r$ on $L^{1,\infty}(\mathcal{X}, \nu)$, which is a simple corollary of Lemma 8.2.3(ii) and $\tilde{N}_r(f) \lesssim \mathcal{M}_{r,5}(f)$. Thus,

$$\inf\{1, \tilde{N}(|T_\epsilon f|^r)\} \in L^{\frac{2}{r}}(\mathcal{X}, \nu).$$

From this, Lemmas 8.3.3 and 8.3.2(ii), together with Corollary 7.1.21, it follows that, for all $p \in [2, \infty)$ and $f \in L^\infty_b(\mathcal{X}, \nu)$ satisfying that $\int_{\mathcal{X}} f(x)\, d\nu(x) = 0$,

$$\begin{aligned}
&\sup_{t\in(0,\infty)} t^p \nu(\{x \in \mathcal{X} : |T_\epsilon f(x)| > t\})\\
&\quad\le \sup_{t\in(0,\infty)} t^p \nu\left(\left\{x \in \mathcal{X} : \tilde{N}_r(T_\epsilon f)(x) > t\right\}\right)\\
&\quad= \sup_{s\in(0,\infty)} s^{\frac{p}{r}} \nu\left(\left\{x \in \mathcal{X} : \tilde{N}(|T_\epsilon f|^r)(x) > s\right\}\right)\\
&\quad\lesssim \sup_{s\in(0,\infty)} s^{\frac{p}{r}} \nu\left(\left\{x \in \mathcal{X} : \mathcal{M}^\sharp(|T_\epsilon f|^r)(x) > s\right\}\right)\\
&\quad\sim \sup_{t\in(0,\infty)} t^p \nu\left(\left\{x \in \mathcal{X} : \mathcal{M}^\sharp_r(T_\epsilon f)(x) > t\right\}\right)\\
&\quad\lesssim \|f\|^p_{L^p(\mathcal{X},\nu)}.
\end{aligned} \tag{8.3.18}$$

Using a density argument, we see that T_ϵ is bounded from $L^p(\mathcal{X}, \nu)$ to $L^{p,\infty}(\mathcal{X}, \nu)$ for all $p \in [2, \infty)$ with the bound independent of ϵ, which, together with the uniform boundedness of T_ϵ from $H^1(\mathcal{X}, \nu)$ to $L^{1,\infty}(\mathcal{X}, \nu)$ and Theorem 8.1.2, then completes the proof of Theorem 8.3.4. □

From Theorems 8.3.4 and 8.2.1, we immediately deduce the following conclusion.

Corollary 8.3.5. *Let $\mathcal{X}$ be separable and T a Calderón–Zygmund operator as in* (8.1.25) *with kernel K satisfying* (8.1.23) *and* (8.1.24). *If $\nu(\mathcal{X}) = \infty$, then the fact that T is bounded on $L^2(\mathcal{X}, \nu)$ is equivalent to each of the following statements:*

(i) *T is bounded on $L^p(\mathcal{X}, \nu)$ for some $p \in (1, \infty)$;*
(ii) *T is bounded from $L^1(\mathcal{X}, \nu)$ to $L^{1,\infty}(\mathcal{X}, \nu)$;*
(iii) *T is bounded from $H^1(\mathcal{X}, \nu)$ to $L^{1,\infty}(\mathcal{X}, \nu)$.*

8.4 Boundedness of Calderón–Zygmund Operators: Equivalent Characterizations III

In this section, based on the results in Sect. 8.3, we prove that, if $\mathcal{X}$ is separable and $\nu(\mathcal{X}) = \infty$, then, for the Calderón–Zygmund operator, its boundedness on $L^p(\mathcal{X}, \nu)$ with $p \in (1, \infty)$ is equivalent to its boundedness from $H^1(\mathcal{X}, \nu)$ into $L^{1,\infty}(\mathcal{X}, \nu)$, or from $L^\infty(\mathcal{X}, \nu)$ into RBMO $(\mathcal{X}, \nu)$ or some other estimates, which is stated as follows.

Theorem 8.4.1. *Let $\rho \in (1, \infty)$, K be a ν-locally integrable function mapping $(\mathcal{X} \times \mathcal{X})\backslash\Delta$ to $\mathbb{C}$ which satisfies* (8.1.23) *and* (8.1.24)*, and T be a Calderón–Zygmund operator as in* (8.1.25)*. If $\nu(\mathcal{X}) = \infty$, then the following seven statements are equivalent:*

(i) *T is bounded from $H^1(\mathcal{X}, \nu)$ into $L^1(\mathcal{X}, \nu)$;*
(ii) *T is bounded from $H^1(\mathcal{X}, \nu)$ into $L^{1,\infty}(\mathcal{X}, \nu)$;*
(iii) *for some $\theta \in (0, \infty)$, there exists a positive constant C such that, for all $\epsilon, t \in (0, \infty)$, balls B and bounded functions f with* $\mathrm{supp}\, f \subset B$,

$$\nu(\{x \in B : |T_\epsilon(f)(x)| > t\}) \le Ct^{-\theta}\nu(\rho B)\|f\|^\theta_{L^\infty(\mathcal{X},\nu)};$$

(iv) *for some $\sigma \in (0, 1)$, there exists a positive constant C such that, for all $\epsilon \in (0, \infty)$, balls B and bounded functions f with* $\mathrm{supp}\, f \subset B$,

$$\frac{1}{\nu(\rho B)} \int_B |T_\epsilon(f)(x)|^\sigma \, d\nu(x) \le C\|f\|^\sigma_{L^\infty(\mathcal{X},\nu)};$$

(v) *T is bounded from $L^\infty(\mathcal{X}, \nu)$ into* RBMO $(\mathcal{X}, \nu)$*;*
(vi) *T is bounded on $L^p(\mathcal{X}, \nu)$ for some $p \in (1, \infty)$;*
(vii) *T is bounded on $L^p(\mathcal{X}, \nu)$ for all $p \in (1, \infty)$.*

Proof. By Theorem 8.3.4, Corollary 8.3.5 and Lemma 8.3.1, we know that

$$\text{(i)} \iff \text{(ii)} \iff \text{(vi)} \quad \text{and} \quad \text{(ii)} \Longrightarrow \text{(iii)} \Longrightarrow \text{(iv)}.$$

Now we prove the implication that

$$\text{(iv)} \Longrightarrow \text{(v)} \Longrightarrow \text{(i)},$$

which implies the statements (i)–(vi) are equivalent. We then finally show that (vi) $\iff$ (vii).

(iv)$\Longrightarrow$(v). For simplicity, assume that (iv) is true with $\rho = 3/2$. To show (v), it suffices to prove that, for all $\epsilon \in (0, \infty)$ and bounded functions f with bounded support,

$$\|T_\epsilon(f)\|_{\mathrm{RBMO}(\mathcal{X},\nu)} \lesssim \|f\|_{L^\infty(\mathcal{X},\nu)}.$$

It follows, from Lemma 8.3.2, that $|T_\epsilon(f)|^\sigma \in \mathrm{RBMO}(\mathcal{X}, \nu)$, where $\sigma \in (0, 1)$. By the John–Nirenberg inequality, we know that $T_\epsilon f$ is ν-locally integrable. For each fixed ball B, let

$$h_B := m_B\left(T_\epsilon\left(f\chi_{\mathcal{X}\setminus\frac{\varrho+1}{2}B}\right)\right).$$

Then, by some arguments similar to those used in the proof of Lemma 8.3.2, we conclude that, for all balls B,

$$\frac{1}{\nu(\varrho B)}\int_B |T_\epsilon(f)(x) - h_B|^\sigma\, d\nu(x) \lesssim \|f\|^\sigma_{L^\infty(\mathcal{X},\nu)} \tag{8.4.1}$$

and, for all balls B and S with $B \subset S$,

$$|h_B - h_S| \lesssim [1 + \delta(B, S)]\|f\|_{L^\infty(\mathcal{X},\nu)}. \tag{8.4.2}$$

From (7.3.10), (8.4.1), (8.4.2) and (7.3.2), it follows that, for all balls B,

$$\begin{aligned}
&\int_B \left|T_\epsilon(f)(y) - \alpha_{\tilde{B}^{6\varrho^2}}(T_\epsilon(f))\right|^\sigma d\nu(y)\\
&\quad\le \int_B |T_\epsilon(f)(y) - h_B|^\sigma\, d\nu(y) + \left|h_B - h_{\tilde{B}^{6\varrho^2}}\right|^\sigma \nu(B)\\
&\qquad + \left|h_{\tilde{B}^{6\varrho^2}} - \alpha_{\tilde{B}^{6\varrho^2}}(T_\epsilon(f))\right|^\sigma \nu(B)\\
&\quad\lesssim \nu(\varrho B)\|f\|^\sigma_{L^\infty(\mathcal{X},\nu)} + \left[1 + \delta\left(B, \tilde{B}^{6\varrho^2}\right)\right]^\sigma \nu(B)\|f\|^\sigma_{L^\infty(\mathcal{X},\nu)}\\
&\qquad + \left|\alpha_{\tilde{B}^{6\varrho^2}}(T_\epsilon(f) - h_{\tilde{B}^{6\varrho^2}})\right|^\sigma \nu(B)\\
&\quad\lesssim \nu(\varrho B)\|f\|^\sigma_{L^\infty(\mathcal{X},\nu)} + \left(m^\varrho_{0,s;\,\tilde{B}^{6\varrho^2}}\left[T_\epsilon(f) - h_{\tilde{B}^{6\varrho^2}}\right]\right)^\sigma \nu(B)\\
&\quad\lesssim \nu(\varrho B)\|f\|^\sigma_{L^\infty(\mathcal{X},\nu)} + \frac{\nu(B)}{\nu(\varrho\tilde{B}^{6\varrho^2})}\int_{\tilde{B}^{6\varrho^2}} \left|T_\epsilon(f)(y) - h_{\tilde{B}^{6\varrho^2}}\right|^\sigma d\nu(y)\\
&\quad\lesssim \nu(\varrho B)\|f\|^\sigma_{L^\infty(\mathcal{X},\nu)}
\end{aligned}$$

and, for any two $(6\varrho^2, \beta_{6\varrho^2})$-doubling balls $B \subset S$,

$$\begin{aligned}
&|\alpha_B(T_\epsilon(f)) - \alpha_S(T_\epsilon(f))|\\
&\quad\le |\alpha_B(T_\epsilon(f)) - h_B| + |h_B - h_S| + |\alpha_S(T_\epsilon(f)) - h_S|\\
&\quad\lesssim |\alpha_B(T_\epsilon(f) - h_B)| + [1 + \delta(B, S)]\|f\|_{L^\infty(\mathcal{X},\nu)}\\
&\qquad + |\alpha_S(T_\epsilon(f) - h_S)|\\
&\quad\lesssim m^\varrho_{0,s;\,B}(T_\epsilon(f) - h_B) + [1 + \delta(B, S)]\|f\|_{L^\infty(\mathcal{X},\nu)}
\end{aligned}$$

$$
\begin{aligned}
&+m^{\varrho}_{0,s;\,S}\,(T_\epsilon(f)-h_S)\\
&\lesssim \left[\frac{1}{\nu(\varrho B)}\int_B |T_\epsilon(f)(y)-h_B|^\sigma\,d\nu(y)\right]^{1/\sigma}\\
&\quad+[1+\delta(B,\,S)]\|f\|_{L^\infty(\mathcal{X},\nu)}\\
&\quad+\left[\frac{1}{\nu(\varrho S)}\int_S |T_\epsilon(f)(y)-h_S|^\sigma\,d\nu(y)\right]^{1/\sigma}\\
&\lesssim [1+\delta(B,\,S)]\|f\|_{L^\infty(\mathcal{X},\nu)}.
\end{aligned}
$$

By Corollary 7.3.7 with $\varphi(t):=t^\sigma$ for all $t\in[0,\infty)$, we see that

$$\|T_\epsilon(f)\|_{\mathrm{RBMO}(\mathcal{X},\nu)}\lesssim \|f\|_{L^\infty(\mathcal{X},\nu)}.$$

Thus, (v) holds true.

(v)$\Longrightarrow$(i). We first claim that, for all $\epsilon\in(0,\,\infty)$, balls B and bounded functions f with support contained in B,

$$\int_B |T_\epsilon(f)(x)|\,d\nu(x)\lesssim \nu(2B)\|f\|_{L^\infty(\mathcal{X},\nu)}. \tag{8.4.3}$$

We consider the following two cases for r_B.

Case I. $r_B\le \operatorname{diam}(\operatorname{supp}\nu)/40$. In this case, choose $\rho=2$ and $\varrho=1$ in Lemma 7.3.6. From the hypothesis and Lemma 7.3.6, it follows that, for all $\epsilon\in(0,\,\infty)$,

$$\int_B \left|T_\epsilon(f)(x)-m_{\widetilde{B^6}}(T_\epsilon(f))\right|\,d\nu(x)\lesssim \nu(2B)\|f\|_{L^\infty(\mathcal{X},\nu)},$$

where, for any ball $B\subset\mathcal{X}$, $\widetilde{B^6}$ denotes the *smallest* $(6,\,\beta_6)$*-doubling ball* of the form 6^jB with $j\in\mathbb{Z}_+$. Hence, in this case, the proof of (8.4.3) is reduced to showing

$$\left|m_{\widetilde{B^6}}(T_\epsilon(f))\right|\lesssim \nu(2B)\|f\|_{L^\infty(\mathcal{X},\nu)}. \tag{8.4.4}$$

Let S be the *smallest ball* of the form 6^jB such that $\nu(6^jB\setminus 2B)>0$ with $j\in\mathbb{N}$. Thus,

$$\nu(6^{-1}S\setminus 2B)=0\quad\text{and}\quad \nu(S\setminus 2B)>0.$$

This leads to

$$\nu\left(S\setminus\left(6^{-1}S\bigcup 2B\right)\right)>0\quad\text{and}\quad \widetilde{B^6}\subset\widetilde{S^6}.$$

By this and Lemma 7.1.13, we choose $x_0 \in S \setminus (6^{-1}S \cup 2B))$ such that the ball centered at x_0 with the radius $6^{-k}r_S$ for some integer $k \geq 2$ is $(6, \beta_6)$-doubling. Let B_0 be the *largest ball* of this form. Then it is easy to show that $B_0 \subset 2S$ and $d(B_0, B) \geq r_B/2$. Observe that

$$\delta(B, 2S) \lesssim 1 \quad \text{and} \quad \delta(B_0, 2S) \lesssim 1,$$

which imply that

$$\delta(B, \widetilde{(2S)^6}) \lesssim 1 \quad \text{and} \quad \delta(B_0, \widetilde{(2S)^6}) \lesssim 1.$$

Therefore, via Lemma 7.3.6, we conclude that

$$\begin{aligned}
&\left|m_{B_0}(T_\epsilon(f)) - m_{\widetilde{B^6}}(T_\epsilon(f))\right| \\
&\leq \left|m_{B_0}(T_\epsilon(f)) - m_{\widetilde{(2S)^6}}(T_\epsilon(f))\right| + \left|m_{\widetilde{(2S)^6}}(T_\epsilon(f)) - m_{\widetilde{B^6}}(T_\epsilon(f))\right| \\
&\leq \left[2 + \delta\left(B_0, \widetilde{(2S)^6}\right) + \delta\left(\widetilde{B^6}, \widetilde{(2S)^6}\right)\right] \|T_\epsilon(f)\|_{\mathrm{RBMO}(\mathcal{X},\nu)} \\
&\lesssim \|f\|_{L^\infty(\mathcal{X},\nu)},
\end{aligned}$$

which implies that, to prove (8.4.4), it suffices to show

$$|m_{B_0}(T_\epsilon(f))| \lesssim \|f\|_{L^\infty(\mathcal{X},\nu)}. \tag{8.4.5}$$

Notice that, for all $y \in B_0$ and $z \in B$, $d(y, z) \geq r_B/2$ and hence

$$d(x_B, y) \leq d(x_B, z) + d(z, y) \lesssim d(z, y).$$

From this, (8.1.23) and (7.1.3), it follows that, for all $y \in B_0$,

$$\begin{aligned}
|T_\epsilon(f)(y)| &\lesssim \int_B \frac{|f(z)|}{\lambda(y, d(y, z))}\, d\nu(z) \\
&\lesssim \int_B \frac{|f(z)|}{\lambda(x_B, d(y, z))}\, d\nu(z) \\
&\lesssim \frac{\nu(B)}{\lambda(x_B, r_B)} \|f\|_{L^\infty(\mathcal{X},\nu)} \\
&\lesssim \|f\|_{L^\infty(\mathcal{X},\nu)},
\end{aligned}$$

which implies (8.4.5). Therefore, (8.4.3) holds true in this case.

Case II. $r_B > \operatorname{diam}(\operatorname{supp} \nu)/40$. In this case, without loss of generality, we may assume $r_B \leq 8 \operatorname{diam}(\operatorname{supp} \nu)$. Then $B \cap \operatorname{supp} \nu$ is covered by balls $\{B_j\}_{j=1}^M$ with radius $r_B/400$, where $M \in \mathbb{N}$. For $j \in \{1, \ldots, M\}$, define

$$a_j := \frac{\chi_{B_j}}{\sum_{k=1}^M \chi_{B_k}} f.$$

Since (8.4.3) holds true if we replace B by $2B_j$, which supports the function a_j, we then see that

$$\begin{aligned}\int_B |T_\epsilon(f)(x)|\,d\nu(x) &\le \sum_{j=1}^M \left[\int_{B\setminus 2B_j} |T_\epsilon(a_j)(x)|\,d\nu(x) + \int_{2B_j} \cdots\right]\\ &\lesssim \sum_{j=1}^M \|a_j\|_{L^\infty(\mathcal{X},\nu)}\left[\nu(B)+\nu(4B_j)\right]\\ &\lesssim \|f\|_{L^\infty(\mathcal{X},\nu)}\nu(2B).\end{aligned}$$

Thus, the claim (8.4.3) also holds true in this case.

Now based on the claim (8.4.3), we prove (i). Take $\rho = 4$ and $p = \infty$ in Definition 7.4.1. It suffices to show that, for all $(\infty, 1)_\lambda$-atomic blocks b,

$$\|T_\epsilon(b)\|_{L^1(\mathcal{X},\nu)} \lesssim |b|_{H^{1,\infty}_{\mathrm{atb}}(\mathcal{X},\nu)}. \tag{8.4.6}$$

Let

$$b := \sum_{j=1}^2 \kappa_j a_j$$

be a $(\infty, 1)_\lambda$-atomic block, where, for any $j \in \{1, 2\}$,

$$\operatorname{supp} a_j \subset B_j \subset B$$

for some B_j and B as in Definition 7.4.1. Write

$$\begin{aligned}\|T_\epsilon(b)\|_{L^1(\mathcal{X},\nu)} &= \int_{2B} |T_\epsilon(b)(x)|\,d\nu(x) + \int_{\mathcal{X}\setminus(2B)} \cdots\\ &\le \sum_{j=1}^2 |\kappa_j| \int_{2B_j} |T_\epsilon(a_j)(x)|\,d\nu(x) + \sum_{j=1}^2 |\kappa_j| \int_{2B\setminus 2B_j} \cdots\\ &\quad + \int_{\mathcal{X}\setminus(2B)} |T_\epsilon(b)(x)|\,d\nu(x)\\ &=: \sum_{i=1}^3 \mathrm{F}_i.\end{aligned}$$

It follows, from (8.4.3), that

$$\mathrm{F}_1 \lesssim \sum_{j=1}^{2} |\kappa_j| \nu(4B_j) \|a\|_{L^\infty(\mathcal{X},\nu)} \lesssim \sum_{j=1}^{2} |\kappa_j|.$$

On the other hand, by (8.1.23) and (7.1.3), we conclude that

$$\begin{aligned}
\mathrm{F}_2 &\lesssim \sum_{j=1}^{2} |\kappa_j| \int_{(2B)\setminus(2B_j)} \int_{B_j} \frac{|a_j(y)|}{\lambda(x, d(x,y))}\, d\nu(y)\, d\nu(x) \\
&\lesssim \sum_{j=1}^{2} |\kappa_j| \int_{(2B)\setminus(2B_j)} \int_{B_j} \frac{|a_j(y)|}{\lambda(x_{B_j}, d(x, x_{B_j}))}\, d\nu(y)\, d\nu(x) \\
&\lesssim \sum_{j=1}^{2} |\kappa_j| \delta\left(B_j, B\right) \|a_j\|_{L^1(\mathcal{X},\nu)} \\
&\lesssim \sum_{j=1}^{2} |\kappa_j|.
\end{aligned}$$

It remains to estimate F_3. We consider the following two cases.

Case (i). $\epsilon \in (0, r_B)$. In this case, it is easy to show that, for all $x \in \mathcal{X} \setminus (2B)$ and $y \in B$, $d(x, y) < \epsilon$. Thus, by the vanishing moment of b with (8.1.24) and (7.1.3), we easily see that

$$\begin{aligned}
\mathrm{F}_3 &\lesssim \int_{\mathcal{X}\setminus(2B)} \int_{\mathcal{X}} |K(x,y) - K(x, x_B)||b(y)|\, d\nu(y)\, d\nu(x) \\
&\lesssim \int_{\mathcal{X}\setminus(2B)} \int_{\mathcal{X}} \left[\frac{d(y, x_B)}{d(x, x_B)}\right]^\tau \frac{|b(y)|}{\lambda(x, d(x,y))}\, d\nu(y)\, d\nu(x) \\
&\lesssim \sum_{j=1}^{2} |\kappa_j| \int_{\mathcal{X}\setminus(2B)} \int_{B_j} \left[\frac{r_B}{d(x, x_B)}\right]^\tau \frac{|a_j(y)|}{\lambda(x_B, d(x, x_B))}\, d\nu(y)\, d\nu(x) \\
&\lesssim \sum_{j=1}^{2} |\kappa_j| \sum_{k=1}^{\infty} \int_{(2^{k+1}B)\setminus(2^k B)} \left[\frac{r_B}{2^k r_B}\right]^\tau \frac{d\nu(x)}{\lambda(x_B, d(x, x_B))} \\
&\lesssim \sum_{j=1}^{2} |\kappa_j|.
\end{aligned}$$

Case (ii). $\epsilon \in [r_B, \infty)$. In this case, we first write

$$\begin{aligned}
\mathrm{F}_3 &\le \int_{d(x, x_B)>r_B+\epsilon} \int_{d(x, y)>\epsilon} |K(x, y) - K(x, x_B)||b(y)|\, d\nu(y)\, d\nu(x) \\
&\quad + \int_{2r_B \le d(x, x_B) \le r_B+\epsilon} \int_{d(x, y)>\epsilon} |K(x, y)||b(y)|\, d\nu(y)\, d\nu(x) \\
&=: \mathrm{F}_{3,1} + \mathrm{F}_{3,2}.
\end{aligned}$$

For the term $\mathrm{F}_{3,1}$, notice that, for all $x \in \mathcal{X}$ such that $d(x, x_B) \le r_B + \epsilon$ and all $y \in B$, $d(x, y) > \epsilon$. Thus, by the same argument as that used in the proof of Case (i), we conclude that

$$\mathrm{F}_{3,1} \lesssim \sum_{j=1}^{2} |\kappa_j|.$$

On the other hand, from (8.1.23), (7.1.1) and (7.1.3), it follows that

$$\begin{aligned}
\mathrm{F}_{3,2} &\lesssim \sum_{j=1}^{2} |\kappa_j| \int_{d(x, x_B)\le 2\epsilon} \int_{d(x, y)>\epsilon} \frac{1}{\lambda(x, d(x, y))} |a_j(y)|\, d\nu(y) d\nu(x) \\
&\lesssim \sum_{j=1}^{2} |\kappa_j| \int_{d(x, x_B)\le 2\epsilon} \frac{1}{\lambda(x_B, \epsilon)}\, d\nu(x) \|a_j\|_{L^1(\mathcal{X}, \nu)} \\
&\lesssim \sum_{j=1}^{2} |\kappa_j|.
\end{aligned}$$

Combining the estimates for $\mathrm{F}_{3,1}$ and $\mathrm{F}_{3,2}$, we see that

$$\mathrm{F}_3 \lesssim \sum_{j=1}^{2} |\kappa_j|$$

in this case.

The estimates for F_1, F_2 and F_3 imply (8.4.6). Thus, (v)$\Longrightarrow$(i), which further implies that (i)–(vi) are equivalent.

(vi)$\Longleftrightarrow$(vii). The implication (vii)$\Longrightarrow$(vi) is obvious. Let us prove (vi)$\Longrightarrow$(vii). Indeed, based on the equivalence of (i)–(vi) proved above, we conclude that, if (vi) holds true, then T is bounded from $H^1(\mathcal{X}, \nu)$ into $L^1(\mathcal{X}, \nu)$ and from $L^\infty(\mathcal{X}, \nu)$ into $\mathrm{RBMO}(\mathcal{X}, \nu)$. By Theorem 8.1.4, we see that T is bounded for all $p \in (1, \infty)$. This means that (vii) holds true. Thus, we prove that (vi)$\Longleftrightarrow$(vii), which completes the proof of Theorem 8.4.1. □

From Theorem 8.4.1, we further deduce the following result.

Corollary 8.4.2. *Let K be a ν-locally integrable function mapping $(\mathcal{X} \times \mathcal{X})\setminus\Delta$ to $\mathbb{C}$ which satisfies* (8.1.23) *and* (8.1.24), *and T a Calderón–Zygmund operator as in* (8.1.25). *Let Φ be a Young function such that, for all $t_1, t_2 \in [0, \infty)$,*

$$\Phi(t_1t_2) \leq C\Phi(t_1)\Phi(t_2)$$

and, for some $\sigma \in (0, 1)$,

$$\int_0^\infty \Phi\left(\frac{1}{t}\right) t^{\sigma-1}\, dt < \infty.$$

If there exists a positive constant C such that, for all $\epsilon, t \in (0, \infty)$ and bounded functions f with bounded support,

$$\nu(\{x \in \mathcal{X} : |T_\epsilon(f)(x)| > t\}) \leq C \int_{\mathcal{X}} \Phi\left(\frac{|f(x)|}{t}\right) d\nu(x),$$

then T is bounded on $L^p(\mathcal{X}, \nu)$ for all $p \in (1, \infty)$.

Proof. By Theorem 8.4.1, the proof of Corollary 8.4.2 is reduced to proving that Theorem 8.4.1(iv) is true. Let B be a fixed ball and f a bounded function with support contained in B. By the homogeneity of $\|\cdot\|_{\mathrm{RBMO}(\mathcal{X},\nu)}$, we may assume that $\|f\|_{L^\infty(\mathcal{X},\nu)} = 1$. By some trivial computation, we easily see that, for any $\sigma \in (0, 1)$,

$$\begin{aligned}\int_B |T_\epsilon(f)(x)|^\sigma\, d\nu(x) &= \sigma \int_0^1 t^{\sigma-1}\nu(\{x \in B : |T_\epsilon(f)(x)| > t\})\, dt \\ &\quad +\sigma \int_1^\infty t^{\sigma-1}\nu(\{x \in B : |T_\epsilon(f)(x)| > t\})\, dt \\ &\lesssim \nu(B) + \int_1^\infty t^{\sigma-1}\Phi\left(\frac{1}{t}\right) dt \int_B \Phi(|f(x)|)\, d\nu(x) \\ &\lesssim \nu(B),\end{aligned}$$

which completes the proof of Corollary 8.4.2. □

We remark that, if we let

$$\Phi(t) := t\log^\gamma(2+t)$$

with $\gamma \in [1, \infty)$ for all $t \in [0, \infty)$, then, by Corollary 8.4.2, we see that, if T is of weak type $(L\log^\gamma L(\mathcal{X}, \nu), L^1(\mathcal{X}, \nu))$, namely, there exists a positive constant C such that, for all $\epsilon, t \in (0, \infty)$ and bounded functions f with bounded support,

$$\nu(\{x \in \mathcal{X} : |T_\epsilon(f)(x)| > t\}) \leq C \int_{\mathcal{X}} \frac{|f(x)|}{t} \log^\gamma \left(e + \frac{|f(x)|}{t}\right) d\nu(x),$$

then T is also bounded on $L^p(\mathcal{X}, \nu)$ for all $p \in (1, \infty)$.

8.5 The Molecular Characterization of Hardy Spaces $H^1(\mathcal{X}, \nu)$ and $\tilde{H}^1(\mathcal{X}, \nu)$

The main purpose of this section is to establish a suitable molecular characterization of $H_{\rm atb}^{1,p}(\mathcal{X}, \nu)$. To be precise, let $p \in (1, \infty]$. We first introduce a version of the atomic Hardy space, $\tilde{H}_{\rm atb}^{1,p}(\mathcal{X}, \nu)$, which is a subspace of the atomic Hardy space $H_{\rm atb}^{1,p}(\mathcal{X}, \nu)$ (see Definition 8.5.3 below), via the discrete coefficient $\tilde{\delta}_{B,S}^{(\rho)}$ for $\rho \in (1, \infty)$ and balls $B \subset S$ of $\mathcal{X}$. Then, we establish the corresponding molecular characterization of $\tilde{H}_{\rm atb}^{1,p}(\mathcal{X}, \nu)$ via a constructive way, which is new even for the Euclidean space $\mathbb{R}^D$ endowed with the D-dimensional Lebesgue measure m. A similar molecular characterization of the Hardy space $H_{\rm atb}^{1,p}(\mathcal{X}, \nu)$ also holds true. As an application, we obtain the boundedness of the Calderón–Zygmund operator on $\tilde{H}_{\rm atb}^{1,p}(\mathcal{X}, \nu)$ as well as on its corresponding dual space $\widetilde{\rm RBMO}(\mathcal{X}, \nu)$, which are new even for spaces of homogeneous type. Moreover, we give a sufficient condition to guarantee that $\tilde{H}_{\rm atb}^{1,p}(\mathcal{X}, \nu)$ coincides with $H_{\rm atb}^{1,p}(\mathcal{X}, \nu)$, and show that this sufficient condition is also necessary in some sense via an example.

We point out that the motivation for introducing the space $\tilde{H}_{\rm atb}^{1,p}(\mathcal{X}, \nu)$ is that, according to the present method used in this section, we can only show that the Calderón–Zygmund operator in a non-homogeneous space $(\mathcal{X}, d, \nu)$ is bounded on $\tilde{H}_{\rm atb}^{1,p}(\mathcal{X}, \nu)$. Our method does not work for the boundedness of the Calderón–Zygmund operators on the space $H_{\rm atb}^{1,p}(\mathcal{X}, \nu)$. Moreover, it is still unclear whether $H_{\rm atb}^{1,p}(\mathcal{X}, \nu)$ always coincides with $\tilde{H}_{\rm atb}^{1,p}(\mathcal{X}, \nu)$ or not so far.

Before we introduce the new atomic Hardy space, we first need a notion of the discrete coefficient $\tilde{\delta}_{B,S}^{(\rho)}$ for $\rho \in (1, \infty)$ and balls $B \subset S$ of $\mathcal{X}$.

Definition 8.5.1. For any two balls $B \subset S$ and $\rho \in (1, \infty)$, let

$$\tilde{\delta}_{B,S}^{(\rho)} := 1 + \sum_{k=1}^{N_{B,S}^{(\rho)}} \frac{\nu(\rho^k B)}{\lambda(x_B, \rho^k r_B)},$$

where x_B is the center of the ball B, r_B and r_S respectively denote the *radii* of B and S, and $N_{B,S}^{(\rho)}$ is the *smallest integer* satisfying $\rho^{N_{B,S}^{(\rho)}} r_B \geq r_S$.

Remark 8.5.2. When $(\mathcal{X}, d, \nu) := (\mathbb{R}^D, |\cdot|, \mu)$ with μ as in (0.0.1), it is easy to see that, for any $\rho \in (1, \infty)$,

$$1 + \delta(B, S) \sim \tilde{\delta}_{B,S}^{(\rho)}. \tag{8.5.1}$$

For a general non-homogeneous space $(\mathcal{X}, d, \nu)$, obviously,

$$1 + \delta(B, S) \lesssim \tilde{\delta}_{B,S}^{(\rho)}$$

for any $\rho \in (1, \infty)$ and all balls $B \subset S$ of $\mathcal{X}$. On the other hand, for a given $\rho \in (1, \infty)$, in general, (8.5.1) is not true.

Now we introduce the new atomic Hardy space associated with $\tilde{\delta}_{B,S}^{(\rho)}$ for $\rho \in (1, \infty)$ and balls $B \subset S$ of $\mathcal{X}$.

Definition 8.5.3. Let $\rho \in (1, \infty)$, $p \in (1, \infty]$ and $\gamma \in [1, \infty)$. A function $b \in L^1(\mathcal{X}, \nu)$ is called a $(p, \gamma, \rho)_\lambda$-*atomic block* if

(i) there exists a ball B such that $\operatorname{supp} b \subset B$;
(ii)

$$\int_{\mathcal{X}} b(x)\, d\nu(x) = 0;$$

(iii) for any $j \in \{1, 2\}$, there exist a function a_j, supported on a ball $B_j \subset B$, and a number $\lambda_j \in \mathbb{C}$ such that

$$b = \lambda_1 a_1 + \lambda_2 a_2 \quad \text{and} \quad \|a_j\|_{L^p(\mathcal{X},\nu)} \le [\nu(\rho B_j)]^{1/p-1} \left[\tilde{\delta}_{B_j,B}^{(\rho)}\right]^{-\gamma}.$$

Moreover, let

$$|b|_{\tilde{H}_{\mathrm{atb},\rho}^{1,p,\gamma}(\mathcal{X},\nu)} := |\lambda_1| + |\lambda_2|.$$

A function $f \in L^1(\mathcal{X}, \nu)$ is said to belong to the *atomic Hardy space* $\tilde{H}_{\mathrm{atb},\rho}^{1,p,\gamma}(\mathcal{X}, \nu)$, if there exist $(p, \gamma, \rho)_\lambda$-atomic blocks $\{b_i\}_{i=1}^\infty$ such that

$$f = \sum_{i=1}^\infty b_i$$

in $L^1(\mathcal{X}, \nu)$ and

$$\sum_{i=1}^\infty |b_i|_{\tilde{H}_{\mathrm{atb},\rho}^{1,p,\gamma}(\mathcal{X},\nu)} < \infty.$$

The $\tilde{H}_{\mathrm{atb},\rho}^{1,p,\gamma}(\mathcal{X}, \nu)$ *norm* of f is defined by

$$\|f\|_{\tilde{H}^{1,p,\gamma}_{\mathrm{atb},\rho}(\mathcal{X},\nu)} := \inf\left\{\sum_{i=1}^{\infty} |b_i|_{\tilde{H}^{1,p,\gamma}_{\mathrm{atb},\rho}(\mathcal{X},\nu)}\right\},$$

where the infimum is taken over all the possible decompositions of f as above.

Remark 8.5.4. (i) When $(\mathcal{X}, d, \nu) := (\mathbb{R}^D, |\cdot|, \mu)$ with μ as in (0.0.1), by (8.5.1), we see that $\tilde{H}^{1,p,\gamma}_{\mathrm{atb},\rho}(\mathcal{X}, \nu)$ becomes the atomic Hardy space $H^{1,p}_{\mathrm{atb}}(\mu)$ in Sect. 3.2. For general non-homogeneous spaces, if we replace $\tilde{\delta}^{(\rho)}_{B,S}$ by $1+\delta(B, S)$ in Definition 8.5.1, then $\tilde{H}^{1,p,\gamma}_{\mathrm{atb},\rho}(\mathcal{X}, \nu)$ becomes the atomic Hardy space $H^{1,p}_{\mathrm{atb}}(\mathcal{X}, \nu)$ in Sect. 7.4. Obviously, for $\rho \in (1, \infty)$, $p \in (1, \infty]$ and $\gamma \in [1, \infty)$, we always have

$$\tilde{H}^{1,p,\gamma}_{\mathrm{atb},\rho}(\mathcal{X}, \nu) \subset H^{1,p}_{\mathrm{atb}}(\mathcal{X}, \nu).$$

(ii) By some arguments completely analogous to those used in Sect. 7.4, we conclude that, for each $p \in (1, \infty]$, the new atomic Hardy space $\tilde{H}^{1,p,\gamma}_{\mathrm{atb},\rho}(\mathcal{X}, \nu)$ is independent of the choices of ρ and γ and that, for all $p \in (1, \infty)$, the spaces

$$\tilde{H}^{1,p,\gamma}_{\mathrm{atb},\rho}(\mathcal{X}, \nu) \quad \text{and} \quad \tilde{H}^{1,\infty,\gamma}_{\mathrm{atb},\rho}(\mathcal{X}, \nu)$$

coincide with equivalent norms. Thus, in what follows, we *denote $\tilde{H}^{1,p,\gamma}_{\mathrm{atb},\rho}(\mathcal{X}, \nu)$ simply by $\tilde{H}^{1,p}_{\mathrm{atb}}(\mathcal{X}, \nu)$.*

(iii) Let $(\mathcal{X}, d, \nu)$ be a space of homogeneous type in the sense of Coifman and Weiss [18, 19] with ν as in (7.1.2) and $H^{1,p}(\mathcal{X}, \nu)$, the atomic Hardy space as in [19] with $p \in (1, \infty]$. Then, it is easy to see that

$$H^{1,p}(\mathcal{X}, \nu) \subset \tilde{H}^{1,p}_{\mathrm{atb}}(\mathcal{X}, \nu) \subset H^{1,p}_{\mathrm{atb}}(\mathcal{X}, \nu).$$

Moreover, if $\nu(\mathcal{X}) = \infty$, by Proposition 7.2.3 and Theorem 7.4.8, together with Lemma 7.4.7, we know that

$$H^{1,p}(\mathcal{X}, \nu) = \tilde{H}^{1,p}_{\mathrm{atb}}(\mathcal{X}, \nu) = H^{1,p}_{\mathrm{atb}}(\mathcal{X}, \nu)$$

and the same are true for their dual spaces. However, for a general doubling measure ν with $\nu(\mathcal{X}) < \infty$, it may happen that

$$H^{1,p}(\mathcal{X}, \nu) \subsetneqq \tilde{H}^{1,p}_{\mathrm{atb}}(\mathcal{X}, \nu) \subset H^{1,p}_{\mathrm{atb}}(\mathcal{X}, \nu).^5$$

We now introduce the notion of the molecular Hardy space.

[5] See [134, p. 317, lines 15 to 16] and [131, p. 125, Example 5.6].

Definition 8.5.5. Let $\rho \in (1,\infty)$, $p \in (1,\infty]$, $\gamma \in [1,\infty)$ and $\epsilon \in (0,\infty)$. A function $b \in L^1(\mathcal{X}, \nu)$ is called a $(p, \gamma, \epsilon, \rho)_\lambda$*-molecular block* if

(i)

$$\int_{\mathcal{X}} b(x)\, d\nu(x) = 0;$$

(ii) there exist some ball B and some constants $\tilde{M}$, $M \in \mathbb{N}$ such that, for all $k \in \mathbb{Z}_+$ and $j \in \{0, 1, \ldots, M_k\}$ with $M_k := \tilde{M}$ when $k = 0$ and $M_k := M$ when $k \in \mathbb{N}$, there exist functions $m_{k,j}$ supported on some balls $B_{k,j} \subset U_k(B)$ for all $k \in \mathbb{Z}_+$, where

$$U_0(B) := \rho^2 B \quad \text{and} \quad U_k(B) := \rho^{k+2} B \setminus \rho^{k-2} B \quad \text{with} \quad k \in \mathbb{N},$$

and $\lambda_{k,j} \in \mathbb{C}$ such that

$$b = \sum_{k=0}^{\infty} \sum_{j=1}^{M_k} \lambda_{k,j} m_{k,j},$$

$$\|m_{k,j}\|_{L^p(\mathcal{X},\nu)} \le \rho^{-k\epsilon} \left[\nu(\rho B_{k,j})\right]^{1/p-1} \left[\tilde{\delta}^{(\rho)}_{B_{k,j},\, \rho^{k+2} B}\right]^{-\gamma} \tag{8.5.2}$$

and

$$|b|_{\tilde{H}^{1,p,\gamma,\epsilon}_{\mathrm{mb},\rho}(\mathcal{X},\nu)} := \sum_{k=0}^{\infty} \sum_{j=1}^{M_k} |\lambda_{k,j}| < \infty.$$

A function f is said to belong to the *molecular Hardy space* $\tilde{H}^{1,p,\gamma,\epsilon}_{\mathrm{mb},\rho}(\mathcal{X}, \nu)$ if there exist $(p, \gamma, \epsilon, \rho)_\lambda$-molecular blocks $\{b_i\}_{i=1}^{\infty}$ such that

$$f = \sum_{i=1}^{\infty} b_i$$

in $L^1(\mathcal{X}, \nu)$ and

$$\sum_{i=1}^{\infty} |b_i|_{\tilde{H}^{1,p,\gamma,\epsilon}_{\mathrm{mb},\rho}(\mathcal{X},\nu)} < \infty.$$

The $\tilde{H}^{1,p,\gamma,\epsilon}_{\mathrm{mb},\rho}(\mathcal{X}, \nu)$ *norm* of f is defined by

$$\|f\|_{\tilde{H}^{1,p,\gamma,\epsilon}_{\mathrm{mb},\rho}(\mathcal{X},\nu)} := \inf\left\{\sum_{i=1}^{\infty} |b_i|_{\tilde{H}^{1,p,\gamma,\epsilon}_{\mathrm{mb},\rho}(\mathcal{X},\nu)}\right\},$$

where the infimum is taken over all the possible decompositions of f as above.

Now we give the first main result of this section as follows.

Theorem 8.5.6. *Let $\rho \in (1, \infty)$, $p \in (1, \infty]$, $\gamma \in [1, \infty)$ and $\epsilon \in (0, \infty)$. Then the spaces $\tilde{H}^{1,p,\gamma}_{\mathrm{atb},\rho}(\mathcal{X}, \nu)$ and $\tilde{H}^{1,p,\gamma,\epsilon}_{\mathrm{mb},\rho}(\mathcal{X}, \nu)$ coincide with equivalent norms.*

Remark 8.5.7. Theorem 8.5.6 is new even when $(\mathcal{X}, d, \nu) := (\mathbb{R}^D, |\cdot|, \mu)$ with μ being the D-dimensional Lebesgue measure m, since $\tilde{H}^{1,p,\gamma}_{\mathrm{atb},\rho}(\mathcal{X}, \nu)$ may be strictly bigger than the classical Hardy space $H^1(\mathbb{R}^D)$ (see Remark 8.5.4(ii)).

Now we state another main result of this section, namely, the Calderón–Zygmund operator T is bounded on $\tilde{H}^{1,p}_{\mathrm{atb}}(\mathcal{X}, \nu)$ if T is bounded on $L^2(\mathcal{X}, \nu)$ and $T^*1 = 0$, where, by $T^*1 = 0$, we mean that, for any $b \in L^{\infty}_{b,0}(\mathcal{X}, \nu)$,

$$\int_{\mathcal{X}} Tb(x)\, d\nu(x) = 0.$$

Theorem 8.5.8. *Let $p \in (1, \infty)$. Suppose that T is a Calderón–Zygmund operator as in* (8.1.25) *with kernel K satisfying* (8.1.23) *and* (8.1.24) *which is bounded on $L^2(\mathcal{X}, \nu)$ and $T^*1 = 0$. Then there exists a positive constant C such that, for all $f \in \tilde{H}^{1,p}_{\mathrm{atb}}(\mathcal{X}, \nu)$,*

$$Tf \in \tilde{H}^{1,p}_{\mathrm{atb}}(\mathcal{X}, \nu) \quad \text{and} \quad \|Tf\|_{\tilde{H}^{1,p}_{\mathrm{atb}}(\mathcal{X},\nu)} \le C\|f\|_{\tilde{H}^{1,p}_{\mathrm{atb}}(\mathcal{X},\nu)}.$$

This section is organized as follows.

Section 8.5.1 is mainly devoted to the proofs of Theorems 8.5.6 and 8.5.8. The key ingredient for the proof of Theorem 8.5.6 is the notion of the $(p, \gamma, \epsilon, \rho)_\lambda$-molecular blocks, which is totally different from those in the Euclidean spaces or spaces of homogeneous type (see [125] and [19]); meanwhile, similar to the classical case, any $(p, \gamma, \rho)_\lambda$-atomic block is automatically a $(p, \gamma, \epsilon, \rho)_\lambda$-molecular block for any $\epsilon \in (0, \infty)$. By borrowing some ideas from the proof of [19, Theorem C] and using basic properties of $\tilde{\delta}^{(\rho)}_{B,S}$ (see Lemma 8.5.9 below), for any fixed $(p, \gamma, \epsilon, \rho)_\lambda$-molecular block b, we construct some $(p, \gamma, \rho)_\lambda$-atomic blocks and $(\infty, \gamma, \rho)_\lambda$-atomic blocks such that their sum is just b, which further implies that $b \in \tilde{H}^{1,p,\gamma}_{\mathrm{atb},\rho}(\mathcal{X}, \nu)$ and the norm of b in $\tilde{H}^{1,p,\gamma}_{\mathrm{atb},\rho}(\mathcal{X}, \nu)$ is controlled by a positive constant multiple of $|b|_{\tilde{H}^{1,p,\gamma,\epsilon}_{\mathrm{mb},\rho}(\mathcal{X},\nu)}$. The desired conclusion of Theorem 8.5.6 is then deduced. A similar molecular characterization of $H^{1,p}_{\mathrm{atb}}(\mathcal{X}, \nu)$ also holds true; see Remark 8.5.11(ii). As an application of Theorem 8.5.6, we prove Theorem 8.5.8 by showing that the Calderón–Zygmund operator T maps atomic blocks into some molecular blocks. One novelty for the proof of Theorem 8.5.8 is to decompose the image of an atomic block under T into a "small part" and a "large part", and further to decompose these two parts by different scales. Basic properties of $\tilde{\delta}^{(\rho)}_{B,S}$ and the geometrically doubling condition play important roles in the estimates of these both parts; in particular, when estimating the "small part", we need to use the discrete coefficient $\tilde{\delta}_{B,S}^{(\rho)}$ and its properties, which, according to the present method used in this section, cannot be replaced by $1 + \delta(B, S)$ and its properties. Using the duality and Theorem 8.5.8, in Corollary 8.5.14 below, we further obtain the boundedness of the Calderón–Zygmund operator on $\widetilde{\mathrm{RBMO}}(\mathcal{X}, \nu)$, the corresponding dual space of $\tilde{H}_{\mathrm{atb}}^{1,p}(\mathcal{X}, \nu)$.

In Sect. 8.5.2, we study the relations between

$$\tilde{H}_{\mathrm{atb}}^{1,p}(\mathcal{X}, \nu) \text{ in this section and } H_{\mathrm{atb}}^{1,p}(\mathcal{X}, \nu)$$

in Sect. 7.4. To this end, we first introduce the so-called weak reverse doubling condition (see Definition 8.5.15 below), under which we show (8.5.1) for all $\rho \in (1, \infty)$ and balls $B \subset S$ of $\mathcal{X}$ (see Theorem 8.5.17 below). Applying Theorem 8.5.17, we further show, in Corollary 8.5.18 below, that $\tilde{H}_{\mathrm{atb}}^{1,p}(\mathcal{X}, \nu)$ and $\widetilde{\mathrm{RBMO}}(\mathcal{X}, \nu)$ coincide, respectively, with $H_{\mathrm{atb}}^{1,p}(\mathcal{X}, \nu)$ and $\mathrm{RBMO}\,(\mathcal{X}, \nu)$ in Sect. 7.2, when a non-homogeneous space $(\mathcal{X}, d, \nu)$ has the domination function satisfying the weak reverse doubling condition. It turns out that there exists a large class of metric measure spaces with dominating functions satisfying the weak reverse doubling condition (see Example 8.5.16 and (i) and (ii) of Remark 8.5.19 below). On the other hand, we also present an example (see Example 8.5.20 below), which does not satisfy the weak reverse doubling condition and

$$1 + \delta(B, S) \nsim \tilde{\delta}_{B,S}^{(\rho)}$$

for some $\rho \in (1, \infty)$. In this sense, the weak reverse doubling condition in Theorem 8.5.17 is also necessary.

8.5.1 *Proofs of Theorems 8.5.6 and 8.5.8*

This section is mainly devoted to the proofs of Theorems 8.5.6 and 8.5.8. We begin with some useful properties of $\tilde{\delta}_{B,S}^{\rho}$. The proof is similar to that of Lemma 7.1.16 and omitted.

Lemma 8.5.9. (i) *For any $\rho \in [1, \infty)$, there exists a positive constant $C_{(\rho)}$, depending on ρ, such that,*

$$\text{for all balls } B \subset R \subset S,\ \tilde{\delta}_{B,R}^{(\rho)} \le C_{(\rho)} \tilde{\delta}_{B,S}^{(\rho)}.$$

(ii) *For any $\alpha \in [1, \infty)$ and $\rho \in [1, \infty)$, there exists a positive constant $C_{(\alpha,\rho)}$, depending on α and ρ, such that,*

$$\text{for all balls } B \subset S \text{ with } r_S \le \alpha r_B,\ \tilde{\delta}^{(\rho)}_{B,S} \le C_{(\alpha,\rho)}.$$

(iii) *For any* $\alpha \in [1,\infty)$ *and* $\rho \in [1,\infty)$*, there exists a positive constant* $C_{(\alpha,\rho)}$*, depending on* α *and* ρ*, such that,*

$$\text{for all balls } B,\ \tilde{\delta}^{(\rho)}_{B,\tilde{B}^\alpha} \le C_{(\alpha,\rho)}.$$

(iv) *For any* $\rho \in [1,\infty)$*, there exists a positive constant* $c_{(\rho)}$*, depending on* ρ*, such that,*

$$\text{for all balls } B \subset R \subset S,\ \tilde{\delta}^{(\rho)}_{B,S} \le \tilde{\delta}^{(\rho)}_{B,R} + c_{(\rho)}\tilde{\delta}^{(\rho)}_{R,S}.$$

(v) *For any* $\rho \in [1,\infty)$*, there exists a positive constant* $\tilde{c}_{(\rho)}$*, depending on* ρ*, such that,*

$$\text{for all balls } B \subset R \subset S,\ \tilde{\delta}^{(\rho)}_{R,S} \le \tilde{c}_{(\rho)}\tilde{\delta}^{(\rho)}_{B,S}.$$

Before we prove Theorems 8.5.6 and 8.5.8, we present the following two important properties of the molecular Hardy space whose proofs are analogous to that of Proposition 7.4.2. We omit the details.

Proposition 8.5.10. *Let* $\rho \in (1,\infty)$*,* $p \in (1,\infty]$*,* $\gamma \in [1,\infty)$ *and* $\epsilon \in (0,\infty)$*. Then the following conclusions hold true:*

(i) $\tilde{H}^{1,p,\gamma,\epsilon}_{\mathrm{mb},\rho}(\mathcal{X},\nu)$ *is a Banach space;*
(ii) *for all* $f \in \tilde{H}^{1,p,\gamma,\epsilon}_{\mathrm{mb},\rho}(\mathcal{X},\nu)$*,*

$$\tilde{H}^{1,p,\gamma,\epsilon}_{\mathrm{mb},\rho}(\mathcal{X},\nu) \subset L^1(\mathcal{X},\nu) \quad \text{and} \quad \|f\|_{L^1(\mathcal{X},\nu)} \le \|f\|_{\tilde{H}^{1,p,\gamma,\epsilon}_{\mathrm{mb},\rho}(\mathcal{X},\nu)}.$$

Proof of Theorem 8.5.6. By similarity, we only consider the case $p \in (1,\infty)$. Let b be a $(p,\gamma,\rho)_\lambda$-atomic block. It is easy to see that b is a $(p,\gamma,\epsilon,\rho)_\lambda$-molecular block for any $\epsilon \in (0,\infty)$. Moreover, for all $f \in \tilde{H}^{1,p,\gamma}_{\mathrm{atb},\rho}(\mathcal{X},\nu)$,

$$\tilde{H}^{1,p,\gamma}_{\mathrm{atb},\rho}(\mathcal{X},\nu) \subset \tilde{H}^{1,p,\gamma,\epsilon}_{\mathrm{mb},\rho}(\mathcal{X},\nu) \quad \text{and} \quad \|f\|_{\tilde{H}^{1,p,\gamma,\epsilon}_{\mathrm{mb},\rho}(\mathcal{X},\nu)} \le \|f\|_{\tilde{H}^{1,p,\gamma}_{\mathrm{atb},\rho}(\mathcal{X},\nu)}.$$

Now we show the converse via a constructive way. By some standard arguments,[6] it suffices to prove that any $(p,\gamma,\epsilon,\rho)_\lambda$-molecular block b can be decomposed into some $(p,\gamma,\rho)_\lambda$-atomic blocks and $(\infty,\gamma,\rho)_\lambda$-atomic blocks, and

$$\|b\|_{\tilde{H}^{1,p}_{\mathrm{atb}}(\mathcal{X},\nu)} \lesssim |b|_{\tilde{H}^{1,p,\gamma,\epsilon}_{\mathrm{mb},\rho}(\mathcal{X},\nu)}.$$

[6]See, for example, [19].

Indeed, for any $(p, \gamma, \epsilon, \rho)_\lambda$-molecular block b, by Definition 8.5.5, we know that

$$b = \sum_{k=0}^{\infty} \sum_{j=1}^{M_k} \lambda_{k,j} m_{k,j} =: \sum_{k=0}^{\infty} b_k, \tag{8.5.3}$$

where, for any $k \in \mathbb{Z}_+$ and $j \in \{1, \ldots, M_k\}$, $\lambda_{k,j} \in \mathbb{C}$ and

$$\operatorname{supp}(m_{k,j}) \subset B_{k,j} \subset U_k(B)$$

with the same notation as in Definition 8.5.5. Let

$$B_{k+2}^\rho := \rho^{k+2} B \quad \text{and} \quad \tilde{B}_{k+2}^\rho := \widetilde{\rho^{k+2} B}^\rho, \ k \in \mathbb{Z}_+.$$

By (8.5.3), we further see that

$$\begin{aligned} b &= \sum_{k=0}^{\infty} \left[b_k - \frac{\chi_{\tilde{B}_{k+2}^\rho}}{\nu(\tilde{B}_{k+2}^\rho)} \int_{\mathcal{X}} b_k(y)\, d\nu(y) \right] + \sum_{k=0}^{\infty} \frac{\chi_{\tilde{B}_{k+2}^\rho}}{\nu(\tilde{B}_{k+2}^\rho)} \int_{\mathcal{X}} b_k(y)\, d\nu(y) \\ &= \sum_{k=0}^{\infty} \sum_{j=1}^{M_k} \lambda_{k,j} \left[m_{k,j} - \frac{\chi_{\tilde{B}_{k+2}^\rho}}{\nu(\tilde{B}_{k+2}^\rho)} \int_{B_{k,j}} m_{k,j}(y)\, d\nu(y) \right] \\ &\quad + \sum_{k=0}^{\infty} \frac{\chi_{\tilde{B}_{k+2}^\rho}}{\nu(\tilde{B}_{k+2}^\rho)} \int_{\mathcal{X}} b_k(y)\, d\nu(y) \\ &=: \sum_{k=0}^{\infty} \sum_{j=1}^{M_k} b_{k,j} + \sum_{k=0}^{\infty} \chi_k \tilde{M}_k \\ &=: \mathrm{I} + \mathrm{II}, \end{aligned}$$

where, for all $k \in \mathbb{Z}_+$ and $j \in \{1, \ldots, M_k\}$,

$$b_{k,j} := \lambda_{k,j} \left[m_{k,j} - \frac{\chi_{\tilde{B}_{k+2}^\rho}}{\nu(\tilde{B}_{k+2}^\rho)} \int_{B_{k,j}} m_{k,j}(y)\, d\nu(y) \right], \ \chi_k := \frac{\chi_{\tilde{B}_{k+2}^\rho}}{\nu(\tilde{B}_{k+2}^\rho)}$$

and

$$\tilde{M}_k := \int_{\mathcal{X}} b_k(y)\, d\nu(y).$$

To estimate I, we first show that, for any $k \in \mathbb{Z}_+$ and $j \in \{1, \ldots, M_k\}$, $b_{k,j}$ is a $(p, \gamma, \rho)_\lambda$-atomic block. Noticing that

$$\operatorname{supp}(b_{k,j}) \subset \widetilde{B}^{\rho}_{k+2} \quad \text{and} \quad \int_{\mathcal{X}} b_{k,j}(y)\, d\nu(y) = 0,$$

it only needs to show that $b_{k,j}$ satisfies Definition 8.5.3(iii). To this end, we further decompose $b_{k,j}$ into

$$\begin{aligned} b_{k,j} &= \lambda_{k,j}\left[m_{k,j} - \frac{\chi_{B_{k,j}}}{\nu(\widetilde{B}^{\rho}_{k+2})}\int_{B_{k,j}} m_{k,j}(y)\, d\nu(y)\right] \\ &\quad + (-\lambda_{k,j})\frac{\chi_{\widetilde{B}^{\rho}_{k+2}\setminus B_{k,j}}}{\nu(\widetilde{B}^{\rho}_{k+2})}\int_{B_{k,j}} m_{k,j}(y)\, d\nu(y) \\ &=: A^{(1)}_{k,j} + A^{(2)}_{k,j}. \end{aligned}$$

By the Hölder inequality, (8.5.2), (iv) and (iii) of Lemma 8.5.9, we know that

$$\begin{aligned} &\left\|A^{(1)}_{k,j}\right\|_{L^p(\mathcal{X},\nu)} \\ &\le |\lambda_{k,j}|\left\{\|m_{k,j}\|_{L^p(\mathcal{X},\nu)} + \frac{[\nu(B_{k,j})]^{1/p}}{\nu(\widetilde{B}^{\rho}_{k+2})}\left|\int_{B_{k,j}} m_{k,j}(y)\, d\nu(y)\right|\right\} \\ &\le |\lambda_{k,j}|\left\{\|m_{k,j}\|_{L^p(\mathcal{X},\nu)} + \frac{[\nu(B_{k,j})]^{1/p}[\nu(B_{k,j})]^{1/p'}}{\nu(\widetilde{B}^{\rho}_{k+2})}\|m_{k,j}\|_{L^p(\mathcal{X},\nu)}\right\} \\ &\lesssim |\lambda_{k,j}|\|m_{k,j}\|_{L^p(\mathcal{X},\nu)} \\ &\lesssim |\lambda_{k,j}|\rho^{-k\epsilon}[\nu(\rho B_{k,j})]^{1/p-1}\left[\widetilde{\delta}^{(\rho)}_{B_{k,j},B^{\rho}_{k+2}}\right]^{-\gamma} \\ &\le c_5|\lambda_{k,j}|\rho^{-k\epsilon}[\nu(\rho B_{k,j})]^{1/p-1}\left[\widetilde{\delta}^{(\rho)}_{B_{k,j},\widetilde{B}^{\rho}_{k+2}}\right]^{-\gamma}, \end{aligned}$$

where c_5 is a positive constant independent of k and j. Let

$$\nu^{(1)}_{k,j} := c_5|\lambda_{k,j}|\rho^{-k\epsilon} \quad \text{and} \quad a^{(1)}_{k,j} := \frac{1}{\nu^{(1)}_{k,j}}A^{(1)}_{k,j}.$$

Then

$$A^{(1)}_{k,j} = \nu^{(1)}_{k,j}a^{(1)}_{k,j}, \ \operatorname{supp}(a^{(1)}_{k,j}) \subset B_{k,j} \subset \widetilde{B}^{\rho}_{k+2}$$

and

$$\left\|a^{(1)}_{k,j}\right\|_{L^p(\mathcal{X},\nu)} \le [\nu(\rho B_{k,j})]^{1/p-1}\left[\widetilde{\delta}^{(\rho)}_{B_{k,j},\widetilde{B}^{\rho}_{k+2}}\right]^{-\gamma}.$$

From the Hölder inequality, (8.5.2), the fact that $\widetilde{\delta}^{(\rho)}_{B_{k,j}, B^{\rho}_{k+2}} \geq 1$, the (ρ, β_ρ)-doubling property of $\tilde{B}^{\rho}_{k+2}$ and Lemma 8.5.9(ii), it follows that

$$\begin{aligned}\left\|A^{(2)}_{k,j}\right\|_{L^p(\mathcal{X},\nu)} &= |\lambda_{k,j}| \frac{[\nu(\tilde{B}^{\rho}_{k+2} \setminus B_{k,j})]^{1/p}}{\nu(\tilde{B}^{\rho}_{k+2})} \left| \int_{B_{k,j}} m_{k,j}(y)\, d\nu(y) \right| \\ &\leq |\lambda_{k,j}| \left[\nu\left(\tilde{B}^{\rho}_{k+2}\right)\right]^{1/p-1} [\nu(B_{k,j})]^{1/p'} \|m_{k,j}\|_{L^p(\mathcal{X},\nu)} \\ &\lesssim |\lambda_{k,j}| \rho^{-k\epsilon} \left[\nu\left(\rho \tilde{B}^{\rho}_{k+2}\right)\right]^{1/p-1} \\ &\leq c_6 |\lambda_{k,j}| \rho^{-k\epsilon} \left[\nu\left(\rho \tilde{B}^{\rho}_{k+2}\right)\right]^{1/p-1} \left[\widetilde{\delta}^{(\rho)}_{\tilde{B}^{\rho}_{k+2}, \tilde{B}^{\rho}_{k+2}}\right]^{-\gamma},\end{aligned}$$

where c_6 is a positive constant independent of k and j. Let

$$\nu^{(2)}_{k,j} := c_6 |\lambda_{k,j}| \rho^{-k\epsilon} \quad \text{and} \quad a^{(2)}_{k,j} := \frac{1}{\nu^{(2)}_{k,j}} A^{(2)}_{k,j}.$$

Then

$$A^{(2)}_{k,j} = \nu^{(2)}_{k,j} a^{(2)}_{k,j}, \ \operatorname{supp}(a^{(2)}_{k,j}) \subset \tilde{B}^{\rho}_{k+2}$$

and

$$\left\|a^{(2)}_{k,j}\right\|_{L^p(\mathcal{X},\nu)} \leq \left[\nu\left(\rho \tilde{B}^{\rho}_{k+2}\right)\right]^{1/p-1} \left[\widetilde{\delta}_{\tilde{B}^{\rho}_{k+2}, \tilde{B}^{\rho}_{k+2}}\right]^{-\gamma}.$$

Thus,

$$b_{k,j} = \nu^{(1)}_{k,j} a^{(2)}_{k,j} + \nu^{(2)}_{k,j} a^{(2)}_{k,j}$$

is a $(p, \gamma, \rho)_\lambda$-atomic block and

$$|b_{k,j}| \lesssim |\lambda_{k,j}| \rho^{-k\epsilon}.$$

Moreover, we conclude that

$$\|\mathrm{I}\|_{\tilde{H}^{1,p}_{\mathrm{atb}}(\mathcal{X},\nu)} \lesssim \sum_{k=0}^{\infty} \sum_{j=1}^{M_k} |\lambda_{k,j}| \rho^{-k\epsilon} \lesssim \sum_{k=0}^{\infty} \sum_{j=1}^{M_k} |\lambda_{k,j}| \sim |b|_{\tilde{H}^{1,p,\gamma,\epsilon}_{\mathrm{mb},\rho}(\mathcal{X},\nu)}.$$

Now we turn to estimate II. Let

$$N_k := \sum_{i=k}^{\infty} \tilde{M}_i.$$

Since

$$\int_{\mathcal{X}} b(y)\,d\nu(y) = 0,$$

it follows that

$$\begin{aligned}
\sum_{k=0}^{\infty} \chi_k \tilde{M}_k &= \sum_{k=0}^{\infty} \chi_k (N_k - N_{k+1}) \\
&= \sum_{k=0}^{\infty} (\chi_{k+1} - \chi_k) N_{k+1} + \chi_0 N_0 \\
&= \sum_{k=0}^{\infty} (\chi_{k+1} - \chi_k) N_{k+1} \\
&= \sum_{k=0}^{\infty} \sum_{i=k+1}^{\infty} \sum_{j=1}^{M} \lambda_{i,j} (\chi_{k+1} - \chi_k) \int_{B_{i,j}} m_{i,j}(y)\,d\nu(y) \\
&=: \sum_{k=0}^{\infty} \sum_{i=k+1}^{\infty} \sum_{j=1}^{M} b_{k,j,i}.
\end{aligned}$$

Now we prove that, for any $k \in \mathbb{Z}_+$, $i \in \{k+1, k+2, \ldots\}$ and $j \in \{1, \ldots, M\}$, $b_{k,j,i}$ is an $(\infty, \gamma, \rho)_\lambda$-atomic block. Observing that

$$\operatorname{supp}(b_{k,j,i}) \subset \tilde{B}^{\rho}_{k+3} \quad \text{and} \quad \int_{\mathcal{X}} b_{k,j,i}(y)\,d\nu(y) = 0,$$

we still need to show that $b_{k,j,i}$ satisfies Definition 8.5.3(iii).

To this end, we further write

$$\begin{aligned}
b_{k,j,i} &= \lambda_{i,j} \chi_{k+1} \int_{B_{i,j}} m_{i,j}(y)\,d\nu(y) + (-\lambda_{i,j}) \chi_k \int_{B_{i,j}} m_{i,j}(y)\,d\nu(y) \\
&=: A^{(1)}_{k,j,i} + A^{(2)}_{k,j,i}.
\end{aligned}$$

From the Hölder inequality, (8.5.2), the fact that $\tilde{\delta}^{(\rho)}_{B_{i,j}, B^{\rho}_{i+2}} \geq 1$ and Lemma 8.5.9(ii), we deduce that

$$\begin{aligned}\left\|A_{k,j,i}^{(1)}\right\|_{L^\infty(\mathcal{X},\nu)} &\le |\lambda_{i,j}|\frac{[\nu(B_{i,j})]^{1/p'}}{\nu(\tilde{B}_{k+3}^{\rho})}\|m_{i,j}\|_{L^p(\mathcal{X},\nu)} \\ &\le |\lambda_{i,j}|\frac{[\nu(B_{i,j})]^{1/p'}}{\nu(\tilde{B}_{k+3}^{\rho})}\rho^{-i\epsilon}[\nu(\rho B_{i,j})]^{1/p-1}\left[\tilde{\delta}_{B_{i,j},B_{i+2}^{\rho}}^{(\rho)}\right]^{-\gamma} \\ &\lesssim |\lambda_{i,j}|\rho^{-i\epsilon}\left[\nu\left(\rho\tilde{B}_{k+3}^{\rho}\right)\right]^{-1} \\ &\le c_7|\lambda_{i,j}|\rho^{-i\epsilon}\left[\nu\left(\rho\tilde{B}_{k+3}^{\rho}\right)\right]^{-1}\left[\tilde{\delta}_{\tilde{B}_{k+3}^{\rho},\tilde{B}_{k+3}^{\rho}}^{(\rho)}\right]^{-\gamma},\end{aligned}$$

where c_7 is a positive constant independent of k and j. Let

$$\nu_{k,j,i}^{(1)} := c_7|\lambda_{i,j}|\rho^{-i\epsilon} \quad \text{and} \quad a_{k,j,i}^{(1)} := \frac{1}{\nu_{k,j,i}^{(1)}}A_{k,j,i}^{(1)}.$$

Then we see that

$$A_{k,j,i}^{(1)} = \nu_{k,j,i}^{(1)}a_{k,j,i}^{(1)}, \ \operatorname{supp}(a_{k,j,i}^{(1)}) \subset \tilde{B}_{k+3}^{\rho}$$

and

$$\left\|a_{k,j,i}^{(1)}\right\|_{L^\infty(\mathcal{X},\nu)} \le \left[\nu\left(\rho\tilde{B}_{k+3}^{\rho}\right)\right]^{-1}\left[\tilde{\delta}_{\tilde{B}_{k+3}^{\rho},\tilde{B}_{k+3}^{\rho}}^{(\rho)}\right]^{-\gamma}.$$

By an argument similar to that used in the estimate for $A_{k,j,i}^{(1)}$, we conclude that

$$\left\|A_{k,j,i}^{(2)}\right\|_{L^\infty(\mathcal{X},\nu)} \le c_8|\lambda_{i,j}|\rho^{-i\epsilon}\left[\nu\left(\rho\tilde{B}_{k+2}^{\rho}\right)\right]^{-1}\left[\tilde{\delta}_{\tilde{B}_{k+2}^{\rho},\tilde{B}_{k+3}^{\rho}}^{(\rho)}\right]^{-\gamma}.$$

Let

$$\nu_{k,j,i}^{(2)} := c_8|\lambda_{i,j}|\rho^{-i\epsilon} \quad \text{and} \quad a_{k,j,i}^{(2)} := \frac{1}{\nu_{k,j,i}^{(2)}}A_{k,j,i}^{(2)}.$$

Then

$$A_{k,j,i}^{(2)} = \nu_{k,j,i}^{(2)}a_{k,j,i}^{(2)}, \ \operatorname{supp}\left(a_{k,j,i}^{(2)}\right) \subset \tilde{B}_{k+2}^{\rho} \subset \tilde{B}_{k+3}^{\rho}$$

and

$$\left\|a_{k,j,i}^{(2)}\right\|_{L^\infty(\mathcal{X},\nu)} \le \left[\nu\left(\rho\tilde{B}_{k+2}^{\rho}\right)\right]^{-1}\left[\tilde{\delta}_{\tilde{B}_{k+2}^{\rho},\tilde{B}_{k+3}}^{(\rho)}\right]^{-\gamma}.$$

Thus,

$$b_{k,j,i} = \nu_{k,j,i}^{(2)} a_{k,j,i}^{(2)} + \nu_{k,j,i}^{(2)} a_{k,j,i}^{(2)}$$

is an $(\infty, \gamma, \rho)_\lambda$-atomic block and

$$|b_{k,j,i}|_{\tilde{H}^{1,\infty,\gamma}_{\mathrm{atb},\rho}(\mathcal{X},\nu)} \lesssim |\lambda_{i,j}| \rho^{-i\epsilon}.$$

Moreover, we have

$$\begin{aligned}
\|\mathrm{II}\|_{\tilde{H}^{1,p}_{\mathrm{atb}}(\mathcal{X},\nu)} &\lesssim \sum_{k=0}^{\infty} \sum_{i=k+1}^{\infty} \sum_{j=1}^{M} |b_{k,j,i}|_{\tilde{H}^{1,\infty,\gamma}_{\mathrm{atb},\rho}(\mathcal{X},\nu)} \\
&\lesssim \sum_{j=1}^{M} \sum_{k=0}^{\infty} \sum_{i=k+1}^{\infty} |\lambda_{i,j}| \rho^{-i\epsilon} \\
&\sim \sum_{j=1}^{M} \sum_{i=1}^{\infty} \sum_{k=0}^{i-1} |\lambda_{i,j}| \rho^{-i\epsilon} \\
&\sim \sum_{j=1}^{M} \sum_{i=1}^{\infty} |\lambda_{i,j}| i \rho^{-i\epsilon} \\
&\lesssim |b|_{\tilde{H}^{1,p,\gamma,\epsilon}_{\mathrm{mb},\rho}(\mathcal{X},\nu)}.
\end{aligned}$$

Combining the estimates for I and II, we see that $b \in \tilde{H}^{1,p}_{\mathrm{atb}}(\mathcal{X}, \nu)$ and

$$\|b\|_{\tilde{H}^{1,p}_{\mathrm{atb}}(\mathcal{X},\nu)} \lesssim |b|_{\tilde{H}^{1,p,\gamma,\epsilon}_{\mathrm{mb},\rho}(\mathcal{X},\nu)},$$

which further implies that

$$\tilde{H}^{1,p,\gamma,\epsilon}_{\mathrm{mb},\rho}(\mathcal{X}, \nu) \subset \tilde{H}^{1,p}_{\mathrm{atb}}(\mathcal{X}, \nu)$$

and, for all $f \in \tilde{H}^{1,p,\gamma,\epsilon}_{\mathrm{mb},\rho}(\mathcal{X}, \nu)$

$$\|f\|_{\tilde{H}^{1,p}_{\mathrm{atb}}(\mathcal{X},\nu)} \lesssim \|f\|_{\tilde{H}^{1,p,\gamma,\epsilon}_{\mathrm{mb},\rho}(\mathcal{X},\nu)}.$$

This finishes the proof of Theorem 8.5.6. □

Remark 8.5.11. (i) As a consequence of Theorem 8.5.6 and Remark 8.5.4, we see that the space $\tilde{H}^{1,p,\gamma,\epsilon}_{\mathrm{mb},\rho}(\mathcal{X}, \nu)$ is independent of the choices of the parameters p, ρ, γ and ϵ. In what follows, we *denote the molecular Hardy space* $\tilde{H}^{1,p,\gamma,\epsilon}_{\mathrm{mb},\rho}(\mathcal{X}, \nu)$ *simply by* $\tilde{H}^{1,p}_{\mathrm{mb}}(\mathcal{X}, \nu)$ *or* $\tilde{H}^1(\mathcal{X}, \nu)$.

(ii) By Lemma 7.1.16 and an argument similar to that used in the proof of Theorem 8.5.6, the conclusion of Theorem 8.5.6 is still valid with $\tilde{\delta}_{B,S}^{(\rho)}$ in Definitions 8.5.3 and 8.5.5 replaced by $1+\delta(B, S)$. That is, there also exists a similar molecular characterization of $H^1(\mathcal{X}, \nu)$.

Now we are ready to prove Theorem 8.5.8.

Proof of Theorem 8.5.8. By Remarks 8.5.4(ii) and 8.5.11(i), without loss of generality, we may assume that $p = 2 = \gamma$ and $\rho = 4$ in Definition 8.5.3, and $\rho = 2 = p$, $\gamma = 1$ and $\epsilon = \tau/2$ in Definition 8.5.5, where τ is as in (8.1.24). Observe that Theorem 8.1.5 is still valid with $H_{\rm atb}^{1,p}(\mathcal{X}, \nu)$ replaced by $\tilde{H}_{{\rm atb},\rho}^{1,p,\gamma}(\mathcal{X}, \nu)$. By this and Proposition 8.5.10, we see that, to show Theorem 8.5.8, it suffices to prove that the Calderón–Zygmund operator T maps a $(2,2,4)_\lambda$-atomic block into a $(2,1,\tau/2,2)_\lambda$-molecular block, where τ is as in (8.1.24).

Indeed, let b be a $(2,2,4)_\lambda$-atomic block. Then

$$b := \sum_{j=1}^{2} \lambda_j a_j,$$

where, for any $j \in \{1,2\}$,

$$\operatorname{supp}(a_j) \subset B_j \subset B$$

for some balls B_j and B as in Definition 8.5.3. Let $B_0 := 8B$. We write

$$Tb = Tb\chi_{B_0} + \sum_{k=1}^{\infty} Tb\chi_{2^k B_0 \setminus 2^{k-1} B_0} =: A_1 + A_2.$$

We first estimate A_1. Since $B_j \subset B$, we have $2B_j \subset 8B = B_0$. Let $N_j := N_{2B_j,B_0}^{(2)}$. We further decompose

$$\begin{aligned} A_1 &= \sum_{j=1}^{2} \lambda_j T a_j \chi_{2B_j} + \sum_{j=1}^{2} \sum_{i=1}^{N_j-2} \lambda_j T a_j \chi_{2^{i+1}B_j \setminus 2^i B_j} + \sum_{j=1}^{2} \lambda_j T a_j \chi_{B_0 \setminus 2^{N_j-1} B_j} \\ &=: A_{1,1} + A_{1,2} + A_{1,3}. \end{aligned}$$

To estimate $A_{1,1}$, by Definition 8.5.3(iii), the boundedness of T on $L^2(\mathcal{X}, \nu)$, (v), (iv) and (ii) of Lemma 8.5.9 and the fact that $\tilde{\delta}_{2B_j,B_0}^{(2)} \ge 1$, we see that, for any $j \in \{1,2\}$,

$$\begin{aligned} \|Ta_j\chi_{2B_j}\|_{L^2(\mathcal{X},\nu)} &\lesssim \|a_j\|_{L^2(\mathcal{X},\nu)} \\ &\lesssim [\nu(4B_j)]^{-1/2} \left[\tilde{\delta}_{B_j,B}^{(2)}\right]^{-2} \end{aligned}$$

$$\begin{aligned}
&\lesssim [\nu(4B_j)]^{-1/2}\left[\tilde{\delta}^{(2)}_{2B_j,8B}\right]^{-2}\\
&\le c_9[\nu(4B_j)]^{-1/2}\left[\tilde{\delta}^{(2)}_{2B_j,4B_0}\right]^{-1},
\end{aligned}$$

where c_9 is a positive constant independent of a_j and j. Let

$$\sigma_{j,1} := c_9\lambda_j \quad \text{and} \quad n_{j,1} := c_9^{-1}Ta_j\chi_{2B_j}.$$

Then

$$A_{1,1} = \sum_{j=1}^{2}\sigma_{j,1}n_{j,1}, \ \operatorname{supp}(n_{j,1}) \subset 2B_j \subset B_0$$

and

$$\|n_{j,1}\|_{L^2(\mathcal{X},\nu)} \le [\nu(2\times 2B_j)]^{-1/2}\left[\tilde{\delta}^{(2)}_{2B_j,4B_0}\right]^{-1}.$$

For $A_{1,3}$, since $B_0 \subset 2^{N_j+3}B_j$, we have $r_{B_0} \sim r_{2^{N_j-1}B_j}$. For any $j \in \{1,2\}$, let x_j and r_j be the center and the radius of B_j, respectively. By (8.1.23), the Hölder inequality, Definition 8.5.3(iii), the fact that $\tilde{\delta}^{(2)}_{B_j,B} \ge 1$, $B_0 \subset 2^{N_j+3}B_j$ and (ii) of Lemma 8.5.9, we obtain

$$\begin{aligned}
&\left\|Ta_j\chi_{B_0\setminus 2^{N_j-1}B_j}\right\|_{L^2(\mathcal{X},\nu)}\\
&\le \left\{\int_{8B\setminus 2^{N_j-1}B_j}\left[\int_{B_j}\frac{|a_j(y)|}{\lambda(x,d(x,y))}\,d\nu(y)\right]^2 d\nu(x)\right\}^{1/2}\\
&\lesssim \frac{[\nu(8B\setminus 2^{N_j-1}B_j)]^{1/2}}{\lambda(x_j,2^{N_j-2}r_j)}[\nu(B_j)]^{1/2}\|a_j\|_{L^2(\mathcal{X},\nu)}\\
&\lesssim [\nu(16B)]^{-1/2}\left[\tilde{\delta}^{(2)}_{B_j,B}\right]^{-2}\\
&\le c_{10}[\nu(2B_0)]^{-1/2}\left[\tilde{\delta}^{(2)}_{B_0,4B_0}\right]^{-1},
\end{aligned}$$

where c_{10} is a positive constant independent of a_j and j. Let

$$\sigma_{j,3} := c_{10}\lambda_j \quad \text{and} \quad n_{j,3} := c_{10}^{-1}Ta_j\chi_{B_0\setminus 2^{N_j-1}B_j}.$$

Then

$$A_{1,3} = \sum_{j=1}^{2} \sigma_{j,3} n_{j,3},$$

$$\operatorname{supp}(n_{j,3}) \subset 8B = B_0 \quad \text{and} \quad \left\|n_{j,3}\right\|_{L^2(\mathcal{X},\nu)} \le [\nu(2B_0)]^{-1/2}\left[\tilde{\delta}^{(2)}_{B_0,4B_0}\right]^{-1}.$$

We now estimate $A_{1,2}$. By (8.1.23), Definition 8.5.3(iii), the Hölder inequality, and (v), (iv) and (ii) of Lemma 8.5.9, we conclude that

$$\begin{aligned}
&\left\|(Ta_j)\,\chi_{2^{i+1}B_j\setminus 2^i B_j}\right\|_{L^2(\mathcal{X},\nu)}\\
&\le \left\{\int_{2^{i+1}B_j\setminus 2^i B_j}\left[\int_{B_j}\frac{|a_j(y)|}{\lambda(x,d(x,y))}\,d\nu(y)\right]^2 d\nu(x)\right\}^{1/2}\\
&\lesssim \left\{\int_{2^{i+1}B_j\setminus 2^i B_j}\left[\int_{B_j}\frac{|a_j(y)|}{\lambda(x_j,d(x,x_j))}\,d\nu(y)\right]^2 d\nu(x)\right\}^{1/2}\\
&\lesssim \left[\lambda\left(x_j,2^i r_j\right)\right]^{-1/2}\|a_j\|_{L^1(\mathcal{X},\nu)}\left[\int_{2^{i+1}B_j\setminus 2^i B_j}\frac{1}{\lambda\left(x_j,2^i r_j\right)}\,d\nu(x)\right]^{1/2}\\
&\lesssim \left[\lambda\left(x_j,2^i r_j\right)\right]^{-1/2}\|a_j\|_{L^2(\mathcal{X},\nu)}[\nu(B_j)]^{1/2}\left[\frac{\nu\left(2^{i+1}B_j\right)}{\lambda\left(x_j,2^i r_j\right)}\right]^{1/2}\\
&\le c_{11}\frac{\nu\left(2^{i+2}B_j\right)}{\lambda\left(x_j,2^i r_j\right)}\left[\tilde{\delta}^{(2)}_{B_j,B}\right]^{-1}\left[\nu\left(2^{i+2}B_j\right)\right]^{-1/2}\left[\tilde{\delta}^{(2)}_{2^{i+1}B_j,4B_0}\right]^{-1},
\end{aligned}$$

where c_{11} is a positive constant independent of a_j, j and i. Let

$$\sigma^{(i)}_{j,2} := c_{11}\lambda_j\frac{\nu(2^{i+2}B_j)}{\lambda(x_j,2^i r_j)}\left[\tilde{\delta}^{(2)}_{B_j,B}\right]^{-1}$$

and

$$n^{(i)}_{j,2} := \left[c_{11}\frac{\nu\left(2^{i+2}B_j\right)}{\lambda\left(x_j,2^i r_j\right)}\right]^{-1}\tilde{\delta}^{(2)}_{B_j,B}\,Ta_j\,\chi_{2^{i+1}B_j\setminus 2^i B_j}.$$

Then

$$A_{1,2} = \sum_{j=1}^{2}\sum_{i=1}^{N_j-2}\sigma^{(i)}_{j,2}n^{(i)}_{j,2},\quad \operatorname{supp}(n^{(i)}_{j,2}) \subset 2^{i+1}B_j \subset 4B_0$$

and

$$\left\| n_{j,2}^{(i)} \right\|_{L^2(\mathcal{X},\nu)} \leq \left[\nu\left(2 \times 2^{i+1} B_j\right)\right]^{-1/2} \left[\tilde{\delta}^{(2)}_{2^{i+1}B_j, 4B_0}\right]^{-1}.$$

Now we turn to estimate A_2. For any $k \in \mathbb{N}$, by the geometrically doubling condition, there exists a ball covering $\{B_{k,j}\}_{j=1}^{M_0}$, with uniform radius $2^{k-2} r_{B_0}$, of

$$\tilde{U}_k(B_0) := 2^k B_0 \setminus 2^{k-1} B_0$$

such that the cardinality $M_0 \leq N_0 4^n$. Without loss of generality, we may assume that the centers of the balls in the covering belong to $\tilde{U}_k(B_0)$.

Let

$$L_{k,1} := B_{k,1},\ C_{k,l} := B_{k,l} \setminus \bigcup_{m=1}^{l-1} B_{k,m},\ l \in \{2, \ldots, M_0\}$$

and

$$D_{k,l} := C_{k,l} \cap \tilde{U}_k(B_0)$$

for all $l \in \{1, \ldots, M_0\}$. Then we know that $\{D_{k,l}\}_{l=1}^{M_0}$ is pairwise disjoint,

$$\tilde{U}_k(B_0) = \bigcup_{l=1}^{M_0} D_{k,l}$$

and, for any $l \in \{1, \ldots, M_0\}$,

$$D_{k,l} \subset B_{k,l} \subset U_k(B_0) := 2^{k+2} B_0 \setminus 2^{k-2} B_0.$$

Thus,

$$A_2 = \sum_{k=1}^{\infty} Tb \sum_{l=1}^{M_0} \chi_{D_{k,l}} = \sum_{k=1}^{\infty} \sum_{l=1}^{M_0} Tb\chi_{D_{k,l}}.$$

From the fact

$$\int_{\mathcal{X}} b(y)\, d\nu(y) = 0,$$

(8.1.24), the Hölder inequality, Definition 8.5.3(iii), the fact that $\tilde{\delta}^{(2)}_{B_j,B} \geq 1$ and (ii) of Lemma 8.5.9, it follows that, for any $k \in \{1, 2, \ldots\}$ and $l \in \{1, \ldots, M_0\}$,

$$
\begin{aligned}
&\|Tb\chi_{D_{k,l}}\|_{L^2(\mathcal{X},\nu)}\\
&\le \left\{\int_{D_{k,l}}\left[\int_B |b(y)||K(x,y)-K(x,x_B)|\,d\nu(y)\right]^2 d\nu(x)\right\}^{1/2}\\
&\lesssim \left\{\int_{D_{k,l}}\left[\int_B |b(y)|\frac{[d(y,x_B)]^\tau}{[d(x,x_B)]^\tau\lambda(x_B,d(x,x_B))}\,d\nu(y)\right]^2 d\nu(x)\right\}^{1/2}\\
&\lesssim \frac{r_B^\tau[\nu(D_{k,l})]^{1/2}}{\lambda\left(x_B,2^k r_B\right)\left(2^k r_B\right)^\tau}\int_B |b(y)|\,d\nu(y)\\
&\lesssim 2^{-k\tau}\left[\nu\left(2^{k+2}B\right)\right]^{-1/2}\sum_{j=1}^2 |\lambda_j|[\nu(B_j)]^{1/2}\|a_j\|_{L^2(\mathcal{X},\nu)}\\
&\lesssim 2^{-k\tau}\left[\nu\left(2^{k+2}B\right)\right]^{-1/2}\sum_{j=1}^2 |\lambda_j|\\
&\le c_{12}2^{-k\tau/2}2^{-k\tau/2}\sum_{j=1}^2 |\lambda_j|[\nu(2B_{k,l})]^{-1/2}\left[\tilde{\delta}^{(2)}_{B_{k,l},2^{k+2}B_0}\right]^{-1},
\end{aligned}
$$

where c_{12} is a positive constant independent of b and k. Let

$$
\lambda_{k,l} := c_{12}2^{-k\tau/2}\sum_{j=1}^2 |\lambda_j| \quad \text{and} \quad m_{k,l} := \lambda_{k,l}^{-1}Tb\chi_{D_{k,l}}.
$$

Then

$$
A_2 = \sum_{k=1}^\infty\sum_{l=1}^{M_0}\lambda_{k,l}m_{k,l},\ \operatorname{supp}(m_{k,l}) \subset B_{k,l} \subset \tilde{U}_k(B_0)
$$

and

$$
\|m_{k,l}\|_{L^2(\mathcal{X},\nu)} \le 2^{-k\tau/2}[\nu(2B_{k,l})]^{-1/2}\left[\tilde{\delta}^{(2)}_{B_{k,l},2^{k+1}B_0}\right]^{-1}.
$$

Combining the estimates of A_1 and A_2, we see that Tb is a $(2,1,\tau/2,2)_\lambda$-molecular block and

$$
|Tb|_{\tilde{H}^{1,p}_{\mathrm{mb}}(\mathcal{X},\nu)} = \sum_{j=1}^2 |\sigma_{j,1}| + \sum_{j=1}^2\sum_{i=1}^{N_j-1}\left|\sigma_{j,2}^{(i)}\right| + \sum_{j=1}^2 |\sigma_{j,3}| + \sum_{k=1}^\infty\sum_{l=1}^{M_0}|\lambda_{k,l}|
$$

$$\begin{aligned}
&\lesssim \sum_{j=1}^{2}|\lambda_j| + \sum_{j=1}^{2}\sum_{i=1}^{N_j-2}|\lambda_j|\frac{\nu(2^{i+2}B_j)}{\lambda(x_j,2^i r_j)}\left[\tilde{\delta}^{(2)}_{B_j,B}\right]^{-1}\\
&\quad+\sum_{k=1}^{\infty}\sum_{l=1}^{M_0}2^{-k\tau/2}\sum_{j=1}^{2}|\lambda_j|\\
&\lesssim \sum_{j=1}^{2}|\lambda_j| + \sum_{k=1}^{\infty}2^{-k\tau/2}M_0\sum_{j=1}^{2}|\lambda_j|\\
&\lesssim \sum_{j=1}^{2}|\lambda_j|\\
&\sim \|b\|_{\tilde{H}^{1,p}_{\mathrm{atb}}(\mathcal{X},\nu)},
\end{aligned}$$

which completes the proof of Theorem 8.5.8. □

As a corollary of Theorem 8.5.8, we now consider the boundedness of Calderón–Zygmund operators on the regularized BMO space, $\widetilde{\mathrm{RBMO}}_{\rho,\gamma}(\mathcal{X},\nu)$, associated with $\tilde{\delta}^{(\rho)}_{B,S}$ for $\rho\in(1,\infty)$ and balls $B\subset S$ of $\mathcal{X}$.

Definition 8.5.12. Let $\rho\in(1,\infty)$ and $\gamma\in[1,\infty)$. A function $f\in L^1_{\mathrm{loc}}(\mathcal{X},\nu)$ is said to be in the *space* $\widetilde{\mathrm{RBMO}}_{\rho,\gamma}(\mathcal{X},\nu)$ if there exist a positive constant C and, for any ball $B\subset\mathcal{X}$, a number f_B such that

$$\frac{1}{\nu(\rho B)}\int_B |f(x)-f_B|\,d\nu(x)\le C \tag{8.5.4}$$

and, for any two balls $B\subset B_1$,

$$|f_B-f_{B_1}|\le C\left[\tilde{\delta}^{(\rho)}_{B,B_1}\right]^{\gamma}. \tag{8.5.5}$$

The infimum of all the positive constants C satisfying both (8.5.4) and (8.5.5) is defined to be the $\widetilde{\mathrm{RBMO}}_{\rho,\gamma}(\mathcal{X},\nu)$ *norm* of f and denoted by $\|f\|_{\widetilde{\mathrm{RBMO}}_{\rho,\gamma}(\mathcal{X},\nu)}$.

Remark 8.5.13. (i) By some arguments similar to those used in Sect. 7.2, we conclude that the space $\widetilde{\mathrm{RBMO}}_{\rho,\gamma}(\mathcal{X},\nu)$ is independent of the choices of $\rho\in(1,\infty)$ and $\gamma\in[1,\infty)$. In what follows, we *denote* $\widetilde{\mathrm{RBMO}}_{\rho,\gamma}(\mathcal{X},\nu)$ *simply by* $\widetilde{\mathrm{RBMO}}(\mathcal{X},\nu)$.

(ii) When $(\mathcal{X},d,\nu):=(\mathbb{R}^D,|\cdot|,\mu)$ with μ as in (0.0.1), by (8.5.1), we see that $\widetilde{\mathrm{RBMO}}(\mathcal{X},\nu)$ becomes the regularized BMO space, RBMO (μ), in Sect. 3.1. For general non-homogeneous metric measure spaces, if we replace $\tilde{\delta}^{(\rho)}_{B,S}$ by $1+\delta(B,S)$ in Definition 8.5.12, then $\widetilde{\mathrm{RBMO}}(\mathcal{X},\nu)$ becomes the space

RBMO $(\mathcal{X}, \nu)$ in Sect. 7.2. Obviously, for $\rho \in (1, \infty)$ and $\gamma \in [1, \infty)$,

$$\mathrm{RBMO}(\mathcal{X}, \nu) \subset \widetilde{\mathrm{RBMO}}(\mathcal{X}, \nu).$$

However, it is still unclear whether we always have

$$\mathrm{RBMO}(\mathcal{X}, \nu) = \widetilde{\mathrm{RBMO}}(\mathcal{X}, \nu)$$

or not.

(iii) Let $\rho \in (1, \infty)$, $p \in (1, \infty]$ and $\gamma \in [1, \infty)$. Then, by an argument similar to that used in Sect. 7.4, we have

$$\left(\tilde{H}^{1,p,\gamma}_{\mathrm{atb},\rho}(\mathcal{X}, \nu)\right)^* = \widetilde{\mathrm{RBMO}}(\mathcal{X}, \nu).$$

By Theorem 8.5.8 and Remark 8.5.4(iii), we have the following conclusion.

Corollary 8.5.14. *Let T be as in Theorem 8.5.8 and T^* the adjoint operator of T. Then there exists a positive constant C such that, for all $f \in \widetilde{\mathrm{RBMO}}(\mathcal{X}, \nu)$,*

$$T^* f \in \widetilde{\mathrm{RBMO}}(\mathcal{X}, \nu) \quad \text{and} \quad \|T^* f\|_{\widetilde{\mathrm{RBMO}}(\mathcal{X}, \nu)} \le C\|f\|_{\widetilde{\mathrm{RBMO}}(\mathcal{X}, \nu)}.$$

8.5.2 Relations Between $\tilde{H}^{1,p}_{\mathrm{atb}}(\mathcal{X}, \nu)$ and $H^{1,p}_{\mathrm{atb}}(\mathcal{X}, \nu)$

In this subsection, we study the relations between $\tilde{H}^{1,p}_{\mathrm{atb}}(\mathcal{X}, \nu)$ in Sect. 8.5 and $H^{1,p}_{\mathrm{atb}}(\mathcal{X}, \nu)$ in Sect. 7.4, or between $\widetilde{\mathrm{RBMO}}(\mathcal{X}, \nu)$ and RBMO $(\mathcal{X}, \nu)$ in Sect. 7.2. In what follows, we give a sufficient condition to guarantee (8.5.1) for all balls $B \subset S$ of $\mathcal{X}$ and $\rho \in (1, \infty)$. We first introduce the definition of the weak reverse doubling condition for the dominating function λ.

Definition 8.5.15. The dominating function λ as in Definition 7.1.1 is said to satisfy the *weak reverse doubling condition* if, for all

$$r \in (0, 2\,\mathrm{diam}\,(\mathcal{X})) \quad \text{and} \quad a \in [1, 2\,\mathrm{diam}\,(\mathcal{X})/r),$$

there exists a number $C_{(a)} \in [1, \infty)$, depending only on a and $\mathcal{X}$, such that, for all $x \in \mathcal{X}$,

$$\lambda(x, ar) \ge C_{(a)}\lambda(x, r) \tag{8.5.6}$$

and

$$\sum_{k=1}^{\vartheta_a} \frac{1}{C_{(a^k)}} < \infty, \tag{8.5.7}$$

where ϑ_a is the *smallest integer* such that

$$\vartheta_a > \log_a(2\,\mathrm{diam}\,(\mathcal{X})/r) \quad \text{if} \quad \mathrm{diam}\,(\mathcal{X}) < \infty$$

and

$$\vartheta_a := \infty \quad \text{if} \quad \mathrm{diam}\,(\mathcal{X}) = \infty.$$

We then show that there exists a large class of spaces with dominating functions satisfying the weak reverse doubling condition.

Example 8.5.16. Let $(\mathcal{X}, d, \nu)$ be a connected metric measure space with a doubling measure ν. Then, by (7.1.2), it is easy to see that there exists $n \in (0,\infty)$, depending only on $C_{(\nu)}$ as in (7.1.2), such that, for all $x \in \mathcal{X}$, $a \in [1,\infty)$ and $r \in (0,\infty)$,

$$\nu(B(x,ar)) \lesssim a^n \nu(B(x,r)). \tag{8.5.8}$$

Let $\beta \in (n,\infty)$. We now claim that the *minimal dominating function* $F_\beta(x,r)$,[7] defined by setting, for all $x \subset \mathcal{X}$ and $r \in (0,\infty)$,

$$F_\beta(x,r) := \beta \int_1^\infty \frac{\nu(B(x,\,sr))}{s^{\beta+1}}\,ds$$

satisfies the weak reverse doubling condition.

To this end, we first show that, for any fixed $\beta \in (n,\infty)$, F_β is a dominating function over $(\mathcal{X}, d, \nu)$ as in Definition 7.1.1. Indeed, from (8.5.8), we easily deduce that, when $\beta \in (n,\infty)$, then, for all $x \in \mathcal{X}$ and $r \in (0,\infty)$, it holds true that $F_\beta(x,r) < \infty$ and, for any given $x \in \mathcal{X}$, $F_\beta(x,r)$ is increasing on $r \in (0,\infty)$. Also, for all $x \in \mathcal{X}$ and $r \in (0,\infty)$, it is easy to see that

$$\begin{aligned} F_\beta(x,2r) &= \beta \int_1^\infty \frac{\nu(B(x,2sr))}{s^{\beta+1}}\,ds \\ &= \beta 2^\beta \int_2^\infty \frac{\nu(B(x,sr))}{s^{\beta+1}}\,ds \\ &\le 2^\beta F_\beta(x,r). \end{aligned}$$

Moreover, for all $x \in \mathcal{X}$ and $r \in (0,\infty)$, we have

---

[7] See [126].

$$\begin{aligned} F_\beta(x,r) &= \beta \int_1^\infty \frac{\nu(B(x,2sr))}{s^{\beta+1}}\, ds \\ &\geq \beta \int_1^\infty \frac{\nu(B(x,r))}{s^{\beta+1}}\, ds \\ &= \nu(B(x,r)). \end{aligned}$$

Thus, F_β is a dominating function over $(\mathcal{X}, d, \nu)$ as in Definition 7.1.1.

On the other hand, recall the known fact that connected metric spaces satisfy the following *uniformly perfect condition*[8]: there exists a constant $a_0 \in [1, \infty)$ such that, for all $x \in \mathcal{X}$ and $r \in (0, \operatorname{diam}(\mathcal{X})/a_0)$,

$$B(x, a_0 r) \setminus B(x,r) \neq \emptyset. \tag{8.5.9}$$

Now we claim that $F_\beta(x,r)$ satisfies the weak reverse doubling condition with

$$C_{(a)} \sim a^m \quad \text{for some} \quad m \in (0,n].$$

We consider the following two cases for $\operatorname{diam}(\mathcal{X})$.

Case (i) $\operatorname{diam}(\mathcal{X}) = \infty$. By (8.5.9) and [156, Proposition 2.1], we conclude that there exists a number $m \in (0,n]$ such that, for all $x \in \mathcal{X}$, $r \in (0,\infty)$ and $a \in [1,\infty)$,

$$\begin{aligned} F_\beta(x,ar) &= \beta \int_1^\infty \frac{\nu(B(x,\, asr))}{s^{\beta+1}}\, ds \\ &\gtrsim \beta a^m \int_1^\infty \frac{\nu(B(x,\, sr))}{s^{\beta+1}}\, ds \\ &\sim a^m F_\beta(x,r). \end{aligned}$$

Then the claim holds true in this case.

Case (ii) $\operatorname{diam}(\mathcal{X}) < \infty$. By (8.5.9) and [156, Proposition 2.1] again, we see that there exist constants $\tilde{C} \in (0,1]$ and $m \in (0,n]$ such that, for all $x \in \mathcal{X}$, $r \in (0, 2\operatorname{diam}(\mathcal{X}))$ and $a \in [1, 2\operatorname{diam}(\mathcal{X})/r)$,

$$\begin{aligned} F_\beta(x,ar) &= \beta a^\beta \int_a^\infty \frac{\nu(B(x,\, sr))}{s^{\beta+1}}\, ds \\ &= \beta a^\beta \left[\int_1^\infty \frac{\nu(B(x,\, sr))}{s^{\beta+1}}\, ds - \int_1^a \frac{\nu(B(x,\, sr))}{s^{\beta+1}}\, ds \right] \\ &\geq \beta a^\beta \left[\int_1^\infty \frac{\nu(B(x,\, sr))}{s^{\beta+1}}\, ds - \frac{1}{\tilde{C} a^m} \int_1^a \frac{\nu(B(x,\, asr))}{s^{\beta+1}}\, ds \right] \end{aligned}$$

[8]See, for example, [49, p. 88] and [156, (2.1)].

$$\geq a^\beta F_\beta(x,r) - \frac{a^{\beta-m}}{\tilde{C}} F_\beta(x,ar),$$

which further implies that

$$F_\beta(x,ar) \geq \frac{a^\beta}{1+a^{\beta-m}/\tilde{C}} F_\beta(x,r) \gtrsim a^m F_\beta(x,r).$$

This finishes the proof of Case (ii) and hence our claim.

The following conclusion gives an application of the weak reverse doubling condition.

Theorem 8.5.17. *Let $\rho \in (1,\infty)$. If the dominating function λ satisfies the weak reverse doubling condition, then* (8.5.1) *holds true for any two balls $B \subset S$.*

Proof. By Remark 8.5.2, we see that, for $\rho \in (1,\infty)$ and all balls $B \subset S$ of $\mathcal{X}$,

$$1+\delta(B,S) \lesssim \tilde{\delta}^{(\rho)}_{B,S}.$$

Conversely, if $\operatorname{diam}(\mathcal{X}) = \infty$, then, by Definition 7.1.1, (8.5.6) and (8.5.7), we conclude that, for $\rho \in (1,\infty)$ and all balls $B \subset S$ of $\mathcal{X}$,

$$\begin{aligned}
\tilde{\delta}^{(\rho)}_{B,S} &= 1+\sum_{k=1}^{N^{(\rho)}_{B,S}} \left[\frac{\nu(\rho^k B\setminus B)}{\lambda(x_B,\rho^k r_B)} + \frac{\nu(B)}{\lambda(x_B,\rho^k r_B)}\right]\\
&\leq 1+\sum_{k=1}^{N^{(\rho)}_{B,S}} \left[\sum_{j=1}^{k} \int_{\rho^j B\setminus \rho^{j-1}B} \frac{1}{\lambda(x_B,\rho^{k-j}d(x,x_B))}\,d\nu(x) + \frac{1}{C_{(\rho^k)}}\right]\\
&\lesssim 1+\sum_{k=1}^{N^{(\rho)}_{B,S}} \sum_{j=1}^{k} \frac{1}{C_{(\rho^{k-j})}} \int_{\rho^j B\setminus \rho^{j-1}B} \frac{1}{\lambda(x_B,d(x,x_B))}\,d\nu(x)\\
&\sim 1+\sum_{j=1}^{N^{(\rho)}_{B,S}} \sum_{k=j}^{N^{(\rho)}_{B,S}} \frac{1}{C_{(\rho^{k-j})}} \int_{\rho^j B\setminus \rho^{j-1}B} \frac{1}{\lambda(x_B,d(x,x_B))}\,d\nu(x)\\
&\lesssim 1+\sum_{j=1}^{N^{(\rho)}_{B,S}} \int_{\rho^j B\setminus \rho^{j-1}B} \frac{1}{\lambda(x_B,d(x,x_B))}\,d\nu(x)\\
&\lesssim 1+\delta(B,S).
\end{aligned}$$

If $\operatorname{diam}(\mathcal{X}) < \infty$, then we may assume that, for any ball $B \subset \mathcal{X}$, r_B, the radius of B, is less than $2\operatorname{diam}(\mathcal{X})$. In this case, for any balls $B \subset S$, by the fact that

$r_S < 2\,\mathrm{diam}\,(\mathcal{X})$ and the choice of $N^{(\rho)}_{B,S}$, we see that, for all $k \in \{1, \ldots, N^{(\rho)}_{B,S} - 1\}$, it holds true that

$$\rho^k r_B < 2\,\mathrm{diam}\,(\mathcal{X}).$$

Then, by this and (7.1.1), arguing as in the case that $\mathrm{diam}\,(\mathcal{X}) = \infty$, we conclude that

$$\begin{aligned}\tilde{\delta}^{(\rho)}_{B,S} &\le 2 + \sum_{k=1}^{N^{(\rho)}_{B,S}-1} \left[\frac{\nu\,(\rho^k B)}{\lambda\,(x_B, \rho^k r_B)}\right] \\ &\lesssim 1 + \sum_{j=1}^{N^{(\rho)}_{B,S}} \int_{\rho^j B \setminus \rho^{j-1} B} \frac{1}{\lambda(x_B, d(x, x_B))}\, d\nu(x) \\ &\lesssim 1 + \delta(B, S),\end{aligned}$$

which completes the proof of Theorem 8.5.17. □

From Theorems 8.5.17, 8.5.6 and 8.5.8, we deduce the following conclusions.

Corollary 8.5.18. *Let $(\mathcal{X}, d, \nu)$ be a non-homogeneous space with the dominating function satisfying the weak reverse doubling condition. Then*

(i)

$$\tilde{H}^{1,p}_{\mathrm{atb}}(\mathcal{X}, \nu) = \tilde{H}^{1,p}_{\mathrm{mb}}(\mathcal{X}, \nu) = H^{1,p}_{\mathrm{atb}}(\mathcal{X}, \nu)$$

and

$$\widetilde{\mathrm{RBMO}}(\mathcal{X}, \nu) = \mathrm{RBMO}\,(\mathcal{X}, \nu);$$

(ii) *if T is as in Theorem 8.5.8, then T is bounded on $H^{1,p}_{\mathrm{atb}}(\mathcal{X}, \nu)$;*
(iii) *if T is as in Theorem 8.5.8 and T^* the adjoint operator of T, then T^* is bounded on* RBMO $(\mathcal{X}, \nu)$.

Remark 8.5.19. (i) If $(\mathcal{X}, d, \nu) := (\mathbb{R}^D, |\cdot|, \mu)$ with μ as in (0.0.1) and

$$\lambda(x, r) := C_0 r^n \quad \text{for all} \quad x \in \mathbb{R}^D \quad \text{and} \quad r \in [0, \infty),$$

where $n \in (0, D]$, then the weak reverse condition holds true automatically in this case and hence all the conclusions of Corollary 8.5.18 are true in this case.
(ii) If $(\mathcal{X}, d, \nu)$ is a space of homogeneous type in the sense of Coifman and Weiss with ν as in (7.1.2), and $\mathcal{X}$ connected, then

$$\lambda(x, r) := \nu(B(x, r)) \quad \text{for all} \quad x \in \mathcal{X} \text{ and } \quad r \in [0, 2\,\mathrm{diam}\,(\mathcal{X}))$$

satisfies the weak reverse condition and hence all the conclusions of Corollary 8.5.18 are true in this case.

(iii) If $(\mathcal{X}, d, \nu)$ is a non-homogeneous space without the dominating function satisfying the weak reverse doubling condition, then it is still unclear whether

$$\tilde{H}^1(\mathcal{X}, \nu) = H^1(\mathcal{X}, \nu) \quad \text{and} \quad \widetilde{\mathrm{RBMO}}(\mathcal{X}, \nu) = \mathrm{RBMO}\,(\mathcal{X}, \nu)$$

or not.

Next, we show that the weak reverse doubling condition is necessary to guarantee (8.5.1) for all balls $B \subset S$ in the sense that there exists a large class of spaces which do not satisfy the weak reverse doubling condition and

$$1 + \delta(B, S) \nsim \tilde{\delta}_{B,S}^{(\rho)} \tag{8.5.10}$$

for some balls $B \subset S$ and $\rho \in (1, \infty)$.

Example 8.5.20. Assume that ν is a finite Borel measure on a metric space $(\mathcal{X}, d)$ such that $\mathrm{diam}\,(\mathcal{X}) = \infty$. We show that the minimal dominating function $F_\beta(x, r)$ in Example 8.5.16 is a dominating function over $(\mathcal{X}, d, \nu)$, which satisfies (8.5.6) but not (8.5.7), and (8.5.10) holds true for some balls $B \subset S$ and $\rho \in (1, \infty)$.

We first show that, for any fixed $\beta \in (0, \infty)$, F_β is a dominating function over $(\mathcal{X}, d, \nu)$ as in Definition 7.1.1. As in Example 8.5.16, instead of $\beta \in (n, \infty)$, by $\beta \in (0, \infty)$ together with $\nu(\mathcal{X}) < \infty$, we see that, for all $\beta \in (0, \infty)$, $x \in \mathcal{X}$ and $r \in (0, \infty)$, it holds true that

$$F_\beta(x, r) \le \beta \int_1^\infty \frac{\nu(\mathcal{X})}{s^{\beta+1}}\, ds = \nu(\mathcal{X}) < \infty. \tag{8.5.11}$$

Using this and the same argument as that used in the proof of Example 8.5.16, we then conclude that F_β, for all $\beta \in (0, \infty)$, is a dominating function over $(\mathcal{X}, d, \nu)$ as in Definition 7.1.1.

Now we claim that F_β satisfies (8.5.6) but not (8.5.7). Indeed, we show that F_β satisfies (8.5.6) with $C_{(a)} = 1$ for all $a \in [1, \infty)$, $x \in \mathcal{X}$ and $r \in (0, \infty)$. To this end, for all $a \in [1, \infty)$, $x \in \mathcal{X}$ and $r \in (0, \infty)$, we have

$$\frac{F_\beta(x, ar)}{F_\beta(x, r)} \ge 1.$$

It remains to prove that, for all $a \in [1, \infty)$,

$$\inf_{x \in \mathcal{X},\, r \in (0,\infty)} \frac{F_\beta(x, ar)}{F_\beta(x, r)} \le 1.$$

Let $x := x_0 \in \mathcal{X}$. By the finiteness of ν, the monotone convergence theorem and the assumption that $\mathrm{diam}\,(\mathcal{X}) = \infty$, we see that, for all $a \in [1, \infty)$,

$$\inf_{x\in\mathcal{X}, r\in(0,\infty)} \frac{F_\beta(x,ar)}{F_\beta(x,r)} \le \lim_{r\to\infty} \frac{F_\beta(x_0,ar)}{F_\beta(x_0,r)} = \frac{\nu(\mathcal{X})}{\nu(\mathcal{X})} = 1.$$

Thus, F_β satisfies (8.5.6) with $C_{(a)} = 1$ for all $a \in [1,\infty)$, $x \in \mathcal{X}$ and $r \in (0, 2\,\mathrm{diam}\,(\mathcal{X}))$. Obviously, (8.5.7) fails in this case.

Finally, we show that (8.5.10) holds true for some balls $B \subset S$ and $\rho \in (1,\infty)$. To this end, fix a ball $B \subset \mathcal{X}$. Let $B_N := \rho^N B$ for all $N \in \mathbb{N}$. Then $N^{(\rho)}_{B,B_N} = N$ and, by (7.1.1) and the finiteness of ν, we know that

$$\begin{aligned}1 + \delta(B, B_N) &= 1 + \int_{2\rho^N B\setminus B} \frac{1}{F_\beta(x_B, d(x,x_B))}\,d\nu(x)\\ &\lesssim 1 + \frac{1}{F_\beta(x_B, r_B)}\\ &\lesssim 1 + \frac{1}{\nu(B)}.\end{aligned}$$

On the other hand, by the fact that

$$\lim_{k\to\infty} \nu(\rho^k B) = \nu(\mathcal{X}) > \nu(\mathcal{X})/2,$$

we see that, there exists some $K \in \mathbb{N}$ such that, for all $k > K$,

$$\nu(\rho^k B) > \nu(\mathcal{X})/2 > 0.$$

By this and (8.5.11), we obtain

$$\frac{\nu(\rho^k B)}{F_\beta(x_B, \rho^k r_B)} \ge \frac{\nu(\rho^k B)}{\nu(\mathcal{X})} > \frac{1}{2} \text{ for all } k > K.$$

Consequently,

$$\frac{\tilde{\delta}^{(\rho)}_{B,B_N}}{1+\delta(B, B_N)} \gtrsim \left[1 + \sum_{k=1}^{N} \frac{\nu(\rho^k B)}{F_\beta\left(x_B, \rho^k r_B\right)}\right]\left[1 + \frac{1}{\nu(B)}\right]^{-1} \to \infty,$$

as $N \to \infty$, which implies (8.5.10). Thus, our claim holds true.

8.6 Weighted Estimates for the Local Sharp Maximal Operator

In this section, we introduce variants of the John–Strömberg maximal operator and the John–Strömberg sharp maximal operator on non-homogeneous metric measure spaces, and then establish some weighted norm inequalities with $A_\infty^\varrho(\mathcal{X}, \nu)$ weights (see Definition 8.6.1 below) related to these maximal operators, where $\varrho \in [1, \infty)$.

For all balls $B \subset S \subset \mathcal{X}$, let

$$\tilde{\delta}_{B,S}^{(30\varrho)} := 1 + \sum_{k=1}^{N_{B,S}} \frac{\mu((30\varrho)^k B)}{\lambda(x_B, (30\varrho)^k r_B)}, \tag{8.6.1}$$

where $N_{B,S}$ is the *smallest integer* j such that $(30\varrho)^j r_B \geq r_S$.

Let $s \in (0, 1)$, $\sigma \in [1, \infty)$ and $\varrho \in [1, \infty)$. For any fixed ball B and ν-measurable function f, define $m_{0,s;B}^{\sigma,\varrho}(f)$ by setting

$$m_{0,s;B}^{\sigma,\varrho}(f) := \inf\{t \in (0, \infty) : \nu(\{y \in B : |f(y)| > t\}) < s\nu(\sigma\varrho B)\}$$

when $\nu(B) > 0$, and letting $m_{0,s;B}^{\sigma,\varrho}(f) := 0$ when $\nu(B) = 0$. For any ν-measurable function f, the *John–Strömberg maximal operator* $M_{0,s}^{\sigma,\varrho}$ is defined by setting, for all $x \in \mathcal{X}$,

$$M_{0,s}^{\sigma,\varrho}(f)(x) := \sup_{B \ni x,\, B\ (30\varrho,\, \beta_{30\varrho})-\text{doubling}} m_{0,s;B}^{\sigma,\varrho}(f),$$

and the *John–Strömberg sharp maximal function* $M_{0,s}^{\sigma,\varrho;\sharp}$ is defined by setting, for all $x \in \mathcal{X}$,

$$M_{0,s}^{\sigma,\varrho;\sharp}(f)(x) := \sup_{B \ni x} m_{0,s;B}^{\sigma,\varrho}\left[f - \alpha_{\tilde{B}^{30\varrho}}(f)\right] + \sup_{\substack{x \in B \subset S \\ B,\, S\ (30\varrho,\, \beta_{30\varrho})-\text{doubling}}} \frac{|\alpha_B(f) - \alpha_S(f)|}{\tilde{\delta}_{B,S}^{(30\varrho)}}.$$

It is easy to show that, for all balls B containing x and all $\epsilon \in (0, \infty)$,

$$\nu\left(\left\{y \in B : |f(y) - \alpha_{\tilde{B}^{30\varrho}}(f)| > M_{0,s}^{\sigma,\varrho;\sharp}(f)(x) + \epsilon\right\}\right) < s\nu(\sigma\varrho B). \tag{8.6.2}$$

Definition 8.6.1. Let $\varrho \in [1, \infty)$, $p \in (1, \infty)$ and $p' := p/(p-1)$. A nonnegative ν-measurable function u is called an $A_p^\varrho(\mathcal{X}, \nu)$ *weight* if there exists a positive constant C such that, for all balls $B \subset \mathcal{X}$,

$$\left[\frac{1}{\nu(\varrho B)} \int_B u(x)\, d\nu(x)\right] \left[\frac{1}{\nu(\varrho B)} \int_B u(x)^{1-p'}\, d\nu(x)\right]^{p-1} \leq C.$$

Also, a weight u is called an $A_1^\varrho(\mathcal{X}, \nu)$ *weight* if there exists a positive constant C such that, for all balls $B \subset \mathcal{X}$,

$$\frac{1}{\nu(\varrho B)} \int_B u(x)\, d\nu(x) \leq C \inf_{y \in B} u(y).$$

As in the classical setting, let

$$A_{\infty}^{\varrho}(\mathcal{X}, \nu) := \bigcup_{p=1}^{\infty} A_p^{\varrho}(\mathcal{X}, \nu).$$

When $\varrho = 1$, we *denote* $A_p^{\varrho}(\mathcal{X}, \nu)$, $A_1^{\varrho}(\mathcal{X}, \nu)$ *and* $A_{\infty}^{\varrho}(\mathcal{X}, \nu)$ *simply by* $A_p(\mathcal{X}, \nu)$, $A_1(\mathcal{X}, \nu)$ *and* $A_{\infty}(\mathcal{X}, \nu)$, respectively.

Our main result in this section is the following theorem.

Theorem 8.6.2. *Let* $\varrho \in [1, \infty)$, $\sigma \in [1, 30]$, $s_1 \in (0, \beta_{30\varrho}^{-1}/4)$, $p \in (0, \infty)$ *and* $u \in A_{\infty}^{\varrho}(\mathcal{X}, \nu)$. *Then there exist a constant* $\tilde{C}_3 \in (0, 1)$, *depending on* s_1 *and* u, *and a positive constant* C *such that, for any* $s_2 \in (0, \tilde{C}_3 s_1)$,

(i) *if* $\nu(\mathcal{X}) = \infty$, $f \in L^{p_0, \infty}(\mathcal{X}, \nu)$ *with* $p_0 \in (0, \infty)$ *and, for all* $R \in (0, \infty)$,

$$\sup_{t\in(0,R)} t^p u(\{x \in \mathcal{X} : |f(x)| > t\}) < \infty,$$

then $\left\|M_{0,s_1}^{\sigma,\varrho}(f)\right\|_{L^{p,\infty}(\mathcal{X},u)} \le C \left\|M_{0,s_2}^{\sigma,\varrho;\sharp}(f)\right\|_{L^{p,\infty}(\mathcal{X},u)}$;

(ii) *if* $\nu(\mathcal{X}) < \infty$ *and* $f \in L^{p_0,\infty}(\mathcal{X}, \nu)$ *with* $p_0 \in (0, \infty)$, *then*

$$\begin{aligned}&\left\|M_{0,s_1}^{\sigma,\varrho}(f)\right\|_{L^{p,\infty}(\mathcal{X},u)}\\ &\quad\le C \left\|M_{0,s_2}^{\sigma,\varrho;\sharp}(f)\right\|_{L^{p,\infty}(\mathcal{X},u)} + Cu(\mathcal{X})(s_1\nu(\mathcal{X}))^{-p/p_0}\|f\|_{L^{p_0,\infty}(\mathcal{X},\nu)}^p.\end{aligned}$$

To prove Theorem 8.6.2, we need to recall some technical lemmas and establish some preliminary results.

For any $\varrho \in [1, \infty)$, the *doubling maximal operator* $\tilde{\mathcal{N}}$ is defined by setting, for all $x \in \mathcal{X}$,

$$\tilde{\mathcal{N}}(f)(x) := \sup_{\substack{B\ni x\\ B\ (30\varrho,\ \beta_{30\varrho})-\text{doubling}}} \frac{1}{\nu(B)} \int_B |f(y)|\,d\nu(y). \tag{8.6.3}$$

Then we have the following result.

Lemma 8.6.3. *Let* $\varrho \in [1, \infty)$ *and* $\eta \in [5\varrho, \infty)$. *Let* $\mathcal{M}_\eta$ *and* $\tilde{\mathcal{N}}$ *be the operators defined by* (7.1.8) *and* (8.6.3), *respectively. Then, for any* $p \in [1, \infty)$ *and* $u \in A_p^{\varrho}(\mathcal{X}, \nu)$, $\mathcal{M}_\eta$ *and* $\tilde{\mathcal{N}}$ *are bounded from* $L^p(\mathcal{X}, u)$ *to* $L^{p,\infty}(\mathcal{X}, u)$.

Proof. It is easy to see that, for all $\eta \in (1, 30\varrho]$, $\tilde{\mathcal{N}}(f) \lesssim \mathcal{M}_\eta(f)$. Hence, to prove Lemma 8.6.3, it suffices to show that $\mathcal{M}_\eta$ with $\eta \in [5\varrho, \infty)$ is bounded from $L^p(\mathcal{X}, u)$ into $L^{p,\infty}(\mathcal{X}, u)$ for $u \in A_p^{\varrho}(\mathcal{X}, \nu)$ with $p \in [1, \infty)$.

For all $R \in (0, \infty)$ and $\eta \in (1, \infty)$, the *operator* $\mathcal{M}_\eta^R$ is defined as in (7.1.9). For all $t \in (0, \infty)$, let

$$E_R(t) := \left\{x \in \mathcal{X} : \mathcal{M}_\eta^R(f)(x) > t\right\}.$$

It is easy to see that, for any $x \in E_R(t)$, there exists a ball $B(z_x, r_x)$ such that $x \in B(z_x, r_x)$ and

$$\frac{1}{\nu(B(z_x, \eta r_x))} \int_{B(z_x, r_x)} |f(y)| \, d\nu(y) > t.$$

By Lemma 7.1.17, we conclude that there exist disjoint balls $\{B(z_{x_i}, r_{x_i})\}_i$ such that

$$\left[\bigcup_{x \in E_R(t)} B(z_x, r_x)\right] \subset \left[\bigcup_i B(z_{x_i}, 5r_{x_i})\right],$$

which, together with the Hölder inequality and the $A_p^\varrho(\mathcal{X}, \nu)$ condition, implies that

$$\begin{aligned}
&\int_{E_R(t)} u(y) \, d\nu(y) \\
&\quad \le \sum_i \int_{B(z_{x_i}, 5r_{x_i})} u(y) \, d\nu(y) \\
&\quad \le \frac{1}{t^p} \sum_i \left[\frac{1}{\nu(B(z_{x_i}, \eta r_{x_i}))} \int_{B(z_{x_i}, r_{x_i})} |f(y)| \, d\nu(y)\right]^p \int_{B(z_{x_i}, 5r_{x_i})} u(y) \, d\nu(y) \\
&\quad \lesssim \frac{1}{t^p} \sum_i \frac{1}{[\nu(B(z_{x_i}, \eta r_{x_i}))]^p} \int_{B(z_{x_i}, r_{x_i})} |f(y)|^p u(y) \, d\nu(y) \\
&\qquad \times \left\{\int_{B(z_{x_i}, r_{x_i})} [u(y)]^{-p'/p} \, d\nu(y)\right\}^{p/p'} \int_{B(z_{x_i}, 5r_{x_i})} u(y) \, d\nu(y) \\
&\quad \le \frac{1}{t^p} \sum_i \frac{[\nu(B(z_{x_i}, 5\varrho r_{x_i}))]^p}{[\nu(B(z_{x_i}, \eta r_{x_i}))]^p} \int_{B(z_{x_i}, r_{x_i})} |f(y)|^p u(y) \, d\nu(y) \\
&\quad \lesssim \frac{1}{t^p} \int_{\mathcal{X}} |f(y)|^p u(y) \, d\nu(y).
\end{aligned}$$

Letting $R \to \infty$, we then conclude the desired result, which completes the proof of Lemma 8.6.3. □

As an application of Lemma 8.6.3, we obtain the following useful conclusions. In what follows, for any ν-measurable set E and $u \in L^1_{\mathrm{loc}}(\mathcal{X}, \nu)$, let

$$u(E) := \int_E u(x) \, d\nu(x).$$

Lemma 8.6.4. *Let ϱ, $p \in [1, \infty)$, $u \in A_p^{\varrho}(\mathcal{X}, \nu)$ and $\eta \in [5\varrho, \infty)$. Then there exist constants $\tilde{C}_4$, $\tilde{C}_5 \in \mathbb{N}$ such that,*

(i) *for any ball B and ν-measurable set $E \subset B$,*

$$\frac{u(E)}{u(B)} \geq \tilde{C}_4^{-1} \left[\frac{\nu(E)}{\nu(\eta B)}\right]^p;$$

(ii) *for any $(30\varrho, \beta_{30\varrho})$-doubling ball B and ν-measurable set $E \subset B$,*

$$\frac{u(E)}{u(B)} \geq \tilde{C}_5^{-1} \left[\frac{\nu(E)}{\nu(B)}\right]^p;$$

(iii) *for any $(30\varrho, \beta_{30\varrho})$-doubling ball B and ν-measurable set $E \subset B$,*

$$\frac{u(E)}{u(B)} \leq 1 - \tilde{C}_5^{-1} \left[1 - \frac{\nu(E)}{\nu(B)}\right]^p.$$

Proof. It is easy to see that (iii) is an easy consequence of (ii) with E replaced by $B \setminus E$, and (ii) follows from (i). Thus, it suffices to prove (i).

Observe that, for any ball B and ν-measurable set $E \subset B$,

$$\inf_{x \in B} \mathcal{M}_\eta(\chi_E)(x) \geq \frac{\nu(E)}{\nu(\eta B)}.$$

By Lemma 8.6.3, we see that there exists a constant $\tilde{C}_4 \in \mathbb{N}$ such that, for all $t \in (0, \infty)$,

$$u\left(\{x \in \mathcal{X} : \mathcal{M}_\eta(\chi_E)(x) > t\}\right) \leq \tilde{C}_4 t^{-p} \int_{\mathcal{X}} |\chi_E(x)|^p u(x)\, d\nu(x).$$

Therefore, for all $t \in (0, \nu(E)/\nu(\eta B))$,

$$u(B) \leq u\left(\{x \in \mathcal{X} : \mathcal{M}_\eta(\chi_E)(x) > t\}\right) \leq \tilde{C}_4 t^{-p} u(E).$$

From this we deduce that, for all $t \in (0, \nu(E)/\nu(\eta B))$, it holds true that

$$\frac{u(E)}{u(B)} \geq \tilde{C}_4^{-1} t^p.$$

Letting $t \to \nu(E)/\nu(\eta B)$, we obtain (i), which completes the proof of (i) and hence Lemma 8.6.4. □

Lemma 8.6.5. *Let ϱ, $p \in [1, \infty)$, $\sigma \in [1, 30]$ and $s \in (0, \beta_{30\varrho}^{-1})$. Then, for all ν-measurable functions f and $t \in (0, \infty)$,*

(i) $\{x \in \mathcal{X} : |f(x)| > t\} \subset \{x \in \mathcal{X} : M_{0,s}^{\sigma,\varrho}(f)(x) \geq t\} \bigcup E$ *with* $\nu(E) = 0$;

(ii) *for* $u \in A_p^{\varrho}(\mathcal{X}, \nu)$, *there exists a positive constant* C, *independent of* f *and* t, *such that*

$$u\left(\left\{x \in \mathcal{X} : M_{0,s}^{\sigma,\varrho}(f)(x) > t\right\}\right) \le Cs^{-p}u(\{x \in \mathcal{X} : |f(x)| > t\}).$$

Proof. It is easy to see that

$$\begin{aligned}\{x \in \mathcal{X} : |f(x)| > t\} &= \left\{x \in \mathcal{X} : \chi_{\{y \in \mathcal{X}:\, |f(y)|>t\}}(x) = 1\right\} \\ &\subset \left\{x \in \mathcal{X} : \tilde{\mathcal{N}}(\chi_{\{y \in \mathcal{X}:\, |f(y)|>t\}})(x) > \beta_{30\varrho}s\right\} \bigcup E,\end{aligned}$$

where $\nu(E) = 0$. From Corollary 7.1.21, it follows that, if $x \in \mathcal{X}$ satisfies that

$$\tilde{\mathcal{N}}\left(\chi_{\{y \in \mathcal{X}:\, |f(y)|>t\}}\right)(x) > \beta_{30\varrho}s,$$

then there exists a $(30\varrho, \beta_{30\varrho})$-doubling ball B containing x such that

$$\frac{1}{\nu(B)}\int_B \chi_{\{y \in \mathcal{X}:\, |f(y)|>t\}}(y)\, d\nu(y) > \beta_{30\varrho}s.$$

This means that

$$\nu(\{y \in B : |f(y)| > t\}) > s\beta_{30\varrho}\nu(B).$$

Notice that $\sigma \in [1, 30]$ and hence

$$\nu(\sigma\varrho B) \le \nu(30\varrho B) \le \beta_{30\varrho}\nu(B).$$

Thus,

$$\nu(\{y \in B : |f(y)| > t\}) > s\nu(\sigma\varrho B).$$

Therefore, $m_{0,s;\,B}^{\sigma,\varrho}(f) \ge t$ and hence $M_{0,s}^{\sigma,\varrho}(f)(x) \ge t$, which implies (i).

Now we turn to prove (ii). For all $R \in (0, \infty)$ and $x \in \mathcal{X}$, set

$$M_{0,s}^{\sigma,\varrho,R}(f)(x) := \sup_{\substack{B \ni x,\ r_B < R,\\ B\ (30\varrho,\, \beta_{30\varrho})-\text{doubling}}} m_{0,s;\,B}^{\sigma,\varrho}(f).$$

Given any $t \in (0, \infty)$, let

$$F_R(t) := \left\{x \in \mathcal{X} : M_{0,s}^{\sigma,\varrho,R}(f)(x) > t\right\}.$$

Then, for any $x \in F_R(t)$, there exists a $(30\varrho, \beta_{30\varrho})$-doubling ball B_x such that $x \in B_x$, $r_{B_x} < R$ and

$$\nu(\{y \in B_x : |f(y)| > t\}) \geq s\nu(\sigma\varrho B_x) \gtrsim s\nu(30\varrho B_x).$$

Via Lemma 8.6.4(i) with $\eta = 5\varrho$, we easily conclude that

$$u(\{y \in B_x : |f(y)| > t\}) \gtrsim s^p u(6B_x).$$

From Lemma 7.1.17, it follows that there exist disjoint balls $\{B_{x_i}\}_i$ such that

$$F_R(t) \subset \bigcup_i 5B_{x_i}.$$

Thus,

$$\begin{aligned} u(F_R(t)) &\leq \sum_i u(5B_{x_i}) \\ &\lesssim s^{-p} \sum_i u(\{y \in B_{x_i} : |f(y)| > t\}) \\ &\lesssim s^{-p} u(\{y \in \mathcal{X} : |f(y)| > t\}). \end{aligned}$$

The conclusion (ii) follows by letting $R \to \infty$, which completes the proof of Lemma 8.6.5. □

Similar to Lemma 5.1.1, we have the following lemma, whose proof is omitted.

Lemma 8.6.6. *Let $\varrho \in [1, \infty)$, $\sigma \in [1, 30]$, $s \in (0, \beta_{30\varrho}^{-1}/4)$ and B be a $(30\varrho, \beta_{30\varrho})$-doubling ball with $\nu(B) \neq 0$. For any constant $c \in \mathbb{C}$ and ν-measurable function f,*

$$\left| m_{0,s;B}^{\sigma,\varrho}(f) - |c| \right| \leq m_{0,s;B}^{\sigma,\varrho}(f - c).$$

Lemma 8.6.7. *Let $\varrho \in [1, \infty)$, $\sigma \in [1, 30]$, $s \in (0, \beta_{30\varrho}^{-1}/4)$ and B be a $(30\varrho, \beta_{30\varrho})$-doubling ball. Then, for any ν-measurable real-valued function f,*

$$|\alpha_B(f)| \leq m_{0,s;B}^{\sigma,\varrho}(f).$$

The proof of Lemma 8.6.7 is similar to that of Lemma 5.1.2. We omit the details here.

Lemma 8.6.8. *Let $\varrho, p \in [1, \infty)$, $\sigma \in [1, 30]$, $s_1 \in (0, \beta_{30\varrho}^{-1}/4)$ and $u \in A_p^\varrho(\mathcal{X}, \nu)$. Then there exists a constant $\tilde{C}_6 \in (1, \infty)$ such that, for all $s_2 \in (0, \tilde{C}_6^{-1} s_1)$, $\gamma \in (0, \infty)$ and real-valued functions $f \in L^{p_0,\infty}(\mathcal{X}, \nu)$ with some $p_0 \in (0, \infty)$,*

$$u\left(\left\{x \in \mathcal{X} : M_{0,s_1}^{\sigma,\varrho}(f)(x) > (1+\gamma)t,\ M_{0,s_2}^{\sigma,\varrho;\sharp}(f)(x) \le \theta_2\gamma t\right\}\right)$$
$$\le \theta_1 u\left(\left\{x \in \mathcal{X} : M_{0,s_1}^{\sigma,\varrho}(f)(x) > t\right\}\right),$$

provided that

(i) $\nu(\mathcal{X}) = \infty$ *and* $t \in (0, \infty)$, *or*
(ii) $\nu(\mathcal{X}) < \infty$ *and*

$$t > t_f := [s_1\nu(\mathcal{X})]^{-1/p_0}\|f\|_{L^{p_0,\infty}(\mathcal{X},\nu)},$$

where $\theta_1, \theta_2 \in (0, 1)$ *are constants depending on* ϱ *and* ν.

Proof. For s_1 and p_0 as in Lemma 8.6.8, let $t_f := 0$ if $\nu(\mathcal{X}) = \infty$, and

$$t_f := [s_1\nu(\mathcal{X})]^{-1/p_0}\|f\|_{L^{p_0,\infty}(\mathcal{X},\nu)}$$

if $\nu(\mathcal{X}) < \infty$. For σ, ϱ, s_1 and s_2 as in Lemma 8.6.8, any fixed $t > t_f$ and $\gamma \in (0, \infty)$, set

$$\Omega_t := \left\{x \in \mathcal{X} : M_{0,s_1}^{\sigma,\varrho}(f)(x) > t\right\}$$

and

$$G_t := \left\{x \in \mathcal{X} : M_{0,s_1}^{\sigma,\varrho}(f)(x) > (1+\gamma)t,\ M_{0,s_2}^{\sigma,\varrho;\sharp}(f)(x) \le \theta_2\gamma t\right\},$$

where θ_2 is a positive constant which is determined later. Notice that, if $t > t_f$, then

$$\nu(\{y \in \mathcal{X} : |f(y)| > t\}) \le \frac{\|f\|_{L^{p_0,\infty}(\mathcal{X},\nu)}^{p_0}}{t^{p_0}} < s_1\nu(\mathcal{X}).$$

Thus,

$$\frac{1}{\nu(\mathcal{X})}\int_{\{y\in\mathcal{X}:\ |f(y)|>t\}} d\nu(y) < s_1.$$

This means that, for all $t > t_f$ and all $x \in \mathcal{X}$,

$$\lim_{\substack{I\ni x,\, r(I)\to\infty \\ I\ (30\varrho,\,\beta_{30\varrho})-\text{doubling}}} \frac{1}{\nu(I)}\int_{\{y\in I:\ |f(y)|>t\}} d\nu(y) < s_1,$$

which implies that, for all $(30\varrho, \beta_{30\varrho})$-doubling balls I containing x with the radius large enough,

$$m_{0,s_1;I}^{\sigma,\varrho}(f) \le t. \tag{8.6.4}$$

On the other hand, for each fixed $x \in G_t$, there exists a $(30\varrho, \beta_{30\varrho})$-doubling ball B containing x such that

$$m_{0,s_1;B}^{\sigma,\varrho}(f) > (1+\gamma/2)t.$$

From this and (8.6.4), it follows that, among these $(30\varrho, \beta_{30\varrho})$-doubling balls, there exists a $(30\varrho, \beta_{30\varrho})$-doubling ball B_x, which has almost maximal radius in the sense that, if some $(30\varrho, \beta_{30\varrho})$-doubling ball I contains x and has radius no less than $30\varrho r_{B_x}$, then

$$m_{0,s_1;I}^{\sigma,\varrho}(f) \le (1+\gamma/2)t.$$

Let R_x be the *ball centered at x with radius* $30\varrho r_{B_x}$, and

S_x the smallest $((30\varrho)^3, \beta_{30\varrho})-$doubling ball in the form $(30\varrho)^{3j} R_x$ with $j \in \mathbb{Z}_+$.

Then, by Lemma 7.1.14, we easily see that S_x, $30\varrho S_x$ and $(30\varrho)^2 S_x$ are all $(30\varrho, \beta_{30\varrho})$-doubling balls. From (iv), (ii) and (iii) of Lemma 8.5.9, it follows that there exists a positive constant $\tilde{C}$, depending only on ϱ and ν, such that

$$\tilde{\delta}_{B_x, 30\varrho S_x}^{(30\varrho)} \le \tilde{C}.$$

Thus, by Lemma 8.6.6 and $s_2 < s_1$, we conclude that

$$\begin{aligned}
&\left|m_{0,s_1;B_x}^{\sigma,\varrho}(f) - m_{0,s_1;30\varrho S_x}^{\sigma,\varrho}(f)\right| \\
&\quad \le \left|m_{0,s_1;B_x}^{\sigma,\varrho}(f) - |\alpha_{B_x}(f)|\right| + \left||\alpha_{B_x}(f)| - |\alpha_{30\varrho S_x}(f)|\right| \\
&\qquad + \left||\alpha_{30\varrho S_x}(f)| - m_{0,s_1;30\varrho S_x}^{\sigma,\varrho}(f)\right| \\
&\quad \le m_{0,s_1;B_x}^{\sigma,\varrho}(f - \alpha_{B_x}(f)) + |\alpha_{B_x}(f) - \alpha_{30\varrho S_x}(f)| \\
&\qquad + m_{0,s_1;30\varrho S_x}^{\sigma,\varrho}\left(f - \alpha_{30\varrho S_x}(f)\right) \\
&\quad \le 3\tilde{\delta}_{B_x, 30\varrho S_x}^{(30\varrho)} \inf_{y\in B_x} M_{0,s_1}^{\sigma,\varrho;\sharp}(f)(y) \\
&\quad \le 3\tilde{\delta}_{B_x, 30\varrho S_x}^{(30\varrho)} M_{0,s_2}^{\sigma,\varrho;\sharp}(f)(x) \\
&\quad \le \tilde{C}_7\theta_2\gamma t,
\end{aligned}$$

where $\tilde{C}_7 := 3\tilde{C}$ is a positive constant, depending on ϱ and ν. If we choose $\theta_2 \in (0, \frac{1}{2\tilde{C}_7})$, we easily see that $m_{0,s_1;30\varrho S_x}^{\sigma,\varrho}(f) > t$ and hence $30\varrho S_x \subset \Omega_t$.

We consider the following two cases.

Case (i) $\sup_{x\in G_t} r_{S_x} < \infty$. In this case, by Lemma 7.1.17, we conclude that there exist disjoint balls $\{S_{x_i}\}_i$ such that

$$G_t \subset \bigcup_i 5S_{x_i} \subset \bigcup_i 30\varrho S_{x_i} =: \bigcup_i W_{x_i}. \quad (8.6.5)$$

We claim that there exists a positive constant $\tilde{C}_8$ such that

$$\nu\left(W_{x_i} \bigcap G_t\right) \le \tilde{C}_8 s_1^{-1} s_2 \nu(W_{x_i}). \quad (8.6.6)$$

For all $y \in W_{x_i} \cap G_t$ and all $(30\varrho, \beta_{30\varrho})$-doubling balls $B \ni y$ satisfying

$$m_{0,s_1;B}^{\sigma,\varrho}(f) > (1+\gamma)t,$$

we have $r_B \le r_{W_{x_i}}/8$. Otherwise, if $r_B > r_{W_{x_i}}/8$, then $B_{x_i} \subset W_{x_i} \subset 18B$ and

$$\begin{aligned}
&\left|m_{0,s_1;B}^{\sigma,\varrho}(f) - m_{0,s_1;\widetilde{18B}^{30\varrho}}^{\sigma,\varrho}(f)\right| \\
&\quad \le \left|m_{0,s_1;B}^{\sigma,\varrho}(f) - |\alpha_B(f)|\right| + \left|\alpha_B(f) - \alpha_{\widetilde{18B}^{30\varrho}}(f)\right| \\
&\qquad + \left|\left|\alpha_{\widetilde{18B}^{30\varrho}}(f)\right| - m_{0,s_1;\widetilde{18B}^{30\varrho}}^{\sigma,\varrho}(f)\right| \\
&\quad \le 3\tilde{\delta}_{B,\widetilde{18B}^{30\varrho}}^{(30\varrho)} \inf_{z\in B} M_{0,s_1}^{\sigma,\varrho;\sharp}(f)(z) \\
&\quad \le 3\tilde{\delta}_{B,\widetilde{18B}^{30\varrho}}^{(30\varrho)} M_{0,s_2}^{\sigma,\varrho;\sharp}(f)(y) \\
&\quad \le \tilde{C}_9 \theta_2 \gamma t,
\end{aligned}$$

where $\tilde{C}_9 \in (1,\infty)$ is a constant, depending on ϱ and ν, such that

$$3\tilde{\delta}_{B,\widetilde{18B}^{30\varrho}}^{(30\varrho)} \le \tilde{C}_9.$$

Choose $\theta_2 := 1/(2\tilde{C}_7 + 2\tilde{C}_9)$. Then,

$$\begin{aligned}
m_{0,s_1;\widetilde{18B}^{30\varrho}}^{\sigma,\varrho}(f) &\ge m_{0,s_1;B}^{\sigma,\varrho}(f) - \left|m_{0,s_1;B}^{\sigma,\varrho}(f) - m_{0,s_1;\widetilde{18B}^{30\varrho}}^{\sigma,\varrho}(f)\right| \\
&> (1+\gamma)t - \tilde{C}_9\theta\gamma t \\
&> (1+\gamma/2)t,
\end{aligned}$$

which contradicts the fact that B_{x_i} is the chosen maximal $(30\varrho, \beta_{30\varrho})$-doubling ball satisfying

$$m_{0,s_1;B}^{\sigma,\varrho}(f) > (1+\gamma/2)t$$

with $B \ni x_i$. Therefore, $r_B \le r_{W_{x_i}}/8$ and hence $B \subset \frac{5}{4}W_{x_i}$. By this, together with the fact $m_{0,s_1;B}^{\sigma,\varrho}(f) > (1+\gamma)t$, we conclude that

$$m_{0,s_1;B}^{\sigma,\varrho}\left(f\chi_{\frac{5}{4}W_{x_i}}\right) > (1+\gamma)t. \tag{8.6.7}$$

On the other hand, from Lemma 8.6.7, it follows that

$$\left|\alpha_{\widetilde{\frac{5}{4}W_{x_i}}^{30\varrho}}(f)\right| \le m^{\sigma,\varrho}_{0,s_1;\widetilde{\frac{5}{4}W_{x_i}}^{30\varrho}}(f) \le (1+\gamma/2)t. \tag{8.6.8}$$

Combining the estimates (8.6.7) and (8.6.8), we obtain

$$\begin{aligned}
&m_{0,s_1;B}^{\sigma,\varrho}\left(\left[f-\alpha_{\widetilde{\frac{5}{4}W_{x_i}}^{30\varrho}}(f)\right]\chi_{\frac{5}{4}W_{x_i}}\right)\\
&\quad\ge \left|m_{0,s_1;B}^{\sigma,\varrho}\left(f\chi_{\frac{5}{4}W_{x_i}}\right)-\left|\alpha_{\widetilde{\frac{5}{4}W_{x_i}}^{30\varrho}}(f)\right|\right|\\
&\quad> \frac{\gamma t}{2}.
\end{aligned}$$

This means that

$$\left(W_{x_i}\bigcap G_t\right)\subset\left\{y\in\mathcal{X}:\ M_{0,s_1}^{\sigma,\varrho}\left(\left[f-\alpha_{\widetilde{\frac{5}{4}W_{x_i}}^{30\varrho}}(f)\right]\chi_{\frac{5}{4}W_{x_i}}\right)(y)>\frac{\gamma t}{2}\right\}.$$

Recall that

$$\theta_2 := 1/(2\tilde{C}_7+2\tilde{C}_9) < 1/4.$$

From this, (ii) of Lemma 8.6.5, (8.6.2) and the fact that W_{x_i} and $30\varrho W_{x_i}$ are $(30\varrho, \beta_{30\varrho})$-doubling balls, we deduce that there exist positive constants ζ and $\tilde{C}_8$ such that

$$\begin{aligned}
&\nu\left(W_{x_i}\bigcap G_t\right)\\
&\quad\le \nu\left(\left\{y\in\mathcal{X}:\ M_{0,s_1}^{\sigma,\varrho}\left(\left[f-\alpha_{\widetilde{\frac{5}{4}W_{x_i}}^{30\varrho}}(f)\right]\chi_{\frac{5}{4}W_{x_i}}\right)(y)>\frac{\gamma t}{2}\right\}\right)\\
&\quad\lesssim s^{-1}\nu\left(\left\{y\in\frac{5}{4}W_{x_i}:\ \left|f(y)-\alpha_{\widetilde{\frac{5}{4}W_{x_i}}^{30\varrho}}(f)\right|>2M_{0,s_2}^{\sigma,\varrho;\sharp}(f)(x_i)+\zeta\right\}\right)\\
&\quad\lesssim s_1^{-1}s_2\nu\left(\frac{5}{4}\sigma\varrho W_{x_i}\right)\\
&\quad\lesssim s_1^{-1}s_2\nu\left((30\varrho)^2W_{x_i}\right)\\
&\quad\le \tilde{C}_8 s_1^{-1}s_2\nu\left(W_{x_i}\right).
\end{aligned}$$

Thus, the claim (8.6.6) holds true. By this, (iii) and (i) of Lemma 8.6.4, we know that

$$\begin{aligned}
&u\left(W_{x_i}\bigcap G_t\right)\\
&\le \left[1-\tilde{C}_5^{-1}(1-\tilde{C}_8 s_1^{-1}s_2)^p\right]u(W_{x_i})\\
&\le \tilde{C}_4(\beta_{30\varrho})^{2p}\left[1-\tilde{C}_5^{-1}(1-\tilde{C}_8 s_1^{-1}s_2)^p\right]u(S_{x_i}). \qquad (8.6.9)
\end{aligned}$$

Let

$$\tilde{C}_6 := \tilde{C}_8\left[1-\left(\tilde{C}_5-\frac{1}{(\beta_{30\varrho})^p}\right)^{1/p}\right]^{-1}$$

and

$$\theta_1 := \tilde{C}_4(\beta_{30\varrho})^{2p}[1-\tilde{C}_5^{-1}(1-\tilde{C}_8 s_1^{-1}s_2)^p].$$

From the facts that $30\varrho S_{x_i}\subset\Omega_t$ for all $i\in\mathbb{N}$ and that $\{S_{x_i}\}_i$ are pairwise disjoint, (8.6.5) and (8.6.9), we deduce that

$$u(G_t)\le\sum_i u\left(W_{x_i}\bigcap G_t\right)\le\sum_i\theta_1 u\left(S_{x_i}\right)\le\theta_1 u(\Omega_t),$$

which is the desired estimate.

Case (ii) $\sup_{x\in G_t} r_{S_x}=\infty$. In this case, we fix $z_0\in\mathcal{X}$. For any fixed $R\in(0,\infty)$, let

$$G_{t,R} := G_t\bigcap B(z_0,\,R).$$

Thus, it is easy to see that there exists some $S_{x_0}\in\{S_x\}_{x\in G_t}$ such that $G_{t,R}\subset S_{x_0}$. For the ball S_{x_0}, repeating the process of the proof in Case (i), we easily see that

$$u(G_{t,R})\le\theta_1 u(\Omega_t).$$

Letting $R\to\infty$, then the above estimate implies that

$$u(G_t)\le\theta_1 u(\Omega_t),$$

which completes the proof of Lemma 8.6.8. □

Lemma 8.6.9. *Let* $\varrho\in[1,\infty)$, $\sigma\in[1,30]$ *and* $s\in(0,\beta_{30\varrho}^{-1}/4)$. *Then, for any* ν*-locally integrable function* f *and* $x\in\mathcal{X}$,

$$M_{0,s}^{\sigma,\varrho;\sharp}(|f|)(x)\le 8M_{0,s}^{\sigma,\varrho;\sharp}(f)(x).$$

The proof of Lemma 8.6.9 is similar to that of Lemma 5.1.3, the details being omitted.

Now we turn to the proof of Theorem 8.6.2.

Proof of Theorem 8.6.2. By Lemma 8.6.9, we may assume that f is real-valued. We consider the following two cases.

Case (i) $\nu(\mathcal{X}) = \infty$, $f \in L^{p_0,\infty}(\mathcal{X}, \nu)$ with $p_0 \in (0, \infty)$ and, for all $R \in (0, \infty)$,

$$\sup_{t\in(0,R)} t^p u(\{x \in \mathcal{X} : |f(x)| > t\}) < \infty.$$

In this case, from Lemma 8.6.8, it follows that, for all $\gamma \in (0, \infty)$ and $t \in (0, \infty)$,

$$\begin{aligned} u\left(\{x \in \mathcal{X} : M_{0,s_1}^{\sigma,\varrho}(f)(x) > (1+\gamma)t\}\right) &\le \theta_1 u\left(\{x \in \mathcal{X} : M_{0,s_1}^{\sigma,\varrho}(f)(x) > t\}\right) \\ &\quad + u\left(\left\{x \in \mathcal{X} : M_{0,s_2}^{\sigma,\varrho;\sharp}(f)(x) > \theta_2\gamma t\right\}\right), \end{aligned}$$

where $\theta_1 \in (0, 1)$ depends only on ϱ and ν. Consequently,

$$\begin{aligned} (1+\gamma)^p t^p u\left(\{x \in \mathcal{X} : M_{0,s_1}^{\sigma,\varrho}(f)(x) > (1+\gamma)t\}\right) &\le \theta_1(1+\gamma)^p t^p u\left(\{x \in \mathcal{X} : M_{0,s_1}^{\sigma,\varrho}(f)(x) > t\}\right) \\ &\quad + (1+\gamma)^p t^p u\left(\left\{x \in \mathcal{X} : M_{0,s_2}^{\sigma,\varrho;\sharp}(f)(x) > \theta_2\gamma t\right\}\right). \end{aligned}$$

Taking the supremum in the last inequality, we know that, for all $R \in (0, \infty)$,

$$\begin{aligned} \sup_{t\in(0,(1+\gamma)R)} t^p u\left(\{x \in \mathcal{X} : M_{0,s_1}^{\sigma,\varrho}(f)(x) > t\}\right) &\le \theta_1(1+\gamma)^p \sup_{t\in(0,R)} t^p u\left(\{x \in \mathcal{X} : M_{0,s_1}^{\sigma,\varrho}(f)(x) > t\}\right) \\ &\quad + \left(\frac{1+\gamma}{\theta_2\gamma}\right)^p \sup_{t\in(0,\infty)} t^p u\left(\left\{x \in \mathcal{X} : M_{0,s_2}^{\sigma,\varrho;\sharp}(f)(x) > t\right\}\right). \end{aligned}$$

From Lemma 8.6.5(ii), it then follows that

$$\begin{aligned} \sup_{t\in(0,R)} t^p u\left(\{x \in \mathcal{X} : M_{0,s_1}^{\sigma,\varrho}(f)(x) > t\}\right) \lesssim \sup_{t\in(0,R)} t^p u(\{x \in \mathcal{X} : |f(x)| > t\}). \end{aligned}$$

Thus, our hypotheses guarantee that, in this case,

$$\sup_{t\in(0,R)} t^p u\left(\{x \in \mathcal{X} : M_{0,s_1}^{\sigma,\varrho}(f)(x) > t\}\right) < \infty.$$

Choosing $\gamma \in (0, 1)$ small enough such that $(1+\gamma)^p\theta_1 < 1$, we see that, when $\nu(\mathcal{X}) = \infty$,

$$\sup_{t\in(0,R)} t^p u\left(\{x \in \mathcal{X} : M_{0,s_1}^{\sigma,\varrho}(f)(x) > t\}\right)$$
$$\lesssim \sup_{t\in(0,\infty)} t^p u\left(\left\{x \in \mathcal{X} : M_{0,s_2}^{\sigma,\varrho;\sharp}(f)(x) > t\right\}\right).$$

Letting $R \to \infty$, we then obtain (i).

Case (ii) $\nu(\mathcal{X}) < \infty$ and $f \in L^{p_0,\infty}(\mathcal{X}, \nu)$ with $p_0 \in (0, \infty)$. In this case, by another application of Lemma 8.6.8, we conclude that, for all $R > t_f$ and $\gamma \in (0, \infty)$,

$$\begin{aligned}
&\sup_{t\in(0,(1+\gamma)R)} t^p u\left(\{x \in \mathcal{X} : M_{0,s_1}^{\sigma,\varrho}(f)(x) > t\}\right)\\
&\quad\le \sup_{t\in((1+\gamma)t_f,(1+\gamma)R)} t^p u\left(\{x \in \mathcal{X} : M_{0,s_1}^{\sigma,\varrho}(f)(x) > t\}\right)\\
&\qquad+ \sup_{t\in(0,(1+\gamma)t_f]} t^p u\left(\{x \in \mathcal{X} : M_{0,s_1}^{\sigma,\varrho}(f)(x) > t\}\right)\\
&\quad\le (1+\gamma)^p \sup_{t\in(t_f,R)} t^p u\left(\{x \in \mathcal{X} : M_{0,s_1}^{\sigma,\varrho}(f)(x) > (1+\gamma)t\}\right)\\
&\qquad+ (1+\gamma)^p t_f^p u(\mathcal{X})\\
&\quad\le (1+\gamma)^p\theta_1 \sup_{t\in(t_f,R)} t^p u\left(\{x \in \mathcal{X} : M_{0,s_1}^{\sigma,\varrho}(f)(x) > t\}\right)\\
&\qquad+ \left(\frac{1+\gamma}{\theta_2\gamma}\right)^p \sup_{t\in(0,\infty)} t^p u\left(\left\{x \in \mathcal{X} : M_{0,s_2}^{\sigma,\varrho;\sharp}(f)(x) > t\right\}\right)\\
&\qquad+ (1+\gamma)^p t_f^p u(\mathcal{X}).
\end{aligned}$$

Since $\nu(\mathcal{X}) < \infty$ implies $u(\mathcal{X}) < \infty$, by choosing $\gamma \in (0, 1)$ small enough such that $(1+\gamma)^p\theta_1 < 1$, we see that, when $\nu(\mathcal{X}) < \infty$,

$$\sup_{t\in(0,R)} t^p u\left(\{x \in \mathcal{X} : M_{0,s_1}^{\sigma,\varrho}(f)(x) > t\}\right)$$
$$\lesssim \sup_{t\in(0,\infty)} t^p u\left(\left\{x \in \mathcal{X} : M_{0,s_2}^{\sigma,\varrho;\sharp}(f)(x) > t\right\}\right) + t_f^p u(\mathcal{X}).$$

Taking $R \to \infty$, we then obtain (ii), which completes the proof of Theorem 8.6.2. $\square$

8.7 Multilinear Commutators of Calderón–Zygmund Operators on Orlicz Spaces

Let Φ be a *convex Orlicz function* on $[0, \infty)$, namely, a convex increasing function satisfying $\Phi(0) = 0$, $\Phi(t) > 0$ for all $t \in (0, \infty)$ and $\Phi(t) \to \infty$ as $t \to \infty$. Let[9]

$$a_\Phi := \inf_{t\in(0,\infty)} \frac{t\Phi'(t)}{\Phi(t)} \quad \text{and} \quad b_\Phi := \sup_{t\in(0,\infty)} \frac{t\Phi'(t)}{\Phi(t)}. \tag{8.7.1}$$

The *Orlicz space* $L^\Phi(\mathcal{X}, \nu)$ is defined to be the *space* of all measurable functions f on $(\mathcal{X}, d, \nu)$ such that

$$\int_{\mathcal{X}} \Phi(|f(x)|)\, d\nu(x) < \infty;$$

moreover, for any $f \in L^\Phi(\mathcal{X}, \nu)$, its *Luxemburg norm* in $L^\Phi(\mathcal{X}, \nu)$ is defined by

$$\|f\|_{L^\Phi(\mathcal{X},\nu)} := \inf\left\{t \in (0, \infty) : \int_{\mathcal{X}} \Phi(|f(x)|/t)\, d\nu(x) \le 1\right\}.$$

For any sequence $\vec{b} := (b_1, \ldots, b_k)$ of functions, the *multilinear commutator* $T_{\vec{b}}$ of the Calderón–Zygmund operator T and $\vec{b}$ is defined by setting, for all suitable functions f and $x \in \mathcal{X}$,

$$T_{\vec{b}} f(x) := [b_k, [b_{k-1}, \cdots, [b_1, T] \cdots]] f(x), \tag{8.7.2}$$

where

$$[b_1, T] f(x) := b_1(x) Tf(x) - T(b_1 f)(x). \tag{8.7.3}$$

The first main result of this section is the following boundedness of multilinear commutators on Orlicz spaces.

Theorem 8.7.1. *Let $k \in \mathbb{N}$, $b_i \in \mathrm{RBMO}\,(\mathcal{X}, \nu)$ for all $i \in \{1, \ldots, k\}$, and Φ be a convex Orlicz function satisfying that*

$$1 < a_\Phi \le b_\Phi < \infty.$$

Assume that T is a Calderón–Zygmund operator which is bounded on $L^2(\mathcal{X}, \nu)$. Then the multilinear commutator $T_{\vec{b}}$ in (8.7.2) is bounded on Orlicz spaces $L^\Phi(\mathcal{X}, \nu)$, namely, there exists a positive constant C such that, for all $f \in L^\Phi(\mathcal{X}, \nu)$,

[9] See [93] for more properties of a_Φ and b_Φ.

$$\|T_{\vec{b}} f\|_{L^\Phi(\mathcal{X},\nu)} \le C\|b_1\|_{\mathrm{RBMO}(\mathcal{X},\nu)} \cdots \|b_k\|_{\mathrm{RBMO}(\mathcal{X},\nu)} \|f\|_{L^\Phi(\mathcal{X},\nu)}.$$

The proof of Theorem 8.7.1 is given in Sect. 8.7.2.

Remark 8.7.2. We remark that there exist non-trivial convex Orlicz functions satisfying the assumptions of Theorem 8.7.1. For example, if

$$\Phi_2(t) := t^p \ln(e+t)$$

for all $t \in [0,\infty)$ with $p \in (1,\infty)$, then

$$1 < p = a_{\Phi_2} \le b_{\Phi_2} < \infty;$$

if

$$\Phi_3(t) := t^p / \ln(e+t)$$

for all $t \in [0,\infty)$ with $p \in (2,\infty)$, then

$$1 < a_{\Phi_3} \le b_{\Phi_3} = p < \infty.$$

The endpoint counterpart of Theorem 8.7.1 is also considered in this section. To this end, we first recall the following Orlicz type function space $\mathrm{Osc}_{\exp L^r}(\mathcal{X},\,\nu)$.

Definition 8.7.3. For $r \in [1,\infty)$, a function $f \in L^1_{\mathrm{loc}}(\mathcal{X},\nu)$ is said to belong to the *space* $\mathrm{Osc}_{\exp L^r}(\mathcal{X},\,\nu)$ if there exists a positive constant $\tilde{C}$ such that,

(i) for all balls B,

$$\|f - m_{\tilde{B}^6}(f)\|_{\exp L^r,\, B,\, \nu/\mu(2B)} := \inf\left\{\lambda \in (0,\infty) : \frac{1}{\nu(2B)}\int_B \exp\left(\frac{|f - m_{\tilde{B}^6}(f)|}{\lambda}\right)^r d\nu \le 2\right\} \le \tilde{C};$$

(ii) for all $(6,\beta_6)$-doubling balls $B \subset S$,

$$|m_B(f) - m_S(f)| \le \tilde{C}[1 + \delta(B,S)].$$

The $\mathrm{Osc}_{\exp L^r}(\mathcal{X},\,\nu)$ *norm* of f, $\|f\|_{\mathrm{Osc}_{\exp L^r}(\mathcal{X},\nu)}$, is then defined to be the infimum of all positive constants $\tilde{C}$ satisfying (i) and (ii).

Remark 8.7.4. Obviously, for any $r \in [1,\infty)$,

$$\mathrm{Osc}_{\exp L^r}(\mathcal{X},\,\nu) \subset \mathrm{RBMO}\,(\mathcal{X},\,\nu).$$

Moreover, from Corollary 7.2.12(i), it follows that

$$\mathrm{Osc}_{\exp L^1}(\mathcal{X}, \nu) = \mathrm{RBMO}(\mathcal{X}, \nu).$$

Now we state another main result of this section, whose proof is given in Sect. 8.7.3.

Theorem 8.7.5. *Let $k \in \mathbb{N}$, $r_i \in [1,\infty)$ and $b_i \in \mathrm{Osc}_{\exp L^{r_i}}(\mathcal{X}, \nu)$ for $i \in \{1,\ldots,k\}$. Let T and $T_{\vec{b}}$ be as in* (8.1.25) *and* (8.7.2)*, respectively. If T is bounded on $L^2(\mathcal{X}, \nu)$, then there exists a positive constant C such that, for all $t \in (0,\infty)$ and all bounded functions f with bounded support,*

$$\begin{aligned}
&\nu\left(\{x \in \mathcal{X} : |T_{\vec{b}} f(x)| > t\}\right)\\
&\quad \le C\Phi_{1/r}\left(\|b_1\|_{\mathrm{Osc}_{\exp L^{r_1}}(\mathcal{X},\nu)} \cdots \|b_k\|_{\mathrm{Osc}_{\exp L^{r_k}}(\mathcal{X},\nu)}\right)\\
&\quad\quad \times \int_{\mathcal{X}} \Phi_{1/r}\left(\frac{|f(y)|}{t}\right) d\nu(y),
\end{aligned}$$

where

$$1/r := 1/r_1 + \cdots + 1/r_k$$

and, for all $t \in (0,\infty)$ and $s \in (0,\infty)$,

$$\Phi_s(t) := t\log^s(2+t).$$

8.7.1 An Interpolation Theorem

In this subsection, we establish an interpolation theorem of Orlicz spaces, which plays a key role in the proof of Theorem 8.7.1. We begin with some properties of the indices a_Φ and b_Φ.

Proposition 8.7.6. *Let Φ be a convex Orlicz function on $[0,\infty)$, a_Φ and b_Φ as in* (8.7.1)*. Then the following hold true:*

(i) *If $b_\Phi < \infty$, then Φ satisfies the ∇_2* condition, *namely, there exists a positive constant C such that, for all $t \in (0,\infty)$,*

$$\Phi(2t) \le C\Phi(t);$$

(ii) *If $b_\Phi < \infty$, then $\Phi(t)/t^{b_\Phi}$ is decreasing for $t \in (0,\infty)$. Moreover, for any given $\lambda \in [0,1]$ and $t \in (0,\infty)$,*

$$\Phi(\lambda t) \geq \lambda^{b_\Phi} \Phi(t);$$

(iii) $\Phi(t)/t^{a_\Phi}$ *is increasing for* $t \in (0, \infty)$. *Moreover, for any given* $\lambda \in [1, \infty)$ *and* $t \in (0, \infty)$,

$$\Phi(\lambda t) \geq \lambda^{a_\Phi} \Phi(t);$$

(iv) *Let*

$$1 < p < a_\Phi \leq b_\Phi < q < \infty.$$

Then

$$\lim_{t \to 0} \frac{\Phi(t)}{t^p} = 0 \quad \text{and} \quad \lim_{t \to \infty} \frac{\Phi(t)}{t^q} = 0.$$

Proof. (i) By $b_\Phi < \infty$, we know that, for any $t \in (0, \infty)$,

$$\frac{\Phi'(t)}{\Phi(t)} \leq \frac{b_\Phi}{t};$$

moreover, by the fact that any convex function on $[0, \infty)$ is absolutely continuous on every finite closed intervals of $[0, \infty)$, we see that

$$\log \frac{\Phi(2t)}{\Phi(t)} = \int_t^{2t} \frac{\Phi'(s)}{\Phi(s)}\, ds \leq \int_t^{2t} \frac{b_\Phi}{s}\, ds = b_\Phi \log 2.$$

Thus, we see that, for any $t \in (0, \infty)$,

$$\Phi(2t) \leq 2^{b_\Phi} \Phi(t).$$

This shows (i).

(ii) For any given $t_1, t_2 \in (0, \infty)$, $t_1 \leq t_2$, by the fundamental theorem of calculus, we see that

$$\frac{\Phi(t_2)}{t_2^{b_\Phi}} - \frac{\Phi(t_1)}{t_1^{b_\Phi}} = \int_{t_1}^{t_2} \frac{s\Phi'(s) - b_\Phi \Phi(s)}{s^{b_\Phi + 1}}\, ds \leq 0.$$

Then $\Phi(t)/t^{b_\Phi}$ is decreasing for $t \in (0, \infty)$. Specially, for $\lambda \in [0, 1]$ and $t \in (0, \infty)$, it holds true that

$$\frac{\Phi(t)}{t^{b_\Phi}} \leq \frac{\Phi(\lambda t)}{(\lambda t)^{b_\Phi}},$$

that is, $\Phi(\lambda t) \geq \lambda^{b_\Phi} \Phi(t)$, which completes the proof of (ii).

(iii) The proof of (iii) is similar to (ii), the details being omitted.

(iv) For $t \in (0, 1]$, since $\frac{\Phi(t)}{t^{a_\Phi}}$ is increasing on t, we then see that

$$\frac{\Phi(t)}{t^{a_\Phi}} \leq \Phi(1) < \infty.$$

This, combined with $a_\Phi > p$, implies that

$$\lim_{t\to 0} \frac{\Phi(t)}{t^p} = \lim_{t\to 0} t^{a_\Phi - p} \frac{\Phi(t)}{t^{a_\Phi}} = 0.$$

For $t \in [1, \infty)$, since $\frac{\Phi(t)}{t^{a_\Phi}}$ is decreasing on t, it follows that

$$\frac{\Phi(t)}{t^{b_\Phi}} \leq \Phi(1) < \infty.$$

This, together with $b_\Phi < q$, further implies that

$$\lim_{t\to \infty} \frac{\Phi(t)}{t^q} = \lim_{t\to \infty} t^{b_\Phi - q} \frac{\Phi(t)}{t^{b_\Phi}} = 0,$$

which completes the proof of (iv) and hence Proposition 8.7.6. □

In what follows, for a convex Orlicz function $\Phi : [0, \infty) \to [0, \infty)$, its *inverse* Φ^{-1} is defined by setting, for all $t \in [0, \infty)$,

$$\Phi^{-1}(t) := \inf\{s \in (0, \infty) : \Phi(s) > t\}.$$

With these conclusions, we establish the following interpolation theorem.

Theorem 8.7.7. *Let* $\alpha \in [0, 1)$, $p_i, q_i \in (0, \infty)$ *satisfy*

$$1/q_i = 1/p_i - \alpha$$

for $i \in \{1, 2\}$, $p_1 < p_2$ *and* T *be a sublinear operator of weak type* (p_i, q_i) *for* $i \in \{1, 2\}$. *Then* T *is bounded from* $L^\Phi(\mathcal{X}, \nu)$ *to* $L^\Psi(\mathcal{X}, \nu)$, *where* Φ *and* Ψ *are convex Orlicz functions satisfying the following conditions:*

$$1 < p_1 < a_\Phi \leq b_\Phi < p_2 < \infty,\ 1 < q_1 < a_\Psi \leq b_\Psi < q_2 < \infty$$

and, for all $t \in (0, \infty)$,

$$\Psi^{-1}(t) = \Phi^{-1}(t)t^{-\alpha}.$$

Proof. First, we show that

$$L^\Phi(\mathcal{X}, \nu) \subset L^{p_1}(\mathcal{X}, \nu) + L^{p_2}(\mathcal{X}, \nu).$$

To this end, for any given $t \in (0, \infty)$, we decompose $f \in L^{\Phi}(\mathcal{X}, \nu)$ as

$$f(x) = f(x)\chi_{\{x\in\mathcal{X}:\ |f(x)|>t\}}(x) + f(x)\chi_{\{x\in\mathcal{X}:\ |f(x)|\le t\}}(x) =: f^t(x) + f_t(x)$$

for all $x \in \mathcal{X}$. For the sake of simplicity, we assume that $f \not\equiv 0$ on $\mathcal{X}$. Then we claim that $f^t \in L^{p_1}(\mathcal{X}, \nu)$ and $f_t \in L^{p_2}(\mathcal{X}, \nu)$. Indeed, by (i) and (iii) of Proposition 8.7.6, there exists a positive constant $C_{(t)}$, depending on t, such that, for all $x \in \mathcal{X}$ satisfying $|f(x)| > t$,

$$\left[\frac{|f(x)|}{t}\right]^{a_\Phi} \le \frac{\Phi(|f(x)|/t)}{\Phi(1)} \le C_{(t)}\frac{\Phi(|f(x)|)}{\Phi(1)},$$

which, together with $p_1 < a_\Phi$, implies that

$$\begin{aligned}
&\int_{\mathcal{X}} |f^t(x)|^{p_1}\,d\nu(x)\\
&\quad= \int_{\{x\in\mathcal{X}:\ |f(x)|>t\}} |f(x)|^{p_1}\,d\nu(x)\\
&\quad\le \int_{\{x\in\mathcal{X}:\ |f(x)|>t\}} \frac{|f(x)|^{a_\Phi-p_1}}{t^{a_\Phi-p_1}}|f(x)|^{p_1}\,d\nu(x)\\
&\quad\le C_{(t)}\frac{t^{p_1}}{\Phi(1)}\int_{\mathcal{X}} \Phi(|f(x)|)\,d\nu(x)\\
&\quad< \infty,
\end{aligned}$$

namely, $f^t \in L^{p_1}(\mathcal{X}, \nu)$.

Now we show $f_t \in L^{p_2}(\mathcal{X}, \nu)$. By (i) and (ii) of Proposition 8.7.6, there exists a positive constant $C_{(t)}$, depending on t, such that, for all $x \in \mathcal{X}$ satisfying $|f(x)| \le t$,

$$\left[\frac{|f(x)|}{t}\right]^{b_\Phi} \le \frac{\Phi(|f(x)|/t)}{\Phi(1)} \le C_{(t)}\frac{\Phi(|f(x)|)}{\Phi(1)}.$$

This, combined with $b_\Phi < p_2$, implies that

$$\begin{aligned}
\int_{\mathcal{X}} |f_t(x)|^{p_2}\,d\nu(x) &= \int_{\{x\in\mathcal{X}:\ |f(x)|\le t\}} |f(x)|^{p_2}\,d\nu(x)\\
&\le t^{p_2-b_\Phi}\int_{\{x\in\mathcal{X}:\ |f(x)|\le t\}} |f(x)|^{b_\Phi}\,d\nu(x)\\
&\le C_{(t)}\frac{t^{p_2}}{\Phi(1)}\int_{\mathcal{X}} \Phi(|f(x)|)\,d\nu(x)\\
&< \infty,
\end{aligned}$$

namely, $f_t \in L^{p_2}(\mathcal{X}, \nu)$, which proves the previous claim, and hence

$$L^{\Phi}(\mathcal{X}, \nu) \subset L^{p_1}(\mathcal{X}, \nu) + L^{p_2}(\mathcal{X}, \nu).$$

Next we show that T is bounded from $L^{\Phi}(\mathcal{X}, \nu)$ to $L^{\Psi}(\mathcal{X}, \nu)$. To this end, let u be a function on $[0, \infty)$ satisfying

$$u^{-1}(t) = \Psi^{-1}(\Phi(t))$$

for all $t \in [0, \infty)$. Then u^{-1} is nondecreasing function defined on $[0, \infty)$ such that $u^{-1}(t) \to 0$ as $t \to 0$ and $u^{-1}(t) \to \infty$ as $t \to \infty$. We also let

$$\sigma(f,t) := \nu\left(\{x \in \mathcal{X} : |f(x)| > t\}\right).$$

Then, by the layer cake representation,[10] we see that

$$\begin{aligned}
\int_{\mathcal{X}} \Psi(|Tf(x)|)\, d\nu(x) &= \int_0^{\infty} \sigma(Tf,t)\, d\Psi(t) \\
&\le \int_0^{\infty} \sigma(Tf^{u(t)}, t/2)\, d\Psi(t) + \int_0^{\infty} \sigma(Tf_{u(t)}, t/2)\, d\Psi(t) \\
&=: \mathrm{I} + \mathrm{II}.
\end{aligned}$$

Since T is of weak type (p_1, q_1), we then see that

$$\sigma(Tf^{u(t)}, t/2) \lesssim \left(\frac{2}{t}\right)^{q_1} \|f^{u(t)}\|_{L^{p_1}(\mathcal{X},\nu)}^{q_1},$$

which, together with $p_1 < q_1$ and the Minkowski inequality, implies that

$$\begin{aligned}
\mathrm{I}^{p_1/q_1} &\lesssim \left\{\int_0^{\infty}\left[\int_{\mathcal{X}} t^{-p_1}|f(x)|^{p_1}\chi_{\{x\in\mathcal{X}:\,|f(x)|>u(t)\}}(x)\, d\nu(x)\right]^{q_1/p_1} d\Psi(t)\right\}^{p_1/q_1} \\
&\lesssim \int_{\mathcal{X}}\left[\int_0^{\infty} t^{-q_1}|f(x)|^{q_1}\chi_{\{x\in\mathcal{X}:\,|f(x)|>u(t)\}}(t)\, d\Psi(t)\right]^{p_1/q_1} d\nu(x) \\
&\sim \int_{\mathcal{X}} |f(x)|^{p_1}\left[\int_0^{u^{-1}(|f(x)|)} t^{-q_1}\, d\Psi(t)\right]^{p_1/q_1} d\nu(x).
\end{aligned} \tag{8.7.4}$$

By integration by parts, together with $u^{-1}(t) \to 0$ as $t \to 0$, (iii) and (iv) of Proposition 8.7.6, we conclude that

[10] See [83, Theorem 1.13].

$$\begin{aligned}
&\int_0^{u^{-1}(|f(x)|)} \frac{1}{t^{q_1}}\,d\Psi(t)\\
&\quad= \frac{\Psi(u^{-1}(|f(x)|))}{[u^{-1}(|f(x)|)]^{q_1}} + q_1\int_0^{u^{-1}(|f(x)|)} \frac{\Psi(t)}{t^{q_1+1}}\,dt\\
&\quad\le \frac{\Psi(u^{-1}(|f(x)|))}{[u^{-1}(|f(x)|)]^{q_1}} + q_1\int_0^{u^{-1}(|f(x)|)} \frac{\Psi(u^{-1}(|f(x)|))}{t^{q_1+1}}\left[\frac{t}{u^{-1}(|f(x)|)}\right]^{a_\Psi} dt\\
&\quad= \frac{a_\Psi}{a_\Psi-q_1}\frac{\Psi(u^{-1}(|f(x)|))}{[u^{-1}(|f(x)|)]^{q_1}}\\
&\quad\lesssim \frac{\Phi(|f(x)|)}{[u^{-1}(|f(x)|)]^{q_1}}\\
&\quad\lesssim \frac{\Phi(|f(x)|)}{|f(x)|^{q_1}}[\Phi(|f(x)|)]^{q_1\alpha}\\
&\quad\sim \frac{[\Phi(|f(x)|)]^{q_1/p_1}}{|f(x)|^{q_1}},
\end{aligned} \tag{8.7.5}$$

where the second and the third inequalities to the last one depend on the facts that, for any $t\in(0,\infty)$,

$$\Psi(\Psi^{-1}(t))\le t,\ \Psi^{-1}(t)=\Phi^{-1}(t)t^{-\alpha}\quad\text{and}\quad \Phi^{-1}(\Phi(t))\ge t.$$

Combining (8.7.4) and (8.7.5), we conclude that

$$\mathrm{I}\lesssim\left[\int_{\mathcal{X}}\Phi(|f(x)|)\,d\nu(x)\right]^{q_1/p_1}.$$

By a method similar to the estimate for I, we also see that

$$\mathrm{II}\lesssim\left[\int_{\mathcal{X}}\Phi(|f(x)|)\,d\nu(x)\right]^{q_2/p_2}.$$

Combining the estimates for I and II, we further conclude that

$$\int_{\mathcal{X}}\Psi(|Tf(x)|)\,d\nu(x)\lesssim\left[\int_{\mathcal{X}}\Phi(|f(x)|)\,d\nu(x)\right]^{q_1/p_1}+\left[\int_{\mathcal{X}}\Phi(|f(x)|)\,d\nu(x)\right]^{q_2/p_2}.$$

By a standard argument, we then know that T is bounded from $L^{\Phi}(\mathcal{X},\nu)$ into $L^{\Psi}(\mathcal{X},\nu)$, which completes the proof of Theorem 8.7.7. □

In Theorem 8.7.7, if we take $\alpha=0$, we then immediately obtain the following conclusion. We omit the details.

Corollary 8.7.8. *Let T be a sublinear operator of weak type (p, p) for any $p \in (1, \infty)$. Then T is bounded on $L^{\Phi}(\mathcal{X}, \nu)$, where Φ is a convex Orlicz function on $[0, \infty)$ satisfying that*

$$1 < a_{\Phi} \le b_{\Phi} < \infty.$$

8.7.2 Proof of Theorem 8.7.1

In this section, we show Theorem 8.7.1. To begin with, we introduce the *sharp maximal operator* $\tilde{\mathcal{M}}^{\sharp}$ associated with the coefficient $\tilde{\delta}^{(6)}_{B,S}$ as in (8.6.1) with 30ϱ replaced by 6.

Definition 8.7.9. For all $f \in L^1_{loc}(\mathcal{X}, \nu)$ and $x \in \mathcal{X}$, let

$$\tilde{\mathcal{M}}^{\sharp}(f)(x) := \sup_{B \ni x} \frac{1}{\nu(6B)} \int_B \left| f(x) - m_{\tilde{B}^6}(f) \right| d\nu(x) + \sup_{(B,S) \in \Delta_x} \frac{|m_B(f) - m_S(f)|}{\tilde{\delta}^{(6)}_{B,S}},$$

where

$$\Delta_x := \{(B, S) : \ x \in B \subset S \text{ and } B, S \text{ are } (6, \beta_6) - \text{doubling balls}\}.$$

Remark 8.7.10. The sharp maximal operator $\tilde{\mathcal{M}}^{\sharp}$ has the following useful properties:

(i) By the fact that, for all $(6, \beta_6)$-doubling balls $B \subset S$,

$$1 + \delta(B, S) \lesssim \tilde{\delta}^{(6)}_{B,S},$$

we easily see that, for all $x \in \mathcal{X}$,

$$\tilde{\mathcal{M}}^{\sharp}(f)(x) \lesssim \mathcal{M}^{\sharp}(f)(x);$$

(ii) From (i), together with the corresponding properties of $\mathcal{M}^{\sharp}$, we deduce that $\tilde{\mathcal{M}}^{\sharp}$ is of weak type $(1, 1)$ and bounded on $L^p(\mathcal{X}, \nu)$ for all $p \in (1, \infty)$;

(iii) For all $x \in \mathcal{X}$,

$$\tilde{\mathcal{M}}^{\sharp}|f|(x) \le 5\beta_6 \tilde{\mathcal{M}}^{\sharp}(f)(x).$$

The following lemma improves Lemma 8.1.3 by Remark 8.7.10(i).

Lemma 8.7.11. *Let $f \in L^1_{\mathrm{loc}}(\mathcal{X}, \nu)$ satisfying that*

$$\int_{\mathcal{X}} f(x)\,d\nu(x) = 0 \quad \text{when} \quad \|\nu\| < \infty.$$

Assume that, for some $p \in (1, \infty)$,

$$\inf\{1, \tilde{N}(f)\} \in L^p(\mathcal{X}, \nu).$$

Then there exists a positive constant C, independent of f, such that

$$\left\|\tilde{N}(f)\right\|_{L^p(\mathcal{X},\nu)} \le C \left\|\tilde{\mathcal{M}}^\sharp(f)\right\|_{L^p(\mathcal{X},\nu)}.$$

Proof. By Lemma 8.5.9 and Remark 8.7.10, repeating the argument used in the proof of Lemma 8.1.3, we obtain the desired conclusion. We omit the details, which completes the proof of Lemma 8.7.11. □

We now establish the boundedness of commutators of Calderón–Zygmund operators with $\mathrm{RBMO}(\mathcal{X}, \nu)$ functions on $L^p(\mathcal{X}, \nu)$ for all $p \in [5, \infty)$. To this end, let $r \in (0, \infty)$, $\rho \in (1, \infty)$ and the maximal operator $\mathcal{M}_{r,\rho}(f)$ be as in (8.3.9). It is easy to see that, for any $p \in (1, \infty)$ and $r \in (1, p)$, $\mathcal{M}_{r,\rho}(f)$ is bounded on $L^p(\mathcal{X}, \nu)$.

Theorem 8.7.12. *Let $b \in \mathrm{RBMO}\,(\mathcal{X}, \nu)$ and T be a Calderón–Zygmund operator which is bounded on $L^2(\mathcal{X}, \nu)$. Then the commutator $[b, T]$ as in* (8.7.3) *is bounded on $L^p(\mathcal{X}, \nu)$ for all $p \in (1, \infty)$.*

Proof. To show Theorem 8.7.12, it suffices to show that, for all $f \in L^p(\mathcal{X}, \nu)$ with $p \in (1, \infty)$ and $x \in \mathcal{X}$,

$$\begin{aligned}
&\tilde{\mathcal{M}}^\sharp([b, T]f)(x)\\
&\quad\lesssim \|b\|_{\mathrm{RBMO}(\mathcal{X},\nu)} \left[\mathcal{M}_{r,5} f(x) + \mathcal{M}_{r,6}(Tf)(x) + T^\sharp f(x)\right],
\end{aligned} \tag{8.7.6}$$

where $T^\sharp$ is as in Sect. 8.2. We assume (8.7.6) for the moment and then show that $[b, T]$ is bounded on $L^p(\mathcal{X}, \nu)$ for all $p \in (1, \infty)$. Indeed, by Theorem 8.2.17(i), $T^\sharp$ is bounded on $L^p(\mathcal{X}, \nu)$ for all $p \in (1, \infty)$. This fact, together with (8.7.6), the boundedness of $\mathcal{M}_{r,5}$ on $L^p(\mathcal{X}, \nu)$ and Theorem 8.2.1, implies that $\tilde{\mathcal{M}}^\sharp([b, T])$ is bounded on $L^p(\mathcal{X}, \nu)$ for all $p \in (1, \infty)$. By a standard argument and a limit argument, without loss of generality, we may assume that b is a bounded function, which, together with the boundedness of $\tilde{N}$ and T on $L^p(\mathcal{X}, \nu)$, implies that

$$\inf\{1, \tilde{N}([b, T]f)\} \in L^p(\mathcal{X}, \nu)$$

if $f \in L^p(\mathcal{X}, \nu)$. We now consider two cases for $\|\nu\|$.

Case (i) $\|\nu\| = \infty$. In this case, applying the Lebesgue differentiation theorem and Lemma 8.7.11, we know that $[b, T]$ is bounded on $L^p(\mathcal{X}, \nu)$ for all $p \in (1, \infty)$.

Case (ii) $\|\nu\| < \infty$. In this case, by Corollary 7.2.12(ii) and the Lebesgue dominated convergence theorem, we see that, for all $r \in (1, \infty)$,

$$\left[\frac{1}{\nu(\mathcal{X})} \int_{\mathcal{X}} |b(x) - m_{\mathcal{X}}(b)|^r \, d\nu(x)\right]^{1/r} \lesssim \|b\|_{\mathrm{RBMO}(\mathcal{X},\nu)}, \tag{8.7.7}$$

where

$$m_{\mathcal{X}}(b) := \frac{1}{\nu(\mathcal{X})} \int_{\mathcal{X}} b(y) \, d\nu(y).$$

Write

$$\tilde{N}([b, T]f) \le \tilde{N}([b, T]f - m_{\mathcal{X}}([b, T]f)) + |m_{\mathcal{X}}([b, T]f)|.$$

Notice that

$$\int_{\mathcal{X}} \{[b, T]f(x) - m_{\mathcal{X}}([b, T]f)\} \, d\nu(x) = 0.$$

Then, by Lemma 8.7.11, the fact that

$$\tilde{\mathcal{M}}^{\#}([b, T]f - m_{\mathcal{X}}([b, T]f)) = \tilde{\mathcal{M}}^{\#}([b, T]f)$$

and the boundedness of $\tilde{\mathcal{M}}^{\sharp}([b, T])$ on $L^p(\mathcal{X}, \nu)$ for all $p \in (1, \infty)$, we see that

$$\begin{aligned}
&\left\|\tilde{N}([b, T]f - m_{\mathcal{X}}([b, T]f))\right\|_{L^p(\mathcal{X},\nu)} \\
&\quad \lesssim \left\|\tilde{\mathcal{M}}^{\sharp}([b, T]f - m_{\mathcal{X}}([b, T]f))\right\|_{L^p(\mathcal{X},\nu)} \\
&\quad \sim \left\|\tilde{\mathcal{M}}^{\sharp}([b, T]f)\right\|_{L^p(\mathcal{X},\nu)} \\
&\quad \lesssim \|f\|_{L^p(\mathcal{X},\nu)}.
\end{aligned}$$

For the term $|m_{\mathcal{X}}([b, T]f)|$, we further write

$$|[b, T]f| \le |(b - m_{\mathcal{X}}(b))Tf| + |T((b - m_{\mathcal{X}}(b))f)|,$$

which, together with the Hölder inequality, (8.7.7) and the boundedness of T on $L^q(\mathcal{X}, \nu)$ for all $q \in (1, p]$, further implies that

$$\|m_{\mathcal{X}}([b, T]f)\|_{L^p(\mathcal{X},\nu)} \lesssim \|f\|_{L^p(\mathcal{X},\nu)}.$$

Thus, $[b, T]$ is also bounded on $L^p(\mathcal{X}, \nu)$ for all $p \in (1, \infty)$ in this case.

Now we prove (8.7.6). By $b \in \mathrm{RBMO}(\mathcal{X}, \nu)$ and Definition 7.2.1, there exists a family of numbers, $\{b_B\}_B$, satisfying that, for all balls B,

$$\int_B |b(x) - b_B|\,d\nu(x) \le 2\nu(6B)\|b\|_{\mathrm{RBMO}(\mathcal{X},\nu)}$$

and, for all balls $B \subset S$,

$$|b_B - b_S| \le 2[1 + \delta(B, S)]\|b\|_{\mathrm{RBMO}(\mathcal{X},\nu)}.$$

For all balls B, let

$$h_B := m_B(T((b - b_B)f\chi_{\mathcal{X}\setminus(6/5)B})).$$

Next we show that, for all $x \in \mathcal{X}$ and balls B with $B \ni x$,

$$\begin{aligned}\frac{1}{\nu(6B)}\int_B |[b,T]f(y) - h_B|\,d\nu(y) \\ \lesssim \|b\|_{\mathrm{RBMO}(\mathcal{X},\nu)}[\mathcal{M}_{r,5}f(x) + \mathcal{M}_{r,\rho}(Tf)(x)]\end{aligned} \tag{8.7.8}$$

and, for all $x \in B \subset S$,

$$|h_B - h_S| \lesssim \|b\|_{\mathrm{RBMO}(\mathcal{X},\nu)}\left[\mathcal{M}_{r,5}f(x) + T^\sharp f(x)\right][1 + \delta(B,S)]\tilde{\delta}^{(6)}_{B,S}. \tag{8.7.9}$$

The proof of (8.7.8) is analogous to that of (5.6.6) with a slight modification, the details being omitted.

To prove (8.7.9), for two balls $B \subset S$, let $N := 1 + N_{B,S}$. Then we control $|h_B - h_S|$ by the following five terms:

$$\begin{aligned}|h_B - h_S| &\le |m_B(T((b - b_B)f\chi_{6B\setminus(6/5)B}))| + |m_B(T((b_B - b_S)f\chi_{\mathcal{X}\setminus 6B}))| \\ &\quad + |m_B(T((b - b_S)f\chi_{6^N B\setminus 6B}))| + |m_B(T((b - b_S)f\chi_{\mathcal{X}\setminus 6^N B})) \\ &\quad - m_S(T((b - b_S)f\chi_{\mathcal{X}\setminus 6^N B}))| + |m_S(T((b - b_S)f\chi_{6^N B\setminus(6/5)S}))| \\ &=: \mathrm{M}_1 + \mathrm{M}_2 + \mathrm{M}_3 + \mathrm{M}_4 + \mathrm{M}_5.\end{aligned}$$

By a slight modified argument similar to that used in the proof of Theorem 5.6.4, we conclude that, for all $x \in \mathcal{X}$,

$$\mathrm{M}_1 + \mathrm{M}_4 + \mathrm{M}_5 \lesssim \|b\|_{\mathrm{RBMO}(\mathcal{X},\nu)}\mathcal{M}_{r,5}f(x),$$

$$\mathrm{M}_2 \lesssim [1 + \delta(B,S)]\|b\|_{\mathrm{RBMO}(\mathcal{X},\nu)}\left[T^\sharp f(x) + \mathcal{M}_{r,5}f(x)\right]$$

and

$$\mathrm{M}_3 \lesssim [1+\delta(B,S)]\tilde{\delta}_{B,S}^{(6)}\|b\|_{\mathrm{RBMO}(\mathcal{X},\nu)}\mathcal{M}_{r,5}f(x),$$

which further implies (8.7.9).

Observe that Lemmas 7.2.6 and 7.2.7 also hold true for $\tilde{\delta}_{B,S}^{(6)}$. Then, by this, together with an argument similar to the proof of Theorem 5.6.4, the fact that

$$1+\delta(B,S) \lesssim \tilde{\delta}_{B,S}^{(6)},$$

(8.7.8) and (8.7.9), we obtain (8.7.6). This finishes the proof of Theorem 8.7.12. □

For $k \in \mathbb{N}$ and $i \in \{1,\ldots,k\}$, the *family of all finite subsets* $\sigma := \{\sigma(1),\ldots,\sigma(i)\}$ *of* $\{1,\ldots,k\}$ *with* i *different elements* is denoted by C_i^k. For any $\sigma \in C_i^k$, the *complementary sequence* $\tilde{\sigma}$ is given by

$$\tilde{\sigma} := \{1,\ldots,k\}\setminus\sigma.$$

For any

$$\sigma := \{\sigma(1),\ldots,\sigma(i)\} \in C_i^k$$

and k-tuple $r := (r_1,\ldots,r_k)$, we write that

$$1/r_\sigma := 1/r_{\sigma(1)} + \cdots + 1/r_{\sigma(i)} \quad \text{and} \quad 1/r_{\tilde{\sigma}} := 1/r - 1/r_\sigma,$$

where

$$1/r := 1/r_1 + \cdots + 1/r_k.$$

Let $\vec{b} := (b_1,\ldots,b_k)$ be a *finite family of locally integrable functions*. For all $i \in \{1,\ldots,k\}$ and

$$\sigma := \{\sigma(1),\ldots,\sigma(i)\} \in C_i^k,$$

we let

$$b_\sigma := b_{\sigma(1)}\cdots b_{\sigma(i)},\ \vec{b}_\sigma := (b_{\sigma(1)},\ldots,b_{\sigma(i)}),$$

$$\left\|\vec{b}_\sigma\right\|_{\mathrm{RBMO}(\mathcal{X},\nu)} := \|b_{\sigma(1)}\|_{\mathrm{RBMO}(\mathcal{X},\nu)}\cdots\|b_{\sigma(i)}\|_{\mathrm{RBMO}(\mathcal{X},\nu)}$$

and, for any $y,z \in \mathcal{X}$ and any ball B in $\mathcal{X}$,

$$\left[m_{\tilde{B}^6}(b) - b(z)\right]_\sigma := \left[m_{\tilde{B}^6}(b_{\sigma(1)}) - b_{\sigma(1)}(z)\right] \cdots \left[m_{\tilde{B}^6}(b_{\sigma(i)}) - b_{\sigma(i)}(z)\right].$$

For any $\vec{b} := (b_1, \ldots, b_k)$, we simply write

$$\|\vec{b}\|_{\mathrm{RBMO}(\mathcal{X},\nu)} := \|b_1\|_{\mathrm{RBMO}(\mathcal{X},\nu)} \cdots \|b_k\|_{\mathrm{RBMO}(\mathcal{X},\nu)}.$$

For any $\sigma \in C_i^k$, we let

$$T_{\vec{b}_\sigma} := [b_{\sigma(i)}, \cdots, [b_{\sigma(1)}, T] \cdots].$$

In particular, when $\sigma := \{1, \ldots, k\}$, $T_{\vec{b}_\sigma}$ coincides with $T_{\vec{b}}$ as in (8.7.2).

Now we turn to the proof of Theorem 8.7.1.

Proof of Theorem 8.7.1. To prove Theorem 8.7.1, by Corollary 8.7.8, it suffices to prove that $T_{\vec{b}}$ is bounded on $L^p(\mathcal{X}, \nu)$ for all $p \in (1, \infty)$. We show this by induction on k.

By Theorem 8.7.12, the conclusion is valid for $k = 1$. Now assume that $k \geq 2$ is an integer and, for any $i \in \{1, \ldots, k-1\}$ and any subset $\sigma := \{\sigma(1), \ldots, \sigma(i)\}$ of $\{1, \ldots, k\}$, $T_{\vec{b}_\sigma}$ is bounded on $L^\Phi(\mathcal{X}, \nu)$.

The case that $\|\nu\| < \infty$ can be proved by a way similar to the proof of Theorem 8.7.12, the details being omitted. Thus, without loss of generality, we may assume that $\|\nu\| = \infty$. Let $p \in (1, \infty)$. We first claim that, for all $r \in (1, \infty)$, $f \in L^p(\mathcal{X}, \nu)$, and $x \in \mathcal{X}$,

$$\tilde{\mathcal{M}}^\#(T_{\vec{b}} f)(x) \lesssim \|\vec{b}\|_{\mathrm{RBMO}(\mathcal{X},\nu)} \left[\mathcal{M}_{r,\rho}(Tf)(x) + \mathcal{M}_{r,5} f(x)\right]$$
$$+ \sum_{i=1}^{k-1} \sum_{\sigma \in C_i^k} \|\vec{b}_\sigma\|_{\mathrm{RBMO}(\mathcal{X},\nu)} \mathcal{M}_{r,\rho}\left(T_{\vec{b}_{\bar{\sigma}}} f\right)(x). \quad (8.7.10)$$

Once (8.7.10) is proved, by an argument similar to that used in the proof of Theorem 8.7.12, we conclude that, for all $p \in (1, \infty)$ and $f \in L^p(\mathcal{X}, \nu)$,

$$\begin{aligned}
\|T_{\vec{b}} f\|_{L^p(\mathcal{X},\nu)} &\leq \left\|\tilde{N}\left(T_{\vec{b}} f\right)\right\|_{L^p(\mathcal{X},\nu)} \\
&\lesssim \left\|\tilde{\mathcal{M}}^\sharp\left(T_{\vec{b}} f\right)\right\|_{L^p(\mathcal{X},\nu)} \\
&\lesssim \|\vec{b}\|_{\mathrm{RBMO}(\mathcal{X},\nu)} \left[\|\mathcal{M}_{r,\rho}(Tf)\|_{L^p(\mathcal{X},\nu)} + \|\mathcal{M}_{r,5}(f)\|_{L^p(\mathcal{X},\nu)}\right] \\
&\quad + \sum_{i=1}^{k-1} \sum_{\sigma \in C_i^k} \|\vec{b}_\sigma\|_{\mathrm{RBMO}(\mathcal{X},\nu)} \left\|\mathcal{M}_{r,\rho}\left(T_{\vec{b}_{\bar{\sigma}}} f\right)\right\|_{L^p(\mathcal{X},\nu)} \\
&\lesssim \|\vec{b}\|_{\mathrm{RBMO}(\mathcal{X},\nu)} \left(\|Tf\|_{L^p(\mathcal{X},\nu)} + \|f\|_{L^p(\mathcal{X},\nu)}\right)
\end{aligned}$$

$$+\sum_{i=1}^{k-1}\sum_{\sigma\in C_i^k}\|\vec{b}_\sigma\|_{\mathrm{RBMO}(\mathcal{X},\nu)}\left\|T_{\vec{b}_{\tilde{\sigma}}}f\right\|_{L^p(\mathcal{X},\nu)}$$

$$\lesssim \|\vec{b}\|_{\mathrm{RBMO}(\mathcal{X},\nu)}\|f\|_{L^p(\mathcal{X},\nu)},$$

which is desired.

To prove (8.7.10), by the homogeneity of RBMO $(\mathcal{X}, \nu)$, we may assume that $\|b_i\|_{\mathrm{RBMO}(\mathcal{X},\nu)} = 1$ for all $i \in \{1, \ldots, k\}$. Then it suffices to show that, for all $x \in \mathcal{X}$ and balls B with $B \ni x$,

$$\frac{1}{\nu(6B)}\int_B |T_{\vec{b}}f(y) - h_B|\,d\nu(y)$$
$$\lesssim \mathcal{M}_{r,5}f(x) + \mathcal{M}_{r,\rho}(Tf)(x) + \sum_{i=1}^{k-1}\sum_{\sigma\in C_i^k}\mathcal{M}_{r,\rho}(T_{\vec{b}_{\tilde{\sigma}}}f)(x) \quad (8.7.11)$$

and, for an arbitrary ball B, a $(6, \beta_6)$-doubling ball S with $B \subset S$ and $x \in B$,

$$|h_B - h_S| \lesssim \left[\tilde{\delta}_{B,S}^{(6)}\right]^{k+1}[\mathcal{M}_{r,5}f(x) + \mathcal{M}_{r,\rho}(Tf)(x)]$$
$$+\left[\tilde{\delta}_{B,S}^{(6)}\right]^{k+1}\sum_{i=1}^{k-1}\sum_{\sigma\in C_i^k}\mathcal{M}_{r,\rho}\left(T_{\vec{b}_{\tilde{\sigma}}}f\right)(x), \quad (8.7.12)$$

where

$$h_B := m_B(T([(m_{\tilde{B}^6}(b_1) - b_1)\cdots(m_{\tilde{B}^6}(b_k) - b_k)]f\chi_{\mathcal{X}\setminus\frac{6}{5}B}))$$

and

$$h_S := m_S(T([(m_S(b_1) - b_1)\cdots(m_S(b_k) - b_k)]f\chi_{\mathcal{X}\setminus\frac{6}{5}S})).$$

Let us first prove (8.7.11). With the aid of the formula that, for all $y, z \in \mathcal{X}$,

$$\prod_{i=1}^{k}[m_{\tilde{B}^6}(b_i) - b_i(z)] = \sum_{i=0}^{k}\sum_{\sigma\in C_i^k}[b(y) - b(z)]_{\tilde{\sigma}}[m_{\tilde{B}^6}(b) - b(y)]_\sigma, \quad (8.7.13)$$

where, if $i = 0$, we let

$$\tilde{\sigma} := \{1, \ldots, k\},\ \sigma = \emptyset \quad \text{and} \quad [m_{\tilde{B}^6}(b) - b(y)]_\emptyset = 1,$$

it is easy to prove that, for all $y \in \mathcal{X}$,

$$T_{\vec{b}}f(y) = T\left(\prod_{i=1}^{k}[m_{\tilde{B}^6}(b_i) - b_i]f\right)(y) - \sum_{i=1}^{k}\sum_{\sigma\in C_i^k}[m_{\tilde{B}^6}(b) - b(y)]_\sigma T_{\vec{b}_{\bar{\sigma}}}f(y),$$

where, if $i = k$, $T_{\vec{b}_{\bar{\sigma}}}f := Tf$. Therefore, for all balls $B \ni x$,

$$\begin{aligned}
&\frac{1}{\nu(6B)}\int_B \left|T_{\vec{b}}f(y) - h_B\right| d\nu(y)\\
&\le \frac{1}{\nu(6B)}\int_B \left|T\left(\prod_{i=1}^{k}[m_{\tilde{B}^6}(b_i) - b_i]f\chi_{\frac{6}{5}B}\right)(y)\right| d\nu(y)\\
&\quad+\sum_{i=1}^{k}\sum_{\sigma\in C_i^k}\frac{1}{\nu(6B)}\int_B \left|[m_{\tilde{B}^6}(b) - b(y)]_\sigma\right|\left|T_{\vec{b}_{\bar{\sigma}}}f(y)\right| d\nu(y)\\
&\quad+\frac{1}{\nu(6B)}\int_B \left|T\left(\prod_{i=1}^{k}[m_{\tilde{B}^6}(b_i) - b_i]f\chi_{\mathcal{X}\setminus\frac{6}{5}B}\right)(y) - h_B\right| d\nu(y)\\
&=: \mathrm{I}_1 + \mathrm{I}_2 + \mathrm{I}_3.
\end{aligned}$$

From the Hölder inequality and Corollary 7.2.12(ii), it follows that, for all $q \in (1,\infty)$,

$$\int_{\frac{6}{5}B}\prod_{i=1}^{k}|b_i(y) - m_{\tilde{B}^6}(b_i)|^q\, d\nu(y) \lesssim \nu(6B). \tag{8.7.14}$$

Take $s := \sqrt{r}$ and write

$$b_i(y) - m_{\tilde{B}^6}(b_i) = b_i(y) - m_{\widetilde{\frac{6}{5}B}^6}(b_i) + m_{\widetilde{\frac{6}{5}B}^6}(b_i) - m_{\tilde{B}^6}(b_i)$$

for $i \in \{1,\ldots,k\}$. By the Hölder inequality, the boundedness of T on $L^s(\mathcal{X},\nu)$ for $s \in (1,\infty)$, and (8.7.14), we conclude that, for all $x \in B$,

$$\begin{aligned}
\mathrm{I}_1 &\le \frac{\nu(B)^{\frac{1}{s'}}}{\nu(6B)}\left\|T\left(\prod_{i=1}^{k}[m_{\tilde{B}^6}(b_i) - b_i]f\chi_{\frac{6}{5}B}\right)\right\|_{L^s(\mathcal{X},\nu)}\\
&\lesssim \frac{\nu(B)^{1-\frac{1}{s}}}{\nu(6B)}\left\|\prod_{i=1}^{k}[m_{\tilde{B}^6}(b_i) - b_i]f\chi_{\frac{6}{5}B}\right\|_{L^s(\mathcal{X},\nu)}\\
&\lesssim \frac{1}{\nu(6B)^{\frac{1}{s}}}\left\{\int_{\frac{6}{5}B}\prod_{i=1}^{k}|b_i(y) - m_{\tilde{B}^6}(b_i)|^{ss'}\, d\nu(y)\right\}^{\frac{1}{ss'}}\left\{\int_{\frac{6}{5}B}|f(y)|^r\, d\nu(y)\right\}^{\frac{1}{r}}\\
&\lesssim \mathcal{M}_{r,5}f(x).
\end{aligned}$$

For I_2, by (8.7.14), we see that, for all $x \in B$,

$$\mathrm{I}_2 \leq \sum_{i=1}^{k} \sum_{\sigma \in C_i^k} \left\{ \frac{1}{\nu(6B)} \int_B \left| \left[b(y) - m_{\widetilde{B}^6}(b) \right]_\sigma \right|^{r'} d\nu(y) \right\}^{\frac{1}{r'}}$$

$$\times \left\{ \frac{1}{\nu(6B)} \int_B \left| T_{\vec{b}_{\bar{\sigma}}} f(y) \right|^r d\nu(y) \right\}^{\frac{1}{r}}$$

$$\lesssim \sum_{i=1}^{k} \sum_{\sigma \in C_i^k} \mathcal{M}_{r,\rho} \left(T_{\vec{b}_{\bar{\sigma}}} f \right)(x).$$

To estimate I_3, we need to calculate the difference

$$\left| T \left(\prod_{i=1}^{k} \left[m_{\widetilde{B}^6}(b_i) - b_i \right] f \chi_{\mathcal{X} \setminus \frac{6}{5}B} \right)(y) - h_B \right|$$

for all $y \in B$. By (8.1.24), (7.1.1), (8.7.13), (8.7.14), Lemma 7.1.16, Proposition 7.2.9, the Hölder inequality and Corollary 7.2.12(ii), we see that, for $y, y_1, x \in B$,

$$\left| T \left(\prod_{i=1}^{k} \left[m_{\widetilde{B}^6}(b_i) - b_i \right] f \chi_{\mathcal{X} \setminus \frac{6}{5}B} \right)(y) - T \left(\prod_{i=1}^{k} \left[m_{\widetilde{B}^6}(b_i) - b_i \right] f \chi_{\mathcal{X} \setminus \frac{6}{5}B} \right)(y_1) \right|$$

$$\lesssim \int_{\mathcal{X} \setminus \frac{6}{5}B} \frac{d(y, y_1)^\tau}{d(y, z)^\tau \lambda(y, d(y, z))} \prod_{i=1}^{k} \left| b_i(z) - m_{\widetilde{B}^6}(b_i) \right| |f(z)| \, d\nu(z)$$

$$\lesssim \sum_{j=1}^{\infty} \int_{2^j \frac{6}{5}B \setminus 2^{j-1} \frac{6}{5}B} 2^{-j\tau} \frac{1}{\lambda(y, 2^j 6 r_B)} \prod_{i=1}^{k} \left(\left| b_i(z) - m_{\widetilde{2^j \frac{6}{5}B}^6}(b_i) \right| \right.$$

$$\left. + \left| m_{\widetilde{2^j \frac{6}{5}B}^6}(b_i) - m_{\widetilde{B}^6}(b_i) \right| \right) |f(z)| \, d\nu(z)$$

$$\lesssim \sum_{j=1}^{\infty} \sum_{i=0}^{k} \sum_{\sigma \in C_i^k} \frac{2^{-j\tau} j^{k-i}}{\nu(2^j 6B)} \int_{2^j \frac{6}{5}B} \left| \left[b(z) - m_{\widetilde{2^j \frac{6}{5}B}^6}(b) \right]_\sigma \right| |f(z)| \, d\nu(z)$$

$$\lesssim \sum_{j=1}^{\infty} 2^{-j\tau} j^{k-i} \mathcal{M}_{r,5} f(x)$$

$$\lesssim \mathcal{M}_{r,5} f(x),$$

where, in the third to the last inequality, we have used Lemma 7.1.16 and Proposition 7.2.9 to conclude that, for all $i \in \{1, \ldots, k\}$,

$$\left|m_{\widetilde{2^j\frac{6}{5}B}^6}(b_i) - m_{\widetilde{B}^6}(b_i)\right| \lesssim \left[1 + \delta\left(B, 2^j\frac{6}{5}B\right)\right] \lesssim j.$$

From the above estimate and the choice of h_B, we deduce that, for all $x,\ y \in B$,

$$\begin{aligned}
&\left|T\left(\prod_{i=1}^{k}\left[m_{\widetilde{B}^6}(b_i) - b_i\right] f\chi_{\mathcal{X}\setminus\frac{6}{5}B}\right)(y) - h_B\right| \\
&\leq \frac{1}{\nu(B)}\int_B \left|T\left(\prod_{i=1}^{k}\left[m_{\widetilde{B}^6}(b_i) - b_i\right] f\chi_{\mathcal{X}\setminus\frac{6}{5}B}\right)(y)\right. \\
&\quad \left. -T\left(\prod_{i=1}^{k}\left[m_{\widetilde{B}^6}(b_i) - b_i\right] f\chi_{\mathcal{X}\setminus\frac{6}{5}B}\right)(y_1)\right| d\nu(y_1) \\
&\lesssim \mathcal{M}_{r,5}f(x)
\end{aligned}$$

and hence, for all $x \in B$,

$$\mathrm{I}_3 \lesssim \mathcal{M}_{r,5}f(x).$$

Combining the estimates for I_1, I_2 and I_3, we then obtain (8.7.11).

Next we prove (8.7.12). Let B be an arbitrary ball and S a $(6, \beta_6)$-doubling ball in $\mathcal{X}$ such that $x \in B \subset S$. Denote $N_{B,S} + 1$ simply by N. Write

$$\begin{aligned}
&|h_B - h_S| \\
&= \left|m_B\left[T\left(\prod_{i=1}^{k}\left[m_{\widetilde{B}^6}(b_i) - b_i\right] f\chi_{\mathcal{X}\setminus\frac{6}{5}B}\right)\right]\right. \\
&\quad \left. -m_S\left[T\left(\prod_{i=1}^{k}\left[m_S(b_i) - b_i\right] f\chi_{\mathcal{X}\setminus\frac{6}{5}S}\right)\right]\right| \\
&\leq \left|m_B\left[T\left(\prod_{i=1}^{k}\left[m_{\widetilde{B}^6}(b_i) - b_i\right] f\chi_{\mathcal{X}\setminus 6^N B}\right)\right]\right. \\
&\quad \left. -m_S\left[T\left(\prod_{i=1}^{k}\left[m_{\widetilde{B}^6}(b_i) - b_i\right] f\chi_{\mathcal{X}\setminus 6^N B}\right)\right]\right| \\
&\quad +\left|m_S\left[T\left(\prod_{i=1}^{k}\left[m_{\widetilde{B}^6}(b_i) - b_i\right] f\chi_{\mathcal{X}\setminus 6^N B}\right)\right]\right. \\
&\quad \left. -m_S\left[T\left(\prod_{i=1}^{k}\left[m_S(b_i) - b_i\right] f\chi_{\mathcal{X}\setminus 6^N B}\right)\right]\right|
\end{aligned}$$

$$+\left|m_B\left[T\left(\prod_{i=1}^{k}[m_{\tilde{B}^6}(b_i)-b_i]\,f\chi_{6^N B\setminus\frac{6}{5}B}\right)\right]\right|$$

$$+\left|m_S\left[T\left(\prod_{i=1}^{k}[m_S(b_i)-b_i]\,f\chi_{6^N B\setminus\frac{6}{5}S}\right)\right]\right|$$

$$=: \mathrm{L}_1+\mathrm{L}_2+\mathrm{L}_3+\mathrm{L}_4.$$

By an estimate similar to that for I_3, together with

$$1+\delta(B,S)\lesssim \tilde{\delta}_{B,S}^{(6)},$$

we see that, for all $x\in B$,

$$\mathrm{L}_1\lesssim\left[\tilde{\delta}_{B,S}^{(6)}\right]^k \mathcal{M}_{r,5}f(x).$$

To estimate L_2, from (8.7.13) and Proposition 7.2.9, we deduce that, for all $y\in S$,

$$\left|T\left(\prod_{i=1}^{k}[m_S(b_i)-b_i]\,f\chi_{\mathcal{X}\setminus 6^N B}\right)(y)-T\left(\prod_{i=1}^{k}[m_{\tilde{B}^6}(b_i)-b_i]\,f\chi_{\mathcal{X}\setminus 6^N B}\right)(y)\right|$$

$$=\left|T\left(\prod_{i=1}^{k}[m_S(b_i)-b_i]\,f\chi_{\mathcal{X}\setminus 6^N B}\right)(y)\right.$$

$$\left.-\sum_{i=0}^{k}\sum_{\sigma\in C_i^k}[m_{\tilde{B}^6}(b)-m_S(b)]_{\tilde{\sigma}}\,T\left([m_S(b)-b]_\sigma\, f\chi_{\mathcal{X}\setminus 6^N B}\right)(y)\right|$$

$$\lesssim\sum_{i=0}^{k-1}\sum_{\sigma\in C_i^k}[1+\delta(B,S)]^{k-i}\left|T\left([m_S(b)-b]_\sigma\, f\chi_{\mathcal{X}\setminus 6^N B}\right)(y)\right|$$

$$\lesssim\sum_{i=0}^{k-1}\sum_{\sigma\in C_i^k}[1+\delta(B,S)]^{k-i}\left\{\left|T\left([m_S(b)-b]_\sigma\, f\right)(y)\right|\right.$$

$$\left.+\left|T\left([m_S(b)-b]_\sigma\, f\chi_{6^N B}\right)(y)\right|\right\}$$

$$\lesssim \sum_{i=0}^{k-1} \sum_{\sigma \in C_i^k} [1 + \delta(B,S)]^{k-i} \left\{ \sum_{j=0}^{i} \sum_{\eta \in C_j^i} |[m_S(b) - b(y)]_{\eta'}| \left| T_{\vec{b}_\eta} f(y) \right| \right.$$

$$\left. + \left| T\left([m_S(b) - b]_\sigma f \chi_{6^N B \setminus \frac{6}{5} S}\right)(y) \right| + \left| T\left([m_S(b) - b]_\sigma f \chi_{\frac{6}{5} S}\right)(y) \right| \right\}.$$

Applying the Hölder inequality, the fact that S is $(6, \beta_6)$-doubling, and Corollary 7.2.12(ii), we see that, for all $x \in B$,

$$\frac{1}{\nu(S)} \int_S |[b(y) - m_S(b)]_{\eta'}| \left| T_{\vec{b}_\eta} f(y) \right| d\nu(y) \lesssim \mathcal{M}_{r,\rho}(T_{\vec{b}_\eta} f)(x). \tag{8.7.15}$$

Moreover, from Corollary 7.2.12(ii) and (8.1.23), it follows that, for all $y \in S$,

$$\begin{aligned}
&\left| T\left([m_S(b) - b]_\sigma f \chi_{6^N B \setminus \frac{6}{5} S}\right)(y) \right| \\
&\quad \le \int_{6^N B \setminus \frac{6}{5} S} |K(y,z)| \, |[m_S(b) - b(z)]_\sigma| \, |f(z)| \, d\nu(z) \\
&\quad \lesssim \frac{1}{\lambda(x, r_S)} \int_{6^N B \setminus \frac{6}{5} S} |[m_S(b) - b(z)]_\sigma| \, |f(z)| \, d\nu(z),
\end{aligned}$$

where r_S denotes the radius of the ball S. By the Hölder inequality, the fact that $6^{N-2} r_B < r_S$, (7.1.1) and Corollary 7.2.12(ii), we further have

$$\begin{aligned}
&\frac{1}{\lambda(x, r_S)} \int_{6^N B \setminus \frac{6}{5} S} |[m_S(b) - b(z)]_\sigma| \, |f(z)| \, d\nu(z) \\
&\quad \lesssim \left[\frac{1}{\lambda(x, r_S)} \int_{6^N B \setminus \frac{6}{5} S} |[m_S(b) - b(z)]_\sigma|^{r'} \, d\nu(z) \right]^{1/r'} \\
&\qquad \times \left[\frac{1}{\lambda(x, r_S)} \int_{6^N B \setminus \frac{6}{5} S} |f(z)|^{r} \, d\nu(z) \right]^{1/r} \\
&\quad \lesssim \left[\frac{1}{\nu(6^{N+1} B)} \int_{6^N B} |[m_S(b) - b(z)]_\sigma|^{r'} \, d\nu(z) \right]^{1/r'} \mathcal{M}_{r,5} f(x) \\
&\quad \lesssim \mathcal{M}_{r,5} f(x).
\end{aligned}$$

Taking the mean over $y \in S$, we obtain

$$m_S \left[\left| T\left([m_S(b) - b]_\sigma f \chi_{6^N B \setminus \frac{6}{5} S}\right) \right| \right] \lesssim \mathcal{M}_{r,5} f(x). \tag{8.7.16}$$

By an argument similar to the estimate for I_1, we see that, for all $x \in B$,

$$m_S\left[\left|T\left([m_S(b)-b]_\sigma\, f\chi_{\frac{6}{5}S}\right)\right|\right] \lesssim \mathcal{M}_{r,5}f(x). \tag{8.7.17}$$

Noticing that

$$1+\delta(B,S) \lesssim \tilde{\delta}^{(6)}_{B,S}$$

and combining (8.7.15), (8.7.16) and (8.7.17), we then conclude that, for all $x \in B$,

$$\mathrm{L}_2 \lesssim \left[\tilde{\delta}^{(6)}_{B,S}\right]^k \left\{\sum_{i=1}^k \sum_{\sigma\in C_i^k} \mathcal{M}_{r,\rho}\left(T_{\vec{b}_\sigma} f\right)(x) + \mathcal{M}_{r,5}f(x)\right\}.$$

Now we deal with L_3. By (8.1.23), the Hölder inequality and Corollary 7.2.12(ii), we see that, for all $x,\ y \in B$,

$$\begin{aligned}
&\left|T\left(\prod_{i=1}^k [m_{\tilde{B}^6}(b_i)-b_i]\, f\chi_{6^N B\setminus\frac{6}{5}B}\right)(y)\right| \\
&\quad\lesssim \sum_{j=2}^{N-1} \frac{1}{\lambda(y,6^{j-1}r_B)} \int_{6^j B\setminus 6^{j-1}B} \prod_{i=1}^k \left|b_i(z)-m_{\tilde{B}^6}(b_i)\right| |f(z)|\,d\nu(z) \\
&\qquad+\frac{1}{\lambda(y,r_B)} \int_{6B\setminus\frac{6}{5}B} \prod_{i=1}^k \left|b_i(z)-m_{\tilde{B}^6}(b_i)\right| |f(z)|\,d\nu(z) \\
&\quad\lesssim \sum_{j=2}^{N-1} \frac{1}{\lambda(y,6^{j-1}r_B)} \left\{\int_{6^j B} \prod_{i=1}^k \left|b_i(z)-m_{\tilde{B}^6}(b_i)\right|^{r'} d\nu(z)\right\}^{\frac{1}{r'}} \\
&\qquad\times\left\{\int_{6^j B} |f(z)|^r\,d\nu(z)\right\}^{\frac{1}{r}} \\
&\qquad+\frac{1}{\lambda(y,r_B)} \left\{\int_{6B\setminus\frac{6}{5}B} \prod_{i=1}^k \left|b_i(z)-m_{\tilde{B}^6}(b_i)\right|^{r'} d\nu(z)\right\}^{\frac{1}{r'}} \\
&\qquad\times\left\{\int_{6B} |f(z)|^r\,d\nu(z)\right\}^{\frac{1}{r}} \\
&\quad\lesssim \sum_{j=2}^{N-1} \frac{1}{\lambda(y,6^{j-1}r_B)} \left\{\int_{6^j B} \prod_{i=1}^k \left[\left|b_i(z)-m_{\widetilde{6^j B}^6}(b_i)\right|\right.\right. \\
&\qquad\left.\left.+\left|m_{\widetilde{6^j B}^6}(b_i)-m_{\tilde{B}^6}(b_i)\right|\right]^{r'} d\nu(z)\right\}^{\frac{1}{r'}} \left\{\int_{6^j B} |f(z)|^r\,d\nu(z)\right\}^{\frac{1}{r}}
\end{aligned}$$

$$+\mathcal{M}_{r,5}f(x)$$

$$\lesssim [1+\delta(B,S)]^k \sum_{j=2}^{N-1} \frac{\nu(5\times 6^j B)}{\lambda(y, 5\times 6^j B)} \mathcal{M}_{r,5}f(x) + \mathcal{M}_{r,5}f(x),$$

where the last inequality follows from an argument similar to the estimate of I_2. Taking the mean over $y \in B$, we see that, for all $x \in B$,

$$L_3 \lesssim [1+\delta(B,S)]^k \tilde{\delta}_{B,S}^{(6)} \mathcal{M}_{r,5}f(x) \lesssim \left[\tilde{\delta}_{B,S}^{(6)}\right]^{k+1} \mathcal{M}_{r,5}f(x).$$

Finally, we estimate L_4. From (8.1.23), the Hölder inequality and Corollary 7.2.12(ii), we deduce that, for all $y \in S$ and $x \in B \subset S$,

$$\begin{aligned}
&\left| T\left(\prod_{i=1}^k [m_S(b_i) - b_i]\, f\chi_{6^N B\setminus \frac{6}{5}S}\right)(y)\right| \\
&\quad \lesssim \frac{1}{\lambda(y, r_S)} \int_{6^N B\setminus \frac{6}{5}S} \prod_{i=1}^k |b_i(z) - m_S(b_i)|\, |f(z)|\, d\nu(z) \\
&\quad \lesssim \left\{\frac{1}{\lambda(y, r_S)} \int_{6^N B} \prod_{i=1}^k |b_i(z) - m_S(b_i)|^{r'}\, d\nu(z)\right\}^{\frac{1}{r'}} \\
&\qquad \times \left\{\frac{1}{\lambda(y, r_S)} \int_{6^N B} |f(z)|^r\, d\nu(z)\right\}^{\frac{1}{r}} \\
&\quad \lesssim \mathcal{M}_{r,5}f(x).
\end{aligned}$$

Therefore, for all $x \in B$,

$$L_4 \lesssim \mathcal{M}_{r,5}f(x).$$

Combining L_1, L_2, L_3 and L_4, we then obtain (8.7.12) and hence complete the proof of Theorem 8.7.1. □

8.7.3 *Proof of Theorem 8.7.5*

In what follows, for any $k \in \mathbb{N}$ and $i \in \{1,\ldots,k\}$, let C_i^k be as in Sect. 8.7.2. For all k-tuples $r := (r_1,\ldots,r_k)$ and $\sigma := \{\sigma(1),\ldots,\sigma(i)\} \in C_i^k$, let $\vec{b}_\sigma := (b_{\sigma(1)},\ldots,b_{\sigma(i)})$,

$$\|\vec{b}_\sigma\|_{\mathrm{Osc}_{\exp L^{r_\sigma}}(\mathcal{X},\nu)} := \|b_{\sigma(1)}\|_{\mathrm{Osc}_{\exp L^{r_{\sigma(1)}}}(\mathcal{X},\nu)} \cdots \|b_{\sigma(i)}\|_{\mathrm{Osc}_{\exp L^{r_{\sigma(i)}}}(\mathcal{X},\nu)}$$

and, for a finite family $\vec{b} := (b_1, \ldots, b_k)$ of locally integrable functions, let

$$\|\vec{b}\|_{\mathrm{Osc}_{\exp L^r}(\mathcal{X},\nu)} := \|b_1\|_{\mathrm{Osc}_{\exp L^{r_1}}(\mathcal{X},\nu)} \cdots \|b_k\|_{\mathrm{Osc}_{\exp L^{r_k}}(\mathcal{X},\nu)}.$$

Proof of Theorem 8.7.5. Without loss of generality, we may assume that, for all $i \in \{1, \ldots, k\}$,

$$\|b_i\|_{\mathrm{Osc}_{\exp L^{r_i}}(\mathcal{X},\nu)} = 1.$$

We prove the theorem in two cases: $k = 1$ and $k > 1$.

Case (i) $k = 1$. For any given bounded function f with bounded support and

$$\lambda > \beta_6 \|f\|_{L^1(\mathcal{X},\nu)} / \|\nu\|,$$

applying the Calderón–Zygmund decomposition to f at level λ, we see that, with the same notation as in Theorem 8.1.1, $f = g + h$, where

$$g := f\chi_{\mathcal{X}\setminus \cup_j 6B_j} + \sum_j \varphi_j \quad \text{and} \quad h := \sum_j (\omega_j f - \varphi_j).$$

Recall that $|g(x)| \lesssim \lambda$ for ν-almost every $x \in \mathcal{X}$ and

$$\|g\|^2_{L^2(\mathcal{X},\nu)} \lesssim \lambda \|f\|_{L^1(\mathcal{X},\nu)},$$

which, together with Theorem 8.7.12, further imply that

$$\nu(\{x \in \mathcal{X} : |T_b g(x)| > \lambda\}) \lesssim \lambda^{-2} \|g\|^2_{L^2(\mathcal{X},\nu)} \lesssim \lambda^{-1} \int_{\mathcal{X}} |f(y)|\, d\nu(y),$$

where $T_b := T_{b_1}$. On the other hand, by (a) of Theorem 8.1.1, we see that

$$\nu\left(\bigcup_j 6^2 B_j\right) \lesssim \frac{1}{\lambda} \int_{\mathcal{X}} |f(x)|\, d\nu(x).$$

Therefore, the proof of Theorem 8.7.5 in Case (i) can be reduced to proving that

$$\nu\left(\left\{x \in \mathcal{X} \setminus \bigcup_j 6^2 B_j : |T_b h(x)| > \lambda\right\}\right)$$
$$\lesssim \int_{\mathcal{X}} \frac{|f(y)|}{\lambda} \log^{1/r}\left(2 + \frac{|f(y)|}{\lambda}\right) d\nu(y).$$

To see this, for all j and $x \in \mathcal{X}$, let

$$b_j(x) := b(x) - m_{\tilde{B}_j^6}(b) \quad \text{and} \quad h_j(x) := \omega_j(x) f(x) - \varphi_j(x),$$

and write

$$T_b h(x) = \sum_j b_j(x) T h_j(x) - \sum_j T(b_j h_j)(x) =: \mathrm{I}(x) + \mathrm{II}(x).$$

For the term $\mathrm{II}(x)$, by Theorem 8.2.1, we know that T is of weak type (1,1) and hence

$$\begin{aligned}
&\nu(\{x \in \mathcal{X} : |\mathrm{II}(x)| > \lambda\}) \\
&\lesssim \lambda^{-1} \sum_j \int_{\mathcal{X}} |b_j(y) h_j(y)| \, d\nu(y) \\
&\lesssim \lambda^{-1} \sum_j \int_{6B_j} \left| b(y) - m_{\tilde{B}_j^6}(b) \right| |f(y)| \omega_j(y) \, d\nu(y) \\
&\quad + \lambda^{-1} \sum_j \|\varphi_j\|_{L^\infty(\mathcal{X},\nu)} \int_{S_j} \left| b(y) - m_{\tilde{B}_j^6}(b) \right| \, d\nu(y) \\
&=: \mathrm{E} + \mathrm{F}.
\end{aligned}$$

It follows, from Proposition 7.2.9, that

$$\begin{aligned}
&\int_{S_j} \left| b(y) - m_{\widetilde{B}_j^6}(b) \right| \, d\nu(y) \\
&\leq \int_{S_j} |b(y) - m_{S_j}(b)| \, d\nu(y) + \nu(S_j) \left| m_{\widetilde{B}_j^6}(b) - m_{S_j}(b) \right| \\
&\lesssim \nu(6S_j) + \nu(S_j) \\
&\lesssim \nu(S_j).
\end{aligned}$$

This further implies that

$$\mathrm{F} \lesssim \lambda^{-1} \sum_j \|\varphi_j\|_{L^\infty(\mathcal{X},\nu)} \nu(S_j) \lesssim \lambda^{-1} \int_{\mathcal{X}} |f(y)| \, d\nu(y).$$

Observe that Lemma 5.6.1 still holds true in the present setting. Then, from Lemma 5.6.1,

$$\|b_j\|_{\mathrm{Osc}_{\exp L^{r_j}}(\mathcal{X},\nu)} = 1 \text{ for } j \in \{1, \ldots, k\},$$

(8.1.1) and (8.1.4), it follows that

$$\begin{aligned}
\mathrm{E} &\lesssim \lambda^{-1} \sum_j \nu(12B_j) \| f\omega_j \|_{L^1(\mathcal{X},\nu)} \|b_j\|_{\exp L^r, 6B_j, \nu/\nu(12B_j)} \\
&\lesssim \lambda^{-1} \sum_j \nu(12B_j) \| f\omega_j \|_{L^1(\mathcal{X},\nu)} \\
&\lesssim \lambda^{-1} \sum_j \nu(12B_j) \\
&\quad \times \inf_{t\in(0,\infty)} \left\{ t + \frac{t}{\nu(12B_j)} \int_{6B_j} \frac{|f(y)|\omega_j(y)}{t} \right. \\
&\quad \left. \times \log^{1/r}\left(2 + \frac{|f(y)|\omega_j(y)}{t}\right) d\nu(y) \right\} \\
&\lesssim \int_{\mathcal{X}} \frac{|f(y)|}{\lambda} \log^{1/r}\left(2 + \frac{|f(x)|}{\lambda}\right) d\nu(y).
\end{aligned}$$

Thus, we conclude that

$$\nu\left(\{x \in \mathcal{X} : |\mathrm{II}(x)| > \lambda\}\right) \lesssim \int_{\mathcal{X}} \frac{|f(y)|}{\lambda} \log^{1/r}\left(2 + \frac{|f(y)|}{\lambda}\right) d\nu(y).$$

Now we turn to $\mathrm{I}(x)$. Let x_j be the center of B_j. Since $\operatorname{supp} h_j \subset S_j$, using (8.1.24) and

$$\int_{\mathcal{X}} h_j(x)\, d\nu(x) = 0,$$

we write

$$\begin{aligned}
&\int_{\mathcal{X}\setminus\bigcup_j 6^2B_j} |\mathrm{I}(x)|\, d\nu(x) \\
&\quad \lesssim \sum_j r_{S_j}^{\tau} \int_{\mathcal{X}} |h_j(y)|\, d\nu(y) \int_{\mathcal{X}\setminus 2S_j} \frac{|b_j(x)|}{d(x,x_j)^{\tau}\lambda(x_j, d(x,x_j))}\, d\nu(x) \\
&\qquad + \sum_j \int_{2S_j\setminus 6^2B_j} |b_j(x)||T(\omega_j f)(x)|\, d\nu(x) \\
&\qquad + \sum_j \int_{2S_j} |b_j(x)||T(\varphi_j)(x)|\, d\nu(x) \\
&\quad =: \mathrm{G} + \mathrm{H} + \mathrm{J}
\end{aligned}$$

Using (7.1.3), Lemma 7.1.16 and Proposition 7.2.9, we see that

$$\begin{aligned}
&\int_{\mathcal{X}\setminus 2S_j} \frac{|b_j(x)|}{d(x,x_j)^\tau \lambda(x_j, d(x,x_j))}\,d\nu(x)\\
&\lesssim \sum_{k=1}^{\infty} \left(2^k r_{S_j}\right)^{-\tau} \frac{1}{\lambda(x_j, 2^k r_{S_j})} \int_{2^{k+1}S_j} \left|b(x) - m_{\widetilde{2^{k+1}S_j}^6}(b)\right| d\nu(x)\\
&\quad + \sum_{k=1}^{\infty} \left(2^k r_{S_j}\right)^{-\tau} \frac{\nu(2^{k+1}S_j)}{\lambda(x_j, 2^k r_{S_j})} \left|m_{\widetilde{B_j}^6}(b) - m_{\widetilde{2^{k+1}S_j}^6}(b)\right|\\
&\lesssim \sum_{k=1}^{\infty} \left(2^k r_{S_j}\right)^{-\tau} \frac{\nu(2^{k+2}S_j)}{\lambda(x_j, 2^k r_{S_j})}\\
&\quad + \sum_{k=1}^{\infty} \left[1 + \delta\left(\widetilde{B_j}^6, \widetilde{2^{k+1}S_j}^6\right)\right] \left(2^k r_{S_j}\right)^{-\tau} \frac{\nu(2^{k+1}S_j)}{\lambda(x_j, 2^k r_{S_j})}\\
&\lesssim r_{S_j}^{-\tau},
\end{aligned}$$

where the last inequality follows from an argument similar to I_3. From this, together with (8.1.4), we then deduce that

$$\mathrm{G} \lesssim \int_{\mathcal{X}} |f(y)|\,d\nu(y).$$

On the other hand, by the Hölder inequality, the boundedness of T on $L^2(\mathcal{X}, \nu)$, Proposition 7.2.9, the fact that S_j is $(6, \beta_6)$-doubling and (8.1.6), we conclude that

$$\begin{aligned}
\mathrm{J} &\lesssim \sum_j \int_{2S_j} \left|b(x) - m_{\widetilde{2S_j}^6}(b)\right| |T(\varphi_j)(x)|\,d\nu(x)\\
&\quad + \sum_j \left|m_{\widetilde{B_j}^6}(b) - m_{\widetilde{2S_j}^6}(b)\right| \int_{2S_j} |T(\varphi_j)(x)|\,d\nu(x)\\
&\lesssim \sum_j \left(\int_{2S_j} \left|b(x) - m_{\widetilde{2S_j}^6}(b)\right|^2 d\nu(x)\right)^{1/2} \|T\varphi_j\|_{L^2(\mathcal{X},\nu)}\\
&\quad + \sum_j [\nu(2S_j)]^{1/2} \|T\varphi_j\|_{L^2(\mathcal{X},\nu)} \left|m_{\widetilde{B_j}^6}(b) - m_{\widetilde{2S_j}^6}(b)\right|\\
&\lesssim \sum_j [\nu(4S_j)]^{1/2} \|T\varphi_j\|_{L^2(\mathcal{X},\nu)} \left(1 + \left|m_{\widetilde{B_j}^6}(b) - m_{\widetilde{2S_j}^6}(b)\right|\right)\\
&\lesssim \sum_j [\nu(4S_j)]^{1/2} \|T\varphi_j\|_{L^2(\mathcal{X},\nu)}\\
&\lesssim \int_{\mathcal{X}} |f(y)|\,d\nu(y).
\end{aligned}$$

To estimate H, observe that, by (8.1.23), for all $x \in 2S_j \setminus 6^2 B_j$,

$$|T(\omega_j f)(x)| \lesssim \frac{1}{\lambda(x_j, d(x, x_j))} \int_{\mathcal{X}} \omega_j(y)|f(y)|\, d\nu(y).$$

Therefore, by Proposition 7.2.9,

$$\begin{aligned}
\mathrm{H} &\lesssim \sum_j \int_{2S_j \setminus S_j} \frac{|b_j(x)|}{\lambda(x_j, d(x, x_j))}\, d\nu(x) \int_{\mathcal{X}} \omega_j(y)|f(y)|\, d\nu(y)\\
&\lesssim \sum_j \frac{\nu(4S_j)}{\lambda(x_j, r_{S_j})} \int_{\mathcal{X}} \omega_j(y)|f(y)|\, d\nu(y)\\
&\quad + \sum_j \sum_{k=0}^{N-1} \frac{1}{\lambda(x_j, (3\times 6^2)^k r_{B_j})}\\
&\quad \times \int_{(3\times 6^2)^{k+1} B_j \setminus (3\times 6^2)^k B_j} \left| b(x) - m_{\widetilde{(3\times 6^2)^{k+1} B_j}^6}(b) \right| d\nu(x)\\
&\quad \times \int_{\mathcal{X}} \omega_j(y)|f(y)|\, d\nu(y)\\
&\quad + \sum_j \sum_{k=0}^{N-1} \frac{\nu((3\times 6^2)^{k+1} B_j)}{\lambda(x_j, (3\times 6^2)^k r_{B_j})} \left| m_{\widetilde{B_j}^6}(b) - m_{\widetilde{(3\times 6^2)^{k+1} B_j}^6}(b) \right|\\
&\quad \times \int_{\mathcal{X}} \omega_j(y)|f(y)|\, d\nu(y),
\end{aligned}$$

where N satisfies $S_j = (3 \times 6^2)^N B_j$. Obviously, for all $k \in \{0, \ldots, N-1\}$,

$$(3 \times 6^2)^k B_j \subset S_j$$

and hence

$$\left| m_{\widetilde{B_j}^6}(b) - m_{\widetilde{(3\times 6^2)^{k+1} B_j}^6}(b) \right| \lesssim 1 + \delta(B_j, (3\times 6^2)^{k+1} B_j) \lesssim 1 + \delta(B_j, S_j) \lesssim 1.$$

Consequently, by the fact that S_j is the smallest $(3\times 6^2, C_{(\lambda)}^{\log_2(3\times 6^2)+1})$-doubling ball of type $(3 \times 6^2)^i B_j$ with $i \in \mathbb{N}$, (7.1.1), Proposition 7.2.9 and an argument similar to that used in the proof of Lemma 7.1.16(iii), together with (8.1.4), we see that

$$\mathrm{H} \lesssim \sum_j \int_{\mathcal{X}} \omega_j(y)|f(y)|\, d\nu(y)$$

$$+\sum_j \sum_{k=0}^{N-1} \frac{\nu((3\times 6^2)^{k+2}B_j)}{\lambda(x_j,(3\times 6^2)^k r_{B_j})} \int_{\mathcal{X}} \omega_j(y)|f(y)|\,d\nu(y)$$

$$+\sum_j \sum_{k=0}^{N-1} \frac{\nu((3\times 6^2)^{k+1}B_j)}{\lambda(x_j,(3\times 6^2)^k r_{B_j})} \int_{\mathcal{X}} \omega_j(y)|f(y)|\,d\nu(y)$$

$$\lesssim \sum_j \int_{\mathcal{X}} \omega_j(y)|f(y)|\,d\nu(y)$$

$$+\sum_j \sum_{k=0}^{N-1} \frac{\nu((3\times 6^2)^{k}B_j)}{\lambda(x_j,(3\times 6^2)^k r_{B_j})} \int_{\mathcal{X}} \omega_j(y)|f(y)|\,d\nu(y)$$

$$\lesssim \int_{\mathcal{X}} |f(y)|\,d\nu(y).$$

Combining the estimates for G, H and J above, we then conclude that

$$\int_{\mathcal{X}\setminus\bigcup_j 6^2 B_j} |\mathrm{I}(x)|\,d\nu(x) \lesssim \int_{\mathcal{X}} |f(y)|\,d\nu(y),$$

which implies the desired conclusion and hence completes the proof of Theorem 8.7.5 in the case that $k=1$.

Case (ii) $k\geq 2$. The proof of this case is completely similar to that of Theorem 5.6.8. We omit the details, which completes the proof of Theorem 8.7.5. □

8.8 Weighted Boundedness of Multilinear Calderón–Zygmund Operators

In this section, we use Theorem 8.6.2 to establish weighted norm inequalities with weights satisfying the multilinear $A_{\vec{p}}(\mathcal{X},\nu)$ condition for the multilinear Calderón–Zygmund operators (see Definition 8.8.1 below for $A_{\vec{p}}(\mathcal{X},\nu)$ condition). We start with the definition of multilinear Calderón–Zygmund operators.

Let m be a positive integer,

$$\Delta_{m+1} := \{(x,\ldots,x) : x\in\mathcal{X}\}$$

and $K(x,y_1,\ldots,y_m)$ a ν-locally integrable function mapping from

$$(\mathcal{X}\times\cdots\times\mathcal{X})\setminus\Delta_{m+1} \text{ to } \mathbb{C},$$

which satisfies the *size condition* that there exists a positive constant C such that, for all $x, y_1, \ldots, y_m \in \mathcal{X}$ with $x \neq y_j$ for some j,

$$|K(x, y_1, \ldots, y_m)| \le C \frac{1}{[\sum_{i=1}^m \lambda(x, d(x, y_i))]^m}, \tag{8.8.1}$$

and the *regularity condition* that there exists some positive constants $\tilde{\tau}$ and C such that, for all $x, \tilde{x}, y_1, \ldots, y_m \in \mathcal{X}$ with $\max\{d(x, y_1), \ldots, d(x, y_m)\} \ge 2d(x, \tilde{x})$,

$$\begin{aligned} &|K(x, y_1, \ldots, y_m) - K(\tilde{x}, y_1, \ldots, y_m)| \\ &\quad \le C \frac{[d(x, \tilde{x})]^{\tilde{\tau}}}{[\sum_{i=1}^m d(x, y_i)]^{\tilde{\tau}} [\sum_{i=1}^m \lambda(x, d(x, y_i))]^m}. \end{aligned} \tag{8.8.2}$$

Throughout this section, a *multilinear operator* T associated with kernel K is assumed to be bounded from $L^1(\mathcal{X}, \nu) \times \cdots \times L^1(\mathcal{X}, \nu)$ into $L^{1/m, \infty}(\mathcal{X}, \nu)$ and to satisfy that, for all bounded functions $f_1, \ldots, f_m$ with bounded support and ν-almost every $x \in \mathcal{X} \setminus (\cap_{j=1}^m \operatorname{supp} f_j)$,

$$\begin{aligned} T(f_1, \ldots, f_m)(x) := &\int_{\mathcal{X}} \cdots \int_{\mathcal{X}} K(x, y_1, \ldots, y_m) \\ &\times f_1(y_1) \cdots f_m(y_m)\, d\nu(y_1) \cdots d\nu(y_m). \end{aligned} \tag{8.8.3}$$

When $m = 1$, the operator T defined by (8.8.3) is just the Calderón–Zygmund operator on non-homogeneous metric measure spaces. In this section, we establish some weighted estimates for the operator T on non-homogeneous metric measure spaces. In what follows, we *always assume* $m = 2$ for brevity.

Definition 8.8.1. Let $\varrho \in [1, \infty)$, $\vec{P} := (p_1, p_2)$ with $p_1, p_2 \in [1, \infty)$ and

$$1/p = 1/p_1 + 1/p_2.$$

A vector-valued weight $\vec{w} := (w_1, w_2)$ is said to belong to $A_{\vec{P}}^{\varrho}(\mathcal{X}, \nu)$ if w_1, w_2 are nonnegative ν-measurable functions and there exists a positive constant C such that, for all balls $B \subset \mathcal{X}$,

$$\left[\frac{1}{\nu(\varrho B)} \int_B v_{\vec{w}}(x)\, d\nu(x)\right] \prod_{j=1}^2 \left\{\frac{1}{\nu(\varrho B)} \int_B [w_j(x)]^{1-p_j'}\, d\nu(x)\right\}^{p/p_j'} \le C,$$

where above and in what follows, for all $x \in \mathcal{X}$,

$$v_{\vec{w}}(x) := \prod_{j=1}^2 [w_j(x)]^{p/p_j}.$$

When $p_j = 1$,

$$\left\{\frac{1}{\nu(\varrho B)}\int_B [w_j(x)]^{1-p_j'}\,d\nu(x)\right\}^{1/p_j'}$$

is understood as $(\inf_B w_j)^{-1}$ for $j \in \{1, 2\}$.

By the Hölder inequality, we can easily show that $\upsilon_{\vec{w}} \in A^{\varrho}_{2p}(\mathcal{X}, \nu)$, if $\vec{w} \in A^{\varrho}_{\vec{P}}(\mathcal{X}, \nu)$ with all notation as in Definition 8.8.1.

Our main result of this section is stated as follows.

Theorem 8.8.2. *Let K be a ν-locally integrable function mapping $(\mathcal{X}\times\mathcal{X}\times\mathcal{X})\setminus\Delta_3$ to $\mathbb{C}$ which satisfies* (8.8.1) *and* (8.8.2) *with $m = 2$ and the operator T as in* (8.8.3). *Then, for all $\vec{P} := (p_1, p_2)$ with $p_1, p_2 \in [1, \infty)$ and*

$$1/p = 1/p_1 + 1/p_2 \quad \text{and} \quad \vec{w} := (w_1, w_2) \in A^{\varrho}_{\vec{P}}(\mathcal{X}, \nu)$$

with $\varrho \in [1, \infty)$, there exists a positive constant C such that, for all bounded functions f_1, f_2 with bounded support,

$$\|T(f_1, f_2)\|_{L^{p,\infty}(\mathcal{X}, \upsilon_{\vec{w}})} \le C\|f_1\|_{L^{p_1}(\mathcal{X}, w_1)}\|f_2\|_{L^{p_2}(\mathcal{X}, w_2)}.$$

To prove Theorem 8.8.2, we consider the multilinear maximal operator as follows. Let $\eta \in (1, \infty)$. The *multilinear maximal operator* $\tilde{\mathcal{M}}_\eta$ is defined by setting, for all $x \in \mathcal{X}$,

$$\tilde{\mathcal{M}}_\eta(f_1, f_2)(x) := \sup_{B\ni x}\prod_{i=1}^{2}\frac{1}{\nu(\eta B)}\int_B |f_i(y_i)|\,d\nu(y_i). \tag{8.8.4}$$

By some arguments similar to those used in the proof of Lemma 8.6.3, we easily obtain the following weighted estimate for the maximal operator $\tilde{\mathcal{M}}_\eta$.

Lemma 8.8.3. *Let $\varrho \in [1, \infty)$, $\eta \in [5\varrho, \infty)$ and $\tilde{\mathcal{M}}_\eta$ be the operator defined by* (8.8.4). *For $\vec{P} := (p_1, p_2)$, with $p_1, p_2 \in [1, \infty)$ and $1/p = 1/p_1 + 1/p_2$, and*

$$\vec{w} := (w_1, w_2) \in A^{\varrho}_{\vec{P}}(\mathcal{X}, \nu),$$

the maximal operator $\tilde{\mathcal{M}}_\eta$ is bounded

$$\text{from } L^{p_1}(\mathcal{X}, w_1) \times L^{p_2}(\mathcal{X}, w_2) \text{ to } L^{p,\infty}(\mathcal{X}, \upsilon_{\vec{w}}).$$

Notice that, for any weight w, the space $L^{p,\infty}(\mathcal{X}, w)$ with $p \in (0, \infty)$ is a quasi-Banach space and the intersection of the space of ν-simple functions with $L^q(\mathcal{X}, w)$

is dense in $L^q(\mathcal{X}, w)$ with $q \in [1, \infty)$.[11] Thus, Theorem 8.8.2 is a consequence of Lemma 8.8.3 and the following conclusion.

Theorem 8.8.4. *Let K be a ν-locally integrable function mapping $(\mathcal{X}\times\mathcal{X}\times\mathcal{X})\setminus\Delta_3$ to $\mathbb{C}$ which satisfies* (8.8.1) *and* (8.8.2) *with $m = 2$ and the operator T as in* (8.8.3). *Then, for $\varrho \in [1, \infty)$ and $u \in A_{2p}^{\varrho}(\mathcal{X}, \nu)$ with $p \in [1/2, \infty)$, there exists a positive constant C such that, for all bounded functions f_1 and f_2 with bounded support,*

$$\|T(f_1, f_2)\|_{L^{p,\infty}(\mathcal{X},u)} \le C \left\|\tilde{\mathcal{M}}_\eta(f_1, f_2)\right\|_{L^{p,\infty}(\mathcal{X},u)},$$

where $\eta \in (5\varrho, 30\varrho)$ is a constant.

To prove Theorem 8.8.4, we first establish an inequality for the John–Strömberg sharp maximal operator $M_{0,s}^{\sigma,\varrho;\sharp}$ and a variant, $M_r^{\sigma,\varrho;\sharp}$, of the sharp maximal operator $\mathcal{M}^\sharp$. Let $r \in (0, \infty)$, $\sigma \in (1, \infty)$ and $\varrho \in [1, \infty)$. The *operator* $M_r^{\sigma,\varrho;\sharp}$ is defined by setting, for all $x \in \mathcal{X}$,

$$\begin{aligned} M_r^{\sigma,\varrho;\sharp}(f)(x) := &\sup_{B\ni x}\left[\frac{1}{\nu(\sigma\varrho B)}\int_B \left|f(y) - \alpha_{\tilde{B}^{30\varrho}}(f)\right|^r d\nu(y)\right]^{1/r} \\ &+ \sup_{\substack{x\in B\subset S\\ B,\,S\ (30\varrho,\,\beta_{30\varrho})-\text{doubling}}} \frac{|\alpha_B(f) - \alpha_S(f)|}{\tilde{\delta}_{B,S}^{(30\varrho)}}. \end{aligned}$$

Then, it is easy to show that, for all $x \in \mathcal{X}$,

$$M_{0,s}^{\sigma,\varrho;\sharp}(f)(x) \le s^{-1/r} M_r^{\sigma,\varrho;\sharp}(f)(x). \tag{8.8.5}$$

The following pointwise estimate plays an important role in the proof of Theorem 8.8.4.

Lemma 8.8.5. *Let K be a ν-locally integrable function mapping*

$$\text{from } (\mathcal{X}\times\mathcal{X}\times\mathcal{X})\setminus\Delta_3 \text{ to } \mathbb{C}$$

which satisfies (8.8.1) *and* (8.8.2) *with $m = 2$ and the operator T as in* (8.8.3). *Then, for any $\varrho \in [1, \infty)$, $\sigma \in (1, 30]$, $\sigma_1 \in (1, \sigma)$ and $r \in (0, 1/2)$, there exists a positive constant C such that, for all bounded functions f_1, f_2 with bounded support and $x \in \mathcal{X}$,*

$$M_r^{\sigma,\varrho;\sharp}[T(f_1, f_2)](x) \le C\tilde{\mathcal{M}}_{\frac{\sigma}{\sigma_1}\varrho}(f_1, f_2)(x). \tag{8.8.6}$$

Proof. For each ball B and bounded functions f_1 and f_2 with bounded support, let

[11] See [4, Lemma 3.4].

$$h_B := m_B\left(T\left(f_1\chi_{\mathcal{X}\setminus\sigma_1 B},\, f_2\chi_{\sigma_1 B}\right) + T\left(f_1\chi_{\sigma_1 B},\, f_2\chi_{\mathcal{X}\setminus\sigma_1 B}\right)\right.$$
$$\left.+T\left(f_1\chi_{\mathcal{X}\setminus\sigma_1 B},\, f_2\chi_{\mathcal{X}\setminus\sigma_1 B}\right)\right).$$

Notice that, for any ball B, $c \in \mathbb{C}$ and ν-measurable function h,

$$\alpha_B(h) - c = \alpha_B(h - c).$$

From this fact and Lemma 8.6.7, it follows that, for all balls B and $r \in (0,\, 1/2)$,

$$\begin{aligned}
&\int_B \left|T(f_1,\, f_2)(y) - \alpha_{\tilde{B}^{30\varrho}}(T(f_1,\, f_2))\right|^r\, d\nu(y)\\
&\quad\le \int_B |T(f_1,\, f_2)(y) - h_B|^r\, d\nu(y) + \left|h_B - h_{\tilde{B}^{30\varrho}}\right|^r \nu(B)\\
&\qquad + \left|h_{\tilde{B}^{30\varrho}} - \alpha_{\tilde{B}^{30\varrho}}(T(f_1,\, f_2))\right|^r \nu(B)\\
&\quad\le \int_B |T(f_1,\, f_2)(y) - h_B|^r\, d\nu(y) + \left|h_B - h_{\tilde{B}^{30\varrho}}\right|^r \nu(B)\\
&\qquad + \left|\alpha_{\tilde{B}^{30\varrho}}(T(f_1,\, f_2) - h_{\tilde{B}^{30\varrho}})\right|^r \nu(B)\\
&\quad\le \int_B |T(f_1,\, f_2)(y) - h_B|^r\, d\nu(y) + \left|h_B - h_{\tilde{B}^{30\varrho}}\right|^r \nu(B)\\
&\qquad + \left[m^{\sigma,\,\varrho}_{0,\,s;\,\tilde{B}^{30\varrho}}\left(T(f_1,\, f_2) - h_{\tilde{B}^{30\varrho}}\right)\right]^r \nu(B)\\
&\quad\lesssim \int_B |T(f_1,\, f_2)(y) - h_B|^r\, d\nu(y) + \left|h_B - h_{\tilde{B}^{30\varrho}}\right|^r \nu(B)\\
&\qquad + \frac{\nu(B)}{\nu(\tilde{B}^{30\varrho})}\int_{\tilde{B}^{30\varrho}} \left|T(f_1,\, f_2)(y) - h_{\tilde{B}^{30\varrho}}\right|^r\, d\nu(y)
\end{aligned}$$

and that, for any two $(30\varrho,\, \beta_{30\varrho})$-doubling balls $B \subset S$,

$$\begin{aligned}
&|\alpha_B(T(f_1,\, f_2)) - \alpha_S(T(f_1,\, f_2))|\\
&\quad\le |\alpha_B(T(f_1,\, f_2)) - h_B| + |h_B - h_S| + |\alpha_S(T(f_1,\, f_2)) - h_S|\\
&\quad\le |\alpha_B(T(f_1,\, f_2) - h_B)| + |h_B - h_S| + |\alpha_S(T(f_1,\, f_2) - h_S)|\\
&\quad\le m^{\sigma,\,\varrho}_{0,\,s;\,B}(T(f_1,\, f_2) - h_B) + |h_B - h_S| + m^{\sigma,\,\varrho}_{0,\,s;\,S}(T(f_1,\, f_2) - h_S)\\
&\quad\lesssim \left[\frac{1}{\nu(\sigma\varrho B)}\int_B |T(f_1,\, f_2)(y) - h_B|^r\, d\nu(y)\right]^{1/r} + |h_B - h_S|\\
&\qquad + \left[\frac{1}{\nu(\sigma\varrho S)}\int_S |T(f_1,\, f_2)(y) - h_S|^r\, d\nu(y)\right]^{1/r}.
\end{aligned}$$

These estimates imply that, to prove (8.8.6), it suffices to show that, for any ball B,

$$\left\{\frac{1}{\nu(\sigma\varrho B)}\int_B |T(f_1, f_2)(y) - h_B|^r\, d\nu(y)\right\}^{1/r} \lesssim \inf_{x\in B} \tilde{\mathcal{M}}_{\frac{\sigma}{\sigma_1}\varrho}(f_1, f_2)(x) \quad (8.8.7)$$

and that, for all balls $B \subset S$ with S being $(30\varrho, \beta_{30\varrho})$-doubling,

$$|h_B - h_S| \lesssim \tilde{\delta}_{B,S}^{(30\varrho)} \inf_{x\in B} \tilde{\mathcal{M}}_{\frac{\sigma}{\sigma_1}\varrho}(f_1, f_2)(x). \quad (8.8.8)$$

Let us first prove (8.8.7). Write

$$\begin{aligned}
&\frac{1}{\nu(\sigma\varrho B)}\int_B |T(f_1, f_2)(y) - h_B|^r\, d\nu(y)\\
&\leq \frac{1}{\nu(\sigma\varrho B)}\int_B |T(f_1\chi_{\sigma_1 B}, f_2\chi_{\sigma_1 B})(y)|^r\, d\nu(y)\\
&\quad+\frac{1}{\nu(\sigma\varrho B)}\frac{1}{\nu(B)}\int_B\int_B |T(f_1\chi_{\sigma_1 B}, f_2\chi_{\mathcal{X}\setminus\sigma_1 B})(y)\\
&\quad-T(f_1\chi_{\sigma_1 B}, f_2\chi_{\mathcal{X}\setminus\sigma_1 B})(z)|^r\, d\nu(z)\, d\nu(y)\\
&\quad+\frac{1}{\nu(\sigma\varrho B)}\frac{1}{\nu(B)}\int_B\int_B |T(f_1\chi_{\mathcal{X}\setminus\sigma_1 B}, f_2\chi_{\sigma_1 B})(y)\\
&\quad-T(f_1\chi_{\mathcal{X}\setminus\sigma_1 B}, f_2\chi_{\sigma_1 B})(z)|^r\, d\nu(z)\, d\nu(y)\\
&\quad+\frac{1}{\nu(\sigma\varrho B)}\frac{1}{\nu(B)}\int_B\int_B |T(f_1\chi_{\mathcal{X}\setminus\sigma_1 B}, f_2\chi_{\mathcal{X}\setminus\sigma_1 B})(y)\\
&\quad-T(f_1\chi_{\mathcal{X}\setminus\sigma_1 B}, f_2\chi_{\mathcal{X}\setminus\sigma_1 B})(z)|^r\, d\nu(z)\, d\nu(y)\\
&=: \mathrm{H}_1 + \mathrm{H}_2 + \mathrm{H}_3 + \mathrm{H}_4.
\end{aligned}$$

The Kolmogorov inequality implies that

$$\mathrm{H}_1 \lesssim \frac{\nu(B)^{1-2r}}{\nu(\sigma\varrho B)}\|f_1\chi_{\sigma_1 B}\|_{L^1(\mathcal{X},\nu)}^r\|f_2\chi_{\sigma_1 B}\|_{L^1(\mathcal{X},\nu)}^r \lesssim \left[\inf_{x\in B}\tilde{\mathcal{M}}_{\frac{\sigma}{\sigma_1}\varrho}(f_1, f_2)(x)\right]^r.$$

To estimate H_2, by (8.8.2) and (7.1.1), we first see that, for all $y, z \in B$,

$$\begin{aligned}
&\left|T\left(f_1\chi_{\sigma_1 B}, f_2\chi_{\mathcal{X}\setminus\sigma_1 B}\right)(y) - T\left(f_1\chi_{\sigma_1 B}, f_2\chi_{\mathcal{X}\setminus\sigma_1 B}\right)(z)\right|\\
&\lesssim \int_{\sigma_1 B}\int_{\mathcal{X}\setminus\sigma_1 B}\frac{|f_1(u_1)f_2(u_2)|}{[d(y, u_1) + d(y, u_2)]^{\tilde{\tau}}[\lambda(y, d(y, u_1)) + \lambda(y, d(y, u_2))]^2}\\
&\quad\times [r_B]^{\tilde{\tau}}\, d\nu(u_2)\, d\nu(u_1)
\end{aligned}$$

$$\lesssim \sum_{k=1}^{\infty} \int_{\sigma_1 B} \int_{(\sigma_1)^{k+1} B \setminus (\sigma_1)^k B} \frac{|f_1(u_1) f_2(u_2)| [r_B]^{\tilde{\tau}}}{[(\sigma_1)^k r_B]^{\tilde{\tau}} [\lambda(x_B, (\sigma_1)^k r_B)]^2} \, d\nu(u_2) \, d\nu(u_1)$$

$$\lesssim \inf_{x \in B} \tilde{\mathcal{M}}_{\frac{\sigma}{\sigma_1} \varrho}(f_1, f_2)(x),$$

which implies that

$$\mathrm{H}_2 \lesssim \left[\inf_{x \in B} \tilde{\mathcal{M}}_{\frac{\sigma}{\sigma_1} \varrho}(f_1, f_2)(x) \right]^r.$$

By an argument similar to the estimate for H_2, we conclude that

$$\mathrm{H}_3 \lesssim \left[\inf_{x \in B} \tilde{\mathcal{M}}_{\frac{\sigma}{\sigma_1} \varrho}(f_1, f_2)(x) \right]^r.$$

For the term H_4, from (8.8.2) and (7.1.1), it follows that, for all $y, z \in B$,

$$\left| T\left(f_1 \chi_{\mathcal{X} \setminus \sigma\sigma_1 B}, f_2 \chi_{\mathcal{X} \setminus \sigma\sigma_1 B}\right)(y) - T\left(f_1 \chi_{\mathcal{X} \setminus \sigma\sigma_1 B}, f_2 \chi_{\mathcal{X} \setminus \sigma\sigma_1 B}\right)(z) \right|$$

$$\lesssim \sum_{k=1}^{\infty} \int_{[(\sigma_1)^{k+1} B]^2 \setminus [(\sigma_1)^k B]^2} \frac{|f_1(u_1) f_2(u_2)| [r_B]^{\tilde{\tau}}}{[(\sigma_1)^k r_B]^{\tilde{\tau}} [\lambda(x_B, (\sigma_1)^k r_B)]^2} \, d\nu(u_1) \, d\nu(u_2)$$

$$\lesssim \inf_{x \in B} \mathcal{M}_{\frac{\sigma}{\sigma_1} \varrho}(f_1, f_2)(x),$$

where we used the notation $E^2 := E \times E$ for any set $E \subset \mathcal{X}$. Thus,

$$\mathrm{H}_4 \lesssim \left[\inf_{x \in B} \tilde{\mathcal{M}}_{\frac{\sigma}{\sigma_1} \varrho}(f_1, f_2)(x) \right]^r.$$

Combining the estimates for H_i with $i \in \{1, 2, 3, 4\}$, we then obtain (8.8.7).

Now we show (8.8.8) for chosen h_B and h_S. Denote the *smallest positive integer* N such that $\sigma_1 S \subset (30\varrho)^N B$ simply by N_1. We first write

$$\begin{aligned}
&|h_B - h_S| \\
&\quad \le \left| m_B\left(T\left(f_1 \chi_{(30\varrho)^{N_1} B \setminus \sigma_1 B}, f_2 \chi_{\sigma_1 B}\right)\right) \right| \\
&\qquad + \left| m_B\left(T\left(f_1 \chi_{\sigma_1 B}, f_2 \chi_{(30\varrho)^{N_1} B \setminus \sigma_1 B}\right)\right) \right| \\
&\qquad + \left| m_B\left(T\left(f_1 \chi_{(30\varrho)^{N_1} B \setminus \sigma_1 B}, f_2 \chi_{(30\varrho)^{N_1} B \setminus \sigma_1 B}\right)\right) \right| \\
&\qquad + \left| m_B\left(T\left(f_1 \chi_{\mathcal{X} \setminus (30\varrho)^{N_1} B}, f_2 \chi_{(30\varrho)^{N_1} B}\right)\right) \right. \\
&\qquad\quad \left. - m_S\left(T\left(f_1 \chi_{\mathcal{X} \setminus (30\varrho)^{N_1} B}, f_2 \chi_{(30\varrho)^{N_1} B}\right)\right) \right| \\
&\qquad + \left| m_B\left(T\left(f_1 \chi_{(30\varrho)^{N_1} B}, f_2 \chi_{\mathcal{X} \setminus (30\varrho)^{N_1} B}\right)\right) \right.
\end{aligned}$$

$$
\begin{aligned}
&\quad -m_S\left(\left[T\left(f_1\chi_{(30\varrho)^{N_1}B},\, f_2\chi_{\mathcal{X}\setminus(30\varrho)^{N_1}B}\right)\right)\right| \\
&+\left|m_B\left(T\left(f_1\chi_{\mathcal{X}\setminus(30\varrho)^{N_1}B},\, f_2\chi_{\mathcal{X}\setminus(30\varrho)^{N_1}B}\right)\right)\right. \\
&\quad \left.-m_S\left(T\left(f_1\chi_{\mathcal{X}\setminus(30\varrho)^{N_1}B},\, f_2\chi_{\mathcal{X}\setminus(30\varrho)^{N_1}B}\right)\right)\right| \\
&+\left|m_S\left(T\left(f_1\chi_{(30\varrho)^{N_1}B\setminus\sigma_1 S},\, f_2\chi_{\sigma_1 S}\right)\right)\right| \\
&+\left|m_S\left(T\left(f_1\chi_{\sigma_1 S},\, f_2\chi_{(30\varrho)^{N_1}B\setminus\sigma_1 S}\right)\right)\right| \\
&+\left|m_S\left(T\left(f_1\chi_{(30\varrho)^{N_1}B\setminus\sigma_1 S},\, f_2\chi_{(30\varrho)^{N_1}B\setminus\sigma_1 S}\right)\right)\right| \\
&=: \sum_{i=1}^{9} \mathrm{I}_i.
\end{aligned}
$$

It follows, from (8.8.1) and (7.1.1), that, for all $y \in B$,

$$
\begin{aligned}
&\left|T\left(f_1\chi_{(30\varrho)^{N_1}B\setminus\sigma_1 B},\, f_2\chi_{\sigma_1 B}\right)(y)\right| \\
&\lesssim \sum_{k=1}^{N_1-1}\int_{(30\varrho)^{k+1}B\setminus(30\varrho)^k B}\int_{\sigma_1 B}\frac{|f_1(z_1)f_2(z_2)|}{[\lambda(y,\, d(y,\, z_1))+\lambda(y,\, d(y,\, z_2))]^2}\,d\nu(z_2)\,d\nu(z_1) \\
&\quad +\int_{30\varrho B\setminus\sigma_1 B}\int_{\sigma_1 B}\frac{|f_1(z_1)f_2(z_2)|}{[\lambda(y,\, d(y,\, z_1))+\lambda(y,\, d(y,\, z_2))]^2}\,d\nu(z_2)\,d\nu(z_1) \\
&\lesssim \sum_{k=1}^{N_1-1}\frac{[\nu((30\varrho)^{k+2}B)]^2}{[\lambda(x_B,\, (30\varrho)^{k+2}r_B)]^2}\frac{1}{[\nu((30\varrho)^{k+2}B)]^2} \\
&\quad \times\int_{(30\varrho)^{k+1}B}\int_{(30\varrho)^{k+1}B}|f_1(z_1)f_2(z_2)|\,d\nu(z_1)\,d\nu(z_2) \\
&\quad +\frac{[\nu(30\varrho B)]^2}{[\lambda(x_B,\, 30\varrho r_B)]^2}\frac{1}{[\nu(30\varrho B)]^2}\int_{30\varrho B}\int_{30\varrho B}|f_1(z_1)f_2(z_2)|\,d\nu(z_1)\,d\nu(z_2) \\
&\lesssim \tilde{\delta}_{B,S}^{(30\varrho)}\inf_{x\in B}\tilde{\mathcal{M}}_{\frac{\varrho}{\sigma_1}\varrho}(f_1,\, f_2)(x).
\end{aligned}
$$

Thus,

$$
\mathrm{I}_1 \lesssim \tilde{\delta}_{B,S}^{(30\varrho)}\inf_{x\in B}\tilde{\mathcal{M}}_{\frac{\varrho}{\sigma_1}\varrho}(f_1,\, f_2)(x).
$$

By an argument similar to the estimate for I_1, we see that

$$
\mathrm{I}_2 \lesssim \tilde{\delta}_{B,S}^{(30\varrho)}\inf_{x\in B}\tilde{\mathcal{M}}_{\frac{\varrho}{\sigma_1}\varrho}(f_1,\, f_2)(x),
$$

and

$$
\mathrm{I}_7+\mathrm{I}_8 \lesssim \inf_{x\in B}\tilde{\mathcal{M}}_{\frac{\varrho}{\sigma_1}\varrho}(f_1,\, f_2)(x).
$$

On the other hand, for all $y \in B$, we find that

$$
\begin{aligned}
&\left|T\left(f_1\chi_{(30\varrho)^{N_1}B\setminus\sigma_1 B},\, f_2\chi_{(30\varrho)^{N_1}B\setminus\sigma_1 B}\right)(y)\right| \\
&\quad\lesssim \sum_{k=1}^{N_1-1} \frac{[\nu((30\varrho)^{k+1}B)]^2}{[\lambda(x_B,(30\varrho)^{k+1}r_B)]^2}\frac{1}{[\nu((30\varrho)^{k+1}B)]^2} \\
&\quad\quad\times\int_{[(30\varrho)^{k+1}B]^2\setminus[(30\varrho)^k B]^2} |f_1(z_1)f_2(z_2)|\,d\nu(z_1)\,d\nu(z_2) \\
&\quad\quad+\frac{[\nu(30\varrho B)]^2}{[\lambda(x_B,30\varrho r_B)]^2}\frac{1}{[\nu(30\varrho B)]^2}\int_{(30\varrho B)^2\setminus(\sigma_1 B)^2} |f_1(z_1)f_2(z_2)|\,d\nu(z_1)\,d\nu(z_2) \\
&\quad\lesssim \tilde{\delta}_{B,S}^{(30\varrho)} \inf_{x\in B} \tilde{\mathcal{M}}_{\frac{\varrho}{\sigma_1}\varrho}(f_1,f_2)(x),
\end{aligned}
$$

which implies that

$$
\mathrm{I}_3 \lesssim \tilde{\delta}_{B,S}^{(30\varrho)} \inf_{x\in B} \tilde{\mathcal{M}}_{\frac{\varrho}{\sigma_1}\varrho}(f_1,f_2)(x).
$$

Analogously,

$$
\mathrm{I}_9 \lesssim \inf_{x\in B} \tilde{\mathcal{M}}_{\frac{\varrho}{\sigma_1}\varrho}(f_1,f_2)(x).
$$

From a familiar argument similar to the estimate for H_2 involving the regularity condition (8.8.2), we deduce that

$$
\mathrm{I}_4 + \mathrm{I}_5 \lesssim \inf_{x\in B} \tilde{\mathcal{M}}_{\frac{\varrho}{\sigma_1}\varrho}(f_1,f_2)(x).
$$

Finally, by some estimates similar to that for H_4, we conclude that

$$
\mathrm{I}_6 \lesssim \inf_{x\in B} \tilde{\mathcal{M}}_{\frac{\varrho}{\sigma_1}\varrho}(f_1,f_2)(x).
$$

Combining the estimates for I_i with $i \in \{1,\ldots,9\}$, we obtain (8.8.8), which completes the proof of Lemma 8.8.5. □

Proof of Theorem 8.8.4. We consider the following two cases.

Case (i) $\nu(\mathcal{X}) = \infty$. In this case, we first prove that, for all $R \in (0,\infty)$,

$$
\sup_{0<t<R} t^p u(\{x \in \mathcal{X} : |T(f_1,f_2)(x)| > t\}) < \infty. \tag{8.8.9}
$$

Fix $x_0 \in \mathcal{X}$. Let $l \in (1,\infty)$ be large enough such that $\cup_{i=1}^2 \operatorname{supp} f_i$ is contained in the ball $B(x_0,l)$. Then,

$$\sup_{0<t<R} t^p u(\{x \in B(x_0, 2l) : |T(f_1, f_2)(x)| > t\})$$
$$\le R^p u(B(x_0, 2l)) < \infty. \tag{8.8.10}$$

On the other hand, by (8.8.1), it is easy to show that, for all $x \in \mathcal{X} \setminus B(x_0, 2l)$,

$$\begin{aligned}|T(f_1, f_2)(x)| &\le \int_{\mathcal{X}}\int_{\mathcal{X}} |K(x, y_1, y_2) f_1(y_1) f_2(y_2)|\, d\nu(y_1)\, d\nu(y_2)\\ &\lesssim \frac{1}{[\lambda(x, d(x, x_0)) + \lambda(x, d(x, x_0))]^2} \prod_{i=1}^{2} \|f_i\|_{L^1(\mathcal{X},\nu)}\\ &= \frac{\tilde{C}_{10}}{[\lambda(x_0, d(x, x_0))]^2},\end{aligned} \tag{8.8.11}$$

where $\tilde{C}_{10}$ is a positive constant depending on f_1, f_2. Recall that $\nu(\mathcal{X}) = \infty$. Thus, for $x_0 \in \mathcal{X}$,

$$\lim_{r\to\infty} \lambda(x_0, r) \ge \lim_{r\to\infty} \nu(B(x_0, r)) = \infty,$$

which implies that, for any $t \in (0, \infty)$, there exists $r_t \in (0, \infty)$ such that

$$[\lambda(x_0, r_t)]^2 \ge \frac{\tilde{C}_{10}}{t}.$$

If there exists $\tilde{t} \in (0, \infty)$ such that, for any $r \in (0, \infty)$,

$$[\lambda(x_0, r)]^2 \ge \frac{\tilde{C}_{10}}{\tilde{t}},$$

then, for all $t \in (\tilde{t}, \infty)$ and $r \in (0, \infty)$,

$$[\lambda(x_0, r)]^2 \ge \frac{\tilde{C}_{10}}{t}.$$

Let

$$t_* := \inf\left\{\tilde{t} \in (0, \infty) : [\lambda(x_0, r)]^2 \ge \frac{\tilde{C}_{10}}{\tilde{t}} \quad \text{holds true for any} \quad r \in (0, \infty)\right\}.$$

If there does not exist $\tilde{t} \in (0, \infty)$ such that, for any $r \in (0, \infty)$,

$$[\lambda(x_0, r)]^2 \ge \frac{\tilde{C}_{10}}{\tilde{t}},$$

let $t_* = \infty$. Moreover, if $t_* \in (0, \infty)$, then, for any $t \in (t_*, \infty)$ and $x \in \mathcal{X}$ satisfying

$$\frac{\tilde{C}_{10}}{[\lambda(x_0, d(x, x_0))]^2} > t, \tag{8.8.12}$$

we conclude that $d(x, x_0) = 0$ and hence $x = x_0$. Thus, for any $t_* \in (0, \infty]$ and $t \in (0, t_*)$, there exists $r_t \in (0, \infty)$ such that

$$[\lambda(x_0, r_t)]^2 \geq \frac{\tilde{C}_{10}}{t} \quad \text{and} \quad [\lambda(x_0, r_t/2)]^2 < \frac{\tilde{C}_{10}}{t}. \tag{8.8.13}$$

This further implies that, for all $x \in \mathcal{X}$ such that (8.8.12) holds true, we have $d(x, x_0) < r_t$. On the other hand, notice that, for all $x \in \mathcal{X} \setminus B(x_0, 2l)$,

$$\frac{1}{[\lambda(x_0, d(x, x_0))]^2} \leq \frac{1}{[\lambda(x_0, l)]^2}.$$

This implies that, for $t > \tilde{C}_{10}/[\lambda(x_0, l)]^2$, there does not exist any point $x \in \mathcal{X} \setminus B(x_0, 2l)$ satisfying $|T(f_1, f_2)(x)| > t$.

Thus, by (8.8.11), (7.1.1), Lemma 8.6.4(i) with $u \in A_{2p}^{\varrho}(\mathcal{X}, \nu)$, and (8.8.13), we see that, if $t_* \leq \tilde{C}_{10}/[\lambda(x_0, l)]^2$, then

$$\begin{aligned}
&\sup_{t\in(0,\infty)} t^p u(\{x \in \mathcal{X} \setminus B(x_0, 2l) : |T(f_1, f_2)(x)| > t\}) \\
&\quad = \sup_{t\in(0,\tilde{C}_{10}/[\lambda(x_0,l)]^2]} t^p u(\{x \in \mathcal{X} \setminus B(x_0, 2l) : |T(f_1, f_2)(x)| > t\}) \\
&\quad \leq \sup_{t\in(0,\tilde{C}_{10}/[\lambda(x_0,l)]^2]} t^p u\left(\left\{x \in \mathcal{X} : \frac{\tilde{C}_{10}}{[\lambda(x_0, d(x, x_0))]^2} > t\right\}\right) \\
&\quad \leq \sup_{t\in(0,t_*]} t^p u(B(x_0, r_t)) + \sup_{t\in(t_*,\tilde{C}_{10}/[\lambda(x_0,l)]^2]} t^p u(\{x_0\}) \\
&\quad \lesssim 1 + \sup_{t\in(0,t_*],\, r_t\in(0,l]} t^p u(B(x_0, r_t)) + \sup_{t\in(0,t_*],\, r_t\in(l,\infty)} t^p u(B(x_0, r_t)) \\
&\quad \lesssim 1 + \sup_{t\in(0,t_*],\, r_t\in(l,\infty)} t^p u(B(x_0, l)) \left[\frac{\nu(B(x_0, 5\varrho r_t))}{\nu(B(x_0, l))}\right]^{2p} \\
&\quad \lesssim 1 + u(B(x_0, l)) \left[\frac{1}{\nu(B(x_0, l))}\right]^{2p} \sup_{t\in(0,t_*],\, r_t\in(l,\infty)} t^p [\lambda(x_0, r_t/2)]^{2p} \\
&\quad \lesssim 1 + u(B(x_0, l)) \left[\frac{1}{\nu(B(x_0, l))}\right]^{2p} \\
&\quad < \infty.
\end{aligned}$$

Similarly, if $t_* > \tilde{C}_{10}/[\lambda(x_0, l)]^2$, then

$$\begin{aligned}
&\sup_{t\in(0,\infty)} t^p u(\{x \in \mathcal{X} \setminus B(x_0, 2l) : |T(f_1, f_2)(x)| > t\}) \\
&\le \sup_{t\in(0,\tilde{C}_{10}/[\lambda(x_0,l)]^2]} t^p u\left(\left\{x \in \mathcal{X} : \frac{\tilde{C}_{10}}{[\lambda(x_0, d(x, x_0))]^2} > t\right\}\right) \\
&\lesssim \sup_{t\in(0,\tilde{C}_{10}/[\lambda(x_0,l)]^2],\, r_t\in(0,l]} t^p u(B(x_0, r_t)) \\
&\quad + \sup_{t\in(0,\tilde{C}_{10}/[\lambda(x_0,l)]^2],\, r_t\in(l,\infty)} t^p u(B(x_0, r_t)) \\
&\lesssim 1 + u(B(x_0, l))\left[\frac{1}{\nu(B(x_0, l))}\right]^{2p} \\
&< \infty.
\end{aligned}$$

Thus, we show that

$$\sup_{t\in(0,\infty)} t^p u(\{x \in \mathcal{X} \setminus B(x_0, 2l) : |T(f_1, f_2)(x)| > t\}) < \infty,$$

which, along with (8.8.10), implies (8.8.9).

Now we conclude the proof of Theorem 8.8.4 in this case. Choose σ and σ_1 such that $\sigma \in (1, 30]$ and $\sigma_1 \in (1, \sigma/5)$. Denote $\frac{\sigma}{\sigma_1}\varrho$ simply by η. Recalling that T is bounded from $L^1(\mathcal{X}, \nu) \times L^1(\mathcal{X}, \nu)$ into $L^{1/2,\infty}(\mathcal{X}, \nu)$, we choose $p_0 = 1/2$ in Theorem 8.6.2(i). Therefore, the desired conclusion of Theorem 8.8.4 follows from Lemma 8.6.5(i), (8.8.9), Theorem 8.6.2(i) with $s_1 = \beta_{30\varrho}^{-1}/5$ and $p_0 = 1/2$, (8.8.5) and Lemma 8.8.5.

Case (ii) $\nu(\mathcal{X}) < \infty$. In this case, choose $p_0 = 1/2$. Then, by the boundedness of T from $L^1(\mathcal{X}, \nu) \times L^1(\mathcal{X}, \nu)$ into $L^{1/2,\infty}(\mathcal{X}, \nu)$, we see that, for $u \in A_{2p}^{\varrho}(\mathcal{X}, \nu)$,

$$\begin{aligned}
&u(\mathcal{X})[\nu(\mathcal{X})]^{-2p} \|T(f_1, f_2)\|^p_{L^{1/2,\infty}(\mathcal{X},\nu)} \\
&\lesssim u(\mathcal{X})[\nu(\mathcal{X})]^{-2p} \prod_{i=1}^{2} \|f_i\|^p_{L^1(\mathcal{X},\nu)} \\
&\lesssim u(\mathcal{X})\left[\inf_{x\in\mathcal{X}} \tilde{\mathcal{M}}_\eta(f_1, f_2)(x)\right]^p \\
&\lesssim \sup_{t\in(0,\infty)} t^p u\left(\{x \in \mathcal{X} : \tilde{\mathcal{M}}_\eta(f_1, f_2)(x) > t\}\right),
\end{aligned}$$

where, in the second inequality, we used the fact that

$$\begin{aligned}
\prod_{i=1}^{2} \frac{1}{\nu(\mathcal{X})} \int_{\mathcal{X}} |f_i(y)|\, d\nu(y) &= \lim_{r_B\to\infty} \prod_{i=1}^{2} \frac{1}{\nu(\eta B)} \int_B |f_i(y)|\, d\nu(y) \\
&\le \inf_{x\in\mathcal{X}} \tilde{\mathcal{M}}_\eta(f_1, f_2)(x).
\end{aligned}$$

Thus, the desired conclusion of Theorem 8.8.4 again follows from Lemma 8.6.5(i), (8.8.9), Theorem 8.6.2(i) with $s_1 = \beta_{30_Q}^{-1}/5$ and $p_0 = 1/2$, (8.8.5) and Lemma 8.8.5, which completes the proof of Theorem 8.8.4. □

8.9 Notes

- Theorem 8.1.1 was proved by Bui and Duong in [9]. Using Theorem 8.1.1, they further proved that if T is a Calderón–Zygmund operator associated with kernel K satisfying (8.1.23) and (8.1.24) such that T is bounded on $L^2(\mathcal{X}, \nu)$, then T is also bounded on $L^p(\mathcal{X}, \nu)$ for all $p \in (1, \infty)$.
- Theorem 8.1.2 was obtained by Liu et al. in [91] and Theorem 8.1.4 was proved by Tolsa in [131] when $(\mathcal{X}, d, \nu) := (\mathbb{R}^D, |\cdot|, \mu)$ with μ satisfying the polynomial growth condition (0.0.1) and in [9] for a general metric measure space $(\mathcal{X}, d, \nu)$.
- Theorems 8.1.5 and 8.1.6 were established by Da. Yang and Do. Yang in [155] when $(\mathcal{X}, d, \nu) := (\mathbb{R}^D, |\cdot|, \mu)$ with μ satisfying the polynomial growth condition (0.0.1) and by Hytönen et al. in [71] for a general metric measure space $(\mathcal{X}, d, \nu)$.
- Example 8.1.7 was given by Volberg and Wick in [147].
- Theorems 8.2.1, 8.2.9 and 8.2.12 were proved by Nazarov et al. in [103] in the case that ν is a non-negative Borel measure on the separable space $\mathcal{X}$ such that ν satisfies the polynomial growth condition, and by Hytönen et al. [69] for a general metric measure space $(\mathcal{X}, d, \nu)$ satisfying the upper doubling condition and the geometrically doubling condition. In [69], $(\mathcal{X}, d, \nu)$ was assumed to be separable, which is superfluous.
- Theorem 8.3.4 was established by Fu et al. in [27] for $(\mathcal{X}, d, \nu) := (\mathbb{R}^D, |\cdot|, \mu)$ with μ satisfying the polynomial growth condition (0.0.1) and by Liu et al. in [91] for a general metric measure spaces $(\mathcal{X}, d, \nu)$. It is still unclear whether the conclusions of Theorem 8.3.4 and Corollary 8.3.5 hold true or not, if we replace the coefficient $1 + \delta(B, S)$ by its discrete counterpart $\tilde{\delta}_{B,S}^{(\rho)}$. Precisely, since for any balls $B \subset S$ with $\mu(2S \setminus B) = 0$, it is unclear whether there exists a positive constant C, independent of B and S, such that $\tilde{\delta}_{B,S}^{(\rho)} \le C$ or not, the method used in the proof of Lemma 8.3.1 does not apply to $\tilde{\delta}_{B,S}^{(\rho)}$.
- Theorem 8.4.1 was obtained by Hu et al. in [63] for $(\mathcal{X}, d, \nu) := (\mathbb{R}^D, |\cdot|, \mu)$ with μ satisfying the polynomial growth condition (0.0.1) and by Hu et al. in [61] for a general metric measure spaces $(\mathcal{X}, d, \nu)$. Similar to Theorem 8.3.4, the results in Theorem 8.4.1 and Corollary 8.4.2 are also unknown when $1+\delta(B, S)$ is replaced by $\tilde{\delta}_{B,S}^{(\rho)}$.
- Let T and $T^\sharp$ be the operators associated with kernel K as in (8.1.25) and (8.2.1), respectively, where K satisfies the size condition (8.1.23) and the Hörmander condition (8.3.1). Liu, Meng and Yang [90] showed that the boundedness of $T^\sharp$ on $L^{p_0}(\mathcal{X}, \nu)$ for some $p_0 \in (1, \infty)$ is equivalent to that of $T^\sharp$ from $L^1(\mathcal{X}, \nu)$ into $L^{1,\infty}(\mathcal{X}, \nu)$, and to that of $T^\sharp$ on $L^p(\mathcal{X}, \nu)$ for all $p \in (1, \infty)$.

Furthermore, they showed that if T is bounded on $L^2(\mathcal{X}, \nu)$, then $T^\sharp$ is also bounded on $L^p(\mathcal{X}, \nu)$ for all $p \in (1, \infty)$ and from $L^1(\mathcal{X}, \nu)$ to $L^{1,\infty}(\mathcal{X}, \nu)$.

- We remark that, among the results related to $1 + \delta(B, S)$ or $\tilde{\delta}_{B,S}^{(\rho)}$ in Chaps. 7 and 8, all conclusions, from Theorems 7.2.11, 7.3.2, 7.4.8, 8.1.1, 8.1.2, 8.1.4, 8.1.5, 8.1.6, 8.5.6, 8.6.2, 8.7.1, 8.7.5 and 8.7.12, hold true for both $1+\delta(B, S)$ and $\tilde{\delta}_{B,S}^{(\rho)}$. On the other hand, Theorems 8.3.4 and 8.4.1 only hold true for $1+\delta(B, S)$, while Theorem 8.5.8 is only established for $\tilde{\delta}_{B,S}^{(\rho)}$. To be precise, let $B \subset S$ be two balls with $\mu(2S \setminus B) = 0$. Then we see that

$$1 + \delta(B, S) = 1, \tag{8.9.1}$$

which played an important role in the proofs of Theorems 8.3.4 and 8.4.1. However, (8.9.1) is not true for $\tilde{\delta}_{B,S}^{(\rho)}$ and it is unclear whether there exists a positive constant C, independent of B and S, such that $\tilde{\delta}_{B,S}^{(\rho)} \le C$ or not. Thus, it is unknown whether the conclusions of Theorems 8.3.4 and 8.4.1 are true or not for $\tilde{\delta}_{B,S}^{(\rho)}$. Also, the discrete form of $\tilde{\delta}_{B,S}^{(\rho)}$ plays an important role in the proof of Theorem 8.5.8. It is still unknown in general whether the conclusion of Theorem 8.5.8 holds true for $1 + \delta(B, S)$ or not.
- Theorems 8.5.6, 8.5.8 and 8.5.17 were proved by Fu et al. in [28]. When $(\mathcal{X}, d, \nu) := (\mathbb{R}^D, |\cdot|, \mu)$ with μ as in (0.0.1), Theorem 8.5.8 was proved by Chen et al. in [11, Theorem 1] in terms of the maximal function characterization, established by Tolsa [134], of the Hardy space $H_{\rm atb}^{1,p}(\mathcal{X}, \nu)$. However, even when $(\mathcal{X}, d, \nu)$ is a space of homogeneous type, Theorem 8.5.8 is also new, since $\tilde{H}_{\rm atb}^{1,p}(\mathcal{X}, \nu)$ may be different from the atomic Hardy space $H^{1,p}(\mathcal{X}, \nu)$ introduced by Coifman and Weiss in [19].
- Examples 8.5.16 and 8.5.20 were given by Fu et al. in [28].
- Let T be the Calderón–Zygmund operator associated with kernel K which satisfies (8.1.23) and (8.1.24). An interesting *problem* is that, under what assumptions, T is bounded on $H^1(\mathcal{X}, \nu)$.
- In [147], Volberg and Wick established the $T(1)$ theorem for Bergman-type operators. In [70], Hytönen and Martikainen further obtained a version of the $T(b)$ theorem in the setting of $(\mathcal{X}, d, \nu)$. Recently, Tan and Li [126] established the Littlewood–Paley theory and also obtained the $T(1)$ theorem in the setting of $(\mathcal{X}, d, \nu)$.
- The $A_p^\varrho(\mathcal{X}, \nu)$ weights of Muckenhoupt type in the setting of $\mathbb{R}^D$ with the measure as in (0.0.1) were first introduced by Orobitg and Pérez [106] for $\varrho = 1$ and by Komori [76] for $\varrho \in [1, \infty)$. We remark that the reverse Hölder inequality, the fact that $u \in A_1(\mathcal{X}, \nu)$ implies $u \in L_{\rm loc}^{1+\sigma}(\mathcal{X}, \nu)$ with some $\sigma \in (1, \infty)$, and some other important properties enjoyed by A_p weights in the classical Euclidean space $(\mathbb{R}^D, |\cdot|, dx)$ may not be true in the non-homogeneous setting. Indeed, when $(\mathcal{X}, d, \nu) := (\mathbb{R}^D, |\cdot|, \mu)$ with μ as in (0.0.1), Orobitg and Pérez [106] have already explicitly pointed out this.
- Theorem 8.6.2 was proved by Hu and Yang in [64] when

$$(\mathcal{X}, d, \nu) := (\mathbb{R}^D, |\cdot|, \mu)$$

with μ satisfying the polynomial growth condition (0.0.1) and by Hu et al. in [62] for a general metric measure spaces $(\mathcal{X}, d, \nu)$.

- Let $\Phi_1(t) := t^p$ for all $t \in (0, \infty)$ with $p \in (1, \infty)$. Then Φ_1 is a convex Orlicz function, with $a_{\Phi_1} = b_{\Phi_1} = p \in (1, \infty)$, and $L^{\Phi_1}(\mathcal{X}, \nu) = L^p(\mathcal{X}, \nu)$. In this case, Theorem 8.7.1 was proved by Bui and Duong in [9, Theorem 7.6] when $k = 1$ under the additional assumption that there exists $m \in (0, \infty)$ such that, for all $x \in \mathcal{X}$ and $a, r \in (0, \infty)$,

$$\lambda(x, ar) = a^m \lambda(x, r), \tag{8.9.2}$$

and by Fu et al. in [30] for general $k \in \mathbb{N}$ and Φ without the assumption (8.9.2); moreover, when $(\mathcal{X}, d, \nu) := (\mathbb{R}^D, |\cdot|, \mu)$ with μ as in (0.0.1), then Theorem 8.7.1 is just [54, Theorem 2] obtained by Hu et al.
- Theorem 8.7.5 was proved in [54] when $(\mathcal{X}, d, \nu) := (\mathbb{R}^D, |\cdot|, \mu)$ with μ as in (0.0.1) and in [30] for a general metric measure space $(\mathcal{X}, d, \nu)$.
- Let T be the operator associated with kernel K as in (8.1.25), where K satisfies (8.1.23) and (8.1.24). Bui [8] showed that, if T is bounded on $L^2(\mathcal{X}, \nu)$, then the maximal commutator generated by RBMO $(\mathcal{X}, \nu)$ functions is bounded on $L^p(\mathcal{X}, \nu)$ for $p \in (1, \infty)$.
- When $m \geq 2$ and $(\mathcal{X}, d, \nu) := (\mathbb{R}^D, |\cdot|, dx)$, the operator defined by (8.8.3) is just the multilinear Calderón–Zygmund operator introduced by Coifman and Meyer in [16]. The study of multilinear Calderón–Zygmund operators is motivated not only by a mere quest to generalize the theory of the classical Calderón–Zygmund operators but also by their natural appearance in analysis; see Grafakos and Torres [44] and Lacey and Thiele [77] for some backgrounds and motivations. For more results on multilinear Calderón–Zygmund operators, see, for example, [42, 43, 45, 150] and the references therein.
- Theorem 8.8.2 was obtained by Hu et al. in [62]. As is well known, in the case that $(\mathcal{X}, d, \nu) := (\mathbb{R}^D, |\cdot|, dx)$, the classical bilinear Calderón–Zygmund operator is bounded from $L^1(\mathbb{R}^D) \times L^1(\mathbb{R}^D)$ into $L^{1/2,\infty}(\mathbb{R}^D)$, if T is bounded from $L^{q_1}(\mathbb{R}^D) \times L^{q_2}(\mathbb{R}^D)$ into $L^{q,\infty}(\mathbb{R}^D)$ for some $q_1, q_2 \in (1, \infty)$ and $q \in (0, \infty)$ with $1/q = 1/q_1 + 1/q_2$, and there exist positive constants C and τ such that, for all $x, y, z, \tilde{y} \in \mathbb{R}^D$ with $\max\{|x - y|, |x - z|\} \geq 2|y - \tilde{y}|$,

$$\begin{aligned}&|K(x, y, z) - K(x, \tilde{y}, z)| + |K(x, z, y) - K(x, z, \tilde{y})|\\&\quad \leq C \frac{|y - \tilde{y}|^\tau}{(|x - y| + |x - z|)^{2n+\tau}};\end{aligned}$$

see [44]. However, in the case of non-homogeneous metric measure spaces, it is still unknown whether T is bounded from $L^1(\mathcal{X}, \nu) \times L^1(\mathcal{X}, \nu)$ into $L^{1/2,\infty}(\mathcal{X}, \nu)$ or not, if we only assume that T is bounded from $L^{q_1}(\mathcal{X}, \nu) \times L^{q_2}(\mathcal{X}, \nu)$ into $L^{q,\infty}(\mathcal{X}, \nu)$ and that there exist positive constants C and τ such that, for all $x, y, z, \tilde{y} \in \mathcal{X}$ with $\max\{d(x, y), d(x, z)\} \geq 2d(y, \tilde{y})$,

$$|K(x, y, z) - K(x, \tilde{y}, z)| + |K(x, z, y) - K(x, z, \tilde{y})| \\ \leq C \frac{[d(y, \tilde{y})]^\tau}{[d(x, y) + d(x, z)]^\tau [\lambda(x, d(x, y)) + \lambda(x, d(x, z))]^2},$$

due to the weak growth assumption of the considered measure ν. Even when $(\mathcal{X}, d, \nu) := (\mathbb{R}^D, |\cdot|, \mu)$ with μ as in (0.0.1), this is also unknown. On the other hand, in Theorem 8.8.2, if we only assume that T is bounded from $L^{q_1}(\mathcal{X}, \nu) \times L^{q_2}(\mathcal{X}, \nu)$ into $L^{q,\infty}(\mathcal{X}, \nu)$ for some $q_1, q_2 \in (1, \infty)$ and q with $1/q = 1/q_1 + 1/q_2$, we conclude that the ranges of the indices p_1 and p_2 in Theorem 8.8.2 are $p_1 \in [q_1, \infty)$ and $p_2 \in [q_2, \infty)$, which are narrower than Theorem 8.8.2, and $\vec{w}$ belongs to some smaller weight class than Theorem 8.8.2. For these reasons, we assume, in Theorem 8.8.2, that T is bounded from

$$L^1(\mathcal{X}, \nu) \times L^1(\mathcal{X}, \nu) \quad \text{to} \quad L^{1/2,\infty}(\mathcal{X}, \nu)$$

and, moreover, that the kernel K satisfies the regularity condition only on the first variable; see [62] for the details.

- If $(\mathcal{X}, d, \nu)$ is the classical Euclidean space, Lerner et al. [82] established a better result than Theorem 8.8.2, that is, the multilinear Calderón–Zygmund operator T is bounded from

$$L^{p_1}(\mathcal{X}, w_1) \times \cdots \times L^{p_m}(\mathcal{X}, w_m) \quad \text{to} \quad L^p(\mathcal{X}, v_{\vec{w}})$$

with $(w_1, \ldots, w_m) \in A^{\varrho}_{\vec{P}}(\mathcal{X}, \nu)$, $p_1, \ldots, p_m \in (1, \infty)$ and

$$1/p = 1/p_1 + \cdots + 1/p_m.$$

However, it is still unknown whether their result is true or not when $(\mathcal{X}, d, \nu)$ is just a non-homogeneous metric measure space. Indeed, it seems that the argument in [82] is not valid in the present setting. Checking the argument used in [82], we see that a priori estimate that, for $w \in A_\infty(\mathbb{R}^D)$, $p \in (0, \infty)$ and bounded functions f_1, f_2 with bounded support, $T(f_1, f_2) \in L^p(w)$, is necessary in their proof. However, without the reverse Hölder inequality, the priori $T(f_1, f_2) \in L^p(w)$ cannot be obtained directly.

- Theorem 8.8.4 was obtained by Hu et al. in [62]. When

$$(\mathcal{X}, d, \nu) := (\mathbb{R}^D, |\cdot|, dx),$$

Lerner et al. [82, Corollary 3.8] proved that, for all $u \in A_\infty(\mathbb{R}^D)$ and $p \in (0, \infty)$, there exists a positive constant C such that, for all bounded functions f_1 and f_2 with bounded support,

$$\|T(f_1, f_2)\|_{L^p(u)} \leq C \left\|\tilde{\mathcal{M}}(f_1, f_2)\right\|_{L^p(u)}, \tag{8.9.3}$$

where $\tilde{\mathcal{M}}$ is the multilinear maximal operator on $\mathbb{R}^D$. Applying the argument in [82], we also see, in $(\mathbb{R}^D, |\cdot|, dx)$, that there exists a positive constant C such that, for all bounded functions f_1 and f_2 with bounded support,

$$\|T(f_1, f_2)\|_{L^{p,\infty}(u)} \le C \left\|\tilde{\mathcal{M}}(f_1, f_2)\right\|_{L^{p,\infty}(u)}, \tag{8.9.4}$$

provided that $p \in (0, \infty)$ and $u \in A_\infty(\mathbb{R}^D)$. Observe that, in $(\mathbb{R}^D, |\cdot|, dx)$, $u \in A_\infty(\mathbb{R}^D)$ implies that u satisfies the reverse Hölder inequality and hence $u \in L^q_{\rm loc}(\mathbb{R}^D)$ for some $q \in (1, \infty)$ sufficiently close to 1, from which, it further follows that, for $u \in A_\infty(\mathbb{R}^D)$, $p \in (0, \infty)$ and all bounded functions f_1 and f_2 with bounded support,

$$\begin{cases} \tilde{\mathcal{M}}(f_1, f_2) \in L^p(u) \Longrightarrow T(f_1, f_2) \in L^p(u), \\ \tilde{\mathcal{M}}(f_1, f_2) \in L^{p,\infty}(u) \Longrightarrow T(f_1, f_2) \in L^{p,\infty}(u). \end{cases} \tag{8.9.5}$$

This, via the sharp function estimate that, for $r \in (0, 1/2)$, there exists a positive constant C such that, for all bounded functions f_1 and f_2 with bounded support and $x \in \mathbb{R}^D$,

$$M_r^\sharp(T(f_1, f_2))(x) \le C\tilde{\mathcal{M}}(f_1, f_2)(x)$$

leads to (8.9.3) and (8.9.4). However, for the non-homogeneous metric measure space, a weight $u \in A_\infty(\mathcal{X}, \nu)$ does not imply $u \in L^q_{\rm loc}(\mathcal{X}, \nu)$ for some $q \in (1, \infty)$, and it is unclear whether (8.9.5) is true or not when $u \in A_\infty(\mathcal{X}, \nu)$. Also, it is unknown whether Theorem 8.8.4 is true or not if only assuming that $u \in A_\infty(\mathcal{X}, \nu)$.

- Let $\eta, \rho \in (1, \infty)$. A real-valued function $f \in L^1_{\rm loc}(\mathcal{X}, \nu)$ is said to be in the *space* $\mathrm{RBLO}(\mathcal{X}, \nu)$ if there exists a nonnegative constant C such that, for all balls B,

$$\frac{1}{\nu(\eta B)} \int_B \left[f(y) - \operatorname*{ess\,inf}_{B^\rho} f \right] d\nu(y) \le C \tag{8.9.6}$$

and, for all (ρ, β_ρ)-doubling balls $B \subset S$,

$$\operatorname*{ess\,inf}_{B} f - \operatorname*{ess\,inf}_{S} f \le C[1 + \delta(B, S)]. \tag{8.9.7}$$

Moreover, the $\mathrm{RBLO}(\mathcal{X}, \nu)$ *norm* of f is defined to be the minimal constant C as above and denoted by $\|f\|_{\mathrm{RBLO}(\mathcal{X},\nu)}$. In [87], Lin and Yang introduced the space $\mathrm{RBLO}(\mathcal{X}, \nu)$, which is a proper subset of RBMO $(\mathcal{X}, \nu)$. Some equivalent characterizations of $\mathrm{RBLO}(\mathcal{X}, \nu)$ and the boundedness of maximal Calderón–Zygmund operators from $L^\infty(\mathcal{X}, \nu)$ to $\mathrm{RBLO}(\mathcal{X}, \nu)$ were also established in [87]. All these conclusions aforementioned hold true when $1 + \delta(B, S)$ is replaced by $\tilde{\delta}^{(\rho)}_{B,S}$.

- Let $(\mathcal{X}, d, \nu)$ be a space of homogeneous type and $\lambda(x, r) := \nu(B(x, r))$ for all $x \in \mathcal{X}$ and $r \in (0, \infty)$. A real-valued function $f \in L^1_{\rm loc}(\mathcal{X}, \nu)$ is said to be in the *space* $\mathrm{BLO}(\mathcal{X}, \nu)$ if there exists a non-negative constant C such that, for all balls B,

$$\frac{1}{\nu(B)} \int_B \left[f(y) - \operatorname*{ess\,inf}_B f \right] d\nu(y) \le C. \tag{8.9.8}$$

The $\mathrm{BLO}(\mathcal{X}, \nu)$-*norm* of f is defined to be the minimal constant C as in (8.9.8) and denoted by $\|f\|_{\mathrm{BLO}(\mathcal{X}, \nu)}$. It is obvious that

$$\mathrm{BLO}(\mathcal{X}, \nu) \subset \mathrm{BMO}(\mathcal{X}, \nu) \tag{8.9.9}$$

and

$$\|f\|_{\mathrm{BMO}(\mathcal{X}, \nu)} \lesssim \|f\|_{\mathrm{BLO}(\mathcal{X}, \nu)} \text{ for all } f \in \mathrm{BLO}(\mathcal{X}, \nu),$$

where $\mathrm{BMO}(\mathcal{X}, \nu)$ is as in Proposition 7.2.3.

- In [29], Fu et al. showed that, when $(\mathcal{X}, d, \nu)$ is a space of homogeneous type and $\nu(\mathcal{X}) = \infty$, $\mathrm{RBLO}(\mathcal{X}, \nu)$ and $\mathrm{BLO}(\mathcal{X}, \nu)$ coincide with equivalent norms. Indeed, it is obvious that

$$\mathrm{RBLO}(\mathcal{X}, \nu) \subset \mathrm{BLO}(\mathcal{X}, \nu)$$

and

$$\|f\|_{\mathrm{BLO}(\mathcal{X}, \nu)} \le \|f\|_{\mathrm{RBLO}(\mathcal{X}, \nu)} \text{ for all } f \in \mathrm{RBLO}(\mathcal{X}, \nu).$$

Conversely, let $f \in \mathrm{BLO}(\mathcal{X}, \nu)$. By (8.9.9) and Proposition 7.2.3, we see that, for all balls $B \subset S$,

$$\begin{aligned}
&\operatorname*{ess\,inf}_B f - \operatorname*{ess\,inf}_S f\\
&\quad \le \operatorname*{ess\,inf}_B f - m_B(f) + |m_B(f) - f_B| + |f_B - f_S| + |f_S - m_S(f)|\\
&\qquad + m_S(f) - \operatorname*{ess\,inf}_S f\\
&\quad \le \frac{1}{\nu(B)} \int_B |f(y) - f_B|\, d\nu(y) + |f_B - f_S|\\
&\qquad + \frac{1}{\nu(S)} \int_S |f(y) - f_S|\, d\nu(y)\\
&\qquad + \frac{1}{\nu(S)} \int_S \left[f(y) - \operatorname*{ess\,inf}_S f \right] d\nu(y)\\
&\quad \lesssim \|f\|_{\mathrm{BLO}(\mathcal{X}, \nu)} + [1 + \delta(B, S)] \|f\|_{\mathrm{BMO}(\mathcal{X}, \nu)}\\
&\quad \lesssim \|f\|_{\mathrm{BLO}(\mathcal{X}, \nu)},
\end{aligned}$$

where, for all balls B, f_B is as in Definition 7.2.1 and

$$m_B(f) := \frac{1}{\nu(B)} \int_B f(x)\,d\nu(x).$$

This shows that (8.9.7) holds true. We then obtain

$$\mathrm{BLO}(\mathcal{X},\, \nu) \subset \mathrm{RBLO}(\mathcal{X},\, \nu)$$

and

$$\|f\|_{\mathrm{RBLO}(\mathcal{X},\nu)} \lesssim \|f\|_{\mathrm{BLO}(\mathcal{X},\nu)} \text{ for all } f \in \mathrm{BLO}(\mathcal{X},\, \nu).$$

Thus, $\mathrm{RBLO}(\mathcal{X},\, \nu)$ and $\mathrm{BLO}(\mathcal{X}, \nu)$ coincide with equivalent norms, which completes the proof.

- In [29], Fu et al. showed that, when $(\mathcal{X}, d, \nu)$ is a space of homogeneous type and $\nu(\mathcal{X}) < \infty$, it may happen that $\mathrm{BLO}(\mathcal{X}, \nu) \subsetneqq \mathrm{RBLO}(\mathcal{X},\, \nu)$. To see this, let $(\mathcal{X}, d, \nu) := (\mathbb{R}^2, |\cdot|, \mu)$ with μ the two-dimensional Lebesgue measure restricted to the unit ball $B(0,1)$, and $\lambda(x,r) := r$ for all $x \in \mathbb{R}^2$ and $r \in (0,\infty)$. This measure is doubling and $\mu(\mathbb{R}^2) < \infty$. Now we claim that

$$\mathrm{RBLO}(\mathbb{R}^2,\, \mu) = L^\infty(\mathbb{R}^2,\, \mu)/\mathbb{C}$$

(the *space of* $L^\infty(\mathbb{R}^2,\, \mu)$ *modulo constant functions*) with equivalent norms. Indeed, by an argument similar to that used in Example 3.1.14, we see that

$$\mathrm{RBMO}(\mathbb{R}^2,\, \mu) = L^\infty(\mathbb{R}^2,\, \mu)/\mathbb{C}$$

with equivalent norms. Then, from the fact that

$$L^\infty(\mathbb{R}^2,\, \mu)/\mathbb{C} \subset \mathrm{RBLO}(\mathbb{R}^2,\, \mu) \subset \mathrm{RBMO}(\mathbb{R}^2,\, \mu),$$

the claim follows. On the other hand, it is not difficult to see that

$$-(\log|x|)\chi_{\{x\in\mathbb{R}^2:\ 0<|x|<1\}}(x) \in \left(\mathrm{BLO}(\mathbb{R}^2, \mu) \setminus L^\infty(\mathbb{R}^2, \mu)/\mathbb{C}\right).$$ [12]

We further conclude that, in this case,

$$\mathrm{RBLO}(\mathbb{R}^2,\, \mu) = L^\infty(\mathbb{R}^2,\, \mu)/\mathbb{C} \subsetneqq \mathrm{BLO}(\mathbb{R}^2,\, \mu),$$

which completes the example.

[12] See Zhou [158].

- Let $\eta, \rho \in (1, \infty)$. A real-valued function $f \in L^1_{\rm loc}(\mathcal{X}, \nu)$ is said to be in the *space* $\widetilde{\rm RBLO}(\mathcal{X}, \nu)$ if it satisfies (8.9.6) and (8.9.7) with $1 + \delta(B, S)$ replaced by $\tilde{\delta}^{(\rho)}_{B,S}$. Moreover, the $\widetilde{\rm RBLO}(\mathcal{X}, \nu)$*-norm* of f is defined to be the minimal constant C as in (8.9.6) and (8.9.7) with $1 + \delta(B, S)$ replaced by $\tilde{\delta}^{(\rho)}_{B,S}$ and denoted by $\|f\|_{\widetilde{\rm RBLO}(\mathcal{X},\nu)}$. Fu et al. [29] introduced the space $\widetilde{\rm RBLO}(\mathcal{X}, \nu)$ and showed that $\widetilde{\rm RBLO}(\mathcal{X}, \nu)$ is independent of the choices of the constants η and $\rho \in (1, \infty)$, and $\widetilde{\rm RBLO}(\mathcal{X}, \nu) \subset \widetilde{\rm RBMO}(\mathcal{X}, \nu)$.
- Fu et al. [29] showed that, when $(\mathcal{X}, d, \nu)$ is a space of homogeneous type, if $\nu(\mathcal{X}) = \infty$, $\widetilde{\rm RBLO}(\mathcal{X}, \nu)$ and ${\rm BLO}(\mathcal{X}, \nu)$ coincide with equivalent norms and, if $\nu(\mathcal{X}) < \infty$, it may happen that $\widetilde{\rm RBLO}(\mathcal{X}, \nu) \subsetneqq {\rm BLO}(\mathcal{X}, \nu)$. For any $\rho, \eta \in (1, \infty)$, it is obvious that ${\rm RBLO}(\mathcal{X}, \nu) \subset \widetilde{\rm RBLO}(\mathcal{X}, \nu)$. Moreover, let $(\mathcal{X}, d, \nu)$ be a non-homogeneous space with the dominating function satisfying the weak reverse doubling condition. Fu et al. [29] showed that ${\rm RBLO}(\mathcal{X}, \nu)$ and $\widetilde{\rm RBLO}(\mathcal{X}, \nu)$ coincide with equivalent norms. However, if $(\mathcal{X}, d, \nu)$ is a non-homogeneous space without the dominating function satisfying the weak reverse doubling condition, then it is still unclear whether ${\rm RBLO}(\mathcal{X}, \nu)$ and $\widetilde{\rm RBLO}(\mathcal{X}, \nu)$ coincide with equivalent norms or not.
- Let $\alpha \in (0, 1)$. A function $K_\alpha \in L^1_{\rm loc}((\mathcal{X} \times \mathcal{X}) \setminus \{(x, y) : x = y\})$ is called a *generalized fractional integral kernel*, if there exists a positive constant $C_{(K_\alpha)}$, depending on K_α, such that, for all $x, y \in \mathcal{X}$ with $x \neq y$,

$$|K_\alpha(x, y)| \le C_{(K_\alpha)} \frac{1}{[\lambda(x, d(x, y))]^{1-\alpha}} \tag{8.9.10}$$

and that there exist positive constants τ and $c_{(K_\alpha)}$ such that, for all $x, \tilde{x}, y \in \mathcal{X}$ with $d(x, y) \ge c_{(K_\alpha)} d(x, \tilde{x})$,

$$\begin{aligned} &|K_\alpha(x, y) - K_\alpha(\tilde{x}, y)| + |K_\alpha(y, x) - K_\alpha(y, \tilde{x})| \\ &\quad \le C_{(K_\alpha)} \frac{[d(x, \tilde{x})]^\tau}{[d(x, y)]^\tau [\lambda(x, d(x, y))]^{1-\alpha}}. \end{aligned} \tag{8.9.11}$$

A linear operator T_α is called a *generalized fractional integral* with kernel K_α satisfying (8.9.10) and (8.9.11) if, for all $f \in L^\infty(\mathcal{X}, \nu)$ with bounded support and $x \notin {\rm supp}\, f$,

$$T_\alpha f(x) := \int_{\mathcal{X}} K_\alpha(x, y) f(y)\, d\nu(y). \tag{8.9.12}$$

In [31], Fu et al. introduced T_α with $\alpha \in (0, 1)$ and showed that the following statements are equivalent:

(i) T_α is bounded from $L^p(\mathcal{X}, \nu)$ into $L^q(\mathcal{X}, \nu)$ for all $p \in (1, 1/\alpha)$ and $1/q = 1/p - \alpha$;

(ii) T_α is bounded from $L^1(\mathcal{X}, \nu)$ into $L^{1/(1-\alpha),\infty}(\mathcal{X}, \nu)$;
(iii) There exists a positive constant C such that, for all $f \in L^{1/\alpha}(\mathcal{X}, \nu)$ with $T_\alpha f$ being finite almost everywhere, $\|T_\alpha f\|_{\mathrm{RBMO}(\mathcal{X},\nu)} \le C\|f\|_{L^{1/\alpha}(\mathcal{X},\nu)}$;
(iv) T_α is bounded from $H^1(\mathcal{X}, \nu)$ into $L^{1/(1-\alpha)}(\mathcal{X}, \nu)$;
(v) T_α is bounded from $H^1(\mathcal{X}, \nu)$ into $L^{1/(1-\alpha),\infty}(\mathcal{X}, \nu)$.

The equivalence of these statements are unknown for $\tilde{\delta}^{(\rho)}_{B,S}$ instead of $1+\delta(B, S)$. These results are all true when $(\mathcal{X}, d, \nu) := (\mathbb{R}^D, |\cdot|, \mu)$ with μ as in (0.0.1); see Hu et al. [60]. However, for a general metric measure space $(\mathcal{X}, d, \nu)$, no conclusion of these results is known to be true. In [31], a specific example of the generalized fractional integrals, which is a natural variant of Bergman-type operators was introduced as follows. Let $(\mathcal{X}, d, \nu) := (\mathbb{B}_{2D}, d, \nu)$ be as in Example 8.1.7. If $\alpha \in (0, 1)$, then the kernel

$$K_{m,\alpha}(x, y) := (1 - \bar{x} \cdot y)^{-m(1-\alpha)}, \quad x, y \in \overline{\mathbb{B}}_{2D}$$

satisfies the conditions (8.9.10) and (8.9.11). Thus, when $\alpha \in (0, 1)$, the fractional integral $T_{m,\alpha}$, associated with $K_{m,\alpha}$, is an example of the generalized fractional integrals. Recall that, when $\alpha = 0$, then the operator $T_{m,0}$, associated with $K_{m,0}$, is just the so-called "Bergman-type" operator in Example 8.1.7. For all $\alpha \in (0, 1)$, $f \in L^\infty_b(\mathcal{X}, \nu)$ and $x \in \mathcal{X}$, the *fractional integral* $I_\alpha f(x)$ is defined by

$$I_\alpha f(x) := \int_{\mathcal{X}} \frac{f(y)}{[\lambda(y, d(x, y))]^{1-\alpha}}\, d\nu(y).$$

Obviously, the integral kernel $K_\alpha(x, y) = \frac{1}{[\lambda(y, d(x, y))]^{1-\alpha}}$ satisfies (8.9.10) and (8.9.11). Under the additional assumption (8.9.2), Fu, Yuan and Yang [31] further showed that all the conclusions of the aforementioned equivalent statements for T_α hold true for I_α.

- Let K be a locally integrable function on $(\mathcal{X}\times\mathcal{X})\setminus\{(x, x) : x \in \mathcal{X}\}$. Assume that there exists a positive constant C satisfying (8.1.23) and that, for all $y, \tilde{y} \in \mathcal{X}$,

$$\int_{d(x, y)\ge 2d(y, \tilde{y})} [|K(y, x) - K(\tilde{y}, x)| + |K(x, y) - K(x, \tilde{y})|] \frac{1}{d(x, y)}\, d\nu(x) \le C.$$

The *Marcinkiewicz integral* $M(f)$ associated to the above kernel K is defined by setting, for all $x \in \mathcal{X}$,

$$M(f)(x) := \left[\int_0^\infty \left|\int_{d(x, y)<t} K(x, y) f(y)\, d\nu(y)\right|^2 \frac{dt}{t^3}\right]^{1/2}.$$

In [88], Lin and Yang proved that the $L^p(\mathcal{X}, \nu)$ boundedness with $p \in (1, \infty)$ of the Marcinkiewicz integral is equivalent to either of its boundedness from

$L^1(\mathcal{X}, \nu)$ into $L^{1,\infty}(\mathcal{X}, \nu)$ or from the atomic Hardy space $H^1(\mathcal{X}, \nu)$ into $L^1(\mathcal{X}, \nu)$. The equivalence of these statements is still unknown when $1+\delta(B, S)$ is replaced by $\tilde{\delta}^{(\rho)}_{B,S}$. Comparing with the corresponding result in [52], this result makes an essential improvement. Moreover, Lin and Yang showed that, if $M(f)$ is bounded form $H^1(\mathcal{X}, \nu)$ into $L^1(\mathcal{X}, \nu)$, then it is also bounded from $L^\infty(\mathcal{X}, \nu)$ into the space RBLO$(\mathcal{X}, \nu)$ and, conversely, if the Marcinkiewicz integral is bounded from $L^\infty_b(\mathcal{X}, \nu)$ (the set of all $L^\infty(\mathcal{X}, \nu)$ functions with bounded support) into the space RBMO$(\mathcal{X}, \nu)$, then it is also bounded from the finite atomic Hardy space $H^{1,\infty}_{\rm fin}(\mathcal{X}, \nu)$ into $L^1(\mathcal{X}, \nu)$. The aforementioned results are also unknown when $1 + \delta(B, S)$ is replaced by $\tilde{\delta}^{(\rho)}_{B,S}$.

References

1. P. Auscher, T. Hytönen, Orthonormal bases of regular wavelets in spaces of homogeneous type. Appl. Comput. Harmon. Anal. **34**, 266–296 (2013)
2. C. Bennett, Another characterization of *BLO*. Proc. Am. Math. Soc. **85**, 552–556 (1982)
3. C. Bennett, R.A. DeVore, R. Sharpley, Weak-L^{∞} and *BMO*. Ann. of Math. (2) **113**, 601–611 (1981)
4. C. Bennett, R. Sharpley, *Interpolation of Operators*. Pure and Application Mathematics, vol. 129 (Academic Press, Boston, MA, 1988)
5. A.S. Besicovitch, A general form of the covering principle and relative differentiation of additive functions. Proc. Cambridge Philos. Soc. **41**, 103–110 (1945)
6. N. Bourbaki, *Topological vector spaces, Chaps. 1–5, Elements of Mathematics (Berlin)* (Springer, Berlin, 1987)
7. M. Bramanti, Singular integrals in nonhomogeneous spaces: L^2 and L^p continuity from Hölder estimates. Rev. Mat. Iberoam. **26**, 347–366 (2010)
8. T.A. Bui, Boundedness of maximal operators and maximal commutators on non-homogeneous spaces. In: *Proceedings of AMSI International Conference on Harmonic Analysis and Applications*, Macquarie University, Australia (to appear)
9. T.A. Bui, X.T. Duong, Hardy spaces, regularized BMO spaces and the boundedness of Calderón–Zygmund operators on non-homogeneous spaces. J. Geom. Anal. **23**, 895–932 (2013)
10. W. Chen, Y. Lai, Boundedness of fractional integrals in Hardy spaces with non-doubling measure. Anal. Theory Appl. **22**, 195–200 (2006)
11. W. Chen, Y. Meng, D. Yang, Calderón–Zygmund operators on Hardy spaces without the doubling condition. Proc. Am. Math. Soc. **133**, 2671–2680 (2005)
12. W. Chen, C. Miao, Vector valued commutators on non-homogeneous spaces. Taiwanese J. Math. **11**, 1127–1141 (2007)
13. W. Chen, B. Zhao, Weighted estimates for commutators on nonhomogeneous spaces. J. Inequal. Appl. **Art. ID 89396**, 14 pp. (2006)
14. M. Christ, A $T(b)$ theorem with remarks on analytic capacity and the Cauchy integral. Colloq. Math. **60/61**, 601–628 (1990)
15. R.R. Coifman, A real variable characterization of H^p. Studia Math. **51**, 269–274 (1974)
16. R.R. Coifman, Y. Meyer, On commutators of singular integrals and bilinear singular integrals. Trans. Am. Math. Soc. **212**, 315–331 (1975)
17. R.R. Coifman, R. Rochberg, Another characterization of BMO. Proc. Am. Math. Soc. **79**, 249–254 (1980)
18. R.R. Coifman, G. Weiss, *Analyse Harmonique Non-commutative sur Certains Spaces Homogènes*. Lecture Notes in Mathematics, vol. 242 (Springer, Berlin/New York, 1971)

D. Yang et al., *The Hardy Space H^1 with Non-doubling Measures and Their Applications*, Lecture Notes in Mathematics 2084, DOI 10.1007/978-3-319-00825-7,

19. R.R. Coifman, G. Weiss, Extensions of Hardy spaces and their use in analysis. Bull. Am. Math. Soc. **83**, 569–645 (1977)
20. G. David, Opérateurs intégraux singuliers sur certaines courbes du plan complexe. Ann. Sci. École Norm. Sup. (4) **17**, 157–189 (1984)
21. G. David, Opérateurs d'intégrale singulière sur les surfaces régulières. Ann. Sci. École Norm. Sup. (4) **21**, 225–258 (1988)
22. G. David, Unrectifiable 1-sets have vanishing analytic capacity. Rev. Mat. Iberoam. **14**, 369–479 (1998)
23. M. de Guzmán, *Differentiation of Integrals in* $\mathbb{R}^n$. Lecture Notes in Mathematics, vol. 481 (Springer, Belin/New York, 1975)
24. D. Deng, Y. Han, *Harmonic Analysis on Spaces of Homogeneous Type*. Lecture Notes in Mathematics, vol. 1966 (Springer, Berlin, 2009)
25. D. Deng, Y. Han, D. Yang, Besov spaces with non-doubling measures. Trans. Am. Math. Soc. **358**, 2965–3001 (2006)
26. C. Fefferman, E.M. Stein, H^p spaces of several variables. Acta Math. **129**, 137–193 (1972)
27. X. Fu, G. Hu, D. Yang, A remark on the boundedness of Calderón–Zygmund operators in non-homogeneous spaces. Acta Math. Sin. (Engl. Ser.) **23**, 449–456 (2007)
28. X. Fu, Da. Yang, Do. Yang, Molecular Hardy spaces on non-homogeneous metric measure spaces, Submitted
29. X. Fu, Da. Yang, Do. Yang, The Hardy space H^1 on non-homogeneous spaces and its applications—a survey. Eurasian Math. J. (to appear)
30. X. Fu, D. Yang, W. Yuan, Boundedness of multilinear commutators of Calderón–Zygmund operators on Orlicz spaces over non-homogeneous spaces. Taiwanese J. Math. **16**, 2203–2238 (2012)
31. X. Fu, D. Yang, W. Yuan, Generalized fractional integrals and their commutators over non-homogeneous spaces, Submitted
32. J. García-Cuerva, A.E. Gatto, Boundedness properties of fractional integral operators associated to non-doubling measures. Studia Math. **162**, 245–261 (2004)
33. J. García-Cuerva, A.E. Gatto, Lipschitz spaces and Calderón–Zygmund operators associated to non-doubling measures. Publ. Mat. **49**, 285–296 (2005)
34. J. García-Cuerva, J.M. Martell, Weighted inequalities and vector-valued Calderón–Zygmund operators on non-homogeneous spaces. Publ. Mat. **44**, 613–640 (2000)
35. J. García-Cuerva, J.M. Martell, On the existence of principal values for the Cauchy integral on weighted Lebesgue spaces for non-doubling measures. J. Fourier Anal. Appl. **7**, 469–487 (2001)
36. J. García-Cuerva, J.M. Martell, Two-weight norm inequalities for maximal operators and fractional integrals on non-homogeneous spaces. Indiana Univ. Math. J. **50**, 1241–1280 (2001)
37. J. Garnett, *Analytic Capacity and Measure*. Lecture Notes in Mathematics, vol. 297 (Springer, Berlin/New York, 1972)
38. D. Goldberg, A local version of real Hardy spaces. Duke Math. J. **46**, 27–42 (1979)
39. L. Grafakos, Estimates for maximal singular integrals. Colloq. Math. **96**, 167–177 (2003)
40. L. Grafakos, *Classical Fourier Analysis* (Springer, New York, 2008)
41. L. Grafakos, *Modern Fourier Analysis* (Springer, New York, 2009)
42. L. Grafakos, N. Kalton, Multilinear Calderón–Zygmund operators on Hardy spaces. Collect. Math. **52**, 169–179 (2001)
43. L. Grafakos, N. Kalton, Some remarks on multilinear maps and interpolation. Math. Ann. **319**, 151–180 (2001)
44. L. Grafakos, R.H. Torres, Multilinear Calderón–Zygmund theory. Adv. Math. **165**, 124–164 (2002)
45. L. Grafakos, R.H. Torres, Maximal operator and weighted norm inequalities for multilinear singular integrals. Indiana Univ. Math. J. **51**, 1261–1276 (2002)
46. H. Gunawan, Y. Sawano, I. Sihwaningrum, Fractional integral operators in nonhomogeneous spaces. Bull. Aust. Math. Soc. **80**, 324–334 (2009)

47. Y. Han, D. Müller, D. Yang, A theory of Besov and Triebel-Lizorkin spaces on metric measure spaces modeled on Carnot-Carathéodory spaces. Abstr. Appl. Anal. **Art. ID 893409**, 250 pp. (2008)
48. Y. Han, D. Yang, Triebel-Lizorkin spaces with non-doubling measures. Studia Math. **162**, 105–140 (2004)
49. J. Heinonen, *Lectures on Analysis on Metric Spaces* (Springer, New York, 2001)
50. G. Hu, J. Lian, H. Wu, An interpolation theorem related to the Hardy space with non-doubling measure. Taiwanese J. Math. **13**, 1609–1622 (2009)
51. G. Hu, S. Liang, Another characterization of the Hardy space with non doubling measures. Math. Nachr. **279**, 1797–1807 (2006)
52. G. Hu, H. Lin, D. Yang, Marcinkiewicz integrals with non-doubling measures. Integr. Equat. Operat. Theory **58**, 205–238 (2007)
53. G. Hu, Y. Meng, D. Yang, Multilinear commutators for fractional integrals in non-homogeneous spaces. Publ. Mat. **48**, 335–367 (2004)
54. G. Hu, Y. Meng, D. Yang, Multilinear commutators of singular integrals with non doubling measures. Integr. Equat. Operat. Theory **51**, 235–255 (2005)
55. G. Hu, Y. Meng, D. Yang, New atomic characterization of H^1 space with non-doubling measures and its applications. Math. Proc. Cambridge Philos. Soc. **138**, 151–171 (2005)
56. G. Hu, Y. Meng, D. Yang, Endpoint estimate for maximal commutators with non-doubling measures. Acta Math. Sci. Ser. B Engl. Ed. **26**, 271–280 (2006)
57. G. Hu, Y. Meng, D. Yang, Estimates for maximal singular integral operators in non-homogeneous spaces. Proc. Roy. Soc. Edinb. Sect. A **136**, 351–364 (2006)
58. G. Hu, Y. Meng, D. Yang, Endpoint estimates for maximal commutators in non-homogeneous spaces. J. Korean Math. Soc. **44**, 809–822 (2007)
59. G. Hu, Y. Meng, D. Yang, Boundedness of some maximal commutators in Hardy-type spaces with non-doubling measures. Acta Math. Sin. (Engl. Ser.) **23**, 1129–1148 (2007)
60. G. Hu, Y. Meng, D. Yang, Boundedness of Riesz potentials in nonhomogeneous spaces. Acta Math. Sci. Ser. B Engl. Ed. **28**, 371–382 (2008)
61. G. Hu, Y. Meng, D. Yang, A new characterization of regularized BMO spaces on non-homogeneous spaces and its applications. Ann. Acad. Sci. Fenn. Math. **38**, 3–27 (2013)
62. G. Hu, Y. Meng, D. Yang, Weighted norm inequalities for multilinear Calderón–Zygmund operators on non-homogeneous metric measure spaces. Forum Math. (2012). Doi: 10.1515/forum-2011-0042
63. G. Hu, X. Wang, D. Yang, A new characterization for regular BMO with non-doubling measures. Proc. Edinb. Math. Soc. **51**, 155–170 (2008)
64. G. Hu, D. Yang, Weighted norm inequalities for maximal singular integrals with nondoubling measures. Studia Math. **187**, 101–123 (2008)
65. G. Hu, Da. Yang, Do. Yang, Weighted estimates for commutators of singular integral operators with non-doubling measures. Sci. China Ser. A **50**, 1621–1641 (2007)
66. G. Hu, Da. Yang, Do. Yang, h^1, bmo, blo and Littlewood–Paley g-functions with non-doubling measures. Rev. Mat. Iberoam. **25**, 595–667 (2009)
67. G. Hu, Da. Yang, Do. Yang, A new characterization of RBMO (μ) by John–Strömberg sharp maximal functions. Czechoslovak Math. J. **59**, 159–171 (2009)
68. T. Hytönen, A framework for non-homogeneous analysis on metric spaces, and the RBMO space of Tolsa. Publ. Mat. **54**, 485–504 (2010)
69. T. Hytönen, S. Liu, Da. Yang, Do. Yang, Boundedness of Calderón–Zygmund operators on non-homogeneous metric measure spaces. Canad. J. Math. **64**, 892–923 (2012)
70. T. Hytönen, H. Martikainen, Non-homogeneous Tb theorem and random dyadic cubes on metric measure spaces. J. Geom. Anal. **22**, 1071–1107 (2012)
71. T. Hytönen, Da. Yang, Do. Yang, The Hardy space H^1 on non-homogeneous metric spaces. Math. Proc. Camb. Phil. Soc. **153**, 9–31 (2012)
72. Y. Jiang, Spaces of type *BLO* for non-doubling measures. Proc. Am. Math. Soc. **133**, 2101–2107 (2005)

73. F. John, Quasi-isometric mappings. In: *1965 Seminari 1962/63 Anal. Alg. Geom. e Topol.*, vol. 2, Ist. Naz. Alta Mat., pp. 462–473, Ediz. Cremonese, Rome
74. F. John, L. Nirenberg, On functions of bounded mean oscillation. Comm. Pure Appl. Math. **14**, 415–426 (1961)
75. J.-L. Journé, *Calderón–Zygmund Operators, Pseudodifferential Operators and the Cauchy Integral of Calderón*. Lecture Notes in Mathematics, vol. 994 (Springer, Berlin, 1983)
76. Y. Komori, Weighted estimates for operators generated by maximal functions on nonhomogeneous spaces. Georgian Math. J. **12**, 121–130 (2005)
77. M. Lacey, C. Thiele, L^p estimates on the bilinear Hilbert transform for $2 < p < \infty$. Ann. of Math. (2) **146**, 693–724 (1997)
78. R.H. Latter, A characterization of $H^p(\mathbb{R}^n)$ in terms of atoms. Studia Math. **62**, 93–101 (1978)
79. A.K. Lerner, On the John–Strömberg characterization of BMO for nondoubling measures. Real Anal. Exchange **28**, 649–660 (2002/03)
80. A.K. Lerner, Weighted norm inequalities for the local sharp maximal function. J. Fourier Anal. Appl. **10**, 465–474 (2004)
81. A.K. Lerner, Weighted rearrangement inequalities for local sharp maximal functions. Trans. Am. Math. Soc. **357**, 2445–2465 (2005)
82. A.K. Lerner, S. Ombrosi, C. Pérez, R.H. Torres, R. Trujillo-González, New maximal functions and multiple weights for the multilinear Calderón–Zygmund theory. Adv. Math. **220**, 1222–1264 (2009)
83. E. Lieb, M. Loss, *Analysis* (American Mathematical Society, Providence, RI, 2001)
84. H. Lin, Y. Meng, Boundedness of parametrized Littlewood–Paley operators with nondoubling measures. J. Inequal. Appl. **Art. ID 141379**, 25 pp. (2008)
85. H. Lin, Y. Meng, Maximal multilinear Calderón–Zygmund operators with non-doubling measures. Acta Math. Hungar. **124**, 263–287 (2009)
86. H. Lin, Y. Meng, D. Yang, Weighted estimates for commutators of multilinear Calderón–Zygmund operators with non-doubling measures. Acta Math. Sci. Ser. B Engl. Ed. **30**, 1–18 (2010)
87. H. Lin, D. Yang, Spaces of type BLO on non-homogeneous metric measure spaces. Front. Math. China **6**, 271–292 (2011)
88. H. Lin, D. Yang, Equivalent boundedness of Marcinkiewicz integrals on non-homogeneous metric measure spaces. Sci. China Math. (to appear)
89. L. Liu, Da. Yang, Do. Yang, Atomic Hardy-type spaces between H^1 and L^1 on metric spaces with non-doubling measures. Acta Math. Sin. (Engl. Ser.) **27**, 2445–2468 (2011)
90. S. Liu, Y. Meng, D. Yang, Boundedness of maximal Calderón–Zygmund operators on non-homogeneous metric measure spaces. Proc. Roy. Soc. Edinb. Sect. A (to appear)
91. S. Liu, Da. Yang, Do. Yang, Boundedness of Calderón–Zygmund operators on non-homogeneous metric measure spaces: equivalent characterizations. J. Math. Anal. Appl. **386**, 258–272 (2012)
92. J. Luukkainen, E. Saksman, Every complete doubling metric space carries a doubling measure. Proc. Am. Math. Soc. **126**, 531–534 (1998)
93. L. Maligranda, *Indices and Interpolation*. Dissertationes Math. (Rozprawy Mat.) **234**, 49 pp. (1985)
94. J. Mateu, P. Mattila, A. Nicolau, J. Orobitg, BMO for nondoubling measures. Duke Math. J. **102**, 533–565 (2000)
95. Y. Meng, Multilinear Calderón–Zygmund operators on the product of Lebesgue spaces with non-doubling measures. J. Math. Anal. Appl. **335**, 314–331 (2007)
96. Y. Meng, D. Yang, Multilinear commutators of Calderón–Zygmund operators on Hardy-type spaces with non-doubling measures. J. Math. Anal. Appl. **317**, 228–244 (2006)
97. Y. Meng, D. Yang, Boundedness of commutators with Lipschitz functions in nonhomogeneous spaces. Taiwanese J. Math. **10**, 1443–1464 (2006)
98. A.P. Morse, Perfect blankets. Trans. Am. Math. Soc. **61**, 418–442 (1947)
99. E. Nakai, Construction of an atomic decomposition for functions with compact support. J. Math. Anal. Appl. **313**, 730–737 (2006)

100. E. Nakai, A generalization of Hardy spaces H^p by using atoms. Acta Math. Sin. (Engl. Ser.) **24**, 1243–1268 (2008)
101. E. Nakai, Y. Sawano, Hardy spaces with variable exponents and generalized Campanato spaces. J. Funct. Anal. **262**, 3665–3748 (2012)
102. F. Nazarov, S. Treil, A. Volberg, Cauchy integral and Calderón–Zygmund operators on nonhomogeneous spaces. Int. Math. Res. Not. **15**, 703–726 (1997)
103. F. Nazarov, S. Treil, A. Volberg, Weak type estimates and Cotlar inequalities for Calderón–Zygmund operators on nonhomogeneous spaces. Int. Math. Res. Not. **9**, 463–487 (1998)
104. F. Nazarov, S. Treil, A. Volberg, Accretive system Tb-theorems on nonhomogeneous spaces. Duke Math. J. **113**, 259–312 (2002)
105. F. Nazarov, S. Treil, A. Volberg, The Tb-theorem on non-homogeneous spaces. Acta Math. **190**, 151–239 (2003)
106. J. Orobitg, C. Pérez, A_p weights for nondoubling measures in $\mathbb{R}^n$ and applications. Trans. Am. Math. Soc. **354**, 2013–2033 (2002)
107. W. Ou, The natural maximal operator on BMO. Proc. Am. Math. Soc. **129**, 2919–2921 (2001)
108. C. Pérez, R. Trujillo-González, Sharp weighted estimates for multilinear commutators. J. Lond. Math. Soc. (2) **65**, 672–692 (2002)
109. M. Rao, Z. Ren, *Theory of Orlicz Spaces* (Marcel Dekker, New York, 1991)
110. W. Rudin, *Functional Analysis* (McGraw-Hill Book, New York/Düsseldorf/Johannesburg, 1973)
111. W. Rudin, *Real and Complex Analysis* (McGraw-Hill Book, New York, 1987)
112. Y. Sawano, Sharp estimates of the modified Hardy-Littlewood maximal operator on the nonhomogeneous space via covering lemmas. Hokkaido Math. J. **34**, 435–458 (2005)
113. Y. Sawano, A vector-valued sharp maximal inequality on Morrey spaces with non-doubling measures. Georgian Math. J. **13**, 153–172 (2006)
114. Y. Sawano, Generalized Morrey spaces for non-doubling measures. Nonlinear Differ. Equat. Appl. **15**, 413–425 (2008)
115. Y. Sawano, S. Shirai, Compact commutators on Morrey spaces with non-doubling measures. Georgian Math. J. **15**, 353–376 (2008)
116. Y. Sawano, H. Tanaka, Morrey spaces for non-doubling measures. Acta Math. Sin. (Engl. Ser.) **21**, 1535–1544 (2005)
117. Y. Sawano, H. Tanaka, Equivalent norms for the Morrey spaces with non-doubling measures. Far East J. Math. Sci. **22**, 387–404 (2006)
118. Y. Sawano, H. Tanaka, The John-Nirenberg type inequality for non-doubling measures. Studia Math. **181**, 153–170 (2007)
119. Y. Sawano, H. Tanaka, Sharp maximal inequalities and commutators on Morrey spaces with non-doubling measures. Taiwanese J. Math. **11**, 1091–1112 (2007)
120. Y. Sawano, H. Tanaka, Predual spaces of Morrey spaces with non-doubling measures. Tokyo J. Math. **32**, 471–486 (2009)
121. E.M. Stein, *Harmonic Analysis: Real-variable Methods, Orthogonality, and Oscillatory Integrals* (Princeton University Press, Princeton, NJ, 1993)
122. E.M. Stein, G. Weiss, On the theory of harmonic functions of several variables. I. The theory of H^p-spaces. Acta Math. **103**, 25–62 (1960)
123. J.O. Strömberg, Bounded mean oscillation with Orlicz norms and duality of Hardy spaces. Indiana Univ. Math. J. **28**, 511–544 (1979)
124. J.O. Strömberg, A. Torchinsky, *Weighted Hardy Spaces*. Lecture Notes in Mathematics, vol. 1381 (Springer, Berlin, 1989)
125. M.H. Taibleson, G. Weiss, *The Molecular Characterization of Certain Hardy Spaces, Representation Theorems for Hardy Spaces*. Astérisque, vol. 77 (Soc. Math. France, Paris, 1980), pp. 67–149
126. C. Tan, J. Li, Littlewood–Paley theory on metric measure spaces with non doubling measures and its applications. Sci. China Math. (to appear)
127. E. Tchoundja, Carleson measures for the generalized Bergman spaces via a T(1)-type theorem. Ark. Mat. **46**, 377–406 (2008)

128. Y. Terasawa, Outer measures and weak type (1,1) estimates of Hardy-Littlewood maximal operators. J. Inequal. Appl. **Art. ID 15063**, 13 pp. (2006)
129. X. Tolsa, L^2-boundedness of the Cauchy integral operator for continuous measures. Duke Math. J. **98**, 269–304 (1999)
130. X. Tolsa, A $T(1)$ theorem for non-doubling measures with atoms. Proc. Lond. Math. Soc. (3) **82**, 195–228 (2001)
131. X. Tolsa, BMO, H^1 and Calderón–Zygmund operators for non doubling measures. Math. Ann. **319**, 89–149 (2001)
132. X. Tolsa, Littlewood–Paley theory and the $T(1)$ theorem with non-doubling measures. Adv. Math. **164**, 57–116 (2001)
133. X. Tolsa, A proof of the weak $(1,1)$ inequality for singular integrals with non doubling measures based on a Calderón–Zygmund decomposition. Publ. Mat. **45**, 163–174 (2001)
134. X. Tolsa, The space H^1 for nondoubling measures in terms of a grand maximal operator. Trans. Am. Math. Soc. **355**, 315–348 (2003)
135. X. Tolsa, Characterization of the atomic space H^1 for non doubling measures in terms of a grand maximal operator. http://mat.uab.cat/~xtolsa/hh.pdf
136. X. Tolsa, Painlevé's problem and the semiadditivity of analytic capacity. Acta Math. **190**, 105–149 (2003)
137. X. Tolsa, The semiadditivity of continuous analytic capacity and the inner boundary conjecture. Am. J. Math. **126**, 523–567 (2004)
138. X. Tolsa, Analytic capacity and Calderón–Zygmund theory with non doubling measures. In: *Seminar of Mathematical Analysis, Colecc. Abierta*, vol. 71 (University Sevilla Secr. Publ., Seville, 2004), pp. 239–271
139. X. Tolsa, Bilipschitz maps, analytic capacity, and the Cauchy integral. Ann. of Math. (2) **162**, 1243–1304 (2005)
140. X. Tolsa, Painlevé's problem and analytic capacity. Collect. Math. (**Extra**) 89–125 (2006)
141. X. Tolsa, Weighted norm inequalities for Calderón–Zygmund operators without doubling conditions. Publ. Mat. **51**, 397–456 (2007)
142. X. Tolsa, Calderón–Zygmund theory with non doubling measures. In: *NAFSA 9-Nonlinear Analysis, Function Spaces and Applications*, Vol. 9. ed. by Jiří Rákosník. Proceedings of the international school held in Třešt, September 11–17, 2010 (Academy of Sciences of the Czech Republic, Institute of Mathematics, Praha, 2011), pp. 217–260
143. X. Tolsa, *Analytic Capacity, the Cauchy Transform, and Non-homogeneous Calderón–Zygmund Theory*. Progress in Mathematics, vol. 307 (Birkhäuser/Springer Basel AG, Basel, 2014)
144. J. Verdera, On the $T(1)$-theorem for the Cauchy integral. Ark. Mat. **38**, 183–199 (2000)
145. J. Verdera, The fall of the doubling condition in Calderón–Zygmund theory. Publ. Mat. (**Extra**) 275–292 (2002)
146. A. Volberg, *Calderón–Zygmund Capacities and Operators on Nonhomogeneous Spaces*. CBMS Regional Conference Series in Mathematics, vol. 100 (Amer. Math. Soc., Providence, RI, 2003)
147. A. Volberg, B.D. Wick, Bergman-type singular operators and the characterization of Carleson measures for Besov-Sobolev spaces on the complex ball. Am. J. Math. **134**, 949–992 (2012)
148. J. Wu, Hausdorff dimension and doubling measures on metric spaces. Proc. Am. Math. Soc. **126**, 1453–1459 (1998)
149. J. Xu, Boundedness of multilinear singular integrals for non-doubling measures. J. Math. Anal. Appl. **327**, 471–480 (2007)
150. J. Xu, Boundedness in Lebesgue spaces for commutators of multilinear singular integrals and RBMO functions with non-doubling measures. Sci. China Ser. A **50**, 361–376 (2007)
151. D. Yang, Local Hardy and BMO spaces on non-homogeneous spaces. J. Aust. Math. Soc. **79**, 149–182 (2005)
152. Da. Yang, Do. Yang, Uniform boundedness for approximations of the identity with non-doubling measures. J. Inequal. Appl. **Art. ID 19574**, 25 pp. (2007)

153. Da. Yang, Do. Yang, Endpoint estimates for homogeneous Littlewood–Paley g-functions with non-doubling measures. J. Funct. Spaces Appl. **7**, 187–207 (2009)
154. Da. Yang, Do. Yang, BMO-estimates for maximal operators via approximations of the identity with non-doubling measures. Canad. J. Math. **62**, 1419–1434 (2010)
155. Da. Yang, Do. Yang, Boundedness of linear operators via atoms on Hardy spaces with non-doubling measures. Georgian Math. J. **18**, 377–397 (2011)
156. D. Yang, Y. Zhou, New properties of Besov and Triebel-Lizorkin spaces on RD-spaces. Manuscripta Math. **134**, 59–90 (2011)
157. K. Yosida, *Functional Analysis* (Springer, Berlin, 1995)
158. Y. Zhou, Some endpoint estimates for local Littlewood–Paley operators. (Chinese) Beijing Shifan Daxue Xuebao **44**, 577–580 (2008)

Index

D. Yang et al., *The Hardy Space H^1 with Non-doubling Measures and Their Applications*, Lecture Notes in Mathematics 2084, DOI 10.1007/978-3-319-00825-7,

LECTURE NOTES IN MATHEMATICS

Springer

Edited by J.-M. Morel, B. Teissier; P.K. Maini

Editorial Policy (for the publication of monographs)

1. Lecture Notes aim to report new developments in all areas of mathematics and their applications - quickly, informally and at a high level. Mathematical texts analysing new developments in modelling and numerical simulation are welcome.
   Monograph manuscripts should be reasonably self-contained and rounded off. Thus they may, and often will, present not only results of the author but also related work by other people. They may be based on specialised lecture courses. Furthermore, the manuscripts should provide sufficient motivation, examples and applications. This clearly distinguishes Lecture Notes from journal articles or technical reports which normally are very concise. Articles intended for a journal but too long to be accepted by most journals, usually do not have this "lecture notes" character. For similar reasons it is unusual for doctoral theses to be accepted for the Lecture Notes series, though habilitation theses may be appropriate.

2. Manuscripts should be submitted either online at www.editorialmanager.com/lnm to Springer's mathematics editorial in Heidelberg, or to one of the series editors. In general, manuscripts will be sent out to 2 external referees for evaluation. If a decision cannot yet be reached on the basis of the first 2 reports, further referees may be contacted: The author will be informed of this. A final decision to publish can be made only on the basis of the complete manuscript, however a refereeing process leading to a preliminary decision can be based on a pre-final or incomplete manuscript. The strict minimum amount of material that will be considered should include a detailed outline describing the planned contents of each chapter, a bibliography and several sample chapters.
   Authors should be aware that incomplete or insufficiently close to final manuscripts almost always result in longer refereeing times and nevertheless unclear referees' recommendations, making further refereeing of a final draft necessary.
   Authors should also be aware that parallel submission of their manuscript to another publisher while under consideration for LNM will in general lead to immediate rejection.

3. Manuscripts should in general be submitted in English. Final manuscripts should contain at least 100 pages of mathematical text and should always include

   – a table of contents;
   – an informative introduction, with adequate motivation and perhaps some historical remarks: it should be accessible to a reader not intimately familiar with the topic treated;
   – a subject index: as a rule this is genuinely helpful for the reader.

   For evaluation purposes, manuscripts may be submitted in print or electronic form (print form is still preferred by most referees), in the latter case preferably as pdf- or zipped psfiles. Lecture Notes volumes are, as a rule, printed digitally from the authors' files. To ensure best results, authors are asked to use the LaTeX2e style files available from Springer's web-server at:

   ftp://ftp.springer.de/pub/tex/latex/svmonot1/ (for monographs) and
   ftp://ftp.springer.de/pub/tex/latex/svmultt1/ (for summer schools/tutorials).

Additional technical instructions, if necessary, are available on request from lnm@springer.com.

4. Careful preparation of the manuscripts will help keep production time short besides ensuring satisfactory appearance of the finished book in print and online. After acceptance of the manuscript authors will be asked to prepare the final LaTeX source files and also the corresponding dvi-, pdf- or zipped ps-file. The LaTeX source files are essential for producing the full-text online version of the book (see http://www.springerlink.com/openurl.asp?genre=journal&issn=0075-8434 for the existing online volumes of LNM). The actual production of a Lecture Notes volume takes approximately 12 weeks.
5. Authors receive a total of 50 free copies of their volume, but no royalties. They are entitled to a discount of 33.3 % on the price of Springer books purchased for their personal use, if ordering directly from Springer.
6. Commitment to publish is made by letter of intent rather than by signing a formal contract. Springer-Verlag secures the copyright for each volume. Authors are free to reuse material contained in their LNM volumes in later publications: a brief written (or e-mail) request for formal permission is sufficient.

Addresses:

Professor J.-M. Morel, CMLA,
École Normale Supérieure de Cachan,
61 Avenue du Président Wilson, 94235 Cachan Cedex, France
E-mail: morel@cmla.ens-cachan.fr

Professor B. Teissier, Institut Mathématique de Jussieu,
UMR 7586 du CNRS, Équipe "Géométrie et Dynamique",
175 rue du Chevaleret
75013 Paris, France
E-mail: teissier@math.jussieu.fr

For the "Mathematical Biosciences Subseries" of LNM:

Professor P. K. Maini, Center for Mathematical Biology,
Mathematical Institute, 24-29 St Giles,
Oxford OX1 3LP, UK
E-mail : maini@maths.ox.ac.uk

Springer, Mathematics Editorial, Tiergartenstr. 17,
69121 Heidelberg, Germany,
Tel.: +49 (6221) 4876-8259

Fax: +49 (6221) 4876-8259
E-mail: lnm@springer.com